"Be Cool and Do Mischief"

Francis Marion's Orderly Book 1775 - 1782

"Be Cool and Do Mischief"

Francis Marion's Orderly Book 1775 - 1782

Patrick O'Kelley

Be Cool and Do Mischief: Francis Marion's Orderly Book 1775 - 1782
by Patrick O'Kelley

ISBN 978-1-956904-29-1

Cover: Francis Marion by Bryant White, wwwbbwhite.com

Printed in the United States of America

Published by Blacksmith LLC
Fayetteville, NC

www.Blacksmithpublishing.com

Direct inquiries and/or orders to the above web address.

To Betsy Puckett
"The little red headed fifer"
2nd South Carolina Regiment
1971 – 2004

Preface

Each company within a regiment was ordered to maintain an orderly book, while the regimental adjutant kept his own book. Each Continental regiment at any moment was maintaining about ten orderly books. If you expand that for each regiment serving in the army at any given point in time, then count the additional books kept at the brigade, division, wing, departmental and headquarters, it's clear that a very large number of books were being maintained on any given day.

The process for maintaining these orderly books would be for brigade adjutants to gather at the office of the adjutant general each morning and the general orders for the day would be read. The brigade adjutants transcribed them into their books. Those brigade adjutants would then receive any division orders and return to their respective brigades. They would then receive any brigade orders for the day, and the regimental adjutants would be read the general orders and any divisional and brigade orders for the day. This process of dictation and transcription was then repeated by the regimental adjutants passing along all daily orders to their companies' adjutants, orderly sergeants or other sergeants.

For any particular date the content of all books would be essentially identical in content, excepting only for the addition of unique division, brigade or regimental orders. There would be differences though. Transcribing from orders being read was a challenge, yielding differences between books in phrasings. By the time the procedure got down to the regimental adjutant reading to nine company sergeants who may not be totally literate, the results can become quite different. Spelling varied considerably and much of the spelling may have been done phonetically. If a sergeant missed a sentence or phrase during the dictation he could not stop the adjutant to catch up.

In 1780 Major General Robert Howe made an inspection of a number of orderly books being maintained at West Point and in the Highlands Department, and he described what he had seen as being an "incomprehensible, incoherent parcel of stuff." A very great percentage of orderly books were transcribed by sergeants doing clerical tasks, not authored by the officers who had issued the orders. Many officers did not physically write their own official correspondence and they had men, usually sergeants, from within their companies acting as aides or secretaries writing those letters. The officers simply signed them.

The idea that officers, especially those of high rank, authored orderly books is a myth which came about through the descendant families, donating sets of books to historical societies and libraries. Those books became associated by the library or institution accession process and records with the officer and created the impression that those books were literally written by him. A major example of this is the collection of the Library of Congress, which is cataloged on the basis of the surname of the supposed officer-author. Almost always those books were written for those officers by a sergeant.

There are some long runs of orderly books from the Revolutionary War, the best run kept by a single individual are those at the Library of Congress, attributed to William Torrey, Adjutant of the 2nd Massachusetts Regiment. They run, in 23 volumes, from Sept 1777 to May 1783 with gaps. There are also good runs for the 8th Massachusetts Regiment, and 6th Massachusetts Regiment. However they cannot rival the 2nd South Carolina orderly book, that has daily entries from 1775 to 1782.

Foreword

When I first read the "Francis Marion" orderly book I was amazed at the amount of information that it contained. I transcribed a few small parts that interested me, and over the years I continued that transcription. I was always amazed that no one had attempted to put the whole orderly book into a single published work. In 1999 I began the task of transcribing the whole book, and finding out what happened to all of the men mentioned in the book. Much of the information from the men listed in the footnotes comes from Bobby Moss's book on the South Carolina Patriots. Other information comes from muster rolls or from family histories.

In the first edition of this book I put all the footnotes at the end, but there were so many, that people told me that it was tiring needing to turn back to the end of the book and then go back to the original page. So, in this second edition I have put all the footnotes on the same page as the subject. However, this makes the first part of the book very "footnote heavy", due to introducing everyone who is mentioned. The footnotes become less common as the book continues.

William Dobein James rode with Francis Marion's partisans when he was 16 years old. Many years later he would write a history of Marion, and his book is an excellent source, since he witnessed much of what he wrote about. In his foreword he writes "On the monument erected by the Greeks at Thermopylae, the names of Leonidas and his three hundred men were not inscribed, because it was thought impossible to imagine they could ever be forgotten." He wrote that passage because he sought forgiveness from his former comrades, in case he left any out. I have attempted to track down every single soldier mentioned in this book, but it is only a fraction of the men who served with Marion. The rest are lost to history at this time.

I wish to thank those who assisted me in compiling this information, Doyle Harper, Paul Burke, and Ron Vido of the recreated 2nd South Carolina Regiment. Contributing artists are Daniel Bode, Jeff Saeger and Dan Clement and Bryant White who painted the cover art. Finally, the second edition of this book has a new name. The original name was "Unwaried Patience and Fortitude". This book, "Be Cool and Do Mischief" comes from Miller Roach, a fellow historian.

Patrick O'Kelley
Barbecue Township, North Carolina
August 2006
Second edition 2022

Introduction

Francis Marion was one of the most versatile commanders in the Revolutionary War. He was able to adapt to whatever was thrown his way, while still exceeding what was required of him. Marion was not the picture of was considered to be the "hero". He was short, and at the height of his military career he was middle aged and walked with a limp.

There are towns, counties, and parks named after Francis Marion in over 30 States. He has been the subject of many books, television series and movies, from Disney's "Swamp Fox" to Mel Gibson's "The Patriot". His impact on the history of the United States will never be fully measured, but without his efforts and sacrifices the fate of the nation would be in doubt.

Marion was born in 1732 at Goatfield Plantation in St. John's Parish, Berkeley County. His grandparents were Huguenots who had fled France with the Edict of Nantes. They settled along the Santee River, north of Charleston. When Marion was five years old his family moved near the English school at Georgetown. Francis grew up in that port town and dreamed of becoming a sailor. At the age of 15 he signed on to a schooner sailing for the West Indies. Unfortunately on the return trip a whale attacked the ship and sank it. The crew escaped in an open boat, and Marion had to suffer under the tropical sun for six days. Their only food was a small dog who had swam to the boat when the ship sank. Two crewmen died while they were adrift, but they were able to reach land on the sixth day. After that cruise Francis gave up on the sea, and decided to become a planter at his parent's plantation.

At the end of the French and Indian war the Cherokees threatened the frontier settlements in South Carolina. Governor William Henry Lyttleton increased the size of the military and Captain John Postell came to the low country to recruit a company of Provincials among the Huguenots. In 1756, at the age of 25, Marion and his brother Gabriel joined the St. John's militia company. When the Cherokee War broke out three years later the two brothers were experienced militiaman. Gabriel was given a captain's commission and he recruited a troop of cavalry, but Governor Lyttleton had made a treaty with the Cherokees, so the two brothers were demobilized. Gabriel Marion moved his family to Belle Isle, in St. Stephen's Parish, while Francis Marion moved up the Santee River to his brother Job's plantation.

The peace did not last and the Cherokees struck again. Colonel Archibald Montgomery and his 77th Highlander Regiment marched to Fort Prince George. Colonel Montgomery had some South Carolina militia with him when he marched for the Little Tennessee Valley. These men were known as "the Buffs" since their uniform was blue coats faced with buff colored cloth. Near the Cherokee town of Echoe, Montgomery's force was ambushed. Montgomery was able to drive the Indians back, but he suffered heavy losses.

Lieutenant Governor William Bull requested reinforcements from Lord Amherst. Lieutenant Colonel James Grant arrived in 1761 with 1,200 regulars. Lieutenant Governor Bull authorized Colonels Thomas Middleton, Henry Laurens and Major John Moultrie to recruit a regiment. This South Carolina Provincial Regiment was to consist of ten companies of 100 men. Francis Marion was commissioned a first lieutenant in Captain William Moultrie's light infantry company of this regiment. The uniform of this regiment was a blue coat, with scarlet facings "made up in the same manner as that of the Light Infantry of His Majesty's Royal Scots." They wore a light infantry-style cap, with a crescent on the turned up brim. The family seal of William Bull consisted of a crescent, and this new regiment adopted the crescent as their symbol to wear on their headgear. This uniform mirrors the one that Marion and his regiment would wear in the Revolutionary War.

Colonel Grant arrived at Fort Prince George in late May, 1761 and followed the same route that Montgomery had taken into Cherokee territory. With Grant and Middleton's forces were some Chickasaw and Catawba Indians, and a few companies of the "Buffs" that brought their numbers up to 2,600 men.

When Grant's army arrived at the same location where Montgomery had been ambushed the year before, they found the Cherokees waiting. Grant chose Lieutenant Francis Marion to drive off the ambushers. He detached 30 men for Marion to use and they moved from tree to tree, advancing up the pass. Marion began to suffer casualties, but he continued forward, leading the men on. By the time the main column had caught up to Marion he had suffered 21 men dead or wounded out of the 30. The army continued to fight uphill until the Indians finally broke. Grant had his men burn the town of Echoe, then turned his army loose on the other Cherokee settlements. Fifteen towns were burned and all the crops in the area were cut down. Finally the Cherokee chief, Attakullakulla came to him at Keowee and asked for peace.

Upon the end of the Cherokee war Lieutenant Francis Marion returned to being a planter, but now he was looked upon as a war hero and had gained the respect of the people. Moultrie wrote that Marion "was an active, brave and hardy soldier, and an excellent partisan officer." Little did Moultrie realize how excellent a partisan Marion would be.

Marion settled down for the next ten years. He bought the Pond Bluff plantation four miles below Eutaw Springs in 1773. The people of the area elected Francis Marion to represent them in Charleston when the situation with England became critical. When they learned that the colony of Massachussetts had fired upon British soldiers on April 19, 1775, they swore they would protect South Carolina. The Continental Congress requested that the state raise two regiments of infantry and one of cavalry. On June 12th, 1775 the Provincial Congress began electing the officers of these two new regiments, and this orderly book begins shortly afterwards.

Francis
Marion

**"Do not make too free with your cannon.
Cool and do mischief."**

**- Governor John Rutledge to General William Moultrie
Battle of Fort Sullivan - June 1776**

Contents

Patrick O'Kelley

[1]

[1] Image by Bryant White, ©2006, www.bbwhite.com

1775

General Orders by Col° Wm Moultrie[2] Tuesday June 20 1775

Officers of the First and 2^d regiments to give in the dates of their commissions to Adjutant Dellient [3] of the 2^d Regiment[4]

The recruits of the 1st & 2^d Regiment to be Quarter'd at the new barracks [5] & put into Messes 6 to each mess and to be drilled morning & Evening

Officers of the 1st & 2nd Regiment to be at the new Barracks at 6 oClock in the morning & 6 OClock in the Afternoon to See the roll Call'd over & attend the drilling of the men

Quartermaster to each Regiiment to procure tin kettles for each mess & few axes for cutting wood & draw the Rations at the rate of 1 ^lb beef & 1 ^lb bread or flour p^r man - some salt to be provided

Adjutants to apply to Mr Weyman [6] who will deliver out Arms for the recruits- Officers to send their recruits to the adjutant of their different Regim^t - A return to be made to the Commanding Officer of the Number of Recruits in the different Regiments

---

[2] William Moultrie was born on 23 November 1730 in Charleston, S.C. He served as a captain of Provincial Troops in the Cherokee War and was a colonel of militia regiment of horse in 1774. On 17 June 1775 he became the colonel of the 2nd South Carolina Regiment, and he was the commander at the battle of Fort Sullivan. He was promoted to brigadier general on 16 September 1776 and resigned from being the colonel of the 2nd South Carolina Regiment on 28 October 1776. On 3 September 1777 the South Carolina line was made into two brigades. Moultrie was made the commander of the 1st Brigade, which consisted of the 1st, 3rd, and 6th South Carolina Regiments. The 2nd Brigade consisted of the 2nd and 5th South Carolina Regiments. He commanded the forces at Port Royal Island and at the Savannah River when General Lincoln invaded Georgia in 1779. He was in the siege of Charleston and was taken prisoner at the fall of Charleston. He was exchanged for General John Burgoyne in 1782. On 15 October 1782 he was promoted to major general. He was elected as Governor of South Carolina in 1785, and again in 1791. He died on September 27, 1805 at the age of 75.

[3] Andrew D'Ellient (Dellient) served as an adjutant in the 2nd South Carolina Regiment during 1776. He was appointed as a brigade major under General Moultrie on 28 August 1777. He was made a prisoner at the fall of Charleston in May 1780.

[4] The officers for the three new regiments were chosen from upper class of Charleston society. Christopher Gadsden wrote that they were " most of them, Gentlemen of considerable fortunes with us who have enter'd into the service merely from Principle and to promote and give credit to the cause." The selection for the officers was a social competition that insured the support of the rich, propertied men of Charleston from the very start of the Revolution. Much of the selection also depended on their experience in the Cherokee War fifteen years earlier.

[5] The "new" enlisted barracks were built in 1757 and was constructed of brick. They were 300 feet long and 80 feet wide. They had forty-two brick chimneys, wood floors and could house 1,000 men. The officer's building was 180 feet long and 24 feet wide. It had twenty-one apartments, each of which could house six officers. These were located approximately near Rutledge and Cannon Streets in Charleston today.

[6] Possibly Edward Weyman, who later served as a lieutenant and captain in the Charleston Battalion of Artillery in 1779-1781. He was also a marshal on the Admiralty Board. He dug the graves for 271 men, attended their funerals and supplied fifty coffins. Whenever an officer was referred to as "Mr." he was normally a volunteer gentleman, or a cadet. They were treated as officers, but had not been commissioned. They usually came from a higher status in society that would prevent them from wanting to be privates.

Regimental Orders by Col° Moultrie
Every Officer to provide himself with a blue cloth coatee Land & Cuff.d with Scarlet Cloth & Lind with Scarlet- White buttons & white waistcoat & breeches (a pattern may be seen at M^{r}. Freezwans) also a cap and blk Feather

---

General Orders by Col° Wm Moultrie June 21 1775
Lieutenant Col Isaac Huger [7] of the 1st Regiment and Lieutenant Col Isaac Motte [8] of the 2d Regiment are Lieutenant Colos in the Provincial Troops & to be obeyd as Such-

Major Owen Roberts [9] of the 1st Regiment & Major Alexa McIntosh [10] of the second Regiment are Majors In the Provincial Service & to be obeyed as Such- Captains Charles C. Pinkney,[11] William

---

[7] Colonel Isaac Huger was born on 19 March 1742 and was the brother to Benjamin Huger (Huger is pronounced "Yew Gee"). He was a planter but in 1761 he served in the militia during the Cherokee War. He became a lieutenant colonel in the 1st South Carolina Regiment on 17 June 1775. On 16 September 1776 he became a colonel in the 5th South Carolina Regiment (1st South Carolina Rifles). On 3 September 1777 the South Carolina line was made into two brigades. The 1st Brigade consisted of the 1st, 3rd, and 6th South Carolina Regiments. Huger was made commander of the 2nd Brigade, which consisted of the 2nd and 5th South Carolina Regiments. On 9 January 1779 he became a brigadier general in the Continental Army. He was wounded at the battle of Stono Ferry on 20 June 1779. He was at the siege of Savannah. He was also wounded in the battle of Guilford Courthouse on 15 March 1781 and participated at the battle of Hobkirk's Hill in April 1781. He died in 1797 at the age of 55.

[8] Isaac Motte was born on 8 December 1738. When he was 16 years old he was appointed an ensign in the 60th Royal American Regiment. He served in Canada during the French and Indian War. He was elected to the Royal Assembly in South Carolina in 1772. He was commissioned a lieutenant colonel in the 2nd South Carolina Regiment and was the second in command during the battle of Sullivan's Island. He was promoted to the rank of colonel of the 2nd South Carolina Regiment on 16 September 1776 when William Moultrie was promoted to brigadier general. He resigned his commission and his command of the 2nd South Carolina Regiment on 19 September 1778 so that he could serve in the Privy Council. He was taken prisoner at the fall of Charleston and exchanged on 14 June 1781. He was commissioned a brevet brigade general on 30 September 1783. He voted to ratify the Constitution. He died on 8 May 1795 at the age of 57.

[9] Owen Roberts was born in 1720. He was commissioned as a captain in the South Carolina Provincial troops and served in Middleton's Regiment during the Cherokee War of 1760. After the Cherokee War he served as a captian in the Charlestown Provincial Artillery and was the commissioner of fortifications. He was commissioned as a major on 17 June 1775 in the 1st South Carolina Regiment. On 14 November 1775 he became a lieutenant colonel in the 4th South Carolina Regiment (Artillery). He was promoted to a colonel on 16 September 1776. He loaned the state government £5,000 in 1777. For comparison a qualified brickmason would make £70 in a year. He was court-martialed on 7 July 1777 for not being at his post at Fort Moultrie when the Marquis de LaFayette visited. He was acquitted of the charge because the court said that commanding officers would normally visit the town "whenever inclination or neccessit should prompt him" without asking for permission from a superior officer. He was killed at the battle of Stono Ferry on 20 June 1779.

[10] Alexander McIntosh had become a major in the 2nd South Carolina Regiment on 17 June 1775. He was promoted to lieutenant colonel in the 5th South Carolina Regiment (1st South Carolina Rifles) during July 1775. He died in 1780.

[11] Charles Cotesworth Pinckney was born on 14 February 1746. He was educated in England as a lawyer. He returned to South Carolina and was a member of the General Assembly. On 17 June 1775 he was commissioned a captain in the 1st South Carolina Regiment. He served under General Moultrie in the defense of Fort Sullivan. On 16 September 1776 he became a lieutenant colonel. He was promoted to colonel on 23 November 1776. He was in the battles of Fort Sullivan, Brandywine, the siege of Savannah, and the siege of

Cattle,[12] Thomas Lynch [13], John Barnwell,[14] Adam McDonald,[15] Benjamin Cattle,[16] Edmund Hyrn,[17] William Scott,[18] Rogers Saunders,[19] Thos Pinkney[20] are Captains in the First Regiment of Provincial Troops & to be obeyd as such-

---

Charleston. He was taken prisoner at the fall of Charleston and exchanged in February 1782. On 3 November 1783 he was made a brevet general and on 19 July 1798 he became a major general in the U.S. Army. He was discharged on 15 June 1800. In 1796 he was appointed Minister to France. He ran for Vice President in 1800, and ran for President in 1804 and 1808. He died on 16 August 1825 at the age of 79.

[12] William Cattell was born in 1747. He became a captain in the 1st South Carolina Regiment on 17 June 1775 and was promoted to major in the 3rd South Carolina (Ranger) Regiment in May 1776. On 16 September 1776 he became a lieutenant colonel. He was taken prisoner at the fall of Charleston. He died in May of 1787.

[13] Thomas Lynch was born on 5 August 1749 in Prince George Winyah Parish, South Carolina. He was educated as a lawyer in England and returned to South Carolina in 1772 to become a planter. He served as a Representative to the First and Second Provincial Congress. He was elected to the Second Continental Congress in 1775 and was the second youngest member to sign the Declaration of Independence at 27 years old. He had been elected a captain in the 1st South Carolina Regiment in 1775 and served at Fort Johnson. He resigned his commission in April 1776. He contracted fever during recruiting duty in North Carolina and never fully recovered his health. He was killed when his ship sank in December 1779 while he was traveling to France to seek a cure.

[14] John Barnwell was born in 1748. He was elected captain in the 1st Regiment prior to him being commissioned on 17 June 1775. He was on the expedition to capture gun powder at Blood Point in July 1775. Full details of this incident are described in "*Nothing but Blood and Slaughter, Volume One*" by Patrick O'Kelley.

[15] Adam McDonald was elected as a captain in the 1st South Carolina Regiment on 17 June 1775. He was promoted to major on 16 September 1776. He resigned on 22 April 1777. He was killed in 1777 or possibly in December 1778.

[16] Benjamin Cattell was born on 13 July 1751. He was elected a captain in the 1st South Carolina Regiment on 17 June 1775. He resigned on 20 July 1778. In 1779 he returned to the service and was an aide to William Moultrie. He was captured at Charleston in May 1780 and remained in town as a prisoner. While he was a prisoner, his estate was sequestered and his family banished. He died in 1782.

[17] Edmund Massingbird Hyrne became a major on 12 May 1779. He served as deputy adjutant general of the Southern Department fomr 17 November 1778 until the end of the war. He was wounded in Charleston on 30 March 1778 during the fighting at Gibbes Plantation. Full details of this action are described in "*Nothing but Blood and Slaughter, Volume Two*" by Patrick O'Kelley. In 1781 he became an aide de camp to General Nathanael Greene. He was in the battle of Eutaw Springs and was in charge of the exchange of prisoners on the American side during 1781. He was the 1st Major in Maham's cavalry (South Carolina State Legion) on 4 August 1782. He resigned his commission from the State Legion on 18 September 1782. He became a brevet lieutenant colonel on 30 September 1783.

[18] William Scott was born in 1725. He became a captain in the 1st South Carolina Regiment on 17 June 1775. He was promoted to major on 24 April 1777 when Major McDonald resigned. He was promoted to lieutenant colonel under Colonel Pinckney on 13 August 1778 after Lieutenant Colonel William Cattle died. During the siege of Charleston he was placed in command of Fort Moultrie along with 118 men of the 1st South Carolina Regiment and 98 Charleston militiamen. After a two day naval bombardment he surrendered Fort Moultrie on 7 May 1780. He was exchanged in June 1781 and served until the close of the war. He died on 4 May 1807

[19] Roger Parker Sanders (Saunders) enlisted in the 1st South Carolina Regiment on 17 June 1775 as a captain. He resigned on 8 October 1778.

[20] Thomas Pinckney was born on 23 October 1750. He became a major on 1 May 1778. In 1779 he was aide de camp to General Benjamin Lincoln. On 3 August 1780 he became aide de camp to General Horatio Gates and was wounded, and taken prisoner at the battle of Camden. He was exchanged in December 1780 and served until the close of the war.

Captains Bernerd Elliot[21], Francis Marion, Daniel Horry,[22] Francis Huger,[23] William Mason,[24] James McDonald,[25] Peter Horry,[26] Nichs Eveliegh,[27] Isaac Harleston,[28] Charles Motte,[29] are Captains in the Second Regiment as Provincials & to be obeyd as Such-

---

[21] Barnard Elliot was born in South Carolina and was educated in England. He returned to America in 1766 and served on the Royal Council before the Revolution. He was elected a captain in the 2nd South Carolina Regiment on 17 June 1775. He was promoted to Major of the 4th South Carolina Regiment on 14 November 1775. He was in charge of constructing a portion of Fort Johnson in December 1775. On 16 September 1776 he was promoted to lieutenant colonel and was the commander of Fort Johnson in 1777. He died in Charleston on 25 October 1778. There were two other officers named Barnard Elliot in the 4th South Carolina Regiment, all of who were related; Captain Barnard Elliot, and Private Barnard Elliot.

[22] Daniel Horry, Jr. was a rice planter and a ship owner (Horry is pronounced Orr-ee). He served in the Royal Assembly during the 1760's. He was commissioned a captain in the 2nd South Carolina Regiment in 1775. On 6 June 1776 he resigned his commission in the 2nd South Carolina so that he could become a lieutenant colonel in the Craven County militia. Later in 1776 he was a lieutenant colonel of an independent corps of cavalry, the South Carolina Dragoons. He was promoted to colonel in 1778. He was in the battle of Stono Ferry and was attached to Pulaski's Legion in the siege of Savannah. He was captured at Monck's Corner in 1780, but took parole with the British afterwards. He did this to possibly preserve his estate from destruction. His plantation located on the Santee River is the site of the present day Hampton Plantation located near McClellanville, South Carolina.

[23] Francis Huger became a captain in the 2nd South Carolina Regiment on 17 June 1775. In addition he was lieutenant colonel and deputy quartermaster-general of the Southern Department from September 1777 until he resigned in October 1778. He died on 18 August 1811.

[24] William Mason became a captain in the 2nd South Carolina Regiment on 17 June 1775. He was in the regiment until 1779, afterwards he served in the militia until 1782.

[25] James McDonald became a captain in the 2nd South Carolina Regiment on 17 June 1775. He was at Fort Moultrie on 28 June 1776. He resigned on 3 June 1777 to become a captain in the South Carolina Dragoons.

[26] Peter Horry was born in 1747. He was elected as a captain in the 2nd South Carolina Regiment on 17 June 1775. He became a major on 16 September 1776. He was promoted to lieutenant colonel in the 5th South Carolina Regiment. After the fall of Charleston he was in Marion's partisans and was in command of Horry's South Carolina Regiment. This was a unit made up of infantry and cavalry from Marion's partisans. He was wounded at the battle of Eutaw Springs on 8 September 1781, but served until the end of the war. He died on 28 February 1815 at the age of 68.

[27] Nicholas Eveleigh was born in 1748. He was educated in Scotland and received a commission as a captain in the 2nd South Carolina Regiment on 17 June 1775. He was present at the battle of Fort Sullivan on 28 June 1776. He married Mary Shubrick. He became a colonel and deputy adjutant-general for South Carolina and Georgia on 24 May 1777. He was appointed as the adjutant general of the South Carolina Regiments on 11 August 1777. He resigned his commission on 24 August 1778. He was the Comptroller of the U.S. Treasury in 1789. He died on 16 April 1791 at the age of 43.

[28] Isaac Child Harleton (Harleston) was born on 9 October 1745. He was elected as a captain in the 2nd South Carolina Regiment on 17 June 1775 (with his Continental commission on 4 November 1775). After the battle of Fort Moultrie his unit entered Continental service. He resigned his commission in the 2nd South Carolina Regiment on 9 August 1778 and transferred to the 6th South Carolina Regiment. He was promoted major on 30 December 1778. When the five regiments were consolidated into two units during February 1780 he was transferred to the 2nd South Carolina Regiment. He was captured at the fall of Charleston but served until the end of the war. He died on 10 January 1798 at the age of 53.

[29] Charles Motte was the fifteenth child out of nineteen born to Jacob Motte. He served in the Charlestown militia as an officer, and was commissioned a captain in the 2nd South Carolina Regiment on 17 June 1775 (with his Continental commission on 4 November 1775). He was at Fort Sullivan on 28 June 1776. He was

Lieutenant John Mouat,[30] Thomas Elliot,[31] Glen Drayton,[32] Richard Singleton,[33] John Vanderhorst,[34] Alexander M^c^Queen,[35] Ben^j^. Dickinson,[36] Joseph Poor, Richard Armstrong,[37] & James Ladson[38] are Lieut^s^ In the 1st Regiment & to be obeyd as Such

---

promoted to Major in September 1779. He was killed at Savannah during the fateful assault on the Spring Hill Redoubt on 9 October 1779.

[30] John Mouatt remained a lieutenant in the 1st South Carolina until he was captured in Charleston in May 1780. He was a prisoner at St. Augustine until 1 July 1781.

[31] Thomas Elliott was promoted to captain in 1777 and remained in service until April 1782.

[32] Glenn Drayton was commissioned a lieutenant in the 1st South Carolina Regiment on 17 June 1775. He was promoted to captain and resigned in August 1779, possibly due to the failed attack at Stono Ferry.

[33] Richard Singleton later served as a colonel in the militia from St. Mark's Parish and was at the fall of Charleston in May 1780.

[34] John Vanderhorst was a first lieutenant on 17 June 1775 in the 2nd South Carolina Regiment. He was promoted to captain in 1777. He was promoted to major on 9 October 1779 after the assault on the Spring Hill Redoubt at Savannah. From August 1780 to December 1782 he was in the partisans under General Francis Marion.

[35] Alexander McQueen was a first lieutenant in the 1st South Carolina.

[36] Benjamin Dickenson was commissioned a first lieutenant in the 1st South Carolina Regiment on 17 June 1775.

[37] Possibly Robert Armstrong, who was born in 1731.

[38] James Ladson was born on 7 July 1753. He was promoted to captain in the 4th South Carolina Artillery on 4 November 1775. He was a brigade major under General Howe in 1777 and was appointed brigade major of the 1st Regiment on 1 May 1779. He resigned on 8 November 1779 after the siege of Savannah. He was a lieutenant colonel in the Colleton County militia until 1781.

Lieut[s] Richard Shubrick,[39] John Allen Walter,[40] William Oliphant,[41] Thomas Moultrie,[42] Thomas Lessesne,[43] Richard Fuller, William Charnock,[44] Anthony Ashby,[45] John Blake,[46] & James Peronneau [47] are Lieutenants in the Second Regiment of Provincial Troops & are to be Obeyed as Such-

---

Regimental Orders by Col Moultrie 22[d] June
Captain Eveliegh & Cap[t] Motte with L[t] Tho[s] Moultrie, James Perronneau, & William Moultrie,[48] to remain in Town & to take charge of the recruits that may be sent to the Regiment from the country. Are also to pick up what Recruits they can ab[t] Town-

---

[39] Richard Shubrick was born in 1751. On 17 June 1775 he became a first lieutenant in the 2nd South Carolina Regiment. He was promoted to captain on 24 May 1776. He was at the battle of Fort Sullivan on 28 June 1776. He died on 8 November 1777.

[40] John Allen Walter resigned on 22 September 1775. He later became the assistant commissary of supplies

[41] William Oliphant became a lieutenant in the 2nd South Carolina Regiment on 17 June 1775. He was promoted to captain on 24 February 1776. He was in the battle of Fort Sullivan on 28 June 1776. He resigned on 20 October 1777.

[42] Thomas Moultrie was born on 24 October 1740. He was commissioned a lieutenant in the 2nd South Carolina Regiment on 17 June 1775. On 2 October 1776 he was promoted to captain and took command of Captain Huger's company on 22 October 1776. He was killed during the siege of Charleston on 24 April 1780.

[43] Thomas Lesesne became a first lieutenant in the 2nd South Carolina Regiment on 17 June 1775. He was in the battle of Fort Sullivan and on 13 August 1776 there was a court of inquiry into whether he showed cowardice during the battle. The court found that he was "was somewhat confused during the Action of the 28th June, But acquit him of Cowardice". He was promoted to captain on 22 November 1776 and took command of Captain Horry's company (the Officer's list says 16 September 1776). This company became the light infantry company. He resigned on 15 August 1779. He later served as a captain in General Francis Marion's partisans.

[44] William Charnock served as a lieutenant in the 2nd South Carolina Regiment on 17 June 1775. He was promoted to captain on 29 August 1776 and he took over Captain Eveleigh's company. He resigned on 25 November 1778.

[45] Anthony Ashby became a first lieutenant in the 2nd South Carolina Regiment on 17 June 1775. He was promoted to captain on 1 December 1775. He resigned on 16 February 1778. After becoming a captain in the militia he was wounded and taken prisoner at McKay's Trading Post in Augusta, Georgia on 18 September 1780. Colonel Thomas Brown hanged him and 12 others. A full description of this incident is in "*Nothing but Blood and Slaughter, Volume Two*" by Patrick O'Kelley.

[46] John Blake was commissioned a Lieutenant in the 2nd South Carolina Regiment on 21 June 1775. He was promoted to captain on 23 April 1776. He resigned on 25 April 1778 after his men were killed on board the Frigate *Randolph.* The fight of the *Randolph* is depicted in "*Nothing but Blood and Slaughter, Volume One*" by Patrick O'Kelley. He served as a Captain in the Charleston Militia in 1780, then served in the militia during 1781-1783.

[47] James Perronneau was commissioned a 2nd Lieutenant in the 2nd South Carolina Regiment on 21 June 1775.

[48] William Moultrie, Jr. was born on 8 August 1752. He was promoted to 1st Lieutenant on 6 June 1776. He was at the battle of Fort Moultrie on 28 June 1776. On 6 August 1776 he resigned his commission. He was appointed as the aid de camp to his father, Brigadier General William Moultrie, on 7 November 1776 with the rank of captain. He was taken prisoner at the fall of Charleston in May 1780.

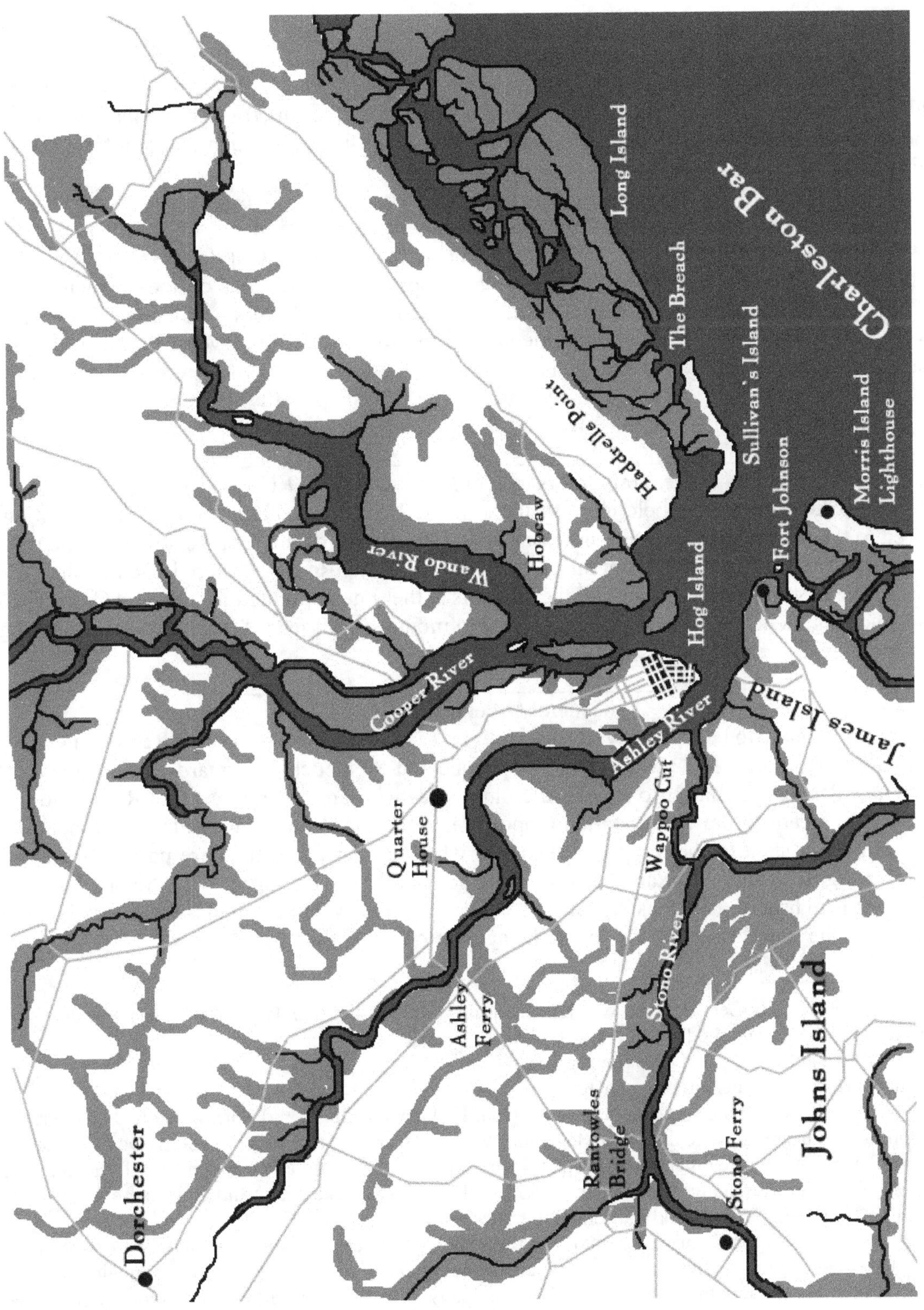
Long Island
Charleston Bar
The Breach
Sullivan's Island
Morris Island
Lighthouse
Haddrells Point
Fort Johnson
Hobcaw
Wando River
Hog Island
Cooper River
Ashley River
James Island
Wappoo Cut
Quarter
House
Stono River
Ashley
Ferry
Johns Island
Rantowles
Bridge
Stono Ferry
Dorchester

Regimental Orders by L[t] Col Motte 25 June

The Commanding Officer at the Barracks not to Suffer Either non Commissioned Officer or Soldier to be absent without leaf to be carefull he does not give Liberty to more than two at a time to be from the Barracks & to report to the Commanding Officer all such as do not appear at roll calling

---

General Orders by Colonel Moultrie 28 June

The quarter Master of the 1st & 2[d] Regiment are to apply to Col Jervais [49] or M[r] Andrew Williamson [50] who have contracted to Supply the troops with provisions on the following manner one Pound good beef p[r] Day or one pound fresh pork or 12 [oz] Salt pork, one pound wheat flour or one pound shipbread or ½ p[t] rice, ½ pt vinegar p[r] week when they are served with Fresh Provisions & one pound blk pepper pr year if to be had-

Soldiers found Drunk on Duty will certainly be punishd- Especially the sergeants. who Ought to sett good examples for the men

---

General Orders by Col Moultrie June 29 1775

All the Subaltern [51] officers belonging to the 1st & 2[d] Regiments that are not gone Recruiting are ordered to take their Quarters at the new Barracks the officers of the first Regiment on the right, & the Officers of the Second Regiment on the left side of the long Room- Ordered also that the Military Law be read morning & Evening to all the Recruits at the time appointed for the Roll Call- The first & Second Regiments of S[o] Carolina are order'd to mount Guard from 8 OClock in the morning till the same time the next day as usual till further orders- [52]

General Orders June 30 1775

The 1st & 2[d] Regem[t] to beat the Revellie at daylight each Regiment to mount a Regimental guard of 1 Serjeant, 1 Corporal, & 12 Men. the Troop to beat at 8 OClock the guard to assemble on their regimental Parades & March to relieve the old guard, who are to draw up in a Rank Entire. the Relieving Sergeant to draw up his men Opposite the old guard. the Sergeants to advance to each other, & the Sergeant of the old guard tells him what number of centries are posted & the orders. The relieving Sergeant is to examine every thing he is charg'd with particularly Prisoners-

Retreat is to beat at 6 OClock & Tattoo at 9 OClock after which time no Soldier is to be allowed to go out of the Barracks & all Soldiers to be at Quarters at that time-[53]

---

[49] Possibly John Lewis Gervais, who was born in 1741. He was the deputy paymaster general in 1778. He married May Sinclair and died on 18 August 1798.

[50] Andrew Williamson was a civilian contractor.

[51] A Subaltern was a lieutenant or an ensign.

[52] The troops garrisoned in Charleston had Town Guard, Magazine Guard and Barracks Guard. Each regiment guarded its own barracks. The different Magazines were the Cumberland Street Magazine and Colonel Henry Laurens Brick House. The other places of guard were the General Hospital; Gadsden Wharf; the house of the President, Governor Rutledge; the State House; General Howe's Headquarters; and at Broughton's, Littleton's, Craven's, Granville's, Elliot's and Laurens Batteries.

[53] There were four company formations a day in the life of an 18th century soldier. These were reveille, troop, noon and retreat. At reveille and noon the soldiers formed in the company street of the camp without arms and the roll would be called by the 1st Sergeant of the company. The men formed with arms for troop and retreat,

Regimental Orders- The quarter master is to provide some pails & a wheel Barrow for the Regiment, to keep the Barracks clean

Saturday Evening

Return of 1st July by order of Lt James Peronneau

Richard Freetop was put under arrest for quitting his post as Centry a 6 OClock, leaving his gun & going to sleep in the room-

John Gardner was discharged for his age & Infirmities By order of the Commanding Officer

General Orders 2d July

As there are a number of raw, undisciplind Soldiers at the Barracks, who are riotous & Disorderly, orderd that one Captain 3 Subalterns from Each Regiment do reside at the Barracks beginning tomorrow-

Order'd that no person be admitted into the Barracks yard but those who are brought by an officer, or those who are bringing Provisions to Sell to the Officers or soldiers

No more officers to be sent out of Town recruitg.

Mr Ramage is appointed Subtler to the Barracks.[54]

order'd that the subtler do not let any soldier have more than two Gills rum, in one day & that to be mix'd with as much water. Any Soldier who shall appear drunk shall not be allowed any Liquor by the Subtler no Liquor to be allowed to be allowed to the men before Troop Beating or after Retreat beating-

A Sergeant & 3 men to go to the contractors every morning from each Regiment for bread-

General Orders Monday Evening.

Order'd that Major Owen Roberts, Capt N. Eveliegh, Thos Pinkney Edmund Hyrne, & Lieut Thomas Moultrie, Attend to a Court Martial to be held tomorrow morng at 8 OClock at the long Room, in the new Barracks on a tryal of James Blenchfield [55] Soldier in the 2d Company of 2d Regt for wanting of Respect & abusing Capt C. Motte Commanding Officer at the Barracks Orderd also that the witness appear at the Same time

---

and then were marched to the color line to form with the regiment. The last roll call each day was held when the men were in their tents, or their beds in the barracks. This was held 15 minutes after tattoo and was conducted by the sergeants.

[54] A sutler was the 18th century version of a military post exchange or PX. The sutler sold items that the soldier could not get issued to him. This could be simple items such as needle, thread, paper and pencils, but the main thing that sutlers sold were food and drink. They would sell cookies, cakes, nuts and a variety of alcoholic beverages. Charles Ramadge, the sutler, later served as the adjutant of the Charles Town Militia.

[55] James Blenchfield was in the 2nd South Carolina Regiment. On 5 July 1775 he was supposed to be whipped at the halberds for mutiny towards Captain Motte, but he was given a reprieve. On 29 September 1775 he was supposed to be court-martialed for being drunk on duty, but he was remanded to another date since he was too drunk to appear before the court. He re-enlisted in the 2nd South Carolina Regiment again on 4 November 1775. The initial enlistments for the war were for 6 months and many did not think the war would last that long. Most of the men enlisted in May 1775, so when that enlistement ran out they reinlisted into a Continental regiment in November 1775. The second enlistment was for 30 months. On 7 July 1777 he received 50 lashes for quiting his post.

General Orders July 5 1775

Order'd that the 1st & 2d So Carolina Regiments of Foot use the manual Exercise as it is directed by a book printed for the use of this Colony as agreed to & Established By the field Officers in the Service of this Colony with Instructions for young officers by Colo Wolf [56]

James Blenchfield private man in the Second Regt who was Confind for mutiny towards Captain Motte & tryd by a Court Martial the 4th Inst. & condemn'd to be whipt was brought to the halbert to receive his Condemnation was Reprievd by order of the Colonel-[57]

General orders

Officers of the 1st & 2d Regiments to provide themselves with Tents Lt & Majors Tents to be 10 feet ridge pole & 8 ft upright pole Captains & Subalterns Tents 8 ft ridge pole & 7 ft upright pole. the two Subalterns of each Company to one Tent

It is order'd that nobody whatsoever shall ride in or out of the Barracks

---

General Orders

One Captain to be Appointed for the Day beginning by the Eldest Captain of the 1st regt. who is to make his report every morning to the Commanding Officer on the Parade-

One Subaltern Officer to be appointed for the day beginning by the Eldest of the 2d Regiment who is to See orders kept in the Barracks & visit the Soldiers Quarters frequently & not allow a person to sell rum to the men (except the Subtler) & make his report every morning to the Captain of the day on Parade-

The Captain of the Day & Subaltern on Duty to be relievd every morning at 8 OClock. the Captain to be relievd by the Eldest of the first Regt & So on till the Whole has taken their turn on Duty-

No Horses to be admitted in the Barracks yard

General Orders 20 July

A Return of the arms receiv'd from the Publick by the 1st & 2d Regts to be made to the Commanding Officer & Also what Arms are Wanted for each Regiment [58] Order'd that 1 Subaltern (the next on duty) 1 Sergeant 1 Corporal & 13 Privates from the 1st & 1 Sergt, 1 Corporal & 12 Privates from the 2d Regiment to Parade at Retreat Beating & mount Guard at the old Magazine[59] that proper Centries be posted about the old Magazine. this guard to be relieved tomorrow at retreat beating, by 1 Subn next on duty from the 2d & 1 Sergt. 1 Corpl. & 12 Privates- One Sergt. 1 Corp. & 13 Privates from the 1st regt - this guard to be continued till Countermanded.

---

56 This was the "Instructions to Young Officers" by Major General James Wolfe that was published in 1768.

57 The halberd was the traditional symbol of the non-commissioned officer. It resembled an ax on the end of a long pole. Though they were not carried in battle anymore, their symbolism was still apparent. If a soldier was "reprimanded at the halberds" he was stripped down, placed in the stocks and punished in front of the company.

58 The musket shortage was so acute that notices appeared in the South Carolina Gazette. The notice stated that "anyone who had any public muskets should immediately turn them in to John Calvert, town armorer."

59 The magazine was built in 1758 and was designed to hold 40,000 pounds of gunpowder. It was 33 feet long and the walls were two feet thick.

Proper Centries to be posted & relieved & the usual time. 1 Orderly Sergeant to attend the Commanding Officer every Day beginning by the first

Parole Liberty [60]

The Magazine guard to Parade the grand Parade at 6 OClock. to march off & relieve the old guard. the Reliev^{g} Officer is to draw up his Men Oposite the old Guard two deep or a Rank Entire (or as the old guard may be Drawn up) then the guard to Rest. the Officers advance to one another Paying the Compliment of the Hutt. & Officer of the old guard tells him what number of Centries are posted & his orders &^{c} the Sergeants. Corpol^{s}. & Drummers of the old guard. at the same time deliver their orders to those of the new guard. after which both Guards Shoulder. & the Corporal of the new Guard numbers off the men & draws out the number of Centries to be posted & march with the Corporal of the old guard to relieve his officer of the guard to have the Roll called frequently & Exercise his guard Morning & Evening All guards to turn out with rested arms to the Command^{g} Officer once a Day & Centries to rest to the Field Officers. No Centry to quit his Arms or Sett down at his Post & not to allow any Horse near his post. all Centries to Challenge after dark. when they challenge & are answer'd rounds or Patroles pass rounds. or Patroles or Stand rounds & Call the Serg^{t} of the guard to turn out the guard, after which he is not to suffer the rounds to Advance till orderd by his Officer. as Soon as the guard is turnd out The officer is to send 1 Serg^{t} & 4 men to relieve the Rounds when they get within 6 or 8 paces the Sergeant is to Challenge briskly- who comes there- they Answer- Rounds- the Sergt again challenges- <u>What Rounds</u>-they Answer- Grand Rounds - or Visiting rounds-[61] the Sergt answers- <u>Stand Rounds</u> advance Sergt with the Patrole- upon which the Serg^{t} of the escort advances alone & gives the Serg^{t} the Parole (of the guard) in his Ear. & with such caution as not to be heard- after receiving the Parole he orders the Serg^{t} to Return to his Escort & Leaving his men to Keep the round from advancing. goes to his Officer & Delivers him the Parole he received from the Serg^{t} which the officer finding Right sends the Serg^{t} back to his Men & calls out- <u>advance Rounds</u>. Upon which the Serg^{t} of the guard orders his four men to wheel back from the Center & form a lane thro' which the officer who goes the rounds is to pass the escort remaining where they where & goes up to the officer who gives the Parole to the grand Rounds. but visiting rounds are to geve the Parole to the officer of the guard. the Captain of the Day to visit the guard & take a Corporal from either guard at Barracks & two men from each guard to go the Rounds with him- The Centries at the magazine are to admit the officers or Sergeants for the Artillery Company within the wall of the magazine or any person bringing an order or note from them If any Centries are taken Sick on his post he must send word to the guard that he may be relieved

---

[60] The parole was the password of the day. Sometimes in the orderly book they had a countersign listed. When on guard and an individual approached, the guard would have the person say the parole. If the person approached and asked for the parole first, the guard would ask for the countersign. The paroles that they used tended to show what were the current events of the day or the talk of the town.

[61] The officers of a guard would visit their guard posts to determine if they were being run correctly. A "visiting round" was when the officer of the guard post, usually a lieutenant, would check to see if his sentries are posted and all is in order. A "grand round" was when the officer of the day, usually a captain, would check all the guard posts. When an officer checked on visiting rounds he would tell this to the sentry, and the parole would then be given by the officer to the sentry. When an officer checked grand rounds, he would also announce this and the parole was given by the sentry to the officer.

General Orders 21st July

Parole Charles Town

Order'd that the guard from the 1st & 2d Regts do mount guard at the old magazine beginning tomorrow at 8 OClock in the morning Precisely.[62] therefore the two Divisions shall meet together at the grand parade at half past Seven oClock in the morning to be joind in Readiness to march to their guard. orderd also that the Sergt & Corps of each Compy in both regiments take a Survey of the Private men to see if their Arms & Cloaths are as Clean as Can be Expected

---

General Orders July 23d

Parole Boston

The Sub. Officers of the 1st & 2d Regts of Foot that are not on duty are orderd to Attend the drilling of the men of their different Companys morning & Eveng at the usual time

All officers on their return from recruiting are Immediately to wait on the commanding officer-

The Adjutant of the 1st & 2d Regts of foot are orderd to acquaint the day before, the officer that is to be on Duty the next day in the Regimental Orders-

All the Officers of the 1st & 2d Regiments that have advanced money for Recruiting are required to give their Accts to the commanding Officer-

---

Parole America 24th July

Parole Cambden 25th July

Gave furlow for a month to James Campbell,[63] Wm Evans-[64] & Patrick McCan[65] to go in search of deserters

Parole Chatham 26 July

General Orders 27th

Parole Effingham

The first & Second So Carolina Regiments of Foot are Ordered to Turn out their men every morning at 5 OClock to be drilld till 7. Also in the afternoon from 5 till 7 OClock. the roll to be calld particularly every Day at 7 in the morning & at 5 in the afternoon-

---

[62] In July the Council of Safety ordered the regulars to assume the mission of guarding the town's powder supply, which was being stored in the old magazine on Queen Street. The guard would consist of 1 Sergeant, 1 Corporal and 13 Privates from the 1st South Carolina Regiment and 1 Sergeant, 1 Corporal and 12 Privates from the 2nd South Carolina Regiment.

[63] James Campbell enlisted in the 2nd South Carolina Regiment on 4 November 1775. On September 29th, 1775 he received 39 lashes for sleeping on guard duty. He was appointed a regimental barber on 24 January 1778. On 8 July 1778 he was promoted to corporal and on 18 November 1778 he was promoted to sergeant. He was court martialed for drunkenness and demoted to private on 26 March 1779. He was in Captain Hall's company in 1779.

[64] William Evans enlisted in the 2nd South Carolina Regiment on 4 November 1775. He reinlisted on 1 February 1777. On 6 February 1777 he was given 50 lashes for being absent without leave. He was discharged on 30 June 1777.

[65] Patrick McCann reenlisted in the 2nd South Carolina Regiment on 17 February 1777. He also served as a post rider under Lieutenant Colonel Samuel Hammond in 1780-1781.

The quarter master of each Regiment to take care that the Chimneys in the Barracks to be kept clean-

Ordered that the Rules & articles for the Government of the military Forces of this Province be read once a month to each Regiment

Every Guard to Report to the Captain of the Day after they are relievd. in which report is to be mentioned the Parole the number of Centries. the time of the Rounds going & whatever happens extra ordinary. the Officer signs it Specifying his Rank & Regiment & their Reports to be deliverd to the Captain of the Day who delivers them to the Commanding officer with the parole-[66]

Parole Granby 28 July

General Orders in the Evening

Orderd that each Company in the 1st & 2d So Carolina Regts of Foot turn out their men with their Blankets. Bedding Arms & Cloathing every Saturday at nine OClock. if a fair day. if not the next fair Day to see that their Arms & Cloaths are kept clean & to air them

Parole Rockingham 29th July

General Orders by Col Moultrie

Parole Gloucester 30th

Parole Cumberland 31st

Order'd that the 1st & 2d South Carolina Regts of Foot do read tomorrow morning at 6 OClock the Rules & articles for the Government of the Military Forces of this Province. & that the Officers be present at the Same-

Order'd the two Sergts Guard of the 1st & 2d Regts be Joined & one Subaltern to take the Command. this guard to be mounted between the pump & the gate five tents to be pitchd for that purpose. this guard to be as a Barrack guard & to Send 1 Corporal & 2 men between every Relief after Tatoo beat round the Barrack yard to prevent the men sleeping out of their barracks at night.[67] beginning tomorrow no orderly Subaltern to be appointed till future orders

Parole Manchester 1st Aug

---

[66] The first sergeant of each company would fill out his forms at reveille. This would include information about his men; who was present, who was absent and who was sick in the hospital. The sergeant would then present the form to the company commander for review, and then both of the men would sign the form. Around 10:00 the duty drummer would beat "First Sergeant's call" and the 1st Sergeants of each company would bring their forms to the adjutant's tent. The adjutant would then use these forms to make his regimental muster form. He would then give this form over to the regimental commander for signing. Around the same time the brigade duty drummer would beat "adjutant's call" and all the adjutants of the different regiments would bring their forms to the brigade major. They would also bring their orderly books so that they could write down the day's orders. The Brigade Major would read to the adjutants the orders of the day from the Brigade and General officers and have the adjutant's write down the orders in their books.

[67] In "A Classical Dictionary of the Vulgar Tongue" written in 1785 it lists that Tatoo is the "beat of the drum, or signal for soldiers to go to their quarters, and a direction to the sutlers to close the tap, and draw no more liquor for them; it is generally beat at nine in summer and eight in winter." Tatto became known as Taptoo.

General Orders 2^{d}
Parole Richmond

Orderd that all Horses that are in the Barracks be Sent away & for the Future none are to be admitted after this Day into the yard-

Mr Charles Ramage Subtler to the 1st & 2^{d} Regiments Shall have no demand paid for what he may Entrust to the Said Regiment-

General Orders
Parole Asaple 3^{d} Augst

The Officers on Duty for the future to wear Sashes

Parole Torrington 4th

General Orders
Parole Saville 5th

Order'd that 4 men be Employ'd from each Regiment every Day under the Command of a Corporal by the Direction of a Quarter Master. to keep the Barracks yard Clean & in good order-

---

Parole Abington 6th

Parole Scarborough 7th
Regimental Orders

Orderd that every company give a return of the Absent or Deserted

General Orders-

Order'd that all officers of the 1st & 2^{d} Reg^{ts} attend Tomorrow at the parade at 7 OClock-

The Adjutant of the 2^{d} Regiment is to give the discription of the men that are absent without leave or deserted-

Parole Radnor 8th Augst

Order'd the quarter master of Each Regiment to have 3 Private places on the green for the use of the men of both regiments. that may be mov'd from place to place when necessary-

Order'd that each Comp'y furnish the Pay Master of the 2^{d} Regiment with a pay Bill for next monday which must be deliverd this Saturday

General Orders
Parole Warren 9th

Order'd that a Court Martial be held tomorrow morning at 9 OClock- for the Tryal of several prisoners of the 1st & 2^{d} Reg^{ts} Confind for absence without Leave & other Misbehaviour. as Shall be presented before the Court- Orderd that all Evidences attend

President
Captain Francis Marion.

Captain Adam McDonald	Nicholas Everly
Lt John Vanderhorst	Richd Fuller

Results of the court martial on the following Tryals

John Burke.[68] of Cap Wm Scotts Comp^y confind for Drunkinness & Insolent behaviour. was by opinion of the Court Martial found guilty of the above & adjudged to Receive 50 stripes on his bare back. but by mercy of the Colonel had 20 omitted & rec^d the Rest

Gabriel Martin-

of Capt F Hugers Comp^y Confin'd for absence without leave by Capt Motte was by opinion of the court Martial found guilty of the above & condemmed to Receive 200 Lashes on his bare Back. but recommended to mercy. by the goodness of the Commanding Officer 150 were Omitted NB [69] Reprievd on condition of his paying £10 for bringing him back

Will Miller [70]

of Captain Scotts Comp^y confind by L^t McQueen for Disobediance & Insolence was found guilty of the Above by his own Confession & by the opinion of the court adjudged to Receive 100 stripes on his bare back but by the mercy of the Commanding officer was Repreivd

General Orders 10th Aug

Parole Craven

Order'd that no guns be Loaded by the guards mounted at the Magazine or Barracks without Necessity or further orders. & that the Sub^s & Serg^ts that are mounted do take great care that the number of Cartridges deliverd the men be redeliverd at their Return to the Parade to the person appointed as also strictly to Examine their Arms-

Orderd also that the roll be Calld at 5 OClock in the morning instead of 7- Any person apprehending a Deserter from Either Regiment shall Receive £10 army reward & all reasonable charges

Orderd also that 1st & 2^d Reg^ts of Foot do give in an Exact return of all deserters & the place they were Inlisted-

The officers of each Regiment to provide themselves with gorgets [71]

---

[68] John Burke had enlisted in the 1st South Carolina Regiment. On 9 August 1775 he was supposed to receive 50 lashes on his bare back for drunkenness and insolent behavior, however he only received 20 lashes due to the mercy of the colonel.

[69] This is Latin and means *Nota Bene*. It means "mark well" and is used to designate something important.

[70] William Miller enlisted in the 2nd South Carolina Regiment in 1775 in Captain Scott's company. On 9 August 1775 he was confined by Lieutenant McQueen for disobedience and insolence and he was found guilty by his own confession. He was sentenced to receive 100 lashes on his bare back, but he was reprieved by the commanding officer. He reenlisted on 4 November 1775 under Captain Daniel Mazyck. He was in the battle of Fort Sullivan. On 13 February 1777 he received 50 lashes for misbehavior, but instead he was reprimanded and pardoned. He was later in the militia and was at the fall of Charleston in May 1780.

[71] A gorget (pronounced gore-jay) was a metal plate in the shape of a half moon that officers wore about their necks. The plate may be made of gold, or brass, but the most common were made of silver. Some had symbols of their units engraved or embossed upon them. There are some myths of the gorget that tell of its use as armor against neck attacks, but this is not what it was used for. The gorget was an identification symbol for an officer and was the last vestige of armor from the days of knighthood. The 1st South Carolina gorget for officers had a symbol of crossed scimitars with the words "Ultima Ratio" in a scroll under the swords. The gorget of the 2nd South Carolina had a symbol of a "pile" of colors, muskets and a drum, and underneath are the words "Libertas, Potior, Vita" in a scroll.

Parole Ponsonby 11th Augst

Parole Spencer 12 -do-

On Monday next being Pay day an Officer from each Compy is to attend at the Paying of the men

Orderd that the Captain of the Day do frequently visit the Hospitals to see that nothing is wanting for the sick & that proper Necessaries are provided for them

Parole Exeter 13th

Parole Stanhope 14

General orders

Orderd that 2 Captains 4 Subalterns 6 Sergeants 6 Corporal & 100 Rank & file to hold themselves in Readiness to march- 1 Cap. 2 Sub. 3 Sergts & 3 Corps. 50 rank & file from each Regiment the first Regiment to march this Evening. the Captn will receive his Orders from the Commanding Officer. The Party of 2d Regt to remain in Readiness till further orders each party to be Served with 12 Rounds of Cartridges- 3 good flints. a Cartridge Box. a gun & Bayonet-

Officers for this Service are-

From the 1st Regt	From the 2d Regt
Cap. Wm. Cattle	Cap. Francis Marion
Lt Armstrong	Lt Oliphant
Lt Ladson	Lt Mazyck [72]

To Captn William Cattell
of the first Regiment of Foot – By Col. Wm Moultrie

You are to proceed with all expedition with the men under your command to Beaufort. Port Royal there to assist in Defending that place & take Charge of the Publick Powder- on your march you are to take care to Quarter your Men in some convenient Place or Plantation & to be careful that they do not injure the Inhabitants on their March- when you come to Port Royal you are to Quarter your Men in some Convenient Place altogether if you can; taking care to have proper Centries & that your men do not leave their quarters upon any pretence whatsoever without your leave, you should move of at 4 OClock in the morning & march till 9 OClock. halt till 5 in the afternoon then march till 7. if it be Cloudy weather you may March later in the ~~Evening~~ morning & earlier in the Evening. If you should receive accounts from Captn Barnwell of all things being well at Port Royal & that they will have no Occasion for your Assistance you must return to Town with all Convenient Speed taking care to keep strict Discipline & good order

---

On 9 July 1775 a joint South Carolina and Georgia naval force captured 16,000 pounds of powder from a Royal Navy supply ship. The powder was divided between the two colonies and 5,000 pounds went to South Carolina. South Carolina's powder was

[72] Daniel Mazyck was born in 1747. He was commissioned a second lieutenant in the 2nd South Carolina Regiment during 1775. He was promoted to first lieutenant during January 1777. He was promoted to captain on 13 July 1777 and took charge of Captain McDonald's company when McDonald resigned. He was captured at the fall of Charleston but was exchanged during June of 1781. Afterwards he served under General Francis Marion.

taken to Tucker's Island where 4,000 pounds were put on board a schooner and delivered to the Congress in Philadelphia. The powder was used in the siege of Boston and the invasion of Canada. The other 1,000 pounds and a supply of "salt-petre, sulphur, blankets and plains" were escorted back to Charlestown. At the same time there was a report that the British out of St. Augustine were going to mount an expedition to take back the powder and military stores. The men in the detail mentioned in the order were used to escort the powder and protect it from attack. No attack occurred.[73]

Parole Archer 15th

Ordered that General Orders be given on the parade to both Regiments at the same time

---

Parole Stamford 16th

Parole Port Royal 17th

Ordered that a Sergeant in each Company do bring to the Surgeon a List of the Sick-

Regimental Orders-

The officers of each comp^y to take all of their men that are not on Duty or Sick and exercise them every morn^g. & the Serjeant Major to take out all the Serj^ts that are of Duty & not Sick & Drill them-

The grand Squad to be formed only in the after^n.

Orderd that the Quarter Masters of each Reg^t do take care & visit the Suttlers Quarters frequently & to examine the Quality of his Liquors. & See that he does not impose on the men in his Prices or by selling bad Liquor-

Parole Congress 18 Aug^st 1775

General Orders

Ordered that 1 Captn 1 Corporal 1 Drummer & 13 private men from the 2^n Reg^t 1 Sub: 1 Serg^t & 12 private men from the 1st Reg^t do turn out at 6 OClock this Even^g. to reinforce the Magazine Guard & there to stay till further Orders

Parole Cholmondly 19th

Parole Windsor 20th

Parole Carolina 21st

Orderd that all Recruiting Officers deliver their Attestations to the Commanding Officer

1775 Aug 22nd

Parole Saluda

Parole Cambridge

Ordered That the parties that were under marching orders is discharged to do Duty as usual-

---

[73] Patrick O'Kelley, *"Nothing but Blood and Slaughter" Military Operations and Order of Battle of the Revolutionary War in the Carolinas, Volume One 1771-1779,* (Booklocker.com, Inc 2003), p. 35.

Regimental orders
To Serg[t] Grey of the 2nd Reg[t] [74]
You are Ordered with Corporal M[c]Leland & Ten privates to march 10 Miles above Moncks Corner to attend 3 Wagons that are under you Charge & See that nothing is taken Out or put in the Said Wagons, When you arrive at the place you are to go to you'll Return to the head Quarters &[c] &[c] [75]

24th Parole Newport
Order'd That every Company do appoint 2 men of each Company as pioners or Workmen for doing the Labor that Shall be required & they Shall be exempted from Guard-

Parole New York
25th Orderd that all the Sergeants & Corporals of The first & Second Reg[ts]: are to be Drill'd together once a Day on the parade

26th Parole Philadelphia

27th Parole Maryland

28th Parole Virginia
The Rev[d]. M[r] Purcel [76] will perform divine Service tomorrow Morning at 7 OClock at the New Barracks Orderd that all Officers & Soldiers do attend the Soldiers are to take care that they appear Clean & Decent with their hir Comb.[77] any Soldier who Shall appear Drunk will Certainly be punished

29th Parole Brunswick
Order'd that the guard be Releived at 9 OClock at which time all deliquents under Sentence are to receive their punishment the Roll to be Called at half past four OClock

30th Parole Georgia

---

[74] William Gray was a sergeant in the 2nd South Carolina Regiment. He was at the battle of Fort Moultrie on 28 June 1776. He was discharged on 12 August 1776.

[75] William McCleland enlisted in the 2nd South Carolina Regiment on 4 November 1775. He was promoted to sergeant on 30 April 1776 in Captain Motte's company.

[76] Henry Purcell was born in 1740. He was commissioned in the 2nd South Carolina Regiment on 8 May 1776 as the regimental chaplain. He was the deputy judge advocate general for South Carolina and Georgia from 3 April 1778. He also became the South Carolina brigade chaplain under General Isaac Huger from April 1779 to July 1781. He died on 24 March 1802 at the age of 62.

[77] The commander constantly had to order the men to conduct personal hygiene. One observer noted that some recruits "when at home, their female relations put them to wash their hands and faces, and keep themselves neat and clean, but being absent from such monitors; through an indolent heedless turn of mind, they have neglected the means of health, and have grown filthy and poisoned by their constitution by nastiness."

31st Parole Mecklinburgh

Order'd that the first & Second Regts each do form one Granadier Company of 50 Men each & that the Captains have the Choice of the men from their Respective Regts after the Granadier Compy is formed,[78]

Ordered that the officers do take the men they Recruited to their Own Companies & if any Should have more than 50 privates then so many as is above their Compliment must be Draughted out of their whole number to be thrown into those Companies are to be Drawn for the Supernumery Men [79]

Regimental Orders

Capt Barnard Elliott is appointed Capt of the Granadier Company [80]

Sept 1st Parole General Ward

2nd Regimental Orders

Ordered that the Sergeants of each Compy do Call the Roll after Tatoo beating to See that none has absented since the Roll at half past four OClock in the afternoon and if any are found out off the Way to take down his name to be reported to the officers of the company to bring the Defaulters to punishment & the Serjeants failing in so doing Shall be punished accordingly

---

3d Parole General Lee

General Orders

As the Blue house Tavern opposite the Barrack yard is a Great Nuisance to the good Order & Discipline of the Soldiers, Order'd that if any Noncommissioned Officers or Soldiers do presume to go into that House Without leave of the Capt of the Day or the Officer of the Barrack Guard that Such noncommissioned officer or Soldier be immediately confined & they may Depend on Being Punished

Regimental Orders

Orderd that the Officers of every Company do See their men Keep themselves Clean decent with their Hair combd & Dressed in a Soldier like manner They that have long Trowzers to have them made into Breeches

4th Parole General Putnam

For Duty tomorrow Capt Barnard Elliott

Order'd that a Court Martial do Sett this morning to try Such prisoners as may be Brought before the court to Sett at Eleven OClock all evidences is Desired to appear-

---

78 The commanders of the grenadier companies were allowed to choose men from all the other companies. The grenadiers would be an elite hand picked force available for the most difficult missions. South Carolina was the only state to have grenadiers in their continental regiments until the Congressional reorganization in late 1779.

79 A good portion of the men in the 1st and 2nd South Carolina Regiments were sailors who were out of work. Two British warships blocked the harbor and nothing could get in or out. Moultrie wrote, "The sailors had no alternative, they must either inlist or starve, as the men of war had completely blocked up the port; the sailors were discharged from the vessels."

80 Bernard Elliot was commissioned a captain in the 2nd South Carolina Regiment in 1775.

President
Capt Thomas Lynch
Lieut Jno Blake   Lieut Armstrong
Lieut Wm Charnock   Lieut McQueen

The officers of each Regiment to provide themselves with gorgets

1st South Carolina Gorget   2nd South Carolina Gorget of Alexander Hume

5th

Parole Newton
General Orders

A Return to be made immediately of all the Arms that have been Received by the 1st & 2nd Regiments and what number is fitt for service in each Regiment, According to the Sentence of Yesterdays Court Martial the undermentioned Prisoners received their Punishment, James Reid charged with stabing Daniel Munrowe with fix'd Bayonet,[81] was found Guilty of the above Charge & Sentenced by the Court to Receive 100 Lashes But Received only 50 by order of the Commanding Officer [82]

Conrad Fitner Confined for Drunkenness was Sentenced to Receive 39 Lashes which was executed [83]

---

[81] James Reid (Reed) re-enlisted in the 2nd South Carolina Regiment on 4 November 1775 and re-enlisted again on 1 November 1779 in Captain Thomas Hall's company. On 12 April 1779 he was appointed the gunner's mate to Fort Sullivan. He was at the siege of Savannah.

[82] Daniel Munroe was in the 1st South Carolina and reinlisted in that unit on 4 November 1775. He was discharged on 1 January 1776.

[83] Conrad Fitner served as a fifer in the 2nd South Carolina Regiment during 1775. On 5 September 1775 he received 39 lashes for drunkenness. A month later on 3 October 1775 he was remanded for lack of evidence on an unnamed charge. On 18 May 1776 he received 150 lashes for an unnamed charge. On 14 July 1776 he

Richard Kersley Confined by Lieu^t Elliott for Sleeping on his post when Centry & Suffering his arms to be taken away was Sentenced to receive 200 Lashes, But had 100 Remitted & Received the other [84]

6th Parole Burke

General Orders The quarter master of Each of the Regiments to draw provissions for the women at the proportion of two to a Company those that are the most orderly & have Children are to be prefered, Orderd that not more than Twenty Men of Each Reg^t do have leave of absence and not then 20 days furlow to be given to any soldier at one time[85]

7th Parole Hillsborough

8th Parole Eugene

9th Parole Quebec

According to the Sentence of the Court Martial of 7th Instant [86] John Gregg [87]/ Corporal/ was turned in the Ranks John White [88] Ditto was Ditto
Unas Dart a private was Released after 3 days Confinement. Sergt McDowell reduced to a private according to Sentence[89]

---

was supposed to receive 200 lashes for "riot", but he was pardoned due to "sore Leggs". On 16 August 1776 he was supposed to receive 200 lashes for being absent from duty, but only received 120 on account of a bad leg. On 24 November 1777 he received 50 lashes for misbehavior. He was supposed to receive 20 lashes on 20 May 1778 for abusing the fife major, but they were remitted. He was under Captain Thomas Hall in 1779. On 26 March 1779 he received 25 lashes on the bare back with a cat of nine tails for disobedience of orders. On 19 May 1779 he received 25 lashes for leaving ranks. He was at the siege of Savannah.

[84] Richard Kersley reinlisted in the 2nd South Carolina on 4 November 1775.

[85] Women and children followed all the different armies at the time of the 18th century, and became known as campfollowers. Most were families of the soldiers and a soldier would split his rations with them. Some were refugees, mingling with the army for protection. Some were there to earn money. As with all armies in history there was an illegal trade of gambling, alcohol and prostitution, but there was also legitimate businesses such as laundry. An "authorized" woman of the army was one that did such jobs as the laundry. One of the myths of the time was that the women were nurses, but there were very few women who were used as nurses during the Revolutionary War.

[86] The term "Instant" was used to clarify dates. If someone in the 18th century wrote "the 7th Instant", it meant that it was on the 7th of the current month. If the word "ultimo" is used it meant the prior month.

[87] John Gregg served in the 2nd South Carolina Regiment. On 9 September 1775 he was returned to the ranks, which meant being reduced to private. He became a conductor of wagons in 1780 and was in the militia in 1781.

[88] John White served in the 2nd South Carolina Regiment. On 9 September 1775 he was reduced to private. In 1778 he was in Jacob Shubrick's company. He received 100 lashes on 7 February 1778 for sleeping on his post. When Captain Shubrick died his company was known as the 2nd Vacant Company since it did not have a captain to command it. After the 2nd Vacant Company was disbanded on 5 October 1778 John White was assigned to Captain Charnock's company. He was later under Captain Adrian Proveaux and served till 1780

[89] John McDowell re-enlisted in the 2nd South Carolina Regiment on 4 November 1775. He received 39 lashes and was mult a weeks pay on 14 December 1776 for drunkenness. He received 50 lashes on 7 June 1777 for

10th Parole Montreal

11th Parole General Schuyler

General Orders

Ordered that if any Soldier is Seen in Town after Dark without leave in Writing from any officer he may Depend upon being Severely punished

12th Parole Cherokee

For Duty tomorrow Capt McDonald

General Orders

Orderd that a court martial do Set this morning at 11 OClock to Try Such prisoners as Shall be Brought before them Witnesses to attend

President

Capt Lynch

Lieut Shubrick [90] Lieut Mouatt

Lieut Harlston [91] Lieut Doharty [92]

A Court of Inquiry to Sett tomorrow morning at 10 OClock to Examine into the conduct of Capt Chas Motte when in town last Sunday Night wh a party of men on an Information from Some of the Inhabitants Wittness Mr Bentham,

President

Major Alexr McIntosh

Capt Marion Capt Saunders

Capt D. Horry Capt Thos Pinckney

13th Parole North Hampton

For Guard tomorrow Lt Perroneau & Lieutanant Harleston According to the Sentence of the Court Martial Patrick Johnston of Capt Lynch's Company in the 1st Regiment received 20 Stripes for not appearing decent and absent from Roll calling received his punishment[93]

---

refusing to do his duty. He received 50 lashes on 30 September 1777 for being drunk on guard. On 30 June 1778 he was promoted to sergeant. He was in Captain Harleston's company in early 1779. On 20 January 1779 he was reduced in the ranks for repeated drunkiness when on guard in town. Francis Marion approved the sentence, but limited the punishment. In 1779 he was under Captain Peter Gray at the siege of Savannah

[90] Jacob Shubrick became a second lieutenant in the 1st South Carolina Regiment on 17 June 1775. On 10 May 1776 he became a first lieutenant in the 2nd South Carolina Regiment and was in the battle of Fort Sullivan on 28 June 1776. He was transferred to Captain Lesesne's company on 16 July 1777. He was promoted to captain on 21 October 1777 when Captain Oliphant resigned. He died on 27 April 1778.

[91] John Harleston was commissioned a lieutenant in the 2nd South Carolina Regiment in August 1775. He resigned his commission in the 2nd South Carolina on 27 January 1777.

[92] I am unable to determine what unit James Doharty was in. He was promoted to captain and was killed near Beaufort in 1779.

[93] Patrick Johnson reenlisted in the 1st South Carolina Regiment on 4 November 1775.

Daniel Lockhart of Cap[t] Hugers Company in the 2nd Regiment for absence & Drunkness received according to Sentance 39 lashes.[94] Bunker Thring of Capt Mottes Company in 2nd Regiment for abusive language and absence at Night- Received 39 lashes,[95] Tho[s] Shores[96] Ja[s] Anderson[97] W[m] Huper Jones[98] John Cody[99] John Reily[100] & Tho[s] Green[101] for Small offenses were dismissed

---

[94] Daniel Lockhart reinlisted in the 2nd South Carolina on 4 November 1775. He died on 1 July 1776.

[95] Bunker Thring was in the 2nd South Carolina in Captain Motte's company. On 13 September 1775 he received 39 lashes for abusive language and being absent at night. On 4 September 1776 he received 100 lashes for an unnamed offense. On 25 November 1776 he was mult a weeks pay for drunkenness and striking Sergeant Simpson. On 13 December 1776 he received 100 lashes for neglect of duty. On 1 December 1777 he was confined for three day on bread and water, and received 50 lashes for being absent from Guard duty. He was in Captain Theus company in the 1st South Carolina Regiment during 1780.

[96] Thomas Shoore (Shors) was in the 2nd South Carolina Regiment. On 13 September 1775 he was reprimanded for "small offenses". On 2 August 1777 he was supposed to receive 50 lashes for being absent from roll call and being drunk, but he was pardoned. He was in Captain John Blake's company in 1778.

[97] James Anderson reenlisted in the 2nd South Carolina Regiment on 4 November 1775 as a corporal. He was promoted to sergeant on 22 July 1777 in Captain Ashby's company. He was demoted to private on 11 December 1777 for "scandalous and infamous behavior." On 19 January 1778 he was sentenced to spend eight days in the "Black Hole" for being absent without leave. He had guard every other day while he was in the Black Hole. He was reprimanded on 17 April 1778 for being absent without leave. He was discharged on 22 July 1778. He served in Marion's Partisans after the fall of Charleston. He was wounded on 8 May 1781. He served 230 days under Lieutenant John Piercey.

[98] William Huper Jones reenlisted as a corporal in the 2nd South Carolina Regiment on 4 November 1775 under Captain Charles Motte. He was reprimanded on 24 June 1777 for being absent from roll call. He was found guilty of an unnamed offense on 14 October 1777, but he asked for a pardon and was granted it. He was reduced to private on 13 January 1778, but he was promoted to corporal again on 20 July 1778 in Captain Moultrie's company. He was confined for neglect of duty and reprimanded on 20 November 1778.

[99] John Caddy is also spelled Jean Caddet, John Cadet, Cadday and Cade. On 16 November 1775 he was sentenced to 200 lashes for absenting himself all Night from Guard, but fifty of the lashes were remitted. On 15 December 1775 he was sentenced to 300 lashes for some unnamed offense. On 10 May 1776 he was sentenced to 100 lashes for an unnamed offense, but it was remitted. On 19 May 1776 he received 150 lashes for drunkeness. On 27 May 1776 he received 250 lashes for drunkiness. On 16 July 1776 he was sentenced to receive 500 Lashes and be drummed out of the regiment for being absent without leave, however his sentence was remitted and he only received 200 lashes. He was supposed to receive 50 lashes for desertion, but he was pardoned. He re-enlisted on 22 May 1777 in Captain Shubrick's company. After Captain Shubrick's death he served in the 2nd Vacant Company until it was disbanded on 5 October 1778. He was then assigned to Lieutenant Colonel Marion's Company. He was under Captain Adrian Proveaux in 1779. He deserted on 10 February 1779 then returned and was court martialed on 9 April 1779. He received 100 lashes on the bare back with switches four different times for this offense. He was in the assault on the Spring Hill Redoubt with Captain Mason on 9 October 1779 and deserted shortly afterwards on 15 October.

[100] John Reiley reinlisted in the 1st South Carolina on 4 November 1775. He was discharged on 24 August 1778. He served in the militia under Captain William Dawkins from 20 February to 26 August 1779. He was a sergeant under Captain Thomas Jones during 1780 and 1781. He was a private horsemen from 1 January to 10 June 1783 under Captain James Kelly. He lost a horse, bridle and saddle at the Battle of Fishing Creek in 1780.

[101] Thomas Green reinlisted in the 1st South Carolina Regiment on 4 November 1775. He enlisted in the 3rd South Carolina (Ranger) Regiment on 27 March 1779. He was omitted from the rolls during August 1779. He continued to serve in the militia from 1780-1782 under Colonel Hopkins.

w^t a Reprimand and the rest Remanded [102]

According to the Court of Inquiry concerning Capt Mottes Conduct, The Charge brought ags^t him is groundless & malicious that he has Behaved himself as a good Officer & do acquit him with Honour

September 14th 1775
Parole Iceland

For guard tomorrow, Capt. Motte
General Orders at 4 OClock in the afternoon
Orderd that Capt Cot^s. Pinckney Capt Barnard Elliotts & Capt Fra^s Marions Companies be immediately Compleated to 50 Men Each from their respective Corps & to hold themselves in Readiness to march in Three Hours-
Colonel Isaac Motte is appointed for this command & will Receive his Orders from the Commanding Officer Each Capt to take his own officers if possible if not the next in Rank every Soldier to be provided with proper Arms Cartridge Boxes 12 Rounds 3 flints his Blanketts and provisions NB The detachment went away about 12 OClock [103]

Fort Johnson was Charlestown's principal fort, built in 1747. It was located on James Island in Charlestown harbor. The Royal Sloop of War *Tamar* lay in Rebellion Road and even though the *Tamar* was an old ship, it still had the ability to destroy the town if it wished. The Carolinians had no artillery that could reach the ship. On board the ship was the Royal Governor of South Carolina, Lord William Campbell, who had returned to South Carolina after an absence of two years. It was learned that the Governor was expecting a ship from England, the Sloop of War *Scorpion* along with the transport ship *Palliser*. The cargo ship was "to receive on board the cannon and other ordinance stores at Fort Johnson."

The South Carolinians became alarmed because the guns at Fort Johnson made up a significant part of its defense and the fort was a primary storage site for a large quantity of round shot belonging to the town's heavy artillery. On September 7th another British ship, the armed sloop *Cherokee*, appeared in the harbor. The Council of Safety thought this ship was the vanguard of a British fleet sent to punish the South Carolinians. Until September the Carolinians had been careful not to provoke the British ship in its harbor, but by the middle of the month the Patriots felt confident that they could take Fort Johnson. The Council of Safety gave orders to Colonel William Moultrie that he need to "detach one hundred and fifty men under such command as you shall judge proper for the seizure."

The Governor had watched the activity in town and deduced that the Carolinians were going to seize the fort, the last position still controlled by the English. Governor Campbell ordered his secretary, Alexander Innes, to take the sailors on board the *Tamar* and dismount the cannon at Fort Johnson. The sailors had labored through the night dismounting and spiking the guns and left the fort only two hours before the arrival of Motte and his men. By dawn only part of Motte's force was ashore. Elliot and Pinckney's Grenadier companies were ashore, but Marion's Light Infantry Company had not arrived yet. Motte decided to attack with the men he had. Once they reached the fort they found the gates were wide open. Lieutenant Mouatt rushed the fort and quickly captured Gunner Walker

---

[102] If a soldier was remanded it meant that his court date was changed to another time to try to collect more evidence. If there was no more evidence then the charges were dropped.

[103] The full details of this incident is depicted in "*Nothing but Blood and Slaughter, Volume One*" by Patrick O'Kelley.

and five sailors from the *Tamar*, who put up no resistance. Inside the fort all the cannons had been thrown from their carriages, however they had not been spiked.[104]

Septr 15th 1775
Parole Dublin

For guard tomorrow Lt Richd Fuller Lt Gray [105]

General Orders

Orderd that Capt Cattle Capt Adam McDonald and Capt Barnwells Comp^{y} of the First Regiment Capt Peter Horry & Capt Hugers Companys of the 2nd Regiment are to hold themselves in Readiness to march the Companies to be Compleated to fifty Men Each to have 12 Rounds 3 flints and as wel Armed as possible, Major Roberts to Command this party & will Receive his Orders from the Commanding Officer

Orders to Major Roberts of the South Carolina Reg^{t} of Foot

Sir

You are to proceed with your party to Gadsdens Wharf where you will find Two Schooners ready to take on Board your Party with them

you are to proceed to fort Johnston, on James Island on y^{r} Arrival there, you are to Send an Officer to Lieu^{t} Co^{l} Motte & acquaint him of y^{r} arrival then march to the fort & put yourselves under his Command your are not to Lett any Boats obstruct your passage-

To Lieut Col^{o} Motte

I have Sent Major Owen Roberts with Two hundred & fifty Men to Reinforce your part. You are to Defend the fort from all parties that may attempt to land, But if the Man of War should Attack the fort & you find you cannot make Any Stand against her you are to With Draw your men to some place of Safety out of the reach of her Guns But you are to Take not to suffer Any parties to Land with any interest to Damage the fort I will provide you with every thing Necessary

After the capture of Fort Johnson the Carolinians feared an attack from the two Royal Navy ships in Charleston Harbor. On September 15th Motte ordered 250 men of the 1st South Carolina Regiment under the command of Major Owen Roberts to reinforce Fort Johnson. Captain Thomas Heyward with a detachment of Charlestown artillery had three cannon immediately mounted.[106]

---

[104] O'Kelley, *NBBAS, Volume One* pp. 41-46.

[105] Henry Gray was appointed a lieutenant in the 2nd South Carolina in 1775. He was commissioned a second lieutenant on 15 June 1776. He was wounded at Sullivan's Island on 28 June 1776. He was promoted to 1st Lieutenant on 7 August 1776 in Captain Motte's company. He resigned his Continental commission and became a paymaster of the 2nd South Carolina Regiment on 15 December 1777 after the death of Thomas Evance. He was promoted to captain during 1778 and was wounded at the siege of Savannah in 1779. He was taken prisoner at the fall of Charleston. He died on 20 July 1824.

[106] O'Kelley, *NBBAS, Volume One* p. 46.

Septr 16th 1775
Parole Corke

General Orders, Orderd that every officer of the Two South Carolina Regiments Keep Themselves in readiness to the first call & if any alarm is given that all Officers Repair to the Barracks immediately & Lieu^{t} Mazyck to go to the fort [107]

Septr 17th 1775
Parole Port Arlingtyon

General Orders, Orderd That the Sick of the 1st & 2nd Regiments be immediately Removed to Mount Pleasant Boat are provided to Carry them Over

A Surgeons Mate from Each Reg^{t} to go with them & take such Medicines as may be Wanted
Capt Mottes Comp^{y} to go Over & Carry 10 Tents to incamp with Near the Hospital-
Ordered that the Q^{r} Master do provide all Necessaries

Septr 18th 1775
Parole Fort Johnson

For Duty Capt Harleston Lieu^{t} Ashby-

Septr 19th 1775
Parole Mount Pleasant

For Guard tomorrow Lieu^{t} Thos Lessesne

Septr 20th 1775
Parole Unanimity

For Duty tomorrow Capt Mason- A Court Martial is Ordered to be held tomorrow Morning at 10 OClock to Try Such prisoners as May be Brought before them Witness to attend

President Capt Mason

Capt Ed^{m} Hyrne Capt Saunders
Lt Anth^{y} Ashby Lieu^{t} Smith [108]

NB Capt Hyrne Being younger than Captain Mason, Capt Mason was president

Septr 21st 1775
Parole Oxford

For guard tomorrow Lieu^{t} Blake

Septr 22nd 1775
Parole Tower Hill

For Duty tomorrow Capt James McDonald

---

107 All troops in Charlestown were to be in readiness to repel an attack from the British warships.

108 Press Smith had been commissioned as a lieutenant in the 2nd South Carolina Regiment on 17 June 1775. He died on 2 October 1777.

According to the Sentence of the last Court Martial Edwd Waugh alias Walker for Threatening to join the Kings Troops, was sentenced to Receive 50 Lashes which he Received accordingly,[109]

Philip Stapleton for sleep^{g} on his post Received 25 Lashes
the Rest was dismissed Except Wm Simpson who is Kept on Bread and Water until tomorrow[110]

Septr 23 1775
Parole Ninety Six

For Duty tomorrow Lt Oliphant Lt Harleston
Regimental Orders Capt Ja^{s} McDonald was Orderd to the fort in the Room of Capt Francis Huger who is Sick-

Septr 24th
Parole Enore

For Duty Capt. Daniel Horry

Septr 25th
Parole Pacolett

For Guard tomorrow Lieu^{t} Hall & Lt Lessesne Inspector of the armourey-[111]

General Orders, A Subaltern Officer from the 1st & 2nd Reg^{t} Be appointed to Inspect the Armourers of Both Reg^{ts}. Ever day By Turns-

---

[109] Edward Walker (Waugh) was charged with drunk and not appearing for guard duty on 29 September 1775, but the charges were remanded due to lack of evidence. He broke out of confinement on 7 October 1775 and received 50 lashes. He was later promoted to corporal. He was captured by the enemy and returned from captivity on 1 September 1779. He continued to serve as a corporal in the 1st South Carolina under Captain Charles Skirving in 1780.

[110] William Simpson served in the 2nd South Carolina Regiment. He was promoted to quartermaster sergeant on 11 January 1776. On 16 March 1776 he was found guilty of abuse to a sergeant and received 100 lashes. On 20 August 1776 he received 100 lashes for drunkenness and neglect of duty. On 25 November 1776 he received 99 lashes and was mult 14 days pay for being drunk on guard and stabbing "three Negroes". He deserted and was returned by William Kelly on 29 November 1776, for which he received 50 lashes. He received 50 lashes for breaking confinement on 4 December 1776. On 6 February 1777 he received an unknown amount of lashes for being drunk on duty. On 12 May 1777 he received 100 lashes for being drunk on guard duty. On 30 August 1777 he was reprimanded "at the Halberts" for abusing Private Edward Fry. He was supposed to receive 50 lashes on 22 September 1777 for breaking confinement, but he was pardoned. He received 75 lashes on 23 October 1777 for helping a prisoner escape. He received 50 lashes for an unnamed offense on 27 October 1777. He was promoted to sergeant in 1778, and then appointed Quarter Master Sergeant of the regiment on 18 July 1778. On 5 October 1778 he was made an orderly at the Grand Hospital. On 27 October 1778 he was given 50 lashes for stealing potatoes. On 12 November 1778 he was ordered to do double duty for a week for neglect of duty. On 24 November 1778 he received 75 lashes for disobedience of orders. He was in Captain Harleston's company in 1778. On 5 January 1779 he received 100 lashes on the bare back with a cat of nine tails for quitting his post while on sentry duty. During 1779 he was under Captain Peter Gray. He was demoted to private in on 9 August 1779 for embezzling clothes from the regiment with Sergeant Newman. His pay was stopped for the loss of the clothing, and he was discharged from the service. He was listed as being at the siege of Savannah in Captain Gray's company so he may have been allowed to return after the British invaded South Carolina. He has the notorius achievement of being the most punished soldier in the 2nd South Carolina Regiment.

[111] John Hall became the regimental quartermaster of the 2nd South Carolina on 1 July 1776. He resigned in December 1776 and joined the British. He was captured and hanged for treason at Ninety Six on 17 April 1779.

Sept[r] 26th 1775
Parole Fair Forrest

For Duty tomorrow Cap[t] Saunders-

Sept[r] 27th 1775
Parole George Town

For duty tomorrow Cap[t] Mason for guard L[t] Fuller

Ordered that One Subaltern Officer from the 1st Reg[t] do go Over to Mount pleasant to See that the sick have proper care taken of Them

This officer to be Relieved every week

Sept[r] 28th
Parole New Markett

For Guard tomorrow Lieu[t] Lessesne

General Orders

Ordered that a General Court Martial do sett this afternoon at 10 OClock to Try such prisoners as Shall be Brought before them

President
Cap[t] Nicholas Everleigh

L[t] Ashby L[t] Turner [112]
L[t] Hall L[t] Godfrey [113]

Evidences to attend at the Same Time NB Cap[t] Mason was member in the Room of Lieu[t] Hall

Sept[r] 29th
Parole Dorchester

For Guard tomorrow L[t] Ashby NB if the first Reg[t] does not give one-

General Orders

Ordered That a Court Martial do Sett tomorrow Morning at 7 OClock at M[r] Ramages Tavern On the Trial of Rich[d] Rogers where M[r] Roger Smith is desired to attend Rogers is a private in Cap[t]. D. Horrys Company- [114]

President
Cap[t] William Scott

Lieu[t] Blake Lieu[t] Shingleton [115]
Lieu[t] Everleigh Lieu[t] Armstrong

---

[112] George Turner was born in 1738. He was commissioned a second lieutenant in the 1st South Carolina Regiment on 17 June 1775. He was promoted to first lieutenant on 16 May 1776. He was promoted to captain on 24 April 1777 when Captain Scott was promoted to major. He was promoted to major and served as aide-de-camp to General Howe on 22 November 1778. He was taken prisoner at the fall of Charleston and served as deputy commissary of prisoners. He was exchanged and became a brevet-major on 30 September 1783. After the war he moved to Kentucky, Ohio, and then Pennsylvania and died on 16 March 1804.

[113] Richard Godfrey was a lieutenant in the 2nd South Carolina Regiment. He served as a forage master and then a captain in 1779-1780. He died in 1817.

[114] Richard Rogers reenlisted in the 2nd South Carolina Regiment on 4 November 1775. On 1 April 1776 he received 100 lashes for forging a pass to go to town with Colonel Moultrie's name. He was killed at the battle of Fort Sullivan on 28 June 1776 while serving under Captain Ashby.

[115] Richard Singleton was a lieutenant in the 2nd South Carolina Regiment from 17 June 1775. He was also a colonel in the St. Mark's Parish militia. He was at the fall of Charleston in May 1780.

Lieu[t] Blake to inspect the Armourers

according to the Sentence of the Court Martial

James Campbell of Cap[t] Dan[l] Horrys Comp[y] in the 2[d] Reg[t]. confined by L[t] Godfrey for sleeping on his post Received 39 Stripes at 9 OClock

W[m] Wheeler for being Drunk on Guard Confined by ditto Rece[d] 50 Lashes[116]

David Stuart Corn[s] Syllavan, Ja[s] Blenchfield & Ed[w] Waugh were Remanded being Drunk & incapable of appearing[117]

John Ansty for sleeping on his Post Received 100 Lashes,[118] Ja[s]. Hooper of Captain M[c]Donalds Company Confind by Lieu[t] Hall for being absent from Duty was Remanded to the fort[119] James Sherwood Remanded to another Tryal[120] Samuel Shaw of Cap[t] Cha[s] C.Pinckneys Comp[y] Confind for Absence from his Guard & Insolence Received 50 Lashes[121]

Sept[r] 30th

For Duty tomorrow Captain Eveleigh

General Orders- Orderd that 1 Cap[t], 2 Subalterns 2 Serg[ts]. 2 Corp[ls] & 40 privates do immediately hold Themselves in Readiness Cap[t] Isaac Harleston L[t] Turner & L[t] Jo[s] Elliott are with the men in Readiness since last Night[122]

---

[116] William Wheeler died in the General Hospital in 1776.

[117] David Stewart (Stuart) served in the 2nd South Carolina Regiment. He was appointed a regimental barber on 24 January 1778. He was confined for 14 days on 4 May 1778 for being absent without leave. He also had do guard duty every other day while he was confined. He was promoted to sergeant in 1778. He was demoted to corporal and served under Captain Thomas Dunbar in 1779. He was reduced to private on 14 February 1779 for drunkenness on guard duty. He was with Dunbar's light infantry at the siege of Savannah.

[118] John Anstey reenlisted in the 2nd South Carolina on 4 November 1775. He died on 26 May 1776.

[119] James Hooper enlisted in the 1st South Carolina Regiment in 1775. He was confined to the guardhouse for beating his wife on 9 December 1775. He then enlisted in the 2nd South Carolina Regiment on 5 August 1776. On 17 September 1776 he received 200 lashes for drunkenness and neglect of duty. On 3 October 1776 him and Sam Shaw was remanded for an unnamed offense for lack of evidence. On 7 October he and Sam Shaw both received 100 lashes for breaking out of confinement on September 28th. On 20 November 1776 he received 60 lashes for being absent without leave. On 12 April 1777 he was reprimanded for misbehavior. On 24 June 1777 he received 25 lashes for being absent from roll call. On 22 July 1777 he received 50 lashes for beating Francis Simpson's wife. On 22 August 1777 he was charged with being absent without leave, but he was acquitted due to lack of evidence. On 6 September 1777 he was reprimanded for being drunk. He received 100 lashes on 23 September 1777 for riotous behavior. He received 100 lashes on 1 May 1778 for being drunk and bad behavior. On 15 February 1779 he was supposed to receive 100 lashes with switches for rioting and drunkenness in town, but he was pardoned.

[120] James Sherwood served in the 1st South Carolina under Captain Adam McDonald. On 12 October 1775 he received 100 lashes for abusing Sergeant Johannes. Two days later he received another 100 lashes for threatening to murder the same sergeant.

[121] Samuel Shaw later served in the 3rd South Carolina (Ranger) Regiment under Captain George Liddell in 1779. After the fall of Charleston he served in the militia under Colonel Brandon.

[122] Joseph Elliott became a second lieutenant in the 1st South Carolina on 17 June 1775. He was promoted to first lieutenant on 1 November 1776. He was made adjutant to the expedition raised to relieve Fort McIntosh on 15 March 1777. He was promoted to captain on 16 November 1778 after the death of Captain Joor on the frigate *Randolph*. He was wounded and taken prisoner at the siege of Charleston on 12 May 1780.

October 1st 1775
Parole Williamsburgh.

For the Guard to morrow Lt Lessesne The party that were order'd to hold themselves in readiness last night are discharged...

General Orders

Order'd that a Court Martial do sett to morrow at 11 oClock in the morning to try all Prisoners that shall be brought before them; All witness to attend. President Captn N. Eveleigh
Members Lts Jno Farr, Jno Harleston, Elliott & Godfrey [123]

1775 October 2nd
Parole Rockingham. Regimental Orders
Order'd that Captn Harleston do go over to Fort Johnston to take the Charge of Captn Hugers Company and Lt Lessesne to relieve Lt Shubrick of the same Company who is sick

NB. Lieut Lessene was order'd to the Fort & Lt Blake took the Guard. Adjutant Massey [124] of the First Regt is order'd to the Fort.[125]

3rd Parole Montesque. Captn Dl. Horry for the Day to morrow
For Guard to morrow Lt Geo Eveleigh

According to the Order of the two last Court Martials the Sentence was executed

Richard Rodgers for saying he would not fight the Kings Troops receiv'd 200 Lashes

Hugh McGuire for Drunkness & Absence from Roll Call receiv'd 50 Lashes,[126] Serjeant John Ross for abusive Language was reduc'd to the ranks.[127] Edwd Waugh Saml Shaw & Hooper remanded for want of Evidences. also Jams Sheerwood & Conrad Fitner.

Timothy Sullivant is confin'd in the Black Hole according to the Sentence on Bread & Water. [128]

4th Parole Montmorencie. For to morrow officers from the 1st Regt.
Order'd that a Court Martial do sit tomorrow at 11 oClock

---

123 John Farr was a second lieutenant in the 2nd South Carolina Regiment on 17 June 1775. He was later in the militia of Colonel Waters.

124 William Massey was born in 1743. He served as the adjutant for the 1st South Carolina Regiment. He later served as a lieutenant colonel and deputy muster-master general for South Carolina from 20 October 1777 to 1780. He was taken prisoner at the surrender of Charleston and taken to St. Augustine. He died in 1841 at the age of 98.

125 Since the seizure of Fort Johnson on September 18th Lieutenant Colonel Motte had not been given orders on what to do next. He wrote to the council of safety "I have received no instructions since the first day I came here, and how to act with respect to the Men of War now lying near Sullivant's Island." A ship would leave Charlestown every day at 7:00 a.m. to bring food and supplies, but no orders. The rank and file at Fort Johnson turned the ship into a smuggling vessel that would bring in illegal alchohol every day. Some of the men would also use the ship to go into to town. Motte ordered that the ship be inspected every day when it arrived and when it left.

126 Hugh McGuire reeinlisted in the 1st South Carolina on 4 November 1775 under Captain Thomas Lynch

127 John Ross enlisted in the 1st South Carolina in 1775 as a sergeant. He was reduced in rank on 3 October 1775 for abusive language. He reinlisted on 4 November 1775 as a sergeant. He died on 7 April 1777.

128 The Black Hole appears to have been an underground jail at Fort Sullivan.

General Orders

That if any Soldier be found in Town belonging to Col° Mottes Detachment without a pass from Col° Motte or the Cap[tn] of the Company to which he belongs, that he be immediately takend up & confin'd at the Barracks

---

1775 Octob[r] 5th Parole Somersett. For the Guard to morrow L[t]. Blake

The Cap[tn] from the first Regim[t]. President of the Court Martial Cap[tn] Edmund Hyrne; Members L[t] Blake L[t] Eveleigh L[t] Armstrong & L[t] Godfrey go over to relieve L[t] Hext of Cap[tn] W[m] Cattells Comp[y][129]

6th Parole James Island. For Guard to Morrow Cap[tn] Mason

Daniel M[c]Kenzie [130] Joseph Reives [131]& John Still [132] who deserted last Friday from Cap[tn] P. Horry's Company were brought back

---

7th Parole Ashley River For Guard to morrow L[t] Hall

General Orders

Order'd that one Subaltern, 2 Serjeants & 25 Private Men do hold themselves in Readiness and to parade at the Old Barracks by 5 o'Clock this Afternoon, where they will see Cap[tn] Cockran Powder receiver,[133] who will conduct them on board a Schooner & there they are to take Charge of Ten Thousand lb Weight of powder & to see it deliver'd to the Committee of Dorchester and safely lodge in the Magazine there & then the Party to return to Charles Town by land. Cap[tn] Cochran Powder receiver will attend the Party, L[t] Armstrong of the 1st Regiment is appointed Commander. According

---

[129] William Hext (Hixt) served as a 2nd Lieutenant in the 1st South Carolina Regiment from November 1776. He was promoted to 1st Lieutenant on 6 November 1778. On 1 May 1779 he was promoted to captain. He was wounded at Stono Ferry and was taken prisoner at the fall of Charleston. He was still a prisoner at Haddrell's Point in November 1780. He died in May 1790.

[130] Daniel McKenzie reenlisted in the 2nd South Carolina Regiment on 4 November 1775. He was discharged on 8 July 1778.

[131] Joseph Reeves reenlisted in the 2nd South Carolina Regiment on 4 November 1775. He received 100 lashes with switches on three different days on 4 December 1777 for being absent without leave. He was supposed to receive 99 lashes on 11 December 1777 for taking £27 out of Captain Charnock's portmanteau (leather suitcase), however the lashes were remitted and his pay was stopped until the money was repaid. He was also supposed to receive 90 lashes for stealing Captain Charnock's horse, saddle, bridle and spurs. One half of his pay was stopped until he paid the £139 for the items. The 90 lashes were withheld with the condition that he reenlist for the duration of the war. He was under Captain Richard Mason in 1779 at the siege of Savannah.

[132] John Steele served in the 2nd South Carolina Regiment under Captain Blake. He received 50 lashes on 18 September 1777 for being drunk. He received 100 lashes on 30 September 1777 for being drunk on guard. He was under Captain Thomas Moultrie at the siege of Savannah in 1779. After the fall of Charleston he was in the light dragoons under Lieutenant Colonel John Thomas and General Sumter.

[133] Robert Cochrane (Cockran) served as a captain and ordinance storekeeper of Charleston. In the middle of the town was the armory, also known as Cochrane's Magazine, that stored the weapons of the military and over 10,000 pounds of powder.

to the Order the Detachm[t] went wth 12 Round, 3 Flint 13 Men from the first Reg[t] & 12_from the 2nd D[o]- [134]

1775 Octob[r] 7th According to the Sentence of the last Court Martial John Hooper & Sam Shaw receiv'd 100 Lashes each for breaking out of Confinement the 28th September [135]
Edward Waugh for the same receiv'd 50 Lashes
John Anstey receiv'd also 50 Lashes for abuse, Serjeant Farshaw was reduc'd to the ranks accordingly. Theophilus Thorper was Confin'd according to the Sentence--- [136]
Ja[s] Chester – }
Laur[es] Murray [137] } remanded for want of Evidence

8th Parole Ticondoroga, Officer for the next Day from y[e] 1st Reg[t]
Order'd that a General Return be made to morrow of what Arms are fit for Service, & also what Arms are carried off by deserters & how many Bayonets are wanted & to examine all the old Pouches (& be nice in the Examination) & make a report to morrow...........

Regimental Orders

Order'd that the quarter Master do make a Return of the Stores & what Number of Cloaths, Shoes &[c] - are given to the Men............

Gen. Orders

Order'd that all Officers that have not been on recruiting Service do hold themselves in readiness, they will receive their recruiting Orders very shortly. A Court Martial to sit to morrow Morning, All Evidences to attend.

1775 Octob[r] 8th A General Court Martial to be held on Thursday the 12th Instant next over at Fort Johnston

Major Roberts President four Capt[ns] & two Subalterns from the first Regiment and four Captains & two Subalterns from the Second Reg[t] to be Members of the Court Martial, Cap[tn] W[m] Mason is appointed Judge Advocate to the said court. All Evidences to have Notice to attend
A Report of the Proceedings & Determinations of the General Court Martial to be made to the Provincial Congress or the Council of Safety. in due Time, A Copy of this Order to be sent to Col: Motte

For the Court Martial to sitt today President Cap[tn] Nich[s]. Eveleigh, Members L[ts] Hall, Jac[b]. Shubrick, Ladson & Eveleigh

---

9th Parole Warsaw For Guard tomorrow L[t] Asby
As L[t]: Shubrick is indispos'd L[t]. Eveleigh is appointed in his Room L[t]: John Allen Walter having resign'd his Commission in the 2nd Reg[t]: is no longer an Officer in the Provincial Troops, and is not

---

[134] Moultrie wrote, "It thought prudent to send a part of our ammunition, ordinance, stores and public records to be lodged at Dorchester, and to build fortifications round that town." The gunpowder in Charleston was moved to Dorchester so that a British raiding party from the warships in the harbor could not capture it.
[135] This was actually James Hooper.
[136] Theophilus Thorpe served in the 1st Regiment under Captain Adam McDonald.
[137] Lawrence Murray reenlisted in the 1st South Carolina Regiment on 4 November 1775 and again on 1 February 1780. He served under Captain William Cattell.

to be obey'd as such. L[t]: John Far having resign'd his Commission in the 2nd Regiment is no longer an Officer in the Provincial Troops, and is not to be obey'd as such.

According to the Sentence of the last Court Martial Lawrence Murray of the 1st Reg[t] rec[d] 100 Lashes

---

1775 Octob[r] 10th Parole Albany. For Duty tomorrow Cap[tn]. D. Horry
Order'd that 2 Capt[ns]. & 1 Subaltern from the first Reg[t]. & 2 Capt[ns]. & 1 Subaltern from the 2nd Regiment, to be Members of the General Court Martial Viz. to join 4 Captains & 2 subalterns at Fort Johnston, Assistant Members, Cap[tn] Scott, Cap[tn] Heyrne, Cap[tn]. D. Horry, Cap[tn] Eveleigh, L[t]. Ladson, L[t]. Ashby

---

11th Parole Paris, For Guard tomorrow L[t]. Harleston.
Order'd that a Court Martial do Set this Morning at 10 o'Clock to try all Prisoners that shall be brought before them & all Evidences to attend. President. Cap[tn] James M[c]Donald Members L[t]. Jos[h]. Elliott L[t]. Tho[s] Hall L[t]. Hughes L[t]. Harleston
Rich[d] Baker is appointed a L[t]. in the 2nd S[o]. Carolina Regiment of Foot & is to be obey'd as such. [138]

12th Parole Pitsburgh, For duty tomorrow Cap[tn]. Harleston
According to the Sentence of the last Court Martial
Jos[h] Jackson for Disobedience of Orders was reprimanded at the Halbard.[139] Richard Trimble (Corporal) for Neglect of Duty was reduc'd to a Private.[140] Tho[s] Hamilton & James Winter discharg'd for want of Evidence.
Archibald Love for being absent from his guard receiv'd 50 Lashes according to his Sentence[141]
Timothy Sullivant for Desertion was sent to Fort Johnston for his Trial. James Sherwood for abuse of Serjeant Johannes [142] & damning his officer receiv'd 100 Lashes according to Sentence; Dennison Chester [143] & Ja[s]. Martin [144] who were sentenc'd for desertion by the Court are kept in Confinement, the said Court having no right to judge Deserters.

---

[138] Richard Bohun Baker was born in November 1755. He was commissioned as a 2nd Lieutenant in the 2nd South Carolina Regiment on 12 October 1775. He was promoted to first lieutenant on 22 November 1776 (also listed as 16 September 1776) in Captain Blake's company. He assumed command of Blake's company on 25 April 1778 when Captain Blake resigned. He was promoted to captain-lieutenant of the Lieutenant Colonel's company on 5 October 1778. He resigned his commission in January 1780. When Charleston fell he was taken prisoner and exchanged during July 1781. He served in General Marion's partisans until the close of the war. He died on 8 November 1837 at the age of 82.

[139] Joseph Jackson reenlisted in the 1st South Carolina Regiment on 4 November 1776.

[140] Richard Trimble served in the 1st South Carolina Regiment under Captain Thomas Lynch.

[141] Archibald Love reenlisted on 4 November 1775 in the 1st South Carolina under Captain Charles Pinckney.

[142] Peter Johannas reenlisted in the 1st South Carolina on 4 November 1775 as a sergeant. He was promoted to sergeant major on 13 January 1778. He was discharged on 27 July 1778.

[143] Tinson Chesson served in the 1st South Carolina under Pinckney. He was court martialed on 19 January 1778 for disorderly conduct during the Great Fire of 1778. He was discharged on 21 September 1778.

[144] This is not James Martin, but is his brother, John Martin. John served the time in place of his brother, and in all the records John is referred to as James. John (aka James) reenlisted as John on 12 November 1778 in the 1st South Carolina.

Captn Wm. Scott, Edmd. Hyrne & Lt. Jas Ladson of the 1st Regiment & Captn Dan. Horry & Nichols Eveleigh & Lt Anthy Ashby of the 2nd Regiment are appointed Members of the General Court Martial to be held at Fort Johnston this day & with such other Officers as shall be appointed by the commanding Officer there & to try all such Prisoners as Shall be brought before them.

13th Parole New England. For Guard tomorrow Lt. Blake
Order'd that a Court Martial do sitt this morning at 11 oClock to try such Prisoners as shall be brought before them. All Evidences to attend. President Captn R. Saunders Lts. Singleton, Elliott, Blake & Hall.

Order'd that a Court of Inquiry do sitt for inquiring into the management of Wm Graham Quarter Master to the 2nd So.Carolina Regiment of foot concerning his Business as shall be presented by a copy of the Information by the Contractr & him. [145]

President
Major McKintosh
Captn D Horry Captn Mason
Captn Eveleigh Captn Harleston

\- 14 - Parole Bourdeaux. For duty to morrow Capt.
According to the Sentence of the Court Martial Wm. Wheeler for Drunkness & Absence receiv'd 200 Lashes
Dennison Chessire for attempting to desert receiv'd 300 Lashes
James Martin for . . . do . . . do. . . . 200 Do
James Sherwood for threatning to Murther Serjt. 100 Do
Jno Hay & Wm Smith for Absence without Leave 200 Do ea
Samuel William too sick to be brought to the Halberts. [146]

\- 15 Parole Calais
For Guard to morrow Lt Hall.
Order'd that Lt. Blake do go to Fort Johnston for a few Days to relieve Lt Oliphant.

\- 16 Parole Dover
For duty to morrow Capt
One Soldier to be appointed to wait on Messrs Dun & Booth to be sent on Errands & c - by them.

1775 Octobr 17th Parole Portsmouth
General Orders The General Court Martial of which Major Roberts was President is dissolved.

---

145 William Graham resigned as quartermaster on 22 October 1775. He reenlisted in the 1st South Carolina Regiment on 4 November 1775. He became the quartermaster sergeant again on 31 December 1776.

146 Living in camp at Fort Johnson were 179 officers and men who were repairing the fort's outer works, and building a battery on a small hill to the southwest. The three companies there were Captain Daniel Horry's Grenadier company, and Charles Motte's and Nicholas Eveleigh's line companies. Soldiers in Fort Johnson would slip into Charlestown illegally for some rest and relaxation. The men being punished here were caught and made an example of so that the other men would curtail their desertions. Some court penalties were conducted within the barracks area to reinforce the image of military justice.

General Orders

Order'd that Lt Neile have leave of Absence from the Fort for 3 or 4 days, & that a Lieut from the first Regt. go down to the Fort this afternoon to relieve him.[147] Samuel Williams of the first Regt receiv'd his 500 Lashes this morning according to the Sentence of the 13th Inst.

18th

Parole Plymouth

For Duty to morrow Captn D. Horry.

Order'd that Major McIntoshs do go over to Fort Johnston to relieve Majr Roberts, who is to relieve Colo. Motte that has leave of Absence for a few Days

Order'd that Lt Hall do go over to the Fort to relieve Lt Dubose who is sick[148]

19th

Parole Inverness

For Guard to Morrow Lt. Moultrie

Order'd that a Court Martial do sett this Morning at 11 oClock to try such Prisoners as shall be brought before them. Evidences to attend.

President Captn D. Horry, Members Lts. Ladson, T. Elliott, Thos Moultrie, Jas Perroneau

General Orders Order'd that at roll call Evening & Morning the Officers of each Compy in both Regiments do attend the drilling of their Compys and also to see that their men keep their Arms clean & appear on the Parade as decent as possible.

20th

Parole Norfolk

For Duty to morrow from the 1st Regimt.

Genl. Orders

A Return to be made immediately of all Seamen & how many are now in the two regts. with their Names & if they are willing to change the Service by acting on board the Colony Schooners where their pay will be considerably advanc'd [149]

One Captn from the 1st Regt. to Fort Johnston today to relieve Captn Thos Pinckney.

According to the Sentence of the last Court Martial D. Munro for absence without Leave receiv'd 100 Lashes[150] Timothy Crimor for repeated Drunkiness receiv'd 50 Do[151]

---

[147] Philip Neyle became a 2nd Lieutenant in the 2nd South Carolina on 17 June 1775. He was promoted to 1st Lieutenant in 1776. He was aide de camp to General Moultrie and was killed at Charleston in 1780.

[148] Isaac Dubose was commissioned as a 2nd Lieutenant in the 2nd South Carolina Regiment on 17 June, 1775. He was promoted to 1st Lieutenant in 6 September 1776 in Captain Oliphant's company. He transferred as a Captain to South Carolina Dragoons in 1779. After the fall of Charleston he served as a lieutenant and then a captain under Peter Horry and Colonel Maham. He was promoted to 2nd Lieutenant on 6 May 1782 and was in 1st Cavalry troop in Maham's Dragoons in 1782. After the war he moved to Georgia.

[149] The Royal Navy sloop of war *Tamar* still was anchored in Rebellion Roads, along with the sloop *Cherokee*. The Carolinians knew that they would need both fixed defenses and some type of naval force. In September the 12-gun schooner *Defence* had been put into commission. However all the sailors in Charlestown had enlisted in the regiments because they were out of work due to the blockade. The Provincial Congress needed to know how many sailors there were in the regiments so that they may possibly be able to outfit the *Defense*.

[150] Daniel Munroe.

[151] Timothy Cremer reenlisted in the 1st South Carolina on 4 November 1775. He was on board the frigate *Randolph* on 12 January 1778 and was killed when it blew up on 8 March 1778.

Additional Orders

Order'd that Cap^{tn} Edmd Hyrne do go over to Fort Johnston with his C^{o}. & to encamp on the Spott where the new Battery is to be built near Fort Johnston

Cap^{tn} Hyrne must apply to Maj^{r} Roberts for tents for his Comp^{y}

21st Parole Normondy

For Duty to morrow L^{t}. Ladson, for Guard D.

General Orders

The oldest officer in rank is to be Commander in the Absence of his superior Officers & to see that all Orders are executed. The Men exchang'd by the Differ^{t}. Comp^{ys}. are to be return'd to their respective Cap^{tns}.

22nd Parole Denmark

For Guard to morrow L^{t}. Blake. for Duty Cap^{tn} Harleston

NB. the 20th Ins^{t} the Key of the Store was deliver'd me by Order of Mr Graham having resign'd, Serj^{t} Burgess is the attendant[152]

The Subalterns of the Second Reg^{t} being order'd recruiting & few left that are sick the Officer for Guard must be taken from either

---

1775 Parole Norway

Octr 23rd For duty to morrow from the 1st Reg^{t}:

For Guard L^{t}. Blake in the room of L^{t} McQueen

Order'd that the paymaster do advance the officers who are going recruiting one Months Pay.

24 Parole Sweden

For Guard tomorrow L^{t}. Ladson for duty L^{t}. Eveleigh

25 Parole Augusta

For Guard tomorrow L^{t}. Theus, for duty Cap^{tn} Mason

General Orders. Mr Joseph Loyd [153] is appointed Quarter Master of the 2nd: S^{o}: Carolina Regiment of Foot & is to be obey'd as such

26 Parole Savannah

For duty to morrow L^{t}. Ashley, for guard L^{t}. Eveleigh

---

[152] John Burgess enlisted in the 2nd South Carolina Regiment on 4 November 1775. On 4 April 1776 he was found guilty of Neglect of Duty and was reduced to private. On 19 April 1776 he was promoted back to sergeant in McDonald's company. On 26 July 1776 he was reprimanded for an unnamed offense. On 17 January 1777 he transferred to Captain Moultrie's company as a sergeant. On 16 June 1777 he was confined for being absent without leave and found guilty, but the other sergeants in the regiment came to his defense and he was released. On 24 June 1777 he was charged with letting a prisoner escape, but he was acquitted. He was appointed quartermaster sergeant to the deputy quartermaster general on 13 November 1777.

[153] Joseph Lloyd was a sergeant in the 2nd South Carolina Regiment. He was made gunner of Fort Johnson after it had been captured. He later served as the adjutant in the 2nd Dragoons under Lieutenant Colonel Charles Myddleton during 1781. He was in the battle of Bacon's Bridge.

General Orders

Order'd that 1 Serg[t]. 1 Corp[l]. & 12 Privates do mount guard on the Exchange Wharf and take charge of the cannon there that no one shall molest them, they must assist in putting them on board a Schooner to morrow

NB They went from the Barracks at Sun Sett

27 ____________________________Parole Pedee______________________________________

For duty & guard from the first Regiment

General Orders

Order'd that 2 Subalterns (L[t]. Ashley of the 2[d]. & L[t]. M[c]Queen of the 1st Reg[t].) 2 Sergeants, 2 Corporals, and 20 Privates do immediately march to Dorchester, and apply there to the Commissioners appointed to carry on the Lines of Fortification round the Town, who will shew the Officers where the Publick Gunpowder, Ordinance Stores & Records are lodg'd, which the Detachment is to take charge of and have them properly guarded, untill reliev'd by the Companies of Rangers [154] order'd for that Service then to return to Town and make a proper Report, The Contractor is to provide Provisions for this Detachment,

____________________________After Orders ______________________________________

All Volunteers from the 1st. & 2[d]. S[o]. Carolina Reg[t]. that are inlisted for Soldiers who are willing to enter on board the Schooner Defence, commanded by Capt Tuff [155] may go if agreeable to their Officers.[156]

N:B: The Detachment for Dorchester went off at 11 oClock & 16 Men (according to order) went to put the Guns on board the Vessell behind the Exchange Wharf.

______________________Additional Orders at 12 o'Clock ____________________________

L[t]. Gray at Mount Pleasant is order'd to return as soon as possible to Head Quarters. N:B: The Guard order'd to keep watch over the Guns behin'd the Exchange return'd at ½ past Three & another guard was sent at ½ past four to take charge

---

[154] This referred to the 3rd South Carolina (Ranger) Regiment. The Rangers were mounted and were supposed to be armed with rifles. John Nix wrote in his pension "each man furnished his own horse and rifle and was to be paid ten dollars per month for services and in the event we lost our horses or guns in the service we were to be paid for them by the Government." Though they were supposed to be armed with rifles, John Richbourg wrote that they were "armed with guns of such description as could conveniently be procured". The average age of enlisted men in the 3rd South Carolina Rangers was 28, while the company grade officers averaged 30 years old. The oldest man in the regiment was 50 years old. The youngest was 16.

[155] Simon Tufts was the captain of the armed schooner *Defense*. He later became the captain of the *Prosper*. He was also storekeeper and and master of the schooner *Dove*. In 1778 he was in the Charles Town Militia Regiment under Captain James Bentham.

[156] Colonel Motte writes, "It was now though necessary to have some armed schooners for the defence of our harbor and rivers, but it was very difficult to man them without taking the seamen from the first and second regiments, as they had already inlisted all the sailors in port; however it was absolutely necessary as the enemy had a schooner cruising on our bar, and we had information that the men-of-war's boats used to come up to town every night and get intelligence of our proceedings: Therefore, on 27th October, the council of safety ordered that thirty seamen from the first and second regiments be put on board the *Defence* Schooner, commanded by Capt. Tuffs; the schooner was stationed between Fort Johnson, and the town, to intercept the men-of-war's boats."

1775 Octb^{r}. __________ Parole Cooper's River __________
28th. For Guard to morrow L^{t}. Eveleigh. For Duty from y^{e} 1st Reg^{t}.

General Orders

Lt Blake of the 2^{d} Reg^{t} is order'd to go to Fort Johnston to relieve L^{t}. Hall 'till further Orders

29 __________ Parole Ashley River __________
For Guard & Duty from the 1st Regiment

30 __________ Parole West Florida __________
For Duty to morrow Cap: Harleston, For Guard Lt Henry Gray

31 __________ Parole Stono. __________
For Duty & Guard to morrow from the 1st Regim^{t}.

General Orders

All Officers & Soldiers to hold themselves in readiness (both of the 1st. & 2^{d}. Reg^{ts}.) to go over to Fort Johnston tomorrow. The Sick to be left in Town at the Hospital

The Quarter Master to provide a Schooner to be ready at Col^{o}. Gadsden's Wharf to transport the Troops over. The Officers that are Members of the Congress may remain in Town if they Chose to attend.

__________ November the 1st. 1775 __________
__________ Parole Provincial Congress __________
For guard L^{t}. Gray. For Duty Cap: Mason. The Guard was reliev'd this Morning at 7 oClock. NB. At ½ past three this Morning the Regiment join'd the head Quarters (the 2nd. Ins^{t}.)

2^{d}. __________ Parole Victory __________
For Duty to morrow Cap: B. Elliott. For Guard Lt Oliphant

Regimental Orders

Order'd that the Officers of each Company do drill their men from relief of the Guard till 12 o'Clock & in the Afternoon to see them turn out between 3 & 4 o'Clock to be form'd in Battallion & exercise accordingly.

All Taylors are order'd to work together for the Regiment, and if the Cap^{ts}. of any of the Comp^{ys}. choose to have their Men fitted by their Taylors they may take Charge of it, and have Cloth deliver'd for the same. N.B. Part of the Detachment from Dorchester return'd this Evening the rest are coming by water.

3^{d} __________ Parole Success __________
For Duty to morrow Cap: Marion. For Guard Lt Blake
Regim^{tl}. Orders- __________ General Orders __________

Order'd that an orderly Serjeant do wait on Col^{o}: Moultrie from Morning 'till the Time of the setting of the Congress at the State House, where he must attend for Orders 'till dismissed.

Order'd that a man from each Comp^{y}. with a Serjeant do go under Inspection of Cadet De Treville [157] to cut Parlmeta Trees for the Service of the Country 'till reliev'd & to carry Six Days Provision

[157] John Francis De Treville became a captain in the 4th South Carolina Artillery in June 1777. In October 1777 Lieutenant Raphel charged him for an unknown offense, but he was found not guilty due to the charges being

with them, their Arms & two Rounds, some Flints The Quarter Master of the 1st. Reg[t]. is order'd to join his Regiment. N.B. The Serj[t]. of every C[o]. is to make a Return to the Surjeon of all the Sick in the Company to which they belong [158]

4. ________________________________Parole Honour________________________________

For duty to morrow Cap[n]. P. Horry. For Guard L[t]. J. Shubrick

Order'd that nine Men (Sailors) and a Serjeant from the 1st. and 2[d]. Reg[ts]. do go on board the two Pilot Boats which are Comm[d]. by Cap: Visey and Cap: Tho[s]. Smith who they must obey & they will receive double Pay during the Time they are in that Service [159]

5 ________________________________Parole Junius________________________________

For Duty to morrow Cap[n]. Fran[s]. Huger For Guard L[t]. Tho[s]. Dunbar [160]

1775 Nov[r]. ________________________Parole Brutus________________________________

6th . For Duty to morrow Capt[n]. Mason, For Guard L[t]. Rich[d]. Shubrick.

Regimental Orders

The Reg[t]. to be exercis'd every Morning beginning at Eight o'Clock, & in the Afternoon by Companies, every Cap: or Sub: to exercise their own Comp[ys].

Order'd that all Sub[ns]: Officers do reside in Barracks except such as have especial Leave from the com[dg]. Officer. L[t]. Mazyck to do Duty in Cap: Marion's Co & L[t]. Baker to do duty in Cap: Ja[s]. M[c]Daniel's C[o].[161]

7 ________________________________Parole Cæsar________________________________

For Duty to morrow Cap: B[d]. Elliott, For Guard L[t]. Tho[s]. Hall

8. ________________________________Parole Aug: Cæsar________________________________

For Duty to morrow Cap: Marion, For Guard L[t]. Charnock, Barrick Guard L[t]. Gray

---

not based on the truth. Lieutenant Raphel died shortly afterwards. Treville was wounded at Savannah on 9 October 1779. After being taken prisoner at the fall of Charleston he was exchanged on 15 June 1781 and served until the end of the war. He was a brevet major on 30 September 1783. He died in 1790.

[158] On 2 November the regiment only turned out 250 men for its normal 4:00 a.m. muster. The next day Colonel Moultrie decided to inspect the regiment at the morning formation. Upon his arrival Moultrie noticed that there was a large number of men absent. Moultrie went into the barracks and found 80 men still in bed that didn't want to go out in the cold and claimed they were sick. Moultrie had the surgeon examine the soldiers and discovered that 74 out of 80 were faking. The commander of the regiment then ordered each company to make a return of all men who claimed they are sick.

[159] In November the South Carolina Navy consisted of the schooner *Defense*, under the command of Captain Simon Tufts, and the two pilot boats *Hawke* under Captain Joseph Vessey and the *Hibernia* under Captain Thomas Smith. Another ship, the *Comet*, was being fitted out but was not ready yet.

[160] Thomas Dunbar was commissioned a 2nd lieutenant in the 2nd South Carolina on 9 November 1775. He was promoted to 1st Lieutenant on 24 February 1776. He was promoted to captain of the Grenadier Company on 9 November 1777 when Captain Shubrick died. When Captain Lesesne resigned his company was no longer the light infantry company, and Dunbar's Grenadiers became the light infantry company on 10 August 1779. He was taken prisoner at the fall of Charleston in May 1780. He was exchanged in June 1781 and he served until the end of the war. He became a brevet major on 30 September 1783.

[161] James McDonald.

Order'd that One Sub: 1 Serj[t], & 20 Rank & File
Do mount the Barrick Guard begin[g] to morrow

9. ______________________________Parole Dorchester ______________________________

For Duty Cap Mason, For Guard L[ts]. Dubose & Tho[s] Moultrie

General Orders Order'd that Cap: Marions & Hugers Comp[y]. do march immediately to Dorchester each C[o]. to be serv'd with twelve Rounds & Three Flints [pr] man. Cap: Marion will receive his orders from the Commanding Officer. Copy of the orders.

You are to proceed with your own & Cap: Huger's Comp[y]. (with all Expedition) to Dorchester to reinforce the Troops there, you are to take the Command of the whole the Troops there. & take especial care in guarding and defending the Cannon, Gunpowder, Stores & Publick Records & remain at that place 'till further Orders. You are to apply to the Committee at Dorchester for a sufficient Number of Negroes (now in the Publick Service) to remove the Cannon lying near the Water side, to a Spot most safe and convenient to the Fort or Barracks, wth. strict Orders to the Centries not to let any Person handle or go near them without proper Auth[y]. Order'd that the Contractor do provide the Troops going to Dorchester during their Stay there. At half past Seven o'Clock this Even[g]. went away of Cap. Marions Company

1 Capt[n]. 2 Sub: 2 Serj[ts]. 2 Corp: 1 Drum & 29 Priv[ts]. of Capt[n]. Hugers

1 Cap. 2 Sub: 2 Serj[ts]. 2 Corp: 1 Fifer & 26 Priv[ts].

Total of all Ranks Seventy One 71

After hostilities commenced between America and Britain the Cherokee Indians "were disturbed by the state of things in the colonies, they were deprived of their usual supplies and were therefore in a very bad humor." To stop any possible raids by the Cherokees, Henry Laurens, President of the South Carolina Council of Safety, had sent a supply of 1,000 pounds of powder and lead to the Indians for hunting. This powder was escorted by a detachment of Rangers under the command of Lieutenant Thomas Charleton. The Loyalists did not want the Patriots to have the Indians as allies, so Loyalist Captain Patrick Cunningham needed to intercept the shipment. Cunningham delayed the first wagon driven by Moses Cotter at the Congarees, when that wagon moved far ahead of the Ranger escort.

Cunningham asked Cotter what he had in the wagon. Cotter told them it was rum. Sixty Loyalists then rose out of the roadside. Cunningham said, "I order you to stop your waggon in his majesty's name, as I understand you have ammunition for the Indians to kill us, and I am come on purpose to take it in his majesty's name." The Loyalists removed the kegs of powder and cut the lead bars into small pieces with their tomahawks.

When Lieutenant Charleton and the Ranger escort appeared in the distance, Cunningham's men quickly hid in the trees. One of Cunningham's men said, "there comes the liberty caps; damn their liberty caps, we will soon blow them to hell." The Rangers were quickly surrounded, and outnumbered, facing rifles at close range. When one of the Loyalist fired into the air Charleton wisely surrendered. The Loyalists marched off with the ammunition and the Ranger prisoners, who were soon released.

Barnard Elliot, of the 2nd South Carolina, wrote "the Tories in the remoter part of the Province having defeated our militia commanded by Major Williamson and a detachment of the Provincial troops under the command of Major Mayson, and intended marching to Dorchester to take possession of our powder magazine there, a detachment was ordered, under the command of Capt. Marion, to proceed immediately for that place." Elliot was incorrect in his report, since Williamson and Mayson were not defeated, only the Rangers escorting the wagon had been.[162]

By Capt[n]. Marion

________________________Dorchester Novemb[r]. 10th. 1775__________________________
General Orders by Cap. Fran[s]. Marion
__________________________Parole Charles Town__________________________________

A Return of the men under Captain Pervis [163] to be made to morrow at 10 O'Clock to the Comm[dg]. Officer

One Serjeant 1 Corporal & twelve Privates of the Rangers to Mount Guard to morrow at a House in the Rear of the Barracks. Centinels to be plac'd by L[t]. Dubose who is to act a Adjutant of the Detachment. _______No Officer of Soldier to leave the Detachm[t]. without Leave of the comm[dg]. Officer

_______'Tis recommended that good order be kep up by the Troops & whatever Soldier who commits any Misdemeanor may expect to be punished by a Court Martial ____A Serjeant of the Rangers to attend at 10 O'Clock every day for orders

__________________________Regimental Orders ___________________________________

One Subaltern, 1 Serj[t]. 1 Corp[l]. & 18 Privates to mount Guard in the place where the guard is now kept. - L[t]. Mazyck for the Day L[t]. Hall for Guard ____________________

1775 Nov[r]. ___________________Parole Congress ___By Cap[t] Marion_________________________

The Royal Governor of South Carolina, Lord William Campbell, had taken refuge aboard the *Tamar*. The *Tamar* had been reinforced by the British sloop *Cherokee*, and was anchored just out of range of Fort Johnson's guns in an area known as Rebellion Road. Rumors of an impending attack by Royal Navy ships brought about changes by the South Carolinians in the harbors defense. The Carolinians decided to close the Hog Island channel by sinking several old schooners.[164]

The plan was to have the South Carolina Navy schooner *Defense* sink several old ships in Marsh Channel and Hog Island Creek. Thirty-five men from the 2nd South Carolina Regiment on Fort Johnson augmented the crew for this mission and were to act as marines. Fort Johnson was placed on alert so that they would be able to fire upon the British ships in support of the *Defense*. Their rules of engagement were to have the British shoot first so that they would be the aggressors. To keep this mission a secret Lieutenant Colonel Motte was ordered to detain all boats and canoes fishing in the harbor, so any Loyalists would not warn the British.

On the afternoon of November 11th Captain Tufts and his ship, *Defense* proceeded with four hulks in order to sink them in the channel. The *Defense* flew the new blue South

---

[162] O'Kelley, *NBBAS, Volume One* pp. 53-54.

[163] John Purvis (Pervis) was born in 1746 in Scotland. He moved to the Ninety-Six District, South Carolina. He was elected as a captain on 17 June 1775 in the 3rd South Carolina Rangers and was stationed at Fort Charlotte (near Augusta, Georgia). On 18 June 1775 he became an adjutant. He was a major in the Florida Expedition from 1 June to 18 August 1777. He was in the battle of Stono Ferry in 1779. He served 94 days as an adjutant general during 1779. He served as a lieutenant colonel in the militia under Colonel Hicks from 1779 to 1781. He was in the siege of Augusta in 1781. He died on 4 May 1792 at the age of 46.

[164] Hog Island is the location of present-day Patriot's Point.

Carolina flag with the crescent in the upper corner.[165] On board the *Defense* was William Henry Drayton, the President of the Provincial Congress.

The Defense sent out a small boat to sound the bar of the Hog Island Channel, but they were spotted by the Royal Navy. At 4:30 in the afternoon the *Tamar*, commanded by Captain Edward Thornborough, fired six shots at the *Defense*; Tufts dropped anchor and returned fire, sending two shots towards the British ships. The *Tamar* continued to fire, but to no effect. Captain Tufts fired one more shot, then ignored the *Tamar* and commenced to sink the hulks. Captain Tufts sank three of the hulks, but the tide came in. Tufts decided to wait until the tide went out again so that the hulk would sink where it was supposed to be.

At four o'clock in the morning on the 12th the *Tamar* and *Cherokee* had drifted close to the *Defense*. They both fired a broadside at the South Carolina Navy ship and continued to fire until seven o'clock, when Tufts carried the hulk to the proper area and sank it. After completing his mission Tufts retired from the scene. The *Tamar* sent an armed boat to the hulk and set it on fire and towed it to shallower water. Tufts fired a parting shot at the armed boat, but since he was ineffective he quit expending valuable powder.

The *Tamar* had shot over 100 rounds at the *Defense*, but only struck her three times, once in the broadside, once in the counter, and one shot cut some rigging. The British shots were too high and most went between the *Defense's* rigging and struck the land behind her. Fort Johnson fired three shots at the British vessels with her 26-pounders, striking the sails of the British ships. The Charles Town Militia was called out during the naval battle, and remained on station for several hours afterwards, prepared for a very improbably land assault.[166]

11th A guard to be constantly mounted by the Rangers consisting of 2 Serjeants & 12 Privates, & to place Centries round the Church, with Orders that no persons whatever be admitted to mar the Church yard without particular Orders from the comm^dg. Officer Lieut Baker Officer of the day.

__________Regimental Orders__________

For Guard to morrow 1 Serjeant 2 Corp^ls & 18 priv^ts:

The two Companies of the 2^d Reg^t to exercise at 9 O'Clock in the Morning & 3 in the afternoon daily- It is expected the Officers will use their Endeavours to make the Men go clean & their Hair comb'd.-

Dorchester Nov^r: 12th __________Parole Moultrie__________

The Rangers to mount Guard as Yesterday & the Orders given yesterday to be continued. The Serj^ts. of the Guards to give in a report of the Guard he commands to the Officer of the Day.___

The two Companies of Rangers to be under Arms to morrow at 10 oClock to be review'd by the com^dg. Officer___For the Day to morrow Lieutenant Monanghan.[167]__________Regim^t: Orders ___The officers of the 2^d. Reg^t. to examine the Mens Arms & Ammunition & to give strict Orders to the Men that they do not Waste their Cartridges; all those who lose or have their Cartridges damag'd may expect to be punish'd by a Court Martial. No Soldier to load their Guns without Orders from an Officer.

---

165 This flag and the other colors of the *Defense* were made by Ann Holmes, of Charlestown, for £37.

166 O'Kelley, *NBBAS, Volume One* pp. 56-60.

167 David Monaghan was born in 1754. He served as a lieutenant in the 3rd South Carolina Rangers under Captain Imhoff from 26 August 1775 until he resigned on 15 December 1775.

13th ____________________________ Parole Motte ______________________________________

As the Officers of the Rangers are not willing to inspect any Guards but their own, it is order'd for the future that the commdg. Officer of that Core do appoint one of their Subalterns as Officer of the day & that Officer so appointed to report to the commdg Officer.

No Drums to beat on any Acct: till the drums of the 2d. Regt gives the Lead, all Drummers who do not obey this order may expect to be confin'd & try'd by a Court Martial for Disobedience of Orders.-

<u>Regimental Orders</u>

Magazine Guard as yesterday. Lt Mazyck is Officer of the day to day, in the room of Lt Monanghan who is to do duty in his own Core only, & Lt Hall for the day to morrow. Roll Call to be at Troop Beating, & the two Company's of the 2d. Regt to turn out to exercise at 10 o'Clock for the future A Sergt: to superintend the Negroes who work at the cannon.-

Dorchester 1775 _________________ Parole Washington ________________________________

Novr: 14th A Party of Rangers consisting of 1 Serjeant & 6 Privates to hold themselves in readiness to proceed after deserters. the Adjutant will deliver ¼ lb: powder to each man, or 12 Cartridges to be taken from every Man of the 2d. Regt. provided the Rangers has no Lead. Guards as Yesterday.-

________________________________ Regimental Orders ____________________________

Lieut. Baker for the Day to morrow. Orders to Serjt: Little [168] of Captn: Purvis Compy: of Rangers.

You are to take six Men under your comd & proceed after five deserters of the 2d. Regt: if you come up with them you are to order them to lay down their Arms, & if they will not obey, you are to endeavor to take them. but if you find it impracticable. You are to fire on them, & endeavour to conquer them by force of Arms, All Prisoners you take you are to cause to be brought to Dorchester or Charles Town & deliver them to the Guards there. As the Men were Inlisted on little Pee Dee it is probable they are gone that way. & if you should meet any other Deserters from the 1st: & 2d: Regiment, you are to take & proceed against as above. _______ You are to your respective Company's in 15 days from this date. All Prisoners which you may take you receive Ten Pounds pr Man & all reasonable Charges.

<u>General After Orders</u>

Counter Sign Lee. 1 Serjt: & 1 Corp from the 2d. Regt. to join the Magazine Guard immediately & a Guard of 1 Serjt: 1 Corp & 6 Men from Magazine Guard to be posted where Lt Dubose will direct. One Centinel of the Rangers to be taken off the Church & posted at the Foot of the Bridge after dark. & a Centinel of the 2d: Regt: planted in the room of the one taken from the Church. 2 lb. Powder & 2 lb. lead to be given to the Rangers immediately, The Lead to be bought at Mr. Warings.-

Dorchester 1775 ___________________ Parole Putnam ______________________________

Novr. 15th The Rangers to detach after dark a Serjeant & 3 men from their guard to the Foot of the Bridge as last Night & to place 1 Centinel on the Foot of said Bridge with Orders not to let any pass either in or out the Town after Tattoo Beat. without the Countersign, & to give notice to their Guard of all Men who may be stopt in consequence of this order. ___Regimtl: Orders___

Counter Sign Boston.

---

[168] George Liddlet was born in Maryland. He was promoted to sergeant in the 3rd South Carolina Rangers on 19 August 1775.

A Serjeant, a Corporal & 6 Men to be detach'd from the Magazine Guard, as a rear guard. as last night. ___A Regimental Court Martial to be held at 11 O'Clock in the Tent to try such Prisoners as may be brot. before ym. Evidences to attend. Cap. Huger President, Lieut^{s}. Baker & Mazyck Members. __

For the day to morrow L^{t}. Mazyck

16th ______________________Parole Schuler Counter Sign Quebeck.
The detachd Guard at the Bridge to continue 'till further Orders. The comm^{dg}: Officer of the Rangers to give charge that their Men do not waste their Ammunition & no Centinels to load without Orders from their Officer. Lt Shubrick to do duty in Captn. Hugers Company in the room of L^{t}. Hall who is to join the Reg^{t}. at head Quarters.

L^{t}. Shubrick for the day to morrow.-

According to the Sentence of the Court Martial held the 15th Ins^{t}. Patrick Berry [169] for drunkness on Guard was sentenc'd to receive 100 Lashes (Confirm'd)

John Cody for absenting himself all Night from Guard, sentenc'd 200 Lashes, the com^{dg}: Officer was pleased to remit 50 Lashes.-

17th ______________Parole Gadsden Counter Sign Philadelphia
______________________________Regimental Orders______________________________
For the day to morrow L^{t}. Baker. A Serj^{t}. to be sent to see the Cannon plac'd L^{t}. Dubose to direct.

18th ______________Parole Effingham C. Sign Carolina.
The Officers of both Cores to visit their different Guards & Centries thrice in the day & the Same at Night. It is expected all Orders given will be punctually obey'd, & the State of Arms & Ammunition be particularly inspected by the Officers of their Respective Companies.____Regimental Orders ____

For the day to morrow L^{t}. Mazyck.-

______________________Parole Chatham Counter Sign Virginia
The Orders of the 13th Ins^{t}: respecting reporting to the commanding officer, & the Drums to be Particularly observ'd, as a Neglect must be taken Notice of for the future.-

1775 Dorchester By Capt Marion ________Regimental Orders______________________________
Novr 19 For the day to morrow Lt Shubrick. Three men from each Company to parade to morrow at Troop Beating as Fatigue Men & 1 Serj^{t}. to command them, who is to see them cut wood for their respective Companies. Two men to be order'd to morrow to make up Cartridges in the Guard Room.-

20. ____________________Parole Elliott Counter Sign Artillery ____________________
A Return of the Ammunition wanting to compleat 12 Rounds p^{r} man from both Cores to be made to day by 3 O'Clock in the Afternoon to the comm^{dg}: Officer ____Regimental Orders

A Regim: Court Martial to be held to day at 11 OClock in the Presid^{ts}: Tent to try such Prisoners as may be brought before them, Evidences to attend _____
President Cap: Huger, L^{ts}. Shubrick & Baker,

---

[169] Patrick Berry enlisted in the 2nd South Carolina on 4 November 1775. He deserted on 16 December 1776.

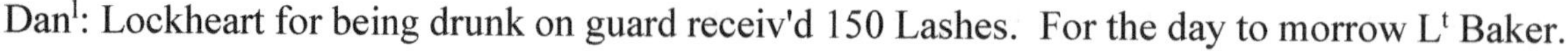

Danl: Lockheart for being drunk on guard receiv'd 150 Lashes. For the day to morrow Lt Baker.

Genl: Orders in Case of an Attack.

On the first Alarm of an Attack being made, the Troops to be immediately under arms, before their different Guards & Barracks & all Officers to attend if the Attack be made on or near the rear guard of the 2^{d} Reg^{t}: the Rangers is immediately to march behind the Church taking their Guard with them (but not to call in their Centries) & there defend the Church 'till call'd in by the commanding officer.

The Magazine guard is to take possession of the Fort under the Command of the Officer of the day, the Centinels to be call'd in & plac'd on the ramparts. no Person to be admitted without knowing him by Parole & Countersign. The Officer must pay particular Attention in defending the Gate. The remainder of the 2^{d}: Regiment to march with the greatest Silence to the rear guard, where that guard is to join them.

If the Attack is made towards the Bridge, the Rangers is to send their Guard to defend the Church wth: their officer of the day, The rest of the Rangers to march to the Foot of the Bridge & Defend that Pass. The 2 Co^{s}: of the 2^{d} Reg^{t}. are also to march to the place attack'd & their Magazine Guard to take possession of the Fort as above. All Officers is to be particularly attentive that no Noise be made & to keep their Men from Hurry & Confusion, All orders will be convey'd to the other Officers (by L^{t}. Dubose) from the commanding Officer.

The reason for the preparation for a possible attack was due to the battle fought at Ninety-Six, the first land battle of the Revolution fought in the South. It came about when Major Andrew Williamson vowed to retake the ammunition captured by Cunningham at Mine Creek, and bring the Tories to justice.

On November 18th Williamson's force of almost 600 South Carolina and Georgia militia marched towards Ninety-Six and erected a square breastwork in the field near the town in two hours. Williamson chose to defend there so that his artillery, three swivel guns would have a better field of fire. This fortified position around a barn became known as Williamson's Fort and was constructed of old fence rails and straw with a rawhide covering. Williamson chose to defend there instead of Ninety-Six so that his artillery of three swivel guns would have a better field of fire and would be used at their maximum potential.

Almost 2,000 Loyalists crossed the Saluda River and arrived at Ninety-Six on November 18th with drums beating and colors flying. The Loyalists took over the town and dug their way into the jail, converting it into a fortified position "from which they fired into the fort, without much exertion". This began the siege of Fort Williamson.

Major Mayson and thirty-seven South Carolina Rangers had reinforced Williamson's force, and they had plenty of supplies but no water. The Loyalists began the siege by demanding that Williamson disperse. Williamson parleyed for some time, delaying any Loyalist assault, but two of his men were seized outside the fort when they went to get water. Williamson ordered his men to fire an opening volley to let the Loyalists know that the parley was over. The ensuing skirmish lasted for two and a half hours, until darkness had made accuracy impossible.

For two days the swivel guns on the fort kept the Loyalists at bay. On November 20th the Loyalists tried to burn the defenders out by setting fire to the grass and fences around the fort, but the grass was too wet. They then attempted to use a "rolling battery," behind which they would approach the barn and set it afire. However the rolling battery caught on fire instead. Williamson's men dug a forty foot well, which brought forth "very good water." They were well fed since they had 38 barrels of flour and 4 steers.

On November 21st both sides agreed to call off the fight. This was just in time for Williamson because the fort's defenders only had forty pounds of powder left and were about to launch a counter-attack to break the siege. The second Treaty of Ninety-Six was signed on 22 November and stated that Major Joseph Robinson would withdraw his troops beyond the Saluda so Williamson could withdraw his troops without being molested.

The Loyalists lost 52 killed and 20 wounded. Twelve of Williamson's men were wounded and one was killed. He was James Birmingham, a South Carolinian, considered as the first Southerner to die in the Revolution for the Patriot cause.[170]

Novr. 21st __________________Parole St. Johns Counter Sign Santee _____________________

The Rangers to receive 3 Rounds Cartridges pr man to be deliver'd by Lt. Dubose this afternoon.-

Regimental Orders- The two Companies of 2d Regt: to receive as many Cartridges as will compleat 12 Rounds pr Man. A Regimental Court Martial to be held at the Presidents Tent to try such prisoners as may be brought before them. Evidences to attend.

Cap: Huger President, Lts. Mazyck & Dubose Members. Samuel Gale [171] & John M. Sharp [172] were Sentenc'd 150 Lashes Each, but by the Intercession of the Officers, the Punishment was remitted. Lt. Mazyck for the Day to morrow.

22d ________________________ Parole Berkley Counter Sign Four ________________________

For the day to morrow Lt. Shubrick.-

23d ________________________ Parole Cambridge C.Sign Seven. _________________________

The four Men of the rangers who were order'd last Night to hold themselves in readiness to proceed after deserters with Lt. Press Smith, & to receive Ammunition to compleat 12 rounds pr man, & to receive 3 Days Bread from the Adjutant.-

No Man to sell any Liquors to the Soldiers without an Order from an Officer, & the Inhabitants to be made acquainted with this Order.-

________________________________Regimental Orders_________________________________

Serjt: Pratt & 3 Men to hold themselves in readiness immediately & to receive 3 days Rations & compleated with 12 rounds pr Man.[173] For the Day tomorrow Lt. Baker. A Regimental Court Martial to be held to day at 11 O'Clock at the presidents Tent to try such Prisoners as may be brought before them; Evidences to attend. Captn. Huger President Lts: Shubrick & Dubose Members. Corporal

---

[170] O'Kelley, *NBBAS, Volume One* pp. 61-65.

[171] Samuel Gale was in Captain Peter Horry's company. On 21 November 1775 he was supposed to receive 150 lashes from an unnamed offense, but it was remitted. On 20 February 1776 he received 100 lashes for neglect of duty while in Francis Marion's company. He was wounded at Fort Sullivan on 28 June 1776. He enlisted in the 4th South Carolina (Artillery) on 15 July 1778..

[172] John Martin Sharp served in the 2nd South Carolina Regiment. On 2 November 1778 he received "50 lashes with a Cat of nine Tails on the Bare Back for Neglect of Duty".

[173] Possible James Pratt, who served in the 3rd South Carolina Rangers.

Ceaton for Neglect of Duty was reduc'd & turn'd in the ranks as a Private. [174] James Grover for the Same Crime reciv'd 100 Lashes[175]

24th ________________________Parole Lexington C.Sign Nine ________________________
______________________________Regimental Orders ______________________________
For the Day to Morrow Lt Mazyck

25th ________________________ Parole Concord C.Sign Thirteen ______________________
Regt: Orders. For the day to morrow Lt. Shubrick.
Gen. after Orders for 23rd Inst: but omitted
The men in both Cores to turn out Immediately & to have their Arms examin'd, The Officers to take particular Care that their Arms are in good Order, no Soldier to absent himself from his Guard or Barracks. on any Account whatever, The Officers to be particularly attentive to their Duty

Novr 26 ___________________ Parole Ninety Six C.Sign Thompson ____________________
All Waggons & Waggoners coming from Chas Towne. this Day to be Stopt 'till Examin'd by an Officer which will be appointed for that purpose.-
______________________________Regimental Orders ______________________________
The Magazine Guard to Consist of 1 Serjeant 2 Corpls: & 15 Privates.. 1 Corporal & 3 Men to be detach'd in the Usual Place. any Soldier going out of the mark'd Lines on the Back of the Town to be Confin'd & punish'd by a Court Martial.-

27th ______________________ Parole Saludy C.Sign Richardson. _____________________
The Orders of Yesterday respecting Waggons & Waggoners coming from Town to be observ'd 'till countermanded.
______________________________Regimental Orders ______________________________
For the Day to morrow Lt. Baker. The Quantity of Provisions recd by the Officer of the day from the Contractor to be given to Lt. Dubose who is to keep an acct of it

28th ______________________Parole Johnson C.Sign Friday_________________________
A General return of the State of each Company of the Rangers & the two Companies of the 2d . Regimt: specifying how many Men Sick, on Command[176], on Duty & for Duty, with the Number of Rifles & Smooth Bore guns, also the Quantity of Powder & Ball they may now have. This return to be made to morrow by 3 O'Clock, that it may be sent to Town

---

[174] John Keaton enlisted in the 2nd South Carolina Regiment on 4 November 1775. On 13 November 1776 he received 80 lashes for theft. During the siege of Charleston in 1780 he was in the militia.

[175] James Grover enlisted in the 2nd South Carolina Regiment on 4 November 1775. On 23 November 1775 he received 100 lashes for neglect of duty. On 20 February 1776 he received 100 lashes again for the same offense. He was absent during the Battle of Fort Sullivan, but he was pardoned. He was promoted to sergeant on 13 July 1778 in Captain Lesesne's company, but he was demoted a month later to private in Captain Moultrie's company on 8 August 1778 for beating a corporal. On 27 November 1778 he received 50 lashes on the bare back with switches for defrauding the sutler and he was put under stoppages to repay the sutler the amount that he swindled from him. He was in Captain Moultrie's company during the siege of Savannah in 1779.

[176] On command meant "on detached service".

Regimental Orders

For the Day to morrow L[t]. Mazyck.

29th Parole St George Counter Sign Monday

Regimental Orders

For the day to morrow Lieut Shubrick.-

30th Parole Richmond Counter Sign Wednesday.

The orders of the 28th Ins[t]: respecting a General State of the Comp[ys]. of the Rangers & Foot now at Dorchester to be made at 12 O'Clock today to the comm[dg]: Officer.

Regimental Orders

A Regimental Court Martial to be held to Day at 11 O'Clock to try such Prisoners as shall be bro[t]. before them L[t]. Mazyck President, L[ts]: Dubose & Baker Members. Thomas Bowen & Christ[r]. Harnom, for Disobedience of Orders were sentenced to 100 Lashes each but it was remitted by the comdg: Officer. [177] Jn[o] Laws [178] for breaking Robert Lance's [179] Leg, but it being prov'd accidental he was released.-

Dorchester General Orders Dec[r]. 1st: 1775 by Cap[t] Marion

Parole Keowee Counter Sign Fifty Nine,

Reg[tl]: Orders, For the day to Morrow L[t]. Mazyck.

2nd Parole Nassaw C.Sign Thirty Seven

Regimental Orders. The Subalterns of the two Companies of 2[d]. Reg[t]. to be on the Parade at roll Call & to visit their Sick twice a day, any Officer who neglects this Order will be taken Notice of & 'tis recommended to them to look over the Orders that has been given to them since the Detachm[t]. came to Dorchester. Lieu[t] Shubrick for the Day to morrow.

3rd Parole Eugene Counter Sign 32.

The Party of Militia that came last Night under Command of Lieu[t]. Clifford to form a guard to morrow of 1 Serjeant & 4 Men to be posted where L[t]. Dubose will direct who will attend at 10 O'Clock for that Purpose. [180]

Regimental Orders.- For the Day to morrow L[t]. Baker.

---

[177] Thomas Bowen enlisted in the 2nd South Carolina Regiment on 4 November 1775. He received 100 lashes for desertion on 18 December 1776. On 23 April 1777 he received 25 lashes for neglect of duty. He reenlisted on 24 September 1777. He received 75 lashes on 16 April 1778 for being drunk on guard duty. He received 35 lashes on 7 July 1777 for being in "liquor" on guard duty. On 15 June 1778 he was supposed to receive 100 lashes for desertion, but he was pardoned. He was under Captain Hall at the siege of Savannah in 1779. After the fall of Charleston he served under Captain John McClure, Hugh Knox and Colonel Lacey.

[178] John Laws enlisted in the 2nd South Carolina on 4 November 1775. On 15 December 1775 he received 100 lashes for some offense.

[179] Robert Lance enlisted in the 2nd South Carolina Regiment on 4 November 1775 as a fifer. On 27 November 1776 he was mult 14 days pay for drunkenness. He was appointed fife major on 16 November 1778. He was under Captain Peter Gray in 1779 during the siege of Savannah..

[180] Charles Clefford was a lieutenant in the Pon Pon Company of the Colleton County Regiment of Foot.

4th Parole St. Michael C.Sign Twenty Nine.

The Party of Militia which came last Night Com^{ded}: by Captn Linnen to join L^{t}. Clifford's Party & to Duty together: [181] All Parties of Militia which may come are to join in one Core & do Duty together. The Eldest Officer present to act as reg^{tl}. Commandant. of the Core, who is to send a Serj^{t}. every day at 10 O'Clock to receive Orders from the Comm^{dg}: Officer of the Whole.

1 Serj^{t}: 1 Corporal & 8 Men from the Militia to form an advanc'd Guard to be posted by L^{t}. Dubose

Orders to be observ'd by all Guards.

You are to stop all arm'd Men which may be coming into Dorchester & not suffer then to pass their Guard 'till Orders are sent by the Comm^{dg}: Officer: on the first Discovery of any arm'd Men; the Officer of the Guard is to send a Man immediately to acquaint the com^{dg}: Officer & to order the arm'd Party to stand; if they refuse, the Guard is to fire 2 Guns, call in their Centries & retreat towards their Comp^{y}. or Reg^{t}. to which they belong.-

All Centries to rest their firelocks when he passes y^{rs}, & shoulder'd Arms to all other Officers. & to be very Vigilant and watchful on their Post. more Particularly at night

Regimental Orders. For the Day to Morrow L^{t}. Mazyck.

5th Parole Berkley Counter Sign Twenty Seven

The Militia to furnish a guard of 1 Serj^{t}: 1 Corp^{l} & 12 Privates every Day for the advanc'd Post & one Subaltern of the Day to visit their Guards & Centries & see they are alert on their Post. no more Drums to beat the revella in the Morning but one from each different Corps & then to take the Lead from the Provincial Drum: all Guards to be relieved at 9 O'Clock & a report of each Guard to be made by the Officer of the Day to their differ^{t}. Command^{rs}: & to the Com^{dg}: Officer of the whole- Captn. Huger Captn. of the Day to go the Grand rounds & visit all the differt. Guards as often as he shall think proper. The Guards when visited by the Captn of the Day to turn out & receive him & obey all orders that may be given by him or those who will succeed him as Captn. of the Day a Verbal Report to be made to the Commandant.-

Reg^{l} Orders. For the Day to morrow L^{t}. Shubrick who is to make his Report to the Captn of the Day.

Decr. 6th Parole Colliton C. Sign Fifteen

The Commandant of the Militia to Send in a return of the Number of Officers & Men in each Detachment & the Names of the Officers now in Dorchester this return to be made by 4 OClock this afternoon

A Guard of the Militia to be mounted consisting of 1 Subaltern, 1 Serj^{t}: 2 Corp^{ls}: & 20 Priv^{ts}: to be posted by L^{t}. Dubose. no Guns to be fir'd within the Guard on any Acc^{t}: whatever without Orders.

For the Day Captn. Peyre Im Hoff [182].-

Reg^{l}. Orders. For the Day Lt Baker. The Subalterns to order and make the men get their Cloaths clean as we may expect to be order'd in a Day or two for Charles Town ________

---

181 Unable to determine Captain Linnen's first name. He served as a captain in the Colleton County Regiment of Foot.

182 John Lewis Peyer im Hoff (Peyerim) was a captain in the 3rd South Carolina Rangers. He mainly served as a recruiting officer in 1775 and 1776. He died in 1776.

7th General Orders. The Subalterns Guard of Militia which mounted today; to detach from his Guard. 1 Corpl. & 4 Privates to relieve the Guard of Rangers at the Bridge & continue that Guard from them 'till further orders.

Officers of the Day to stop all Drums which may beat contrary to Orders, and to take Notice of all such as still continue firing their Guns within the Guards.

For the Day Captn. Linnen. Regl. Orders. A Return of the Cartridges wanting the Number last given to each Man by to morrow 10 O'Clock.________________________

The reason that the militia was called out, and replaced many of the Continentals guarding the town was because of an expedition that became known as the "Snow Campaign". Five hundred Rangers and 4,000 militiamen were sent to the back country of South Carolina to subdue to Loyalists that had dispersed after the first siege of Fort Ninety-Six. This campaign ended at the Battle of the Great Cane Break on 22 December 1775. This is also why there is extra emphasis on not playing any drums or having any indiscriminate firing on the guard posts. Such noises could start a full scale panic within the town, which would interpret it to be a Loyalist attack. [183]

8th Parole Port Royal Countersign Nine.
For the Day Captn. Snipes [184] Regl. Orders For the Day Lt. Shubrick

9th Parole Savanna Counter Sign Six.____________________________________
For the Day Captn. Somers. [185] A Report to be made by the Officer of each Guard to the Captn. of the Day, & the Captn. of the Day to report the Same writg. to the Comdg. Officer

10th by Captn. Marion Regimental Orders Dorchester
For the day Tomorrow Lieut. Baker.-
General Orders. Parole Sunbury Countersign Ogrehee
6 Men to be added tomorrow to the main guard of the Militia & 2 men to the detached guard at the Bridge.[186]

---

[183] Full details of the Snow Campaign are in "*Nothing but Blood and Slaughter, Volume One*".

[184] William Clay Snipes was a captain in the Horse Shoe Company of the Colleton County Regiment of Foot. From 1776 to 1781 he was a captain in the Rangers. In 1781 he was a major in the dragoons under Colonel William Polk and General Sumter. He was also in Marion's partisans under Colonel Peter Horry.

[185] John Sommers was the captain of the Stono Company of the Colleton County Regiment of Militia.

[186] By December 10th the entire 2nd South Carolina Regiment, minus two companies coming back from Dorchester, had left the downtown barracks and established a camp near Fort Johnson. Colonel Motte writes, "We had now a camp on James' Island, near Fort Johnson, of at least five hundred men, well armed, well accoutered, and well clothed with a sufficient number of regular good tents: the field officers and captains, each a tent and marquee, and a tent and marquee for the subalterns of each company. The officers tents were at their own expense: We now began to look, and act like soldiers, and keep up the strict discipline. The men were taught the maual manoevering, and the exercise of the great guns, which made them matrosses as well as infantry; they were as well clothed as troops could be, and made a handsome appearance: we thought it best to form our camp on James' Island, for the benefit of the soldiers healths, and the better situated to keep them from liquor, which it was impossible to do in Charlestown, notwithstanding our strict discipline, and by being there, they were ready to support Fort Johnson, should it be necessary."

For the day to day Cap[t]. Waring [187]

11th Regimental Orders. For the day tomorrow L[t]. Mazyck
General Orders Parole Simons. Count[r] Sign Talbot
No Officer or Soldier to leave Dorchester without permission from the commanding officer. Cap[t]. Smith will order his adjutant to keep an exact account of all that who may leave the command with or without leave- For the day tomorrow L[t] Shubrick
Regimental Orders. A Court Martial to be held at 11 OClock at the presidents Tent to Try such prisoners as may be bro[t]. before them.

Cap[t]. Huger President
L[t] Shubrick {Mem[s]} L[t]. Baker

Prisoners Gasper Mints [188] & Henry Peters [189] who were both Sentenced by the Court to receive 100 lashes each but the commanding officer remitted their Punishments ________

12th General Orders. Parole Cumberland C[r].Sign S[t]. Johns
For the day Cap[t]. Wigfall ________________
Regimental orders. Lieu[t]: Baker for the day tomorrow A Court Martial to be held at 11 OClock at the presidents Tent to try such prisoners as shall be bro[t]. before them.

Lieu[t]. Daniel Mazyck President
L[t]. Dubose } Mem[s].{ L[t] Baker

Prisoner Nicholas Flynn for being absent from roll calling was Sentencd 150 lashes on the bare back ________ [190]

Directions to the Serg[t]. of the advance guard of the Militia

You are not to suffer any armd men ~~who may attempt~~ to come into Dorchester. All such armd men who may attempt to come in you are to stop & Send a man to the Cap[t]. of the day or the commanding officer. Acquainting who they are & what may be their numbers. And if they should attempt to force a pass, you are to fire two guns as an alarm Calling your Centries And retreat Immediately to the main guard and join them the officer of which will have his directions. this order

---

[187] Richard Waring was a captain in the St. George's Company of Militia. He later became the assistant state commissary.

[188] Gasper Mentz enlisted in the 2nd South Carolina on 4 November 1775. He received 50 lashes for neglect of duty on 18 December 1776. He was discharged on 8 July 1778.

[189] Henry Peters enlisted in the 2nd South Carolina on 4 November 1775. He was in Captain Francis Huger's company and was wounded at Fort Moultrie on 28 June 1776. He was discharged on 18 July 1778.

[190] Nicholas Flinn served in the 2nd South Carolina Regiment from 1775 to 1779. On 30 June 1776 he received 200 lashes and was ordered to be dressed in women's clothes for being absent the day of the battle. On 30 August 1776 he was reprimanded for some unnamed offense. On 26 May 1777 he was ordred to return a knife of another soldier after he stole it because the other soldier owed him money. He was supposed to receive 50 lashes on 1 December 1777 for neglect of duty, but it was reduced to being confined two days on nothing but bread and water. In 1778 he was under Captain Thomas Dunbar. He was supposed to receive 50 lashes on 14 May 1778 for striking Corporal Stone, but the sentence was postponed due to him being sick. He received 25 of these lashes on 20 May 1778, and 25 were remitted. On 14 July 1778 he was sentenced to do double duty for 8 days for being drunk. On 30 November 1778 he received 30 lashes on the bare back with a cat of nine tails for selling his regimental blanket. He was in Captain Lesesne's company at the end of 1778. He was in Dunbar's Light Infantry at the siege of Savannah in 1779.

is to be given to the guard which relieve you And no person to be admitted after Tatoo Beating without giving the Countersign-

Directions to the main guard of the Militia. On the first notice of an alarm you are immediately to Parade your men before the Guard house & give the earliest notice to the commanding Officer, & when the advance guard join you, the officer of the Main guard must command the whole & oppose all arm'd force which may attempt passing his guards And keep the post till order'd otherways by the Commandant- this order to be given to the officer who relieve you-

Directions For the Body of the Militia in Case of an alarm ____________

On the first Notice of an alarm every officer who commands a party of men. must immediately (without beat of drum) parade thier men before thier Barracks. And when in order march them to the place near the Market. where the guards are usually paraded. and there join the main Body of Militia. the officers must be particularly Attentive to keep their men from Hurry and Confusion and to march with the greatest silence. no drums to beat without particular order from the commandant.-

13th General Orders Parole Augustin. CrSign St Marks

For the day Capt. Ladsden [191]

Regimental orders. for the day Lt. Mazyck

A Court Martial to be held to day at the presidents Tent at 11 OClock to try such prisoners as may be brought before them Capt. Huger President

Lieut. Dubose} Mems { Lt. Baker

Prisoner James Hendry For neglect of duty was Sentenced 200 lashes on the bare back. and Confirmed-[192]

14th General Orders. Parole Pinsacola CrSign Mississippi

For the day tomorrow Capt Fergason [193]

The Rangers to add 3 men to their guard and place One Centinel by the Cannon by the Waters Edge Tomorrow

Regimental Orders. For the day tomorrow Lt. Shubrick.

Decr. 15th General Orders. Parole Bolingbrook. CrSign David

The Militia to make a return immediately of the Number of Cartridges wanting to compleat 12 rounds pr man and the officers to examine the mens arms and see they are in good order-

The officers of the Rangers to Examine also their Mens arms and see they are fitt for Service- The officers of guards not to let any of their men be absent from their guards particularly after retreat

---

191 Thomas Ladson was a captain in the St. John's Island Company of the Colleton County Regiment of Foot. He died in March 1778.

192 James Henry was born in 1753 in Cumberland County, PA. He enlisted on 4 November 1775 under Captain James Duff and Colonel Sumter while at Tryon County, North Carolina. This enlistment was in the 6th South Carolina Regiment, but he was transferred to the 2nd South Carolina in 1776. He was given 50 lashes on 9 May 1776 for disobedience. He was in the battle of Sullivan's Island on 28 June 1776. He also served under Colonel Neel in the South Carolina militia and under Colonel Graham in the North Carolina militia after the fall of Charleston in 1780. He married Elizabeth Russell on 11 May 1780. He died on 28 February 1841 at the age of 88.

193 David Ferguson was a captain in the Lower Salt Catchers Company of the Colleton County Regiment of Foot. After the fall of Charleston he was in the militia under Colonel Waters.

beating and 'tis expected they wont absent themselves from their Guards but be vigilant and watchful And keep strictly to the orders given

'Tis recommended to the officers of each corps that they see their men lodge their arms at night in such places as they may easily and handily find them without confusion or hurry-

For the day tomorrow Capt. Waring

Regimental Orders, For the Day tomorrow L^t^. Baker

A Court Martial to sitt at 11 OClock at the presidents Tent to try such prisoners as may be bro^t^. before them

Capt. Huger President
L^t^. Mazyck} Memb^r^ { L^t^. Baker

The Court Sentencd John Smith 30 lashes. & confirmd by the Commandant.[194] John Cody Sentenced 300 lashes & confirmd by the Commandant. John Laws Sentenced 100 lashes & likewise Confirmed

15th General After Orders: The Rangers to send at retreat beating this evening 1 Serg^t^. and 3 Men to reinforce The rear guard of 2^d^ Reg^t^. And Capt. Pervis to give orders Immediately to all his Officers to remain with their men to night and not to let any of their men be absent on any acct. whatever. And to send an officer to receive ammunition at 4 OClock this Even^g^ to compleat 12 rounds p^r^ man- The orders formerly given respecting an alarm to be observed-

Regimental Orders- Lieu^t^. Mazyck with one Corporal & 6 men of the first company to hold themselves in readiness with one days provision. ___ Capt. Huger with his company And one subaltern to take Possession of the Church after tatoo Beating- And to remain till further orders, the remaining men of the 1st Comp^y^. to reinforce the Magazine guard & the officer of that guard is the officer of the day. the orders formerly given respecting an alarm to be observd-

16th General Orders. Parole Peterborough C^r^Sign Canterbury
The orders of yesterday to be observed. For the day Capt. Pervis. Regimental Orders_____

For the day L^t^. Shubrick- A man to be sent to the out post of L^t^. Mazyck with 1 days provision Capt. Huger & his company. to take post as last night And the 1st Company to reinforce the Magazine guard & the officer of the day to be officer of that guard

17 General Orders. Parole Marlborough. C^r^Sign Joab
For the day tomorrow Capt Peyre Im Hoff

Regimental orders a man to be sent to L^t^. Mazyck to carry orders this evening __ guards as Usual

___

For the day tomorrow Lieu^t^. Baker-

18th The Detachment of the 2nd Regiment to hold themselves in readiness to march by 10 OClock for Chas town

---

[194] John Smith was in the 2nd South Carolina Regiment. On 15 December 1775 he received 30 lashes for an unnamed offense. On 1 January 1777 he received 100 lashes for being absent without leave. On 10 February 1777 he was remanded for want of evidence for an unnamed offense. However in the same court he received 100 lashes for drunkeness and misbehavior. He was reprimanded at the halberts on 30 September 1777 for rioting and drunkeness. He was in Dunbar's light infantry company at the siege of Savannah. He was appointed a corporal under Captain Henry Lenud and Colonel Peter Horry on 6 October 1781 while he was in Marion's partisans. He was promoted to sergeant on 15 December 1781. He died prior to January 1786.

NB. The fort magazine & the Command in Dorchester was Deliver'd up to Coll. Sam[l]. Elliott at 9 OClock this Day,[195] with Cap[tn] Marions Compliments to the Officers of the Militia (Except Cap[t]. Wigfall) & Returns his thanks for the ready Obediance to all his Orders while he Had the Honour of Commanding in Dorchester, also to there men for there good behavior-

The Detachment march'd out of Dorchester at 11 OClock & gott in Ch[s]. Town at dark-[196]

From December 18th to January 6th there are no entries in the Order Book. There is a blank page in between the entries, denoting a new year had begun. During that time the British built huts on Sullivan's Island for the crews of the British warships in Rebellion Road. The British were planning on living there for the winter months. The Carolinians were no longer going to put up with the Royal Navy ships anymore and decided to build a fortification on Haddrell's Point.

Before building the battery on Haddrell's Point Colonel Moultrie was ordered to clear Sullivan's Island of all English sailors "and Negroes, who are said to have deserted to the enemy." Moultrie was also ordered to make the island uninhabitable for the British by burning the "pest house" that the British were using as a living quarters. On December 19th, Captain Allston's Raccoon Company of Riflemen, consisting of Indian warriors, crossed over to Sullivan's Island and surprised the British and burned the Pest House on Sullivan's Island. They also destroyed some water casks, and burned "Gunner" Walker's house. No white men were casualties, but four Blacks "who would not be taken" were killed. The Indians in the Racoon Company also captured four white men, four women, and three children. Several British sailors hid from the Indians throughout the night and when they were taken off the island in the morning they were fired upon by the Indians. The Indians "were permitted to fire (being all rifle men) and must have killed every sailor what appear'd in the boats."

With Sullivan's Island clear of any enemy the Carolinians could build the battery on Haddrell's Point. The battery was to mount four 18-pounders and a force of 200 men from the regulars would build it. Throughout December 22nd the troops spent their time constructing the 228 fascines that were needed to build the 58-foot long battery. Pinckney's men emplaced the fascines while others positioned the cannon and laid down wooden gun platforms. All through the night and the next day the men quickly worked before the British ships noticed their efforts.

At dawn on December 24th Governor Campbell and the British received an unwelcome Christmas present.[197] A battery with two guns ready for action confronted the hungry

---

[195] Samuel Elliott was the lieutenant colonel of the Colleton County Regiment of Foot. He died on 7 April 1777.

[196] With the British "fleet" in Charlestown harbor increasing to six vessels, the Provinical Congress and Council of Safety knew that they needed professional soldiers manning the harbor defenses. The Council of Safety decided that the militia could safely guard Dorchester, twenty miles inland from Charlestown. Colonel Job Rothmaler and the Prince George county militia regiment assumed the duties at Dorchester. Henry Laurens called Dorchester an unhappy place that needed barracks, sentry boxes and a jail. The Council of Safety suggested that Colonel Rothmaler send young and robust men who can guard against domestic insurrection.

[197] Christmas had only been celebrated in December for a little over 20 years. In 1582 the Catholic church replaced the Julian calendar with the Gregorian calendar, but Protestants kept the old Julian calendar for almost 200 years more because they didn't want the Pope telling them what to do. When England accepted the calendar in 1752 they were eleven days off from the rest of Europe, and so they dropped the eleven days, moving Christmas from January 6th to December 25th. The colonists continued celebrating Christmas on

sailors. The battery had walls that were 24 feet thick and they also had cannons that were larger than those on the ships. The British ships weighed anchor and moved to a position near the Bar where they were out of range from Haddrell's Point. [198]

January 6th for many more years. January 6th is still celebrated in parts of the Carolinas, such as Rodanthe on the Outer Banks, and is known as "Old Christmas".

[198] A detailed account is in "*Nothing but Blood and Slaughter, Volume One*" by Patrick O'Kelley.

1776

Haddrells Point January 6th 1776

Orders by Captm Marion__Parol Chs Town

Lt Baker for the Day __ The Officers of the day to visit the guards twice at Night & as many in the Day

For the Battery guard 1 Sergeant 1 Corporal & 12 men

The Rear and Magazine guard, 1 Corporal & 6 Men Each

The Officers of guards to give strict Orders to the Centinels not to Lett any persons (who does not belong to the Detachment) come near their guard, the Centinels at the Battery to keep a good Lookout next to the water & not lett any boats pass without Orders, & those at the Magazine not to Lett any fire be brought near the Magazine on any Acct. whatever __ any thing which may happen During the night to Acquaint the Commandg. Officer & the officers of the Day Emediately

Roll call to be at 8 OClock in the morning at which time the guards is to be relieved, roll-call 4 oClock in Evening __ the men Left of the first regiment to do Duty part in Captn. Marion & part in Captn Hugers Compy. so as to make the two Compys. Equall in the number of men __ For the Day tomorrow Lieut. Jackson [199] _ the officers of the Day to lodge in the House near the Battery

Orders 7th January 1776 ________ Parole Coll: Moultrie

For the Day tomorrow Lt Baker

Gabriel Marion [200] Cadet,[201] to act as Adjutant to the Detacht. and to be obey'd as such

---

[199] Basil Jackson was commissioned a second lieutenant in the 2nd South Carolina on 2 December 1775. He was at Fort Sullivan on 28 June 1776. That seemed to be enough war for him, and he resigned his commission on 4 July 1776.

[200] Gabriel Marion was Francis Marion's nephew. He entered the 2nd South Carolina as a cadet on 7 January 1776. He was commissioned as a lieutenant in the 2nd South Carolina in 1776. He was at Fort Sullivan on 28 June 1776. He resigned his commission on 24 September 1776. He returned and was in the siege of Savannah and the siege of Charleston. He joined his uncle, Francis Marion, as a partisan in 1780, but was murdered by a Loyalist mulatto named Sweat, who killed him after he had been captured. Sweat was killed by one of Marion's officers the next day after he had been captured. The movie "*The Patriot*" was made in 2000, starring Mel Gibson. Gibson's character "*Benjamin Martin*" is loosely based upon Francis Marion. There is very little about the movie that is historically accurate, but in the movie Gibson's son is played by Heath Ledger. Ledger's character's name was Gabriel Martin, and he was loosely based upon Francis Marion's nephew, Gabriel Marion

[201] Since there were no military academies in the colonies at this time, officers would learn how to do their job by being cadets in a regiment. If they were able to learn the duties and responsibilities of an officer of the regiment, and they showed that they had leadership potential, they would be promoted to second lieutenant.

The troops to turn out at 10 OClock in morning to Exercise the Cannon under the Direction of Lt Mitchel of the artillery[202] __all Officers to attend [203]

Mr Marion to take particular Account of all the tents & Intrenching tools left by the first Regimt. & to apply to Sergt. Redmond for provision, & give a receipt [204]

Orders 8th Jany.________ Parole Coll: Motte

For the Day tomorrow Lt Jackson

A Court Martial to be held today at 11 OClock at the Presidents Apartment, to try such prisoners as shall be brought before them Evidences to Attend

Lieut Baker President Lts. Jackson & Mitchell members

NB. Robert Galloway [205] & Danl Sullivan [206] of Captn. Thos Pinckney Compy. 1st Regt for Disobediance of order was Sentenced to receive 200 lashes, Confirmed & Inflected- Jacob Sadler[207] of the same Compy. & Regt for Disobediance of Orders Sentenced to receive 200 Lashes had 100 remitted

Orders of 9th Jany __Parole McIntosh__

For the Day tomorrow Lt. Baker

Orders 10th January __ Parole Collo. Gadsden [208]

---

[202] The commissioners of fortifications knew that it would take trained gunners to aim and fire the guns located on all the fortifications around the town. The infantry regiments could do the unskilled work of pushing and pulling the artillery, but on 2 November 1775 the Council of Safety authorized a new regiment, the 4th South Carolina Regiment, also known as the Artillery Regiment. The regiment would consist of three companies of 100 men each. The commander of this new regiment was Owen Roberts. The regiment never fought as a single unit, but was detached to man the artillery throughout the town. The artillery barracks was located on Gadsden's Wharf. Gasden's Wharf was located where the US 17 bridge crosses the Cooper River.

[203] William Mitchell was born in 1747 in Virginia. He was commissioned a second lieutenant on 17 June 1775 and was promoted to first lieutenant of the 4th South Carolina Regiment (Artillery) in 1776 after the new regiment was created. He was promoted to captain in 1779. He was court-martialed again on 25 March 1780 with being absent from the camp on the night of 3 March, but he was acquitted. He was killed during the siege of Charleston on 12 May 1780.

[204] Andrew Redmond (Redman) was a corporal and then sergeant in the 1st South Carolina from 4 November 1775 to 1779. He was a military storekeeper from 1780 to 1783.

[205] Robert Galloway enlisted in the 1st South Carolina on 4 November 1775.

[206] Daniel Sullivan was in Thomas Pinckney's company of the 1st South Carolina in 1775. In November 1779 he was in Captain De St. Marie's company and was captured. After capture he enlisted in the British army.

[207] This may be Jacob Sellers who later served in Lieutenant Colonel Myddleton's 2nd Dragoons under Captain Isaac Ross during 1781.

[208] Christopher Gadsden was born in South Carolina on 16 February 1724. He was on the Royal Ship *Aldoborough* at the siege of Louisburg in the French and Indian War. He established the Charlestown Artillery Company in 1756. He participated in the Cherokee War of 1760. He served as a South Carolina Representative during the Stamp Act Crisis and served in the First Continental Congress from 1774 to 1776. On 17 June 1775 he became the colonel of the 1st South Carolina Regiment. He had been the commander of Fort Moultrie and Sullivan's Island. He was promoted to brigadier general in the Continental Army on 16 September 1776. He resigned his Continental commission on 2 October 1777, but was still in the militia and was taken prisoner at the fall of Charleston. He was imprisoned in St. Augustine. He died in Charleston on 28 August 1805 at the

For the Day tomorrow Lt Jackson

The troops to Exercise the small Arms at 4 OClock in after noon for the futer __ the Adjutant to see the wading of the Cannon put up in the storeroom & have all the Cannon shott carry'd in the Battery & piled up

The Centry to prevent any persons from taking them & through about

Orders 11 Jan[y].______ Parole Coll[o]. Huger

For the day tomorrow L[t]. Baker

Sergeant Simpson of Capt[n] Hugers Comp[y]. to act as Quarter-Master Serg[t]. to the Detachment

---

age of 81. The Gadsden flag, that depicts a rattlesnake and states "don't tread on me" is named after Christopher Gadsden, though he may not come up with the design. He presented a copy of this flag to Esek Hopkins, the commander in chief of the Continental Navy, and to the South Carolina Legislature. The South Carolina legislative journal wrote "Col. Gadsden presented to the Congress an elegant standard, such as is to be used by the commander in chief of the American navy; being a yellow field, with a lively representation of a rattle-snake in the middle, in the attitude of going to strike, and these words underneath, "Don't Tread on Me!"

Orders 12th Jany. _______ Parole Majr. Pinkney
For the Day tomorrow Lt. Jackson

Orders 13th Jany. _______ Parole Coll: Roberts
For the Day tomorrow Lt. Baker

Orders 14th Jany._________ Parole Majr. Elliott[209]
For the Day tomorrow Lt. Jackson
Ten men for fatigue tomorrow to work on the Battery under the care of Sergt. Simpson who is to give a gill rum pr Man a day

> The battery they were working on would become Fort Sullivan. The fort would be built by following the guidelines of one of the best military engineers at the time, John Müller. Müller wrote in his books on fortifications, that if ten to twelve feet of earth were added to a fort's walls, it would prove sufficient to stop cannon balls and splinters.[210] The fort on Sullivan's Island was constructed of palmetto logs laid one upon the other in two parallel rows that were sixteen feet apart. The space between the logs were filled with sand. The walls were fifteen feet high to protect the batteries, and longer planks were set up along the walls to stop any musket fire from enemy marines located in ship's tops, firing downwards. Unfortunately Fort Sullivan was unfinished and only had walls on two sides by the time of the battle in June 1776

For Capt Marion **Haddrell Point 1776** [211]

Orders 15th Jany __________Parole Washington
For the Day tomorrow Lt Mazyck

The officers to see the men keep there arms in good order & to Examine there Cartridges once a Day __ any Soldier who lose or wast his Ammunition shall be confin'd & punished __ no Officer or soldier to absent himself from his quarters on any pretence whatever & to hold themselves in

---

[209] The commander of Fort Johnson was Major Bernard Elliot. He had taken command on 5 December 1775. During his first inspection tour he became upset over the lack of progress on the fort. After four months of occupation only seventeen guns were mounted, and the battery on the hill to the west only had five guns. Elliot figured that he needed 248 more men to adequately man the two batteries. He would also need a reserve of 50 men to replace casualties during combat. With his present garrison he could only provide three men per cannon. Each cannon was recommended to have nine men. He also discovered he only had enough shot for each cannon to fire nine rounds. On December 6th, Elliot saw some palmetto logs floating along the beach that were to be used in the batteries. Elliot asked the overseer to get the slaves to pull the wood out of the water. The overseer refused and Elliott drew his sword and slapped him on the cheek. The overseer refused to work on the battery anymore until the Council of Safety could assure his safety. Those assurances did not come and work ceased on the fort. Also "the negroes disappeared and he couldn't get them back."

[210] John Muller, *A Treatise Containing the Practical Part of Fortification*, 1755, University of Michigan.

[211] At this point in the orderly book there will be a single line mentioning an officer and a location, this information lets the reader know who was in charge at the time, and where the unit was located. This also signifies the start of a new page in the orderly book. For example on this line it shows that Captain Francis Marion was in charge of the regiment, and the unit was located at Haddrell's Point. This would normally be at the top of a new page, but I will highlight it in "bold".

readiness with there Arms at a minutes warning, this Order to be made known to the Soldiers by there respective Officers

On January 12th Major Pinckney at James Island reported that he had sighted "two ships & a sloop Northward of the Bar." At seven in the morning of January 12th the Royal Navy frigate *Syren* approached Charlestown harbor. With the *Syren* was the sloop *Raven* towing a dismasted prize ship, the *Rittenhouse*. Not knowing if it was safe or not the captain of the *Syren*, Tobias Furneaux, sent a boat into the harbor to determine if the *Tamar* was still there. The smaller boat did not find any British vessels and turned around to return.

On the return trip the British boat was attacked by the pilot boat *Hibernia*. The British boat fired upon the Pilot boat and pursued it near Sullivan's Island. In the "fort" the 2nd South Carolina Regiment fired two shots at the boat. The British boat quickly left the harbor. The captain of the *Syren*, Tobias Furneaux, did not know what was going on due to the weather being "thick and hazey". When he heard the two shots from the fort on Sullivan's Island, they answered with a shot of their own, thinking that this was the *Tamar* signaling them. Most of the officers of the regular regiments were not in town, so it was with a sigh of relief that they were granted an extra day to prepare for a possible British attack. Major Pinckney reported that due to the wind and tide the British couldn't cross the Bar until January 14th. The South Carolina schooner *Defense* was told to transport the remainder of the 2nd South Carolina Regiment to Sullivan's Island then "do every thing in your power to cover and protect the works intended to be carried on there."

Colonel Moultrie, commanding the men at Haddrell's Point and Sullivan's Island, was reinforced with 82 riflemen from the 3rd South Carolina Rangers under the command of Major James Mayson.[212]

On January 13th Captain Ladson and 55 men of the 1st South Carolina arrived at Sullivan's Island. To make up for the loss of one company of men the Council of Safety told Major Pinckney that he could enlist a number of "trusty able bodied slaves to assist in the defense of the new battery to the westward of Fort Johnson." Three 26-pounders were moved to that location. Anticipating an attempt to take the city, the country militia were called on to march to Charlestown in case the *Syren* was the vanguard for such an attack. A report that the British had sailed south on Christmas Day from New York bolstered this idea of a possible attack.

On January 14th, a "General Alarm" was sounded at 8:00 on Sunday morning ordering the Charlestown militia to report for duty. Hundreds of men reported for duty. On January 15th, Captain Isaac Hayne and 173 militia arrived to help defend Haddrell's Point.[213]

A "damage control" plan was introduced for different points in town. General Orders for January 15th stated "That when the town shall be attacked or alarm'd one subaltern, one sergt. and twenty rank and file from Col. Pinckney's regiment take post at the State House with one of the fire engines and a sufficient number of negroes with fire hooks, axes and

---

[212] James Mayson was born in 1739. He was commissioned a major in the 3rd South Carolina Rangers on 18 June 1775. He was in the siege of Ninety Six in November 1775. He was promoted to lieutenant colonel on 18 May1776. He became a brevet colonel on 30 September 1783.

[213] Isaac Hayne was a captain in the Pon Pon Company of the Colleton County Regiment of Foot in 1775. He was later promoted to colonel and was taken prisoner at the surrender of Charleston. On 8 July 1781 he was taken prisoner when he was doing a raid on the home of Andrew Williamson, "the Benedict Arnold of the South". He was executed by the British in Charleston on 4 August 1781. This execution led to several revenge raids by the Patriots.

ropes and ladders to observe if any fires should break out in town, immediately repair to the place with the party and engine and endeavor to extinguish the fire."

Similar fire details were ordered to take post at Blandford's Corner, Grimke's Corner, Ramadge's corner and Col. Brewton's corner. In addition, the orders stated "If any fire should break out at night in town some persons form the Guard, are to go up in the upper gallery of St. Michael's Church steeple and then hold out a lanthorn on a pole pointing towards the fire." [214]

Fort Sullivan 1776

Orders 16th Jany ________ Parole General Lee
For the Day tomorrow Lt Baker

Orders 17th Jany ________ Parole Genl. Putnam

1 Serg^t. 1 Corp^l. & 9 Privates of Captn Vanderhorst Comp^y. of Militia to relieve the Provincial quarter guard & 1 Corporal & 4 privates of the above Comp^y. to mount a guard at the Magazine at 10 OClock today __ [215]

Adj^t. Marion will give the Officers of Each guard there Orders in Wrighting who is to Deliver those Orders to the next who relieves them.

A Sergeant of the Militia to attend dayly at 10 OClock in Morn^g. for Orders

Officers of the day to visit twice in the day & as many at night there guard, & Centinels & see they are alert & do there duty __ all guards to turn out once a day with rested Arms to the Commd^g. Officer all Centry to rest there arms when he passes them

The Officer of the Day to give in wrighting a report of the guards to the Command^t. when the guards is relieved

No Officer or Private to leave Haddrells without Leave from the Commanding Officer, all who do to be reported __ The Commanding Officer of the Malitia to keep a perticular Account of all there men who do duty __ No Drums to beat without taking the Lead from the Provincial Drum

---

214 O'Kelley, *NBBAS, Volume One*, pp. 77-78.

215 Arnoldus Vanderhorst was born on 21 March 1748. He was a resident of Christ Church. He was a captain in the Berkeley County Regiment of Militia until 1780. He served 91 days with Francis Marion in 1782. He was governor of South Carolina from 1794 to 1796. He died on 29 January 1815 at the age of 67.

For the Malitia guard to day L[t]. Gab[l]. Capers

Regimental Orders

For the day tomorrow L[t] Jackson
Serg[t]. Simpson to keep an Exact account of Those men who work on the Battery who will be paid five Shillings a man per Day

---

Orders 18th Jan[y]: _________ Parole Cambridge
Cap[tn] Quelcho Comp[y] to relieve the Malitia guard tomorrow [216]
A Return from the Different Comp[ys] of Malitia to be made tomorrow, with the Number of Commissioned & non Commissioned officers & privates for duty
For the Malitia guard tomorrow L[t]. Bolton
Regimental Orders For the day tomorrow L[t] Mazyck

By Cap[t] Marion **Haddrell Point 1776**

General Orders, 19th Jan[y]. _______ Parole Boston
The Magazine Guard to be relieved tomorrow by Cap[t] Quelch's Comp[y] __ Quart[r] Guard by Cap[t] Murrells [217]
Officer for the Day tomorrow from Cap[t] Murrells Comp[y]. Lieut[t]. Benj[m]. Villepontaux to do duty in the room of L[t] Mitchell who is relieved [218]

Regimental Orders

For the Day tomorrow L[t] Baker
The Subalterns to take it by turns to Exercise the men every day after roll-call in Aft:Noon

---

Gen[l]. Orders 20th Jan[y]. __ __ __ __ __ Parole Lexington
The Magazine & Quart[r] Guards to be relieved by Capt[n]: Murrell's Comp[y]. tomorrow
Off[r]. of the Day tomorrow L[t] George Irons

Regimental Orders

for the Day tomorrow L[t] Jackson

General Orders 21st Jan[y]. ___________ Parole Massichusett
The Eldest Officer of the Malitia to Order there guard

Regimental Orders

For the Day tomorrow L[t]. Mazyck

Haddrells Point 1776

Gen[l] Orders 22nd Jan[y].________ Parole Rode Island
The Quart[r]. & Magazine Guard to be relieved today by Capt[n]. Vanderhorst, Comp[y]. also by his Comp[y]. tomorrow

---

216 Andrew Quelch was born in 1747. He served as a captain of the Wando Militia Company in 1775 - 1776
217 Possily William Murrell, who was a captain in the Christ Church Company of Militia in 1781
218 Benjamin Villepontoux became the paymaster of the 4th South Carolina Artillery. He was taken prisoner at Charleston in May 1780

For the Day today Lt Capers for tomorrow Lt. Toomes [219]

Regimentl: Orders

For the day tomorrow Lt Mazyck

Genl Orders 23d Jany.________ Parole Providence
Regimental Ordrs   For the day tomorrow Lt Baker

Genl Orders 24th Jany.________ Parole Roxbury
Regimentl. Ordrs: for the day tomorrow Lt Jackson

Genl Orders 25th Jay: __________ Parole Conecticut
Regiml: For the day tomorrow Lt Mazyck

A Court Martiall to be held today at 11 OClock at President Apartment to Try such prisoners as shall be Brought before them Evidences to Attend

Presedt. Lt Mazyck
Lts Baker and Jackson- Members

Genl Orders 26th Jay. ___________ Parole New York
Regimentl Ordrs. For the day tomorrow Lt Baker

A Return of the Arms & Accouterments unfit for service & wanting to be made this after noon at 6 OClock

By Capt Marion **Haddrell Point 1776**

Genl. Orders 27th Jany __________ Parole Albany

Regimental Orders

Lt Thos Moultrie to do duty in the room of Lt Baker & to mount his guard to day, as Lt Baker is relieved

For the Day tomorrow Lt Mazyck

Genl. Orders 28th January ____________ Parole Crown Point
For the day tomorrow Lt. Benjn Villepontaux

Genl Orders 29th Jany.________ Parole Montreal

Regimental Orders

For the Day tomorrow Lt Thos Moultrie

Genl Orders 30th Jany.__________ Parole Quebeck

Regiml. Orders

For the day tomorrow Lt. Mazyck

---

[219] Joshua Toomer enlisted on 20 October 1775 in the St. Helena Militia under Captain John Jenkins. He later was promoted to lieutenant and then captain. In 1782 he was a captain under Colonel Vanderhorst and General Marion.

Gen[l]. Orders 31st Jan[y].______________ Parole Canada
Any Soldiers who shall cut any live Oak trees or Branches or Burn any rails may depend on being severly punished and any Non Commissioned Officer who Should see or know, any Soldier Cutting or loping such trees or Burning rails, & do not Emediately confine him or them may Depend on being punished as if he had commited the offence ___ Regimental Orders

For the Day tomorrow L[t] Benj[n]. Villepontaux

General Orders 1st Feb[y].____________ Parole Niagara
Regiment[l]. Orders

For the Day tomorrow L[t]. Moultrie

A Court Martial to be held at 11 OClock today at the Quarter Masters room to try such prisoners as Shall be brought before them Evidences to Attend

Capt[n] Huger President
L[t]. Tho[s] Moultrie and Mazyck Members

Gen[l] Orders 2nd February ___________ Parole Ohio
Regim: Orders

For the day tomorrow L[t]. Mazyck

General Orders 3[d] Feb[y]. _____________ Parole Pittsburgh
Regt. Ord[rs].

For the day tomorrow L[t]. Benj[n]: Villepountaux

General Orders 4th Feb[y].__________ Parole Orleans
Reg[t]. Ord[rs].

For the Day tomorrow L[t] Moultrie

General Orders 5th Feb[y].__________ Parole Pensylvania
Reg[t]. Ord[rs].

For the day tomorrow L[t]. D. Mazyck

Gen[l]. Orders 6th Feb[y]._________ Parole Philadelphia
Reg: Ord[rs].

For the day tomorrow L[t]. Villepontaux

General Ord[rs]. 7th Feb[y].______________ Parole Maryland
Reg: Ord[rs].

A court martial to be held today at 11 OClock to try such prisoners as shall be brought before them Evidences to attend

Cap[t]. Huger Presid[t]. L[ts]. Moultrie & Mazyck Members

A General Return of the two Comp[ys]. to be made this after noon __ For the Day tomorrow L[t] Mazyck

By Cap[t] Marion **Haddrell Point 1776**

Gen[l] Orders 8th Feb[y]. ___________ Parole Virginia

Regimt[l]: Orders

For the day tomorrow L[t] Tho[s] Moultrie

The Taylers of Each Company to make the Soldiers Cloaths & be Excused all Duty__

Gen[l] Orders 9th Feb[y] ______________ Parole Williamsburgh

Reg[l]: Ord[rs].

For the Day tomorrow Lt Mazyck

A Court Martial to sett to day at 11 OClock to Try such prisoners as shall be Brought before them Evidence to attend

Cap[t]: Huger Presed[t].

L[t] Tho[s] Moultrie & Villepontaux Members

After Orders

The Detachment to turn out this After Noon at 4 OClock to Exercise- a perticular Account taken of those Absent & report them to the Command[g] Officer

NB Mathew Martin Sentenced to 50 Lashes which was remitted [220]

Gen[l] Orders 10th Feb[y].______________ Parole Norfolk

Reg[t] Orders

For the Day tomorrow L[t] Moultrie

Gen[l] Orders 11th Feb[y].___________ Parole Newburn

Reg[t]. Orders

For the Day tomorrow L[t] Villepontaux

By Cap[t] Marion **Haddrells Point __ 1776 __**

Gen[l] Orders 12th Feb[y].________ Parole Wilmington

Reg[t]. Orders

For the day tomorrow L[t] Mazyck

Gen[l] Orders 13th Feb[y]: ___________ Parole Brunswick

Reg[t]. Orders

For the Day tomorrow L[t] Tho[s] Moultrie

A Court of Inquiry to be held today at 11 Clock to Examine in to the Claime of one Cross concerning a Kegg rum which was taken a few days agoe by our Soldiers from a Negroe at Mottes plant[n]: which rum suspected to be sold or Intended to be sold to the soldiers of the Detachment Contrary to Orders

Cap[t]: Huger Presed[t].

L[ts]. Moultrie & Mazyck members

Groves & Jackson to attend as Evidence [221]

---

[220] Matthew Martin was born in 1755. He enlisted in the 2nd South Carolina Regiment on 4 November 1775. On 13 April 1776 he was sentenced to 50 lashes for being drunk and insolent, but it was remitted

[221] Joseph Graves was a lieutenant in the St. David Parish Volunteer Company under Colonel Powell

NB the Court acquit Mr Cross of the Suspecion of selling the rum to the Soldiers against Orders & prove it his property which was Delivered to him by Order of the Commandant:

Genl Orders 14th Feby.______________ Parole Georgetown

Regt. Orders

For the Day tomorrow Lt Mazyck

Lieutt. DeTreville of the Artillery to do duty in room of Lt Benj. Villepontaux who is relieved

After Genl Orders ____________ 14 Feby.

The Quartr. Guard to be Relieved at 1 OClock today by the Provincials and Captn Quelch & his men are Dischargd from doing duty at Haddrells Point __

Regemt. Orders

3 men to reinforce the Battery guard & that guard to place a Centry where the Malitia kept guard, the tents there to be taken down, & stor'd in the Quartr. Mastr. Sergts. room [222]

Orders 15th Feby. _______________ Parole Santee

For the day tomorrow Lieutt. Jackson

For the feuter non but a propper Sergeant is to mount a guard,[223] no Lance Sergeant is ever to command a guard [224]

Orders 16th Feby._____________ Parole St. Johns

For the day tomorrow Lt Moultrie

Sergeant Simpson & ten men with proper tools to go Emediately to pull down the Chimney, at Mottes old place & Batter the Bricks so as to Lay as flatt as possible, to prevent its being a Land_mark for ships which come up [225]

Orders 17th Feby._____________ Parole Mecklenburgh

For the Day tomorrow Lt Mazyck

---

222 By February 1776 Charlestown harbor was one of the most heavily defended cities in America. The Provincials knew that they would need more men to defend all these positions and they voted that 1,050 men of the country militia be required to serve in Charleston. The militamen would be rotated so as to keep them from deserting. Militia units were billeted in the many empty houses that were available in Charelston, since their inhabitants had left for the safety of the countryside. The landlords of these houses drew rations from the Commissary General, Thomas Farr. The militia began to "strip the paper hangings, chop wood upon parlour flours & a thousand such improper acts." They did this because they thought "every house belongs to what is called a Tory & make the House suffer for the imputed Sins of its owner."

223 feuter = future.

224 Simes' Military Medley of 1768 defines a lance-corporal as a private, acting and doing duty as a Corporal, for soldier's pay. It defines a lance-sergeant as a Corporal, acting and doing duty as a Serjeant; though he receives only Corporal's pay.

225 Captain Edward Blake was the commander of the Charlestown pilots. He was ordered by the Provincial Congress to stop up the Marsh and Hog Island Channels. The hulks that were sunk there in November 1775 proved to be ineffective. The swift currents merely cut a new path around the wooden obstructions. Blake sunk a schooner in Marsh Channel in February. This would channelize any enemy ships to pass right by the guns of Fort Sullivan. Captain Blake was also ordered to destroy any landmarks along the coastline that might aid enemy ships in an attempt to cross the Charlestown Bar. Motte's old place was located at present day Mount Pleasant.

It is recommended to the Subalterns who has roll-call every day, to make the men Keep there Arms Clean and in good Order

By Capt Marion **Haddrells Point ____ 1776**

Orders 18th Feb^{y}.__________ Parole Waxsaw
For the day tomorrow Lt Jackson

Orders 19th Feb^{y}: ____________ Parole Saludey
For the day tomorrow Lt Moultrie

Orders 20th Feb^{y}.____________ Parole Keowee
For the day tomorrow Lt Mazyck
A Court Martial to be held today at 11 OClock to try such prisoners as shall be brought before them
Cap^{tn} Huger Presid^{t}.
L^{ts}. Mazyck & Jackson Members
NB. Ja^{s}. Grover of Cap^{tn} Hugers & Saml Gales of Marions Comp^{y} for neglect of duty rec^{d}. 100 Lashes Each.

Orders 21st Feb^{y}.__________ Parole Newcastle
For the day tomorrow Lt Jackson

Orders 22nd Feby. ___________ Parole Eyeorce
For the day tomorrow Lt Thos Moultrie
When the weather will permit the troops to Exercise the cannon as usual

---

Orders 23^{d} Feb^{y}.____________ Parole Carolina
For the Day tomorrow Lt Mazyck

Orders 24th Feby. ____________Parole Windsor
For the day tomorrow L^{t}. Jackson
A Court Martial to be held today at 11 OClock to try such prisoners as shall be brought before them __ Lt Moultrie Presed^{t}. L^{ts} Mazyck and Jackson members NB James Hunter found guilty of absents without Leave Sentenced to 50 Lashes but remitted [226]

By Capt Marion **Haddrells Point ____ 1776**

Orders 25th Feby 1776 ________ Parole Radford
For the Day tomorrow Lt Thos Moultrie

Orders 26th Feby __________Parole Richmond
For the day tomorrow L^{t}. Mazyck

---

[226] James Hunter enlisted in the 2nd South Carolina on 4 November 1775 under Captain Francis Huger. On 24 February 1776 he was supposed to receive 50 lashes for being absent without leave, but it was remitted. On 27 November 1776 he was mult a weeks pay for striking Sergeant Simpson and drunkeness.

Orders 27th Feby __________ Parole Cumberland
For the Day tomorrow Lt. Jackson

Orders 28th Feby __________ Parole Montgomery
For the day tomorrow Lt. Moultrie
A Court Martial to be held today at 11 OClock to try Such prisoners as shall be brought before them, Evidences to Attend

Lt Moultrie President
Lt. Mazyck & Jackson Members

NB The court found Sergt McDonald guilty of Disobidence of Orders & reduct him to a private in the ranks, Confirmed [227]

Orders 29th Feby ____________ Parole Coll. Arnold
For the day tomorrow Lt. Mazyck

Sullivants Island 1776

Captn Marion is appointed to the Majority in 2d Regt and Order'd to take the Command on Sullivants Island, & to Leave the Command at Haddrells to Capt. F. Huger which was done the same day [228]

By Major Marion

Orders 1st March ____ Parole Congress
Counter Sign 22 __
For the Day tomorrow Capt: Motte
Advance Guard Lt. Blake [229] __ Quartr Guard Lt. Baker
Rear guard a Sergeant
All Orders Isued by the Late Commanding Officer to be Obey'd __ A General Return of Each Company and a return of Musketry & Ordinance Stores to be made by 10 OClock tomorrow
Any soldier who will work at the fortifycation shall be Excused all Duty & be paid [230]

---

227 John McDonald reenlisted in the 2nd South Carolina Regiment on 4 November 1775 as a sergeant. On 28 February 1776 he was demoted to private for disobedience of orders. He was promoted back to sergeant in 1776, but on 26 November 1776 he was demoted to corporal. He was promoted to sergeant in Captain Harleston's company on 15 July 1778. He was reprimanded on 26 August 1778 for signing a false return. He re-enlisted on 13 February 1779 in Captain Dunbar's Grenadier Company as a corporal. He was demoted to private a short time later. He was promoted again to sergeant on 12 June 1779. During the assault on the Spring Hill Redoubt at Savannah on 9 October 1779 he was wounded, but saved the regimental colors of the 2nd South Carolina from capture after Lieutenant James Gray had died while planting them there.

228 The 2nd South Carolina Regiment was garrisoned at Sullivan's Island, but detachments still manned the downtown batteries along Charleston's waterfront. These detachments were garrisoned in the town barracks.

229 After Marion assumed command of Fort Sullivan he learned that he commanded two other positions on the island. The Advanced Guard was located at the northern end of Sullivan's Island and was garrisoned by a company of infantry. This position was an observation post for a possible beach-landing sight, known as The Breach. The other position was known as the Quarter Guard. This was a position at the narrowest part of the island east of Fort Sullivan. This was garrisoned by a platoon of infantry and overwatched a second possible beach landing area.

230 There was slow progress on the fort, mainly due to the fact that the enlisted men objected to working side by side with the contracted slaves of Daniel Cannon's work force. The Provincial Congress even authorized

L[t] Hall is to act as Adjutant at this post, all Orders from him as Adj[t] is to be Obeyed.
After Orders
No Oak trees to be cut down any persons cuting such trees may Expect to be punished __ no persons to Sell spirituous Liquors or Beer without a perticular Order from the Commanding Officer[231]

Orders 2nd March ____________ Parole Coll Arnold Count[r]S[n]. 5
For the Day tomorrow Capt M[c]Donald the Advance Guard L[t]. Pimineau
Quart[r]. Guard L[t]. Proveaux [232]

Each Comp[y] to order two men for fatigue every Evening & they to parade when the pioneers march is beat the acting Adjutant to warn a Serg[t]. for Fatigue

---

Orders 3rd March _______ Parole Amarica Count[r]S[n]. Boston
For the ~~Adv~~ Day tomorrow Capt Mason
Advance Guard L[t]. Charnock
Quart[r] Guard L[t]. John Harleston
Rear Guard a Sergeant

The men are to Exercise the Cannon every day from 11 to 12 OClock in the fore noon, & the small arms at four in after noon __ all Officers to attend [233]

By Maj[r]. Marion [234] **Sullivants Island 1776**
Orders 4th March _______ Parole Cannada Count[r]S[n] Quebeck
For the Day tomorrow Capt[n]. Cha[s]. Motte
Advance guard L[t]. Lesesne
Quart[r]. guard L[t]. Gray __ Rear a Sergeant

Orders 5th March _______ Parole Crown Point Count. S[n]. Lake Geo[r]:
For the Day tomorrow Capt: A. Ashby

Daniel Cannon to pay money to any enlisted man who would work in the fort. This all changed when Francis Marion took over. Captain Marion was known as an officer who would not put up with any nonsense.

231 Marion first orders were to increase the men's morale and discipline. To provide his men with shelter from the sun he ordered that the oak trees not be chopped down. The men would also not be able to buy liquor from the unauthorized "sutlers and negroes" on the island. After Marion took command there were numerous court martials by the men who were still used to the old ways. Marion ordered his men to look and act like soldiers. If not they could be guaranteed punishment.

232 Adrian Proveaux was commissioned as a 2nd lieutenant in the 2nd South Carolina in February 1776. He was promoted to 1st Lieutenant on 22 October 1776 in Captain Moultrie's company. In September 1777 he was court martialed for conduct unbecoming an officer while he was on guard. Though this incident happened in July, Proveaux demanded a court of inquiry into the matter, which led to a court martial. He was found guilty and was reprimanded by the Marion in the presence of the officers of the regiment. He was promoted to captain on 27 April 1778. He took command of Jacob Shubrick's company when Shubrick died. Shubrick's company was known as the 2nd Vacant Company since it did not have a captain to command it. The 2nd Vacant Company was disbanded on 5 October 1778 and the men were reorganized into the 10th company, with Captain Proveaux in command. He was taken prisoner at the fall of Charleston..

233 The South Carolina Regiments were one of the few in the war that were trained as infantrymen and as artillerymen. This was necessary in the fortress that was Charleston.

234 Francis Marion had been promoted to Major once he assumed command of Fort Sullivan.

Advance Guard Lt Thos Dunbar
Quarter- Lt. Baker__ Rear a Sergt.

A General Return of the Granidier & Captn Ashbys Compy. to be made tomorrow at 10 OClock_fore nn. _

Orders 6th March _________ Parole Oswego. Count: Sn. 63 __
For the Day tomorrow Captn. Jas. McDonald
Advance Guard Lt. Jno Blake
Quartr Guard Lt. Proveaux _______ Rear a Sergt.

An Officer of a Compy. to see there Sergeants warn there men properly for duty every Evening & see them parraded at troop beating Clean, there Hairs comb'd & there Arms in good order
Any Soldier who has been warn'd for Duty & do not attend punctually when the drums beat must be confin'd, any Sergeant who neglect confining such Soldiers may Expect to be punish't __ Tis Expected that one Commissioned Officer at Least will Attend roll-calling constantly __ this order has been often Issued & tis hop'd the Gentlm. will make the repeat of it unnessiary ~~for the~~ Maurice Fowler Corp: in Capt. Eveleigh Comp. is apointed a Sergeant in sd compy. & is to be Obey'd as Such __[235]

By Majr. Marion **Sullivants Island 1776**

Orders 7th March _______ Parole Orleans ________ CountrSn. Albany
For the Day tomorrow Captn. Wm Mason
Advance Guard Lt. Perreneau
Quarter Lt. Geor Eveleigh __ Rear a Sergt.

Orders 8th March _______ Parole Mexico ______ Count. Sn. York
For the Day tomorrow Captn Motte
Advance guard Lt. Richd. Shubrick
Quarter __ Lt. Thos. Hall __ Rear a Sergt.

Orders 9th March _______ Parole Georgea _____ Count: Sn. St Johns
For the Day tomorrow Capt: Ashby
Advance guard Lt Fuller
Quarter __ Lt. Henry Gray __ Rear Sergeant
A Regimental Court martial to be held today at 11 OC: to try such prisoners as shall be brought before them Evidences to attend __ Captn. Petr. Horry Presedt.
Lts. Fuller, Charnock, Baker & Marion members

Orders 10th March ________ Parole Savanna __CountrSn. 17
For the Day Capt: P. Horry
Advance Guard Lt. Charnock
Quartr Guard Lt. Baker __ Rear a Sergt.

---

[235] Maurice Fowler enlisted in the 2nd South Carolina Regiment on 4 November 1775. He was promoted ot sergeant on 6 March 1776 in Captain Eveleigh's company. He was discharged on 14 July 1778. He enlisted in the 3rd South Carolina (Ranger) Regiment on 19 February 1779 under Captain Felix Warley.

Orders 11th March ________ Parole Fredrica __Count^{r}S^{n}. Talbot

Fpr the Day Capt: Ja^{s}. McDonald

Advance guard L^{t}. Proveaux

Quart^{r}. guard L^{t}. Marion __ Rear a Serg^{t}.

NB. The court martial of the 9th Inst. found Fran^{s}. Ferrell guilty of absents without Leave he rec^{d}. 200 Lashes [236]

Mercy, punishment Remitted

By Maj^{r}. Marion **Sullivants Island 1776**

Orders 12th March ______ Parole Florida ___ Count^{r}S^{n}. 89

For the Day tomorrow Captn. Motte

Advance guard L^{t}. Blake

Quart^{r}. guard L^{t}. Eveleigh __ Rear a Serg^{t}.

Orders 13th March ______ Parole Pensicola _____ Count^{r}S^{n}. London

For the day tomorrow Capt: Ashby

Advance guard L^{t}. Thos. Hall

Quart^{r} G _L^{t}. Perreneau _____ Rear a Serg^{t}.

A Regimental court martial to be held today at 11 OClock to try such prisoners as shall be brought before them, Evi: to attend

Captn ~~Motte~~ Presed^{t} Ja^{s} McDonald

L^{ts}. Shubrick, Hall, Marion & Proveaux memb^{rs}.

Orders 14th March ______ Parole Effingham ______ Count^{r}S^{n}. Paris

For the Day tomorrow Captn. P. Horry

Advance guard L^{t}. H Gray

quart^{r} __ L^{t}. Shubrick __ Rear a aerg^{t}.

Orders 15th March ______ Parole Camden _____ count^{r}S^{n}. Aurther

For the Day tomorrow Captn. Ja^{s}. McDonald

Advance Guard L^{t}. Fuller

Quart^{r} Guard L^{t}. Adrian Proveaux __ Rear a Serg^{t}.

Orders 16th March ______ Parole Cumberland Count: S^{n}. David

For the Day tomorrow Capt. Motte

Advance guard L^{t}. Marion

quart^{r}. G: L^{t}. Fuller __ Rear G: a serg^{t}.

A regimental court martial to be held to day at 11 OC: at the Presed. tent to try such prisoners as shall be brought before them, Evidence to attend

Captn Motte Presed^{t}. L^{ts}. R Shubrick, Fuller, Eveleigh, & H Gray memb^{rs}.

---

[236] Francis Ferrill enlisted in the 2nd South Carolina on 20 November 1775. On 11 March 1776 he was supposed to receive 200 lashes for being absent without leave, but he was given mercy and the sentence was remanded. On 13 December 1776 he was reprimanded for neglect of duty.

NB the above court found Wm Simpson guilty of abuse to a Serg^{t}. & rec^{d} 100 Lashes __ Jno Wheeler for Brak^{g}. a public arm rec^{d} 100 Lashes [237]

By Maj^{r}. Marion **Sullivants Island __ 1776**

Orders 17th March ______ Parole Holland, Count^{r}S^{n}. Orange

For the Day tomorrow Capt: Ashby

Advance guard L^{t}. Blake

Quart^{r}. L^{t}. G. Eveleigh __ Rear a Serg^{t}.

The officers of the day to go frequently to visit the mens Hutts in the Day & in the night also in the sutlers tents to Examine in all Disorders & see the men are orderly, perticularly Examine the sutler that they do not act contrary to orders

---

Orders 18th March ______ Parole Bollinbroke Count^{r}S^{n}. S^{t}. Ge^{o}.

For the Day tomorrow Capt: Pet^{r}. Horry

Advance guard Lt Perreneau

quart^{r}. G: L^{t}. Thos. Hall __ Rear G: a serg^{t}.

Serg^{t}. Edmunds of Captn. Horrys Comp^{y}. to act as Sergeant Maj^{r} till further Orders [238]

Orders, 19th March ________ Parole Digby, Count^{r}.S^{n} Dennis

For the day tomorrow Capt Ja^{s} McDonald

Advance guard L^{t}. R. Shubrick

quart^{r}. Lt Gab^{l} [239] __ Rear G. a serg^{t}

No Searg^{t} or Corporall to Leave the Island without Leave from the Commanding Officer

A Reg: Court martial to be held today at 11 OClock at the Presed^{ts}. tent to try such prisoners as shall be brought before them

Evidence to attend- Capt Jam^{s} McDonald presed^{t}.

L^{ts} Rich. Shubrick, __Fuller __ Charnock & Baker members

---

[237] John Wheeler enlisted in July 1775 in the 2nd South Carolina under Captain Barnard Elliot. On 14 October 1776 he was ordered to receive 500 lashes for striking William Donavan, but only received 400. In 1782 he served 30 days in the militia.

[238] William Edmunds had reenlisted in the 2nd South Carolina Regiment on 4 November 1775. He was made acting sergeant major on 18 March 1776. He was promoted to sergeant major on 14 April 1776. On 29 June 1777 he was tried for disrespecting Captain Charnock. The court decided that his time spent in confinement and the abuse he received there was enough punishment, and he returned to duty. He was severely reprimanded on 12 September 1777 for being absent without leave. On 1 October 1777 he was court-martialed for mutiny, but the charge was reduced to behaving contrary to good order and military discipline. He was demoted to private and received 100 lashes on the bare back with the cat-o-nine tails. The cat-o-nine tails was a leather whip with 9 strands of leather. On 28 November 1777 he became a sergeant again in Captain Lesesne's company and was transferred to Captain Moultrie's company. He was promoted when Sergeant Burgess was appointed the Quarter Master Sergeant. He was discharged on 14 June 1778.

[239] A name was scratched through and "Gab^{l}" was written on top of it, referring to Gabriel Marion.

By Maj[r]. Marion **Sullivants Island 1776**

Orders 20th March ______ Parole Marlborough Count[r].S[n]. Windson

For the Day tomorrow Capt Ashby

Advance Guard L[t]. Baker

quart[r].- L[t] R. Fuller__ Rear a Seargeant

The Long Roll to Beat at 5 OClock in morning the troops to Exercise the Cannon from 5 to 6 OC: in morn[g]: Guards to be reliev'd at 7 OClock AM [240]

Orders 21st March ___ Parole Peterborough Count[r].S[n]. Dover

For the Day tomorrow Capt[n] P. Horry

Advance guard L[t]. Adrian Proveaux

quart[r] __ L[t]. W. Charnock __ Rear a Serg[t].

A Reg[tl]. court martial to be held to day at 11 OClock to try such prisoners as shall be brought before them Evidences to attend

Capt P. Horry Presed[t].

L[ts]. Perrineau, Charnock, H Gray & Proveaux members

Orders 22nd March ___ Parole Wilkes ______ Count[r]S[n] North

For the Day tomorrow Capt[n] P. Horry

Advance guard L[t]. Tho[s] Dunbar

quarter L[t]. Gab[l]. Marion __ Rear a Sergeant

Orders 23[d] March ______ Parole Berwick _____ Count[r].S[n]. Edward

For the Day tomorrow Capt[n]. F. Huger

Advance guard L[t]. G Eveleigh

Quarter__ L[t] Blake __ Rear a Serg[t]

Three men to be added to the Quart[r]. Guard & that guard to direct a Corporal & 6 men to be posted where the adjutant will Direct

NB The court martial 21st Inst: which Capt P Horry was Presed[t]. found Rich[d]. Williamson Guilty of Absents without Leave & Drunkenness on Guard- Sentenced to 150 Lash which he received [241]

---

[240] Francis Marion conducted surprise alarms in the early hours of the morning as part of the ongoing effort to turn his men into a highly effective fighting force.

[241] Richard Williamson was in the 2nd South Carolina Regiment. On 4 September 1776 he received 100 lashes for some offense. On 14 December 1776 he received 39 lashes and mult a weeks pay for drunkenness. On 23 December 1776 he received 99 lashes and mult 14 days pay for disorderly behavior while in Charleston. On 12 April 1777 he was mult 14 days pay for drunkenness. On 6 September 1777 he had to run the gauntlet three times for defrauding Private Hendrickson. He also had his pay deducted until he paid Hendrickson back. When a soldier ran the gauntlet he was stripped to the waist and had to run the length of the regiment. The soldiers of the regiment would stand in two lines and would hit the prisoner with their ramrods or with sticks. On 29 July 1778 he lost his regimental coat and his pay was stopped until it was paid for. On 2 November 1778 he received 50 Lashes on the bare back with a Cat o' Nine Tails for drunkenness on duty. On 4 November 1778 he received 45 Lashes on the bare back with switches for quitting his guard without leave.

By Majr. Marion **Sullivants Island __ 1776**
Orders 24th March ______ Parole Tweed ____Countr.Sn Whales
For the Day tomorrow Captn. Oliphant
Advance guard Lt. Perrineau
Quarter __ Lt. Thos Hall __ Rear a Sergeant

Captn Petr Horry is to take the Command of the Light Infantry Compy Formerly Marions and Captn Wm Oliphant to take the Command of Captn P Horrys

---

Orders 25th March ___ Parole Severn ____Count.Sn Scotland
For the Day tomorrow Captn P. Horry
Advance guard Capt: Oliphant & Lts Shubrick & Gray__ Quartr.__ Lt Isaac Dubose __Rear a Sergt.

One Captn. 1 Subaltern 1 Sergt & 15 privates to be added to the Advance guard tomorrow

---

General Orders by Coll: ~~Moultrie~~ Gadsden

No Soldier to be permitted (unless a most pressing necessety) to go to Chs. town, From Fort Johnson, Sullivants Island and Haddrells Point

---

By Major Marion Orders, 26t March ____ Parole Kent _____ Countr.Sn. 57
For the day tomorrow Captn. Wm Mason
Advance Guard Captn Huger, Lts. Fuller & Baker
Quartr __ Lt Charnock __ Rear a Sergeant
No Officer to Leave his guard on any account whatever

Lieutts. Wm Moultrie & Proveaux to hold themselves in readyness to go a recruiting

By Majr. Marion **Sullivants Island 1776**
Orders 27th March ___ Parole Chester ___ Countr.Sn. 72
For the Day tomorrow Captn. Oliphant
Advance guard Capt P Horry & Lts. Mazyck & Marion
quartr __ Lt Thos Dunbar __ Rear a Sergt.

A Regt court martiall to be held today to try such prisoners as shall be brought before them Evidences to Attend

Lieutt. Jno Blake Presedt
Lts. Mazyck, Dunbar _Hall & Marion Members

NB the court sentenced Adm. Meek to 50 Lashes & Geo. Pratt to be reprimanded __ Ad: Meek remitted [242]

---

[242] Adam Meek was born in 1760 in Cecil County, Maryland. He enlisted in the 2nd South Carolina Regiment on 4 November 1775 and re-enlisted on 1 November 1779. On 6 May 1776 he was supposed to receive 100 lashes for disobedience of orders, but it was remitted. He was absent during the battle of Fort Sullivan on 28 June 1776, so on 30 June he received 200 lashes and was ordered to be dressed in women's petticoats and cap. On 20 August 1776 he received 100 lashes for drunkenness and neglect of duty. On 14 April 1777 he received 39 lashes for being absent from roll call. On 20 July 1777 he received 100 lashes for being drunk on guard and abusing the inhabitants of the town. On 2 August 1777 he received 100 lashes for being drunk on guard. On 8 August 1777 he was accused of stealing, but he was acquitted. On 18 October 1777 he received 100 lashes on two days for sleeping on his guard post. He received 100 lashes on 20 February 1778 for neglect of

---

Orders 28th March ____Parole Yorkshire _ Countr.Sn. 27
For the Day tomorrow Capt: Huger
Advance guard Captn. Ashby & Lts. T. Moultrie & Eveleigh
quartr __Lt. Blake __ Rear a Sergt

Orders 29th March ______ Parole Middlesex ____ Count. Sn. 92
For the Day tomorrow Capt Mason
Advance guard Capt: Oliphant & Lts. Perrineau & Hall
quartr __Lt R. Shubrick __ Rear a sergt.

Orders by Coll: Moultrie 29th March

The Representives of the good people of this Collony having made choice of John Rutledge Esqr.to be there Presedt. & Commander in Chief [243]__ Ordered that all Due Honours & Respect be paid him __ All guards to turn out to receive him with Rested arms if an Officers arms the Drum to Beat a march as he passes __ all Centrys to rest to him, and all Orders from him to be punctually to be Obey'd __ The vice President Coll: Henry Laurence [244] to be received with rested Arms __ [245]

Only two Officers to be allowed to be in town at one time __

no soldiers to have furloughs till further Orders __

---

duty. He received 75 lashes on 16 April 1778 for being drunk on guard and sitting down on his post. After the surrender of Charleston he became a lieutenant in the militia under Colonel Bratton. He also served in the dragoons under Lieutenant Colonel Wade Hampton. He was in the battles of Briar Creek, Rocky Mount, Hanging Rock, Fishing Creek, Fishdam Ford, Blackstock's, Biggin Church and Eutaw Springs. He died on 26 January 1807 at the age of 47.

243 John Rutledge was born in 1739 in Charleston, South Carolina. He became President of South Carolina and commander-in-chief of the South Carolina forces in 1776. He was given dictatorial powers during the war. In late 1776 he delegated his commander in chief status to Major General Charles Lee when the British threatened Charleston. He was elected to Governor of South Carolina in February 1779. When the British captured Charleston in 1780 he ran the government in exile from North Carolina. He died on 18 July 1800.

244 Henry Laurens was born on 24 February 1724. He studied in England to become a merchant. Before the Revolution he had served in the militia a lieutenant and then as a colonel in Middleton's regiment in 1760. He served in the Cherokee Campaign of 1761. In 1771 he sailed for England and spent three years abroad then returned in 1775. He was elected President of the First Provincial Congress in June 1775 and served as the President of the Council of Safety. He was elected Vice President of South Carolina in March 1776, then elected to the Continental Congress in June 1777. He was President of Congress in 1777 and 1778. In 1779 he was named as the commissioner to negotiate the treaty with France, but was captured by the British enroute and imprisoned in the Tower of London. He was released on bond in December 1781 and exchanged for Lord Cornwallis after his capture at Yorktown. His son, Colonel John Laurens, was killed at Combahee Ferry 27 August 1782. Henry Laurens died at his estate, Mepkin, on 8 December 1792 at the age of 68.

245 The Council of Safety was replaced by the General Assembly. John Rutledge and Henry Laurens were sworn in as the new President and Vice President by the General Assembly on March 28th. Charleston celebrated its new Congress and Constitution on April 2nd. A parade was held downtown where the regular and militia units displayed their military skills. The 4th South Carolina (Artillery) Regiment and the Charlestown Artillery Company cheered and fired their pieces in celebration, while the South Carolina Navy vessels discharged their cannons.

A Return of the pay bills to be made & Deliver'd to the Commanding Officers Present every Fryday before pay Day, who is to transmit them to the paymaster with all possible Dispatch

---

By Major Marion

Orders 30th March ___________ Parole Rutledge Count^r.S^n. 13

For the Day tomorrow Captn P. Horry

Advance guard Capt Motte, L^ts Charnock & Gray

Quarter__L^t Fuller__Rear a Sergeant

Orders 31st March ______ Parole Laurence Count^r.S^n. 30.

For the day tomorrow Capt: Ashby

Advance guard Capt: Huger L^ts. Mazyck & Dubose

quart^r.__ Lt Baker __ Rear a Sergeant

Orders 1st April _____ Parole Manchester Count^r.S^n. 31.

For the Day tomorrow Capt. Oliphant

Advance guard Captn Horry L^ts. Dunbar & Marion __ Quarter Lt Jacob Shubrick

A Reg^l. court martial to be held today at 11 Oclock at the presidents tent to try such Prisoners as shall be brought before them Evidences to Attend

Capt: Oliphant President

L^ts Blake __Jacob Shubrick __T. Moultrie & Dunbar members

NB Richd Rogers of Captn Asby Comp^y received 100 Lashes for forging a pass to town with Coll. Moultries name

By Major Marion **Sullivants Island 1776**

Orders 2nd Aprill _____ Parole Milford Count^r.S^n. 33

For the Day tomorrow Captn Motte

Advance guard Ashby L^ts. Blake & R Shubrick

Quarter__ L^t. H Gray __ Rear a Serg^t

A court Martial to be held today at 11 OClock to try such prisoners as shall be brought before them, Evid: to attend

Capt. Motte President

L^ts. R. Shubrick __ Fuller __Lesesne __ T. Moultrie Members

NB moses Therrel for sleep^g. on his post rec^d. 170 Lashes [246]

Edwd Johnson for absents from Roll-Call rec^d 100 Lashes [247]

Joseph Butler for gaming rec^d 50 Lashes [248]

---

[246] Moses Therrell was enlisted by Captain Calvin Spencer to serve in the 2nd South Carolina Regiment. On 6 May 1776 he received 100 lashes for disobedience of orders, but it was remitted. On 11 August 1777 he was to receive 39 lashes for taking a prisoner into a tavern, but he was pardoned. He was in the battle of Beaufort on 3 February 1779. He died on 10 July 1787.

[247] Edward Johnson enlisted in the 2nd South Carolina on 4 November 1775.

[248] Joseph Butler enlisted in the 2nd South Carolina on 4 November 1775 and was discharged on 7 July 1778. He enlisted in the 1st South Carolina Regiment on 6 February 1779 and in 1780 was under Charles Skirving.

Orders 3[d] Ap: ____ Parole Cornwall __ Count[r].S[n]. 66
For the Day tomorrow Capt[n] Mason
Advance guard Capt[n]. Huger & L[ts] R Fuller
& Baker __ quart[r] __ L[t] Dubose __Rear a Sergeant

Orders 4th Ap: ____ Parole Bristol ____ Count[r].S[n] 9
For the Day tomorrow Capt[n] P. Horry
Advance guard Capt P. Horry L[ts] Tho[s] Moultrie
& Baker __ quart[r]__L[t] Dixon__ Rear a sergeant

A Reg[tl]: court martial to be held today at 11 OClock to try such prisoners as shall be brought before them. Evid: to Attend

Capt[n] Oliphant Presed[t].
L[ts] Mazyck- Jacob Shubrick- Baker and Marion Members

NB. The Court found Serg[t] Burges of M[c]Donalds Comp[y]. guilty of Neglect of Duty, he was reduct to a private. Corporal M[c]Collough of Hugers Comp[y]. for neglect of Duty is reduct to the ranks__ Boath confirmed [249]

By Major Marion **Sullivants Island __1776__**

Orders 5th April ________Parole Cantebury____ Count[r]S[n] 96.
For the Day tomorrow Capt[n] Ashby
Advance Guard Capt[n] Motte L[ts] Shubrick & Marion
quart[r] __L[t] Mazyck__Rear a serg[t].

A General return to be made tomorrow at 10 OClock AM, to send to his Excellency the president, this return to be compleat in every respect

---

Orders 6th Ap: ____ Parole S[t] Asaph ____ Count[r].S[n]. 69
For the day tomorrow Capt: Huger
Advance guard Capt. Mason L[ts] Blake & H Gray
quarter __ L[t] Dunbar __Rear a Sergeant

Orders 7th Ap: ____ Parole Oxford ___ Count[r].S[n].58.
For the day tomorrow Capt: Oliphant
Advance guard Capt[n]. M[c]Donald & L[ts] Charnock &
Lesesne __ quart[r]__ L[t] Fuller __ Rear a Serg[t]

Orders 8 Ap: _____ Parole Durham _____ Count[r].S[n]. 87
For the Day tomorrow Capt. Motte

---

[249] William M[c]Cullock (McCullogh) enlisted in the 2nd South Carolina Regiment on 4 November 1775. He was promoted to corporal on 24 February 1777. He received 100 lashes for neglect of duty on 17 April 1777. On 2 August 1777 he received 100 lashes from being absent from guard duty. He was court martialed on 17 November 1778 for losing his musket. He was sentenced to have half his pay taken every pay day until the the sun of £25 was made up. He was promoted to corporal in Captain Moultrie's company on 20 January 1779. He reinlisted as a sergeant on 1 July 1779 in Captain Dunbar's Grenadier company. When Captain Dunbar's company became a light infantry company he was transferred to Captain Hall's company. He was killed in the assault on the Spring Hill Redoubt at Savannah on 9 October 1779.

Advance Guard Capt. P. Horry Lts Baker & Dubose
quartr. __ Lt T Moultrie __ Rear a Sergeant __

Orders 9th Ap: ____ Parole Northampton ____ Countr.Sn.6.
For the Day tomorrow Capt. Mason
Advance GuardCapt. Huger, Lts Mazyck &
Marion __ Battry __Lt. R.. Shubrick __ Rear a sergeant

By Major Marion **Sullivants Island 1776 __**

Orders 10th April _____ Parole Westmorland Countr.Sn.9.
For the Day tomorrow Captn Oliphant
Advance guard Capt: Ashby Lts Dunbar &
H Gray __ Battry __ Lt. Blake __ Rear a Sergt.

Orders 11th Ap: ___ Parole Crafton ___ Countr.Sn 28.
For the Day tomorrow Capt: P. Horry
Advance guard Capt McDonald Lts Fuller & ~~Baker~~
Lesessne __quarter __ Lt Charnock

Orders 12 Ap: ____ Parole Essex _____ Countr.Sn 28.
For the day tomorrow Capt Huger
Advance guard Capt Mason Lts. T Moultrie &
R B Baker __ quarter __ Lt Dubose __ Rear a sergeant

Orders 13 Ap: ____ Parole Exeter ____ Countr.Sn 32 _____
For the day tomorrow Captn Motte
Advance guard Lts. Jacob Shubrick & Marion
quarter __ Lt. Mazyck __ Rear a sergt
A Court Martial to be held today at 11 OClock to try such prisoners as shall be brought before them Evidences to Attend
Captn Motte President
Lts. J Shubrick __ Mazyck __ Dunbar & H Gray members
NB, the court found Mathw Martin of Captn Oliphants Compy. guilty of Drunkeness & Insolence Sentenced to 50 Lashes, which was remitted __

By Major Marion **Sullivants Island 1776 __**

Orders 14th Ap: _____ Parole Brunswick ___ Countr.Sn 15 _____
For the Day tomorrow Captn. Oliphant
Advance guard Lts. Fuller & Charnock ___
Battry __ Lt Dunbar __ Rear Sergt. guard
A Regt. court Martial to be held at 11 OClock to try such prisoners as shall be brought before them Evidences to attend
Capt. Ashby Presedt
Lts Fuller __ Lessesne__ T Moultrie & Baker members

NB, The court found Norman [250] & Carter [251] of Capt Masons Compy. Guilty of Exciting the troops to mutiny Sentenced to receive 500 Lashes Each & to be put on board an armed vessel, Norman recd 500 Lash Carter recd 300 __ Whitsett recd 100 for mutinous words [252]

General Orders by Coll. Moultrie 14th April

Fifty pounds reward will be given to any persons who will take any Deserter from the Carolina troops & deliver them to head quarters of their respective regiments or Company with 2/6 Mileage from the place they were they are taken to the place they are Delivered up to there Officers _____ Any persons who shall harbour or Conceal or ferry over any Deserters shall forfeit the sum of twenty pounds, one half to the Informers __ [253]

Sergt. Edmunds is Apointed Sergt Major of the 2d Regt. Sergt. P. Laforce is apointed Quarter Master Sergt in 1st Regt [254]

One man of a Compy is allowed to go to town but must return in 24 hours

By Coll Moultrie **Sullivants Island 1776**

14 Ap: A General Court martial is to be held tomorrow morning at 10 OClock in fore noon to try such Prisoners as shall be brought before them Evidences to attend

From the 2nd Regiment Major Marion Presedt.

Captns. Jas McDonald __ Huger __ C Motte & Lts Blake Lesesne &

H Gray Members __ Captn Mason Judge Advocate

By Major Marion

Orders 15th Ap: __ Parole Somersett __ Countr.Sn 17

For the Day today Capt P Horry

For the Day tomorrow Captn. Ashby

Advance guard Lts. T Moultrie & Baker

Battery __ Lt Dubose __ Rear a Sergt.

---

[250] William Norman enlisted in the 2nd South Carolina Regiment on 20 November 1775. He was supposed to receive 100 lashes on 19 March 1778 for neglect of duty, but it was remitted. He reinlisted on 1 November 1779. In 1778 he was under Captain Blake and in 1779 he was under Captain Adrian Proveaux.

[251] This may be William Carter who originally enlisted in the 1st South Carolina Regiment, but I am unable to determine if he is the same one. There was another William Carter who was in Captain Blake's company.

[252] John Whitsett served in the 2nd South Carolina Regiment in Captain Blake's Company in 1777 and in the Lieutenant Colonel's Company in 1778. He received a reprimand on 18 September 1777 for being absent from guard. On 17 October 1778 he was sentenced to receive 50 lashes for disrespectful behavior, but was recommended to have mercy by the court.

[253] The General Assembly passed several laws to keep men in the ranks and quit deserting. Any soldier who was killed or disabled by enemy action would have annuities paid to their families every six months for the rest of their lives. Rewards were also posted for anyone who would capture deserters and bring them in. Fines were levied for anyone hiding a deserter.

[254] Peter LaForce enlisted in the 2nd South Carolina Regiment on 4 November 1775. On 14 April 1776 he was made the quartermaster sergeant of the 1st South Carolina Regiment. He died on 26 August 1776 while in the hospital.

Orders 16th Ap. Parole Newcastle Count^{r}.S^{n} Mexico

For the Day tomorrow Capt Oliphant

Advance guard L^{ts}. Mazyck & Marion

Battery __ L^{ts}. Jacob Shubrick __ Rear a Serg^{t}.

A Reg^{tl}. Court martial to be held today to try such prisoners as shall be brought before them Evidences to attend

Capt. Oliphant Presed^{t}.

L^{ts}. Jac: Shubrick __ Dunbar __ Hall & Marion Members

NB, Staley of Capt Marions Comp^{y} Sentenced to 100 Lashes for abuse to Lt Charnock who beged him off __[255]

Orders 17th Ap: ___ Parole Tinmouth ______ Count^{r}.S^{n}. 71.

For the Day tomorrow Capt. P. Horry

Advance guard L^{ts}. Charnock & Baker

Quarter __ L^{ts}. Moultrie __ Rear a Serg^{t}

By Major Marion **Sullivants Island 1776**

After Orders 17th Ap. that L^{t}. Thos. Dunbar do Emediately mount the Battery guard NB which he refus'd

Lt Thos Dunbar you are Ordered to keep your tent for Disobediance of orders and remain under Arrest till further Orders __ NB this Order was Serv'd by the Adjutant __

By Coll: Moultrie 96

Orders 18th Ap: _______ Parole Genl Armstrong. Count^{r}.S^{n}.

For the Day tomorrow Capt. Nich^{l}. Eveleigh

Advance guard L^{ts}. Dunbar & Hall

Battery __ L^{t}. R. Shubrick

made

NB. L^{t}. Thos Dunbar ^ confession & his arrest was taken off

---

Orders 19th Ap: _____ Parole Constitution ____ Count^{r}.S^{n} 92.

For the Day tomorrow Capt. Ashby

Advance guard L^{ts}. Fuller & Mazyck

Battery __ Lt Marion __ Rear a Sergeant

John Burgess of Captn McDonalds and James McDonald of Captn Hugers Comp^{y}. are Apointed Serg^{ts}. in S^{d} Comp^{ys}. & are to be Obey as Such[256]

---

[255] Benjamin Staley enlisted in the 2nd South Carolina Regiment on 1 July 1775 under Captain Bernard Elliot. On 16 April 1776 he was sentenced to 100 Lashes for abuse to Lieutenant Charnock, but Charnock begged to have Staley's sentence dropped. The court ignored Charnock's request and on June 20th, Staley received those 100 lashes.

[256] James McDonald was appointed a sergeant in Captain Huger's company of the 2nd South Carolina Regiment on 19 April 1776. He re-enlisted on 4 January 1778. On 2 April 1778 he was acquitted for theft.

A Regimental court martial to be held tomorrow morning at the Presedt. tent to try such Prisoners as shall be brought before them Evidences to attend __
Capt Pet. Horry Presedt.
Lieuts. Richd Shubrick & Charnock
Lts. Dunbar & Jacob Shubrick __ } Members

NB Jefrey Bails [257] & Edwd Magee [258] was sentenced to 100 lashes Each for Absents without Leave __ alias Desertion & was pardoned

By Coll: Moultrie **Sullivants Island 1776**
Orders 20th Ap ___ Parole Effingham ___ Countr.Sn. 45.
For the Day tomorrow Captn Oliphant
Advance guard Lt Charnock & J. Shubrick
Battery __ Lt Hall __ Rear a Sergt

Orders 21st Ap: ___ Parole Hopkens ____ Countr.Sn. 49.
For the Day tomorrow Captn P Horry
Advance guard Lts. Lesesne & T Moultrie
Battery ____ Lt Grey ___ Rear a Sergt.
out
No Officer to Leave Sullivants Island with ^ Leave of the Commanding Officer

---

Orders by Coll: Gadsden 21st April
The General court martial of which Major Marion was President the 11th Inst: benow & is Disolved
NB. The Sentence of the court martial against the following prisoners __ of Coll: Thompsons Regimt of Rangers [259]
John McKenny of Capt Caldwells Compy to receive 800 lashes
W Cunningham of do .... 200 Lashes [260]
Patrick Forbes of Do 200

---

[257] Jesse Bails enlisted in the 2nd South Carolina on 4 November 1775 and reinlisted on 18 April 1776.
[258] Edward McGee enlisted in the 2nd South Carolina Regiment on 4 November 1775.
[259] Lieutenant Colonel William "Danger" Thomson was born on 16 January 1727 in Pennsylvania. When he was still a child his family moved to South Carolina and settled on the west side of the Congaree River. Thomson had been the Sheriff of the Orangeburgh District and had served in the Provincial Legislature. He served as a lieutenant colonel and commander of the 3rd South Carolina (Ranger) Regiment from 18 June 1775. He was in the Snow Campaign during November 1775. During the battle of Fort Sullivan on 28 June 1776, the Rangers prevented the British regulars from crossing over from Long Island. Thomson became a colonel on 16 September 1776. He was taken prisoner at the fall of Charleston and was on parole until the end of the war. On 30 September 1783 he became a brevet brigadier general. He died on 22 November 1796 at the age of 69.
[260] William Cunningham was born in Virginia in 1756. After the Snow Campaign against the Loyalists he became disillusioned with the Patriot cause. He deserted because he had an arugment with Captain John Caldwell over a horse. Caldwell had Cunningham whipped for disrespect. Cunningham joined the Loyalist militia and would later become "Bloody Bill" Cunningham, one of the most notorious Loyalist partisans.

Sam[l] Huggins of Capt. Pervis Comp[y] [261] 400 [262]
Will Skinner of d[o] ______________________________ [263]
Fludd Mitchell_ d[o] ______________________________100 [264]

The above Persons was Pardoned by his Excellency the Presed[t] Jn[o] Rutledge Esq[r]. Governor S[o] Carolina

By Coll Moultrie **Sullivants Island 1776**

Orders 22[d] Ap. ___ Parole Manley ___ Count[r].S[n]. 69

For the Day tomorrow Capt Eveleigh
Advance guard L[ts]. R Shubrick & Fuller
Battery _ L[t] Marion ____ Rear a Serg[t].

By Capt: Pet[r]. Horry

Orders 24th Ap: ___ Parole Wilkes, Count[r].S[n]. 45

For the Day tomorrow Capt Huger
Advance guard L[ts]. Hall & shubrick
Battery __ L[t] Charnock __Rear a Serg[t].

Orders 25th Ap: _____Parole Marion ____ Count[r].S[n] 12

For the Day tomorrow Capt. M[c]Donald
Advance guard L[ts] Lesesne & H Gray
Fort g __ L[t]. T Moultrie __ Rear a Serg[t].

Orders 26th Ap: ___ Parole Fort Johnson ____ Count[r].S[n]. 33

For the Day tomorrow Capt Motte
Advance guard L[ts]. Dunbar & Mazyck
Fort __ L[t]. Blake __ Rear a Sergeant

The Commanding officers ^of companys^ present to Attest there pay bills & Deliver them at the usual time

A Court Martial to be held today at 11 OClock to try such prisoners as shall be brought before them Evidences to attend

---

261 John Purvis.

262 Samuel Huggins later served in the militia in 1777 while residing in Christ Church parish. He was under Captain Brown and Colonel Sheldon Bull. He was drafted again under Captain Hutchinson and Colonel Bull. He was in the skirmish at the Coosawhatchie River in 1779 when Lieutenant Colonel Laurens was wounded. He was drafted a second time under Captain John Murrell and Colonel Maybank. After Charleston fell he rode with Marion's partisans, under Captain John Armstrong and Colonel Richardson. Huggins was in the skirmish at Parker's Ferry and at the raid on White's Plantation.

263 William Skinner was born in North Carolina. He enlisted in the Rangers on 8 July 1775.

264 Flud Mitchell was born on 10 February 1757 in Brunswick County, Virginia. He enlisted in the Rangers on 12 July 1775 when he lived in the Ninety-Six District. He was in the Battle of Fort Sullivan on 28 June 1776 under Captain Francis Boykin. He reenlisted on 24 July 1776. He reinlisted again under Captain Thomas Jones and fought in an expedition against the Indians under Colonel LeRoy Hammond. He then volunteered in a company under Captain John Rains to act as a spy for Count Pulaski. He lived in Virginia, Georgia and Alabama after the war, married to Sarah Bennett. Mitchell died on 29 December 1839.

Capt. McDonald Presedt __ Lts. Fuller__ Hall__ Dubose & Marion Members
NB Wm Staily for gameing Sentenced 100 Lashes pardoned [265]
Adam Creighton for Drunkenness & Insolence recd. 150 __ [266]

By Coll Moultrie **Sullivants Island 1776**

The Light Infantry Compy. to hold themselves in readyness to march Imediately to be compleated to one Capt. three Subalterns three Sergts. & sixty Rank & file men, to be served with 10 rounds pr man & 2 spare flints __ Captn. Petr Horry will command this Detachment & will receive his Orders from the commandg. Officer __ A surgeon to go with them with necessary medicines.

Lts. T. Dunbar, Jacob Shubrick & Thos Hall to hold themselves in readyness to go with the Detachment

Copy of Orders to Captn Horry

You are to proceed to Santea Creek with your party, & with all Expedition to Bulls Island (taking great care at your Landing not to be surprised) there to reconnoiter & prevent any party from Landing or taking any Cattle or sheep from the Island, you are to repel them force by force, taking great care to keep good Order & Disipline among yr men, you are to send Emediate Intelligence should any thing happen__ if you find the man of war put to sea you are to return with all Expedition to Camp with your party you are to be carefull to keep an Advance guard & flankg party to yr Detachment

NB, the above party took boat at the advance guard at 6 OClock, they carry'd 950 Cartridges 130 flints 1 barrel Biscuit 1 Barr: pork

Lt. Jackson with Sergt. McDonald & 6 privates sett off on Commd in Search of Deserters __

Two Men of War were sighted off Racoon Keys. Peter Horry and his company were to proceed down Santee Creek and reccoinoitre the area. Charlestown had intelligence at this time that the British may attempt to attack the city. After the failed attempt by the British to take Wilmington, North Carolina, British General Clinton had to determine what to do next with his fleet. Clinton had wavered on whether to invade the Chesapeake or to move on to Charlestown. General Howe wanted Clinton to try to take Charleston.

By Coll Moultrie **Sullivants Island 1776**

Orders 27th Ap:______ Parole Bulls Island _____ Countr.Sn. 60.

For the Day tomorrow Capt Ashby
Advance guard 1 Sergt. 14 privates & 1 Corpl.
Fort guard Lt Mazyck __ Rear a Corporals guard

The Sergeants of Each Company to parade there men before there quarters & march them regularly to the grand Parade with there Hairs comb'd & there hands & face clean __ on failure of which such Sergeants shall be confin'd & brought before a Court martial

---

Orders 28th Ap: _____ Parole Port Royal ______ Countr.Sn 70.

For the day tomorrow Captn. Oliphant
Fort guard Lt. Blake
Advc.__ a Sergeant __ Rear a Corporal

---

[265] Unable to determine who William Staily was.

[266] Adam Creighton enlisted in the 2nd South Carolina on 4 November 1775. On 26 May 1777 he was acquitted for letting a prisoner escape. He received 50 lashes on 16 April 1778 for being drunk on guard duty.

All the arms of the Regiment to be sent to the Armorours by Company, beginning by the Granidiers, an Officer of a Comp[y] to attend the armourey frequently in the day to have the arms finished as soon as possible

Jeremiah Early of the Granid[rs]. is Apointed a Serg[t] in Capt Ashby Comp[y]. & is to be Obey'd as such [267]

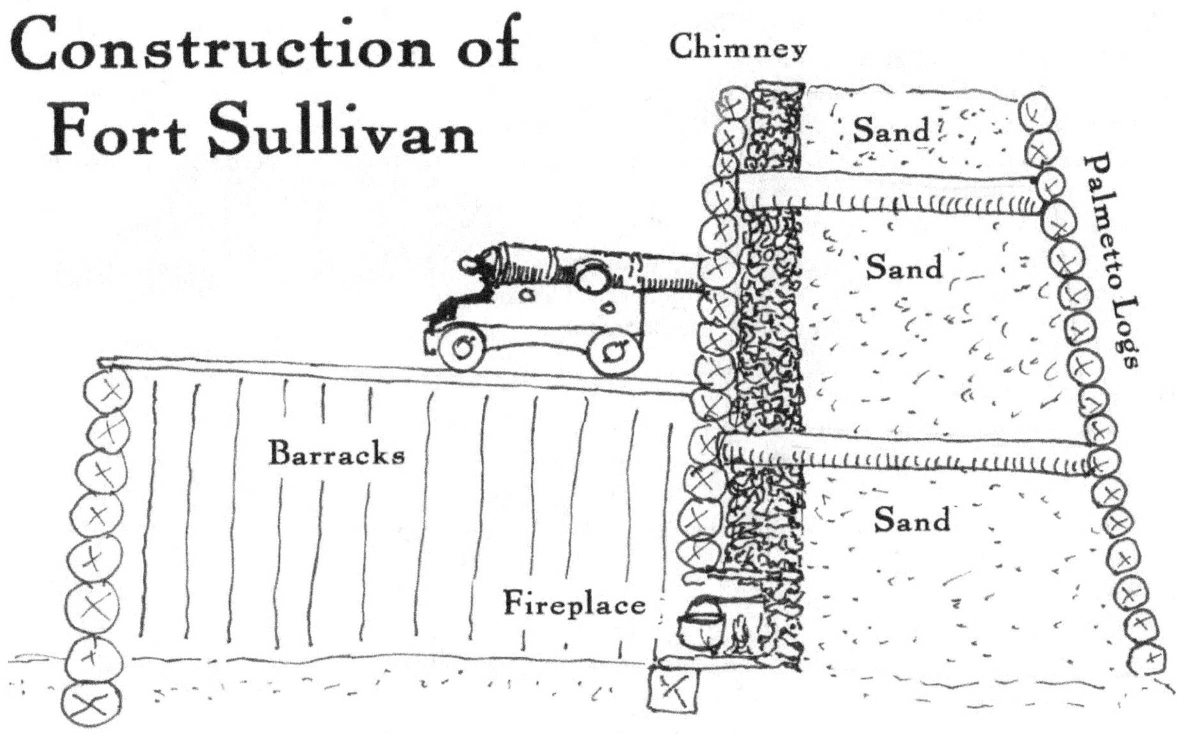

Orders 29th Ap.__ Parole Rutledge__Count[r].S[n] 26.

For the Day tomorrow Capt[n]. Eveleigh

Fort guard L[t]. Fuller

Advance a Sergeant__ Rear a Corporal

NB at 7 OClock this Evening the Detachment from Bulls Island return'd

---

[267] Jeremiah Early enlisted in the 2nd South Carolina Regiment on 4 November 1775. He was promoted to sergeant on 28 April 1776. On 20 June 1776 he was mult a fortnights pay for disobeying the sergeant major. This was to be made by two different payments. He was reduced to private on 5 February 1777. On 13 February 1777 he was reprimanded for misbehavior, but he was pardoned. On 24 June 1777 he was reprimanded for being absent from roll call.

Rogers,[268] armourer mate is Apointed Chief Armourer in the room of John Proby [269]

By Coll Moultrie **Sullivants Island 1776**
Orders 30th Ap:__Parole James Island__ Countr.Sn 20.

For the Day tomorrow Capt McDonald
Fort Guard Lt. R. Shubrick
Advance __ a Sergeant __ Rear a Corporal

Mr Mason who is sutler on this Island for selling Beer is also allowed to sell two gills rum pr man a day & that to be mixed with water & no more____ Any Soldier who Appear in Liquor not to be Allowed any, NO Liquor to be allow'd before troop Beating or after retreat____ the rum to be Sold according to the cost, The Capt of the day to visit all Soldiers Hutts & tents &c to see if any Liquor is concealed or sold Contrary to Orders__ John Conyers of Captn Ashby Compy [270]& Wm Lealand of Motte,[271] are Apointed Sergeant in there respective Compy. & are to be Obey as such_

Orders 1st May__ Parole Geortown__ Countr.Sn. 42.

For the day tomorrow Capt Huger
Fort guard Lt. Charnock
Advance__ a Sergeant__ Rear a Corpl.

By Capt. Pet. Horry
Orders 2d May__ Parole Moultrie __Countr.Sn. 75.

For the Day tomorrow Captn. Motte
Fort guard Lt. Gray
Advance a Sergeant__ Rear a Corpl.

By Major Marion **Sullivants Island 1776**
Orders 3d May __Parole Genl Armstrong__ Countr.Sn. 57.

For the Day tomorrow Capt Ashby

---

268 Nathaniel Rogers was born on 25 July 1755. He enlisted in the 2nd South Carolina Regiment on 20 November 1775. On 29 April 1776 he was made the Chief Armourer. He was under Captain Charles Motte at the siege of Savannah in 1779. He died in 1833 at the age of 78.

269 John Proby enlisted in the 2nd South Carolina Regiment on 4 November 1775. He was wounded at the battle of Fort Sullivan on 28 June 1776 while in Captain Ashby's company. On 29 July 1777 he was made the armourer to the regiment. He was reduced from being armorer on 11 December 1777 when he stole a handvise and had his pay stopped until he paid £20 for the handvise. He was confined for disobedience of orders on 12 January 1779 and he was sentenced to receive 100 lashes on the bare back with a cat o nine tails. He was under Captain Thomas Hall at the siege of Savannah in 1779.

270 John Conyers (Conners) enlisted in the 2nd South Carolina Regiment on 4 November 1775 as a corporal. He was promoted to sergeant on 30 April 1776 in Captain Ashby's company. He was reprimanded on 18 September 1777 for being absent from exercises. He was demoted to private on 22 March 1778 for being absent without leave. He was promoted back to corporal again prior to July 1778. He was in Captain Mason's company in 1779. On 9 February 1779 he was court-martialed for being absent without leave and he was sentenced to do double duty for ten days. On 15 February 1779 he was reduced to private for drunkenness on guard. He received 25 lashes on 18 May 1779 for quitting his rank without leave. He was at the siege of Savannah.

271 William McCleland.

Fort guard L[t] Lesesne
Advance__ a Serg[t]__ Rear a Corp[l].

Orders 4th May__ Parole Hallifax__ Count[r].S[n]. 42
For the Day tomorrow Capt. Oliphant
Fort guard L[t]. Tho[s] Moultrie
Advance__ a sergeant__ Rear a Corp[l].
A Regimental court martial to be held today at 11 OClock at the Presed[ts] tent, to try such prisoners as shall be brought before them__ Evidences to attend
Capt. P. Horry Presed[t].
L[ts]. T Moultrie__ J. Shubrick__Mazyck & Dubose Members
NB this court was countermanded__

Orders 5th May__ Parole Newport __ Count[r].S[n]. 24.
For the Day tomorrow Capt M[c]Donald
Fort guard L[t] Mazyck
Advance a Serg[t] __ Rear a Corp[l].

Orders 6th May__ Parole Rode Island__ Count[r].S[n]. 3.
For the day tomorrow Capt Huger
Fort guard L[t]. Jacob Shubrick
Advance a Sergeant__ Rear a Corp[l].
A Regt[l]. court martial to be held to day to try such prisoners as shall be brought before them Evidences to attend__
Capt[n] Huger Presed[t].
L[ts]. Blake__ J. Shubrick__ Hall & H Gray members
NB David Parsons & Moses Therrel for Disobediance of orders Sentenced to Rec[d] 100 Each__ Remitted also Ad. Meek 100 was remitted[272]

By Coll Moultrie **Sullivants Island 1776**

In February of 1776 Congress decided to make a Southern Department under the command of one major general and six brigadier generals. The overwhelming choice for command of the Southern Department was Major General Charles Lee. The brigadiers who would serve under Lee were John Armstrong, William Thompson, Andrew Lewis, James Moore, William "Lord Stirling" Alexander and Robert Howe. Brigadier General John Armstrong was assigned to the post of Charlestown and arrived on May 3rd. He discovered that the entire South Carolina military had been placed under the command of Governor Rutledge, and Rutledge had been given dictatorial powers that were not answerable to Continental commands. Armstrong also discovered that he did not have any Continental troops to command, since all the South Carolina troops were still Provincial State troops. To remedy this problem the Continental Congress voted on May 18th to send 1,200 Continentals from Virginia to Armstrong.

---

[272] David Parsons enlisted in the 2nd South Carolina Regiment on 4 November 1775. On 22 September 1777 he received 100 lashes for being absent without leave. He was discharged on 2 August 1779.

Orders 6th May __Brigadier Genl. Armstrong being arrived in this Collony he is to be recd by all guards & Centrys in the same manner as the Presedt. & Commander in Chief

The Brigadr will review the Regt. Some day this week tis hoped & Expected that the Officers in the Different Compys will take care to have their men as Clean as possible and Especially there Arms

---

Orders 7th May__ Parole Boltimore__ Countr.Sn 6

For the Day tomorrow Capt. Is. Harleston

Fort Guard Lt Thos Dunbar

Advance__ a Sergt__ Rear a Corporal

Orders 8th May__ Parole Williamsbourgh __Countr.Sn. 7

For the day tomorrow Captn Motte

Fort guard Lt Thos Hall

Advance a Sergt.__ Rear a Corpl.

A R. court martiall to be held today at 11 OClock at the Presedts. tent to try such prisoners as shall be brought before them Evidences to attend

Capt Nichs. Eveleigh Presedt.

Lts Fuller__ Lesesne__Dubose & Baker members

NB Sergt Semple of Capt Horrys Compy. for Disobediance of orders Sentenced to be reduced in the ranks, & receive 300 Lashes, the Commandg. Offr. remitted 300 Lashes [273] Jacob Dunbar of the same Compy sentenced to 250 Lashes Recd 150 [274]

By Major Marion **Sullivants Island 1776**

Orders 9th May__Parole Newburn__Countr.Sn 9.

For the Day tomorrow Capt Ashby

Fort guard Lieut: Jno Blake

Advance a Sergeant__ Rear a Corporal

A R. court martial to be held today at 11 OClock to try such prisoners as shall be brought before them. Evidence to attend

Capt Jas McDonald Presedt.

Lts Charnock__ T Moultrie__Mazyck & Jacob Shubrick members

NB Wm Simpson recd 200, Jno Price 300 Remitted,[275] Timothy Downing recd 150 Lashes [276] __ Jas Hendry recd 50-

---

[273] Unable to determine who James Semple was.

[274] Jacob Dunbar enlisted in the 2nd South Carolina Regiment on 2 July 1775 under Captain Barnard Elliot. He received 100 lashes on his "posterior" on 22 July 1777 for stealing the public's lead. He received 50 lashes on 25 October 1777 for falsely accusing Sergeant Henderson of stealing.

[275] John Price enlisted in the 2nd South Carolina Regiment on 4 November 1775. He reinlisted into the 1st South Carolina Regiment on 6 May 1777.

[276] Timothy Downing served in the 2nd South Carolina Regiment until 1779. On 9 May 1776 he received 150 lashes for disobedience of orders. He was wounded at the Battle of Fort Sullivan on 28 June 1776. On 17 September 1776 he received 100 lashes for drunkenness and neglect of duty. On 23 December 1776 he was reprimanded & Mult 14 days pay for disorderly behavior in the town. He received 100 lashes on 4 August

---

Orders 10th May__ Parole Wilmington__ Count[r].S[n] 11.

For the Day tomorrow Capt Oliphant
Fort guard L[t] Shubrick
Advance a Sergeant__ Rear a corporal

A quarter Guard of one 1 Serg[t]. 1 Corp[l]. & 9 men, every day

All Drumr[s] & fifers to attend the Batallion when Draw'd out, those who Neglect may Depend on being punisht by a court martial__ One drum & fife to be every day on the fort guard & to remain till the guard is relieved

A R. court martial to be held today at 11 OClock to try such prisoners as shall be brought before them Evidences to attend__ Capt Is: Harleston Preseď[t]
L[ts]. Fuller__ Dunbar__ H Gray & Hall members

All reports of Guards & morning reports to be made by 9 OClock in the morning

NB Jn[o] Cody Sentenced to 100 Lashes remitted; Ja[s] Quin rec[d] 100 [277] Serg[t]. C H Simmon mult 14 days pay for neglect of Duty [278] __ [279]

By Major Marion **Sullivants Island 1776**

Orders 11 May__Parole Roxbury__ Count[r].S[n]. 13.

For the Day tomorrow Capt
Fort guard L[t] Fuller
Advance a Sergeant __ Rear a Corp[l].

Orders 12th May__ Parole Beaufort__ Count[r].S[n]. 15

For the day tomorrow Capt. Eveleigh
Fort guard L[t]. Charnock
advance G__ a Serg[t] __ Rear a Corp[l].

1777 for being drunk and absent from guard. In 1778 he was in the Lieutenant Colonel's Company. He received 30 lashes on 19 May 1778 for being drunk. On 27 October 1778 he was sentenced to receive 100 lashes on the bare back with a cat of nine tails for neglect of duty. In 1779 he was under Captain Richard Baker.

277 James Quin enlisted in the 2nd South Carolina Regiment on 7 August 1775 under Captain Barnard Elliot. He reinlisted on 4 November 1776. He was in Captain Lesesne's Company in 1779. On 19 January 1779 he received fifty lashes on his bare back with a cat o nine tails, and sentenced to be put under stoppages for the musket he lost while on guard in town.

278 Charles Howell Simmons was in the 2nd South Carolina Regiment. He was mult 14 days pay for neglect of duty on 10 May 1776. On 24 August 1776 he was reduced in ranks from a sergeant to a private. On 14 October 1776 he was sentenced to receive 200 lashes for threatening to desert, however he was pardoned by the intrusion of Francis Marion. On 13 December 1776 he was tied to the halberds and reprimanded for neglect of duty. On 1 March 1778 the men of the regiment petitioned Francis Marion to allow Charles Simmons to teach them how to read and write. Marion tapped into Simmons ability to be a teacher, and due to this Simmons was exempt from all other duties except making the soldiers literate. He was commissioned as a lieutenant after the fall of Charleston and served seventy-seven days in the militia during 1781 and 1782.

279 Mult means to punish with a fine or forfeiture.

Orders 13th May __ Parole Armstrong__ Count^{r}.S^{n} 25.

For the Day tomorrow Capt Harleston
Fort guard Lt Hall
Adv. guard a Serg^{t}__ Rear a Corp^{l}.

No Officer to be absent from Exercise of the Cannon in the morning__ an Officer to take the Command of a gun & See the men do there Exercise Properly & not to quit there post till the drum beat to arms

No Officer to Appear on the parrade without there Uniform & Sidearms__ The Officers of a Comp^{y} to march there men on the parrade morning & Evening

The Carpenters of the 2nd Reg^{t}. now on the Island may work from Morning till roll-call in the Aft^{r}. noon at which time they must repaire to the parrade Accoutered, to be Exercised

The South Carolinians knew that a large British fleet was only a few days away at the entrance to the Cape Fear River in North Carolina. Charleston's defenders rushed to complete the defenses of the city. Colonel Gadsden rented slaves for a month and obstructions were being built in the channels of Charleston harbor. Six schooners were purchased to be turned into fireships

By Lt Coll Motte [280] **Sullivants Island 1776**

Orders 14th May __Parole Washington __ Count^{r}.S^{n}. 46.

For the Day tomorrow Capt Huger
Fort guard L^{t}. Lesesne
Advance__ a Serg^{t} __ Rear a Corp^{l}.

All Officers Servants are to turn out at Roll-call in After Noon and join there Different Comp^{y}. to be exercised with the Batallion [281]

Orders 15th May__ Parole Prosper __ Count^{r}.S^{n}. 45.

For the day tomorrow Capt Motte
Fort guard L^{t}. H Gray
Advn a Serg^{t}.__ Rear a Corp^{l}.

Lt Jno Blake is Apointed a Captain in the S. Carolina 2nd Reg^{t}. & is to be obey'd as such, Capt. Blake is to take the Command of the Comp^{y}. Late Captn Masons

Mr Richard Mason is Apointed a Second Lieutenant & to be Obey'd as such.[282]

---

[280] Fort Sullivan had increased in size and a rank higher than that of Major Francis Marion was required for the garrison. Lieutenant Colonel Isaac Motte assumed control of the fort on May 14th.

[281] The officer's servants were required to do the drill and military exercises in their assigned companies (the company to which the officer was assigned). These men would also go into battle as soldiers of the 2nd South Carolina Regiment. Most of these servants were Black slaves, but a few were white indentured men.

[282] Richard Mason was appointed a second lieutenant in the 2nd South Carolina Regiment on 15 May 1776 under Captain Harleston. He was in the battle of Fort Sullivan on 28 June 1776. He was promoted to first lieutenant on 27 January 1777 under Captain Charnock. He was promoted to Captain on 25 November 1778 when Captain Charnock resigned. He was captured at the fall of Charleston but continued to serve until the end of the war under General Marion.

L[t] Mason is to join Capt Harleston Comp[y]. M[r]. Pet[r] Gray is Apointed a Second L[t]. and is to be Obey'd as such he is to Join Capt Eveleigh, Comp:[283] L[t] Dunbar is to join Capt Mottes Comp[y]. & to do duty as first Lieut[t]:

A Reg[tl]: court martial to be held tomorrow at 11 OClock to try such prisoners as shall be brought before them Evidences to attend

Capt Blake President
L[ts] Lesesne__Mazyck__ Baker & Pet[r] Gray members

283 Peter Gray was born in 1750. He became a second lieutenant in the 2nd South Carolina Regiment on 15 May 1776 under Captain Eveleigh. He was promotoed to 1st Lieutenant on 6 April 1777 (the Officer's list says 28 January 1777) in Captain Oliphant's company. He was transferred to Captain Charnock's company on 16 July 1777. He was transferred to Captain Motte's company on 16 December 1777. He was promoted to Captain on 30 December 1778. He took command of the company of Captain Harleston on 23 February 1779, when Harleston was promoted to major in the 6th South Carolina Regiment. He was taken prisoner at the fall of Charleston and exchanged in June 1781. He served till the end of the war. He died in 1814 at the age of 64.

L^{t}. Coll Mott **Sullivants Island 1776**

Orders 16th May__ Parole Rutledge__ Count^{r}.S^{n} 9.

For the Day tomorrow Capt. Ashby
Fort guard Lt Thos Moultrie
Advance guard a Serg^{t} __ Rear g __ a Corp^{l}.

For the futer the Cannon are to be exercised every Tuesd^{y}. & fryday, The rest of the week they are to exercise with Small arms by the Commanding Officer of Each Comp^{y}. at the beating the Long-roll __ All Officers to attend and to be carefull that none of there men be Absent but there Servants

The men are not to turn out tomorrow but to make themselves as clean as possible__ The Officer of the Fort guard to visit the Centry, between every relief from Nine OClock at Night till day light

The Officers of Each Comp^{y}. are desired to be very attentive that their men keep there Arms clean & in good Order

The Capt. of the day to visit the Advance guard once a Day, & report any thing Extraordinary to the Commanding Officer______

Orders 17th May__ Parole Effinghan __ Count^{r}.S^{n} Lee

For the Day tomorrow Capt Oliphant
Fort guard Lt Thos Moultrie
Advance__ a Sergeant__ Rear guard a Corp^{l}

The Commanding Officer of each Company are desir'd to size there men properly before they march them on the grand parade [284] __The Command^{g}. Officer Orders that all men off Duty do attend Divine Service today and request that the Officers Do See this Order punctually Complied with The pay bills to be given in today as soon as possible

L^{t}. Coll Mott **Sullivants Island 1776 __**

A Regemental court martial to sett tomorrow morning at the Presidents tent to try such prisoners as shall be brought before them Evidences to Attend

President Capt Pet^{r} Horry
L^{ts}. Mazyck__ H Gray__ Marion & Shubrick Members

Orders 18th May _____ Parole Wilmington _ Count^{r}.S^{n}.

For the day tomorrow Capt Pet^{r} Horry
Fort guard L^{t}. Mazyck
Advance__ a Serg^{t}.__ Rear a Corp^{l}

NB according to the last court martial Conrad Fitner rec^{d} 150 Lashes
N Buk__ Ja^{s} Crawford [285] Corp^{l} Gambell[286] Pardoned

---

[284] When a company is "sized" the shorter men would be on the front row and the taller men on the back row. It also would have the tallest men on the right and the men getting shorter as it goes to the left. This insures that as each two rank "file" of men fire, the rear man will not have to worry about having a taller man in front of him throwing his aim off.

[285] James Crawford enlisted in the 2nd South Carolina Regiment on 4 November 1775. He was found guilty of an unnamed charge but pardoned on 18 May 1776. Unfortunately he died on that same day.

[286] Robert Gammell (Gamble, Gambell) enlisted in the 2nd South Carolina Regiment on 4 November 1775 as a corporal. On 18 May 1776 he was found guilty of some unnamed charge, but he was pardoned. On 30 June

---

Orders 19th May __ Parole Quebeck __ Count^r.S^n Chester
For the Day tomorrow Capt Danl Horry
Fort guard Lt J. Shubrick
Advance__ a Serg^t __ Rear a Corp^l.

Provision Return to be made regularly every day & sign'd by an Officer of a Comp^y. a Sergeant to Attend the Delivery of the provisions & see it properly Divided __ the Non Commissioned Off^rs & soldiers are forbid from Interfering with the workmen and Labourers

The Articles against Mutiny & Desertion to be read tomorrow by an Officer of a Comp^y. & to be carefull all there men are present____The Troops to attend Divine Service this aft^r noon when the Long roll beats

A court martiall to sett tomorrow at 10 OClock AM to try such prisoners as shall be brought before them, Evid: to attend

Capt Danl Horry Presed^t.
L^ts. Fuller__ Charnock__ Dunbar & Hall members

NB Serg^t. Young was order'd by the court today to make consessions to Mr Baldwin for his Behavior to him[287]__ John Cody Rec^d 150 Lashes for Drunkiness

By L^t. Coll Mott **Sullivants Island 1776**

Orders 20th May__ Parole New England__ Count^r.S^n Cork
For the Day tomorrow Capt Eveleigh
Fort Guard L^t. Thos Dunbar
Adv: a Serg^t __ Rear a Serg^t.

By Coll Moultrie

One Captn & one Subaltern to Join the General court martial in town 22nd Inst: of which Major Cattle is Presed^t by Order of Coll Gadsden [288]

By Coll Motte

Orders 21^s May __ Parole Concord __ Count^r.S^n Essen
For the Day tomorrow Capt Motte
Fort guard Lt Marion
Advance guard Lt Fuller

A Return to be made Emediately of the Number of Arms wanting in each company

NB This Evening the Advance guard was reinforced by L^t. Shubrick, 1 Serg^t. & 20 Rank & file _____

1776 he was reprimanded for being absent during the Battle of Fort Sullivan on 28 June 1776. On 14 October 1776 he was reprimanded for striking William Donovan. He was reduced in ranks to a private on 23 April 1777 for neglect of duty. He later was promoted to corporal again, but he was demoted again to private on 15 August 1777 for behaving in a disorderly manner and being out of the barracks. On 16 April 1778 he was sentenced to do double duty for two weeks for neglect of duty. He was in Captain Dunbar's light infantry at the siege of Savannah in 1779. He was also in the militia from 3 November 1780 to 5 January 1782.

[287] John Young was in the 2nd South Carolina Regiment. He lost a leg in the Battle of Fort Sullivan but he survived and became the regimental sutler in 1777.

[288] William Cattel.

British Admiral Sir Peter Parker had sent Major James Moncrieff on a mission to conduct a reconnaissance of Charleston harbor. Moncreiff was to determine if there was a chance of taking the Southern port. On the night of May 21st Moncrieff was able to sound the bar and he determined that a British ship could get over it. He also sketched the half finished fort on Sullivan's Island. Moncrieff landed his boat at Fort Sullivan and went inside the works, sketching the defenses.

The 2nd South Carolina soldiers inside the fort assumed that he was one of their own. It wasn't until Moncrieff returned to his boat that the sentinels discovered who he was and fired upon the British boat. The sentinels fired fifty shots at Moncrieff's party, but did not touch any of them. When Moncrieff returned he told Admiral Parker that Fort Sullivan "was in so unfinished a State as to be Open to a coup-de-main." Parker made the decision to attack Charleston and use it as a base of operations to take back the colonies [289]

Orders 22nd__ Parole Long Island __ Countr.Sn 3.

For the Day tomorrow Capt. Ashby

Advance guard Lt. Charnock

Fort __ Lt Petr Gray __ Rear a Sergt.

It is Expected by the Commdg. Officer that the Officers without their Side Arms & Regimental Caps: the Commandg Officer of Compys. are to apply to the Quarter Master Sergt: for Arms agreeable to the Returns given in yesterday for which they are to give a Receipt.

A court martial to Sett Emediately to try such prisoners as shall be brought before them All Evidences to attend

Captn Eveleigh President

Lts. Lesesne __ J. Shubrick __ Baker & Petr Gray Members

By Lt. Coll Mott **Sullivants Island 1776**

After Orders 22nd May

A Reinforcement of 1 Capt: 2 Subalterns 2 Sergts & 40 Rank & file to Join the Advance guard in Capt Eveleigh & Lts ~~Lesesne~~ & Baker for this command __ NB the reinforcement marched off

---

Orders 23rd May __ Parole Tankerville __ Countr.Sn Liberty

For the day tomorrow Captn. Oliphant

Advance guard Lt Thos Moultrie

Fort __ Lt Lesesne __ Rear a Sergeant-

A Regtl court martial to Sett this morning at 11 OClock to try such prisoners as shall be brought before them. Evid: to attend

Captn. Huger President

Lts. Lesesne __ Shubrick __ Hall & Dunbar Members

Two Subalterns 2 Sergts. 2 Drum & 40 Rank & file for the Advance guard tomorrow __ to be continued every day till Further Orders __ The Capt: of the day to visit the Hutts in & near Camp frequently to Seize Rum or any other spiritous Liquors which he finds in any of them; unless Permission has been granted by the Commandg. Officer

---

289 O'Kelley, *NBBAS, Volume One,* pp. 109-110.

NB the court martial of the 22nd of which Capt Eveleigh was preset Sentenced Wm Boyd to 250 Lashes for striking Sergt Conyers [290]

Orders 24th May __ Parole Geo.Town __ Countr.Sn.

For the Day tomorrow Capt. Petr Horry
Advance guard Lt H Gray
Fort guard Lt Blake __ Rear a Sergt

No Boat or Boats to be suffer'd to pass the fort the Officer of the Fort guard to see this Order strictly complied with [291]

By Lt. Coll Mott **Sullivants Island 1776 __**

The Commanding Officer of each Company to be perticularly carefull that no more then One man of a Company be absent at a time, And that they will Inspect the mens Arms every morning & punish those that are not in good order

The Officers of the Advance guard to make his report in wrighting as soon after he returns to Camp as possible

A court martial to sett Emediately to try such prisoners as shall be brought before them Evid: to attend

Capt Ashby President
Lts Fuller __ H Gray __ Baker & Marion members

Orders 25th May__ Parole Maryland __ Countr.Sn 92.

For the Day tomorrow Captn. Danl. Horry
Advance guard Lt Mazyck
Fort __ Lt Shubrick __ Rear a Sergt

A court martial to sitt Emediately at Presedts tent to try such prisoners as shall be brought before them Evidences to attend

Captn Oliphant President
Lts Dunbar __ J Shubrick __ Hall & Petr Gray members

Orders 26th May __ Parole New Jersey __ Countr.Sn 17

For the Day tomorrow Capt Huger
Advance guard Lt Dunbar
Fort __ Lt G Eveleigh __ Rear a Sergt.

As there is no field Officer of the day at present. the Capt of the day to be Look'd upon as such & to be recd: by going his rounds as Acting in that Character; agreable to the Orders of the 22nd Inst: the Commanding Officers of Companies are perticularly Desired to make there men put there Arms in good Order & keep them so __ Lt Geor Eveleigh is to do Duty in Capt: Ashbys Company till further Orders [292]

---

290 Willliam Boyd enlisted in the 2nd South Carolina Regiment on 4 August 1775, and reinlisted on 4 November 1775 under Captain Barnard Elliot. He was discharged on 21 October 1776.

291 This was done because of British Major Moncrieff's reconnaissance of Fort Sullivan.

292 George Eveleigh became a 2nd Lieutenant on 26 May 1775. He was promoted to 1st Lieutenant on 6 June 1776. He died on 20 May 1777.

By L^{t}. Coll Mott **Sullivants Island 1776 __**

Orders 27th May __ Parole Kent __ Count^{r}.S^{n}. 12.

For the day tomorrow Captn. Ashby
Advance guard L^{t}. Fuller
Fort guard L^{t}. Marion __ Rear a Sergeant

After the Reg^{t}. has turn'd out to exercise in afternoon the Capt of the Day is to take an Escort from the Fort guard & search all Hutts & Sutlers in perticular & all the men fitt for duty he finds lurking there, to confine them & have them reported

this Order to Continue every day __ the Capt. of the day is not to visit the Advance guard till furth^{r} Orders

A Reg^{tl}. court martiall to sitt Emediately &^{ct}.

Capt Blake President
L^{ts}. Fuller __ Lesessne__ T Moultrie & Baker members

NB Jno Cody for drunkinness rec^{d} 250 Lashes __ Thos Walker for Drunkinness Sentenced 200 Lashes rec^{d}. 100[293]

By ~~Coll Moultrie~~ ~~Captn Pt^{r} Horry~~

By Coll Moultrie

Orders 28th May __ Parole Motte __ Count^{r}.S^{n} 32

For the day tomorrow Capt: Oliphant
Advance guard Lt Lesessne
Fort __ Lt Pet^{r} Gray__ Rear a Sergeant

A man in Each Company of the Second Regiment to be Apointed as pioneer by the Commanding officer of Each Comp^{y}. they are to be provided with an Ax & spade & he is to have charge of them & do no other duty in the Reg^{t}. but Pioneer duty-

The Rever^{d} Henry Purcell is Apointed Chaplain to the 2nd S^{o}. Carolina Regiment & to be respected as such

Corporal John Henderson of Capt: Horrys Comp^{y}. is apointed a Serg^{t}. in s^{d}. Comp^{y}. & to be obey as such [294]

By Capt Pet^{r} Horry **Sullivants Island 1776 __**

Orders 29th May __ Parole Brunswick __ Count^{r}.S^{n}. 45.

For the day tomorrow Capt Blake
Advance Guard Lt H. Gray
Fort __ Lt Thos Moultrie __ Rear a Serg^{t}

---

[293] Thomas Walker was in the 2nd South Carolina Regiment until the fall of Charleston in May 1780. Afterwards he was in Colonel Roebuck's regiment of militia and served 310 days on horseback during 1780. He also supplied provisions and sundries. On 27 May 1776 he was sentenced to 200 lashes for drunkiness, but only received 100. On 9 April 1777 he received 100 lashes for "Misbehaviour & insolence to Lt. Eveleigh".

[294] John Henderson enlisted in the 2nd South Carolina Regiment on 4 November 1775. He was promoted to sergeant on 28 May 1776. On 13 November 1776 he was mult 14 days pay for drunkenness. On 4 November 1777 he was reprimanded for an unnamed offense. He was demoted to private on 11 December 1777 for disobedience of orders.

By Lt Coll: Motte

Orders 30th May __ Parole Chelsea __ Countr.Sn. 23.
For the day tomorrow Capt. P. Horry
Advance guard Lt. Baker
Fort __ Lt Mazyck __ Rear a Sergt.

The Commanding Officer Desires that all officers off Duty do attend the Exercise morning & Evening with more punctuallity for the feuter __ the pay bills to be given in tomorrow __ the Officers upon return to camp after have Leave of Absente, are Desir'd to call on the Commandant

No Officers to Leave Camp till the Orders of the day are given out

---

Orders 31st May __ Parole Grafton __ Countr.Sn. 1.
For the day tomorrow Capt Eveleigh
Advance guard Lt Jacob Shubrick
Fort __ Lt Geor Eveleigh __ Rear a Sergt.

A General Return of the Regiment to be given in by tomorrow 12 OClock __ The Articles of War to be read tomorrow & Munday After noon to the Regimt. The Officers are Desired to be very carefull that the men off duty do Attend [295]

Orders 1st June __ Parole Virginia __ Countr.Sn:
For the day tomorrow Capt: Jams McDonald
Advance guard Lt. Dunbar
Fort __ Lt Jackson __ Rear a Sergeant

By Lt Coll: Motte **Sullivants Island 1776 __**

Orders 1st June __ All Centries are to face with shoulder arms to any Officers that may be coming towards them, this Order to be read to the men by an Officer of the Compy. & the Officer of the fort guard is perticularly Desired that this order may be given to the Corporal of his guard before_the relief of the Centry

---

Orders 2nd June __ Parole Hannibal __ Countr.Sn.
For the day tomorrow Capt: Harleston
Advance guard Lt Richd Shubrick
Fort __ Lt Mason
Quarter guard today Lt Marion
Quartr G __ tomorrow Lt P. Gray__ Rear a Sergt __

Orders 3d June __ Parole Scipio __ Countr.Sn. 5.
For the day tomorrow Captn Huger
Advance guard Lt Wm Moultrie
Fort __ Lt Fuller __ Quartr G Lt Hall __ Rear a Sergt

---

[295] The Articles of War are located in Appendix A of this book.

Orders 5th June __ Parole Genl. Lee __ Countr.Sn. 8.
For the Day tomorrow Capt Ashby
Fort guard Lt. Thos Moultrie
Quartr Lt Baker __ Rear Lt Proveaux
The advance guard to be relieved by Coll. Thompson Rangers [296]

Orders 6th June __ Parole Congaree __ Countr.Sn. 2.
For the day tomorrow Capt. Oliphant
Fort guard Lt. Mazyck __Quarter guard Lt Jackson
Rear guard Lt Jacob Shubrick

By Lt Coll: Motte **Sullivants Island 1776 __**

Orders 7th June __ Parole Rutledge__ Countr.Sn 41.
For the day tomorrow Capt Blake
Quarter Lt Mason __ Rear Lt Petr Gray

General Orders 6th by Coll Moultrie

Capt Daniel Horry being under a perticular service as Coll. of the Craven County Militia which will require his presence for some time to prevent the Service suffering he has desir'd to resign his commission as Capt. in 2nd Regt. The President on this account has Accepted his Resignation

Lt Richd Shubrick is Apointed a Captain.

Lts Geor Eveleigh & Wm Moultrie are Apointed 1st Lieuts and are to be Obey'd as such __ Captn Shubrick is to take the command of the Granidier Compy. & Lt Wm Moultrie to do duty in the same

7th June by Coll Moultrie

After Orders, One Captn. 1 Subaltern & 13 Privates to Join the Fort guard Emediately and 1 Sergeant & 4 men for a patrole to patrole constantly during the night round the fort An Officer of the guard to visit the Centries every half hour during the night __ The Officers of the Day to visit the guards & Centries at Least twice in the night

For the Fort Guard Captn Blake & Lt Marion

By Coll: Moultrie [297]

Orders 8th June __ Parole Quebeck __Countr.Sn. 20.
For the day tomorrow Capt Shubrick
Fort guard Captn. McDonald & Lts Charnock & Hall
Quartr. Lt H Gray- Rear Lt Baker

When the Enemy shall Attack this post by Shiping, Coll Thompson with his Regement is to keep a good Look out & to endeavor to prevent any boats from Landing troop on the Island __ If any troops should make good there Landing Coll: Thompson with his Regt & Captn Allstons Compy. or any Other which may Join them, must use there Utmost Endeavor to prevent them from Marching to

[296] When the threat of a British invasion became real, the Rangers were moved from the frontier to the coast.

[297] Colonel Moultrie assumed command of Fort Sullivan on 8 June 1776.

this fort.[298] if he find he cannot stop there march he must Retreat with his Reg^{t}. & any other troops which may hereafter Join him [299]

The Officers of all guards are to be very carefull to Order there Centries not to fire on any boats coming to the Shore with a White flagg, But they are to draw up there men on the Bank ready to receive them & to allow only one man to come ashore, who is to deliver his message to the Officer of the guard who is Emediately to send word to the Commanding Officer __ No guns to be fired in Camp on any account whatever without Leave first Obtain'd from the Commanding Officer

On June 1, 1776 Admiral Parker dropped anchor near Charlestown with fifty vessels, including troop transports. General Clinton did not really have a plan except to seize Sullivan's Island. By June 7th most of the British lighter frigates and troop transports were brought over the Charlestown bar. The British sent a boat with a flag of truce towards Sullivan's Island to attempt to negotiate a surrender, but a sentinel fired upon it. This boat was actually on an intelligence gathering mission to "ascertain if the Americans had erected an abates and the conditions of the defenses."

The next day Colonel William Moultrie sent an officer to the British fleet and explained to them that the sentinel had fired without orders. Major General Clinton was satisfied with the apology and then gave the South Carolina officer a proclamation to deliver to Colonel Moultrie telling him to surrender the town.[300]

Orders 9th June __ Parole Arnold __ Count^{r}.S^{n}. 10.

For the day tomorrow Capt Pet^{r} Horry

Fort Guard Capt Harleston & L^{ts} Mazyck & Proveaux

Quart^{r}. Lt Thos Moultrie __ Rear Lt Jackson

Ordered that all the men do remove there Baggage & lodge within the fort tonight [301]

Orders 10th June __ Parole Geo^{r} town __ Count^{r}.S^{n}. 60.

For the day tomorrow Captn Eveleigh

Fort guard Captn. Huger & L^{ts}. Shubrick & Marion

Quarter __ Lt Dunbar __ Rear L^{t}. Mason

Charles Lee, the son of Major General John Lee, was born in England in 1731. He attended school in England and Switzerland, then he entered his father's regiment as an ensign at the age of 15 in 1747. In 1751, after his father died, he purchased a Lieutenant's

---

298 Captain John Allston is listed in several accounts as John Allen. He commanded the "Raccoon Company." This company was also known as the Foot Ranger Company and consisted of Peedee, Waccamaw, Cheraw, and Catawba Indians. Allston and his Raccoon Company was in the battle of Fort Sullivan in 1776 and in the battle of Stono Ferry in 1779. Allston served in the militia in 1780-1782. After the surrender of Charleston in 1780 he served under Colonel Peter Horry's cavalry with Francis Marion's partisans.

299 Thompson was guarding the northeastern end of Sullivan's Island, known as The Breach..

300 O'Kelley, *NBBAS, Volume One* pp. 125-126.

301 Since the temperature was so hot, many men slept outside, or in tents outside, and used the barracks built into the fort walls as storage for their baggage. When Lee inspected the fort he wrote that the 2nd South Carolina was sheltered "in huts and booths covered in palmetto leaves." When the *Bristol* crossed over the bar the men were ordered to strike their tents, raze their palmetto shelters, and sleep inside the barracks, away from any possible harm that may be caused when the British began shelling the fort.

commission in the 44th Regiment of Foot and was on Braddock's expedition during the French and Indian War. He was in the supply command and was not on Braddock's defeat with other notable leaders, such as George Washington or Daniel Morgan.

He purchased a commission as a captain in 1756 for £900 and went on garrison duty at Fort Johnson near Albany, New York. He was adopted by the Mohawk Indians and received the name Ounewaterika, which meant "boiling water". He "married" the daughter of Seneca chief White Thunder, and they had twins, a girl and a boy.

He was with Lord Howe on patrol around Fort Carrilon (later Ticonderoga) when Howe was killed. He was badly wounded during a bayonet charge during Abercromby's ill planned head-on attack against Fort Carrilon in July of 1758. He "received a severe wound from a musket shot, which passed through his body and broke two of his ribs". He recovered and was with his regiment during the capture of Niagara and Montreal.

He went back to England in 1760 and was appointed as a major in the 103rd Regiment of Foot. There he served with Burgoyne in Porrtugal. He retired on half pay in November 1763 when his regiment was disbanded, so he became a soldier of fortune.

He was in the Polish army where he was the aide de camp to King Stanislaus Poniatowski. He received a medal but he saw no action in Poland. He was promoted to major general in 1767. He spent two years in England, criticizing the government, and then returned to Poland in 1769 to fight the Turks. He also served with the Russian army against the Turks. In 1769 "He left the [Russian] army and crossed the Carpathian Mountains, on his route to try the waters of Buda. In Hungary, he was attacked with a fever that threatened his life."

Lee went to Vienna and then to Italy in 1770, where he killed an officer in a duel, while losing two fingers. He was invalided home to England in 1770. In 1772 he was promoted to brigadier general in the British army, still on half pay.

In 1773 he moved to New York with a letter of introduction from Benjamin Franklin and immersed himself in the revolutionary politics. The wife of a Delaware tavern keeper described him as "strange, nervous man, his cultivated manners and speech so much at odds with his worn leather jerkin, greasy vest, frayed deerskin breeches, old and dirty jackboots and a big broad-brimmed beaver hat crammed down to his pointed nose".

He knew there would be great possibilities for advancement in the colonies and he urged the Patriots to raise an army, one that he would naturally be in command of. Congress appointed him major general on 17 June 1775, subordinate only to George Washington and Artemus Ward. He made Congress promise that they would repay for losses of his property in Britain before he would accept the appointment. He served in the defense of Boston, where one witness wrote "his dirty habits and obscenities gave offense".

He was detached from the army in January 1776 to raise volunteers in Connecticut for the defense of New York City. On 17 February 1776 he was ordered by Congress to take command of the Northern department from General Schuyler, but Congress countermanded the order a month later, and placed him in command of the Southern Department when it was learned the British fleet was off the Carolinas.[302]

[302] O'Kelley, *NBBAS, Volume One,* pg 118.

By Coll Moultrie Sullivants Island 1776

A Letter to Coll Thompson

Sir

Generall Lee has sent down Capt Debram [303] an Engineer to throw up some Breast work at our advance guard,[304] you will therefore show him the most proper place to prevent the Enemy crossing over to us, You will give him all the assistance he may want, Set your men to work they will be allowed 10/ p^r Day & a Gill rum pr man [305]

Y^rs. Wm Moultrie

The Command of all the regular Forces & Militia of this Collony (acting in Conguntion) being Invested his Excellency Genl Lee, All orders from him are to be Obey'd

Captn Mayham [306]& Coutreare's [307] Comp^y. are to act as one body- A return to be made today of the number of men as well Regulars or Militia now on this Island Lt Petr Gray is Apointed Quartr Master to all the troops on the Island till further Order__ all Provision returns to be made to him

Coll Thompson & Coll Sumpter Reg^ts. Capt^ns Allston Maham & Coutreares Comp^y. are to hold themselves in readyness to March [308] __Coll: Thompson will receive his orders from the Commanding Officer

---

303 Military engineer J. Ferdinand S. Debrahm had been the King's Surveyor General of the Southern District and had supervised the work on the town's defenses. Debrahm was ordered to construct a floating bridge from Sullivan's Island to Haddrell's point, but it was not very practical for a withdrawal. There were not enough boats to finish the construction since the distance was over a mile. Another bridge was constructed using empty Hogsheads, placing two planks on each barrel. When the North Carolina Continentals tried to march 200 men over the barrels, the bridge sank.

304 When it was finished the Advance Guard was described as being constructed of "palmetto logs, with merlons on a brick foundation." The bricks were still there in 1850 when they were discovered after the sands had shifted.

305 10 / pr day means that the soldiers would receive an additional 10 shillings a day. If it was written as 10/2 it would mean 10 shillings and 2 pence. 4 farthings = a pence. 12 pence = 1 shilling. 20 shillings = 1 pound. A pence is a penny, but there was no penny coins until after the Revolutionary War. Sometimes money would be written as 10 s 4 d, which means 10 shillings and 4 pence. This is carried on today by the manufacturers of nails. A box of 10 penny nails is labeled 10 d. To give an idea of the cost of things a loaf of bread is 1 pence. A bottle of wine is around 6 shillings. A wool suit of clothes would cost 5 pounds. Five shillings in England, before the war, could buy you 1 gallon of rum, or a copy of Guliver's Travels, or 2 bushels of salt, or 2 pounds of chocolate, or 2 pounds of the best gunpowder, or 4 pounds of candles or soap, or 4 decks of playing cards, or 50 acres of unimproved wilderness land in Virginia.

306 Hezekiah Maham was born on 26 June 1739. He served as a captain in the 1st South Carolina Regiment during 1776. He resigned from the regiment on 8 November 1779, possibly due to the debacle in Savannah. He was promoted to lieutenant colonel of an independent corps of cavalry under Colonel Daniel Horry on 22 June 1781. He was taken prisoner at his home in August of 1782. He died in 1789 at the age of 50.

307 John Couterier was a captain in the militia under Colonel Richardson during 1775. He was also a lieutenant in the dragoons in 1775.

308 Thomas Sumter was born on 14 August 1734 in Virginia. He settled in the Orangeburgh District of South Carolina in 1764. At the beginning of the Revolution he was a captain of a volunteer militia company in St. Mark's Parish. He served as an adjutant for Colonel Richardson in 1775. Impressed by the speed with which the 5th South Carolina Rifles filled their ranks, the Congress authorized a second rifle regiment, the 6th South Carolina Regiment on 27 February 1776. Even though the designation of the regiment was as the second rifle regiment of South Carolina, in reality there were no rifles in the 6th South Carolina. The men were all armed with muskets, and half the regiment did not have any arms at all. Instead of seven companies they could only

These units were being marched to the exposed northern tip of Sullivan's Island. The force consisted of the Thomson's 3rd South Carolina Rangers, Sumter's 6th South Carolina, and Alston, Maham and Coutirier's companies. They were to defend an area known as The Breach. This was an inlet between Long Island and Sullivan's Island. The British had landed on Long Island, opposite the Breach, and set up camp there.

Lee ordered Moultrie to send this force to Long Island and attack the British who landed there. Two field pieces were sent there to cover the possibility of a retreat. Moultrie did not get this order until the afternoon of the 10th and he decided to leave the Rangers in the Advance guard until morning, when they would attempt to cross the Breach. Moultrie did order Thomson to "attack and if possible dislodge the corps of the enemy", but when the British Man-of-War *Bristol* crossed the bar this plan was abandoned.[309]

Orders 11th June __ Parole Haddrells Point __ Countr.Sn. 5.

For the day tomorrow Capt McDonald

Fort guard Capt. Motte Lts. Charnock & Dubose

Quartr. Lt Hall __ Rear Lt Lesessne

General Orders

Coll: Sumpters Regt. Captns Allston Mahams & Coutreares Compys. to hold themselves in readyness to March Emediately to Haddrell point __ Boats will be provided to carry them over [310]

Orders 12th June __ Parole Long Island __ Countr.Sn 4.

For the day tomorrow Captn. Harleston

Fort guard Capt Ashby Lts. Thos Moultrie & Baker

quarter Lt Mazyck __ Rear Proveaux

Orders 13th June __ Parole Fort Johnson __ Countr.Sn. 7.

For the day tomorrow Captn Huger

Fort guard Captn Oliphant Lts. Shubrick & Dunbar

Quartr Lt Jackson __ Rear Lt Marion

All Officers & soldiers are to be in the fort after Tatoo beat & to remain till Revallee Beat __

---

field five. Thomas Sumter became a lieutenant colonel in the 6th South Carolina Regiment on 29 February 1776 and was at Haddrell's Point during the battle of Fort Sullivan. He resigned from the 6th South Carolina Regiment on 23 September 1778 and became a colonel of the South Carolina Militia. After the fall of Charleston he became a brigadier general of the militia. He was in command of the troops at the battle of Fish Dam Ford, Blackstock's Plantation, Hanging Rock, Fort Granby, Quinby Bridge, and Shubrick's Plantation. He was wounded at both Fish Dam Ford and Blackstock's Plantation. After the war Sumter served in both the House and the Senate. Sumter died on 1 June 1832 at the age of 98 after he had taken his customary horse ride for the day. He was the oldest surviving general of the Revolutionary War.

[309] O'Kelley, *NBBAS, Volume One,* pg 143.

[310] Lieutenant Colonel Peter Horry, the temporary commander of Haddrell's Point, was given orders to extend his defenses to guard against a British flank attack from Long Island. Horry was augmented with Colonel Sumter's 6th South Carolina Regiment and 100 men from the 3rd South Carolina Rangers. On June 10th Brigadier General John Armstrong assumed command of the forces on Haddrell's Point.

Orders 14th June __ Parole Camden __ Count^r.S^n. 2.

For the day tomorrow Captn Motte
Fort guard Capt Blak L^ts. Eveleigh & Wm Moultrie
Quart^r. Lt Charnock & Rear Lt Hall

Five men of a Comp^y to turn out emediately to through up some Intrenchments __ they are to be paid & allowed a Gill rum p^r man

A court martial to sett Emediately to try such prisoners as shall be brought before them Evidences to attend

Capt Pet^r Horry Presed^t.
L^ts. Charnock __ Hall __ Lessesne & H Gray members

Orders 15th June __ Parole Chatham __ Count^r.S^n. 12.

For the day tomorrow Capt Ashley
Fort Guard Captn. Shubrick L^ts Lessesne & Dubois
Quarter Lt Baker __ Rear Lt Thos Moultrie

By Coll Moultrie **Sullivants Island 1776 __**

Orders 16th June __ Parole Effingham __ Count^r.S^n. 15.

For the day tomorrow Captn Oliphant
Fort Guard Captn P Horry L^ts. Mason & Jackson
Quart^r. Lt Shubrick __ Rear Lt Marion

Orders to Captn Simon Tuffts on board the Defence [311]

Sir.

As we expect the Enemy to Land on this Island every moment to attack this post you are here by order'd to use your utmost Endeavours to Annoy Destress & Distroy them when they shall come within reach of y^r guns & you are to destroy them in every shape to the utmost of your power __ Y^rs.

W Moultrie __

Orders 17th June __ Parole N. Carolina __ Count^r.S^n 27.

For the day tomorrow Captn Blake
Fort guard Captn Eveleigh L^ts. Wm Moultrie & Mason
Quart^r __ Lt Dunbar __ Rear Lt Proveaux

two Subalterns, one Sergeant & 60 Rank & file to turn out Every day for fatigue

---

Orders 18th June __ Parole Brunswick __ Count^r.S^n. 13.

For the day tomorrow Captn Shubrick
Fort guard Captn McDonald L^ts. Lesessne & Hall
Quart^r. L^t. H. Gray __ Rear Lt Charnock
For Fatigue L^ts Jackson & Marion

---

[311] Moultrie ordered the armed schooner *Defense* to move behind Fort Sullivan to prevent the Advance Guard from being flanked by any smaller British vessels.

Two field pieces to be sent to the Advance guard this Evening NB ~~One 18 pound~~^{r} [312] A Regimt^{l} court martial to sitt at 11 OClock to try such prisoners as shall be brought before them __

Captn Eveleigh President

L^{ts}. Mazyck __ D Dubose __ Shubrick __ Baker Members

By Coll Moultrie **Sullivants Island __1776 __**

Orders 19th June __Parole Ansonborough __ Count^{r}.S^{n}. 1.

For the day tomorrow Captn P Horry

Fort guard Captn Harleston L^{ts}. Thos Moultrie & Proveaux

Quart^{r} L^{t}. Dubose __ Rear Lt Baker

Orders 20th June __ Parole Bulls Island __Count^{r}.S^{n} 6.

For the day tomorrow Captn Eveleigh

Fort guard Captn Huger L^{ts}. Shubrick & Jackson

Quarter __ Lt Wm Moultrie__ Rear Dunbar

Fatigue L^{ts}. Hall & Gray

NB. the court martial 18th Inst of which Captn Eveleigh was Presed^{t} sentence Ben^{j}. Staly 100 Lashes rec^{d} __ Wm Smiley [313] rec^{d} 50 Lash for calling Fickling [314] a murderor

---

[312] These were an 18-pounder and a 6-pounder that was given to Colonel Thomson to use against the British forces camped on Long Island.

[313] William Smiley (Smily) enlisted in the 2nd South Carolina Regiment on 7 August 1775 under Captain Barnard Elliot. On 20 June 1776 he received 50 lashes for calling William Fickling a murderer. He received 60 lashes on 22 July 1777 for quiting his guard. On 15 August 1777 he was reprimanded for quarreling and fighting.

[314] William Fickling enlisted in the 2nd South Carolina Regiment on 18 December 1775. Fickling must have been connected to someone's death, since many men were charged with calling a murderer. He was tried for desertion on 23 September 1777, and was sentenced to receive 100 lashes on 2 October 1777, but was pardoned by General Howe. On 18 October 1777 he received 50 lashes for neglect of duty. He was captured at the fall of Charleston. A man named Fickling was sentenced to death for desertion during the time of the Siege of Savannah. In "The American Revolution, including also the beauties of American history" there is a section entitled "THE SURGEON AND THE GHOST". It states "A circumstance occurred during the encampment of General Lincoln at Perysburg, that from its singularity deserves to be recorded. A soldier named Fickling, by the irregularity of his conduct, long excited the indignation of his comrades, and at length, from repeated efforts to escape to the enemy, had been brought to trial, and condemned to death. It happened that, as he was led to execution, the surgeon-general of the army passed accidentally on his way to his quarters, which were at some distance off. On being tied up to the fatal tree, the removal of the ladder caused the rope to break, and the culprit fell to the ground. This circumstance, to a man of better character, might have proved of advantage; but being universally considered as a miscreant, from whom no good would ever be expected a new rope was sought for, which Lieutenant Hamilton, the adjutant of the First Regiment, a stout and heavy man, essayed by every means, but without effect, to break. Fickling was then haltered, and again, turned off, when, to the astonishment of the bystanders, the rope untwisted, and he fell a second time, uninjured, to the ground. A cry for mercy was now general throughout the ranks, which occasioned Major Ludson, aid-de-camp to General Lincoln, to gallop to head-quarters, to make a representation of facts, which no sooner were stated, than an immediate pardon was granted, accompanied with the order that he should instantaneously be drummed, with every mark of infamy, out of camp, and threatened with instant death if ever he should, at any future time, be found attempting to approach it. In the interim, the surgeon-general had established himself at his quarters, in a distant barn, little doubting but that the catastrophe was at an end, and that Fickling was quietly resting in the

__ David Whyley,[315] Jn° Fleming [316] & Sam[l]. Henry [317] for Disobediance of Orders sentenced 100 Lash each but was remitted

Sergeant Early for Disobeying the Serg[t] Major was mult a fortnights pay to be made by two different payments [318]

Will[m] Low of the Ship Prosper for Disobediance of Orders, Sentenced 100 Lashes remitted

---

Orders 21st June __ Parole Hobkaw __ Count[r].S[n]. 16.

For the day tomorrow Capt[n] Ja[s] M[c]Donald

Fort guard Capt. Motte L[ts] Charnock & Marion

Quarter guard L[t] Hall __ Rear L[t]. Mason

For Fatigue L[ts]. Mazyck & Proveaux

The Capt. of the Day for the futer is to see the fort gates shut every night at retreat beating & open'd at Reville

---

Orders 22nd June __ Parole Fort Johnston __ Count[r].S[n]. 2.

For the day tomorrow Capt[n] Harleston

The Fort guard Capt[n] Ashby L[ts]. Mazyck & Dubose

Quart[r] __ L[t] Lesesne Rear L[t]. H. Gray

For Fatigue L[ts] Charnock & Shubrick

By Coll Moultrie **Sullivants Island __1776 __**

A Regimental court martial to sett Emediately to try such prisoners as shall be brought before them Evidences to attend

Capt. Harleston President

L[ts]. Henry Gray __ T Moultrie __ Wm Moultrie & Dubose Memb[rs]:

---

grave. Midnight was at hand, and he was busily engaged in writing, when hearing the approach of a footstep, he raised his eyes, and saw with astonishment the figure of the man who had, in his opinion, been executed, slowly with haggard countenance approaching towards him. "How! How is this?" Exclaimed the doctor, in great terror. "Whence come you? What do you want with me! Were you not hanged this morning?" "Yes sir," replied the resuscitated man, "I am the wretch you saw going to the gallows, and who was hanged." "Keep your distance," said the doctor, "eat and welcome; but I beg of you, in future, to have a little more consideration, and not intrude so unceremoniously into the apartment of one who had every reason to suppose that you were an inhabitant of the tomb."

315 David Wiley (Whyly, Whily) served in the 2nd South Carolina Regiment under Captain Thomas Dunbar. On 20 June 1776 he was supposed to receive 100 lashes for disobedience of orders, but the sentence was remitted. He was supposed to receive 100 lashes on 2 November 1777 for being absent without leave, but he was pardoned. He received 100 lashes on 23 March 1778 for insolent behavior. He was with Dunbar's light infantry at the siege of Savannah.

316 There were two soldiers named John Fleming in the 2nd South Carolina at this time. One was under Captain Francis Huger and was killed at the battle of Fort Sullivan. The other had enlisted on 4 November 1775 and was in the regiment during the siege of Charleston in 1780. I do not know which one this is.

317 Samuel Henry enlisted in the 2nd South Carolina Regiment on 24 January 1776. On 20 June 1776 he was supposed to receive 100 lashes for disobedience of orders, but the sentence was remitted. He received 50 lashes on 14 May 1778 for being absence from roll call several times. He was beaten and maimed by Private James Grover on 27 October 1778.

318 Jeremiah Early.

NB Part of the Reg[t] ie: 200 men, under Maj[r]. Marion marcht to Advance guard at 5 OClock Aft[r].noon __ Gen[l] Armstrong Order'd them to return __ arrv'd at the fort at 8 OClock at Night

It was unsure on whether the British would attempt to assault the island by landing at the northern tip, or to reduce the fort by naval bombardment. Marion's detachment marched out in the morning when it was suspected that the British were going to land troops, however as soon as it was determined that the British were not, they were ordered to return to Fort Sullivan.

Orders 23[d] June __ Parole Washington __ Count[r].S[n]. 30.

For the day tomorrow Capt. Huger
Fort guard Capt[n]. Oliphant L[ts]. T Moultrie & Baker
Quarter __ L[t] Proveaux __ Rear L[t] Shubrick

NB Sentence against Jeremiah M[c]Guiness for abusing L[t] Proveaux was 200 Lashes but had 50 remitted [319]

The Officer of the Rear guard to make a report to the Commanding Officer Emediately of all the Boats that shall arrive with the cargoes & to whom they are come & for what use

---

Orders 24th June __ Parole Mecklenburgh __ Count[r].S[n]. 25.

For the day tomorrow Capt[n] Motte
Fort guard Capt[n]. Blake L[ts]. W[m] Moultrie & Dunbar
Quarter. L[t] Marion __ Rear L[t] Jackson

For Fatigue L[ts]. Baker & Dubose

Orders 25th June __ Parole Willmington __ Count[r].S[n]. 29.

For the Day tomorrow Capt. Ashby
Fort guard Capt Shubrick L[ts] W[m] Moultrie & H Gray
Quarter L[t] Hall __ Rear Charnock

For Fatigue L[ts]. Shubrick & Baker

On the morning of the 25th the Man of War *Experiment* came near the fort with her gun ports opened to attack, but soon retreated. That afternoon the *Experiment* hoisted her topsails, fired a gun and got under way towards the fort again. Several other ships followed her lead, but a squall came on, shifting the wind and preventing the fleet from coming any closer. The wind also prevented the start of the Battle of Fort Sullivan. The British fleet did approach about a mile closer to Sullivan's Island[320]

---

[319] Jeremiah McGuiness (McGinnis) enlisted in the 2nd South Carolina Regiment on 4 November 1775. On 23 June 1776 he was supposes to receive 200 lashes for abusing Lieutenant Proveaux, but 50 were remitted. On 16 February 1777 he was found guilty of misbehavior and sentenced to receive 50 lashes, but he was pardoned. On 24 April 1777 he was reprimanded at the Halberts for neglect of duty. On 6 May 1777 he was court martialed for some unnamed offense, but he was acquitted. On 25 October 1777 he received 100 lashes for stealing a musket from Alex Simmor, but the lashes was remitted. His pay was also stopped until he paid off the musket. He had his pay stopped on 2 November 1777 for losing a cartridge box. He was with Marion's partisans in 1782.

[320] O'Kelley, *NBBAS, Volume One,* pg 129.

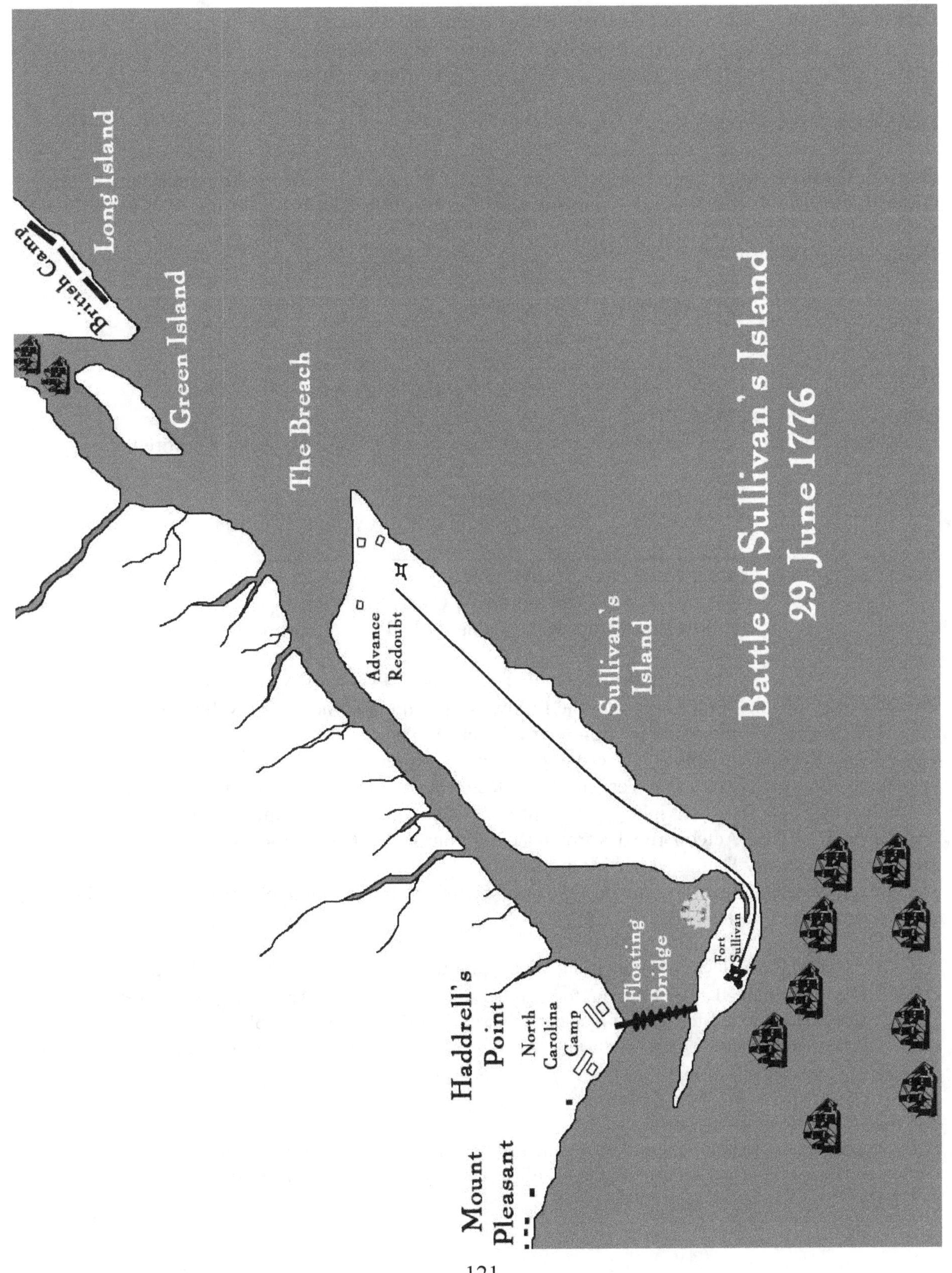
Battle of Sullivan's Island
29 June 1776
Long Island
British Camp
Green Island
The Breach
Advance Redoubt
Sullivan's Island
Floating Bridge
Fort Sullivan
Haddrell's Point
North Carolina Camp
Mount Pleasant

Orders 26th June __ Parole Mecklenborough __ Count^{r}.S^{n}

For the day tomorrow Capt Oliphant

Fort guard Capt P Horry L^{ts} T Moultrie & Dubose

Quarter. Lt Mazyck __ Rear Lesesne

Fatigue L^{ts}. Jackson & Marion

By Coll Moultrie **Sullivants Island 1776 __**

Order'd that the Officers of all working partys where the Engineer Baron Massenborough is Imployed are to follow & Expedite all Orders & Directions which shall be given by him with the Utmost Despatch [321]

Orders 27th June __ Parole Geo^{r} Town Count^{r}.S^{n} 124.

For the day tomorrow Captn Blake

Fort guard Captn Eveleigh L^{ts} Shubrick & Baker

Quart^{r} __ Lt Jackson __ Rear Lt Proveaux

Fatigue L^{ts} Wm Moultrie & H Gray __

A Court Martial to Sett Emediately to try such prisoners as shall be brought before them __ Evidences to attend

Captn Huger Presed^{t}

L^{ts} Shubrick __ Mason __ Dunbar & Proveaux Members

Orders 28th June __ Parole America __ Count^{r}.S^{n}. 25 __

For the day tomorrow Captn. Shubrick

Fort guard Capt Ja^{s} McDonald L^{ts} Hall & Charnock

Quarter guard Lt Dunbar __ Rear L^{t} ~~Proveaux~~ Marion

This day at 10 OClock in the morning the follow^{g} ships attacked the fort at Sullivants Island Viz: Bristol of 50 guns Commanded by Commod^{r}. Sr Pet: Parker

Experiment 50 guns __ Solbay __Active __Acteon __ Syren

Frigates of 28 guns Each The Terrible Bom Ketch & the Spyren of 20 guns, The Engagement continued till 9 OClock at Night __ a continual Cannonade & Bombardment with only Little Intermission __ The Acteon run'd agrown & was burnt the Next Morning, The Bristol lost her mezenmast & greatly Damaged her Mainmast

The Solebay lost her Bowsprit, they all received Considerable Dammage & was Oblige to slip there cable & retreat to 5 fath^{m}. hole __

At 10 o'clock in the morning of the 28th the *Thunder*, a bombship with mounted mortars, anchored away from the fort and began to throw shells. The first 13-inch mortar shell landed in the fort on the magazine, but did not do any considerable damage. After about sixty shells the recoil of the mortars ruptured the heavy beams of the mortar bed. The *Thunder* was out of action.

---

[321] Baron von Massenbourgh is most likely Lieutenant Felix Lewis Massenbach, an engineer from the Maryland Regiment who was attached to the 4th South Carolina Artillery. Baron von Massenbourgh died and was buried with full military honors on 20 August 1777.

Though Fort Sullivan had twenty-six guns, only twelve could be brought to bear on the British fleet and the fort only had 28 rounds for the 26 guns when the battle began. A little before 11 o'clock the South Carolinians fired four or five shots at the *Active* while she approached under sail. Some of the shots struck her but the ship did not seem to regard the fort until it was within about 350 yards of Fort Sullivan.

The British ships dropped anchor and opened fire on the fort with broadside fire. Over 7,000 rounds of cannon and mortar fire hit the fort. Around 3 o'clock the fort ran out of powder. Major Francis Marion left the fort with a small party and received 200 pounds of powder from the Schooner *Defense*, lying in Stop Gap Creek behind the fort. The British thought that fort was surrendering due to the silence. The fort's fire continued at that slow rate until John Rutledge sent 500 more pounds from the city to the island by small boat.

For every fifty shots fired by the British, the tiny fort returned fire with one shot, but the shots were deadly accurate. The British weren't doing any damage to the fort because the spongy palmetto logs absorbed the cannon balls. Injuries normally caused by flying splinters were non-existent. The South Carolina artillery was extremely accurate. The main mast of the *Bristol* had nine 32-pound balls imbedded in it and the mizzenmast had been shot away.

The *Sphinx, Syren*, and *Actaeon* sought to take advantage of the tremendous British broadsides and sailed past the fort, planning to take up a position from where they could attack the fort on its weak side. The ships ran aground on a sandbank known as the Middle Ground. The *Sphinx* cut away her bowsprit because it became fouled in the *Actaeon's* mainmast. By mid-afternoon the *Sphinx* and the *Syren* had refloated, but the *Actaeon* remained stuck in the sandbar that would one day become Fort Sumter.

The battle continued into the night. Flashes lit up the night and the sound of the Carolinians shot crashing through the timbers echoed around the harbor. At half past nine o'clock Admiral Parker finally had enough and withdrew. The *Bristol* alone had fired 1,840 shot from her cannon. At 2 o'clock in the morning of the 29th the British set the grounded *Actaeon* on fire and abandoned it. [322]

By Coll Moultrie **Sullivants Island 1776 __**

Orders 29th Jun__ Parole Unamimity __ Countr.Sn. Second

For the day tomorrow Captn P Horry

Fort guard Captn Harleston Lts Lesesne & Baker

A Return of the Dead & Wounded in the Engagement yesterday to be made the Commanding Officer & if they leave wives or Children & where they are

---

322 Full details of the battle can be read in "*Nothing but Blood and Slaughter, Volume One*".

A Return of the troops under the Command of Coll Wm Moultrie which fought the British fleet in the Fort on Sullivants the 28th June 1776 __ with a List of the Killed & Wounded

Commissioned Off^{rs}						Staff Off^{rs}				Effective Rank & File							
Collonels	Lt Coll:	Majors	Captains	1st Lieut:	2 Lieut:	Adjut:	Qua: Mas	Surgeons	Mates	Sergeants	drums & Fifes	Fit for duty	Sick present	Sick in Hosp:	Recruiting	On furlough	On Comm^{d}
1	1	1	10	7	10	1	1	2	2	24	14	306	17	33		2	

Number of Commissioned & None Commiss: of Officers Killed, Missen & taken Prison: in the Engagement										
Coll:	Lt Coll:	Maj^{r}	Captn	1st Lieut	2 Lieut	Serg^{ts}	Corp	Fifers	Privates	
					0	1	1	0	7	Killed
					1			1	20	wound
					1			1	27	Total 31

Number men Killed of the Artillery			
2nd Reg^{t}	Matross [323]	1 Killed	2 Wounded

Total Killed 10 _ 34 All
Wounded _ 24
a mulatto Boy of Lt Dunbar Killed

NB [324] there was 2 Artillery Officer Captn. Beekman [325] L^{t}. W^{m}. Mitchel & Lt Spencer [326] __ Mitchel was at the Advance guard

Number of Cartridges for Cannon 26 rounds & 20 d^{o} p^{r} musket

---

[323] A matross was a soldier who assists the artillery in firing and moving guns.

[324] The "mullato boy of Lieutenant Dunbar" being in battle was not unusual in the South. The Council of Safety had authorized the military "the use of able bodied negro men" at the batteries and forts of Charlestown "at the rate of two to each gun." This freed up two hundred soldiers for assignments in other positions in Charlestown. The Council also stated that "suitable rewards shall be given to those slaves, who behave well in time of action."

[325] Barnard Beekman was elected a captain in the 4th South Carolina Regiment (Artillery) on 14 November 1775. He was promoted to major on 18 November 1776. He was promoted to lieutenant colonel on 25 October 1778 and promoted to colonel on 20 June 1779. After being taken prisoner at the fall of Charleston on 12 May 1780 he remained a prisoner on parole until the end of the war.

[326] Calvin Spencer was born in 1754. He enlisted early in 1776 as a lieutenant in the 5th South Carolina Regiment. He was in the battle of Fort Sullivan in June 1776 and was made a captain around the same time. He resigned on 10 December 1777 and on the same day was appointed assistant deputy quartermaster general. He was at the siege of Savannah in 1779. In 1780 he was an aide to General Howe and was the quartermaster general. In March of 1781 British dragoons captured him near Georgetown. After a few months imprisonment he was exchanged and became a captain of the marines on a privateer. He married Rebecca Ford in the spring of 1782. He died on 19 January 1801 at the age of 47.

It will always be an honour to the Man that can say he was in the Engagement on Sullivants Island against the British Fleet

By Coll Moultrie **Sullivants Island 1776 __**

Mens Names Killed & Wounded on the 28th June & in What Company

Mens Names & What Comp^y	Killed	Wounded	
Saml Yarbury [327]................ Granidiers	1		In the Artillery Rawley Purdey [348] ___ Killed David Witherspoon [349] & Josiah Niblets [350] Wounded
Josiah Stone [328]			
Saml Gale .... P Horry	....	Wounded	
Serg^t Young [329] }			
Jno Keel [330] ....McDonalds	___ ___	Wounded	
Serg^t McDaniel [331]............. Huger	Killed		
John Long [332] .... Harleston	________	Wounded	
	Killed		
	....	Wounded	
	....	Wounded	
	Killed		
	________	Wounded	
	Killed		
	__Blakes__	Wounded	
		Wounded	

[327] Samuel Yarbury.

[328] Josiah Stone was in Captain Horry's company when he was wounded. On 1 January 1777 he received 100 lashes for being absent without leave. On 21 January 1777 he received 100 lashes for being absent from roll call and drunkiness. On 28 January 1777 he received 50 lashes for losing his cap and being absent. On 2 April 1777 he was placed in confinement, but escaped with James Kelley. On 22 May 1777 he returned from his desertion and was pardoned due to the influence of Captain Fogarty and John Huger. On 27 May 1777 he received 100 lashes for being absent from roll call. On 24 June 1777 he was charged with neglect of duty for selling his regimental coat and stealing a musket. He received 100 lashes, in three days, and was drummed out of the regiment with a halter around his neck.

[329] This is John Young. Major Elliot wrote in a letter to his wife, "Young, the barber, an old artilleryman, who lately enlisted as Sergeant, has lost a leg. Several arms are shot away. Not an officer is wounded." Young survived and became the regimental sutler in 1777.

[330] John Keel was in Captain James McDonald's company during the Battle of Fort Sullivan. He had his leg shot away by a cannon ball, but he survived.

[331] This was James McDonald. Major Elliot wrote, "The expression of a Sergeant McDaniel, after a cannon ball had taken off his shoulder and scouped out his stomach, is worth recording in the annals of America: "Fight on, my brave boys; don't let liberty expire with me to-day!"

[332] John Long enlisted in the 2nd South Carolina Regiment on 4 November 1775 under Captain Isaac Harleston. He was wounded at Fort Moultrie on 28 June 1776, and died from his wounds later that day.

[348] Rawley Purdy was in the 4th South Carolina Artillery Regiment when he was killed at Fort Sullivan

[349] David Witherspoon was a matross attached to the 4th South Carolina Artillery, but he was in Captain Daniel Mazyck's company of the 2nd South Carolina Regiment. He was wounded in the Battle of Fort Sullivan. Later he served 85 days in the militia in 1782. He also supplied beef to the state.

[350] Josiah Niblets served in the 4th South Carolina Artillery and was wounded at the Battle of Fort Moultrie.

Jn° Fleming & D°: Hunter [333] .... Jn° Mason [334] & Wm Jones [335] } Hugers __ Jn° Sayer [336] Henry Peters & Nath[l] Beek [337] L[t] Henry Gray Robert Wade [338] } Mottes ____ Jn° Boxall [339] & Jn° Hickie Luke Fludd [340] & Rich[d] Rogers Ashby __ } Isaac Edwards [341] Jn° Proby, Owen Hinds [342] & Jn° Griffen [343] } Ashby	________	Wounded	

[333] John Hunter enlisted in the 2nd South Carolina on 4 November 1775. He deserted on 12 February 1776. He returned sometime that year and he was killed in the battle.
[334] George Mason enlisted in the 2nd South Carolina Regiment on 4 November 1775 under Captain Francis Huger. He was wounded in the battle.
[335] There were at least eight soldiers named William Jones in the 2nd South Carolina during 1776. One was in Captain Charles Motte's company; one in Captain Bernard Elliot's company; one was William Jones, Sr. in Captain Francis Huger's company, and his son, William Jones, Jr. was in also in the regiment. There was a William Shepherd Jones, a William Skipper Jones, and a William Huper Jones. The one who was wounded, and would die of his wounds on 1 July 1776, was William Jones, Sr. He had enlisted in the 2nd South Carolina Regiment on 15 July 1775.
[336] John Sawyer was in Captain Francis Huger's company at Fort Sullivan when he was wounded, and would die of his wounds later that day.
[337] Nathaniel Beek enlisted in the 2nd South Carolina Regiment on 4 November 1775. He was wounded in the battle and would die from his wounds on 1 July 1776.
[338] Robert Wade (Watt) served in the 2nd South Carolina Regiment under Captain Motte in 1776. His arm was broken during the battle of Fort Sullivan on 28 June 1776. He was promoted to sergeant on 10 July 1778.
[339] John Boxall enlisted in the 2nd South Carolina Regiment on 4 November 1775. He was killed in the battle.
[340] Luke Flood enlisted in the 2nd South Carolina Regiment on 4 November 1775. He was killed in the battle.
[341] Isaac Edwards enlisted in the 2nd South Carolina Regiment on 4 November 1775. He was killed in the battle while serving under Captain Ashby. Flood, Rodgers and Edwards were all killed at the same time when a ball tore into them as they were handspiking an 18-pounder back into battery.
[342] Owen Hinds enlisted in the 2nd South Carolina Regiment on 4 November 1775 under Captain Ashby. He was wounded in the battle. He was discharged on 23 November 1776.
[343] John Griffin (Griffen) enlisted in the 2nd South Carolina Regiment on 4 November 1775 under Captain Ashby and was wounded in the battle. On 14 September 1776 he received 100 lashes for some unnamed offense. On 12 December 1776 he received 100 lashes for neglect of duty. On 5 Februray 1777 he receieved 35 lashes for disobedience of orders. On 25 May 1777 he was charged with beating Mary Burges, but the charges were remitted. In 1778 he was in Captain Hall's company. He was supposed to receive 75 lashes on 14 April 1778 for quitting his guard duty but it was remitted. On 24 September 1778 he was reduced in rank

Jn° Ryan [344] ____________ Blakes ____________ Tim: Downing Tho[s] Smith [345] & B Reeves [346] ____ Jn° Campbell Fife Maj[r] [347] __ __ __ __ __ Jn° Taylor & And[w] Nelson Eveleigh ___			

NB Sam[l] Yarbury has left a wife but parted from her Long before he Inlisted, he has Fath[r] Mother, Broth[r]. & Sisters___

Robert Black [351] has Left a wife & 6 or 7 Children in S[t] Johns Parish [352]

Died of there wounds } Robert Black, W[m] Jones, Nath[l]. Beek of Hugers Comp[y]
John Long of Harlestons [also Jn° Sawyer of Hugers Comp[y]

---

and ordered to run the gauntlet for theft. On 15 November 1778 he was sentenced to run the gauntlet again through the regiment for theft. On 13 December 1778 he was sentenced to receive 50 lashes on the bare back with switches for stealing a musket. He was also put under stoppages until the gun was paid for. After the fall of Charleston he served in the 2nd South Carolina Dragoons under Captain Isaac Ross and Colonel Myddleton

[344] John Ryan enlisted in the 2nd South Carolina Regiment on 4 November 1775 and was in Captain John Blake's company when he was killed in the battle.

[345] Thomas Smith served in the 2nd South Carolina Regiment under Captain John Blake. He had been wounded at the battle of Fort Sullivan on 28 June 1776. He received 100 lashes on 27 January 1778 for being absent without leave.

[346] Benjamin Reeves enlisted in the 2nd South Carolina Regiment on 4 November 1775. He was under Captain Richard Mason in 1778. He was wounded in the battle while he was in Captain John Blake's company. He was discharged on 8 March 1779. He reenlisted on 1 July 1779 and was with Captain Mason's company at the siege of Savannah.

[347] John Campbell enlisted in the 2nd South Carolina Regiment on 21 February 1776 and became a fife major under Captain John Blake. He was wounded in the battle.

[351] Robert Black enlisted in the 2nd South Carolina Regiment on 4 November 1775. He died on July 1st, of his wounds received during the battle. He left behind a wife, Eleanor.

[352] In pre-Revolutionary South Carolina the Church of England dominated the colony. The term "parish" refers to the local Church of England church and all of its members. Everyone within that parish would be considered a member and they would have to pay a tithe, even though someone may attend the Presbyterian or Methodist church in the area. This is one of the main reasons that the 1st Amendment was written, so that the government would not be able to create a church, like England did, and use this church to further government concerns.

By Coll Moultrie **Sullivants Island 1776**

29th June

Coll: Moultrie take this early Opportunity of returning the Officers & men of his Regim[t]. & the Detachment of Artillery his ~~Hardy~~ thanks for their gallant & Soldier Like Behavour in the Engagement of yesterday__It will always be an honour to the Man that can say he was in the Engagement on Sullivants Island against the British Fleet __

Orders 30th June __ Parole Second Regiment CS[n]. 50

For the day tomorrow Capt[n] Eveleigh

Fort guard Capt[n] Huger L[ts]. T Moultrie & Proveaux

Quart[r]. Lt Shubrick __ Rear L[t] Dunbar

Fatigue L[t]. Mazyck & Lesesne

A court martial to sett tomorrow morning to try such prisoners as shall be brought before them __ Evidences to attend __

Capt Horry Presed[t]

L[ts]. Charnock __ Hall __ Lesesne __ & Mazyck Members

His Excellency Jn[o] Rutledge Presed[t]. & Maj[r] Gen[l] Charles Lee Desires there most Perticular & harty thanks to the Brave Officers & Men for their Heroic & gallant Behavour, in the Engagement against the British fleet on the 28th June in the Fort at Sullivants Island

General Lee says no man ever did, & it tis impossible that any ever can behave better & that he will do us Justice in his letters to the Continental Congress

the President

NB His Excellency ^ has sent a Hogshead of Rum to the Regiment as a Present

Order'd that the quarter master do serve a gill rum p[r] man a day while it Lasts

Order'd that the Sergeants of Company do read the thanks of the Presed[t]. & Gen[l] Lee to the men

By Coll Moultrie **Sullivants Island 1776**

Order'd that the Orderly Booke of each Comp[y]. be brought up by the 8th July next & be Delivered to the Command[g] Officers

~~A Court martiall to sett tomorrow to try such prisoners as shall be brought before them~~

NB The Court sentenc'd N. Flinn, Edw[d] George,[353] Jacob Copeland [354] E Barton [355] Adam Meek for absents on the day of Battle to be Dressed in Womens petticoats & Cap & to receive 200 Lashes Each which they rec[d].

John Thompson [356] & Ja[s] Grover was pardoned ____

---

353 Edward George enlisted in the 2nd South Carolina Regiment on 9 May 1776. He served in Captain Shubrick's company in 1777. After Captain Shubrick's death he served in the 2nd Vacant Company until it was disbanded on 5 October 1778. He was then assigned to Captain Mazyck's Company. He was under Captain Adrian Proveaux in 1779.

354 Jacob Copeland enlisted in the 2nd South Carolina Regiment on 4 November 1775. He was absent the day of the Battle of Fort Moultrie, so he was sentenced to receive 200 lashes and was ordered to dress like a woman. He was in Captain Blake's company in 1778.

355 Elias Barton enlisted in the 2nd South Carolina Regiment on 4 November 1775. He died on 27 July 1777.

356 John Thompson served in the 2nd South Carolina Regiment under Captain Richard Mason. He was absent the day of the Battle of Fort Sullivan, but he was pardoned. On 16 September 1776 he was was supposed to

Serg[ts]. Bolliant [357] & Gambol & Drury Abbot [358] prev[y]. reprimanded for the Same crime

After Orders

The Second Regiment to turn out this Aft[rN]Noon to receive his Excellency Gen[l] Lee

NB Gen[l] Lee at the head of the Regiment & part of the Artillery Repeated his former thanks & added he was very glad to do it in person

Maj[r]. Elliotts Lady at the Same time presented the Regiment with a most Elegant p[r] of Collours Embrodered, ~~what was~~ with these Words, Your gallant behavior in Defense of Liberty & your Country Entitles you to the Highest Honours, Accept of these two Standards as Reward Justly due to your Regiment & I make not the Least Doubt under heavens protection you will stand by them as Long as they can wave in the Air of Liberty __ which was Accepted off by Collonels Moultrie & Motte [359]

Orders 1st July __ Parole Concord __ Count[r].S[n]. 12

For the Day tomorrow Capt[n] M[c]Donald

Fort guard Capt[n] Motte L[ts]. Jackson & Mason

Quarter L[t] Marion Rear Lt Dubose

NB M[r] W[m] Logan presents his compliments to Coll. Moultrie the officers & Soldiers of 2[d] Reg[t] & beg there acceptance of a Hogshead of Old antique rum which being scarce at this time may be more acceptable __ [360]

By Coll Moultrie **Sullivants Island 1776**

Orders 2nd July __ Parole Brunswick __ Countr.Sn 25

For the day tomorrow Captn Harleston

For guard Captn Ashby Lts. Lesesne & May

Quarter __ Lt. Hall __ Rear Lt Charnock

---

receive 300 Lashes, but he only received 63 due to his "swooning". On 14 December 1776 he received 39 lashes and was mult 7 days pay for drunkenness and neglect of duty. On 23 December 1776 he received 99 lashes and was mult 14 days pay for disorderly behavior in town. On 28 April 1777 he received 100 lashes for an unnamed offense. He received 50 lashes on 29 September 1777 for being drunk on guard. He received 50 lashes on 11 October 1777 for stealing beef. On 2 June 1778 he was sentenced to be confined for one day and do double duty, but it was remitted. On 14 July 1778 he put under stoppanges for one half of his pay for defrauding a Mr. Cross of a sum of money. He was under Captain Adrian Proveaux during 1779. He was with Captain Mason during the siege of Savannah.

[357] David Boilliet (Boillat) enlisted in the 2nd South Carolina Regiment on 4 November 1775 as a corporal. He was later promoted to sergeant. He was reprimanded on 30 June 1776 for being absent at the Battle of Fort Sullivan. On 13 July 1776 he was reduced to private for neglect of duty. He was reprimanded for neglect of duty on 29 September 1777 and was mult 7 days pay. He was discharged in 1778 and then served under Captain James Bentham in the Charleston Militia Regiment.

[358] Drury Abbott enlisted in the 2nd South Carolina on 4 November 1775. He was reprimanded on 30 June 1776 for being absent at the Battle of Fort Sullivan. He was discharged on 21 September 1778.

[359] The 2nd South Carolina regiment had two colors, one red and one blue. One of these was captured at Savannah and the other at the surrender of Charleston in May 1780. The blue colors were returned to South Carolina from Britain in 1992 and is on display in the South Carolina State Museum in Columbia. Up until 2006 they were the only known Southern regimental colors to exist. In 2006 the two captured colors of Buford's Virginia regiment was brought forward by the descendants of Banastre Tarleton and sold at auction.

[360] William Logan was a merchant in Charleston. He was one of the "seventeen young gentlemen" who started the Charleston Library Society in 1748.

Orders 3^d July __ Parole Magnanemous __ Count^r.S^n. 36

For the day tomorrow Captn Huger

Fort guard Captn Oliphan L^ts Shubrick & Proveaux

Quarter Lt Baker __ Rear Lt Wm Moultrie

A Court martial to sett this morning to try such prisoners as shall be brought before them __ Evidences to Attend

Captn Motte President

L^ts. Thos Moultrie, Proveaux, Baker & Jackson Members

NB James Boyd rec^d. part of his punishm^t. rest remitted

Orders 4th July __ Parole Genl. How__ Countr.Sn. 7.

For the day tomorrow Captn Motte
Fort guard Captn Blake Lts Dunbar & Marion
Quarter Thos Moultrie __ Rear Lt David Adams [361]

Lt. Jackson resigned his Commission & is no more to be obey'd as an Offr

Lt Thos Hall is Apointed a first Lieut:[362] & David Adams a Second Lt. & are to be Obey'd as such __ Lt. Hall to do duty in Capt: Jas McDonalds Compy. & Lt Adams to do duty in Capt: Shubricks

---

Orders 5th July __ Parole Canterbury __Countr.Sn. 57-

For the day tomorrow Captn Ashby
Fort guard Captn Shubrick Lts. Charnock & Lesesne
Quarter Lt Mazyck __ Rear Lt Hall __

By Coll Moultrie **Sullivants Island 1776 __**

Orders 6th July __ Parole Bristol __ Countersign 44.

For the day tomorrow Captn Oliphant
Fort guard Captn P Horry Lts. T Moultrie & Baker
Quarter Lt Charnock ~~Hall~~ __ Rear Lt Hall

A Court martial to sett this morning at 10 OClock to try such prisoners as shall be brought before them __ Evidences to Attend

Captn Ashby President
Lts. Dunbar, Proveaux, Marion & Adams Members

Orders 7th July __ Parole Friends __ CounterSign 11.

For the day tomorrow Captn. Blake
Fort guard Captn. Eveleigh Lts Hall & Adams
Quarter Lt. Marion __ Rear Lt H Gray

NB according to Sentence of Last court martial Sergt Gaultier & others was repremanded at Head of the regimt for dasterly Behavior in the Engagement 28th June [363]

Orders 8th July __ Parole Manchester __ Countr.Sn. 60.

For the day tomorrow Captn Shubrick
Fort guard Captn McDonald Lts T Moultrie & Mazyck
Quarter Lt Marion __ Rear Lt Charnock

---

361 David Adams was commissioned a 2nd Lieutenant in the 2nd South Carolina Regiment in Captain Shubrick's company on 4 July 1776. He resigned his commission on 22 April 1777.

362 Thomas Hall was born on 9 June 1750. He was commissioned a 2nd Lieutenant on 17 June 1775 in the 2nd South Carolina Regiment. He was wounded at Fort Moultrie while serving under Captain William Mason. He was promoted to 1st Lieutenant on 13 June 1776 in Captian Joseph McDonald's company. On 16 February 1778 he was promoted to captain when Captain Ashby resigned his commission. He was in the siege of Savannah in October 1779. He resigned in January 1780 but was captured at the fall of Charleston and then sent to St. Augustine. After being exchanged he served as an aide-de-camp to General St. Clair until the end of the war. He died on 28 August 1814 at the age of 64.

363 Joseph Gaultier enlisted in the 2nd South Carolina Regiment on 14 February 1776. On 7 July 1776 he was reprimanded in front of the regiment for "dasterly Behavior in the Engagement 28th June".

No more than one man of a Company is permitted to go to Town at a time on condition to be back in 24 hours, all passes to be sign'd by a Field Officer __

A court martial to sett tomorrow morning at 10 OClock to try such prisoners as shall be brought before them Evidences to attend

Captn Wm Oliphant Preseď.
L^ts^. Shubrick, Hall, H Gray & Proveaux Members

Orders 9th July __ Parole Beufort __ Count^r^.S^n^. 3.
For the day Captn P Horry
Fort guard Capt: Harleston L^ts^. Mason & Proveaux
Quart^r^ Guard Lt T. Moultrie __ Rear Lesesne

By Coll Motte **Sullivants Island 1776**

Orders 10th July __ Parole Limerick __ Count^r^.S^n^. 80
For the Day tomorrow Captn. Eveleigh
Fort guard Capt Huger L^ts^. Hall & Proveaux
Quarter L^t^. Adams__ Rear L^t^ H Gray

No person presume to ease himself within the fort any who do may depend on being severely punished, this order to be read by an Officer of each Company

The Morning reports to be given in by 8 OClock every morning The men in each Company to clean in and about there Barracks every day, The Officers are desired to see this Order put strictly in execution

A General return to be given in emediately to the Commanding Officer

---

Orders 11th July __ Parole Dorchester __ Count^r^.S^n^. 38.
For the day tomorrow Captn M^c^Donald
Fort guard Captn Motte L^ts^ Mazyck & Baker
Quart^r^ Lt Lesesne __ Rear Lt Marion

The pay bill to be given in tomorrow morning to the Commandant

The Arms of each Company to be inspected this afternoon by there respected Officers & all that are not clean to be mad so emediately

The Commanding Officers expects that all the Officers will be very punctuall in turning out at the beating the Long roll & march there men to the Alarm posts & report such as may be absent

The Quarter master to collect all the fire Buckets & Lodge them in the Fort guard room taking a receipt for the numbers he delivers him

By Coll Motte **Sullivants Island 1776**

Orders 12th July __ Parole Arnold __Count^r^.S^n^. 6.
For the day tomorrow Capt: Harleston
Fort guard Captn. Ashby L^ts^. Charnock & Shubrick
Quarter Lt Lesesne __ Rear Thos Moultrie

A Court martial to sett emediately to try such prisoners as shall be brought before them. Evidences to attend.

Captn Blake Preseď.
L^ts^. Thos Moultrie, Charnock, Lesesne & Eveleigh Members

Order'd that the pilot boat, Commet Captn Turpen [364]& Captn Morgans sloop [365] be suffer'd to pass & repass the fort till further Orders

---

Orders 13th July __ Parole Liberty __Count^{r}.S^{n} Wilkes

For the day tomorrow Captn Motte
Fort guard Capt Oliphant L^{ts}. Hall & P Gray
Quarter Lt Proveaux __ Rear Lt Marion

NB according to Sentence of last court martial Sergt Boillat for neglect of duty was reduct in the Ranks as a private [366]
Jacob Johnson for riot pardon'd __ Conrad Fitner was pardon'd from receiving 200 Lash on account of sore Leggs [367]

Orders 14th July __ Parole Elbert __ Count^{r}.S^{n}. 9.

For the day tomorrow Captn. Motte
Fort guard Captn. Blake L^{ts} H Gray & Adams
Quarter Lt Mazyck __ Rear Lt Baker

The Officers of the Quarter & Rear Guards to be perticular in reporting the hours the rounds visit there respective guards

The firing of Small Arms in or near the Fort has been repeatedly forbid, all those who Disobey this order may depend on being severely punished

By Coll Motte **Sullivants Island 1776**

A R. court martiall to sett tomorrow morning at the Presed^{t}. tent to try such prisoners as shall be brought before them

Captn. Shubrick president
L^{ts} Shubrick, P Gray, Proveaux & Marion members

---

[364] Joseph Turpin was the captain of the *Comet*, a brigantine that had eighteen 6-pounders. The *Comet* had played a key part in the defense of Charlestown by denying British scouting ships access to the harbor, however on May 31st a fierce thunderstorm tore through the harbor and the *Comet* was struck by lightning. The *Comet's* mainmast was destroyed and a sailor was killed. Afterwards the *Comet* could only be used as a floating battery.
[365] Francis Morgan, and his brother Charles, commanded several privateers during the war. On June 16th Francis Morgan's privateer Sloop *Polly* tried to run the gauntlet of British ships and get into Charlestown harbor. The *Polly* was carrying a cargo of 300 barrels of gunpowder, 20 chests of cartridges, several hundred stands of arms and 90 barrels of rum, sugar and gin from St. Eustatia. The *Polly* almost made it into the harbor, but she ran aground near Stono Creek. That night the privateers scuttled the *Polly* and abandoned ship. The British set fire to her and she "blew up with a great Explosion…It would have been much greater but she had five feet of water in her hold, which had damaged a great deal of Powder. On 13 June 1777 Francis and Charles Morgan would raid Castle Harbor in Bermuda.
[366] David Boilliet.
[367] Jacob Johnson was born in Pennsylvania on 8 October 1758. He enlisted in the 2nd South Carolina Regiment on 1 April 1776 and was discharged on 23 July 1777. He re-inlisted in 1778. After the fall of Charleston he served in General Marion's partisans under Captain Goodman and Colonel Baxter. He was in the battles of Quinby Bridge and Eutaw Springs. On 1 March 1783 he enlisted in the 1st South Carolina Regiment. He also served from 1 July 1783 to 15 November 1783. After the war he moved to Kentucky.

Orders 15th July __ Parole Goose Creek __ Countr.Sn. 29.
For the day tomorrow Captn Ashby
Fort guard Captn P Horry Lts Shubrick & Eveleigh
Quarter Lt Charnock __ Rear Lt Lesesne

Orders 16th July __ Parole Washington __ Countr.Sn Camden
For the day tomorrow Captn Oliphant
Fort guard Captn Eveleigh Lts T Moultrie & Marion
Quarter Lt Hall __ Rear Lt Proveaux

NB the sentence of Last court martial Richd Bennet for absents without Leave recd. 200 Lashes,[368] John Cody to receive 500 Lashes & drum'd out of the Regt. The Last remitted & he recd 200 Lashes, Jams McBride pardon from receiving 50 Lashes [369]

Orders 17th June __ Parole Constantinople __ Countr.Sn 5.
For the day tomorrow Captn Blake
Fort guard Captn. McDonald Lts Mazyck & Adams
quarter Lt P Gray __ Rear Lt Marion

A court martial to set Emediately to try such prisoners as shall be brought before them Evidences to attend

Capt Huger Presedt
Lts. Lesesne, Eveleigh, Baker & Adams members

Jno Burtell Corpl of Capt Oliphants Compy [370] & Thos Rybold, of Harlestons are Apointed Sergts. in Sd. Compy.[371] & are to be obey'd as Such__ The priming horns given to Officers 28 June to be returnd to the officer of artillery by 12 OClock tomorrow

---

[368] Richard Bennet enlisted in the 2nd South Carolina Regiment on 4 November 1775. On 14 July 1776 he received 200 lashes for being absent without leave. He deserted with Duncan McFarlin on 2 December 1776.

[369] James McBride was born on 17 August 1756. He enlisted on 4 November 1775. On 16 July 1776 he was supposed to receive 50 lashes due to some unnamed offense, but he was pardoned. On 5 November 1776 he received 50 lashes for being drunk. He was discharged on 4 August 1778. He rejoined the regiment and was in Dunbar's light infantry company at the siege of Savannah. After the fall of Charleston he served in the militia. He died on 11 July 1808 at the age of 52.

[370] John Burtell (Burthell) enlisted on 4 November 1775. On 17 June 1776 he was promoted to sergeant in Captain Oliphant's company. He was discharged on 12 July 1778. He re-enlisted in Captain Adrian Proveaux's company in 1779 and was appointed as a corporal on 15 March 1779. He was transferred to Captain Hall's company and was promoted to sergeant on 18 March 1779. He was in the siege of Savannah.

[371] Thomas Rybold (Raybout, Raybold, Raybould, Royboald) enlisted in the 2nd South Carolina Regiment on 4 November 1775. On 1 July 1776 he was court martialed for disobedience of orders and abusing Mary Sighton, an inhabitant of the town on June 20th. He was acquitted of these charges. On 17 July 1776 he was promoted to a sergeant in Captain Harleston's company. On 11 September 1776 he was mult seven day's pay for an unnamed offense. On 1 July 1777 he was charged with threatening the life of Mary Singleton, but he was acquitted of the charge. On 10 November 1777 he was fined 14 days pay for neglect of duty. He was under Captain Richard Mason in 1778. On 9 April 1778 he was reprimanded for beating Lucy Dunn, who may have been the wife of Private John Dunn. He was reduced to the ranks on 24 August 1778 for drunkenness and impertinence to Lieutenant Mason. On 26 January 1779 he was sentenced to receive 100 lashes on the bare back with a cat of nine tails for neglect of duty. Fifty of those lashes were remitted by Francis Marion.

By Maj^r Marion **Sullivants Island 1776**

Orders 18th July __ Parole Long Island __ Count^r.S^n. 3.
For the day tomorrow Captn Shubrick
Fort guard Captn Motte L^ts Lesesne & Baker
Quart^r L^t H Gray __Rear Charnock

Orders 19th July __ Parole America __ Count^r.S^n. Liberty
For the day tomorrow Captn. P Horry
Fort guard Captn Ashby L^ts. T Moultrie & Proveaux
Quarter Lt Shubrick __ Rear Lt Eveleigh

The Long Roll to beat at 5 OClock in the morning & 5 in the Evening when the Battallion must turn out as usual

---

Orders 20th July __ Parole FairForest __ Count^r.S^n Cherokee
For the day tomorrow Captn Eveleigh
Fort guard Captn Oliphant __ L^ts Hall & Marion
quarter Lt Dunbar__ Rear Lt Mazyck

Orders 21st July __ Parole Cambridge __ Count^r.S^n Oxford
For the day tomorrow Captn. Harleston
Fort guard Captn Horry L^ts. Baker & Charnock
Quarter Lt Adams __Rear L^t H Gray

Order that the Quart^r Master do give a gill rum p^r man every Rainey day while there is rum

Orders by Coll Gadsden Fort Johnston 19 July

In Consequence of an Order from his Excellency the President that a general court martial be held on Wednesday next at ~~at~~ the Barracks on Gadsden Warff Ch^stown for the tryall of Jno Davis of the artillery for Desertion [372] that the 1st 2nd & 4th Regiments do Each furnish two Captains & 2 Subalterns to be members of S^d court

By Maj^r Marion **Sullivants Island 1776**

That Maj^r Elliott sett as President of the same Lt Charles Linin as Judge Advocate [373] & that Captn Motte of 2^d Regim^t. L^ts Detreville & Gilbank of the fourth regem^t. be Summoned as Evidences to Attend & all other Witnesses [374]

On 19 April 1779 he received 30 lashes on the bare back with switches for being absent without leave. He was under Captain Richard Mason at the siege of Savannah.

[372] John Davis enlisted in the 4th South Carolina Artillery Regiment on 14 June 1776. He was court martialed for desertion on 21 July 1776. He was a gunner from 25 February 1778. He was reduced to matross in late 1778.

[373] Charles Lining was born on 26 October 1753. He became a lieutenant in the 1st South Carolina Regiment during May 1776. He was made a brevet captain on 17 November 1778 when Captain Benjamin Cattell resigned. He was taken prisoner at the fall of Charleston in 1780 and exchanged during June 1781. He served until the end of the war. He died on 16 August 1813 at the age of 60.

[374] John Gilbank was a lieutenant in the 4th South Carolina Artillery Regiment. He was later promoted to captain. He was killed during the siege of Charleston when he was a major in the militia.

Regimental Orders by Maj^{r} Marion

Captn Oliphant & Blake L^{ts}. Thos Moultrie & Baker is Apointed Members of the above Genl Court Martial

Two Subalterns 2 Sergeants & 40 Rank & file to Parrade Emediately with two days provisions & 18 Rounds p^{r} Man they will receive there Orders when Compleat

L^{ts} Lesesne & Mason is Apointed for this Command

Orders to Lieut Lesesne

Sir your are to take under your command 1 Sub: 2 Sergeants & 40 Rank & file men with one Surgeons Mate 18 Rounds p^{r} man & two days provisions & proceed Evediately to the Advance guard & Embark your party on board the floating Battery where you will follow all Order given you by a Superior Officer; I'm y^{rs} &^{c}

F Marion Maj^{r} 2^{d} Reg^{t}

NB the detachment marched off at half past 3 OClock aft^{r}noon

The British were able to get most of the transports over the Bar and had them waiting in Spencer's Inlet until the troops could board them. The British soldiers at the lighthouse boarded the transports on July 19th; unfortunately some of the British soldiers were left behind, including almost an entire company of the Royal Highland Emigrants. Some of the Royal Highland Emigrants were able to get off Long Island on the Transport Brig *Glasgow Packet*. They lost thirty-eight men and seven officers who were left on the beach and taken prisoner by the South Carolinians.

The *Glasgow Packet* had been ordered to be the last ship off the Bar since it had some small guns on board that could be used to cover the withdrawal. On July 20th all the ships went over the Bar and got under way, but the *Glasgow Packet* became calmed as she arrived at the Bar, and had to drop her anchor and wait. The brig requested assistance from Sir Peter Parker, and he replied that the *St. Lawrence* would come to their aid. Unfortunately for the brig, it never arrived. On the morning of the 21st the wind picked up and the *Glasgow Packet* was able to get over the Bar, only to get stuck on a shoal.

Captain Porterfield of the *Glasgow Packet* fired signal guns to gain the attention of anyone in the British fleet. He then tried to row his ship off the shoal by having his men row out, drop the anchor, and then pull on the anchor to move the ship. Porterfield sent a rowboat with Captain Alexander Campbell of the Royal Highland Emigrants to alert the British fleet that they were stuck. Campbell arrived at the *Bristol* after rowing twelve miles. Parker sent back a flat-bottomed boat with a kedge anchor to assist in pulling the *Glasgow Packet* off the shoal. He also sent the Schooner *St. Lawrence* to go with the flat-bottomed boat. The *St. Lawrence* fell behind the rowboats because of the lack of wind, but all of the vessels arrived too late.

A South Carolina Row Battery had been sent to annoy the British fleet and heard the signal guns of the *Glasgow Packet*. The Carolinians rowed towards the British brig by using some smaller boats to tow her. The Row Battery, under the command of Lieutenant Pickering of the 2nd South Carolina Regiment, fired upon the *Glasgow Packet* with round ball and grapeshot. Because of the angle of the *Glasgow Packet* being stuck on the shoal, she was unable to bring her guns to bear on the row battery. Captain McNicoll of the Royal Highland Emigrants saw the Row battery approaching and ordered the men to throw their muskets overboard so the Carolinians would not capture them. The South Carolinians boarded the transport and captured her at 4 o'clock in the afternoon. On board were forty-three men of the Royal Highland Emigrants, "6 sailors, and 2 Negro Boatswains." [375]

Orders 22nd July__Parole Glascow __ Countr.Sn Capture

For the Day tomorrow Captn. Motte

Fort guard Captn Eveleigh Lts Eveleigh & Proveaux

quartr. 1 Sergt. 1 Corpl. & 12 men

Rear 1 Sergt. 1 Corpl. & 18 men

The fort guard to be reduct to 33 men Lt Shubrick to mount his guard Emediately in room of Lt Charnock who is Order'd to hold himself in readyness with 14 men to Embark in two Boats for Dewees Island

By Majr Marion **Sullivants Island 1776**

Orders 23d July__ Parole Cherokee__ Countr.Sn. 3

For the day tomorrow Capt Shubrick

Fort guard Captn Harleston Lts Dunbar & Marion

For Quartr & Rear guard a Sergt & 12 men Each

A court martial to sett at 11 OClock at the Presidents tent to try such prisoners as shall be brought before them Evidc. to attend

Captn Harleston Presedt

Lts. Hall, Dunbar, H Gray & Mason Members

Orders 24th July __Parole Congarees __ Countr.Sn 96.

For the day tomorrow Captn Horry

Fort guard Captn Huger Lts. Charnock & Lesesne

Rear Lt Petr Gray ____Quartr a Sergeant

The Commanding officer of Compys. to Exercise there men aftr Roll call in ~~the After Noon~~ morning

After Orders by Coll Gadsden

In Consequence of an Order from his Excellency the President dated at Fort Johnston 22nd July, the general court martial Apointed 19th Inst: to sitt also to try Lieutt Withers [376] of the 5th Regt.[377] for such Crimes & Offences as shall be Laid before them Evidences to attend and to try Nathl Hatch, Durham Sullivan & Saml Moore of Coll Thompsons Regt. for Desertion[378] & also to try Captn Lewis Peyer Im hoff of 3d Regt for acting unbecoming of an Officer in an Affair of mutiny at the relieving of the

---

[375] O'Kelley, *NBBAS, Volume One*, pp. 160-163.

[376] Possibly John Withers. John Withers had been a captain-lieutenant in an Indian company of Foot Rangers in 1776. The Lieutenant Withers of the 5th South Carolina Regiment was court martialed in July 1776 for disobedience of Colonel Huger's orders and other misdemeanors. On 9 January 1777 the court finally convened and they found that he was guilty and recommended that he be pubicly admonished and reprimanded at the head of his regiment. He resigned the next day instead of facing the punishment.

[377] Even though there was hardly enough men to fill three infantry regiments and an artillery regiment, the Congress authorized a 5th South Carolina Regiment (1st South Carolina Rifles) on 24 February 1776. The unit was initially designated as the 5th South Carolina (Rifle) Regiment and the regiment would have seven companies made up of 60 "expert riflemen." Each recruit was to provide himself "a good rifle, shot pouch, and powder horn" and would receive a bounty of £9 and £5 if he supplied his own rifle.

[378] Thomson's Regiment was the 3rd South Carolina Ranger Regiment.

guards, the same President the same Officers & Judge Advocate to compose the court All Evidences to Attend & the court to continue setting with necessary adjournments till the prisoners are try'd

By Maj^{r} Marion **Sullivants Island 1776**

Orders 25th July __ Parole Independince __ CS^{n} Wilkes

For the day tomorrow Capt Eveleigh

Fort guard Captn Shubrick L^{ts} Mazyck & Adams

Rear Lt Shubrick __ quart^{r} a Serg^{t}

A court martial to sett tomorrow at 11 OClock to try such prisoners as shall be brought before them Evid: to Attend

Capt P Horry President

L^{ts}. Hall, Eveleigh, Proveaux & P Gray members

Orders 26th July __ Parole Washington __ Count^{r}.S^{n}. 7.

For the day tomorrow Captn Harleston

Fort guard Captn. P Horry L^{ts}. H Gray & Proveaux

Rear __ Lt Hall __ Quart^{r} Sergeant

It tis recommended to the Officers of the Guard to take perticular Notice of the Centries who sett on there post and the Slovenly Manner the relief marches

One hundred with 4 Sergeants to bring away the Cannon from the Advance guard, horses will be sent to Assist

NB according to last court martial Serg^{ts}. Burgess & Newman was repremanded,[379] Robert Williamson for Absents without Leave rec^{d} 100 Lashes [380]

The Adjutant is to see that all the mens Arms which come on the parrade are clean & those men whose arms are not is to be confined

---

Orders 27th July __ Parole Relief __ Count^{r}.S^{n}. 8.

For the day tomorrow Capt. Huger

Fort guard Captn. Eveleigh L^{ts}. Dunbar & Eveleigh

Rear Lt Marion __ Quart^{r} a Serg^{t}

By Maj^{r} Marion **Sullivants Island 1776**

Orders 28th July __ Parole Cha^{s}Town __ Count^{r}.S^{n} Gadsden

For the day tomorrow Captn Shubrick

Fort guard Captn Harleston L^{ts} Mazyck & Mason

Rear Lt Marion __ Quart^{r}. a Sergeant

---

[379] Hugh Newman enlisted in the 2nd South Carolina Regiment on 4 November 1775 as a sergeant. He was demoted to private on 9 August 1779 for embezzling clothing from the regiment. He was promoted back to sergeant again on 11 February 1779 in Captain Charles Motte's company. He was demoted again on 9 August 1779 to private when he was caught embezzling clothing of the regiment. He also received 100 lashes for that offense. He was killed at Savannah on 9 October 1779.

[380] Robert Williamson was in Captain Oliphant's company of the 2nd South Carolina Regiment in 1776. On 26 July 1776 he received 100 lashes for being absent without leave. On 24 January 1777 he was promoted to sergeant and transferred to Capain Moultrie's company.

The 2nd Regiment to hold themselves in readyness to go to Chs.town tomorrow The Mens Arms to be Clean and in good order all those who disobey this Order may expect to be severly punished

A Return of the Ordinance stores to be made by 10 OClock tomorrow by an Offr. of the Artillery, The Quarter Master make a return of all the Provisions, spears, Intrenchg. tools & Muskett Cartridges now in Garrison, also the number of Waggons carts & Horses now in the Service on the Island this return to be made by 10 OClock

Mr Doughty to make a return of the numbers of White & negroe Carpenters & Laborours now employ'd in this Garrison

The adjutant to acquaint Mr. Doughty of this Order

Orders 29th July __ Parole Girgea __ Countr.Sn. 74

For the day tomorrow Captn P Horry
Fort guard Captn Huger Lts. Shubrick & Adams
Rear Lt. Charnock __ Quartr a sergt.

Orders 30th July __ Parole Port Royal __ Countr.Sn 42.

For the day tomorrow Captn Harleston
Fort guard Captn Shubrick Lts. Hall & H Gray
Rear Guard Lts Proveaux __Quartr a sergeant

By Majr Marion **Sullivants Island 1776**

Orders 31st July __ Parole Geor Town __ Countr.Sn. 14

For the day tomorrow Captn Huger
Fort guard Lts. Dunbar & P Gray
Rear Lt Eveleigh __ quartr. a Sergeant

A Regt Court martial to sett Emediately to try such prisoners as shall be brought before them Evidences to attend

Captn. Huger President
Lts. H Gray, Dunbar, Lesesne & Eveleigh Members

Orders 1st August __ Parole Dorchester __ Countr.Sn 9.

For the day tomorrow Captn Horry
Fort Guard Lts. Charnock & Lesesne
Quartr. a sergt__ Rear Lt Mazyck

The Officer of the day to visit the fort & outguards & all reports to be made him. & he to the Commandg. Offr.

Orders 2nd August __ Parole Talbot __ Countr.Sn. 61.

For the day tomorrow Captn. Harleston
Fort Guard Lts. H Gray & Mason
Rear Lt Shubrick __quartr a Sergt.
Orders by his Excellency the president __

The first Regiment is to relieve the Second on Sullivants Island the Artillery is to go to Fort Johnston & the 2nd Regt. is to come to town forthwith, there Collonels will settle the times of removing so as to make it as Convenient as may be to them & these Regiments

---

Orders 3[d] August__ Parole Alfred __ CoountrS[n]. The Great

For the day tomorrow Capt[n] Huger

Fort guard L[t]. Hall __ Rear a Sergeant

Sollomon Barefield [381] Corp[l]. of Capt[n] Horrys Comp[y]. is Apointed a Serg[t]. in S[d] Comp[y]. & to be obeyd as such

A General Return of the 2nd Regim[t]. to be made a Munday next __

Maj[r] Marion Sullivants Island 1776

Order'd that Capt[ns] Horry, Harleston & Oliphants Company, to proceed & Embark on board Capt[n] Whyley Emediately for Ch[s] town & to Land on Gadsden Warff where Capt Horry will receive orders from Coll: Moultrie[382]__ NB the 3 Comp[y]. Embarkt in Even[g]:

Orders 4th August __ Parole S[o] Carolina __ Count[r].S[n]. 76.

For the day tomorrow L[t] H Gray

Fort guard L[ts] Lesesne & Hall

Quart[r]. a Serg[t] __Rear a Serg[t]

The Sergeants of each Company to take all the men off duty after roll-call in morning & clean all the fort, mens barracks & under the platform, every place to be as clean a possible

The Artillery to make clean in & about there quarters

The Sergeant Major is to see this Order executed__

The Artillery to place a Centry at the Magazine & in Each Cavelare [383] __ The fort guard to Consist in 2 Sub: 1 serg[t]. & 32 R & file

---

Orders 5th August __ Parole Amarica State__ Count[r].S[n] 13 Collony,

For the day tomorrow L[t] Dunbar

Fort guard L[t] Hall

Rear a Serg[t] __NB the quarter guard vacated

Capt[ns] Eveleigh, Ashby, & Shubricks Companies to Embark today on board Capt[n] Whyley & proceed to town, & Land on Gadsden Warff

---

Orders 6th August __ Parole Thursday __ Count[r].S[n] 2.

For the Day tomorrow L[t] Lesesne

Fort Guard L[t]. H Gray __ Rear a serg[t].

---

[381] Solomon Barefield enlisted on 4 November 1775 in the 2nd South Carolina Regiment. On 3 August 1776 he was promoted to sergeant in Captain Horry's Company. He was discharged on 12 March 1777.

[382] Unable to determine if Captain Whyley is a person, or the name of a ship.

[383] A cavalier was a raised portion of a defense in a horseshoe shape, that was 10 to 12 feet higher than the rest of the foritification. They are usually built in the middle of the works, and on a raised up portion of ground, so that the cavalier could command a position that overlooked the whole fortification. In the case of Fort Sullivan this would be the blockhouses at each corner of the fort.

Orders 7th August __ Parole Sullivants Island __ Count^r.S^n Farewell
For the day tomorrow Lt Thos Moultrie
The Remainder of the 2nd Reg^t. to carry all there baggage down the water side tomorrow morning aft^r. roll-call & put Centries over them till the troops Embark't two carts to be ready to carry the Baggage down

By Coll: Moultrie **Ch^stown __ 1776 __**
NB Maj^r. Marion Deliver'd the fort on Sullivants Island to Captn Scott of the first Regiment, & Embark on Thursday 8th Augt. at 7 OClock in Afternoon, & arived next morning in Ch^s town__
By Coll Moultrie Ch^s town

Orders 6th August __ Parole Sparta __ Count^r.S^n. Rome
Lieut Wm Moultrie having with Leave resigned his Commission in the Second Reg^t. is no longer to be Obey'd as an Officer__
An Officer in each company to see roll-called in there Respective companys morning & evening
NB Lt Charnock with 1 Sergeant & 30 Rank & file men from the 2nd Regiment sent to Georgia on Command __

After the battle of Sullivan's Island General Lee turned his attention to the only other place in the South where there were British soldiers, the colony of Florida. Florida had originally been a colony of Spain and had one of the oldest colonial cities in North America, St. Augustine. Spain had allied itself with France during the Seven Years War and gave up the colony when Britain won the war. In the Treaty of 1763 Spain gave up all possessions in Florida, mainly to regain Havana. Florida was divided into two colonies. West Florida contained the settlements of Mobile and Pensacola. East Florida contained St. Augustine. France rewarded its ally by giving Louisiana to Spain in 1763.

The Florida colonies maintained their Loyalty to the King, mainly because all of the colonists were new and had not began to have ideals of independence from a government so far away. If these colonies had swung towards the independence movement the United States would have had fifteen original colonies, instead of thirteen. This detail under the command Lieutenant Charnock may have been an advanced party of the force that would go down in September to "liberate" Florida.

Orders 7th August __ Parole Liberty __ Count^r.S^n Fortitude
One Captn. 3 Subalterns 3 Sergeants 1 Drum & 1 fife & 38 Rank & file men to hold themselves in readyness to march emediately each man to have Eighteen rounds & 3 spare flints
One Subaltern 1 Serg^t. & 24 rank & file men Emediately to releave the Artillery at the Magazine, Lt Proveaux for this service
Captn Huger L^ts. Mazyck, Jacob Shubrick & Baker for the Comm^d. going to Georgia

---

By Lt Coll: Motte
Orders 8th August __ (Parole Rutledge) by Genl How [384]

[384] In February 1776 Congress decided to make a Southern Department under command of one major general and six brigadier generals. The choice for command of the Southern Department was Major General Charles

All the Officers of the different Company, to attend the calling of the roll morning & evening & to report such men as may be absent __ A Captain of the day to be Apointed immediately who is constantly to be near the quarters of the Regem^t. till he is relieved __ all reports to be made to him who will deliver the same to the Command^g. Off^r. by 10 OClock in the morning

The Off^rs. are desired to inform the Adjutant in Wrighting where they are quartered this to be done by tomorrow morning by 10 OClock [385] For the day tomorrow Captn Shubrick

For Guard Lt Mason

By Genl Howe **Ch^s town ____ 1776 __**

Genl Orders __ Rutledge __

Officers of the different detachments order'd to march are to prepare there men to proceed as expeditiously as possible whenever they are ready to inform the Adjutant General. It is expected they will exist themselves upon this Occasion, as upon the Arival of waggons which are momentarily expected they will be ordered to march without the least delay, commanding Off^rs. of each detachment are to report at the head quarters the quantity of ammunition in the possession of their men __ the waggon master to report Emediately to Genl Howe what number of public waggons are in town & as soon as any Arrives to inform him of them __ It is expected the different Corps will in futer attend perticularly to have Adjutants or orderly Officers, at the Orderly house at the State House, as it is impossible to Communicate the orders to the different Corps properly thro any other channel

Coll: Moultries Battalion to Join the town duty in rotation with other troops

NB twelve oclock & evening at sun set to attend for Orders at 8 OClock in the morning one third of the men Appear between the state house while meeting well accutered

---

By Genl Howe

Genl Orders 9 Augt. __ Parole Union __

Field Officer to the day tomorrow Coll: Parsons [386]

Detail for Guard __

2nd Regim^t. 1/3 of the men fit for duty relieving a regimental Guard ____ Coll: Pinckneys Regim^t. 3 Companies

Regimental Orders by ^L^t Coll: Motte

For the day tomorrow Capt P Horry, for guard L^t P Gray

The Off^rs. of respective comp^ys to frequently visit there mens quarters & exert them selves in keeping them in good order

---

Lee. The four brigadiers who would serve under Lee were John Armstrong, William Thompson, Andrew Lewis, James Moore, William "Lord Stirling" Alexander and Robert Howe. Robert Howe was born in 1732. He was elected colonel of the 2nd North Carolina Regiment in September 1775 and was present at the burning of Norfolk, Virginia in January 1776. He served in Charleston during the campaign of 1776. He was promoted to Major General in October 1777. After losing Georgia to the British in 1779 he was replaced by General Benjamin Lincoln. Lincoln would then be in command when Charleston was captured, the greatest defeat in US Army history until the Fall of the Phillippines in WWII. Robert Howe died on 12 November 1785.

385 The officers did not stay in the barracks with the men of the regiment, but instead stayed in town at private residences, such as their own homes or a boarding house.

386 Unable to identify who Colonel Parsons is.

The morning reports to be delivered to Capt[n] of the day who will deliver them when he make his report to the Commanding officer__

Ch[s].town 1776

~~After~~ Gen[l] Orders by Gen[l] Howe 10th Aug[t] Parole Lee CS[n] Liberty

The troops intending to go to the southward are to march early tomorrow morning & are therefore to prepare Accordingly __ Waggons will be Apointed to take all proper baggage & Officers of Companys are to take care that no unnecessary Articles are Carry'd

Th Ammunition return required yesterday is as yett not made nor any reason Assigned why it is not made. A Compliance with that order & a more perticular Observance of those hereafter issued consistance with service will give great pleasure to the Commanding Officer __ It will be necessary for every detachment to be provided with bulletts & shott moulds, Commanding Officers of Comp[ys]. are therefore emmediately to inquire & make report to the head quarters what moulds they have & what they want, & also what other Millitary requisits they stand in need of for there men, or if they are out of town Comm[dg]. Off[r]. of Battalion from which detachments have been taken are desired to be at Head Quart[rs]. at 5 OClock this afternoon where the Captains of those detachments are also to be __ The Surgeon Apointed for this expedition will be carefull to prepare every necessary articles & to have them in readyness as soon as possible as they must be sent with the troops

Regimental Orders by L[t] Coll. Motte
For the day tomorrow Capt[n] Eveleigh
Magazine guard L[t] Eveleigh

The Comm[dg]. Off[rs].of comp[ys]. to fix upon such men as are best able to undergoe the greatest fatigues for the detachment orderd the 7th Inst: __ this Detachment to be provided with three good flints & 18 rounds p[r] man, there arms & pouches to be in the best order & to march this after Noon at 4 OClock

L[t] Coll. Motte **Ch[s].town ____ 1776**

The quartermaster to deliver out the ammunition & provisions as soon as possible & to see that the baggage Waggons be ready in time also to provide a Bullet mould for this detachment

M[r] Adanus Burk is Apointed a Second Lieutenant in the 2nd Reg[t]. & to be Obey'd as such, L[t] Burke is to join Capt[n] Hugers Company & to do duty in the same [387]

A General return to be given in by Munday Morn[g]. 10 OClock. this order to be punctually Complied with

1 Sergeant 1 Corp[l] & 12 privates to relieve the Artillery Guard on Gadsdens Warff at 7 OClock in Morning. The Reg[t]: to be in readyness to take possession of the Barracks at present Occupied by the Reg[t] of Artillery

L[t] Dunbar to do duty in Captn Shubricks Comp[y] till further Orders

After Genl. Orders by Genl Howe

The Detachment of Coll: Moultries Reg[t] to march this after noon The Commisary to furnish the men with two days salt & 2 days fresh provision each man to be furnisht with Eighteen rounds p[r]

---

[387] Ædanus (Adam, Edamus, Adanus) Burke was born in Ireland in 1743. He was appointed a 2nd Lieutenant in Captain Huger's company of the 2nd South Carolina Regiment on 10 August 1776. He transferred to Captain Oliphant's company on 6 September 1776. He was promoted to first lieutenant on 4 August 1777 in Captain Oliphant's company (the Officer's list says 6 May 1777). He was transferred to Captain Oliphant's company on 16 July 1777. He resigned on 22 February 1778. He also served in the Charleston Militia Regiment under Captain James Bentham and Colonel Simons in 1778. He served in the 1st South Carolina Regiment in 1779. He was a captain in the militia during 1780-1782. He died on 30 March 1782 at the age of 39.

man, they are to march till they join Coll Thompsons Reg^{t} & to wait for Orders, if the detachment is not furnisht with Bullet & shott moulds Captn will apply to Mr Calvert for them [388]

The Commisary will Deliver Captn Huger 4 Hogshead rum which is to go with the army this he will put under the charge of a guard Commanded by a carefull Officer the Waggon Master is to furnish waggons to carry it

Captn Huger will apply for three Rheams Cartridge paper & 2000 flints to the ~~to the~~ Publick Store Keeper __

By Lt Coll. Motte **Ch^{s}.town ____ 1776**

A Copy of the Resolves of the Honorable the Continental Congress __ In Congress July 22nd 1776

Resolved that the thanks of the United States of America be given to Maj^{r} Genl Lee Coll Wm Moultrie Coll: Wm Thompson & all Officers & Soldiers under their Command who on the 28th June last repulsed with so much valour the attack which was that day made on the State of S^{o} Carolina by fleet & Army of his Britanic Majesty

That the president transmits the same to Genl Lee Colls. Moultrie & Thompson

By Order of Congress

Jno Hancock Presed^{t}

A Copy of a Lett^{r}. from the H. the Continental Congress to Coll: Wm Moultrie

Philadelphia July 22nd: 1776

Sir

I am extreamly happy to have it in my power to transmit to you by Order of congress the thanks of the United States of America for your patriotic & Spirited Exertions in behalf of Liberty & your country; this Success of our Arms attended with every curcumstances that can add Luster to the Caracter of those who conducted it, will render for ever your name estimable with every friend of America, & posterity will be astonisht when they read, that on the 28th June an inexperienced handful of men under your Command repulsed with Loss & Disgrace a powerful fleet of army of Veteran troops, headed by officers of rank & reputation____ May you go on thus to merit & to receive the gratatude of your Country, and as reward of y^{r}. millitary Service may your name be enrolled on the List of American Worthies on whom posterity will bestow the most greatful & unceasing Applauses

I have the Honour to be with respect

Jno Hancock

President

Lt Coll Motte **Ch^{s}.town 1776**

Orders by General Howe }
11th August__Parole Mifflin __ Count^{r}.S^{n} Franklin
Field Officer tomorrow Coll: Pinkney
From Coll Moultries Reg^{t}. 2 Sub: & 40 privates
From Coll Pinkneys, 2 Companys

Regimental Orders

For the day tomorrow Capt Ja^{s} McDonald

For guard L^{ts} Hall & H Gray

---

388 John Calvert served as a clerk to the Commissioners of the Navy board during 1778 and 1779.

NB Capt[n] Huger sett off yesternight with his detachment for Georgia
M[r] Paul Warley was entered a Cadet in 2nd Reg[t]. the 25th July by Order of Coll: Moultrie[389]

General Orders by G. Howe } 12th August __ Parole Moultrie __ Count[r].S[n]. Success
Field Officer for tomorrow Coll: Parsons
Detail of Guards as yesterday

Regimental Orders by L[t] Coll Motte
For the day tomorrow Capt[n] Motte
For Town guard L[ts] Dunbar & Mason
Barrack L[t] P Gray

The troop to beat at 8 OClock in Morning & tatto at 9 at night __ The Regem[t]. to march this day to Gadsden Warff & take possession of the Barracks thereon

The Barrack guard to consist of 1 Subaltern 1 Serg[t]. 21 rank & file the Major is desired to post the Centries & give such orders as may be necessary for the present __ the Capt[n] of the day is to remain within the Centries of the Barrack guard Day & night till he is relieved & to go his rounds at least once at night.

The roll of each Company & troop & retreat beating and the reg[tl]. orders given the 8th Instant is expected will be Strictly Complyed with __ The Quart[r] Mast[r] to apply to the Quart[r] Master of Artillery for the key of the Stores on Gadsdens Warff, where he is to Lodge all the Stores belonging to the Reg[t].

By L[t] Coll Motte **Ch[s].town 1776**

A court of Enquiry to sett tomorrow morning at 11 OClock in the guard room on Gadsden Warff to enquire into the conduct of L[t] Tho[s] Lesesne on the 28th June last __

Major Marion President

Capt[ns] Horry, Eveleigh, M[c]Donald, & L[t] Tho[s] Moultrie Members

---

Orders 13th August __ Parole Gadsden __ Count[r].S[n]

For the day tomorrow Capt[n] Blake

For guard L[ts] Eveleigh, Hall & P Gray

The Quart[r] Master to provide wood for the reg[t]. emediately & to see it carted up to the Barracks,

The Officer of the Barrack guard to take care that the men do not injure the Barracks by riping any boards belongd to the same __ Anyone found offending may depend on being severely punish'd

L[t] Coll Motte desires all the Officers off duty to meet him on the parrade at Gadsden Warff tomorrow evening at 6 OClock

Ordered, That the Off[rs]. on duty do Appear in their Regementals

---

[389] Paul Warley became a cadet in the 2nd South Carolina Regiment on 11 August 1776. He was promoted to second lieutenant on 24 August 1776 in Captain Motte's company. He resigned on 15 June 1777. He returned as a second lieutenant under Captain Blake on 8 August 1777. He was promoted to first lieutenant under Captain Charnock on 18 July 1778 when Lieutenant Henry Perronneau resigned (the Officer list says 25 April 1778). He was under Captain Richard Mason in 1779. He transferred to Captain Mazyck's company on 18 May 1779.

L[t] Henry Gray is Apointed a first Lieut: & to be Obey as such__ he is to remain Capt[n] Mottes comp[y].

A Court Martial to sett tomorrow morning to try such prisoners as shall be brought before them-

Capt[n] Blake President

L[ts] Lesesne, Dunbar, Mason & P Gray Members

NB the Regemental guard was Augmented 12th Inst: to 1 Serg[t] 2 Corporals & 15 privates

---

Orders 14th August __ Parole Middleton __ Count[r].S[n].

For the day tomorrow Capt: Horry

For guard L[ts] Lesesne, T Moultrie & Proveaux

The Commanding Officer expects that the Capt[n] of the day will comply strictly to the Orders of the 12th Inst:

By L[t] Coll Motte **Ch[s].town ____ 1776**

The Court of enquiry of which Maj[r] Marion was Presed[t]. the 12th Aug[st] last has given it as their Opinion that L[t] Tho[s] Lesesne was somewhat confused during the Action of the 28th June, But acquit him of Cowardice

Ordered that L[t] Lesesne do join & do duty in the Comp[y] which he formerly belong

---

Orders 15th August __ Parole Parsons __ Count[r].S[n].

For the day tomorrow Capt[n]. Eveleigh

For guard L[ts]. Dunbar, Dubose & Mason

No non commissioned Off[r]. or soldiers to leave the Barracks after retreat beating, but upon extraordinary business & with permission of the Capt[n] of the day who is required with the Off[r]. of the guard to see this order complied with

John Wickham of Capt Mottes Comp[y]. is Apointed a Sergeant in Capt[n] Eveleighs Comp[y]. & is to be obeyd as such [390]

A Court martial to sett immediately to try such prisoners as shall be brought before them

Capt[n] P Horry Presed[t]

L[ts] Hall & Adams members __ NB the prisoners was reprimanded & Discharged

---

Orders 16th Aug[t] __Parole

For the day tomorrow Capt[n] M[c]Donald

For guard L[ts]. Eveleigh, Adams & P Gray

Commd[g] Off[rs]. of Comp[ys]. are desired to inspect the mens arms & accutrements at 6 OClock in aft[r].Noon, & order such as may not be in good order to be made so emediately

The Officers are to visit the Barrack rooms frequently & see the men keep them clean [391]

---

[390] John Wickham (Wickom, Wickam) was promoted to sergeant in the 2nd South Carolina Regiment on 15 August 1776. He was reprimanded for neglect of duty on 11 December 1777. He was appointed the sergeant major of the regiment on 16 July 1778. He was appointed an ensign on 6 November 1778 under Captain Peter Gray. He was assigned to Captain Moultrie's company on 17 November 1778. He was promoted to first lieutenant on 1 December 1778 under Captain Harleston, when Lieutenant Mason was promoted to captain. He was made adjutant to the 2nd South Carolina Regiment on 18 December 1778. He was killed at Savannah on 9 October 1779.

[391] Fifteen men were assigned to each room in the barracks.

The Quart[r] Master to have the Chimneys of the Barracks Swept once a week [392]

NB the court martial of yesterday sentenced Conrade fitner for absents 4 days to 200 Lashes - but rec[d] 120 on account of a bad leg __

By L[t] Coll Motte **Ch[s].town 1776**

Orders 17th August __ Parole

For the day tomorrow Capt[n] Motte
For guard L[ts]. Hall, H Gray & Burk

All the men to attend Divine Service tomorrow when the Long roll beats__ the off[rs]. of the Barrack guard to Post Centries emediately so as to prevent Serg[ts]. & privates from going to town with Leave from the Capt[n] of the day who will take care not more than two men of a Comp[y] be absent at a time

The Off[rs]. of duty not to leave town without permission from the comm[dg]. Off[rs]. till they have rec[d]. the orders of the day

A Court martial to be held on Munday next at 11 OClock in the morning to try such prisoners as shall be brought before them

Capt[n] M[c]Donald presed[t].
L[ts]. Henry Gray and Rich[d] Mason Members __

Orders 18th August __ Parole

For the day tomorrow Capt[n] Eveleigh
For guard L[ts]. Lesesne, T Moultrie & Dubose

Orders 19th August __ Parole

For the day tomorrow Capt[n] M[c]Donald
For guard L[ts]. Dunbar, Proveaux & Mason

of a Comp[y].

One Off[r] ^ to perpetually attend the calling of the roll morning & evening & before they are dismissed in the morning see the clean in and about their Barrack room

---

Orders 20th August __ Parole

For the day tomorrow Capt[n] Motte
For guard L[ts]. Eveleigh, Gray & Adams

---

392 John Poag, the Barracks Master warned its inhabitants to guard against the dreadful calamity of fire. Everyone living in the barracks had to have their chimneys swept every fortnight or suffer eviction.

NB the Sentence of the last court martial Wm Simpson rec^d 100 lashes for drunkenness & neglect of duty __ Ad^m Meek rec^d 100 Lashes for the same crime two deserters brought from beaver creek by Luke Pattey [393] Named Henry [394] & Jam^s Smith[395] sent by order Coll: Motte to Fort Johnston

By Lt Coll Motte **Ch^s.town 1776 __**

Orders 21st August __ Parole ....

For the day tomorrow Captn Blake

For guard L^ts. Hall, H Gray & Burk

Orders 22nd August __ Parole

For the day tomorrow Captn Eveleigh

For guard L^ts. Lesesne, T Moultrie & Dubose

An Officer of a Company to take up his quarters Emedeatly in Coll: Laurences tennement [396]

A Court Martial to sett today at 12 OClock to try such prisoners as shall be brought before them Evid to attend

Captn Blake presed^t

L^ts. T Moultrie & Hall members

---

[393] Luke Petty was a captain in the Camden militia under Colonel Joseph Kershaw.

[394] Henry Smith was born on 18 August 1759 in Rockingham County, Virginia. He was drafted while living in York District into the 2nd South Carolina Regiment. During 1777 he enlisted in the militia under Captain McCullough and General Williamson. He was in the battle of Briar Creek, Georgia, and at Rocky Comfort Creek. His brother, James Smith, was wounded at the battle of Stono Ferry and died from his wounds. After the fall of Charleston Henry was with the partisans of General Sumter and Colonel Hampton. He was at the fall of Fort Granby and was in the battle of Blackstock's Plantation. He led General Morgan to the Pacolet River, and then was used as a spy to get information on Loyalist "Bloody Bill" Cunningham at Bush River. He was in the battle of Cowpens and he escorted the prisoners to Rowan County, North Carolina. He also was assigned to pick up the British stragglers and watched the Broad River fords for the approach of Cornwallis. After the war he moved to Georgia. He married Margaret Henning on 10 February 1819. He died on 8 January 1840 at the age of 81.

[395] James Smith served in the 2nd South Carolina Regiment. He deserted from the regiment, but returned on 20 August 1776, where afterwards he was sent to Fort Johnson. On 6 November 1776 he was mult a week's pay for neglect of duty. On 1 January 1777 he received 100 lashes for being absent without leave. On 12 February 1777 he received 100 lashes for being drunk and for misbehavior. On 24 June 1777 or he was reprimanded for being absent from roll call. On 4 August 1777 he received 100 lashes for being drunk and abusing Sergeant Anderson. On 15 August 1777 he was demoted to private for behaving in a disorderly manner and being out of the barracks. On 23 March 1778 he was reprimanded at the halberts for being drunk on guard and attempting to sell his shirt. He was also put on guard every other day for eight days. He was confined for 14 days on 4 May 1778 for being absent without leave and he also had do guard duty every other day while he was confined. On 15 June 1778 he received 100 lashes on the bare back with a cat-o-nine tails for desertion. The next year he once again received 25 lashes on the bare back with a cat of nine tails for willfully abusing his musket on 25 March 1779. He died on board a galley of wounds received at the battle of Stono Ferry on 20 June 1779.

[396] Henry Laurens.

Orders 23[d] August __ Parole …

For the day tomorrow Capt[n] M[c]Donald
for guard L[ts]. Dunbar, Proveaux & Mason

Corporal Monroe of Capt[n] Mottes Comp[y]. is apointed a Sergeant in Capt[n]. Ashbys Comp[y]. & is to be obey'd as such [397]

Orders 24th August __ Parole Moultrie

For the day tomorrow Capt[n] Motte
for guard L[ts]. Eveleigh, P Gray & Adams

NB the last court martial reduced in the ranks Serg[t] Ch. H Simmons

The Reg[t] to attend divine Service tomorrow afternoon at 4 OC: The Off[rs]. of duty to be present; the Capt[n] of the day not to give non-Comm: Off[rs]. & Soldiers leave to go to town aft[r]. 10 OClock M[r] Paul Warley & M[r] John Hart is Apointed Second Lieut[t]. in 2nd Reg[t]. & to be Obey'd as such [398]

L[t] Warley to join Capt[n] Motte Comp[y]. & L[t] Hart Cap[t] Blake & do duty in the same till furth[r]. Orders __

By L[t] Coll Motte **Ch[s].town 1776**

Orders 25th August __Parole Franklen __

For the day tomorrow Capt[n] Blake
for guard L[ts]. Hall, H Gray & Burk

Orders 26th August __ Parole Howe

For the day tomorrow Capt[n] Eveleigh
for guard L[ts]. Lesesne, Warley & Hart

The Guards to be marched from the parrade by their respective Officers.

---

397 Hugh Monroe (Munro, Monrow, Munroe) enlisted in the 2nd South Carolina Regiment on 1 January 1776 in Captain Motte's company. He was promoted to sergeant on 26 August 1776 and transferred to Captain Ashby's company. He was confined for stealing some lead on 8 July 1777, but he was acquitted of the charge. On 15 August 1777 he was reprimanded for "breaking and abusing an inhabitant of the town." He was reprimanded on 18 September 1777 for being absent from exercises. He was fined a half a month's pay for neglect of duty on 9 October 1777. He was demoted to private on 11 December 1777. He received 50 lashes with a cat-o-nine tails on 19 January 1778 for being absent without leave. Two days later he received an additional 50 lashes for forging the commissary general's signature to an affidavit and forging Captain Ashby's name to a pass to town. He was also supposed to be picketted for five minutes, but this was remitted.

398 John Hart was born on 6 March 1758. He was appointed a second lieutenant in the 2nd South Carolina Regiment on 24 August 1776 in Captain Blake's company. He was promoted to 1st Lieutenant on 4 August 1777 (the Officer's list says 21 may 1777) in Captain Ashby's company. When the British threatened Charleston in May 1779 he was assigned to Captain Hall's company. He became a captain on 18 August 1779 and took command of Captain Lesesne's company. He resigned on 19 October 1779 after the failed assault on Savannah. On 28 February 1780 he returned to the 2nd Regiment as a 2nd Lieutenant. He was taken prisoner at the fall of Charleston and exchanged. He became a first lieutenant on 21 November 1781 and served till the end of the war.

The Officers to provide themselves emmediately with the Articles of War for the government of the forces of the state The Articles of War against mutiny & Desertion to be read by an Officer in each Compy. every day in evening this Week

NB Sergt. Laforce died this evening in Hospital

Orders 27th August__ Parole

For the day tomorrow Captn Harleston

for guard Lts. T. Moultrie, Proveaux & Mason

Orders 28th August . . . . Parole . . .

For the day tomorrow Captn Motte

for guards Lts. Dunbar, P Gray & Adams

Captn Nicholas Eveleigh having resigned his commission in the Second Regiment with Leave from his Excellency the President is no longer to be Obey'd as Such . . .

A Court martial to Sett this morning at 11 OClock to try such prisoners as shall be brought before them. Evid: to attend

Captn Harleston President

Lts. Thos Moultrie and Hall Members

Orders 29th August __ Parole

For the day tomorrow Captn Blake

For Magazine guard Lt Eveleigh, Regimental Lt Hall New Barrack Lt Burck

A court martiall to sett this morning at 11 OClock to try such prisoners as shall be brought before them Evidences to attend __

By Lt Coll Motte **Chs.town 1776 _____**

Captn Motte president

Lts. Thos Dunbar & Geor Eveleigh members

NB by last court martial Sergt Laurens was sentenced to receive a privates pay for a fortnight [399]

Orders 30th . . . August . . . Parole

For the day tomorrow Captn Shubrick

Regimental guard Lt. Warley new Barrack Lt H Gray

Magazine guard Lt Hart

Ordered that Commanding Offrs. of all Compys. be very Attentive to all orders __ It is expected all Offrs. will comply strictly to all orders, to prevent disagreeable consequences

One third of the men fit for duty be ordered for guard every day & that they Appear with their arms very clean & in good order. A Court Martial to be held this morning at 11 OClock to try such prisoners as shall be brought before them. Evin: to attend

Capt Blake President

Lts. H Gray and P Gray members

---

399 Unable to determine Sergeant Laurence's (Lawrence) first name. He served in the 2nd South Carolina Regiment until at least 1778. On 29 August 1776 he was sentenced to receive a private's pay for two weeks for an unnamed offense. He was reprimanded on 18 September 1777 for being absent from exercises. He was fined 14 days pay on 30 September 1777 for refusing to do duty.

NB the sentence of court 29th Basil Hollis recd 300 Lashes [400]
the Last court Wm Murphy & N. Flinn was repremanded [401]

Orders 31st August . . . .Parole . . . .

For the day tomorrow Captn Harleston
Regt guard Lt Proveaux __ Magazine Lt T Moultrie
New Barrack Lt Lesesne

A General return to be given in to the Adjutant by Munday morning 9 OClock

A Court Martiall to sett on munday next to try such prisoners as may be brought before them, evidences to Attend

Captn Mott president
Lts. Thos Lesesne & Proveaux members

Orders 1 Septr . . . Parole . . .

For the day tomorrow Captn Motte __ Magazine guard Lt Dunbar __ Regiment. Guard Lt. P. Gray
New Barrack guard Lt. Mason

By Coll Motte **Chs.town 1776**

Orders 2nd Septr . . . Parole . . .

For the day tomorrow Captn Ashby
Magazine guard Lt Eveleigh
Regt. guard Lt Adams __ Barrack G. Lt Burk

The Regimt. to be under Arms tomorrow at 5 OClock after Noon and to be drawn up opposite the Barracks no one to be excused from turning out __ The Commadg. Offr. expects to see the mens Arms in very good order

The Offrs. of New Barrack & Magazine Guard, not to dismis their men before they get to the Regimental parade & for the feuter to make their reports to the Captn of the day

---

Orders 3d Sept . . . Parole . . .

For the day tomorrow Captn Shubrick
Magazine guard Lt Hall
Regimt. Lt. Hart __ New Barracks Lt. H Gray

Lts Lesesne and Warley to proceed on the recruiting Service Emediately, they will call on the commandg Offr. at eleven OClock today for their orders

---

[400] Basil Hollis enlisted in the 2nd South Carolina Regiment on 17 June 1776. On 30 August 1776 he received 300 lashes for an unnamed offense. On 4 September 1776 he received 250 lashes for another unnamed offense. He transferred over to the 5th South Carolina Rifles on 15 October 1776.

[401] William Murphy enlisted in the 2nd South Carolina Regiment on 1 April 1776. On 30 August 1776 he was reprimanded for some unnamed offense. He was promoted to corporal on 23 August 1777. On 24 April 1778 he was appointed as a sergeant under Captain Charles Motte. On 20 January 1779 he was confined for disobedience of orders and insolence. He was reprimanded by the adjutant and had his rank suspended for 15 days. He fought at the siege of Savannah in 1779.

Corp[l]. James ONeal of Capt[n] Blakes comp[y]. is Apointed a Sergeant in S[d] Comp[y]. & to be Obey as such [402]

Orders 4th Sept . . . Parole . . .

For the day tomorrow Capt[n] Harleston
Magazine guard L[t] P. Gray
New Barracks L[t] T Moultrie __ Reg[t]: L[t]. Mason

NB Sentence of last Court Martiall Rich[d]. Williamson rec[d] 100 Lash
Bunker Tring rec[d] 100 Lashes Basil Hollis rec[d] 250 Lashes

A Court martiall to sett this morning at 11 OClock to try such prisoners as shall be brought before them

Capt Ashby president

L[ts]. Mason & Hart members

NB No prisoners were punished

By L[t] Coll Motte **Ch[s].town 1776 __**

Orders 5th Sept . . Parole . . .

For the day tomorrow Capt[n] Shubrick
Magazine guard L[t] Adams
Reg[t]: L[t] Burk __ New Barracks L[t] Eveleigh

L[t] Burke to do duty in Capt[n]. Oliphants Comp[y]. till further orders & L[t] Proveaux to be returned in Capt[n] Hugers Comp[y].

Orders 6th September . . . Parole . . .

For the day tomorrow Capt[n] Horry
Magazine guard L[t] H Gray
Reg[t]. L[t] Hall __ New Barracks L[t] Hart

L[t]. Charnock is Apointed a Capt[n] in Second Reg[t]. & is to be Obey'd as such __ Capt[n]. Charnock to take the Command of the Comp[y] late Capt[n] Eveleigh

L[t] Dubose is Apointed first L[t]. in 2[d] Reg[t]. & is to be obey'd as such __ L[t] Dubose to do duty in Capt[n] Oliphants Comp[y] till further orders

---

Orders 7th Sept . . . Parole . . .

For the day tomorrow Capt[n]. Harleston
Magazine guard L[t]. P Gray
Reg[t]. L[t] T Moultrie __New Barracks L[t]. Mason

---

[402] James O'Neil enlisted in the 2nd South Carolina Regiment on 4 November 1775 as a corporal. He was promoted to sergeant on 3 September 1776. On 16 August 1777 he was sentenced to be demoted to private for refusing to do duty and abusing Sergeant Major Edmunds, however his sentence was pardoned and he was reprimanded instead. Due to that he was reduced in ranks to private on 19 September 1777. He was sentenced to receive punishment for an unnamed offense on 16 October 1777, but some of the punishment was remitted due to the request of Lieutenant Colonel McIntosh. He deserted on 15 December 1777.

Orders 8th Sept . . . Parole . . .

For the day tomorrow Captn Motte
Magazine guard Lt Eveleigh
Regt. Lt Dunbar __ New Barracks Lt. Adams

The Regt to attend divine service this Afternoon; Captn of the day not to give Leave to any one to be Absent from the Barracks after 10 OClock this Morng

By Lt Coll Motte **Chs.town 1776 ____**

Orders 9th Septr - Parole . . .

For the day tomorrow Captn Ashby
Magazine guard Lt. Burke
Regt Lt H Gray __ New Barracks Lt Hall

the Commandg. Offrs. of Compys. to Apply to the Quartr Master this after Noon at 5 OClock for arms & accoutrements wanting to compleat their companys Agreeable to returns given in Last week for which they are to give receipt

General Orders by Genl Lee.

General Lee think it tis his duty before his departure to express the high sense he entertains of the conduct and behavior of the Coll: & Officers of the Severall Battallions of So Carolina boath as Gentlemen and Soldiers & begs leave to Assure them that he thinks himself Obliged to report their Merit to the Continental Congress

---

Orders 10th Septr. . . Parole . . .

For the day tomorrow Captn. Shubrick
Magazine Lt T Moultrie
Regt. Lt. Mason __ New Barracks Lt Hart

A court martial to sett this morning to 10 OClock to try such prisoners as shall be brought before them __ Evid: to attend

Captn Ashby president
Lts Thos Moultrie & Adams Members

Orders 11th Septr. . . Parole . . .

For the Day tomorrow Captn. P. Horry
Magazine Guard Lt. Adams
Regimental Lt. P Gray
New Barracks Lt. Dunbar

NB According to Sentence of Last Court Martial Sergt Raybold & Corporal Salts [403] was Mult Seven days pay

By Lt Coll Motte **Chs.town . . . 1776 __**

Orders 12th Septr. . . . Parole Moultrie

For the day tomorrow Captn Harleston
Magazine guard Lt Thos Hall
Regt. Lt. Eveleigh __ New Barracks Lt Burke

---

403 Unable to determine who Corporal Salts was.

Orders 13th Sept: . . . Parole Lee

For the day tomorrow Captn Motte
Magazine guard Lt. T. Moultrie
Regt. Lt Hart __ New Barracks Lt H Gray

Orders of the 8th Inst: by Genl. Lee

As it is of the Utmost Importance to the common Interest of America, that the So Carolina and Georgia Battalions shoud be compleated as soon as possible; Genl. Lee in Capacity of Continental Commander of the So.thern District gives full authority to the officers of the Sd. So Carolina & Georgia Battallions, to engage in the Service of their respective Regiments any soldier or non Commissioned Officer of the Regemts of Virginia & North Carolina on Condition they pay into the hands of the Collo. or Commandg. Offrs of the Regt. which the Soldiers or non Commissioned Offrs. quit, the bounty which is now allowed to Recruits & Settle the Accounts which Subsist between Such Soldiers or Non Commissioned Offrs. & their Captains__ Drummers & fifers is not allowed to be Engaged [404]

Signed Chs Lee __ Majr Genl. & Commandr in
the Southern District

Lt Coll: Motte desires all the Offrs of duty to meet him on the Parade this Evening.

A Court Martial to sett this morning at 11 OClock to try such prisoners as shall be brought before them evid. to attend

Captn Harleston president

Lts. Dunbar & P Gray members

---

404 As Lee left his command he gave an final order that harmed both the Carolinas. Because there was a shortage of men in the South Carolina and Georgia Continentals Lee allowed some of the North Carolina men to be "translated" into the regiments of those two states. A large number of North Carolina Continentals took advantage of this situation, gaining an additional bounty from a second State. The North Carolina Council of Safety became alarmed at the number of men fleeing to the ranks of South Carolina and ordered Robert Howe to reclaim those men and return all North Carolina troops to their home State. Howe attempted to follow this order, but was stopped by the South Carolina Council who demanded six pounds and five shillings for each man returned. Howe told the North Carolina Council that the North Carolina men should be considered lost. If the North Carolina men were discharged from the South Carolina line they would consider themselves as discharged by all commitments and return home. In 1776 and 1777 a large amount of the men in the Georgia and South Carolina Line were from North Carolina. In 1778 the 2nd Georgia Regiment was almost entirely made up of men from North Carolina. Lee left the Southern Department on 7 October 1776 and returned to Philadelphia, where he was very critical of Washington's tactics and his leadership. His actions during the battle of Trenton made Washington suspect that he would not support him fully, and hoped for his eventual defeat so that Lee would take over. Lee was captured at Basking Ridge, New Jersey on 13 December 1776. George Germain, Secretary of State for the Colonies, ordered Lee to return to England to be tried as a deserter, but General Howe thought Lee had resigned his commission from the British army and did not comply with Germain's order. While a prisoner of the British Lee gave them the plan for ending the rebellion by an offensive operation against Maryland, Pennsylvania and Virginia. The British did not listen to his plan. Lee was exchanged in April 1778, but his actions during the Battle of Monmouth and his conduct afterwards led to his court martial. He almost had a duel with Baron Steuben, but he did duel with South Carolinian John Laurens, who wounded him. After that duel he accepted another challenge by General Wayne. In 1779 Lee heard that Congress was going to dismiss him, and he wrote them a letter so offensive that Congress did dismiss him in January of 1780. He died in Philadelphia in 1782.

By Lt Coll Motte **Chs.town 1776**

Orders 14th Septr. . . . Parole Laurence

For the day tomorrow Captn. Ashby
Magazine guard Lt Eveleigh
Regt. Lt Adams __New Barracks Lt P. Gray

The Long roll to beat every after noon (Sundy excepted) at 5 OClock when the Regt. is to turn out to be exercised by Captn. Motte

Such as Absent themselves without Leave may Depend on being punished severely.

Lt Dunbar to prepare for the Recruiting service without Loss of time

A Court Martial to sett at 11 OClock to try such prisoners as shall be brought before them evidences to Attend

Captn Motte President
Lts. Burke & Hall members

NB by the Court Martiall 13th Inst: Jno Griffin recd 100 lashes

Edward Fry [405] recd 150, Corpl Oaks [406] mult 7 days pay

Orders 15th Sept: . . . Parole Howe

For the day tomorrow Captn. Shubrick
Magazine guard Lt Burke
Regt. Lt Hall __ N Barracks Lt. H Gray

The Regt. to Attend divine service this Aftr.Noon the Captn of the day not to give permission to any being Absent

Genl Orders by Genl. Lee 9th Inst: recd: 14th.

The deputy Adjutant Genl. to send Abstracts of all the General Orders Issued for regulating the troops in the Southern Department to its different States, also review the troops of them & instruct them in the maneuvres agreeable to a plan given for forming a Battalion when two Compys. Act as rifle men armed with spears as well as rifles [407]__also from a regular plan with the Commanding Officers of

---

[405] Edward Fry enlisted in the 2nd South Carolina Regiment on 12 August 1775 under Captain Barnard Elliot. On 14 September 1776 he received 150 lashes for an unnamed offense. On 31 January 1777 he received 50 lashes for an unnamed offense. On 13 May 1777 he received 30 lashes for drunkenness and neglect of duty. In 1779 he was under Captain Thomas Hall at the siege of Savannah.

[406] James Oakes enlisted in the 2nd South Carolina Regiment on 1 July 1776 as a corporal. On 14 September 1776 he was mult a weeks pay for an unnamed offense. On 24 December 1776 he was reduced in rank and received 39 lashes for drunkenness and neglect of duty. In 1777 he was in Captain Blake's company. On 22 August 1777 he received 100 lashes for being absent without leave. In 1778 he was in Captain Baker's company. On 20 November 1778 he was confined for losing his waistcoat and breeches and he was put under stoppages to replace them.

[407] Lee knew that the biggest disadvantage that riflemen had was the lack of a bayonet on their weapon. If an enemy charged with the bayonet, riflemen would have to run away and abandon their positions. On May 10, 1776, when Lee was still in Williamsburg, Virginia, he wrote George Washington, telling him that "I have form'd two Companies of Grenadiers to each Regt arm'd with spears of thirteen feet long -- their Rifles (for they are all Rifle Men) slung over their Shoulders -- their appearance is formidable and the Men are conciliated to the weapon. I am likewise furnishing myself with four ounc'd Rifled Amusets which will carry an infernal distance - the two ounc'd hit a half sheet of Paper at five hundred yards distance."

the different States for receiv^{g}. Returns Agreeable to the Instructions from the Board of War and order any whose duty is to be perticularly attended to as well by the Commanding Officers of the Different States as the Adjutant Generall

Orders 16th Sept^{r}. . Parole Lynch

For the day tomorrow Captn Horry
Magazine Guard L^{t}. Mason
Reg^{t}. L^{t}. T Moultrie __ New Barracks L^{t}. Hart

A Court Martiall to sett this morning at 11 OClock to try such prisoners as shall be brought before them Evidences to Attend __

Captn Shubrick president
L^{ts}. G Eveleigh & T Hall members

Sergeant Wm Fletcher of Captn Hugers Comp^{y}. is Apointed Quarter master Sergeant & is to be Obey'd as such__[408]

NB Sentence of the Court 14th Inst: Corp^{ls}. Steward [409] & Laurence Mitchell [410] was mult a week pay & confin'd within the Centries a week __ Jno Thompson was to receive 300 Lashes but had only 63 occasioned by his Swooning

---

Orders 17th Sept^{r} . . . Parole Armstrong . . .

For the day tomorrow Captn. Ashby
Magazine guard L^{t}. Eveleigh
Reg^{t}. L^{t}. P Gray __ New Barracks L^{t}. Adams

When ever the men Appear under Arms they are always to have their Bayonets & in good Order, those who neglect shall be severly punished __ an Off^{r}. of a Comp^{y}. to make known this order

NB sentence of last court Timothy Downing rec^{d} 100 Lashes __
Jam^{s}. Hooper rec^{d} 200 Lashes boath for drunkenness & neglect of duty

---

408 William Fletcher was born on October 1729. He enlisted in the 2nd South Carolina Regiment on 4 November 1775. He was in Captain Huger's company. On 16 September 1776 he was appointed the quartermaster sergeant of the regiment. The day before his discharge he was court martialed on the charge of misconduct for losing the regimental books. He was found not guilty and discharged on 16 July 1778.

409 Unable to determine which Stewart this is. There were several soldiers in the 2nd South Carolina at this time with the name of Stewart or Stuart, however this one is most likely Alexander Stewart. Alexander was born in Ireland in 1753. He emigrated to America in 1768. He was promoted to sergeant of the 1st Vacant Company commanded by Lieutenant Richard Bohun Baker on 12 July 1778. He received a reprimand on 18 September 1777 for being absent from guard. A week later was reprimanded for stealing a blanket on 29 September 1777. He was confined for eight days on 11 December 1777 for absenting himself from guard. During the eight days he had to pull guard every other day. He served under Captain Adrian Proveaux during 1779. He was reduced in ranks on 7 March 1779 for being "unworthy of that office." He was in the York District militia at the fall of Charleston. After the war he moved to Georgia, where he died in 1810 at the age of 57.

410 Lawrence Mitchell enlisted in the 2nd South Carolina Regiment on 4 December 1775.

Orders 18th Septr... Parole Powell . . .

For the day tomorrow Captn blake
Magazine guard Lt. H Gray
Regimental Lt Burke
New Barracks Lt Hall

By Lt Coll Motte **Chs.town 1776**

Orders 19th Septr . . . Parole Parsons __

For the day tomorrow Captn. Shubrick
Magazine guard Lt. Hart
Regt __Lt. Mason __ New Barracks Lt T. Moultrie

All the Officers to turn out in Aftr.Noon with their arms & Gorgetts. __ the Captns of the day to frequently visit the Armourers shop & if he finds them Idle to report them Emediately __The Commandg Offrs of Compys. to have their mens arms put in Complete Order as soon as possible & to have the pouches properly fitted & marked __ the drummers & fifers to practice every morning after troop beating till 11 OClock this order to be made known to them by the Sergt. Majr. who is to report such as neglect their duty [411]

A Court Martial to sett this morning 11 OClock to try such prisoners as may be brought before them Evid: to attend

Captn. Blake presedt.
Lts. Mason & Hart members

NB Omitted in Orders the 18th Inst:

The Commanding Offrs of Compys. to apply to the quarter master for Oznabergs to make a Hunting Shirt pr. man at the rate of 4 yds each shirt 10/ to be allowed for making,[412] the pattern shirt is in Lt Grays Hands by which all the rest must be made___ the pattern for Sergeants Shirts not yett fixed on [413]

Orders 20th Septr __ Parole Independence
For the day tomorrow Captn Horry
Magazine guard Lt Adams
Regimental Lt Eveleigh __ New Barracks Lt P Gray

Orders 21st Septr... Parole America . . .

For the day tomorrow Captn McDonald
Magazine guard Lt Hall
Regt Lt H Gray__ New Barracks Lt Burke

---

411 Once the regiment returned from Sullivan's Island it was harder to keep the men from going into town, drinking and shirking their duties. These men had just defeated the Royal Navy in one of the most one sided fights of the war and were acting like most young men do after surviving what seemed to be certain defeat. After numerous warnings to have the men maintain order and for the officers to do their duty, Lieutenant Colonel Motte became more strict.

412 A hunting shirt was one of the most common garments in the South at this time, and would be comparable to a pair of blue jeans today. This was a shirt split up the middle with no buttons and merely wrapped around the body and normally had a cape. It was normally not dyed and remained white.

413 Unfortunately the patterns of Lieutenant Gray do not exist at this time.

A Court martiall to sett 11 OClock in morn^{g} to try such prisoners as shall be brought before them

Capt Horry president

L^{ts}. T Moultrie & H Gray members

By Lt Coll Motte **Ch^{s}.town . . . 1776**

Orders 22nd Sept: . . . Parole Pinckney

For the day tomorrow Captn Motte

Magazine guard L^{t}. T Moultrie

Reg^{t}. Barrack Lt Hart __ New Barracks Lt Hall

The Regem^{t} to turn out this after noon half past 4 OC: to Attend Divine Service __ Captn of the day not to Lett any one go from y^{e} Barrack

---

Orders 23^{d} Sept^{r}. . . . Parole Congress

For the day tomorrow Captn. Ashby

Magazine guard L^{t}. P Gray

Reg^{t}. Lt Adams __New Barracks Lt Eveleigh

NB by last court Robert Potts rec^{d} 150 Lashes & mult 14 days pay [414]

Orders 24th Sept: . . . Parole Laurence

For the day tomorrow Captn. Blake

Magazine guard Lt Burke

Reg^{t}. . . . Lt Hall __ New Barrack L^{t} H Gray

Lt Gab^{l} Marion having resigned his Commission in the 2nd Reg^{t} with Leave is no longer to be Obey'd as such

The Commanding Off^{r}. of Comp^{ys}. to apply to Quart^{r} Master for a Blanket for each man present & to keep a list of the mens names that receive them

---

[414] Robert Potts served in an independent company of Rangers under Captain Robert Ellison during 1775 and enlisted in the 2nd South Carolina Regiment on 20 May 1776 under Captain Charnock. On 23 September 1776 he received 150 lashes and was mult 14 days pay for an unnamed offense. On 14 October 1776 he received 100 lashes for selling his blanket. He was also fined for the cost of the blanket. On 21 October 1776 he received 350 lashes for selling his shirt and blanket and his pay was stopped to pay for the shirt and blanket. On 9 November 1776 he was court martialed for offering to inlist in the 5th South Carolina Rifles for the bounty. He was found guilty and sentenced to receive 100 lashes, 50 of which would be on his bare back. General Howe reduced that number to 30 lashes since it was the first time any soldier tried to enlist into another regiment. He also had his pay stopped until he repaid the bounty back to Colonel Huger. On 31 March 1777 he received 100 lashes for being absent without leave and losing his arms and accountrements. He also had his pay stopped until the musket, pouch and bayonet were paid for. On 7 May 1777 he received 99 lashes and he had to run the gauntlet for quitting his post. On 10 June 1777 he was imprisoned for some unnamed offense, and the officer of the guard was warned to make sure that he did not escape, because it would be an "unpardonable neglect of duty." He was sentenced to be shot for desertion on 5 November 1777, but was pardoned by General Howe on the condition that he enlist for the duration of the war. He received 50 lashes on 12 December 1777 for stealing a hat, and he had his pay stopped until he paid the £10 for the hat.

A Court Martial to sett morn^g: 11 OClock to try such prisoners as shall be brought before them Evidences to attend

Captn Ashby Presed^t
L^ts. Hall & Burke members

Orders 25th Sept^r . . . Parole Cattle

For the day tomorrow Captn Shubrick
Magazine guard Lt Mason
Reg^t. . . . L^t T Moultrie __ New Barracks Lt Hart

NB Genl Howe, Coll: Moultrie & Captn Huger returned from Georgia

By Lt Coll Motte **Ch^s.town 1776**

Orders 26th Sept^r . . . Parole Edwards . . .
For the day tomorrow Captn McDonald
Magazine guard L^t P Gray
Reg^t . . a serg^t __ New Barracks Lt Adams

The Regimental Barracks Guard to consist of 1 Serg^t: 1 Corp^l & 12 privates till further Orders

---

Orders 27th Sept: . . . Parole Harwick . . .

For the day tomorrow Captn Motte
Magazine guard L^ts Hall
New Barrack Lt Burke __ Reg^t. a sergeant

NB James Anderson had a furlow for 10 days by Order__

Orders 28th Sept^r . . . Parole Washington . . .

For the day tomorrow Captn Ashby
Magazine Guard L^t H Gray
New Barracks. Lt Hart __ Regl B: a Serg^t.

NB last night arrived from Georgia Lt Jacob Shubrick with 19 men

---

Orders 29th Sept^r . . . Parole Howe

For the day tomorrow Captn Shubrick
Magazine guard L^t. T. Moultrie
New Barracks L^t. Mason __ Reg^t. a Serg^t

Orders by Coll Moultrie The 2^d Reg^t to attend devine Service this aft^r Noon 5 OC:

Orders 30th Sept^r . . . Parole Rutledge

For the day tomorrow Capt Horry
Magazine guard L^t. Adams
New Barracks L^t. P Gray __ Reg^t G. a Serg^t

General Returns to be given in by Command^g. Officers of each Comp^y tomorrow Morning by 10 OClock __

NB Ch^s Loyd [415] Deserted sometime agoe was brought Back by Wm Clark [416] __

---

[415] Charles Lloyd enlisted in the 2nd South Carolina Regiment on 4 November 1775. He deserted from the unit, was captured and returned on 30 September 1776. He was discharged on 13 August 1778.

[416] William Clarke (Clark) enlisted in the 2nd South Carolina Regiment on 4 November 1775. He was supposed to receive 35 lashes for abusing Abner Moses on 9 October 1777, but the sentence was remitted. On 27 July

1778 he was appointed a corporal in Captain Motte's company. He was reduced to private on 19 September 1778 under Captain Thomas Dunbar. On 7 November 1778 he was sentenced to receive 50 lashes for riot and drunkenness, but it was remitted. On 27 November 1778 he received 30 lashes on the bare back with switches for abusing and threatening the life of Corporal Jones. He was in Dunbar's light infantry at the siege of Savannah.

By Coll. Moultrie **Chs.town 1776**

Orders 1st October . . . Parole Armstrong . . . by presedt

For the day tomorrow Captn. McDonald

Magazine guard Lt Burke

New Barracks Lt. Hall __Regt G. a sergt

Orders 2nd Oct. . . . Parole Assembly . . . by G. How

For the day tomorrow Captn Motte

Magazine guard Lt Hart

New Barracks Lt H Gray__ Regt. G.__ a sergt.

One Sergt. & 6 men to be sent on board the ship Clarissia as guard to prevent any of the cargoe &c. being taken out without Orders from Genl. Howe, the presedt. or Coll: Moultrie for this guard Sergt Raybout, which recd his Orders from Majr Marion as above__

Orders 3d Octobr. . . . Parole Pinckney . . . by Presedt

For the day tomorrow Captn. Ashby,

Magazine Guard Lt Thos Moultrie

New Barracks Lt Shubrick __Regt G. a sergeant

NB this Mo. Genl Return is 10 Capt. 18 Lts 27 Sergts. 13 drums & 339 Rank & file men & 12 deserters __

No Offrs. on duty to Appear without his regimental Compleat, his cap not Excepted

A Court Martial to sett this Morng. 11 OC: to try such prisoners as shall be brought before them Evidences to Attend

Captn Motte President

Lts. R Mason & P Gray members __ NB Officers non estandventas [417]

NB This Court Ordered to sett tomorrow

Orders 4th Oct: . . . Parole Liberty . . . by G. Howe

For the day tomorrow Captn Blake

Magazine guard Lt Mason

New Barracks Lt Adams__ Regt G. a sergt.

No Cloathing to be given out till further Orders

By Coll. Moultrie **Chs.town 1776 __**

Genl Orders by Genl Howe 4th Octr 1776

Orders 5th October . . . Parole Huger . . . . by Presedt

For the day tomorrow Captn Shubrick

Magazine guard Lt. Hall

New Barracks__ Lt Burke__Regt G. a sergt.

---

[417] Non Est Adventis. This is latin for "not attending" or "absent".

Orders 6th Octob^{r}. . . . Parole Carolina . . . by G. Howe

For the day tomorrow Captn Horry
Magazine guard L^{t}. Hart
New Barracks L^{t}. H Gray___ Reg^{t}. G. a sergeant

Orders 7th Oct^{r} . . . Parole Parsons . . . by Presed^{t}

For the day tomorrow Captn Motte
Magazine guard Lt Shubrick
New Barracks L^{t}. T. Moultrie__ Reg^{t}. G. a serg^{t}

The men returned from Command to be supply'd with hunting shirts & blankets Emediately

General Orders by Genl Howe

The Adjutant of the Line & the Quart^{r} Master Genl, to attend at head quarters every day at 12 OClock for Orders, this order is not to include the Adj^{ts} on out posts

The Commanding Officers of Corps are Immediately to make a return of their men and at what place they are Stationed, to the Generall in order to Lay them before the honorable House of Assembly

---

Orders 8th Oct^{r} . . . Parole Armstrong . . . by G. Howe

For the day tomorrow Captn. Ashby
Magazine Guard L^{t}. Adams
New Barracks Lt Mason__ Reg^{t}. G. a Serg^{t}.

When the major is not on the Parrade in the Evening the next ~~Off^{r}~~ Eldest Off^{r}. is to Exercise the Battalion

by Maj^{r}. Marion twenty men & a serg^{t}. is Order'd to fetch the boat from Ropers Warff which came from Georgia

By Coll. Moultrie Ch^{s}.town 1776 __

The Captn of the day to visit the different Posts one a day & as many at night & to Wait on the General with the reports after they have been with the Coll: of the Reg^{t}.

Orders 9th Oct^{r} . . . Parole Laurence . . . by the presed^{t}

For the day tomorrow Captn Shubrick
Magazine guard Lt Burke
New Barracks Lt Hall __ Reg^{t}. G. a serg^{t}.

Orders 10th Oct^{r} . . . Parole Gadsden . . . by Genl Howe

For the day tomorrow Captn Shubrick
Magazine guard L^{t}. H Gray
New Barracks Lt Hart __ Reg^{t}. G. __ a Serg^{t}.

Commanding Off^{rs}. of Comp^{ys}. are to examin there mens Accoutrements Arms, & Cloathing one a week a report to the Command^{g}. Officer of the Reg^{t}. every Saturday

---

Orders 11th Oct: . . . Parole Moultrie . . . by Genl Howe

For the day tomorrow Captn. Motte
Magazine Guard Lt T. Moultrie
New Barracks Lt Shubrick __ Reg^{t} G.__ a serg^{t}.

A Court martial to sett at 11 OClock this morning to try such Prisoners as shall be brought before them. Evid: to Attend

Capt[n] Hory President
L[ts]. T. Moultrie & Shubrick members
Gen[l] Orders by Gen[l]. Howe

As Cleanlyness is Essential to the Health of the men Off[rs]. of Comp[ys]. are requested to have the Barracks frequently inspected and whenever tis necessary to Order them Cleaned,, the Quart. Master to see it properly executed__ The Chimnies of barracks & where Soldiers are quartered to be swept without delay , under the Direction of a Sergeant are to perform this duty__ and when there are no camp colour men or not enough of them Orderly men of each Comp[y]. are to be Apointed by Command[g]. Off[rs]. of Comp[ys]. the Quart[r] Mast[r]. to take care that the Chimneys are well swept & regularly Once a fortnight

A Regimental Court Martial to sett tomorrow at 8 OClock in Morn[g]: to try several prisoners of the Second Reg[t] now in the guard House, the Off[rs]. who confind the prisoners are to Order the ~~prisoners~~ Evidence to attend.

Capt[n] Motte Presid[t].
L[ts]. Burke & Adams members

Orders 12th Oct[r]. . . . Parole Middleton . . . by Presid[t]

For the day tomorrow Capt[n] Ashley
Magazine Guard L[t] Adams.
New Barracks L[t] Burke__ Reg[t]. G.__a serg[t].

All Officers present at roll-call in the morning do Order the soldiers to put out their Blankets to Air

---

Orders 13th Oct[r] . . . Parole Widon . . . By Gen[l] Howe

For the day tomorrow Capt[n] Charnock
Magazine Guard L[t] Eveleigh
New Barracks L[t] P Gray ____ Reg[t] G _ a serg[t].

Orders 14th Oct[r] . . . Parole Drayton . . . by Presid[t]

For the day tomorrow Capt[n] Horry
Magazine Guard L[t]. Hall
New Barracks H Gray __ Reg[t] G __ a serg[t]

NB Last court sentence John Wheeler for striking W[m] Donnavan to 500 Lashes rec[d] 400__[418] Corp[l] Gambell repremanded, W[m] Baldwin & Robert Potts for selling their Blankets 100 Lashes each & Mult the price of the Blankets,[419] Sam[l] Henderson for drunkenness 200 rec[d] 100 [420] Ja[s]. Hunt for

---

[418] William Donavan enlisted in the 1st South Carolina Regiment on 4 November 1775.

[419] William Baldwin enlisted in the 2nd South Carolina Regiment on 4 November 1775. On 14 October 1776 he received 100 lashes for selling his blanket and he was fined for the cost of the blanket. He served under Captain Shubrick in 1777. On 9 April 1777 he received 100 lashes for neglect of duty. On 22 October 1777 he received 50 lashes for being drunk on guard. After Captain Shubrick's death he served in the 2nd Vacant Company until it was disbanded on 5 October 1778. He was then assigned to Captain Mazyck's Company. On 11 June 1778 he received 39 lashes for neglect of duty.

[420] Samuel Henderson served in the 2nd South Carolina Regiment during 1775 to 1777 under Captain Thomas Hall. On 14 October 1776 he was sentenced to receive 200 lashes for drunkenness, but only received 100. On

Insolents 50 Lash remitted,[421] Edwd Murphy for quit is guard & Drunkenness 200 remitted 100__[422]
Ch^s: Howel Simmons for threating to desert 200 lashes Pardoned by intersession of Major Marion
By Coll. Moultrie **Ch^s.town 1776 __**
Jebediah Cobbs Corp^l. in Captn Mottes Comp^y. is Apointed a serg^t. in Captn Huger's & is to be obey'd as such [423]

Orders 15th Oct^r . . . Parole Armstrong . . . by Genl Howe
For the day tomorrow Captn Motte
Magazine guard L^t. Mason
New Barracks L^t T Moultrie __ Reg^t G. a serg^t.
Long roll to Beat at half aft^r 4 in Aft^rnoon by Ord^r Maj^r Marion
Genl Orders by General Howe 14th Oct:
Quart^r Masters of Reg^ts to fix upon proper places to build nessesary houses for the men which they are to have done emediately [424] They are to Employ the Camp Coular men for this purpose & where their is non Off^rs. of Comp^ys. to apoint sufficient numbers of Orderly men who are to take their Orders from the quart^r. Master____ the Store Keeper is to furnish the Off^r. of the Magazine Guard with an Ax & that Off^r to give it in charge to the next Relieving Officer . . . .

Orders 16th Oct^r . . . Parole Elliott , , , by the Presid^t.
For the day tomorrow Captn Charnock
Magazine guard L^t. Burke
New Barracks L^t. Adams __ Reg^t G. __ a serg^t.
One Subaltern, 1 Serg^t. all the drums & fifes & 25 Rank & file men to Attend the funeral of Lt Armstrong of the 1st Reg^t at the house of the late Dr Haly at 4 OClock this aft^rnoon [425]
Lt Thos. Hall for this duty

---

7 July 1777 he was pardoned for quitting his post. He received 30 lashes on 9 October 1777 for being absent without leave. A month later on 9 November 1777, he once again received 50 lashes for being absent without leave. He was sentenced to do double duty for the duration of the war on 11 December 1777 after he was convicted of neglect of duty. On 15 July 1778 he was placed under stoppages to replaced his regimental coat that he lost. He was under Captain Lesesne in 1778. On 13 December 1778 he was sentenced to do double duty for four days for being drunk on guard. On 5 January 1779 he received 50 lashes on the bare back with a cat of nine tails for losing the lock of his gun when he was on guard. On 9 February 1779 he received 100 lashes with switches for being absent without leave. On 15 February 1779 he was put under stoppages for losing his regimental clothing. He was in Captain Hall's company at the siege of Savannah.

[421] James Hunt enlisted in the 2nd South Carolina Regiment on 4 November 1775.

[422] Edward Murphy enlisted in the 2nd South Carolina Regiment on 4 November 1775. On 14 October 1776 he was sentenced to receive 200 lashes for being drunk and quitting his guard post, but 100 of those lashes were remitted. He received 100 lashes on 26 August 1777 for being drunk. On 12 September 1777 he received 45 lashes for repeatedly being absent from roll call. He received 100 lashes on 17 April 1778 for being drunk and for neglect of duty, but 50 of those lashes were remitted. On 16 February 1779 he was sentenced to be stopped one months pay for some unnamed offense. The money was to be used for the regimental hospital. He was under Captain Thomas Moultrie at the siege of Savannah in 1779.

[423] Jeremiah Cobb enlisted in the 2nd South Carolina Regiment on 4 November 1775 in Captain Motte's company. He was promoted to sergeant on 14 October 1776 and transferred to Captain Huger's company.

[424] Neccesary Houses were the latrines.

[425] Unable to determine who this Lieuteant Armstrong was.

Orders 17th Oct^{r} . . . Parole Moore . . . by Genl Howe

For the day tomorrow Captn

Magazine guard L^{t} H Gray

New Barracks Lt Hall __Reg^{t} G __ a sergt.

By Coll. Moultrie **Ch^{s}.town 1776 __**

Orders 18th Oct^{r} . . . Parole watch . . . by the president

For the day tomorrow Captn Charnock

Magazine guard L^{t} H Gray

New Barracks Lt Hall __ Reg^{t} G __ a serg^{t}.

Orders 19th Oct^{r} . . . Parole Council . . . by Genl. Howe

For the day tomorrow Captn

Magazine guard L^{t}. Dunbar

New barracks L^{t}. Mason __ Reg^{t} G __ a serg^{t}.

Genl Orders by Genl Howe

Genl Howe is at a loss to know why he has no report of the Proceedings of the Court Martial ordered to sett the 10th Instant and in feuter expects that all proceedings of courts martial Wether General or regimental upon prisoners reported to him are to be laid before him as soon as determined upon

Captn Charnock having reported to the general that the sick in hospital complain that bread is not regularly served them, he expects those who have the charge of this matter will be carefull no feuter complaint may happen & public good exact of all persons the greatest attention to sick Soldiers who devote their lives to the Service of the Common Cause

A detachment from Coll: Hugers Reg^{t} will relieve the guard at the new Barracks tomorrow morning at 8 OC: & continue that duty till further orders

NB Lt Mason to mount the Reg^{t}. Guard by Ord^{r} Maj^{r} Marion

---

By Maj^{r} Marion

Orders 20th Oct^{r}. . . . Parole New York . . . by the presid^{t}

For the day tomorrow Captn Charnock

NB Coll: Moultrie absent Lt Coll Motte sick

Magazine guard Lt Adams, 1 Serg^{t}. 19 rank & file

Regimental Barracks Lt Burk, 1 Serg^{t}. 16 rank & file

The Captn of Day (& when non) the Off^{rs}. of the Reg^{t} guard is to receive all reports of guards, & morning reports & deliver General Howe a report of all guards and such prisoners as are confined for Capital crimes only & Deliver the Commanding Off^{r}. of the Reg^{t}. the Morning & reports of guards mounted by the Second Regim^{t} & the prisoners Confin'd for small crimes only to be try'd by a Regimental court martial

The Off^{rs}. of the day to visit the Barracks & see if they are kept clean, or if the sick are neglected also to inquire if the Quart^{r} Master has the Chemnies swept according to Orders__

The Surgeons mate of the Reg^{t}. to report every day the State of the sick in Barracks, specifying their disorder & complaints and are to send those who may be very sick to the Genl Hospital

Roll call & guards relieved to be at 7 OC: in Morning & Exercise at half past 4 OC: in afternoon

No soldier to mount guard or appear on the parrade without his split shirt & Regimental Leggens

Major Marion is sorry he is obliged to take notice of the great neglect of most of the Officers not Attending their duty more punctually, Roll call in the morning & Exercise in the afternoon is almost intirely neglected to the very great prejudice of the Service __ He beggs leave to remind them that their country has reposed Great trusts in them, which is expected their Honours will not permit them to Neglect

Maj[r] Marion hope they will consider that neglect of Duty in any point is Attened with great Evils not only to the State, but is sure in the end to bring Dishonour on themselves, the Maj[r]. woud rejoice to see the Gentlemen do their duty with Willingness & punctuality, and hope their good sense on serious reflection will prevent the Disagreeable Nesessity of reminding them of their duty in futer__ Divine Service will be at 4 OC. this aft[r]Noon

A Serg[t] & 6 men to go to the Gen[l] Hospital this aft[r].noon to Bury a soldier of the 1st Reg[t] & to take a Cartridge p[r] man to fire over the corps as usual__

By Maj[r] Marion **Ch[s] town 1776**

Orders 21st Oct[r] . . . Parole Carolina . . . by Gen[l]. Howe

For the day tomorrow Capt[n] Horry

Magazine guard L[t] Eveleigh

Reg[tl]. guard L[t] Hall

L[t] Peter Gray & Serg[t]. Fowler to hold themselves in readyness to a recruiting, it is expected they will sett off as soon as possible

---

Orders 22nd Oct[r]. . . . Parole Laurence . . . by the presid[t]

For the day tomorrow Capt[n] Harleston

Magazine guard L[t] Mazyck

Regimental L[t]. H Gray

NB According to the last court James Allwell for selling his shirt & Blanket rec[d] 200 Lashes & stoppages to replace them__[426] Rob[t]. Potts for the same crime rec[d] 350 & stoppages to replace his Blankets & shirt

Lieut[t] Dunbar to hold himself to go a recruiting

Orders by Gen[l] Howe

The General thinks proper to insert the following resolves of the General Assembly of this state respecting the troops in order that every officer may be Acquainted with them

Sept. 20th 1776

Resolved that this House do Acquiese to the resolve of the Continental Congress of the 18th of June & 24th of July last relating to the two Regiments of Infantry the Rangers, Artillery & the two Reg[ts] of Rifflemen in the Service of this State upon the Continental establishment

Resolved that this House will defray the expenses of the difference between the S[o] Carolina Bounty Cloathing & Pay & the Bounty Cloathing & pay allowed to those Regiments respectively by the Act of this State

---

[426] James Allwell enlisted in the 2nd South Carolina Regiment on 26 November 1775. On 27 October 1776 he received 200 lashes for selling his shirt and blanket, and his pay was deducted to cover the loss of the items. He received 25 lashes on 7 November 1777 for neglect of duty. On 15 May 1778 he accidentally shot himself while on guard duty.

15th Octr 1776

That in Order to comply with the recommendation of the Continental Congress to take the most spedy & Effectual measures for Inlisting our Quota of troops they are of Oppinion that a bounty of ten dollars over & above the Continental bounty be given to such non Commissioned Officers & soldiers, who shall inlist during the present war in any of the So Carolina Regiments on the Continental Establishment and that in lieu of the bounty formerly given the Officers for recruits that each recruiting Offrs. be Allowd besides their pay two dollars per Day for each day he is actually in the Service

Your commitee Likewise recomend that the fifth regt of this state shall emediately or as soon as may be, be put on the same establishment with regard to Arms, pay, Bounty, & Cloathing as the two Regiments of Infantry are & that the Sixth shall be likewise be put on the same establishment in the like Respect as soon as the field officers of the last mentioned Regt shall agree thereto

18th Octr. 1776

Resolved that the 3d & 4th Regts. be Augmented to 600 men each & that a Collonel be Appointed to each of Sd Regts.

Resolved that it is there opinion Captn Richbourgs Independent Compy should be Added to the 6th Regt.[427] & that the Artillery Compy of Geor town be augmented to one hundred men & be added together with the artillery Compy. of Beufort to the fourth Regiment

Commanding Offrs of Regemts & Corps will Attend to the above resolutions & immediately order out a sufficient number of Offrs. for the recruiting service in doing which it is recomended to them not to consider whose tour of duty it may be but to apoint such offrs. whose former merit in recruiting or whose quallifications promise the greatest success. The necessity of service must frequently, Supercede common rule and as Success in this kind of undertaking in great measure depend on popular curcomstances, No Offrs. tho it may be his tour of duty ought to think himself injured by not being Order'd on this Occasion and it is hoped no Offr. will conceive themselves Oppressed tho sent out contrary to detail when they consider that at this time they cannot render to their country & the common cause any service more Essential than in Contributing to Compleat those battalions so absolutely necessary for the defence & support of them

The Board of ordinance having Orderd that monthly return Should be ~~given~~ made of every Regt. in the Continental Service in which they are raised the time when & the period the men were Inlisted; the Commandg. Officers are therefore required to made an Emediate & exact return agreable to the Directions above mentioned that the General may be able to comply with the requisition of the Honorable the Board of War & Ordinance

By Majr. Marion | **Chs Town 1776**
Orders 23d Octr . . . | Parole Moultrie . . . | by Genl Howe

[427] Henry Richbourg (Richborough, Richbough) was a captain in the 6th South Carolina Regiment until he resigned on 22 February 1778. After the fall of Charleston he became a turncoat and joined the British, commanding a troop of Loyalist horsemen.

Gen[l] Orders by Gen[l] Howe

All the officers and men of the Continental Army/guards except those now in town & fit for duty are to parrade on fryday morning at 10 OCl: on the green near where Coll: Gadsden formerly lived, the Adjutant Gen[l]. & Brigade Major will attend & train them in the exercise of the pike [428] The Adjutant Gen[l] will direct the adjutant to collect the number of pikes wanting [429]

Complaint having been again made from the Hospital that the sick are not served with Bread, General Howe positively Orders the Comisary to inspect this matter & prevent any cause of feuter complaints And that ignorance of orders may not be pleaded in Excuse, Adjutant Delliant will serve a Copy of this & the former orders upon this subject

Regimental Orders by Maj[r]. Marion

On thursday next at 3 OC. in Aft'N[n]. the 2nd Reg[t] is Order'd parrade that the muster master general may review them No Commissioned non Commissioned Off[rs]. or Soldier to absent himself on any account whatever the Off[rs]. of Comp[ys]. are to make a full return of the number men fit for duty, on duty when on duty sick in G. Hospital or Barracks on Command, recruiting, on furlow, names present & absentees are to be inserted with the particulars of their arms & Accouterments

As Gen[l] Howe has Order'd a review on Fryday it is expected the Off[rs]. will make their men appear as clean as possible especially their Arms

Ordered that the Quarter master do see the Barracks kept clean & the Chemnies swept, as it is reported it has been neglected

---

Orders 24th Oct[r] . . . Parole Pinckney . . . . by the presid[t]

For the day tomorrow Capt[n] Shubrick

Magazine guard L[t] Adams

Regimental L[t] Eveleigh

By Maj[r]. Marion Ch[s] Town 1776

Orders 25th Oct[r] . . . Parole Motte . . . by Gen[l] Howe

For the day tomorrow Capt[n] Charnock

Magazine guard- L[t] Burke

Regimental g __ L[t]. Hall

Orders 26th Oct[r] . . . Parole Marion . . . by the presid[t]

For the day tomorrow Capt[n] Horry

Magazine guard L[t]. Henry Gray

Regimental __ L[t] Hall

L[t] David Adams to hold himself in readiness to go recruiting it is recomended that he be ready if possible to sett out a Munday next

---

428 The brigade majors were staff officers of the brigadier general. They came to headquarters at a set time and the adjutant general dictated the orders. They received from the regimental adjutants all men detailed for guard, inspected their arms and clothing and assigned them to posts by lot. They then returned to their brigades and assembled the regimental adjutants and gave the necessary orders. The adjutants returned to their regiments, assembled the orderly or first sergeants, and gave orders for the companies.

429 The main threat to Fort Sullivan during the battle in June had been the threat of an assault by infantry on the rear of the fort. To repell the enemy was a reserve force of musketmen and men armed with pikes. After the battle the men were trained on use of the pike in case there was ever a need.

Orders 27th Octr ... Parole Pinckney ... by G Howe

For the day tomorrow Captn. Motte
Magazine guard Lt Mason
Regimental Lt. Mazyck

By Genl Howe

As the provisions usually served out to the soldiers when well may not be proper for them that are sick, the Director of the Hospital will Occasionally order the Commisary to purchase for the Indesposed such things as are necessary of which he is to keep an Account The Surgeons of the Hospital will attend to the provision Served to the men under their care & if they find it improper are immediately to make a report to the commanding Offrs. for the time being, that Survey may be order'd upon it

---

Orders 28th Oct^r^... Parole Bullet... by president

For the day tomorrow Captn. Blake
Magazine Guard L^t^. Eveleigh
Regimental Lt Burke

Coll: Moultrie Acquaints the Officers & Soldiers of the Second Regiment that the Honorable the Continental Congress has promoted him to the rank of Brigader General he therefore takes this method of resigning his command as Coll: of the Second Reg^t^. and wishes all health & happiness to the Off^rs^ & Soldiers & hope they will always support that good name they have acquired with so much Bravery and is happy that on his Leaving the Regiment the command devolves on so worthy a Gentleman as Coll: Motte who he is sure will do everything in his power to make the Service agreeable to the Officers & men __ Genl Moultrie tho he has resigned the regiment Yett still he has a Command over them & will look upon them with a partial Eye & assures the officers & Soldiers they shall allways find a friend in him

---

NB from the monthly return 25th Oct^r^
Total 10 Capt^ns^. 10 L^ts^. 1 Adj^t^: 1 quart Mast^r^. 1 Chaplain 2 Surgeons mate
1 Paymaster, 26 Serg^ts^, 13 drums & fifes & 332 Rank & file & recruits
3 dead & 6 Deserted __

Orders 29th Oct^r^.. . . Parole Congress.. . . . by G Howe

For the day tomorrow Captn Charnock
Magazine guard L^t^. Mazyck
Regimental L^t^. H Gray

By Genl Howe

The Honorable the Continental Congress having promoted Col^o^ Christ: Gadsden of the first Reg^t^ & Col^o^: Wm Moultrie of the Second S^o^ Carolina reg^ts^ to the rank of Brigadier Generals in the Army of the United States they are to be obey'd as such

---

By Coll: Motte
Orders 30th Oct^r^ . . . Parole Laurence . . . by presid^t^

For the day tomorrow Captn Horry
Magazine guard L^t^. Hart
Regimental __ L^t^ T Moultrie

Orders 31st Oct^r^. . . . Parole Moore . . . by G. Howe

For the day tomorrow Captn Harleston
Magazine guard Lt Mason
Regimental__ Lt Eveleigh

A monthly return to be given in to the Adjutant today by the Command^g^ Off^rs^: of each Comp^y^. & to enter names of the Off^rs^. with the dates of their Commissions on the Back__ A Court Martial to sett today to try such prisoners as shall be brought before them__

Captn Horry presid^t^
L^ts^ Eveleigh & Burke members

By Coll: Motte **Ch^s Town __1776 __**

Ord^rs. by Genl Howe

A Subaltern & twenty men to relieve the Militia guards opposite the State House tomorrow morning at 8 OC: the men to be taken by detachment from the 2nd & 5th Regiments of S^o Carolina Continental troop in proportion to the number fit for duty in each Corps __ Mr Singleton being Apointed Deputy Muster Master General for the State of S^o Carolina & Georgia is to be respected as such __ this order to be transmitted to the out posts by the Brigade Major__ Adjutant Delliant is requested to furnish Coll Huger with a copy of this Order

---

Orders 1st Nov^r. . . . Parole Oliphant . . . by presed^t

For the day tomorrow Captn Motte

Magazine guard L^t. H Gray

Town guard L^t. Mazyck __ Reg^t. Guard a Serg^t.

No man in future to apply for a furlow till he has first Obtained his Capt^ns. Leave in Wrighting & they are order'd not to grant such permission to more than two men p^r Comp^y. agreeable to the Articles of War

The Guard to be relieved at 8 OC: in morning till further orders

---

Orders 2nd Nov^r . . . Parole Georgia . . . by the presed^t

For the day tomorrow Captn. Shubrick

Magazine guard Lt T. Moultrie

Town G_ L^t. Hart ___ Reg^t G__ a serg^t

Captn Huger having resigned his commission in the Second Regiment with Leave is no longer to be Obey'd as such

The Regiment to Attend devine Service tomorrow aft^r.noon it is expected all the Off^rs. will Attend, the Captn. of the day not give Leave any men absent aft^r. 1 OC:

A Court Martial to sett 11 OC: for the tryal of Prisoners

Captn. Motte president L^ts. T Moultrie & Eveleigh members

By Coll: Motte Ch^s Town 1776

Orders 3^d Novemb^r. . . . Parole Sumpter . . . by Genl Howe

For the day tomorrow Captn Charnock

Magazine Guard L^t. Mason

Town Guard L^t. Eveleigh __Reg^t G __ a serg^t

NB Hamilton for quiting his post & getting drunk rec^d 50 lashes [430]

Orders 4th Nov^r . . . Parole Motte . . . by Genl Howe

For the day tomorrow Captn Harleston

Magazine Guard L^t. Burke

Town G__ L^t. Hall__ Reg^t G__ a Serg^t

---

[430] Charles Hamilton enlisted in the 2nd South Carolina Regiment on 4 December 1775. On 3 November 1776 he received 50 lashes for quitting his post and getting drunk. On 13 November 1776 he received 60 lashes for illegally inlisting in the 5th South Carolina Regiment, and taking the bounty money. His pay was stopped until he paid back the bounty money to Colonel Huger.

A Court martial to Sett this morn^g: 11 OC: to try such prisoners as shall be brought before them

Captn. Charnock President

Lieut^s. Hall & Dubose members

Corp^l Jno Marlow is apoint a serjeant in Granidiers Comp^y. & is to be Obey'd as such [431]

Orders by Genl Howe

The Honble. the Continental Congress having by a resolution of the 20th Septr Last repealed all former Articles of War, & Establish a Sett of Articles for the futer Government of their Army, Commanding Officers of regem^ts. & Corps are as emediately as possible have those Articles read to their men __ the Articles to be had by applying to Mr P. Timothy

---

Orders 5th Nov^r . . . Parole Howe . . . by the presid^t.

For the day tomorrow Captn Shubrick

Magazine guard L^t. Mazyck

Town g__ L^t H Gray__ Reg^t G_ a serg^t

The Continental Articles of War to be read at the Head of the Regiment this Aft^r.Noon at 4 OC: no one to be Absent that is able to stand ____ _____ _____

The Reg^t is to muster nest Thursday 4 OC: in After noon by by the Continental muster master__ The Muster rolls at in the mean time to be gott ready, at the same time the adjutant is to make out a list of the absentees, mentioned Opposite to each mens name the time he is Absent And the Surgeons mate to make out a list of the sick which he is to sign to be deliverd to the muster master

The staff Off^rs. to have notice to attend by the Adj^t.

A Court martial to sett at 11 OC: this morning to try such prisoners as shall be brought before them, Evidences to attend

Captn Harleston presed^t. L^ts. Mazyck & Baker Memb^rs: } this court did not sett

NB Sentence of the court Yesterd^y John McBride for drunkenness rec^d 50 lashes
Jno Fenwick rec^d 150 lashes & to be under stoppages to replace his Blankets and shirt which he sold [432]

---

431 John Marlow enlisted in the 2nd South Carolina Regiment on 4 November 1775 under Captain Thomas Hall. He was promoted to sergeant in the Grenadier company on 4 November 1776. On 28 March 1777 he was reprimanded for misbehavior to Captain Jervey. He was discharged on 11 July 1778, then re-enlisted on 13 February 1779. He fought at the siege of Savannah.

432 John Fenwick enlisted in the 2nd South Carolina Regiment on 20 November 1775. On 2 August 1777 he received 50 lashes for being drunk. On 5 November 1776 he received 150 lashes and had his pay stopped until he paid off the blanket and shirt the he sold. He received 50 lashes on 2 August 1777 for drunkenness. He received 100 lashes a month later on September 18th, for stealing cartridges. He received 50 lashes the next month, on October 14th, for neglect of duty. He was under Captain Richard Baker in 1778. He received 100 lashes with switches on 7 December 1778 for being absent without leave. He was also put under stoppages until he could pay back the watchcoat he lost. On 18 August 1779 he was sentenced to receive 100 lashes with switches for theft and selling his shirt, and his pay was withheld to pay for the shirt. He was in Captain Baker's company in 1779. He served as a private and quartermaster in the militia during 1780-1781.

Orders 6th Novr . . . Parole Pinckney . . . by G. Howe

For the day tomorrow Captn Charnock

Magazine Guard Lt Hart

Town G__ Lt. Baker__Regt G__ a Sergt.

The Muster Rolls to be carried to the Major by an Offr. of a Compy. for his Inspection & direction how to make them out properly__ the Offrs. to see their mens arms are very clean & their Accoutrements are in good order & that they appear as neat as possible tomorrow AftrNoon & to be drawn up as they are on the Muster roll

Jno Newton of Captn Harleston's company is Apointed a Sergeant in the same & to be Obey'd as such [433]

A Court Martial to sett at 12 OClock today &tc.

Captn Shubrick president Lts. T Moultrie & Baker Membrs

NB James Smith & Thos Welch mult a weeks pay for neglect of duty. [434]

Orders by Genl Howe

A Generall court martial to sett tomorrow within the House prescribed by Art: of War for the tryal of Chs: Hamilton of Captn Mottes Compy. for Inlisting in the 5th Regt. Coll: Huger will produce the evidence and Majr. Conner [435] will have the Officers Warn'd according to Detail

---

Orders 7th Novr. . . . Parole Bull . . . by the president

For the day tomorrow Captn Horry

Magazine Guard Lt Mason

Town G__ Lt T Moultrie __ Regt G__ a Sergeant

By Coll: Motte **Chs Town 1776**

The Regemt to be drawn up Opposite the Barrack at half past 3 OC: this aftr.noon The Adjutant to keep an exact Account of all Fines Orderd to be pay'd in his hands.

Ords by Genl Howe__

The General Court martial order'd to sett yesterdy. is for sett and postponed till further orders.__ Captn Wm Moultrie being Appointed Aid de Camp to General Moultrie is to be respected & Obey'd as such

---

433 John Newton enlisted in the 2nd South Carolina Regiment on 2 November 1775. He was promoted to sergeant on 6 November 1776 in Captain Harleston's company. He was demoted to private on 13 April 1778 and discharged four days later. He re-enlisted again on 4 February 1779 under Captain Thomas Dunbar. Prior to the siege of Savannah in 1779 Newton and Jasper teamed up and did missions behind the British lines. Details of these missions are described in "*Nothing but Blood and Slaughter, Volume One*".

434 Thomas Welsh (Welch) served in the 2nd South Carolina Regiment during 1777. On 6 November 1776 he was mult a weeks pay for neglect of duty. On 31 March 1777 he was in confinement for being absent from the Command to Georgia, but the charge was dismissed. He received 100 lashes on 4 August 1777 for being drunk on guard. He received 75 lashes on 15 August 1777 for gambling. He received 50 lashes on 25 October 1777 for beating an inhabitant of the town. A month later he received 50 lashes on 2 November 1777 for beating Private Phillips. He was fined seven days pay on 4 December 1777 for stealing wood. In 1779 he was a sergeant under Captain Thomas Hall. He received 25 lashes on the bare back with a cat of nine tails for willfully abusing his musket on 25 March 1779. He fought at the siege of Savannah.

435 Clement Conyers was a captain in the 5th South Carolina Regiment. After the fall of Charleston he was a captain in Marion's Brigade of Partisans.

Aft^{r}.Orders

One Subaltern 1 Sergeant 2 Corporals & 20 privates to embark this morning on board the Brigg Commet Capt. Allen, with their Blanketts & 20 rounds p^{r} man, the Quarter master to Issue out Ammunition Emediately [436]

NB Lt Mazyck & the above party marched off at 6 OC: this evening to the Exchange warff, but no boats being provided he return'd [437]

Orders 8th Nov^{r}. . . . Parole Rutledge . . . by G. Howe

For the day tomorrow Captn. Harleston

Magazine Guard L^{t}. Eveleigh

Town G__ L^{t}. Burke__ Reg^{t}. G__ a serg^{t}

A court martial was order'd today, but could not get members__

NB Lt Mazyck with his party Embarked on board the Commet at half after ten OClock this morning

---

Orders 9th Nov^{r}. . . . . Parole Congress . . . . by the presed^{t}.

For the day tomorrow Captn. Shubrick

Magazine Lt Hall

Town G__ L^{t} H Gray__ Reg^{t}. G a serg^{t}.

It is Expected that the Off^{rs}. will not Leave their Quarters till they have rec^{d}. the Orders of the day, if they have pressing Occasion to do so, they are desired to leave in Wrighting where they may be found

The court martial Ordered to Sett yesterd^{y}. Captn Horry presed^{t}. & Lieut^{ts}: Hall & Burke members is order'd to sett today

Genl Orders by ~~Howe~~ Genl Howe

The Honorable the continental Congress having been pleased to enter the foregoing resolution in favour of the Off^{rs}. & Soldiers of the army. Commanding Off^{rs} of all Regim^{ts}. & Corps are as immediately as possible to have these read to their men Company by Company

By Coll: Motte **Ch^{s} Town 1776**

In Congress October 7th 1776

Resolved that as a further Encouragement for Gentl^{m}. of abilities to engage as Commissioned Off^{rs}. in the Battalion to be furnished by the several states to serve during the war. their Monthly pay be increased as follows

A Collonel 75 dollars,__ Lt Coll: 60 dollars__ Maj^{rs}: 50, Capt^{ns}. 40, a Lieut: 27__ a Quarter Master 27½__ Adjutant 40__

Octob^{r}. 8th__

---

436 Edward Allen was the captain of the brigantine *Comet*. The *Comet* carried eighteen 6-pounders. On December 22nd the Royal Navy ship *Daphne* fought the *Comet* near Cuba and captured her.

437 Governor Rutledge wrote, "Two armed vessels from St. Augustine infest the coast. A detachment from the Second Regiment is therefore ordered on board the Brigt *Comet*. You will be pleased to give Capt Allen orders to proceed with them in her with the utmost dispatch, on a cruize for a fortnight, scouring the coast from this port to St. Augustine in conjunction with the *Defense*." The detachment consisted of a Lieutenant, a Sergeant, two Corporals and twenty-eight privates of the 2nd South Carolina Regiment.

Resolved that for the futer encouragement of the non Commissioned Offrs & soldiers who shall engage in the service During the war, a suit of Cloath be given each of the sd. non Comm: Offrs. & Soldiers to Consist for the present year of two linning huntg. shirts, two pair of overalls a Hatt a Leathern cap two shirts two pr hose two pr shoes, amounting in the whole to 20 Dollars, or that be paid to each soldier who procure those articles for himself & produce a Certificate from the Captn. of the Company to which he belongs, to the pay master of the regemt.

Extract of the Minutes __ Chs Thompson Sec [438]

A Garrison court Martial to be held tomorrow for the tryal of Robt. Potts of Captn Charnocks Compy in the Regt. of Coll: Motte Confined by Lt Farrar for Offering to Inlist in another corps the Lt will produce the evidence,[439] the Majrs. Brigade will have the Offrs. warned according to detail; & the court to sett according to the Continental Articles of War

Aftr. Regt. Orders

The Resolution of the Honorable the Continental Congress of the 7th Octr last to be read at the Head of the regt. this AftrNoon by the Major

By Coll: Motte **Chs Town 1776**

Orders 10th Novr ... Parole Lynch .... by the presedt

For the day tomorrow Captn Charnock

Magazine Guard Lt. Baker

Town guard Lt. Hart__ Regt. G__ ~~Lt~~ a Sergt

Divine Service will be performed as usual the Regt. to attend

Orders by Genl. Howe

The Garrison court martial when orderd yesterdy the Genl did not recollect it was Sunday that court therefore is to Sett tomorrow Coll: Hugers Regt. to furnish two Subalterns for this duty & those from the 2nd Regt__

Captn. Horry president

Lts. Ths. Moultrie __ Dubose __ Eveleigh & Mason Members

Lt Dubose of 5th Regt __

Orders 11th Novr.... Parole Heyward ... by Genl. Howe

For the day tomorrow Captn Horry

Magazine Guard Lt T. Moultrie

Town guard Lt. Mason__ Regt G__ a sergt

The Commanding Offrs of Compys to give in their pay bills the Fryday preceeding Payday, Such of the Offrs. as are not provided with the Continental Articles of War will apply to Coll: Motte tomorrow fore noon

---

[438] Charles Thompson was the secretary to the Continental Congress. When the Declaration of Independence was adopted by the members of the Continental Congress there were only two signatures placed upon the 24 draft copies. This was of John Hancock, President of Congress, and the other was of Charles Thompson, secretary.

[439] Field Farrar had been a lieutenant in the 3rd South Carolina (Ranger) Regiment since 1775. When the 5th South Carolina Regiment activated he transferred to that unit. He was promoted to captain on 18 December 1778. He was wounded in the siege of Savannah on 9 October 1779. He was taken prisoner at the surrender of Charleston on 12 May 1780, and was paroled later that year. He retired on 1 January 1781. He died in 1796.

Orders 12th Novr . . . Parole Armstrong . . . by the presedt:

For the day tomorrow Captn Harleston

Magazine Guard Lt Burke

Town guard Lt Eveleigh __ Regt G__ a sergt.

A Regtl. court martial to sett 11 OC this morng. & Captn Harleston Lt Burke & sergt Laurence to attend as Evidences against the prisoners

Captn Motte President.

Lts. Thos Hall & H Gray members

By Coll: Motte **Chs Town __ 1776 _**

Orders by Genl Howe

The Garrison court martial which sett yesterdy. for the tryal of Chs. Hamilton & Robt Potts boath of the 2nd Regt. having found them Guilty of the Crimes Laid to their Charge & Sentence the first to receive 100 lashes & the latter 50, on their bare backs the General Approves of & ratify the Sentences, with this reserve that Potts shall receive only 30 lashes & Hamelton 60, the Genl. mitigate the Sentence upon the consideration that they are the first offenders that have been try'd for Inlisting & Indeavouring to inlist in a Battalion while they where soldiers in another, This lenity no Soldiers for the futer is to expect as they may be Certain should they commit this hanious crime that the Sentence of any futer court martial however severe it may be shall be executed to the utmost rigour__ the above sentence to be executed at such time & place at Coll: Motte shall appoint__ the court is dissolved __

Orders 13th Novr . . . Parole Moultrie . . . . by G. Howe

For the day tomorrow Captn Blake

Magazine guard Lt H Gray

Town Guard Lt Hart__ Regimentl. G__ a sergt.

Mr Henry Perreneau Junr. is Apointed a 2nd Lieutt. in the Second So Carolina Regt. of foot, & is to be Obey'd as such___[440] Lt Perreneau is to join & do duty in Captn P Horry"s compt.

NB by Sentence of last Regt Court martial Corpl Campbell for drunkness mult 14 days pay- Sergt. Henderson for the same crime mult 14 days pay__ John Keating for theft recd 80 Lashes [441]

NB According to Sentence of the Last Garrison court martial Robt Potts for offering to inlist in 5th Regemt. recd 30 Lashes & Chs Hamilton for inlisting in Sd Regt. recd 60 lashes & put under Stoppages of 3/9 pr day till 12 dollars he recd. is reimburse Coll: Huger

---

[440] Henry Perronneau became a second lieutenant in the 2nd South Carolina Regiment on 12 November 1776 in Captain Horry's company. He was promoted to first lieutenant on 9 November 1777 in the Grenadier Company. He was transferred to Captain Charnock's company on 19 April 1778. He was court-martialed for disobedience of orders and neglect of duty in July 1778. Colonel Motte was willing to drop the charges if Lieutenant Perronneau would resign his commission. Lieutenant Peronneau resigned on 15 July 1778.

[441] John Keaton.

Orders 14th Novr Parole Moultrie by presedt

For the day tomorrow Captn Charnock
Town guard Lt Baker
Magazine Lt Hart
Regimental a Sergeant Guard

NB this day I brought up my Orderly Book for the first time after being a year behind hand

By Coll: Motte **Chs Town __ 1776**

Orders 15th Novr. . . . . Parole Roberts . . . . by Genl Howe

For the day tomorrow Captn. Horry
Town guard Lt Perreneau
Magazine Lt Mason__ Regtl. G__ a Sergt.

The Arms of each Company to be immediately put in good Order the men to be so employ'd all this After Noon __

The Regiment is not to turn out to exercise this After noon

A R: court martial to sett this morning at 10 OC: to try such prisoners as shall be brought before them Evidences to attend

Captn Horry President
Lieutts. Eveleigh & Perreneau members
After Orders

The releiving Captn of the day be punctual is being at the Barracks, by half past nine OC: in the morning & he observe his Orders with great exactness

---

Orders 16th Novr. . . . . Parole New York . . . by the president

For the day tomorrow Captn Harleston
Town Guard Lt Burke
Magazine Guard Lt Eveleigh__ Regtl. a Sergt.

Orders 17th Novr. . . . . Parole Small wood . . . . by G Howe

For the day tomorrow Captn Horry
Town Guard Lt. H Gray

Regimental Orders

For the Magazine guard Lt Hall__ Rl.G.__ a sergt.

Divine Service as usual this aftr noon, The Officers to attend

---

Orders 18th Novr. . . . . Parole Pinckney . . . . by the presedt

For the day tomorrow Captn. Blake
Town Guard Lt Hart

R. Ordr. Magazine Guard Lt Baker__ Regt G a Sergt

The Quarter Master to make a return this day of the number of Muskets in the store and the condition they are in he is also to deliver to the Sergeant of the guard at the Genl Hospital a watch coat taking a receipt for it

---

Orders 19th Novr . . . . Parole Carolina . . . . by Genl Howe

For the day tomorrow Captn Charnock

Town guard Lt. Perreneau

Regt Ordrs: Magazine guard Lt T Moultrie__ Regt G. a sergt.

A court martial to sett this morning at 11 OClock to try such prisoners as shall be brought before them Evid: to attend

Captn Blake presidt.

Lts Dubose and Mason members

Orders 20th Novr. . . . . Parole Huger . . . . by the presedt.

For the day tomorrow Captn Horry

Town Guard Lt Eveleigh

Regl. Ordrs. For Magazine guard Lt Mason__ Regl. G__ a sergt

Any Offrs. who may have been absent with Leave upon their return to the Regiment they are to call on the Commanding Offr. & Acquaint him there with

NB Sentence of last court martial Jams Greenwood for drunkinness mult 14 days pay,[442]Abrah: Carslik for the same crime mult 7 days pay,[443]__ Jno Clements,[444] Em: Lopez [445]

Zach: Lucas [446] for neglect of duty mult 14 day pay each

Jas Hooper recd 60 Lashes for absence without Leave

---

[442] James Greenwood enlisted in the 2nd South Carolina Regiment on 4 November 1775. On 20 November 1776 he was mult 14 days pay for drunkiness. He received 50 lashes on 25 October 1778 for neglect of duty.

[443] Abraham Kerslick enlisted in the 2nd South Carolina Regiment on 20 November 1775. On 20 November 1776 he was mult 7 days pay for drunkiness. He was promoted to corporal on 7 April 1777. Seven months later was demoted to private on 18 November 1777, but within a year he was promoted to corporal again on 19 October 1778. He was promoted to sergeant on 2 October 1779 in Captain Hall's company.

[444] John Clements served as a drummer in the 2nd South Carolina Regiment under Captain Thomas Hall. On 20 November 1776 he was mult 14 days pay for neglect of duty. On 14 December 1776 he received 39 lashes and was mult 14 days pay for neglect of duty. On 22 January 1777 he was acquitted on the charge of misbehavior in town. On 31 March 1777 he was confined for being absent from the command to Georgia, but the charges were dismissed on 2 April 1777. On 13 May 1777 he received 50 lashes for drunkenness and neglect of duty. On 20 May 1777 he received 100 lashes for neglect of duty. He received 50 lashes on 30 August 1777 for being drunk on guard. Just a few weeks later he received 50 lashes on 12 September 1777 for riotous behavior and for beating his wife. Then a month after that he received 75 lashes on 23 October 1777 selling his blanket. He was under Captain Hall at the siege of Savannah in 1779.

[445] Emanuel Lopez enlisted in the 2nd South Carolina Regiment on 4 November 1775 as a fifer. On 20 November 1776 he was mult 14 days pay for neglect of duty. On 12 May 1777 he received 50 lashes for stealing a shirt, stockings, and handkerchief. He also had to pay £9 to the soldier in the 5th South Carolina Regiment that he stole the items from. He received 35 lashes with switches on 11 December 1777 for having possession of a stolen pocketbook. He was discharged on 18 July 1778.

[446] Zachariah Lucas enlisted in the 2nd South Carolina Regiment on 4 November 1775. On 20 November 1776 he was mult 14 days pay for neglect of duty. He reinlisted in the 6th South Carolina Regiment on 1 August 1779. After Savannah the regiments suffered massive losses due to the assault on the Spring Hill Redoubt and he was consolidated into the 1st South Carolina Regiment in February 1780.

Gen[l] Orders 21st Nov[r]. . . Parole Roberts . . . . . by G Howe
For the day tomorrow Capt[n]
Town Guard L[t]. Hall
Reg[l]. Ord[rs]. For Magazine Guard L[t] Burke__ Reg[l]. G__ a sergeant

By Coll: Motte **Ch[s] Town 1776**
A Regimental court martial to sett this morning at 11 OC: to try such prisoners as shall be brought before them Evid: to attend punctually Capt[n]. Horry prese[d][t].
L[ts]. H Gray and Hart members

Orders G Howe } 22nd Nov[r] . . . Parole Armstrong . . . . by the presid[t]:
For the day tomorrow Capt[n]. Blake
Town Guard L[t]. Hart
Magazine L[t] Perreneau__ Reg[tl]: G__ a Serg[t].

Reg[tl] Ord[rs]. All the Off[rs] of duty to attend the parrade this after noon when the Reg[t]. turns out to Exercise

L[ts]. Lesesne & T Moultrie are Appointed Capt[ns] in the 2nd Reg[t]. & to be Obey'd as such

Capt[n] Lesesne to take the Command of the Late Capt: Horrys Comp[y]. & Capt[n]. Moultrie that of the Late Capt[n] Hugers.

L[ts] Baker & Proveaux are Appointed first L[ts]. & are to be Obey'd as Such, Lieut[t] Baker to Join Capt[n] Blakes comp[y]. & L[t]. Proveaux to join Capt[n] Moultries & to do duty in s[d] Comp[ys]. as 1st Lieut[s]. till further orders

M[r] Will[m]. Galvan is Appointed a Second L[t]. in the 2nd Reg[t]. is to be Obey'd as such & to join & do duty in Capt Ashbys Comp[y] till further Orders [447]

The Orders of the 19th Last August to be strictly Obey'd

---

[447] William Galvan was commissioned as a second lieutenant in the 2nd South Carolina Regiment on 22 November 1776 in Captain Ashby's company. He was promoted to first lieutenant on 21 October 1777 in Captain Lesesne's company. He resigned on 25 July 1778. He later became a captain in 1779 and then became a major and an Inspector of the Continental Army from 12 January 1780 to 26 March 1782. In November 1783 he was acting aide-de-camp to General Washington.

A R. court martial to sett at 12 OC: on munday Next to try such prisoners as shall be brought before them, all Evidences to Attend

NB Capt[n] Blake fell Sick & was relieved by Capt Charnock

By Coll: Motte **Ch[s] Town 1776**

Gen[l] Orders by Gen[l] Howe } 23[d] Nov[r].. . . Parole America . . . G. Howe

For the day tomorrow Capt[n]. Moultrie
Town Guard L[t]. Mason
Magazine L[t] Baker . . . Regiment[l]: Guard a Sergeant

Reg[tl]: Orders The taptoo to beat at 8 OClock till further Orders

M[r] Alexander Keith is Appointed a Second L[t]. in the 2nd Regiment & is to be Obey'd as such; L[t] Keith to join Capt Moultries Comp[y]. & do Duty in the same till further Orders [448]

The Reg[tl]. court martial Ordered to sett on munday next is to Sett this day at 11 OClock in fore noon

Capt[n] Blake president

L[ts]. Dubose & Baker members

Gen[l] Aft[r]: Orders by Gen[l] Howe

In Consequence of the promotions of General Gadsden & Gen[l]. Moultrie, the following promotions takes place in the 1st & 2nd Regiments of the S[o] Carolina continental troops Viz[t]:

Lieut[t]. Coll: Cha[s]. Cotesworth Pinckney of the first Reg: to be Coll: of the same, Maj[r] W[m] Cattle to be Lieut. Coll: & Capt[n]. Adam M[c]Donald to be major __

L[t]. Coll: Isaac Motte of the 2nd Reg[t]: to be Coll: of the same Maj[r]: Francis Marion to be Lieut[t]: Colonel & Capt[n] Pet[r] Horry a Major

Agreable to a resolution of the Hon[ble]: the Gen[l]: Assemble of this state the following promotions takes place in the 3rd & 4th Regiments of the Continental troops in this state Viz[t]:

Lieut: Colonel W[m] Thompson of the 3[d] Reg[t] to be Coll: of the same Maj[r]. James Mayson to be Lieut: Coll: & Capt[n] Sam[l] Wise to be major in the same [449]

By Coll: Motte **Ch[s] Town 1776**

Lieut[t] Coll: Owen Roberts of the 4th Regiment to be Collonel of the same, Major Bernard Elliott to be Lieut Col: Capt[n]. Bernard Beekman Major of the same.

---

[448] Alexander Keith was commissioned a 2nd Lieutenant in the 2nd South Carolina Regiment on 23 November 1776 in Captain Moultrie's company. He resigned his commission in the 2nd South Carolina Regiment on 4 December 1776, and then became a second lieutenant in the 5th South Carolina Regiment in January 1777. On 21 January 1778 he became a first lieutenant.

[449] Samuel Wise was born in 1738 in England. He was commissioned a captain in the 3rd South Carolina (Ranger) Regiment on 17 June 1775. He was elected major on 18 May 1776. He was killed at Savannah on 9 October 1779.

The Ranks of the following Officers of the 1st, 2nd, 3^{d}, & 4th Regiments is as follows

Viz^{t}.__		Majors
Col: Isaac Huger	1st Lieut Col: M^{c}intosh	Benj^{m}. Huger 1st
Coll Isaac Motte	2nd Lieut^{t}. Coll: Ja^{s} Mayson	Pet^{r} Horry 2nd
Coll: Wm Thompson	3^{d} L^{t}. Coll: Barnard Elliott	Ad^{m} McDonald 3^{d}
Col. Owen Roberts	4th Lt Col. Fran^{s} Marion [450]	Saml Wise 4th
Col: Ch^{s} Coteswth. Pinckney	5th Lt Coll Wm Cattle	Bernard Beekman 5th

The following Rations allowed to the several officers on the Staff of the Army of the United States not heretofore Settled

Deputy Quart^{r} Mast^{r} General __ Deputy Pay Master General __ Deputy Commisary Genl __D. Adj^{t}. General __ D. Muster Master Genl __ D. Judge Advocate General ___ Each six rations ~~a piece~~ per Diem.

Chaplains __ Regimental Surgeons __ Each 3 rations, Surgeons Mate 2 rations
Maj^{r} Generals 15 rations __ Brigad^{r}. Generals 12 rations__Colonels 6 d^{o}
Lieut^{t}: Col: 5 rations Maj^{rs}. 4 __ Captn 3 __ Subalterns & staff Off^{rs} 2 Each p^{r} Diem __

The Honorable the Continental Congress have Ordain'd the following Oath to be taken & subscrib'd by all Officers in the Service of the United States who now hold or hereafter shall hold a Commission or Office from Congress Viz^{t}.

I FM do Acknowledge the Thirteen United States of America, Namely New Hampshire, Masachusets Bay, ~~Johnson~~ Rode Island, Connecticut, New York, New Jersey, Pensylvania, Delaware, Maryland, Virginia, North Carolina, S^{o} Carolina and Georgia, to be free Independant & Sovereign States, and Declare that the people thereof owe no allegiance or Obediance to George the third King of Great Britton, & renounce refuse & abjure any Alligeance and Obediance And I do swear that I will to the Utmost of my power, Support, Maintain and Defend the said United States against the S^{d} King George the Third and his heirs Successors, & his and their abetors, Assistant and adherents, and will serve the s^{d}. United States in the Office of L^{t} C__ which I now hold by their Appointment or under their Authority with fidelity and honour, according to the best of my skill & understanding, So Help Me God __

Commanding Officers of all Regiments & Corps are as soon as possible, fix on some day when their men shall be drawn out under Arms & their Officers orderd to Attend in their places, at which time and in the presence of their men, the S^{d} Declaration & Oath shall be subscribed & taken, first by the field Officers and then by all other Officers, and that the S^{d}. Oath may be Administered with proper Solemnity, the Commanding Officers are to apply to his Excellency the president requesting him to appoint some magistrate duly Qualify'd to administer the same and that all Officers who being absent on duty or otherways, may not be able to attend on the day Appointed, are immediately to be and take the S^{d} Declaration and Oath, in the presence of a field officer two Captains and two Subalterns

All guards is to turn out & receive his Excellency the President with Rested arms, also to the General Commanding in Chief__ all other Generals with shoulderd arms__ Centries to rest their firelocks to all field officers

---

[450] This is not correct. Francis Marion did not transfer to the 4th South Carolina, but stayed in the 2nd South Carolina as the lieutenant colonel.

This order is not intended to restrain guards merely Regimental from paying such honours to their own off^{rs}. as the Commanding Off^{rs} of the Reg^{ts}. may direct

The Capt^{ns}. of the day to visit all guards & all reports to be made through him till further orders

By Coll: Motte **Ch^{s} Town 1776**

by the presid^{t}

Genl Orders by Gen Howe } 24th Nov^{r} . . . . Parole Bullock . . . .

For the day tomorrow Captn. Motte

Town guard L^{t}. Burke

Magazine Guard L^{t}. Eveleigh

Reg^{l}. Orders___ Regimental guard a sergeant

The Regiment to attend divine Service this Afternoon as usual__ the Officers are desired to be present_____ _____ _____ _____ _____

Genl. Orders by Genl. Howe

The Main guard to be reinforc'd with six men tomorrow, two Centries to be fixed at his Excellency the Presidents door, the men to reinforce the main Guard to be taken by detachment from 2nd & 5th Regiments in proportion to their strength

Genl. Howe Setts out for georgia to morrow, he strongly recomends to Commanding Officers of Battalions to have their men frequently trained to ye Exercise of spears & to the Soldiers to be attentive in Learning what in course of service may so effectually contribute to their honour & safety, he is Obliged to the Officers of every department for their attention which he with pleasure has observed they pay to their duty & takes this opportunity to express his Approbation of the orderly behavior of the soldiers which he hopes a continuance

The Important & beneficial publick work General Gadsden has undertaken & he is so happyly executing on Sullivants Island requiring all his Attention he has desired to be confind all his Command for the present to this work & the Island__ the Command therfore of the town & out post in the absence of Genl. Howe will devolve on General Moultrie, til Genl Gadsden Chuse to Assume it

Coll: Motte having represented to Genl Howe that James Kelley now under confinement for Desertion has one Curcomstance in his favour which in some measure mitigates his Crime that is that he had surrendered himself to one of his corporals & the Colonel having Compassionatly solicited his pardon,[451] Genl Howe in respect to Col: Motte will for once deviate from a resolution he had fixt never to pardon a deserter and consents to pardon S^{d} Kelly upon this condition that of inlisting in his Battalion During the war & by futer good behavior he promises to attone for the henious Crime he has commited so contrary to all duty & to the Solemn Oath he has taken, he is to do duty in Col: Mottes Reg^{t}. till the arrival of Col: Thompson; Lest this lenity shoud have a bad effect the Genl warn all soldiers against desertion which he now declares he never will again pardon on any condition

This order to be read to the men on parrade

---

[451] James Kelley was in the 3rd South Carolina Ranger Regiment. Though Colonel Motte gave Kelley extraordinary leniency in this instance, Kelley deserted again a few days later with William Simpson of the 2nd South Carolina. He was captured and returned on 29 November 1776. He received 100 lashes for desertion. He was in jail for deserting in April, but he escaped the jail with Josiah Stone.

Orders 25th Novr... Parole Cicely ... by Genl. Moultrie

For the day tomorrow Captn. Motte
for town Guard Lt H Gray
Magazine Lt Hall

Regt Orders__ Regimental Guard a sergt.

Danl Holiday of Captn McDonalds Compy. is Appointed a sergeant in the same & is to be Obeyd as such [452]

A court martial to sett this morning at 12 OClock to try such prisoners as may be brought before them evidences to attend

Captn Harleston Presidt.

Lts Gray & Baker members

NB according to Sentence of Last court.

Wm Simpson for being Drunk on guard & stabing 3 negroes sentenced to receive 99 lashes & Mult 14 day pay__ Jno Crawfor for absents for guard recd. 100 Lashes [453]
Jams Alwell for do recd 50 Lashes & mult 7 days pay, Bunker Tring & Jas Hunter for striking Sergt. Simpson & drunkinness multed each a weeks pay

---

Orders 26th Novr... Parole Moultrie ... by the presedt.

For the day tomorrow Captn Blake
Town guard Lt. Perreneau
Magazine Lt Hart

Regtl. Ordrs. Regimental guard a sergt.

The Orders of the 9th Instant, to be comply'd with, with strictness the commanding Officer hopes the Officers will not give Occasion to repeat it again

The Regtl: court martial Order'd yesterdy. & did not meet for want of members is to sett this Morning at 10 OC: precisely to try such prisoners as shall be brought before them Evidences to attend punctually

Captn Motte president
Lts. Dubose and Baker members

By Coll: Motte **Chs Town 1776**

Orders 27th Novr... Parole Fabius

For the day tomorrow Captn Moultrie
Town guard Lt Baker
Magazine__ Lt Mason

---

452 Daniel Holladay was born in 1752. While residing in the High Hills of the Santee he enlisted when South Carolina's troops were first organized on 4 November 1775. He was promoted to orderly-sergeant in the 2nd South Carolina Regiment on 25 November 1776. He served under Captain James McDonald and was in the battle of Fort Sullivan on 28 June 1776. On 8 August 1777 he was reprimanded for gambling. He was reprimanded on 3 April 1778 for neglect of duty and discharged three days later. He later moved to Alabama and died on 14 February 1837.

453 John Crawford enlisted in the 2nd South Carolina Regiment on 4 November 1775. On 25 November 1776 he received 100 lashes for being absent from guard. On 26 May 1777 he received 50 lashes for beating Mary Burgess. He was reprimanded on 24 June 1777 for being absent from roll call. He was in Captain Hall's company in 1779. He was listed as missing after the assault on the Spring Hill Redoubt in Savannah on 9 October 1779.

Reg^{tl}. Ord^{rs}.   Regimental guard a Serg^{t}.

The Genl Order of the 24th Inst to be read at the head of the Regiment by the Adjutant the first time it turns out

James Kelly of Col^{o}. Thompson Reg^{t}. is hereby orderd to do duty in Captn Charnocks Comp^{y}. till the arival of the regiment he belongs to arrive in town

The Centries of the different guards do always carry their Arms & front any Officer which may Approach them__ this order the Officers of guards to do their utmost to have Strictly complyed with

NB according to last court martial And^{w} Blan of Captn Charnocks Comp^{y}. for riotous behavour mult 14 day pay [454]

Fran^{s} Pickering for being in town at midnight mult 7 days pay [455]

Robert Lance for drunkenness mult 14 days pay

Auger Gaurley for being in town at midnight mult 14 day pay [456]

Orders 28th Nov^{r} . . .   Parole Howe . . .   by the presed^{t}.

For the day tomorrow Captn Harleston

Town guard L^{t}. Eveleigh

Magazine L^{t}. Burke

Reg^{tl} Ord^{rs}.__ Regimental Guard a serg^{t}.

Commanding Off^{rs} of Comp^{ys}. are desired to send in their acc^{ts}. tomorrow by 12 OC: to the Adjutant for making the Hunting shirts for the Regiments__

A court martial to sett tomorrow at 11 OC: to try such prisoners as shall be brought before them Evidences to attend

Presid^{t}. Captn Moultrie

L^{ts} Hall and Burke members

---

454 Andrew Bland (Blan, Le Bland, De Bland, Debland, Deblong) enlisted in the 2nd South Carolina Regiment on 26 October 1776 in Captain Charnock's company and re-enlisted on 31 December 1778. On 27 November 1776 he was mult 14 days pay for riotous behavior. On 14 December 1776 he was reprimanded for threatening two soldiers. On 1 January 1777 he was released from the charges against him due to lack of evidence. On 24 June 1777 he was supposed to receive 99 lashes and be picketed for desertion, but he was pardoned. On 11 August 1777 he received 50 lashes for disobedience of orders. Afterwards he escaped into town and did "riotous behavior." On 12 August 1777 he received 100 lashes, at four different times for this offense. He received 50 lashes on 18 September 1777 for being absent from guard. On 6 March 1779 he was supposed to receive 100 lashes with a cat of nine tails for desertion, but 40 of these were remitted by the commander. He was in Captain Mason's company in 1779. On 4 April 1779 he received 35 lashes on the bare back with a cat of nine tails for being absent without leave.

455 Francis Pickering enlisted in the 2nd South Carolina Regiment on 11 March 1776. On 27 November 1776 he was mult 7 days pay for being in town at midnight.

456 Austin Gourley enlisted in the 2nd South Carolina Regiment on 4 November 1775. He was mult 14 days pay for being in town at midnight with Francis Pickering. He deserted on 2 December 1776, but was returned and was mult 7 days pay.

After Ord^rs. by Genl Moultrie

One Sergeant, 1 Corp: 10 privates from the 2nd and 5th Reg^ts. to march up to the magazine at Pritchards ship yard tomorrow morning there to take post In some Empty house [457] & post 2 Centries on the Magazine & not to lett any persons go into it after the powder is placed there without an order from the president or commanding officer

This guard to take two days provisions with them & to be relieved every forty Eight hours

NB proportion, 1 Serg^t & 6 privates from 2^d Reg^t.
1 Corp^l & 4 privates from 5th d^o.

Orders Genl. Moultrie } 29th Nov^r . . . Parole Capua . . . by Genl Moultrie
For the day tomorrow Captn. Motte
Town Guard L^t. Hall
Magazine__ Lt Hart

For the futer all guards (except Regemental) are to parade on the green before M^r. Garrards House till further Orders

Officers for the Different Guards by detachments are to be appointed in proportion to their numbers fit for duty

NB M^r Ch^s. Coadwell brought two deserters Kelley of 3^d Reg^t.
W^m. Simpson of the 2nd Reg^t.

Orders 30th Nov^r . . . . by Genl Moultrie Parole Marcelus . . . . by Genl Moultrie
For the day tomorrow Captn Blake
Town Guard from Col: Hugers Reg^t. [458]
Magazine L^t H Gray
For Magazine at ShipYard Lt Perreneau & a Corp^l. & 8 men

One Subaltern 1 serg^t. 1 Corp^l. & 15 privates to relieve the guard near the ship yard tomorrow, this guard to be relieved as before described

The Second Reg^t. to remove to the new Barracks on Munday next more wood to be allowed that constant fires may be Kept in the different rooms that they may be the sooner dry'd & mad more wholsome for the men

Regimental Orders by Lt Col: Marion ____

The Quart^r. Master to Appoint the rooms for the new Barracks for the different companies of y^e Reg^t. & furnish a Cord of wood to the men who will be sent there, this to be done Emediatly he is also to provide wagons to carry the mens Bagggage to New Barracks on Munday next

two men out of each company to be sent Emediatly to the New Barracks to take possession of the rooms which the Quart^r. master app'd for their Comp^y. & to keep a constant fire in the rooms to dry them as soon as possible__ NB the room next the S^o. end for store & Guard room the end above the store for the Officers of the Guard

---

[457] This is the magazine at Pritchard's Shipyard. This guard took two days provisions with them and was changed every 48 hours.

[458] 5th South Carolina Regiment.

By L[t] Col: Marion **Ch[s] town 1776**

Orders 1st Decem[r] . . . . Parole Syracuse . . . . by G Moultrie

For the day tomorrow Capt[n].
Town guard L[t]. Mason
Magazine L[t]. Baker

Regem[tl]. Orders

The Regiment to attend devine service this aft[r] noon as usual__

A Regimental court martial to sett tomorrow at 10 Cl: in morning to try such prisoners as shall be brought before them Evidences to attend. Capt[n]: Moultrie president
L[ts]. Hall & Burke __members

The Regiment to hold themselves in readyness to remove to the new barracks tomorrow at 4 OClock in After Noon, at which hour all officers & soldiers not on duty is to attend, the quarter master to have the Baggage Wagons ready to move with the Regiment

One Corporal & 4 men to Remain on Gadsden Warff as a guard to the Battery, and to post one Centry in it

---

Orders 2nd Decem[r] . . . . Parole Gadsden . . . . by the presid[t]
by Gen[l] Moultrie

For the day tomorrow Capt[n] Motte
Town guard L[t] Eveleigh
Magazine L[t]. Burke
Ship Yard L[t]. from 5th Reg[t].

As the Soldiers who have been brought up to the sea Impose themselves on Masters of vessels for Sailers not belonging to the Army every Soldier for the feutar must ware his Regimental Cap when Out of Quarters, that Masters of Vessels may know them to be Soldiers, those Regiment that have no caps to their Uniform, must distinguish themselves by wearing feathers or a piece of Bear skins in their Hatts ~~to distinguish them~~

Regim[tl]. Orders by Coll Motte

A monthly return to be given in by tomorrow 10 OC: to the adjutant

Arnold Drummer in Capt[n] Moultries company is appointed Drum Major to the Reg[t]. & is to be Obey'd as such [459]

By Coll: Motte **Ch[s] Town 1776**

Orders 3[d] Decem[r]. Parole Carthage . . . by G Moultrie

For the day tomorrow Capt[n] Blake
Town guard L[t] Hart
Magazine L[t]. H Gray
Ship Yard L[t] Hall

Reg[tl]. Ord[rs]. For Regimental guard at New Barrack 1 Serg[t]. & 12 men 1 Corporal and 4 men at Battery on Gadsden Warff

---

[459] James Arnold enlisted as a drummer in the 2nd South Carolina Regiment on 29 January 1776 in Captain Moultrie's company. On 14 December 1776 he was reprimanded for some unnamed offense. He was appointed as the drum major on 2 December 1776. He was discharged on 14 July 1778, but he reinlisted on 16 September 1779 when the British had invaded Georgia.

The Regiment to be under Arms this After noon at half past 3 OC: in After noon to be drawn up in front of the Barracks the Commanding Officer expects to see the Arms of the different Companies well cleaned & in good Order & the officers take some pains in sizing their men properly__ No man to be Absent that can possible turn out__ the Officers to appear in their compleat Regimentals, half Gaiters & gorgets the staff Officers to be present, this Order to be sent to them by the Adjutant

The Declaration and Oath to be Subscribed & taken this After noon by all the Officers belonging to the Regiment, that are off duty, Agreable to the Generals Order of the 23[d] of Last Month

The Quart[r] Master with all the Fatigue men to Clean out the Barracks on Gen[l] Gadsden Warff this fore noon and to deliver the Keys to the Warfenger [460]

A Reg[tl]. court martial to sett in the Officers Guard Room this morning at 11 OC: to try such prisoners as shall be brought before them Evidences to Attend

Capt[n] Motte presed[t].

L[ts] H Gray and Hart members

NB Six deserters were brought Back yesterd[y] Aft[r]. Noon by Sergeant Brown [461] Viz: W[m] Maxey,[462] Nesmith,[463] Rowlin Thomas,[464] W[m] Collons,[465] Wilkerson,[466]

---

[460] A warfenger or Wharfinger is a person in charge of a wharf or dock. He handles the administrative details of ship and cargo movement on the dock.

[461] There were numerous men named Brown in the South Carolina regiments, and there were three William Brown's in the 2nd South Carolina alone who were sergeants in 1776. I am unable to determine which Sergeant Brown this is.

[462] William Maxey enlisted in the 2nd South Carolina Regiment on 4 November 1775. He deserted sometime in 1776 and was captured then returned on 3 December 1776. He was pardoned for his offense.

[463] John Nesmith enlisted in the 2nd South Carolina Regiment on 4 November 1775. He deserted sometime in 1776 and was captured then returned on 3 December 1776. He was supposed to receive 100 lashes but it was remanded. On 27 December 1776 he was reprimanded for some unnamed offense. After the fall of Charleston in 1780 he served in Marion's Partisans from 30 December 1780 to 24 March 1782.

[464] Thomas Rawlins (Rowan Thomas, Rowland Thomas) was born on 12 March 1750 in North Carolina. He enlisted in the 2nd South Carolina Regiment on 4 November 1775 under Captain William Mason. He fought in the battle of Sullivan's Island in June 1776. He deserted sometime in 1776 and was captured and returned on 3 December 1776, for which he received 100 lashes six days later. On 24 April 1777 he received 100 lashes for neglect of duty. He served under Captain Shubrick until Shubrick's death. He served in the 2nd Vacant Company until the 2nd Vacant Company was disbanded on 5 October 1778. He was then assigned to Captain Charnock's company. He was in Captain Mason's company at the siege of Savannah. He was later in the militia under Captain William Nettles and Colonel Baxter. During 1780 he was under Captain Joseph Greaves. In 1781 he was under Captain Valentine Rowell.

[465] William Collins enlisted in the 2nd South Carolina Regiment on 4 November 1775. He deserted sometime in 1776 and was captured and returned on 3 December 1776 for which he received 100 lashes six days later with Thomas Rawlins. He deserted again and was returned on 3 May 1777. He was court martialled and he received 99 lashes and he also had to run the gauntlet. He was in the regiment at the fall of Charleston in 1780.

[466] William Wilkinson served in the 2nd South Carolina Regiment from 1776. He deserted sometime in 1776 and was captured and returned on 3 December 1776. In 1778 he was in the 2nd Vacant Company. He deserted but was apprehended in April 1778. When the 2nd Vacant Company was disbanded on 5 October 1778 he was assigned to Captain Lesesne's company. On 15 June 1778 he received 100 lashes on the bare back with a cat-o-nine tails for desertion. In 1779 he was with Captain Dunbar's light infantry at the siege of Savannah. In

Wane & Snow [467]

By Coll: Motte **Ch^s Town 1776**

Orders by G Moultrie

One Sub. 1 serg^t. 1 corp: & 15 privates from the 2nd & 5th Reg^t. to mount guard at the New Barracks & to take charg of the Prisoners of war there

The Captn. of the day to stay constantly at the new barracks unless when he visits the different Guards

---

Orders 4th Decem^r. . . Parole Thompson . . . . by the presed^t

For the day tomorrow Captn Charnock
Magazine Guard Lt Perrineau
Town Guard Lt Dubose
at Ship Yard H Gray __ a Corp^l at Gadsden Warff

R Ord^rs.

Lieut^t Keith having resigned his Commission with Leave is no Longer to be respected & obey'd as an Off^r. in the Reg^t.

Such of the Officers as did not subscribe & take the Oath Yesterday, Ordered by the Genl. to be taken by all the Continental Officers, is to do it this Afternoon, when the Regiment turns out to Exercise - Lieut^t. Col: Marion is Desired to see it done Accordingly

NB by sentence of Last Court Martial Jno Dunn for Absents sentenced to 100 Lashes, pardoned [468] Austin Gourly mult 7 days pay - Shedrich Williamson for Leting a prisoner escape 100 Lashes [469] Pardoned - Wm Simpson for breaking confinement rec^d 50 Lashes - Joseph Siles for Absents Pardoned

Jno Alex^d. Simmons for suspicion for desertion reprimanded [470] Henry Page for absent without Leave received 50 Lashes

---

1781 he was in the 2nd South Carolina Dragoons under Captain Isaac Ross, Lieutenant Colonel Myddleton and General Sumter.

467 William Snow was in the 2nd South Carolina Regiment. He deserted sometime in 1776 and was captured and returned on 3 December 1776 for which he received 50 lashes five days later, with Rawlins and Collins. He later served as a captain of the militia.

468 John Dunn enlisted in the 2nd South Carolina Regiment on 4 December 1775 under Captain Charles Motte. He was in the hospital for wounds on 10 December 1775. These wounds may have been inflicted during the fighting between the South Carolina schooner *Defense* and the Royal Navy sloops of war *Tamar* and *Cherokee*. On 4 December 1776 he was court-martialed for being absent without leave and was sentenced to receive 100 lashes, but he was pardoned.

469 Shadrack Williamson was in the 2nd South Carolina Regiment. He was sentenced to receive 100 lashes on 4 December 1776 for letting a prisoner escape, but he was pardoned. After the fall of Charleston in 1780 he was a lieutenant in Marion's Partisans under Colonel Benton.

470 John Alexander Simmons was in the 2nd South Carolina Regiment, but he had been born in Virginia. On 4 December 1776 he was reprimanded for suspicion of having deserted. He ws sentenced to receive 99 lashes and be picketed for desertion, but he was pardoned on 24 June 1777. After the fall of Charleston he served under Colonel Peter Horry in Marion's Partisans. At some point he was in the North Carolina militia and was wounded in the Battle of Cowpens. He died in 1837.

Orders 5th Decem^r ... Parole Rome .... by G Moultrie

For the day tomorrow Captn. Charnock
Magazine Guard L^t. Baker
Town G __ Lt Shubrick
Prisoners G. at New Barracks Lt Mason

By Lt Col: Marion **Ch^s town 1776**

Agreable to Genl Orders of 2nd Inst for every soldier to wear their Regimental caps - Commanding Officers of Companies is desired to apply to the quart^r Master for caps for their men giving a receipt for the same

Tis expected the Officers will comply with the former Orders if making a weekly return of their mens Arms and Accoutrements & Cloathing, & be particular in confining all who may have lost or made away with any part of them

The Quart^r. Master to make a weekly return to the Barracks Master of the wood wanting for the Regiment -

As there has been Great Irregularity hitherto in serv^g. wood to the men, It is Ordered for the futer that they shall be served with that Article by Companies beginning by the Granideers - Any Soldiers who take any wood brought to the Barracks without Leave shall be tryed by a court martial ~~& Suffer~~ for Disobediance of Orders & Suffer Accordinly -

Non commissioned Officers to see their men regularly served with wood as they shall answer for the Neglect

---

Orders 6th Decem^r .... Parole Laurens .... by the presed^t.

For the day tomorrow Captn Moultrie
Town Guard L^t. Shubrick
Magazine Lt Baker
Ship Guard L^t from 5th Reg^t.

Reg^tl Orders __ Regimental Guard a Sergeant

Jno Cantey Corp^l. of Captn. McDonalds Comp^y is appointed a sergeant in the same & to be Obey'd as such [471]

A Regimental Court Martial to sett tomorrow forenoon at 10 OClock to try such prisoners as shall be brought before them

By Lt Col: Marion **Ch^s town 1776**

Captn T Moultrie president
Lieut^ts. Eveleigh and Hall members

Orders 7th Dec^r ... Parole Alexandria ... by Genl Moultrie

For the day tomorrow Captn Ashby
Town Guard L^t P Gray
Magazine Lt Eveleigh

---

[471] John Cantey enlisted in the 2nd South Carolina Regiment as a corporal on 4 November 1775. He was promoted to sergeant in Captain McDonald's company on 6 December 1776, but was discharged three months later on 28 March 1777.

Reg^{tl}. Ord^{rs}. Regimental Guard in Barracks a serg^{t}. on Warff a Corp^{l}.

The Quarter master to make a return today of the number of Cartridges now made & what Quantity of flints, Balls, Lead and Cartridge paper that are on the store; & to put Six men emediately to make Cartridges, he is to take such men out of the Regem^{t}. as understand making them __ and to have an Amunition Chest made

The Quarter master also is to have the yard before the mens barracks cleaned as fare as the pump, & all round the Barracks also to have Necessary houses made for the men & one for the Officers in most convenient place, its expected this will be done this day

---

Orders 8th Decem^{r}. . . . Parole Washington . . . by the presed^{t}

For the day tomorrow Captn Blake
Town Guard L^{t}. from 5th Reg^{t}.
Magazine L^{t}. Burke
Guard at Pritchard Lt Hall

Reg^{tl}. Ord^{rs}. Regimental Guard a serg^{t}. on the warff a Corp^{l}.

Divine Service as usual, at new Barracks where it is expected the officers will attend

---

Orders 9th Decem^{r}. . . . Parole Antioches . . . by Genl Moultrie

For the day tomorrow Captn Shubrick
Town Guard Lt Dubose
Magazine Lt Hart
Pritchards G__ from the 5th Reg^{t}.

By Coll: Motte **Ch^{s} Town __ 1776**

Genl Ord^{rs}. by Genl Moultrie

One Subaltern 1 Serg^{t}. 1 Corp^{l}. & 15 privates from the 2nd & 5th Regiments to go over to Hobcaw emediately there post themselves at the magazine as a guard to a Quantity of powder Lodged there, not to allow any persons to go into the magazine without an Order from the presed^{t}. or some General Officer __ this Guard to be relieved every week [472]

Reg^{tl}. Orders by Coll Motte __

A court martial to sett tomorrow 10 OC: to try such prisoners as shall be brought before them Evidences to attend

---

472 In 1770 the South Carolina Legislature wanted to build a ferryboat landing at Hobcaw's Point to connect Charleston with the Georgetown post road. Clement Lempriere, a privateer captain who owned the land, refused the Assembly's request. Edward Rutledge made a deal with Lempriere where the Assembly would build a powder magazine at Hobcaw's Point on land bought from Lempriere, in exchange of allowing the ferryboat to be built. Charleston got its ferry, but it also got a magazine placed in a terrible location. The magazine was built in the wilderness not anywhere near any other military installation in Charleston. Any powder stored there would have to be transported 5 miles over very poor roads. To move powder from this magazine to Charleston some unlucky volunteer would have to make a journey in an open boat across the Cooper River, being totally at the mercy of any accurate enemy artillerymen. No fortifications were ever built that were serviced by the powder at Hobcaw Point.

Captn President
Lieut Mazyck and Mason members

NB according to Sentence of Last court martial of the 6th Inst Rowland Thomas for absence without Leave rec^d^ 100 lashes
Wm Wilkinson for d^o^ rec^d^ 50 Lashes __ Wm Collin for d^o^. rec^d^ 100 Lashes
Wm Snow for d^o^ rec^d^. 50 lashes __ W^m^. Maxey for the same pardoned
Jno Nesmith Sentenced for d^o^. to 100 Lashes not yett confirm'd remanded
Christopher Gaymond for neglect of duty reprimanded [473]

Orders 10th Decr by Genl Moultrie } ... Parole Moore ... by the presed^t^
For the day tomorrow Captn Charnock
Town guard Lt Mazyck
Magazine Lt Mason
Pritchards from 5th Reg^t^.
one private from 2nd Reg^t^. & 1 do from 5th to attend Wm Debraham to survey the town __

Aft^r^: Genl Ord^rs^.
One Corporal & six men to go on board Captn Hatters [474] ship now lying in Rebellion road,[475] to bring five prisioners a Shore charged with Mutiny __ the Captn will find a boat to convey them to the road

Reg^t^. Ord^rs^. by Col. Motte

The Officers are desired to lodge their Provincial Commissions in the hands of the Major this fore noon, who is to deliver them to the Commanding Officer of the Regiment __

By Coll: Motte Ch^s^ Town __ 1776
Orders 11th Decem^r^. . . . Parole Hobkaw. . by Genl Moultrie
For the day tomorrow Captn Moultrie
Town guard Lt Perrineau
Magazine Lt P. Gray
Barrack a Serg^t^. on Gadsden Warff a Corp^l^.

Orders 12th Decem^r^. . . . Parole Mathews . . . by the presed^t^
For the day tomorrow Captn Harleston
Town guard from the 5th Reg^t^.
Magazine Lt Eveleigh
Pritchards Shipyard Lt Burke
R Orders The Regiment not to turn out this after noon the men to gett their Arms clean & in good Order

---

473 Christopher Gayman (Gamond) enlisted in the 2nd South Carolina Regiment on 4 November 1775 for the duration of the war. He was reprimanded for neglect of duty on 9 December 1776. He was in Dunbar's light infantry at the siege of Savannah.
474 John Hatter. I do not know what his ship was.
475 The northern passage of the port of Charleston was called Rebellion Road, a large ship anchorage near Haddrell's Point. This name had been given to it in 1744 and had nothing to do with the rebellion in 1776.

A Regimental court martial to sett this fore noon at 11 OC. to try such prisoners as shall be brought before them Evidences to Attend

Capt[n] Moultrie president

L[ts]. Eveleigh and Burke

Orders 13th Decem[r]. . . . Parole Tarentum . . . . by Gen[l] Moultrie

For the day tomorrow Capt[n] Motte

Town Guard Hall. L[t].

Magazine L[t]. Dubose

R Ord[rs]

A court martial to sett this morning at 11 OC: to try such prisoners as shall be brought before them Evidences to Attend

Capt[n] Harleston president

Lieut[s]. Hall and Pereneau memb[rs].

NB According to Last court, Jn[o] Griffin for neglect of duty rec[d] 100 Lashes
Ch[s]. Howel Simmons for d[o]. was tyed to the Habert & reprimanded
Fran[s] Ferrel for d[o] reprimanded __ Jn[o] Sullivant[476] for d[o] rec[d] 50 Lashes
Bunker Tring for d[o] rec[d] 100 Lashes

After Gen[l] Orders by Gen[l] Moultrie

Orderd that the Commanding Officers of the Different Corps of Continental troops to forthwith Transmit to their several Officers now out & those to be sent out on the recruiting service the following resolution of the Hon[ble]. the Continental Congress & that they Inlist their men Agreeable to said Resolutions __

By Coll: Motte **Ch[s] Town 1776 _____**

In Congress Nov[r] 12th. 1776

Resolved that all non-commissioned Officers & Soldiers who do not incline to engage their Service during the Continuance of the present war shall Inlist for three years, unless sooner discharged by Congress shall be Intitled to & receive all such Bounty & payment as are allowed to those who Inlist during the Continuance of the present war. Except the one hundred Acres of Land, which Land is to be granted to those only who inlist without Limitation of time & Each recruiting Officers is requested to provide two distinct Rolls one to sign who Inlist during the War. the other for such who Inlist for three years, if their service shall be so long requested

By John Hancock presed[t]

A copy of the above resolution to be sent to the Different posts

---

Orders 14th Decem[r]. . . . Parole Armstrong . . . . by the presed[t]

For the day tomorrow Capt[n].

Town Guard L[t] P Gray

Magazine L[t] Mazyck

Pritchards Ship yard from 5th Reg[t].

---

476 John Sullivant served in the 2nd South Carolina in 1777-1778 in Captain Charnock's company. He received 50 lashes on 13 December 1776 for neglect of duty. He received 50 lashes on 16 April 1778 for being drunk on guard duty. He also served in the militia as a private from 1778 to 1780 and as a quartermaster sergeant from 20 April to 20 December 1780. During 1784 he served as a constable.

5th South Carolina Regiment

Reg[tl]. Orders

The Pay Bills to be delivered to the Pay master immediately & the Commanding Off[r]. hopes the Officers Commanding each Company will not in futer be remiss on this part of their duty in giving them in Agreable to orders repeatedly given __Orderd that an Officer of each Company be always present till all the men are paid

A court martial to sett this morning at 11 OC: to try such prisoners as shall be brought before them Evidences to Attend ___

Capt[n] Motte president

Lieutenants Eveleigh & Gray

NB according to Sentence of yesterd[ys] court martial Jn[o] Heyrne drum[r] was mult a weeks pay & reprimanded [477] __ Ja[s] Arnold for d[o] mult 14 days pay & reprim[d].

Alexes Simanar [478] & And[w] Blan __ for Threating 2 soldiers __ was reprimanded

Jn[o] Thompson for drunkenness & neglect of duty rec[d]. 39 Lashes & mult 7 day pay

Rich[d] Richardson,[479] Rich[d]. Williamson & Jn[o] M[c]Dowell for Drunkenness rec: 39 lashes Each & mult a weeks pay - Jn[o] Clements drumm[r]. for neglect of duty rec[d] 39 Lashes & mult 1 weeks pay

By Coll: Motte **Ch[s] Town 1776 __**

Orders 15th Decem[r]. . . . Parole Moultrie . . . by Gen[l] Howe

For the day tomorrow Capt[n] Blake

Town guard L[t] from 5th Reg[t].

Magazine L[t] Burke

Barrack Guard L[t] Eveleigh __ Regimental G __ a Serg[t].

Commanding Officers of Battallions & Corps to report Emediately to Gen[l]. Howe what Cartridges they have made up, & what number of Boxes to contain them, that more may be Orderd if Nessesary

Commandants of Forts & Batterys also to report what quantity of Cartridges they have for their Cannon, the Quantity of Ammunition & Military stores they may have in their possession, they will make a list of every thing nessesary for their defence which they have not, that it may be provided.

The Quart[r] Mast[r]. Gen[l]. will immediately report the Publick Waggons in town ___ he is to apply to his Excellency the presid[t]. & request a favour in him to give an Exact State of the Arsenal, which he will bring to head Quart. tomorrow morning - As all Corps & Battallions Aught to be furnished an Ammunition Cart, the Quart[r] Mast[r]. Gen[l]. is therefore imediately to employ proper persons to make them- he is to Apply to Commanding Off[rs]. of each Batallion & Corps for a Model, & be very

---

477 John Hyrne served as a drum major in the 1st South Carolina Regiment from 22 June 1775. He enlisted in the 2nd South Carolina Regiment on 4 November 1775 as a drummer under Captain Peter Gray. On 14 December 1776 he was mult a weeks pay and reprimanded for some unnamed offense. In 1779 he was a drummer in Captain Hall's company. He was confined on 25 March 1779 for disobedience of orders. He was sentenced to receive 25 lashes on his bare back with a cat of nine tails, but he received mercy for his former good behavior. He became a drum major on 30 July 1779. He was in the siege of Savannah under Captain Peter Gray.

478 Alex Simmor (Seymour, Semanar) served in the 2nd South Carolina Regiment. He received a reprimand on 14 December 1776 when he and Andrew Bland threatened two soldiers. He received a reprimand on 18 September 1777 for being absent from guard, but he was pardoned.

479 Richard Richardson enlisted in the 2nd South Carolina Regiment on 20 November 1775. He served as a private and then a sergeant under Captain Thomas Dunbar. He was at the siege of Savannah.

carefull they are well built - As it will probably be nessesary to make a great Numb^r^ of Cartridges & as the Caliber of the Musquets may not be equall
Officers of Companies are to see that every man in their Company is furnished with a former exactly suited to his gun.[480]

A Subaltern 1 Serg^t^. 1 Corp^l^. & 15 priv^t^. from the first Reg^t^. is to relieve the Magazine Guard at Hopkaw on Munday next, this guard to be relieved by the first Reg^t^. till further Orders

NB according to Sentence of last court Robt Cole was reprimanded [481]

Orders 16th Decem^r^. . . . Parole Huger . . . By the presid^t^

For the day tomorrow Captn Shubrick
Town guard Lt Hall
Magazine Lt Hart
Barrack guard L^t^ from 5th Reg^t^.

Reg^l^. Ord^rs^. for Regimental Guard a Sergeant _____ _____

By Coll: Motte **Ch^s^ Town 1776 __**

The Quarter Master to make a return to the Comand^g^ Off^r^. immediately the number of Cartridges musquet Balls & Amunition Chests he has in Store

A Reg^tl^. court martial to set this morning at 11 OC: to try such prisoners as may be brought before them; Evidences to attend punctually

Captn Blake presid^t^.
Lieut^ts^. Hall and Hart members

Orders 17th Decem^r^. . . . Parole Lee . . . . by Genl Howe

by Genl Howe } For the day tomorrow Captn Charnock
Town guard L^t^ from 5th Reg^t^.
Magazine Lt Perreneau
Barrack Guard Lt Baker

Commanding Officers of every Batallion, Corps, fort or Battery Either Continental provencial or Militia if in Actual service are to send some Officers to head Quarters every day at 12 OC: to receive Orders, & all Officers of out posts are immediately upon their coming in to Garrison to leave their names at the house of the Commanding Officer for the time being or otherways Acquaint him of their arival and what time they shall return to their posts

An Emediate report According to the orders of Saturday is Expected at head Quarters __

Complaint having been made that great damage has been done to Brigidier Gadsden Building by the guard at his warff, the Officer of that Guard will be made answerable for any futer mischief done during the time of his guard __ the Captn of the day is to inform the officer posted there of this Order, & be particularly Attentive himself to prevent mischief being done or to detect the persons doing it

All Guards in town not merely Regimental are in futer to be mounted by detachment from the different Battallions in Garrison According to their number __ Battalions of musquetry is to Compleat the number of Cartridges to 50 rounds p^r^ man __

---

[480] A former was a wooden dowel that was used to make the cartridges.
[481] This may be Robert Coleman.

Command^{g}. Off^{rs}. are to be carefull that such men are Selected to make them who will do it properly, & such persons Appointed to superintend them as will prevent the waste or embezelment of Amunition and as the Calibers of the Guns may not be alike formers should be made as already directed to fit each gun; the Cartridges when made to be packed up in Bundles, each Bundle to have some mark to distinguish what gun it belongs to, that they may be served out when wanted with regularity & Expedition they are to be carefully deposited in Boxes __

Genl Moultrie, Col^{s}. Huger, Motte & CC Pinckney are requested to Survey the Cannon in the Arsonal and in town as immediately as possible & report such as are fit to be mounted upon field ~~pieces~~ Carriages

The quart^{r}. Mast^{r}. Genl: is to make Strict inquiry what number of flints there is in the Store for the publick service & to report to the Genl. & to attend every day at head Quart^{rs}. at 12 OC: for Ord^{rs}. or send his Deputy

The Quart^{r} Mast^{r}. of the 2nd Reg^{t} will Endeavour to find a house for the guard at Genl. Gadsden's warff that they may Extend their Centrys as usual without inconvenientcy __ to which if he suceed the guard is to remove

No Adjutant having called for orders yesterd^{y} the General woud have it understood that any futer neglect will be treated with rigour __

A General court martial to sett on Thursday next at 9 OC: at some convenient place in Ch^{s} town for the tryall of all the prisoners brought to them, Maj^{r}. Conner will warn all the Officers, Col: Huger to be president __ if there should not be Off^{rs}. Sufficient in Garrison the numbers wanting is to be furnished from Fort Johnson & Fort Moultrie

Regimental Orders __ Regimental Guard a Serg^{t}.

A Regimental court martial to sett tomorrow morning at 11 OClock to try all prisoners brought before them __ Evidences to attend __Presed^{t} Captn Ashby
Lieut^{s}. Hart and Perreneau members

By Coll: Motte **Ch^{s} Town 1776 __**

Orders 18th Decem^{r} . . . Parole Laurence __________ by the presed^{t}

G. O.
G. H. } For the day tomorrow Captn Moultrie
Magazine guard Lt Mazyck
Town L^{t}. from the 5th Reg^{t}.
Barrack L^{t} P Gray
Ship Yard __ Lt Mason

The Commesary to make an Emediate return to Genl Howe what Quantity of provisions he has on hand & at what place stored

The Genl. is dissatisfyd with the report made from the Arsenal the 14th Inst: as irregular & inexplicit, he remains uninformed as the quantity of cannon shott, & desires immediately to be acquainted with the exact number

For the Genl. Court Martial

From first Reg^{t}. 3 Sub: from 2nd __ 4 Capt^{ns}. 3 Sub: from the 5th __2 subalterns

Reg^{tl}. Orders __

A Reg^{tl}. court martial to sett this forenoon at 11 OC: to try such prisoners as shall be brought before them Evidences to attend

Captn Charnock presedt.
Lts. P Gray & R Mason members.
Members for the Genl. Court martial are Captns. Motte, Ashby, Shubrick & Harleston Lts. Hall H. Gray & Burke

NB According to Last C. Martial sentence E Wood to receive 39 Lashes being sick was remitted [482]__ Thos Bowen for Desertion recd 100 __
C. M. the 17th Inst: Wm Ashford for drunkness & neglect of duty recd. 50 Lashes. [483]
Timothy Green for Insolenses reprimanded [484]__ Gasper mants for neglect of duty 50
Wm Bryan for neglect of duty Acquited [485]
NB tis a pity there is so many court martials on trival crimes

By Col. Motte **Chs Town 1776 __**
Orders 19th Decemr. . . . . Parole Moore . . . . by Genl Howe

G. O. H. }
For the day tomorrow Captn. Blake
Town Guard Lt. Eveleigh
Magazine from 5th Regt.
Barrack Lt Dunbar

Gen Howe with concern Observes a great want of Attention to forms in many of the reports made to him perticularly from the guards, he is far from construing this into an Intentional disrespect to the Commanding Officer tho it bears the Appearance of it __ As the mode of the Army are the result of experience & all its forms has Essense, it behoves an Offr. who wishes to cut any Capital figure in his profession to make himself Acquainted not only with them, but with every other part of Millitary discipline __ an Application to books & reference upon all Occasions to persons of experience is the best way to obtain the Theory of war, a strict attention to duty may furnish the Occasion & will establish the meathod of calling it in to practice

NB the Genl. Court Martial adjourned to 9 OC: tomorrow

Orders 20th Decemr __ Parole Beaufort . . . by presedt

G. H.
For the day tomorrow Captn. Charnock
Town guard Lt from 5th Regt.
Magazine Lt from do
Barrack Lt Dubose
Ship Yard Lt Hart __ Regt. Guard a Sergt.

---

482 Possibly Edward Wood, who was in the militia at the fall of Charleston.

483 William Ashford enlisted under Captain Barnard Elliott on 1 July 1775. He received 50 lashes for drunkenness and neglect of duty of 18 December 1776. He served in Captain Dunbar's Grenadier Company on 11 July 1777. On 4 August 1777 he received 50 lashes for neglect of duty. On 11 June 1778 he received 39 lashes for neglect of duty. He was promoted to Corporal on 3 August 1778.

484 Timothy Green enlisted in the 2nd South Carolina Regiment on 20 November 1775 under Captain Richard Mason. He was reprimanded for insolence on 18 December 1776. On 12 February 1777 he received 100 lashes for drunkenness and misbehavior. He received 35 lashes on 7 July 1777 for being in "liquor" on guard duty. He was appointed a regimental barber on 24 January 1778. He was killed at Savannah during the assault on the Spring Hill Redoubt on 9 October 1779.

485 William Bryan was in the 2nd South Carolina Regiment. He was acquitted of the charge of neglect of duty for sleeping on his post on 18 December 1776. He received 50 lashes for neglect of duty on 2 January 1778.

Orders 21st Decem^{r}. . . . Parole Liberty. . by G Howe

For the day tomorrow Captn Moultrie

Town Guard L^{t} from 5th Reg^{t}.

Magazine Lt Baker

Barrack Guard Lt Mazyck

Regiment^{l}. Ord^{r}.__ Regimental Guard a Sergeant & 9 men

By Col. Motte **Ch^{s} Town 1776**

A Reg^{tl}. court martial to sett at the Barracks at 11 OC: this morn^{g}. to try such prisoners as shall be brought before them Evid. to attend

Captn Charnock president

L^{ts}. Mazyck & Baker members

Genl Orders by Genl Howe

James Kelley of the S^{o} Carolina 3^{d} Reg^{t}. try by a General Court Martial whereof Col^{o}. Huger was president was found guilty of Desertion & Sentenced to receive 100 lashes, the Sentence is approved & Ratifyd & Court is Desolved __ The Sentence of the above court not to be put in execution till the directions of the Hon^{bl}. the Continental Congress can be had upon it According to the 8th Articles of the 14 Sections of the Articles of war

---

Orders 22nd Decem^{r}. . . . . Parole Philadelphia . . . . by presed^{t}

For the day tomorrow Captn Harleston

Town Guard L^{t} from 5th Reg^{t}.

Magazine Lt Geo^{r} Eveleigh

Barrack Lt Burke

Ship Yard L^{t} from 5th Reg^{t}.

One Subaltern 1 Sergeant & 15 privates from the 2nd & 5th Reg^{ts}. to be ready tomorrow morning at 7 OC: to Escort some powder to the Magazine at dorchester & their to remain as a guard till relieved

They are to parrade at the Main guard to receive further Orders, the Officer to call at Head quarters as soon as he brings his party on the parrade

Reg^{tl}. Orders for the command Lt Dunbar, for Reg^{tl}. G a serg^{t}

NB Lt Dunbar with 1 Serg^{t}. 1 Corp: & 15 priv: from the 2nd & 5th Reg^{t}. went at half past Eight for Dorchester

---

G. Orders 23^{d} decem^{r}. . . . Parole Woodford . . . . by G. Howe

For the day tomorrow Captn Motte

Town guard Lt Hall

Magazine L^{t} from 5th Reg^{t}.

Barrack Lt Hart ___ Regim^{t}. Guard a serg^{t}

By Col. Motte **Ch^{s} Town 1776 __**

Regimental Ord^{rs}.

For the feuter no Party or Detachment to be sent from the Reg^{t}. without the Commanding Off^{r}. being first Acquainted with therewith __ the Major is perticularly Order'd to this Order Obey'd _____

All Off[rs]. that has not been on the recruiting service to hold themselves in readyness to proceed On that Duty at the Shortest notice __

A Reg[tl]. court martial to sitt this morn[g]. at 11 OC: to try all prisoners brought before them __ Evid. to Attend

Capt[n] Harleston president
L[ts] Hall and Hart members

NB Sentence of Last court Ben[j]: Wilmot for improper behavior on board the Commet rec[d]. 99 lashes & his prize money to be distributed among the other men of our Detachment [486] __ Rich[d]. Williamson & Jn[o] Thompson for Disorderly behavior in town rec[d] 99 lashes & Mult 14 days pay- And W Adams & Timothy Downing was reprimanded & Mult 14 days pay[487] __ Adam Meek for Absents wth out leave rec[d] 99 lashes & Mult 14 days pay

---

General Orders by G Howe 23[d]

A corporal and 10 men to be aded to the Magazine Guard to taken from the 2nd & 5th Reg[t]. this addition to the guard to continue till further Orders __ At this season when disorder & riot Generally prevails more than Common attention is requisite in All guards, & all off[rs]. of Companys are expected to pay strict regard to the conduct of their men they will be as Little absent from their Barracks as possible & by every means in their power prevent those excesses which may Occasion injury to the Inhabitants & bringing Disgrace upon the army, the General hope the men will by their sobriety & good conduct merit his Approbation, he wishes this the more Anxiously, as severe punishment however painfull to himself will be the certain consequence of Misbehaviour

The Commisary to serve out provisions dayly to John Baltizore a soldier in the Continental Army in Col[o] Habersham Reg[t] [488]

By Col. Motte **Ch[s] Town 1776 __**

G Orders 24th Decem[r]. . . . . Parole Drayton . . . . by the presedent

For the day tomorrow Capt[n]. Ashby
Town guard L[t] from 5th Reg[t].
Magazine L[t]. H Gray
Barrack L[t] from 5th Reg[t].
Ship Yard L[t] Perrineau

Gen[l] Howe Leaves the town tomorrow for a few days, the Command will Devolve as by a former order when he went to Georgia upon Gen[l] Moultrie, unless Gen[l]. Gadsden chuse to Assume it; __ a Centinel to be kept boath day & night at the generals quarters with strict charge to be Vigilent & carefull

---

[486] Benjamin Wilmot was in the 2nd South Carolina Regiment. On 23 December 1776 he received 99 lashes and his prize money was distributed to the rest of the detachment for his conduct while on the *Comet*. He was reprimanded at the Halberts for neglect of duty on 24 April 1777. He received 50 lashes on 12 May 1777 for repeatedly being absent from roll call. On 7 June 1777 he received 100 lashes for trying to get on board a ship.

[487] Unable to find any information on William Adams.

[488] This may be John Habersham, who was not a colonel, but was only a captain at this time. He would be promoted to major on 1 April 1778. Habersham was captured when Savannah fell to the British on 29 December 1778. His unit was the 1st Georgia Regiment. The 1st Georgia was not in Charleston at this time, but the one soldier, John Baltizore, was receiving rations from South Carolina since he was in Charleston.

Regimental Orders _____________ For Regimental Guard a serg[t].

Commanding Officers of Comp[ys]. are to make a return by three OC: this after noon to the Adjutant of the quantity of Osnab[g] each rec[d]. from the quarter Master for the making of Hunting shirts for the men __[489]

When the weather or any thing else prevents the Reg[t] from turning out to Exercise in the After noon, the Roll of each comp[y] is to be called at retreat, in the presents of an Off[r] who is desired to order in Confinement such of the men who may be absent.

The Gen[l] Order of yesterday to be perticularly Attended to by the Off[rs]. who are desired to read the same at the head of each Comp[y]. this Afternoon

NB sentence of last Court Jn[o] Hem was to receive 39 Lashes pardoned [490] Corp: Jam[s] Oakes_for drunkenness & neglect of duty reduced & rec[d]. 39 lashes

---

G. Orders 25th Decem[r]. . . . Parole Trentown . . . by G Moultrie

For the day tomorrow Capt[n] from 5th Reg[t]
Town guard L[t] Dubose
Magazine from 5th Reg[t]
Barrack from 5th Reg[t]

Regimental Guard 2[d] Reg[t] a Serg[t]. __

by Col[o]. Motte **Ch[s] town 1776 __**
by the presed[t]

G. Orders 26th decem[r]
By Gen Moultrie } __ Parole Philadelphia . . .
For the day tomorrow Capt[n] Shubrick
Town guard Sub: from 5th Reg[t]
Magazine L[t] Baker
Barrack__L[t] Mazyck
Ship Yard Sub. from 5th Reg[t].

The Deputy Quart[r] master to make a report immediately of fire wood that is distributed weekly to the troops in Ch[s] Town & in what proportion it is served to the troops__ Reg[t] Guard a Serg[t].

G. Orders 27 Decem[r]. . . . Parole Luxemburg . . . by G Moultrie
by G. Moultrie . . For the day tomorrow Capt[n] Moultrie
Town guard L[t] Eveleigh
Magazine Sub. from 5th Reg[t]
Barrack L[t] Mason__ Gadsden warff a Corp[l]:

M[r] Kaltieson the waggon master to send some horses to Gen[l] Gadsden Warf to draw the field pieces from thence to the new Barracks there to be placed before the guard room door [491]

---

[489] Osnaburg is similar to linen.

[490] Possibly James Hem, who served in the militia of Colonel Roebuck after the fall of Charleston.

[491] Michael Kalteissen was born on 18 June 1729 in Württemberg, Germany. He was living in South Carolina in 1755. He was a wagonmaster in the Cherokee War of 1760-61. He organized the German Fusiliers of Charleston in May 1775 and was elected lieutenant. In February of 1776 he was the commissary of military stores and was appointed wagonmaster later that year. On 29 November 1777 he became the wagonmaster general of the Provincial Army of South Carolina. He would later become the captain of the marines on board the South Carolina Navy frigate *South Carolina.* After the Revolution he was appointed commander of Fort

Reg[t]. orders by Col[o]. Motte

A Subaltern Off[r]. of the day to be immediately Appointed who is to remain constantly at the Barracks till relieved & to receive the morn[g]: Reports of the different Comp[ys]. the Report of the Regimental guard & to prevent any Irregularitys amongst the men __ he is to make this Report in wrighting when relieved to the Commanding Off[r]. of the Regiment __ the Adjutant is to receive the morning & Regim[tl]. Guard reports for this morning

A Regimental court martial to sett forenoon at 11 OC: to try all prisoners brought before them. Evid: to Attend __

Capt[n] Shubrick presid[t]
L[ts]. Eveleigh and Mason members
For the day today L[t] Burke
Regimental guard a Sergeant

NB by sentence of Last court J. Nesmith was reprimanded

by Col[o]. Motte — **Ch[s] town 1776 __**

General Orders G. Moultrie } 28th Decem[r] . . . Parole Carolina . . . — by the presed[t]
For the day tomorrow Capt[n] Harleston
Town guard from 5th Reg[t].
Magazine L[t] Perreneau
Barrack Sub: from 5th Reg[t]
Ship Yard L[t] H Gray __ Gadsden Warff a Corp[l].
Reg[tl] Ord[r]. } L[t] Hall for the day __ Reg[tl]. guard a Serg[t].

G Orders G. Moultrie — 29th decem[r] . . . Parole Bullock . . . — by presed[t].
For the Day tomorrow Capt[n] Motte
Town guard L[t] Mazyck
Magazine Sub: from 5th Reg[t].
Barrack L[t] Baker __ Gadsden Warff a Corp[l].
Reg[tl] Ord[rs]. — L[t] Dubose for the day __ Regimental guard a Sergeant

---

G Orders G Howe } 30th decem[r] . . . Parole Virginia . . . — by the Gen[l]. Howe
For the day tomorrow Capt[n] from 5th Reg[t]
Town guard from 5th Reg[t].
Magazine L[t] Eveleigh
Barrack a Sub: from 5th Reg[t].
Ship Yard L[t] Hart__ Gadsden warf a Corp[l].

A Gen[l]. Court Martial to sett on friday next for the tryal of L[t] Withers of the 5th Regiment put in arrest by Col[o] Huger for disobediance of Orders & other misdemeanours. Coll. Huger will furnish this court with the evidences; Col[o] Motte to be president: if a sufficient Number of Off[rs] should not be in town those wanted to be called from fort Johnson & fort Moultrie; they will also try James

---

Johnson. He was also the founder of the German Friendship Society. He died on 3 November 1807 at the age of 78.

Pliscan of the 4th Reg[t] for desertion; Adjutant Nixon will furnish the evidence [492]__ the Magazine Guard to be reduc't to the usual number of men tomorrow

The court to consist of the following Off[rs]

From the 2nd Reg[t], Capt[ns]. Charnock & Moultrie L[ts]. Baker, Eveleigh Hall & Mason __ L[t] Perreneau to act a Judge Advocate

By Col[o]. Motte **Ch[s] town 1777 __**

Gen[l] Orders 31st Decem[r] . . . Parole Moultrie . . . by presed[t]
G Howe} For the day tomorrow Capt[n] Shubrick
Town Guard L[t] Hall
Magazine Sub from 5th Reg[t].
Barrack L[t] Perreneau __ Gadsden Warf a Corp[l].

James Philan ordered to be tryed Yesterd[y]. by a Gen[l]. Court martial having been improperly Confined by Adjutant Nixon is to be Discharged without tryall the Off[rs] of the guard is therefore to dismiss him R. O __ for Regimental Guard a serg[t] __

---

[492] George Nixon was the adjutant to the 4th South Carolina Artillery.

1777

G. Orders by G. Howe — 1 Jan[y]: ~~1777~~ Parole Mead .... by G. Howe

For the day tomorrow Capt Charnock
Town guard from 5th Reg[t].
Magazine L[t] H Gray
Barrack L[t] Eveleigh
Ship guard Sub: from 5th Reg[t] __Gadsden Warf a Corp:

Great complaints having been made that the Bread furnished to the army is neither made with good flower or well Baked, the commisary will Certainly be made Accountable if this matter is not immediately rectified __ the director of the Hospital complains that pork has been served out to the sick for some days past & that it is very improper for them; that the bread served to the Hospital is exeeding bad & unholsome, The director will therefore apply to the Commisary for flower for the hospital & have the bread made by a Baker under his direction __ he will also order what other provisions he thinks necessary for the sick, which the Commisary is to furnish __ The Commisary will immediately bring before Gen[l] Howe a coppy of his Contract with the publick

The magazine Guard at Pratchards is for the feuter to be relieved weekly __ the party for relief tomorrow will therefore prepare Accordingly __

NB Sentence of last Court Serg[t] Tho[s] Stockwell for absents without leave was reduced & received 100 Lashes __ Josiah Stone & Jn[o] Smith for the same crime rec[d] 100 Lashes
Andrew LeBane was Discharged for want of Evidence [493]

By Col[o]. Motte — **Ch[s] Town __1777** By presed[t]

Gen Orders by G Howe } 2nd January ... Parole Gadsden ...

For the day tomorrow Capt[n] Harleston
Town guard L[t] Burke
Magazine L[t] P Gray
Barrack Sub: from 5th Reg[t] __ Gadsden Warf a Corp[l]:
Reg[tl] Barrack Guard a Sergeant

G. Orders by G Howe } 3[d] Jan[y] .... Parole Roberts by G. H

For the day tomorrow Capt Motte
Town guard Sub: from 5th Reg[t]
Magazine L[t] Dubose
Barrack L[t] Mazyck __ Gadsden Warff a Corp[l]:

If Col[o]. who is out of garrison should not arrive before the setting of the court ordered tomorrow the field officer next for duty is to be warned__Commanding Off[rs]. of out posts when in town will send some persons to head quarters every day at Orderly time to receive Orders __L[t] Henry Perreneau is Appointed Judge Advocate of the Gen[l]. Court martial order'd to try L[t]. Withers, arrested by Col[o] Huger for Disobediance of Orders & other midsdemeanors, Col[o] Huger will furnish the Judge

---

[493] Andrew Bland.

Advocate with the different charge against Lt Withers in wrighting to be laid before the court, he will also find the evidence against the prisoners in support of the Charge

Col° Roberts who is the next field off^r. in the Absence of Col° Motte is to preside in said court

After Genl Orders

A Sergeant & 12 men to take charge of a Magazine near Col° Laurence house, this Guard is to mount immediately __[494]

A Sergeant & 6 men to hold themselves in readyness for a march tomorrow morning early; this duty to be done by Detachment from 2nd & 5th Reg^t ____

Regimental guard a Serg^t & 10 rank & file

NB the above guard mounted at Col° Laurences house at 6 OC even^g

By Col°. Motte **Ch^s Town 1777**

Genl Orders by G. Howe } 4th Jan^y . . . . Parole Moultrie . . . . by presed^t

For the day tomorrow Captn. Ashby

Town guard L^t H Gray

Magazine Lt Hart

Barrack Sub: from 5th Reg^t __ Gadsden Warf a Corp^l.

Guard at Laurences house a Sergeant

All Officers who have recruited men out of the Continental Battalion of North Carolina are to make immediate return to Genl Howe of the numbers & names of the men they have inlisted, the Battalions & Company from which they were taken, & the perticular Officer to whom they returned the Bounty money & no Officer from thence forward to Inlist any men from any North Carolina Battalion or Detachment on any pretence whatsoever, this order to be transmitted to the Out Post.

Regimental guard a sergeant

Genl Orders by G Howe } 5th Jan^y . . . Parole Motte . . . . by G H

For the day tomorrow Captn. from 5th Reg^t

Town guard Sub: from 5th Reg^t

Magazine Sub: from 5th R.

Barracks Lt Pet^r Gray

Laurence House a Serg^t __ Gadsden Warf a Corp^l.

The General Highly Disapproves the loose Disorderly manner in which he observes the relief of Centries are marched up, Officers of Guards Aught to prevent this want of Decipline in the men & the Genl. expects that a soldier in the most trifling manoevure should be made to observe Punctilio's as much as in the greatest; [495] they aught indeed to have impressed upon their men that every duty is important, and tis absolutly incumbent upon all Officers to endeavour to effect this, a negligence in the minor dutys of a soldier will beget negligence in points of greater moment, & all remissness without disgrace upon an army __

Regimental guard a Sergeant __

---

[494] Colonel Henry Laurens.

[495] Definition of Punctilio is "the strict observance of formalities".

By Col°. Motte **Chs Town__ 1777 __**

G Orders 6th Jany. .... Parole Roberts .... by presedt
G Howe For the day tomorrow Captn Shubrick
Town guard Lt Warley
Magazine Sub: from 5th Regt
Barrack Lt Burke
Laurence House a sergt __ Gadsden Warf a corporal
Regimental guard a sergeant

G Orders 7th January .... Parole Pinckney .... by G. H
G Howe For the day tomorrow Captn Lesesne
Town guard Sub: from 5th Regt.
Magazine Lt H Gray
Barracks Lt Hart
Laurence House a Sergeant __ Gadsden Warff a corporal
Regtl. Orders __ Regimental Guard a Sergt.

A Return to be made today to the major of the number of blankets wanting to each company The morning Reports & Regimental Guards to be delivered to the Adj: for this day who is to bring them to the Commandg. Offr. as soon as he receives them

NB at 12 OC. at noon, the sergeant & 6 men ordered the 3d Instant went on board Captn Minnot for port Royall with 6 days provisions

---

G Orders 8th Jany. ... Parole Nash ... by presedt
G Howe For the day tomorrow Captn Harleston
Town guard Lt Mazyck
Magazine Lt Dubose
Barrack Sub: from 5th Regt
Ship guard Lt P. Gray
Guard at Col° Laurence's house a Sergt. __ Gadsden Warf a Corp.

Genl Howe havine Occasion to question severall Sergts. Corpl. & men upon duty respecting some orders issued, found to his surprise they were entirely ignorant of them, the Usage of the Army in general, and the Standing Orders of this particular Garrison injoin that the Soldier should be made Acquainted with every order relative to them it is an Unpardonable Neglegence in Officer of companies not to be carefull that orders are communicated to their men who may otherwise be liable from Ignorance to transgress and be punished; And though Young Gentlemen may suppose this more perticularly the duty of the Adjutant, yett a mind Actuated with proper Zeal for the Service & Emulous in excelling in the noble profession of a soldier will verge beyond the Surcomcribed Limits of men in Duty, & Intertain Ideas more liberal than barely to escape Censure; they will not only fill the measure of their own department, but attentively Observe the Operation in every other, by this means they will learn what to imitate & adopt & what to avoid & in respect to the perticular subject of this order the attention required by it will prevent the ill consequence which might otherwise result from the neglect of Ignorance of those Executive Officers whos immediate duty it may be to communicate orders to the men __

The general court martial Order'd by Lt Withers having Doubts that the Appointment of the Judge Advocate was not authorized, decline proceeding thereupon, They are therefore Desolved and a court of Inquiry composed of the same members that made up the General Court martial is to sett tomorrow, the court will call on Col°. Huger for the Charge & evidence against Lt Withers, what he is commited is worthy of a Genl Court martial __

Regimental guard a Sergeant

Genl Orders by G Howe } 9th Jan^y^ . . . Parole Moore . . . . By G. H.___

For the day tomorrow Captn Motte

Town guard Sub: from 5th Reg^t^.

Magazine Lt Burke

Barrack Lt Warley

Col° Laurences house a Serg^t^ __ Gadsden Warf a Corporal

By Col°. Motte Ch^s^ Town 1777

General Howe waits with impatience for the return orderd to be made him the 4th Instant by the Officers who had recruited men from the North Carolina Battalions, & which aught to be immediately comply'd with; the delay has oblig'd him to detain some despatches that was nessesary to send off, an Officer should consider it as a point of honour to Obey Orders & therefore desdain any neglect or violation of them, a want of this punctual observance may be productive of the worst consequence to the Service, & ruin that cause which soldiers were Ordaind to Support

The court of inquiry, orderd to inquire into the Conduct of Lt Withers of the 5th Regiment report as follows __

The court having maturely inquired of Lt Withers of the 5th Reg^t^: are of Opinion it is such as do not merit a Genl. Court Martial; but are also clearly of opinion his Behavior to Col°. Huger his Commanding Off^r^. has not been altogether consistant with the Carracter of a Gentleman & a good Officer __ they therefore recommend that he be pubicly admonished & reprimanded at the head of the regement to which he belong

The Court appointed is dissolved & the Sentence Approved by the Generall __ Col° Hugers Battalion to parrade tomorrow morning 10 OC: when Lt Withers will attend the Admonition & the reprimand he is to receive according to this Sentence of the Court that sett upon him, it will be sent to the Col°, who will order the Major of his Reg^t^. or in his absents to read it to L^t^. Withers within the hearing of the Battalion; After which he is to be released from his Arrest & to do duty on the Reg^t^. as usual __ ____________

Regimental Orders__ For Regimental guard a sergeant

Commanding Off^rs^. of Comp^ys^. to make a return immediately of the name of each deserter from their comp^ys^. & the places they were Inlisted

Sergeant Brown & 6 men to hold themselves in readyness to go after deserters __ the men to be fixed on by the Sergeant __

By Col°. Motte **Ch^s^ Town 1777 __**

G. Orders by Genl Howe } 10th Jan^y^. . . . Parole Davis . . . by presed

For the day tomorrow Captn. Ashby

Town guard Lt Mason

Magazine Sub: from 5th Reg^t^.

Barrack L^t^. Eveleigh

Guard at Col° Laurence house a Serg[t] __ Gadsden Warf a Corp[l]

L[t]. Withers thinking himself agrieved by the Report made of his conduct Request Leave to resign his Commission which Gen[l]: Howe consenting to receive, he resignes it Accordingly, the orders of yesterday is therefore not to be executed __

Regimental Orders __ for Regimental G __ a Serg[t]:

Ordered that the men allways Appear with their Regimental Caps such as disobey this Order may depend on being severely punished __ Commanding Off[rs]. of Comp[ys] are immediately to apply to the Quarter Master for as many Regimental Caps as they want

A Court Martial to sett this morning at 11 OC: to try such prisoners as shall be brought before them Evidences to Attend __ Capt[n]. Motte president

L[ts]. Henry Gray and Hart members__

G Orders by G. Howe } 11th Jan[y]. . . . . Parole Jones . . . by presed[t]

For the day tomorrow Capt[n]. from 5th Reg[t]

Town Guard Sub: from 5th Reg[t].

Magazine L[t]. Hall

Barrack Sub: from 5th Reg[t].

Guard at Col°. Laurence house a Serg[t]. __ Gadsden Warf a Corp[l].

Regimental Orders __ Reg[tl]. guard a Sergeant

NB according to Last court Vincent de Camp,[496] Jn° Walker, Geo[r] Dauphine Jn° Le Sage & Lewis Couret, all of Capt: Charnocks Comp[y]. for inlisting on board the Commet & taking bounty money, received 99 lashes, & stopage from their remain[g]. bounty money, & their pay till Capt[n] is reimbursed

By Col°. Motte **Ch[s] Town 1777 __**

G Orders by G Howe } 12th Jan[y]. . . . . Parole Harnett . . . by G.H __

For the day tomorrow Capt[n] Shubrick

Town guard L[t]. H Gray

Magazine Sub: from 5th Reg[t].

Barrack L[t] Hart

G. Col° Laurense house a serg[t]. __ Gadsden Warf a Corporal

Commanding Officers of Corps & Battalions are immediately to report to Gen[l]. Howe the names of those Off[rs]. who have inlisted men from the North Carolina Battalions __ Adjutant Deliant will serve the Commanding Off[rs]. of out posts now in town with a Coppy of this Order & of the 4th __ 9th & 12th Instant relative to those Off[rs]. whows name are required __

Reg[tl]. Orders__ for Reg[tl]. Guard a Serg[t].

The Adjutant to receive the morning Reports & that of the Regimental Barrack guard tomorrow & when ever their is a Capt[n]. of the day that does not belong to the Regiment

---

[496] Vincent De Camp enlisted in the 2nd South Carolina Regiment on 5 November 1776. On 11 January 1777 he received 99 lashes for illegally enlisting on board the Brigantine *Comet* and taking the bounty money. His pay was stopped until he repaid Captain Edward Allen of the *Comet* the bounty money. He deserted a few weeks later on 27 January 1777.

G Orders by G Howe } 13th Jan^y^. .... Parole Newborn .... by presed^t^
For the day tomorrow Captn Charnock
Town Guard L^t^. Perreneau
Magazine Lt Mazyck
Barrack Sub: from 5th Reg^t^.
Guard Col° Laurense house a Serg^t^.__ Gadsden Warf a Corp^l^;
R. Ord^rs^. For Regimental Guard a Serg^t^.______

G Orders by G Howe } 14th Jan^y^ .... Parole North Carolina .... by GH __
For the day tomorrow Captn Lessesne
Town guard Sub: from 5th Reg^t^
Magazine Sub: from 5th
Barracks L^t^. Mason
Guard Col° Lawrense's house a serg^t^ __ Gadsden Warf a Corp^l^.
Regem^tl^: Orders __ For Reg^tl^: Guard a Serg^t^ __

The Orders of the 10th Instant to be read at the Head of each Company this Afternoon by a Commissioned Off^r^ __ Such of the Off^rs^. as have not subscribed & taken the Declaration & Oath as ordered by the Honorable the Continental Congress are to do it today at 11 OC: in the forenoon, and to meet at Lt Col°. Marions House; Lt Col° Marion with any two Capt^ns^ & two Subalterns he may fix on to see this order Complyed with __

NB According to the above Ord^r^. Capt^ns^. Ashby & Lessesne & Lt Warley took the Oath in the state House before I. Shadd Esqr. in the presence of Lt Col° Marion, Maj^r^. Horry Lt Hall & those who was sworn __

G Orders by G Howe } 15th Jan^y^..... Parole Philadelphia .... by presed^t^
For the day tomorrow Captn Moultrie
Town Guard L^t^. Eveleigh
Magazine Lt Shubrick
Barrack Sub: from 5th Reg^t^.
Ship Yard Sub: from 5th Reg^t^.
Guard Col° Laurences house a serg^t^ __ Gadsden Warf a Corporal

---

G Orders by G: Howe } 16th January.... Parole Hughes [497]
For the day tomorrow Captn. Harleston
Town Guard L^t^. from 5 Reg^t^.
Barrack Guard L^t^. from d° Col°. Laurence's a Serg^t^
Magazine Guard L^t^. Burk Gadsdens Wharf a Corporal

By the President Parole Nash
For the Day to morrow Captn. Motte, Town Guard L^t^. Hall
Barrack Guard L^t^. Warley Magazine from 5th Reg^t^.
Col° Laurence's a Serg^t^. Gadsdens Wharf a Corporal

---

[497] The handwriting changes on this date, indicating a new orderly was chosen.

John Burgess of Captⁿ. McDonalds Co. is appointed a Sergt in Capt. Moultrie's Co. & is to be obey'd as such.
A Court Martial to sett this morning at 11 o'Clock to try such Prisoners as shall be brought before them Evid: to attend
Capt. Harleston Presedt: Lts. Hall & Warley members __

By General Howe **Charlestown 18th January 1777 ~**

Parole Rutledge
For the Day to morrow Captn. Charnock, magazine Guard Lt. Hart
Town Guard & Barrack Guard from 5th Regt.
Colo. Laurence's a Sergt. Gadsdens Wharf a Corporal

The Articles of War to be read at the head of the Regimt. this Afternoon when it turns out to exercise Lt. Co. Marion is desir'd to read them The Commanding Officer expects that no one will be absent that is able to turn out. The Staff Officers in particular to attend. __ NB. Seven men one of which to act a Corp: are to be added to morrow morning to the Brick House Magazine Guard near Colo. Laurences Gate Subaltern officer is to be warn'd to take charge of this Guard, the men are to be taken by Detachmt from 2d & 5th Regimts: Complaints have been made that the men of that Guard have burnt the Fences The officer of the Guard is to prevt. this for the future if possible & if not, he is to exert himself to detect those who do it & they may depend on being punish'd. __

Mr Righton is to furnish two light rowing Boats for the use of the Troops at Haddrels Point; Adjutant Delliant will give immediate Notice of this. The Boats are to be deliver'd to Colo Francis Nash or the Commanding Officer for the Time being at Haddrels Point who will appoint proper persons to take care of them,[498] The Adjutant will transmit this order to Col: Nash. Gen: Howe goes out of Town for a few days.

By General Howe **Charlestown 18th January 1777 ~**

The Command under his Direction will devolve in his absence upon Co. Huger who will take charge of it & be respected & obey'd accordingly.

---

[498] Francis Nash was from Orange County, North Carolina. He fought against the Regulators at the Battle of Alamance in 1771. He was a representative of Orange County in the 1st Provincial Congress and he represented Halifax County in the 3rd Provincial Congress. He was made the lieutenant colonel of the 1st North Carolina Regiment on 1 September 1775. He was promoted to Colonel on 10 April 1776. He was in Charleston with his regiment during the Battle of Fort Sullivan. He was promoted to brigadier general on 5 February 1777 and commanded the North Carolina Brigade when they marched north to join Washington's army. He was mortally wounded as he led his men into battle at Germantown on 4 October 1777 when "his thigh shattered by a solid shot, and fainting from the loss of blood, was born to a nearby house and lingered only three days." One of his men wrote "we lost our General, Frank Nash, who was killed by the cannon ball which struck his horse behind his right thigh, and passing through, cut off his right thigh, except a small bit of skin on the fore part, which was cut before he was raised, and put in the carriage with him. This gallant and brave officer lived till the next day and died." Nash County, North Carolina is named after him.

By C: Motte / Parole How By Presidt.

For the day to morrow Captn. Shubrick
Town Guard L^{t}. Shubrick Magazine L^{t}. Peter Gray
Barrack Guard L^{t} from 5th Regiment
C^{o} Laurence's House L^{t}. Harleston
Gadsden Wharf a Corporal

Regimental After Orders by Col^{o}. Motte
Lieutenants Baker & Perenneau to hold themselves in readiness to proceed on the recruiting Service on Monday next __ They will call on the comd^{g}. Officer for their Orders at 10 o'Clock next Munday Morning. Gen: After Orders by Gen: Howe

A Centry to be posted both Night & Day at the Quarters of General Howe wth. strict charge to be vigilant __

By C^{o} Motte / Parole Laurens By Gen: Moore[499]

For the day to morrow Captn Charnock
For the Town Guard L^{t}. from 5th Regim^{t}.
Magazine L^{t}. Eveleigh C^{o}. Laurens's L^{t}. Burke
Barrick Guard L^{t}. Warley, Gadsdens Wharf a Corporal

A Court Martial to sett this morning at 11 o'Clock to try such Prisoners as shall be brought before them __ Evid: to attend
Captn Shubrick Pres^{d}. L^{ts}. Burk & Warley Members __

By Gen: Moore / **Charlestown January 20th 1777**

Gen: Orders

The detach'd Situation of Fort Moultrie, Haddrells Point & this Town with Fort Johnston from each other making it necessary that the Command of the Troops should be divided
Gen: Howe will command in Town & at Fort Johnston
Gen Gadsden at Fort Moultrie & Sullivants Island &
Gen: Moultrie the N^{o}. Carolina Troops at Haddrells Point

By C^{o}. Motte __ Parole Horry ____ By the president
January 21st

For the day to morrow Captn Moultrie
Town Guard Lt Dubose, Magazine L^{t}. Mason
C^{o}. Laurences Guard L^{t}. Mazyck
Barrack Guard from 5th Reg^{t} Gadsdens Wharf a Corporal

---

[499] James Moore was a company commander in the French and Indian War. He was in the Regulator War and commanded the North Carolina Artillery during the battle of Alamance in 1771. He was the representative of New Hanover in the North Carolina Provincial Congress in 1775. He was made the colonel of the 1st North Carolina Regiment on 1 September 1775. He commanded the Patriot forces during the Battle of Moore's Creek Bridge, North Carolina on 27 February 1776. He was promoted to brigadier general on 1 March 1776. He commanded the North Carolina Brigade in Charleston during 1776. On April 5, 1777 General James Moore died of "a fit of Gout in his Stomach."

R. O. William Carter of Captn. Blakes Compy is appointed a Sergeant in Captn. Lesesne's Co. & to be obey'd as such. __[500]
N.B. According to the Sentence of the last Court Martial Jno. Clements & Abram Berlin for misbehavior in Town were acquitted for want of Evidence.[501] Thomas Stockwell for want of Evidence remanded. Josiah Stone for absence from Roll call & Drunkeness receiv'd 100 Lashes. Thomas Lewis for carrying a Prisoner abt. at 10 o'Clock at night to a dram shop receiv'd 50 Lashes,[502] Francis Duprin for same Crime but being ignorant of the Language & a new Recruit was acquitted with a severe reprimand ~[503]

By Colo. Motte. / Parole Howe .... by Gen: Moore January 22nd.
For the day to morrow Cap: Motte
Town Guard Lt. Shubrick, Barrack Lt. P. Gray. Pritchards Lt. Harleston
Co. Lawren's & Magazine from 5th Regt. Gadsdens Wharf a Corporal __

By Co. Motte / Charlestown 22d Jany. 1777 ~
R. O. A Court Martial to sett this morning at 11 o'Clock to try such Prisoners as shall be brought before them
Presidt. Captn Blake Lts. Shubrick & Gray members __
Brigade Orders by Gen: Howe 21st Inst.
The Guard at the Battery on Gadsden's Wharf is to be called in to morrow & the Brick House Magazine Guard is to take charge of the Battery.

---

500 William Carter enlisted in the 2nd South Carolina Regiment on 4 November 1775 in Captain Blake's company. He was promoted to sergeant and transferred to Captain Lesesne's company on 21 January 1777. He later was promoted to sergeant major. He was discharged on 20 July 1778.

501 Abraham Berlin (Berlean) enlisted in the 2nd South Carolina Regiment on 4 November 1775. On 21 December 1777 he was charged with misbehavior in town, but was acquitted due to lack of evidence. On 20 February 1777 he received 25 lashes for threatening Sergeant Smith's life. He received such a small amount due to "age and former good behaviour". On 22 August 1777 he was sentenced to receive 40 lashes for being absent without leave, but he was pardoned due to his age and being sick. In 1779 he was under Captain Peter Gray. He deserted in February 1779, but was captured and received 50 lashes. He deserted again on 10 October 1779 after the assault on the Spring Hill Redoubt in Savannah.

502 Thomas Lewis enlisted in the 2nd South Carolina Regiment on 4 November 1775. On 21 January 1777 he received 50 lashes for taking a prisoner to a tavern to have some drinks at 10:00 at night. He received 35 lashes on 7 July 1777 for being in "liquor" on guard duty. On 22 July 1777 he received 60 lashes for quitting his guard and insulting Doctor Page. He was discharged on 16 July 1778.

503 Francis Dupree (Dupuis, Du Pre') enlisted in the 2nd South Carolina Regiment on 19 December 1776. On 21 January 1777 he was supposed to receive 50 lashes for taking a prisoner to a tavern to have some drinks at 10:00 at night, but he was acquitted with a severe reprimand for being "ignorant of the language and a new recruit". On 31 March 1777 he was court martialed for some unnamed offense but he was remanded for want of evidence. On 2 April 1777 he received 30 lashes for abusing and striking Sergeant Coffer. On 7 July 1777 he had to run the gauntlet for stealing a pair of silk stockings from Sergeant Munroe. On 31 July 1777 he received 100 lashes with switches for some unnamed offense.

The Guard posted at Dorchester Magazine to be reliev'd on Thursday next 1 Subaltern 1 Sergeant & 18 men are to be warn'd for this Purpose they are to be taken by Detachment from 2d & 5th Regimts: Lt. Petrie from 5th Regimt to take Charge of this Detachmt. [504]

By Co. Motte. / Parole Constitution __ By the President /

Jany. 23d 1777

For the day to morrow Captn Ashley__
Town Guard Lt from 5th Regimt.
Magazine Guard Lt. Warley, Barrack Lt. Burke
Brickhouse Lt Eveleigh __

R. O. The Surjeons Mate to make a report in writing every Monday Morning of the Sick in the Reg: Hospital to the commanding officer __ NB. According to the Sentence of the last Court Martial, Jno Brown of the Grenadier Compy. for carrying a Prisoner to a dram Shop was reprimanded __ [505]

By Co. Motte. / Parole Gadsden ________________ by Gen: Moore January 24th }

For the day to morrow Captn Blake __
Magazine Guard Lt. Mazyck, Barrack Guard Lt. Dubose
Brickhouse & Town Guard from 5th Regimt.

By Co. Motte / Charlestown 24 January 1777 ~

R. O. Robert Williamson of Captn Oliphants Compy is appointed a Serjeant in Captn. Moultries Compy. & is to be obey'd as such - A Regimental Court Martial to sitt morning 11 o'Clock to try such Prisoners as shall be brought before them Evidences to attend - Captn Moultrie President Lts. Mazyck & Dubose Members - Gen: Orders by General Moore

Generals Howe, Gadsden & Moultrie are desired to order exact returns to be made of the Strength of their respective Commands on Tuesday the 31st Inst. in order for their being transmitted to Congress

Town Brig: Orders. Commanding officers of Battalions in Town & of the Corps of Artillery at Fort Johnston are without fail to make exact returns of the Strength of their respective Corps by 31st of this Inst. . . . .The return to be deliver'd to Major Connor - The Serjeant of the Hospital Guard having reported that several Patients & some of them Prisoners of War have beheav'd in a very improper Manner at the Hospital. The Captn. of the day is to make strict Enquiry & report the particulars to Gen: Howe, The Officer of that Guard will confine such Prisoners who for the future shall behave ill & will certainly suffer the Sentence of a Court Martial, this is to be read to the Patients in the Hospital by the Serjt. of the Guard- The Report of the Town Guard to be made to Gen Howe -

---

[504] There were two officers named Alexander Petrie at this time. One was a lieutenant in the 2nd South Carolina and the other was a lieutenant in the 5th South Carolina. This Alexander Petrie was the officer in the 5th South Carolina Regiment. He was later promoted to captain and resigned on 8 October 1778

[505] John Brown enlisted in the 2nd South Carolina Regiment on 4 November 1775 in the Grenadier Company. On 23 January 1777 he was reprimanded for taking a prisoner to a tavern to have some drinks at 10:00 at night. He received 35 lashes on 7 July 1777 for being in "liquor" on guard duty. He deserted a month later on 11 August 1777

By Col^{o}. Motte / **Charlestown Jany. 25th 1777 ___**

Parole Motte By President

For the day to morrow Captn. Shubrick, Town Guard L^{t}. P. Gray
Magazine from 5th Reg^{t}. Barrack Lt Shubrick Brick house L^{t}. Mason

Gen: Orders by Gen: Moore

Col^{o}. Thompson & Co. Sumpter's Regim^{ts}: are to make each a return of the Strength of their Corps on Thursday 31st Ins^{t}. __

____________________January 26th 1777 ____________________

Parole Moultrie by y^{e}. General

For the day to morrow Captn. Lesesne, Town Guard L^{t} from 5th Reg^{t}.
Magazine Guard L^{t}. Hall, Bark Guard from 5th Reg^{t}:
Brick house L^{t}. Burke, Brigade Orders of 25th

Several Soldiers of the 2^{d}. & 5th S^{o}. Carolina Battallions who are patients in the Gen: Hospital having been reported to Gen: Howe by the Officer of the Hospital Guard for Crimes & Misdemeanors worthy Punishm^{t}. The Director General of the Hospital will examine those delinquents & inform Gen: Howe to morrow whether any or all of y^{e}. are well enough to attend the Tryal & undergo the Sentence of a Court Martial, The Officer of the Guard is in the mean Time to take Care that they do not escape

Some Prisoners in the main Guard having been reported to Gen: Howe for Crimes within the Cognizance of Regim^{t}. Courts Martial the comd^{g}. Officers of the 2^{d} & 5th S^{o}Carolina Battallions to wch. these Men belong will examine into this matter & take measures accordingly __

Brigade Orders. / Ch^{s} Town 26 January 1777 ~

The Crime for which James McDonald of the 2^{d} S^{o}.Carolina Regim^{t}. stands reported to Gen Howe might be made the Subject of a higher Courtt than a reg: Court Martial, but as the Number of Officers in Town are not sufficient for such a Court without breaking in upon Duties more immediately important it is recommended to C^{o}. Motte to order him to be try'd by a reg^{t}. Court Martial

____________________January 27th ____________________

by C^{o}. Motte. / Parole Williamson By Presidt.

For the day to morrow Cap: Moultrie, Town Guard L^{t}. Dubose
Magazine from 5th Reg^{t}: Bark Guard L^{t}. from 5th Reg^{t}.
Brick House L^{t}. Warley -

R.O L^{t}. Harleston having resign'd his Commission with leave, is no longer to be obey'd & respected as an Officer in the Regiment - A Court Martial to sett this Morning at 11 o'Clock to try all Prisoners brought to it Captn Lesesne, L^{ts}: Dubose & Warley Members

By C^{o}. Motte. / **Charles town January 28. 1777 __**

Parole Huger By Gen: Moore

For the day to morrow Captn. Ashley: Town Guard from 5th
Magazine L^{t}. Mason Barrack L^{t}. Baker
Brick House from 5th Reg^{t}:

N. B. According to the Sentence of last Court Martial Jesse McDonald for Suffering the Loss of an apron of a gun to be stolen was reprimanded. - Josiah Stone for loosing his Cap & Absence reciv'd 50 Lashes -

January 29th

Parole Thompson . . . . .By Presidt.

For the day to morrow Cap: Blake, Town Guard Lt. Mazyck
Magazine Lt. Shubrick, Barrack from 5th Regimt.
Brick House Lt. P. Gray, Pritchards Guard from 5th Regt.

R. O. Commandg Officers of Companies are to make out & deliver a monthly return of each Company to the Adjutant to morrow forenoon ____

January 30th

Parole Philadelphia . . . . By the Gen:

For the day to morrow Captn. Shubrick, Town Guard Lt from 5th: Magazine Lt. Dunbar, Brick House from 5th. Barracks Lt. Burke __ R. O. A Court Martial to sett this mnorning at 11 o'Clock to try such Prisoners as shall be brought before them. Presd Cap: Blake Lts.

Dunbar & Burke Members __

By Co Motte. / **Charlestown 31 January 1777 ~**

Parole Trenton __ By President

For the day to morrow Captn. Lessesne, Town Guard Lt. Warley
Magazine from 5th Regt. Brick House Lt. Eveleigh
Barracks Lt. from 5th. R. O. In Case of an Alarm of any kind, the Officers are immediately to repair to the Barracks, when they are to turn out their men wth. the greatest Silence __ The Capt. of the day, or the Adjutant is then to send an Orderly Sergt: to the Commanding officer of the Regt. to acquaint him of the alarm. No one to presume to march the men away 'till Orders are given for that Purpose by the Officer comdg. the Regimt: This Order to be strictly complied with NB. The Sentence of the Court Martial last was executed & Edwd. Fry rec'd 50 Lashes

February the 1st 1777

Parole Drayton .... by the General

For the day to morrow from 5th Regimt: for the Town Guard Lt. Hall, Magazine Lt. Baker, Brickhouse Lt Dubose Barrack Lt from the 5th Regimt.__
R. O. A Regl. Court Martial to sett this morning at 11 o'Clock to try all Prisoners that may be brought to it Captn. Lesesne President Lts: Eveleigh & Baker Members __

By C° Motte. / **Charlestown February 1. 1777 ~**

General Orders.

The following Articles being wanted at Fort Moultrie the Commissary of Stores must immediately procure them & when procur'd deliver them to the comd^g. officer of the Garrison__ Viz^t.

23 Spunges, Staffs & Rammers for 26 Pounders
18 ditto.............d°:................18 d°:
32 ditto.............d°:................12 d°:
11 ditto.............d°:............ .....9 d°:
10 ditto.............d°:.................4 d°:
2 Worms for 20 Pounders
3 d°..........12 d°.
1 d°...........4 d°.
1 Ladle for 18 Pound^r.
11 d°......... 12 d°.
4 d°............9 d°.
4 d°............4 d°.
45 Lints for stocks [508]
100 hand Spikes

14 Lanthrons [506]
6 tann'd Hides
Oakum for Wading
4 Horns for Prickers [507]
4 Iron Crows
4 Fledges

Brigadier Orders

The Case of Accident from Fire, All officers of Battalions & all Soldiers will assemble wth. their arms at the usual Place of Parade & from thence be March'd by the comd^g. officer pres^t. to the place where the fire is they are to act in Conjunction, or be divided in Parties as the comd^g officer finds it necessary, but always w^th: an Officer at their head, they are to consider it as their indispensable Duty to exert themselves to the utmost of their Power to stop the Progress of such a dreadful Calamity, to aid & assist the Inhabitants of the Town in saving their Effects, & to take those Effects if requir'd by the Owner, under their immediate Protection, to guard them securely from being embezel'd & return them uninjur'd to the Proprietors when demanded. The Gen: hardly thinks it possible that there can be a Soldier so base as to add to the Distress of the Inhabitants in a Situation so Loind [509] by secreting any Part of their Property, but if there shou'd exist a writh [510] so dead to every feeling of humanity & so lost to all Sense of common Honesty, he is in Case of Detection to expect no mercy __

Gen: How takes this opportunity to express his Approbation of the Conduct of those Officers & Men whom he found regularly drawn up upon the Accident the Night before last his pleasure upon the Occasion would have been greater had he seen more Officers among them, it will not in his oppinion paliate much less excuse than Ommission, shou'd they plead that no Order like the foregoing had been issu'd because where no regular Alarm Posts are appointed, both Duty & common Sense unite to dictate that the usual Place of Parade is. that. all officers & men should upon any

---

[506] Sometimes Lanterns are listed as Lanthorns. This is because thin cow horn was used in the place of glass, since it would be more durable.

[507] Part of the loading process for artillery was to put a thin brass needle-like object into a hole at the breach of the cannon. The needle would then pierce the powder bag already loaded inside the barrel. The "pricker" would then pour powder into the hole, or he would place a firing tube into the hole, and light it with a slow match to fire the cannon.

[508] The Linstock was the device that held the burning match, or "lint", used to fire the artillery.

[509] Loind is an abbreviated version of the word purloined, or to steal.

[510] Writh – something that causes pain.

Alarm repair, not to know this wou'd reflect upon the understanding of those who wou'd make use of this Plea & not to Comply wth. it is most assuredly a neglect of Duty __
NB. This Order to be read on Parade __

Gen: Orders of 31st Jany. 1777 __

5 Privates of the 5th Reg^{t}. to be added to the Guard at the Gen: Hospital __ Col^{o}. Sumpter is to send 1 Subaltern 2 Serg^{ts}: 2 Corporals & 20 Privates to relieve the Guard at Dorchester & the reliev'd to join their respective Companies. The Barrack Mast^{r}. is to provide a Waggon to carry wood at the magazine near Pritchards Yard __

By C^{o} Motte. / **Charlestown 2nd February 1777 -**

Parole Georgia .... By Presid^{t}.

For the day to morrow Captn Moultrie, Town Guard L^{t}. Jacob Shubrick, Magaz: d^{o}. L^{t} from 5th Regim^{t}. Brick house L^{t}. P. Gray Barks L^{t}. Mason
R. O. The Brigade orders of yesterday to be read at the head of the Regim^{t}. to morrow afternoon by L^{t}. C^{o}. Marion__

______________February 3rd ______________

Parole Huger .... By Gen: Moore

For the day to morrow Cap. Motte, Town Guard L^{t} from 5th: Magazine L^{t}. Burke, Brick house from 5th. Barracks L^{t}. Dunbar. R. O. Col^{o}. Motte is much displeas'd to see several of his officers appear about Town without their Regim^{tl}. Caps contrary to Orders he hopes they will in future be carefull not to be so deficient in their duty as to disregard that Order. __ A Regim^{l}. Court Martial to sett at 11 o'Clock this morn^{g}. to try such Prison^{rs}. as be brought to it Evid: to attend __ Cap. Moultrie Presid^{t}. L^{ts}: Dunbar & Burke members ____
NB According to the Sentence of yesterday's Court W^{m}. Harrington for absence without Leave receiv'd 100 Lashes __[511]

By C^{o} Motte. / **Charlestown Feby 4th 1777 -**

Parole Cherockee, By the Presid^{t}.

For the day tomorrow Cap^{n}. Blake, Town Guard L^{t}. Eveleigh
Magazine D^{o}. L^{t}. Warley, Brickhouse L^{t}. Baker
Barracks L^{t}. from 5th Reg^{t}.

R. O. Presidents of Regim^{l}. Courts Martial are desir'd to report the Proceedings to the comd^{g}: officer of the Reg^{l}: the Same day of the Setting of the Courts

No noncommission'd Officer or Soldier to be absent from the Barracks after retreat beating without leave from the Captn of the day. any seen in Town after that Time & without Permission as above directed. the comd^{g}. Officers orders they be confin'd for disobedience of Orders. That no one may plead Ignorance, the above Order is to be read to the Men by an Officer of each Company every

[511] William Harrington enlisted in the 2nd South Carolina Regiment on 4 November 1775. On 3 February 1777 he received 100 lashes for being absent without leave.

afternoon this Week __ According to the Sentence of last Court, Laome Husband for being out at an unreasonable Hour was reprimanded. [512]

Corp: Roberts for neglect of Duty was reduc'd to the Ranks __[513]

By C^{o} Motte. / **Charlestown February 5th 1777 ~**

Parole Motte __ By the General

For the day to morrow Cap Charnock, Town Guard from 5th Reg^{t}: Magazine L^{t}. Mazyck, Brickhouse L^{t}. P. Gray, Barracks L^{t}. from 5th: Pritchards Magazine L^{t}. Mason R. O. L^{t}. Mason is appointed a 1st. Lieut^{t}. in the Room of L^{t}. Harleston who have resign'd Lieut Mason to remain in Captn Harleston's Comp^{y}. M^{r}. Albert Roux is appointed 2^{d} Lieut in the Reg^{t}. & to be obey'd as such & to join Captn. Moultries Co. till further orders [514]__ A Court Martial to sett this morn^{g}. to try all Prison^{rs}: brought to it Evid: to attend Presid^{t} Captn Blake, L^{ts}: H. Gray & Mason members

______________________________February 6th ______________________________

Parole Brunswick By y^{e} Presid^{t}

For the day to morrow Cap^{n}. from 5th Reg^{t}. Town Guard L^{t}. Dunbar, Magaz: L^{t}. from 5th D^{o}. Brickhouse from D^{o} __ Barracks L^{t}. Shubrick __ N B. According to y^{e} Sentence of last Court Jno Griffin for disobedience of Orders rec'd 35 Lashes, W^{m}. Simpson for Drunkeness on Duty rec'd Nathan, Clark [515] & Wm Evans for absence without Leave receiv'd 50 Lashes Each __ NB By Order of the commanding Officer Serg^{t}. Early was reduc'd to the Ranks __

By C^{o} Motte. / **Charlestown February 7th 1777 ~**

Parole Thompson. By G. Moore

For the day to morrow Captn. Moultrie, Town Guard L^{t}. Burke
Magaz: from 5th. Brickhouse L^{t}. Eveleigh Barks from 5th

---

512 Laomi Husbands enlisted during August 1775 in the 2nd South Carolina Regiment under Captain Barnard Elliot. On 4 February 1777 he was reprimanded for being out at an unreasonable hour. He was publicly reprimanded on 14 October 1777 for striking Corporal William Jones. He also had to publicly apologize to Corporal Jones. He was discharged on 31 July 1778. In his book *Roster of South Carolina Patriots in the American Revolution* author Bobby Moss wrote "He (She)" when describing Laomi Husbands, suggesting that Private Husbands might have been a woman. When I asked Bobby Moss where was his source for this, he wrote, "One of my research colleagues suggested that Husband might be a female. You notice that I put the entry thus: (She). In the mss. to the printer it was (She?) and some how the typesetter left out the question mark. I do not think the soldier was a female."

513 William Roberts had enlisted in the 2nd South Carolina Regiment on 4 February 1776 and had risen in rank. He was reduced to private on 4 February 1777. On 9 October 1777 he was reprimanded for abusing a "Negro" while on guard duty.

514 Albert Roux was commissioned as a second lieutenant in Captain Moultrie's company of the 2nd South Carolina Regiment on 15 December 1776 (the Officer's list says 26 January 1777). He was promoted to first lieutenant on 5 December 1777 in Captain Charnock's company when Lieutenant Henry Gray resigned. He was placed under arrest for disobeying orders on 26 December 1777. He was transferred to Captain Dunbar's company on 19 April 1778. He was promoted to Captain on 17 August 1779 and took command of Captain Motte's company on 17 September 1779 when Motte became a major. He was wounded in the right arm at Savannah on 9 October 1779. He retired on 1 January 1781.

515 Nathaniel Clark (Clarke) enlisted in the 2nd South Carolina Regiment on 4 November 1775 from the Lancaster District. On 6 February 1777 he received 50 lashes for being absent without leave. He survived the war and died on 12 April 1834.

G. O.

The officer for the day to visit Gen[l]. Hospital dayly to see that the Guard conduct themselves properly & to receive the report from the Serjeant & enquire if y[e]. Patients do punctually obey the directions given them

____________________February 8th ____________________

Parole Lee .... By the Presid[t]:

For the day to morrow Capt[n] Jervey [516] of the 5th
Town Guard from the 5th. Magazine L[t]: Dubose
Brickhouse L[t] from 5th Barracks L[t]. Warley __
Regimental Orders,

The Pay Bills of the different Comp[ys]: to be deliver'd to the Pay Master, immediately__ The commanding officer hopes the Officers comd[g]: Comp[ys]: will not give him the Trouble to remind them again in this Part of their Duty __ [517]

All Recruits brought to the Reg[t], are immediately to be conducted to the officer comd[g]: the Reg[t]. & to be approv'd of by no one but himself. The Adjutant to see this order put into Execution __The Qr. Mast[r]. not to deliver any Stores under his Care without an order sign'd by the comd[g]. officer of the Reg[t]. __

By Col[o] Motte. / **Charlestown 8th February 1777 ~**

A Regimental Court Martial to sett at 11 o'Clock this morning to all Prisoners brought to it Evid: to attend

President Capt[n]. Moultrie L[ts]. Dubose & Warley members

NB. Gen Orders. . By Gen: Moore

The orders issued by Gen: Lee on 8th Sep[r]. last by w[ch]. the Troops of N[o]:Carolina & Virginia were inlisted into the Service of this State and of Georgia not having extended to the Inlistm[t]. of Drum[rs]. & Fifers, but on the Contrary forbidding such Inlistm[ts]. The comd[g]. Officers of Battallions in this

---

[516] George Jervey served as a captain in the 5th South Carolina Regiment (1st South Carolina Rifles). He was court martialed for conduct unbecoming an officer on 14 June 1777, but he was acquitted of the charges. He was wounded at Beaufort on 3 February 1779.

[517] A company officer at this time had to be able to run a company with the funds that were given to him, and he also had to make up all losses with that meager pay. If a private deserted with all of his equipment, the officer of the company had to buy new equipment from those funds. Though the rank and file of the company were issued their gear, each officer had to provide for his own uniform and equipment out of their own personal expenses. This is why most officers came from the more wealthy class, since a poorer man would not have been able to afford being an officer. Many officers would take out loans to maintain their company and in the end this would place many of the heroes of the Revolution in deep debt, so bad that they would never be able to recover. Men such as "Light Horse" Harry Lee, son of Robert E. Lee, and George Rogers Clark, would die in debt, with no support from the government that they fought to create. That these men did the job of commanding their units while slowly going into debt shows their commitment to the cause more than any other indicator. In the British army it was not much better and officers had to buy their commissions before they could even get into their regiments. The image of the British officer being from the uppercrust of society, and only being in the army as a hobby until something more responsible came along does not do them, or history, justice. Though there were some extremely rich officers, such as Captain William Leslie, son of the Fifth Earl of Leven, most were barely scraping by on their officer pay, and they had to have the added expense of maintaining their station in society. The image of the British officers being elitist snobs has more basis in the Victorian era British army and did not reflect what the British officer corps of the 18th century was like.

State who have such Drummers & Fifers in their respective Corps are to have them deliver'd to morrow Morning at 11 o'Clock to an officer ~~to an officer~~ who will be sent to the Barracks to receive them __ Capt[n] Blount of the 5th Reg[t] of N[o]:Carolina Troops is order'd to attend for that Purpose [518]

February 9th

Parole Roberts .... By the Presid[t] & say G. Moore

For the day to morrow Cap. Ashby
Town Guard L[t]. Baker
Magazine Guard L[t]. from the 5th Reg[t].
Brickhouse L[t]. Mazyck
Barrack Guard L[t]. from 5th Reg[t].

By C[o] Motte. / Charlestown 10th February 1777 ~

Parole Philadelphia By Presid[t].

For the day to morrow Capt[n] Oliphant, Town Guard Lieut from 5th Reg[t]. Magazine L[t] Shubrick __ Brickhouse L[t]. from 5th reg[t]. Barracks L[t]. P. Gray

R.O. A regimental Court Martial to sett this morn[g] at 11 o'Clock to try all Prison[rs]: brought before it Presid[t] Captain Ashley, members L[ts]. Shubrick & P Gray. NB. The comd[g]. Officers of Companies that have any demands for the making the Split Shirts are desir'd to call on C[o]: Motte for the Paym[t]. of the Same

By Sentence of the Court 8th Ins[t].

Corp: Cidwell [519] was acquitted & Luke Woodward & Jn[o] Smith remanded for want of Evidences —

518 Reading Blount was born at Blount Hall in Pitt County, North Carolina. He was commissioned as a captain in the 3rd North Carolina Regiment on 16 April 1776. He was with his regiment when they marched north to Washington's army in the spring of 1777 and fought in the battles of Brandywine and Germantown. He transferred to the 2nd North Carolina Regiment on 1 June 1778, then returned home when the regiments were reorganized before the Battle of Monmouth. He commanded the New Bern Regiment of Militia at the Battle of Briar Creek in March 1779. He became a major in the 5th North Carolina Regiment and was in the Battle of Stono Ferry, South Carolina in June 1779. He transferred to the 1st North Carolina Regiment on 1 January 1781. He commanded what remained of the North Carolina Continentals after their capture at Charleston in the Battle of Guilford Courthouse on 15 March 1781. He was cited for bravery in that battle. He "gallantly" commanded the 2nd North Carolina Regiment at the Battle of Eutaw Springs, South Carolina on 8 September 1781. This was one of the bloodiest battles of the war, lasting over three hours. The British lost 42% of their force and the Patriots suffered 25% of theirs. The North Carolina Brigade suffered the most with 154 men being killed and wounded. Only two Continental regiment commanders were able to make it out of the battle without being wounded or killed. Blount transferred to the 4th North Carolina Regiment on 1 January 1783 and served until the end of the war. He received a land warrant on 27 October 1783 for his 84 months of Continental service. He died on 12 October 1807.

519 Thomas Kidwell enlisted in the 2nd South Carolina Regiment on 4 November 1775. He was acquitted of some unnamed charge for want of evidence on 10 February 1777. He was promoted to Corporal. He was reprimanded for gambling on 8 August 1777. He was reduced in ranks on 14 October 1777 for striking Sergeant Lewis Coffers. He was promoted to corporal again on 13 July 1778 in Captain Charnock's company. He was sentenced to do the duty of a private for 30 days on 6 October 1778 due to gambling. In 1779 he was in Captain Richard Mason's company. On 27 April 1779 he was court martialed on an unspecified charge by Vincent Maroni, but he was acquitted. He was in Captain Mason's company at the siege of Savannah.

February 11th

Parole, Trentown. By Gen Moultrie

For the day to morrow Captn Pott of the 5th Regt:[520]
Town Guard Lt. Burke, Magaz: Lt. from 5th regt.
Brick house Lt. Dunbar, Barracks Lt. from 5th regt.
R. O. by Colo. Motte.

The Regt. to be in readiness to march at the Shortest Notice The Officers to provide themselves wth. Tents & Field Equipage immediately

February 12th | Gen: Howe

Parole Charleston | By the Presidt

For the day to morrow Captn: Blake, Town Guard Lt from 5th regt.
Magaz: Lt. Warley, Brickhouse Lt. from 5th Barks Lt. Eveleigh
Pritchards Guard Lt from 5th Regt.

By Co Motte. / Charlestown 12 February 1777 ~

R. O. A Regimental Court Martial to sett this morning at 11 o'Clock to try such Prisoners as are brought to it. Evidences to attend at that hour.

Captn Blake Presidt. Lts: Eveleigh & Hall members

N. B. According to the Sentence of last Court Jno Smith & Tim Green for drunkness & misbehaviour receiv'd 100 Lashes Each

G. O. By Gen: Moore 9th this month

Second & Fifth Battallions of this State are to hold themselves in readiness to relieve the No:Carolina Brigade at Haddrells Point on Thursday the 13th Inst. Mr Righton to prepare a Sufficient Number of Boats for the Purpose

Colo. Cannon is immediately to send some careful person to Haddrells Point to serve out Provisions to the No:Carolina Brigade.[521] The Person sent is to continue in Camp 'till the Brigade is reliev'd.

February 13th

Parole By

For the day to morrow Captn Charnock
Town Guard Lt. Hall, Magazine Lt. from 5th
Brickhouse Lt. Baker, Barracks from 5th

By Co Motte. / Charlestown 13 Febry 1777 ~

R. O. Order'd that Captns: McDonald & Oliphant Lts Mazyck, Dunbar & Adams call on Lt. Co. Marion this forenoon at 11 o'Clock & before him take & subscribe the Oath order'd by the Honble: the Continental Congress to be taken by all officers in the Service of the United States

NB. The Detachmt from Dorchester arriv'd last Night

[520] Thomas Potts of the 5th South Carolina Regiment (1st South Carolina Rifles). He served as a 2nd Lieutenant of the First Company of the Prince Frederick's Parish Volunteers in 1775. After the fall of Charleston he served as a captain with Marion's partisans.

[521] Possibly Daniel Cannon, the commander of Captain Daniel Cannon's Volunteers in the Charleston militia.

According to the Sentence of last Court martial Jeremiah Early & Wm. Miller for Misbehavior were reprimanded & pardon'd & Owen Griffin for the same Crime receiv'd 50 Lashes__[522]

G. O. Information having been made to General Howe that Serjt. Ross of the 1st Regt. had insulted the officer of the Genl Hospital Guard while on duty & had otherwise behav'd in a disorderly Manner. Colo. Pinckney is desir'd to enquire into this matter & take Measures accordingly. Sergt: Simpson of C: Mottes Battallion will furnish the Evidence__

By Co Motte. / **Charlestown 14th Feby 1777 ~**

Parole Fortitude By Gen: Howe

For the day to Morrow Captn
Town Guard Lt. from 5th regt. Magaz. Guard Lt Mazyck Barracks Lt. from ~~5th Regt~~ P Gray
Brickhouse Lt from 5th Regt.

R. O. a Regimental Court Martial to sett this morning to try such Prisonrs. as shall be brought before them __
Captn. Charnock Presidt.
Lts: Dunbar & Burk members

General Orders by General Howe __

Gen Howe wth Concern & Surprise has observ'd the frequent Applications made by Gentlemn of the Army for leave to resign their Commissions at this important cricis, where it is difficult to find any Reason sufficient to excuse Men for not endeavouring to get into Service, what can possibly exculpate those who desire to forsake it __ The Freedom of America & all its essential Priveledges are at Prest. the objects of Contest & Compar'd wth. these All private Interests however important, & every darling Inclination, Attachmt. & Simpathy however endearing, & heartfelt; are but futile Considerations __ The prest. Generation & all the the Generations of succeeding ages have the strongest Claim upon a Soldier for every strenuous Endeavour & upmost Effort to preserve & maintain such invaluable Rights & to hand them down to posterity: unimpair'd Difficulty & Distress & Danger are the mediums thro: which this Purpose is to be effected, which every Officer must have been sensible of at the Time he sollicited a Commission; Local Advantages therefore, or temporary Inconveniences are but contemptible pleas for retiremt: an Opinion of an Officer's Spirit & Abilities, a Belief that he wou'd by attention to duty, & by every other means in his Power gain a proper knowledge in his profession, with a firm Perswasion, that he wou'd not forsake the Service, at the very moment he has qualified himself to be useful, must have been what induced his Country to honour him wth a Commission otherwise undoubtedly woud have been granted to those who emulous to serve, were probably possessed of equal Abilities & who by greater Perseverance wou'd have continued to the common cause ~

The Benefit of that Experience they must have obtain'd had not the Interference of those very officers who now desire to leave the Army, deprived them of the opportunity; let therefore those Officers solicitous of resigning, but for one moment consider, that by the superior Confidence plac'd

---

[522] Owen Griffin enlisted in the 2nd South Carolina Regiment on 4 November 1775. On 13 February 1777 he received 50 lashes for misbehavior. On 11 April 1777 was accused of selling another soldier's musket, but was acquitted due to lack of evidence. On 15 August 1777 he was charged with fighting and quarrelling but he was remanded for want of evidence. On 10 March 1778 he received 100 lashes for stealing. He received 100 lashes on 23 March 1778 for being insolent and disobeying orders.

in them by the Country, they were prefer'd to those who wou'd have serv'd it to the last; & let them ask their oun Hearts if withdrawing from their duty at this critical juncture, is not a poor return for the very honorary Preference given them? let them them add to this, the noble & animating Consideration that they are actors upon that glorous Stage where every Incident is to come an historical Fact, & the Gen[l]: perswades himself that they will not by future applications for leave to resign reduce him to the painfull alternative of refusing the Request of Gentl[m]. he respects, or by complying wth. it, of depriving the army of offic[rs]: so capable of doing honour to themselves & rendering Service to the common Cause. __

This Order to be transmitted to the com[dg]. off[r]. at Haddrells Point, Sullivants Island & Fort Johnston, who are to take Care that it is made known to the officers of their Corps __

Adjut[t]. Dellient will inform Gen: Gadsden (if in Town) Gen: Moultrie & all the field officers in Town that Gen: Howe wou'd be glad to speak w[th]: them at his Quarters at 5 o'Clock this afternoon

February 15th 1777

by C[o]. Motte Parole By the Presid[t].

For the dat to morrow Cap: from 5th Reg[t]. Town Guard L[t]. Shubrick, Magazine L[t]. Dunbar, Brick house L[t] from 5th Barrack Guard L[t]. from 5th Reg[t]. __

Charlestown 15th Feby 1777 ~

R. O.

A Court Martial to sett this morning at 10 o'Clock to try such Prisoners as shall be brought before them. Evidences to attend

Presid[t]. Capt[n]. Lessesne, Members L[ts]. Dubose & Mason

February 16th 1777

Parole Assembly By Gen: Howe

For the day tomorrow Capt[n]. Charnock, Town Guard L[t]. from 5th
Magazine Guard L[t]. from 5th. Brickhouse L[t]. Eveleigh
Barracks L[t] Burke NB According to the Sentence of the Court Martial 14th Ins[t]: Dan: Ross for Suspicion of deserting was ty'd to the Post & reprimanded.[523] Jeremiah M[c]Ginnis for misbehaviour was Sentenc'd 50 Lashes but pardon'd __ W[m] Ryan for Absence without Leave reciv'd 500 Lashes [524] __

R. O. By L[t]. C[o]. Marion./ The General to beat to morrow morning 9 o'Clock, The assembly & March at 10. at w[ch]. time the Reg[t]. will march to G. Gadsdens Wharf, there to embark for Haddrells Point, All Off[rs]. & Soldiers off duty to be on the Parade at that Hour & comd[g]. Officers of Companies to give in a field Return to the Adjutant at the Same Time & to order a Serj[t]. & party about the Town for those men who may be missing before the Gen: beat __ Any Soldier who may be found drunk or out of their Barracks at 9 o'Clock in the Morn[g]. may expect to be severely punish'd: it is expected the above Orders will be Punctually complied with As the hour of marching will not be put off __ The

---

[523] Daniel Ross enlisted in the 2nd South Carolina Regiment on 12 January 1776. On 16 February 1777 he was tied to a post and reprimanded for suspicion of desertion. He died on 26 November 1778.

[524] William Ryan enlisted in the 2nd South Carolina Regiment on 24 September 1776. On 16 February 1777 he received 500 lashes for being absent without leave. He reinlisted on 1 November 1777 under Captain Richard Mason. He was at the siege of Savannah.

Q^r^. Mast^r^. to provide 3 waggons to carry the Bags & Stores of the reg^t^. by 6 o'Clock in the Morning & to have all the ammunition &c ready to move wth. the Regim^t^.
Comd^g^; Off^rs^: of Companies are desir'd to have their Mens Rooms clean before the General beat _
The Q^r^. Mast^r^. to see that the Officers, Doctor's & Guard Room likewise his Rooms clean at the same Time __

February 17th 1777 __
Parole Williamson By the Presid^t^.

For the day to morrow Cap: Lessesne, Town Guard L^t^. Dubose
Magazine Guard Lt Baker, Brick house L^t^. from 5th
Barrack Guard L^t^ from 5th G. Orders. The Reg^t^. is order'd to march back to the Barracks, & there remain (holding themselves in readiness) 'till further Orders -
R. O. A Court Martial to sett this morning at 11 o'Clock to try such Prisoners as are brought to it. Evid: to attend

Presid^t^. Captn. Moultrie, members L^ts^ Dunbar & Hall

G. After Orders ~
The Soldiers confin'd in the main guard belonging to the continental Battallion of N^o^: Carolina are to be sent to morrow under a proper Guard at Haddrells Point & deliver'd to the comd^g^ officer at that Place -
Mr Righton, will Furnish a Boat for this Occasion & Adj^t^. Dellient will give Notice & Appoint the Place & Time where & when the Boat is to lie -

G. After Orders Continued } Charlestown February 17th 1777 ~
The Commissary of Stores will apply to his Excellency the Presid^t^ for the Key of Dorchester magazine which he will assist & examine within this 5 days he will take proper steps to prev^t^. the Powder from being damag'd, in doing this the greatest Caution is to be observ'd that no accid^t^. may happen: Carelessness upon such Occasion will be held an inexcusable offence
A report is to be made to the General of the State of the Magazine

February 18th 1777
Parole Virginia By G. Howe

For the day to morrow Captn. Moultrie, Town Guard L^t^. 5th
Magazine L^t^. from 5th. Brickhouse L^t^. Mazyck
Barrack Guard L^t^. Shubrick __
The Commissary of Stores may postpone the Order of Yesterday 'till further orders. __

Kings
Mountain
Charlotte
Cowpens
Cross Creek
New Berne
Waxhaws
Camden
Wilmington
Ninety-Six
Long Cane
Augusta
Savannah
River
Georgetown
Briar
Creek
Stono Ferry
Charleston
Oconee
River
Ogeechee
River
Purisburgh
Port Royal
Savannah
Altamaha
River
Sunbury
Saltilla River
Fort McIntosh
St. Augustine

A General Court Martial to sett in Ch^s^. Town on Monday next at 9 o'Clock in the Morning to try James Barron [525] Hezekial Melipsis [526] Jos: Gordon [527] Jn°. Steel,[528] Jam^s^. Pritchard [529] Thos M^c^Can[530]& Thos. Burkett [531] of the 1st Battallion of S°. Carolina Troops for desertion. Col°. Pinckney will order the Evidences to attend & is to be Presid^t^. of this Court, if there are not officers enough in Town, the members wanting are to be called from Fort Johnston & Fort Moultrie, this court will also try such other prisoners as may be brought before y^e^. __

R. O. by L^t^. C°. Marion

No Officer or Soldier to change the Guard he is appointed for, or get another to take his Tour of duty, without Leave from the comd^g^ officer, as 'tis contrary to Orders & Articles of War __

A Regim^t^. Court Martial to sett to morrow at 11 o'Clock to try such Prisoners as shall be brought before them Evidences to attend. President Cap: Motte
L^ts^: Mason & Warley Members

_______________ February 19th 1777 _______________

Parole Henry By

For the day to morrow Captn. from 5th reg^t^: Town Guard L^t^. Dunbar, Magazine L^t^. Eveleigh, Pritchards L^t^. Burke
Brickhouse from 5th Barrack from 5th

G. O. A Subaltern Officer & 15 men wth. a Serj^t^. & Corp^l^. to be taken by Detachment from 2^d^ & 5th Reg^ts^. are to relieve the Magazine Guard at Hobkaw to Morrow morning, Mr Righton will Furnish Boats for this Occasion __

525 James Barron enlisted in the 1st South Carolina Regiment as a sergeant on 4 November 1775 under Captain Thomas Lynch. He was court martialed for desertion on 18 February 1777. He was a lieutenant in the militia under Captain Jacob Barnett and Colonel Brandon from 12 May 1780 to 1 March 1781. He was also in the light dragoons of Sumter's partisans under Captain Samuel Martin and Lieutenant Colonel William Polk.

526 Ezekiel Malphurs (Zekiel Malpas) enlisted in the 1st South Carolina Regiment on 4 November 1775. He was court martialed for desertion on 18 February 1777. He was captured at the surrender of Charleston in May 1780. He was a corporal after 20 October 1780. He was a servant to Captain Thomas Gadsden while he was a prisoner, however he deserted and enlisted with the British army afterwards.

527 Possibly John Gordon, who enlisted in the 1st South Carolina Regiment on 4 November 1775 under Captain William Scott. He was court martialed for desertion on 18 February 1777.

528 This is a different John Steel than the one that was in the 2nd South Carolina. This John Steel was in the 1st South Carolina Regiment and was court martialed for desertion on 18 February 1777.

529 James Pritchard enlisted in the 1st South Carolina Regiment on 4 November 1775. He served in the 2nd South Carolina Regiment during 1777. He was sentenced to be punished for desertion on 4 August 1777, but he was pardoned since he returned on his own, and because he helped find other deserters.

530 Thomas McCann enlisted in the 1st South Carolina Regiment on 4 November 1775 under Captain Charles C. Pinckney. He was court martialed for desertion on 18 February 1777. He was sentenced to be shot for desertion, but was pardoned by General Howe on 5 November 1777. He was court-martialed again on 2 January 1778 for an unnamed offense.

531 Thomas Burkett enlisted on 4 November 1775 in the 1st South Carolina Regiment. He was court martialed for desertion on 18 February 1777. In 1778 he was in Captain John Blake's company. He was discharged on 3 July 1778.

February 20th 1777 ~
For the day to Morrow Captn. M^{c}.Donald Town Guard L^{t} 5th
Magazine L^{t} from 5th, Brickhouse L^{t}. Hall Barr^{k}. L^{t}. Warley
R. O. A Court Martial to sett this morning at 10 o'Clock to try such Prisoners as shall be brought before them. Ev: to attend
Captn. McDonald Pres^{dt}. Lt Shubrick & P. Gray Members __
By Lt C^{o} Motte. / Charlestown 20th February 1777 -

NB According to the Sentence of last Court, Edwd. Johnston for absence without Leave receiv'd only 50 Lashes on ac^{t}. of his Youth,[532] Abraham Berlin for threatning Serjeant Smith's Life rec'd only 25 Lashes on Acct. of age & former good Behaviour __

February 21. Parole Gates By the Presid^{t}.
For the Day to morrow Captn Oliphant, Town Guard L^{t}. Baker
Magazine Guard L^{t}. Mazyck, Brick house from 5th reg^{t}
Barrack d^{o}. from 5th R. O. A Court Martial to sett to morrow Morning at 11 o'Clock to try such Prisoners as shall be brought before them Evid: to attend __
Presed^{t}. Captn. McDonald L^{ts}. Baker & Mason members

For the Gen: Court Martial on Monday morning Capt^{ns}: Oliphant & Lessesne L^{ts}: J^{c}. Shubrick & Dunbar

---

February 22nd Parole Georgia by Gen Howe
For the day to morrow Captn. Blake Town Guard L^{t}. from 5th
Magazine L^{t}. from 5th. Brickhouse L^{t}. Burke
Barracks L^{t}. Peter Gray

Feby 23^{d} Parole Georgia By the Presid^{t}:
For the day to morrow Captn. Charnock, Town Guard L^{t}. from 5th
Magazine L^{t}. from 5th Brickhouse L^{t}. Gray
Barracks L^{t}. Dubose __ Head Q^{rs}. Ch^{s}. Town 21st Feby 1777
The Situation of the State of Georgia now under actual Invasion rendering it absolutely necessary that a considerable Body of Troops be immediately march'd to their assistance, the following Detachm^{t}: from the several Corps of Continental Troops of this State are to hold themselves in immediate Readiness to march, From Col^{o}. Pinckneys Battallion 2 Capt^{ns}. 4 Subalterns 4 Serg^{ts}: & 100 Rank & File __ From Col^{o}. Mottes Battallion 2 Capt^{ns}. 4 Subalterns 4 Serj^{ts}: & 100 Rank & File, from Col^{o} Roberts Corps of Artillery 1 Captn 3 Subalterns 2 Serg^{ts}. & 50 Rank & File__
From Col^{o}. Huger's Battallion 2 Cap^{ns}: 4 Sub: 4 Serg^{ts} 100 Rank & file, The Detachm^{t}: of Artillery will take charge of the Field Pieces wch. will be commmitted to their Care. Col^{o}. Sumpter wth. his Reg^{t}. is without Delay to march to Purisburgh where he will receive Orders,[533] The Detachment from Pinckneys, Mottes, Roberts, & Hugers are to march under their field officers 1 Col^{o}. 1 L^{t}. C^{o}. & 1 Maj^{r} Col^{o}. Motte ~~L^{t}. C^{o} Marion & Maj^{r} McDonald will com^{d}~~ will command this Detachm^{t}: L^{t}: C^{o}. Marion & Maj^{r} McDonald are also for this Service __ Gen Howe entertains the fullest Confidence

---

[532] Edward Johnston enlisted in the 2nd South Carolina Regiment on 4 November 1775. He received 50 lashes for being absent without leave on 20 February 1777. He was supposed to receive more, but it was reduced on account of his youth.
[533] Purisburgh was located northwest of Savannah and was a major supply point on the Savannah River.

that the Off[rs]: who are to go on this Com[d]. will exert themselves to the utmost to be ready to march wth. all possible Expedition & that the Men will march wth. alacrity & Chearfulness Waggons & everything necessary will be provided __

The Waggon Master is immediately to Furnish Ten Waggons for the use of the Troops & also two Ammunition Waggons, The Commissary will take Care to furnish a Detachm[t]: of 350 men wth. Provisions in this march to Purrisburgh an Ammunition waggon exclusive of these above mention'e to be provided by the Wag[n]. Mast[r]. for the detachment of Artillery __

The Time of march will be appointed as soon as the comd[g]: Off[rs]: of the several Corps from w[ch]. the men are taken report at head quart[r] s. that they are ready __

R. O. That Capt[ns]. Motte & Ashley, L[ts]. Dunbar, Hall, P. Gray, & Burke wth. 4 Serg[ts]. & 100 Rank & file do hold themselves in readiness to march immediately for Georgia the officers & men to parade at 4 o'Clock this afternoon the Q[r]. Mast[r]. to get 50 rounds p[r] man in readiness for this Detachm[t]. & to take particular Care that the Cartridges are such as will fit the different Caliber of the Muskets,[534] he also to get ½ doz flints p[r] man,[535] the Off[rs]. for this detachm[t]. to pick out such men in the Reg[t]: as are able to go through the Service w[ch]. is requir'd of them this is to be done immediately & the men to be warn'd without delay __

R. After Orders __ L[t]. Henry Gray to go in the Com[d]. in the Room of L[t]. Hall who was put out of his Tour of duty. This Order to be made known to L[t]. Gray immediately L[t]. Hall to be a Memb[r]. of y[e]. G. C. Martial in the Room of L[t] Dunbar, who is order'd on Com[d]. __

When Charles Lee conducted his expedition towards Florida in late 1776 he built two forts on the King's Road. These were Fort Howe, built on the site of Fort Barrington on the Altamaha River, and Fort McIntosh, built on the Satilla River. Howes whole force had returned to Charleston by December 1776, but the two forts were manned.

Captain Richard Winn of the 3rd South Carolina Rangers occupied Fort McInstosh and built it up. The fort was "built of split puncheons...the earth thrown up about four feet, and secured by small pines from the ditch upwards---the ditch was only begun upon---the flanks and curtain extended about Sixty paces." The fort was barely completed when the British appeared from Florida. The British were basically a raiding party that had ventured into Georgia for cattle. On February 23rd British Regulars, Indians and Loyalists from Florida began to surround the fort. In Charleston the South Carolina Regiments mobilized to relieve the fort before it had to surrender.

---

534 The problem of different muskets would continue to plague the Continental army until after France declared war on England and became allies of the United States. Until that time the soldiers would carry early model British muskets (aka the 1st model Brown Bess), Spanish muskets (of the 1756 model), Dutch muskets, and early French muskets. Between all of these models the caliber would go from a .62 caliber to an 80 caliber. After France joined with the United States to fight the British they shipped over thousands of French muskets, mainly of the 1763 model.

535 The issuing of extra flints was a sign that there was expected to be a tough fight, or a long expedition. Ammunition could be carried in loose ball or loose powder form, and did not necessarily need cartridges, however a flint may only last a couple of dozen shots and if there was to be a fight, more would be issued to the men.

By L^{t}. C^{o}. Marion **Ch^{s}. Town 24th Feby 1777**

Parole Savannah By the Gen:

For the day to Morrow Captn. McDonald Town Guard Lt Warley
Magazine d^{o}. L^{t}. Baker Brickhouse d^{o}. L^{t}. from 5th. Reg^{t}.
~~Barrack Guard L^{t}.~~ Barrack Guard a Serj^{t}. __

R. O. Commanding Officers of Companies to have their men going on the Command compleated in arms & accoutrim^{ts}. they are to exchange their Musketts if not in order with those that are to stay taking care of ye Exchange that is made in their own Companies & the Arms wanting for those who have not had any to apply to the Q^{r}. Mast^{r}. & give a receipt for the Same likewise what Cloaths are made & apply for Shoes & Blanketts & stop the price out of their Pay if they had them before __ The detachm^{t}: to be ready as soon as possible & when ready to acquaint the comd^{g}. officer immediately _____ Gen: after Orders head Quarters by G. Howe

The Troops order'd for Georgia are to be in readiness by 11 o'Clock to morrow forenoon, when Vessells will be provided for them at a Place that will be appointed convenient. The Detachm^{t}. from Fort Johnston & Fort Moultrie will embark from their Respective posts __ Gen: Moultrie will be so obliging as to give Orders that the Detachm^{t}. from the 2^{d} & 5th Reg^{ts}. be furnish'd wth. a proper Number of Cartridges & flints & every necessary Article for their Reg^{l}. Stores, he will also order those articles of which he has a memorandium & wch. are to compose the Gen: store of this Detachm^{t}: to be in readiness to be put on board the Vessells as soon as possible to wch. he will add whatever he thinks necessary__

Col^{o} Roberts will attend to the necessary Preparation for his Detachm^{t}. who are to take Charge of four field pieces, The Commissary will be at Quarters this afternoon at 4 o'Clock

February 25th. 1777 Parole New York

For the day tomorrow Captn Blake
Town Guard Lieut from 5th Reg^{t}:
Magazine D^{o}. L^{t}. from 5th D^{o}.
Brickhouse Do. L^{t}. Barrak: Guard a Serg^{t}

February 26th. 1777 Parole Port Royal

For the day to morrow Captn. Charnock, Town Guard L^{t}. Mazyck, Magazine Guard Lt Baker Brickhouse L^{t} from 5th Reg^{t}. Pritchards Guard L^{t}. from 5th reg^{t}: Reg^{l}. Guard a Serg^{t}:

Gen: Orders

One Surjeon 1 Surjeons mate to go wth the Detachm^{t}. to Georgia, they will apply to the Director General of the Hospital for such medicine as may be thought necessary, the Remainder of the 2^{d} Reg^{t}. to hold themselves in readiness to go to Fort Moultrie The N^{o}. Carolina Troops are to be in readiness to come to Ch^{s}. Town & Quarter at the new Barracks

Ch^{s}. Town 26th Feby 1777

Orders by Gen. Moultrie

The Troops under Orders for Georgia are to embark this afternoon on board several Schooners provided for that Purpose, the Detachm^{t} from the diff^{t}. reg^{ts}. will each have a vessel Col^{o}. Robert's Detachm^{t}. will be on board the Sloop already allowed them, they will take Charge of the Field Pieces & 10 Barrels Powder to be put on board, one Vessell will be for Provisions & Stores, the Commissary to put on board 4 weeks Provisions of Flour for 600 men & two weeks Provisions wth fifty Bushell of Salt in Barrells, the Comd^{g} Off^{r}. who goes wth. this Detachm^{t}. will apply to the Publick Store Keeper for 4 Reams of Cartridge Paper 1000 w^{t}: Muskett Balls 600 w^{t} Buckshott 2 doz axes 4 doz hoes __ great care must be taken of the Stores & also to apply to his Excellency the Presid^{t}. for an Order for 4000 w^{t}: gun Powder wch. is to be distributed amongst the dift. Detachm^{ts}: on board the Severall Vessels __ An Off^{r} & 20 men to be on board the vessell wth the Stores as soon as all the Troops & stores are on board, the Vessell are to proceed wth. all Expedition to Georgia taking great care to keep together 'till there arrival at the Post __ The comd^{g}: Officer of this Detachm^{t}: will receive his particular Orders in writing __

After orders by Gen: Moultrie Ch^{s}. Town 26 Feby. 1777 ~

1 Surgeon from 5th Reg^{t} & Surgeons mate from 2^{d} reg^{t}: to go wth the Detachm^{t}. to Georgia, they are to apply to the Director Gen: of the Hospital for such medicines as may be thought necessary- The 2^{d} & 5th Reg^{ts}. to hold themselves in readiness to go over to Haddrells Point & the N^{o}. Carolina Troops to be ready to come to C. Town to take their Q^{rs}: at the new Barks. the Q^{r} Mast^{r}. of the 2^{d} Reg^{t} to apply to his Excellency the Presid^{t}: for 6 hogsheads of Rum for the use of the Troops going to Georgia __

________________________Parole Georgia 27th February ____________________________

For the day to Morrow Cap: Potts [536] T. Guard L^{t} from 5th reg^{t}:
Magazine G^{d} L^{t}. from 5th reg^{t}.

G. O. By Gen: Moultrie

The remainder of the 2^{d} & 5th Battallions are to go over to Haddrells Point to morrow, the Guard to be first reliev'd __ An Off^{r}. of each reg^{t} is to apply to Captn. Copithorn [537] for a Schooner __ & a Corp: & 4 men from each Reg^{t}: to sent to ~~him~~ Cap: Copithorn to take Possession of the Schooners he may impress for the Service they are to take 2 days Provisions with them.[538]

____28th February ___________ By L^{t}. Co. Marion

Orders to Captn. on b^{d}. the Schooner

Sir/

You are to take Com^{d}. of the men on Board the Schooner & take great Care the Powder & Stores are put in a Place where it may be safe from Fire or Dampness & you must keep good Order on board ~~your Vessel~~ & make y^{r}. men lodge their arms in such places, as they may readily take them at any Time without Confusion & at a moments warning, You are to use your utmost Endeavour to keep close up to the Vessell wch. carries a Jack [539] at her main topmast head, wch you are not to pass on any acc^{t}. without Orders, & come to an Anchor, or make Sail when you see the above Vessell do so; Should any thing happen Extraordinary at any Time, you are immediately to send & acquaint the comd^{g}. Off^{r}. You are not to suffer any Boats or men to come on board your Vessell without their giving the Parole & Counter Sign: nor suffer any men to go on shore without Orders from the comd^{g}. officer for wch. purpose you are to keep a Guard of 1 Sub: 1 Serj^{t}: & 15 rank & file, Shou'd any Vessell or arm'd Boat appear to be coming towards you, your men must be immediately got ready to

[536] Thomas Potts was a 2nd Lieutenant in the 1st Company of the Prince Frederick's Parish Volunteers, commanded by Captain Benjamin Screven in 1775. He later became a captain in the 5th South Carolina Regiment.

[537] John Copithorn was a naval captain.

[538] John Copithorn impressed four schooners to carry 300 men and their baggage from Haddrell's Point to Georgia. Two of these transports were the *Pendarvis*, owned by William Glen and the *Florentine*, owned by Barnard Elliot. This was the exedition to relieve the siege of Fort McIntosh on the Satilla.

[539] A Jack was a flag that tells what nationality a ship is. The British flag is known as the Union Jack.

oppose them & make all the Sail you possibly can to be near the Vessell wth the Flagg, but shou'd it happen that you are at a Distance in the rear & the Enemy attempt to board you, the greatest Defence must be made, but always send first to the comd^{g}. Officer if practicable to acq^{t}. him wth. the particulars of whatever may present. Every Evening after coming to anchor you are to send for Orders & make a report of whatever may have happened during the Day, to the Command^{r}. of the ~~Reg^{t}~~ Detachm^{t}. You are to be carefull of all the Stores & Provisions & to account for such as may be made use of

---

Orders of L^{t}. C^{o}. Marion of Yesterday
Captn. Ashby wth L^{ts}. Mason & Burke 3 Serg^{ts}. & 60 Rank & file of 2^{d} Reg^{t}. to embark on board Pandarvan's Schooner__

Cap^{n} Vanderhorst two Sub: 3 Serg^{ts}. & 80 rank & file from the 1st Reg^{t} to embark on board M^{r}. Middleton's Schooner

Captn Potts wth the officers & men of the 5th Regim^{t}: to embark on board Col^{o}. Elliott's Schooner

Cap^{n}. Lesessne L^{ts}: Gray & Dunbar wth 1 Serj^{t}: & forty rank & file from the 2^{d} & 1 Serj^{t}. & 20 rank & file from the 1st Reg^{t} to embark on board the fourth Schooner __

The Detachm^{t}. to be in readiness to embark at Eight oClock in the forenoon__ Waggons will be ready to take the Military Stores & Officers Baggage by 6 oClock in the morn^{g}:

By L^{t}. C^{o}. Marion 28th February 1777

Parole Beaufort CounterSign 5

Commanding Officers of the diff^{t} Vessells to be very carefull of their Provisions and Water, they must give out Provisions from the Stores in each vessell where they will find Beef Pork & Ship Bread,[540] a Serj^{t}: to be appointed in each to act as Q^{r}. Mast^{r}. Serj^{t}. who is to keep an acar^{t} acc^{t}. of all Provisions issued __

By L^{t}. C^{o}. Marion 28th Feby. 1777 __ (Copy)
Cap^{n}. White/[541]

You are to take on board the Sloop the Provisions & Stores which are in the Schooner, wch I order you to carry from Prioleau's Wharf along Side the Sloop

This I order to be obey'd without Countermanded by a Gen: Officer, you will as soon as possible get in your men & Stores & join me (Sign'd)

F. Marion L^{t}. C^{o}. 2^{d} Reg^{t}.

Orders 1st March 1777 Parole C. S^{n}.
Sett sail to lay near Fort Johnston & came to an anchor opposite to Mr Lambols Plant^{n}. James Island waiting for the Artillery, Sent Cap^{n}. Lesesne to see if we could get thro: Wappo-Cut; if we cou'd to acquaint the Comd^{t}. & if not to return back immediately & join the Squdron __

---

540 In the 19th century ship's bread would be known as hard tack.

541 Sims White was born in 1738. He served as a quartermaster during July 1775 and became a captain in the Charles Town Artillery on 14 November 1775. He resigned on 12 September 1777, but he was still taken prisoner at the fall of Charleston. He died on 12 August 1799 at the age of 61.

Rec'd Letter from Cap^{n}. Lessesne at 9 o'Clock at night acquainting me that wth. a little Labour we might render it practicable to go thro: the Cutt __ [542]

Chas. town harbour

Orders 2^{d} March 1777 Parole Wappo C. S^{n}. 9

If the artillery joins us in the Beginning of next Flood the whole Squadron must proceed to Wappo Cut to night observing all former Orders __

NB Captn Cogdell [543] / I shall be glad to know the reason you have anchored so great a Distance from the Flagg Schooner contrary to orders. if it was impracticable to come up I expect you woud acquaint me with the Occasion. I expect to sail for Wappo Cut soon to day & hope you will immediately join me

I am Sir &c __ F. Marion

Ch^{s}town harbour —

Orders 3rd March __ Parole Gadsden CounterSign 11

Sett out from Ch^{s}. Town harbour last Night at 8 oClock Captn. Vanderhorst brought me a Letter from Captn. which informing me he had orders to proceed to Beaufort by Sea _ Arrived at Wappo 4 oClock in the afternoon, found the Schooner of Captn. Lesesne in the Cutt __

Wappo Cut

Orders 4th March Parole G Moultrie Counter S^{n} 15.

Any Soldier who molests any of the Inhabitants or takes any thing from them, may depend on being severely punish'd No Person whatever to sell any Spirituous Liquors to any Soldier without a Permit in writing from their officer__ No Officer to give a permit to any man for

---

542 Charlestown was the largest port south of New York City and it was the largest harbor in the South. To get into, or out of the harbor was not an easy matter. An arriving ship had to wait for the right combination of incoming tide and onshore breeze to begin the task of entering the harbor. A ship had to have a local pilot to guide it around the sandbars and shifting underwater obstacles, and many of the pilots in Charlestown were black slaves, sometimes making money on their own time, or freemen. A ship could take one of six channels to get to the city, depending on the draft of the vessel. If the ship were heavily laden it would enter through the main ship channel, however Lawford's channel could be used. Once across the Charlestown sandbar the pilot would tack the ship four miles to the mouth of the harbor, the ship would then make a turn to port in front of Sullivan's Island and proceed towards town. A large sand bar, called the middle bank, divided the harbor into two sections. In the 19th century this middle bank would have Fort Sumter built on it. If a ship went into the southern passage it would pass in front of Fort Johnson on its way to the city of Charlestown. The fort was Charlestown's principal fort, built in 1747. It was located on James Island in Charlestown harbor. The northern passage would pass through Rebellion Road, a large ship anchorage near Haddrell's Point. This name had been given to it in 1744 and had nothing to do with the current rebellion against the British. These passages would take a ship seven hours before a pilot could tie up to the Charlestown docks. Very small craft could get to Charlestown by entering Stono Inlet, proceeding up Wappoo Creek and use the Ashley River to approach the Charlestown docks from the west.

543 George Cogdell served in the 5th South Carolina Regiment (1st South Carolina Rifles). On 17 June 1778 he was court martialed by Colonel Isaac Huger for neglect of duty and disobedience of orders. The court found him not guilty and acquitted him. He died on 5 March 1792.

Liq^{rs} more than ½ p^{t} p^{r} day__ The Soldiers belonging to the diff^{t} Vessells to go on board their respective Vessells by retreat Beating. Any Soldier who shall be found on shore after that time will be treated as a Deserter & suffer accordingly Cap^{ns} Ashby & Lesesne to furnish a Centry each to be planted at the house wch sells Liq^{r}. opposite the Cutt wth Orders not to permit any Soldiers in the house or Suffer Liq^{r}. to be Sold them without a Permit as above __

By L^{t} C^{o}. Marion **5th March 1777** **Parole Gen: How C. S^{n} 4** __
at Wappo Cut _____

No Drums to beat without taking the Lead from the Flagg. All Guards to be reliev'd at 8 o'Clock in the morning__ The Orders of Yesterday to be strictly complied with, particularly that part wch. forbids the men from being on Shore after retreat beating, as a Party will be order'd out after that Time to take up such who may disobey that order & suffer accordingly__ It is expected that a Serg^{t} will be sent to the comd^{g}: officer for Orders every day at 4 o'Clock in the afternoon while the Troops remain at Wappo & those Serg^{ts}: who are sent for orders to wait on board the Flagg till they receive them__

A Court Martial of the Line to sett this morn^{g}. at 11 o'Clock to try such prisoners as may be brought before it Evid: to attend Captn Ashby Presid^{t}: 1 Sub: from 1st Reg^{t}. 1 from 5th: D^{o}. 2 Sub: from 2^{d}. D^{o}. members

NB this Evening we got three Schooners over the Cutt of at 4 o'Clock in the morning the other, we were oblig'd to stay till the Flood made __

Orders 6th March Parole Georgia Counter S^{n}. 6 __
at Wappo River

As we shall be oblig'd to proceed at Night when the Tide serves, the off^{rs}. Commanding the diff^{t}. Vessells are desired to be watchful when the Flagg moves & to keep as Close to her rear as possible __

Orders 7th March Parole St Hellens CounterS^{n}. 8

I am sorry to acquaint the comd^{g} Officers of Vessels that I find it necessary to acquaint them to exert their utmost to forward their different Vessells in this most important Expedition to Georgia & to observe wth. the greatest Strictness all orders issued wth. more Exactness than heretofore. Those Gent^{n}: who have exerted themselves in forwarding the Expedition & observing orders wth. Exactness will merit the Thanks of the comd^{g}. off^{r}.

Orders 8th March Parole Stono- CounterSign 10-

This morning seven o'Clock we sett off & two of the Schooners ran aground half way to Edisto river by this accid^{t}. we lost 2 days__

Orders 9th March Parole Charlestown C. S^{n}. 27.

As all sick Soldiers require the greatest humanity all comd^{g}. officers of vessells are desir'd to take particular Care of their sick & appoint one or two men to attend them whose Business is to administer such medicines & nourishm^{t}: as the Surgeon or mates will order & the officers on Duty is to see it done __ Two G^{ns} Rum is to be drawn from the Publick Stores in each Vessell to be given such

Soldiers as work the Vessells in cold & rainy weather great care must be taken that there is no waste in this article as well as all other Provisions__ When there is occasion for the Vessells to move in the night the Gen: will beat on board the Flagg at wch Time all the Vessells must weigh anchor & the beating of a march will be a Signal that the Flag is under Sail & each vessell is to answer the different Beats as they weigh anchor or are under way__

It is necessary that an Officer should be constantly on deck when the Vessels are under way in order to make the men do their Duty otherways the ~~men~~ Soldiers will not ~~do their duty~~ assist: Six men that are Seamen to be appointed to act as Sailors who are to do no other Duty__

Orders 10th March Parole Col^{o} Pinckney C. S^{n} Sullivants Island

Whenever the Tatto beats it is a Signal for all the Vessels to come to an Anchor__

Orders 11th March Parole Col^{o}. Cattle C. S^{n}. 26.

When any vessells run aground in the rear of the Flagg or cannot keep up wth. her they are immediately to hoist a white Flagg at the fore top mast head & send their Boat immediately to acquaint the comd^{g}. officer with the occasion, shou'd it happen at night that any Vessell by accident cannot follow the Flagg they are to beat the Tattoo as a Signal at night

---

Orders 12th March Parole Col^{o}. Motte C. S^{n}. 29-

Commanding officers of Vessells must not permit any Soldier to go on Shore & such officers who may be permitted to go a Shore must return immediately at retreat beating as we shall proceed as soon as the Tide serves

---

Orders 13th March Parole Fort Lytleton C. S^{n}. 92-

1 Serjeant 1 Corporal & 20 Privates from the Schooner Seaford & 5 Men from the Schooner Susannah to go on board ~~the~~ one of the largest Boats__

1 Serj^{t}: 1 Corp: & 15 Privates from the Mary & 10 Privates from the Susannah to go on Board the 2^{d} largest Georgia Boat__

1 Serj^{t}. 1 Corp: & 14 Privates from the Herculas to go on board the small Boats

These Boats are to keep as close as possible to the Schooners the men were taken out of & come too, close under their Stern when the Schooners come to an Anchor__

Savannah 14th March 1777 ~

Disembark'd the Troops this day at Savannah

NB No orders issued by the Col^{o}.-

By Gen Howe **head Quarters Savannah** **March 15th 1777**

Parole Georgia

For the day to morrow Cap. Ashby__

A main Guard to be mounted to morrow morning at 9 o'Clock consisting of 1 Sub. 2 Serj^{ts}: & 30 Rank & File This Guard to be taken from Col^{o}. Mottes Detachment till further Orders Col^{o} Elbert will fix upon some proper place for this Guard & direct the places at wch. Centries ought to be extended [544] __ Commanding Officers of every Corps in Town will order an Adjutant or some person acting as such to attend at Head Quarters every Morning by 12 o'Clock to receive Orders, they will also make an exact Return to morrow Morning of the Strength of their several Corps __

This order extends only to such Men as they have in Town __

Orders by Col°. Motte

Return to be made as soon as conveniently can be of the Number of Commissioned and non Com^{d}. Officers Drums Fifes & Rank & File from each Regim^{t}. that compose his Detachm^{t}. at the same Time mentioning the State of the arms accoutremts. & ammunition __

The men to be in their Quarters after the Role has been called at Retreat Beating & not to be seen after that Time about the Streets without Permission of a Commissioned Officer__ All Officers are desir'd to visit the mens Quarters and see that they are provided wth. every Necessary that can be procur'd for them. The men are to be made acquainted wth all orders relating to them, by a Commissioned Officer of the Regim^{t}. to wch. they belong. The Guards appointed by L^{t}. Col°. Marion to be continued till further orders__

Lt Elliot of the 1st Regiment is appointed to act as adjutant to the Detachm^{t}. and is to be obey'd as such __

General Orders By Gen How 16th March

Parole M^{c}:Intosh

For the day to morrow Captn White

The Guards to be releiv'd every morning at 8 o'Clock__ officers of Companies to be carefull that the men do no Injury to the Houses when they Barrack or to any other Property of the Inhabitants. Gen: Howe hopes every Soldier will consider himself as ordain'd to protect & defend his fellow Citizens from Injury & Insult; to act contrary to this End will certainly meet wth. punishmt: Captains,

---

544 Colonel Samuel Elbert of the 2nd Georgia Regiment. The Georgia Continental Line was created in January 1776, but only one regiment could be raised from the state's small population. Congress resolved that Georgia could raise two battalions for the defense of the colony of Georgia from the states of Virginia, North Carolina and South Carolina. The recruits were promised bounties of money and clothes, but did not get either. When the Virginia recruits arrived in Georgia they were given "as much Osnabrigs as will make Coats and Kilts." Each soldier was only given 6 cartridges and spanish moss was used as wadding in each cartridge. The soldiers were given loose ball and powder in horns to supplement their ammunition. On February 13, 1778 Congress ordered the four Georgia Battalions to be consolidated into two. Georgia refused to obey, but there were not enough men to fill their battalions. In the Battle of Briar Creek, in March 1779, the Georgia Continentals only had 70 men commanded by Colonel Elbert. They fought a delaying action allowing other soldiers to escape the disastrous battle, but most of them were captured, including Colonel Elbert. Many of the remaining Georgia Continentals were attached to South Carolina units. In April Sergeant Jasper of the 2nd South Carolina conducted a series of clandestine raids across the Savannah river, capturing British soldiers and gathering intelligence. On these raids he took Georgia Continentals who knew the area. During the siege of Savannah the 137 soldiers of the four Georgia battalions were consolidated under Colonel John White and were known as the 2nd Georgia Battalion. After the slaughter of Savannah, the survivors of the four Georgia battalions were officially merged into one unit. At the surrender of Charleston in May 1780 there were only six Georgia officers who were still in service and they became prisoners of war. Elbert remained a prisoner until 1781 when he was released. He was promoted to Brigadier General and served under Washington at the siege of Yorktown in October 1781. When the war ended he attempted to rebuild his mercantile business in Savannah. In 1785 he was elected governor of Georgia. He died at Rae's Hall, his home north of Savannah, in 1788.

Lieut[s]. & Ensigns will be attentive that Orders are communicated to their men that they may not thru Ignorance transgress, or have a Pretext to plead in in Excuse__ [545]

Colonel Sumpters Division will prepare to follow the former Detachm[t]: of his Corps & march the moment the Waggons arrive. The Col[o]. will have a report immediately made to the General what ammunition he wants that it may be order'd___

The Commissary will furnish Col[o] Sumpter with what Provisions he thinks necessary for his march & take care they are sufficiently provided at Sunbury__

The Commissary to attend the Gen[l]. at Head Quarters to day at ½ past 12 o'Clock___

Orders by Col[o]: Motte ___

The whole Detachment off duty to be under Arms every Morning at 9 o'Clock (Sunday only excepted) to be exercis'd by Maj[r]: M[c]Donald___ The Major to fix upon a proper place for Parade. The officers are desir'd to be very punctual in their attendance___

Serjeant Welsh of the 1st: Regim[t]: to act as Q[r] Mast[r] Serj[t]: and Serj[t]. Jasper [546] of the 2[d] to act a Serj[t]: Maj[r]: to the Detachment and are to be obey'd as such ~

The Guards mounted as Barrack Guards to be reduc'd to 1 Corp. & 4 men. Reports of w[ch] & of those Guards on board the Schooners to be made to Maj[r]. M[c]Donald, who is desir'd to report any thing extraordinary to the commanding officer of the Detachm[t]:

Orders by Major M[c]Donald

L[t]. James Mitchell for the Main Guard to morrow [547]__

General Orders by Gen: Howe **17th March 1777 ~**

Parole Elbert

For the day to morrow Cap: Vanderhorst

The Military & other Stores on board the Transports to be examin'd & air'd as immediately as possible & the Vessells properly clear'd out. Col: Marion will appoint the officers & men for this Service & take Care that it is done when the Weather is fair__

---

545 Ensigns was a rank that was under a second lieutenant but were equal in pay. Those that are ensigns are sometimes referred to as a lieutenant's because of the equality of pay.

546 William Jasper enlisted in the 2nd South Carolina Regiment on 7 July 1775 under Captain Barnard Elliot. During the battle of Fort Sullivan on 28 June 1776 the fort's flagstaff was cut by British shot, and toppled to the ground. In the 18th century a lowered flag meant that a fort had surrendered. The regimental colors were also a symbol of honor. Sergeant Jasper jumped up on the ramparts and walked the length of the fort until he came to the colors. He then jumped over the walls of the fort and retrieved the banner, ignoring the rain of shot and shell. Jasper then climbed the walls and tied the flag to an artillery sponge staff and erected it on the walls. He then calmly returned to his gun. He was offered a lieutenant's commission, but he turned it down. He served as a quartermaster sergeant under Captain Thomas Dunbar in 1777. During the siege of Savannah he attempted to plant the 2nd South Carolina's colors on the parapets of the Spring Hill redoubt. He was shot down and as he lay dying, he passed the colors to Lieutenant John Bush, who also fell. The British captured the colors. These colors were returned in 1992 and are periodically on display at the South Carolina State Museum in Columbia.

547 James Mitchell was in the 4th South Carolina Artillery. He was at the siege of Savannah and the siege of Charleston. After the fall of Charleston he was in Sumter's Partisans, with the Fairfield Regiment under Colonel Winn. He fought at Huck's Defeat, Hanging Rock, Blackstock's Plantation, Congaree Fort and Biggin's Bridge. He was severly wounded at Hanging Rock. He died in 1819.

An Exact return of the Military Stores & Provisions receiv'd on board each Transport & also a Return of their pres[t]. State to be made to Col[o]. Marion who is to report to the General__

Doct[r]. Harris is to take Direction of the Hospital & is to regulate it properly, he is (till a Purveyor is appointed) to furnish the Commissary wth a List of such Articles as are necessary for the Sick, who is to proceed accordingly __[548]

Orders By Maj[r] M[c]Donald

Lieut. Gadsden for the main Guard to morrow [549]

Gen: Orders by Gen Howe 18th March

Parole Motte __

The main Guard to be reliev'd to morrow from Col: Elbert's Battallion __

The Detachm[t] under the Command of Col: Motte are to prepare for their return to Charlestown as immediately as possible; they will embark to morrow on board the Transports under the command of Col: Marion who is desir'd to issue the necessary orders__

Orders by L[t]. C[o]. Marion The Transports to be ready to take in the Troops by to morrow 8 o'Clock & hale Close to the wharf__

The Capt[ns]: Officers & Men to go on board the several Vessells that brought them to Savannah All orders from Ch[s]. Town to this Place to be punctually observ'd; The General to beat at 9 o'Clock in the morn[g]: the assemble & march at 10 o"Clock at w[ch]. Hour Col[o]. Mottes Detachm[t]. is to embark as above the Q[r] Mast[r]. Serj[t]: is to apply to the Barrack Master for half a cord of wood for each Vessell w[ch] with water is to be put on Board this Evening or before the General beats to morrow__

Officers Baggage to be on board early in the morning__

Savannah March 19th: **By L[t]. C[o]. Marion**

Parole Cockspear Countersign

all former orders from Charlestown to Savannah to be observ"d, tis expected the Commanding Officers of Vessels will use their utmost Endeavours to keep close up to the Vessell w[ch]: carries the Flagg __

On February 23rd British Regulars, Indians and Loyalists from Florida began to surround Fort McIntosh. For the next 21 days Captain Richard Winn, 52 men of the 3rd South Carolina Rangers and 23 men of the 1st Georgia Regiment held out against a force of 1,200 British troops and artillery. When British Colonel Fuser sent a message to Winn he asked him to look at the British forces arrayed against the fort. Fuser had all his men form up in a single line to show Winn the might against him. He also had three cannons with him.

Winn saw the superior numbers against him, and the additional firepower of the cannons, but he did not trust the Indians. Only after guarantees of their safety from Fuser, did he surrender the fort. The British regulars never fired a shot in this action and only loaded their weapons after the fort was taken.

---

548 John Harris was a physician and surgeon in the militia until 1781.

549 Thomas Gadsden became a first lieutenant in the 1st South Carolina Regiment on 12 May 1776 and was promoted to captain on 6 October 1778 after Captain Roger Parker Saunders resigned. He was appointed Assistant to the Adjutant General of the Southern Department in September 1779. He was killed during the siege of Charleston on 24 April 1780.

Fuser initially had wanted unconditional surrender, but Winn was able to get demands that were favorable to him and his men. The Carolinian's captured rifles were distributed to the Indians. The Georgian's captured muskets were destroyed by fire, along with the fort. Winn's men were able to return to South Carolina. Unfortunately their horses had been turned out into the field when the siege began, and the Indians had captured most of them. Three-fourths of the men had to walk back home.

Upon arriving in South Carolina Winn had to disband his Ranger company, in accordance to the surrender agreement. Once Winn had been exchanged he was appointed the commander of the South Carolina militia Fairfield Regiment. During the siege of Fort McIntosh, the commander of Fort Howe, Colonel Harris, tried to relieve the fort with 300 men, but he was ambushed with a loss of six men killed or wounded.

Lieutenant Colonel Francis Marion and 107 men of the 2nd and 5th South Carolina Regiments, along with four guns of the 4th South Carolina Artillery, and two sloops, hurried to the relief of Fort McIntosh, but they were too late. When Marion learned of the surrender he immediately returned to Charlestown to protect it from possible attack. [550]

Port Royal Bay 20th March
Parole Calliboge Countersign 5

Hilton head 21st March
Parole Genl: Howe Countersign 19 __

Orders 22nd March
Beaufort Parole Beaufort CounterS^{n} Port Royal
The Detachm^{t}. of the 2^{d} & 5th Regim^{ts}: to receive 1 days Salt Provisions to disembark immediately & proceed by Land to Charlestown; the Artillery Stores & c now on board the Sloop. 1 Subaltern 1 Serjeant & 15 men of the artillery to remain on board as a Guard who will receive orders; Likewise a Surgeon to go on board to mind the Sick Soldiers __

Capt. Vanderhorst will put on board the Susannah 1 Subaltern & 20 men & Cap^{n}. White will devide his men between the Mary & Harlequin, 1 Serjeant & 6 men of the 2^{d} to remain on board the Mary __ Capt^{ns}. Ashby, Cogdell & Lesesne will make an exact Return of the Provisions & Military stores now on board their respective Vessells & take a Receipt for the same from the officers Succeeding in the Command of their Vessells __

Captn Ashby will receive his orders as soon as he lands his men __

Sir / Beaufort Port Royal 22nd March 1777 ~
You will take the Command of the Detachm^{t}: of the 2^{d} & 5th Regim^{ts}: & proceed by Land wth all Expedition to Charlestown wch. you will join to their respective Regim^{ts}: upon your arrival __ You are to prevent as much as possible your Men from Stragling from the main Body by placing Centrys by Night round your Camp & not to suffer any Men to go in the Plantations wch. they may come to __ You must be particularly carefull the Soldiers do not Insult or injure any of the Inhabitants of the Country thro: wch. you pass & ~~good~~ Keep good Order__ The Commissary will provide Provisions for the Troops from day to day __ You will give Certificates to the Ferrys wch you pass over

You must also give great charge to the Officers who bring up the rear that they do not Suffer any men to lag behind __

I am yours &c __ Sign'd
Fran^{s}. Marion
L^{t} C^{o}. in 2^{d} Reg^{t} in Con: Service

---

[550] O'Kelley, *NBBAS, Volume One* pp. 182-185.

To
Capt Anthy. Ashby
2^{d} Reg^{t}

Sir/ Beaufort Port Royal 22nd March 1777
You are to proceed on board the Sloop Beaufort commanded by Captn Mercier wth the sick of Col^{o} Mottes Detachm^{t}. & 15 men of the 4th Reg^{t}: as a Guard, & proceed from hence wth all Expedition by Sea to Charlestown, delivering the Men of the 1st Regiment at Fort Moultrie, the 4th at Fort Johnston & the 2^{d} & 5th with the Remainder of the Provisions in Charlestown__

You are to be very carefull of all your Provisions & make an exact Return of what is expended & that wch remains to the commanding officer in Charlestown.

Shou'd you be pursued by an Enemy, you are not to suffer yourself to be taken, but rather run your vessel on Shore, if by that Means you can save ~~your~~ the Troops from being ~~kill'd~~ taken __ Doct^{r}. Marshall [551] will attend the Sick, & in all Things you must conduct yourself wth Prudence & good Discipline to the Service & advantage of the United States

To
Lt Wm Mitchell
4th Regim^{t}

I am yours &c
Francis Marion
L^{t} C^{o} of 2^{d} Reg^{t} in Con: Service

Musqetoo Creek Orders 23rd March Parole Block Island CounterSign 43
Commanding Officers of Vessells will draw 3 Gal Rum out of the Publick Stores & give occasionally to the men who work the vessell

By L^{t}. C^{o}. Marion at Block Island 24th March 1777
Parole Col^{o} Roberts CounterSign 34

At Deadman's Cut. Parole Elbert C. S^{n}. 7 25th March

At Wappoo 26th March Parole Ch^{s} Town CSign 8.
Officers commanding Vessells are to make an exact return of all the Provisions made use of during the voyage to & from Georgia & the Provisions remaining also ammunition & Military Stores &c. this return to be given in to the Commanding Officer as soon as all the Vessells arrive in Ch^{s} Town harbour__

Ch^{s}. Town harbour 27th March Parole CS^{n}
Lt Col: Marion returns his Thanks to the officers for their Endeavours to forward their vessells in this most tedious voyage to & from Georgia and cannot but make known his satisfaction in seeing the officers & men live wth so much unanimity wch. remands to their honour & at all Times will be productive of the happiest Consequences in the Service of their Country & begs that the men may be made acquainted wth his appreciation of their good Behaviour __

[551] Francis Marshall served as a doctor in the militia until 1782.

Copy

Sir/

You will proceed wth the Detachment of the S^{o} Car Regim^{ts}: in your Command to Fort Moultrie & there deliver the men. The Military Stores, Provisions &c you will order to be deliverd to the Publick store Keeper and Commissary in Charlestown taking a recp^{t}. for the Same __

I am yours & c Sign'd

To Lt Gadsden
1st Reg^{t}:

Franc^{s}. Marion
L^{t}. C^{o}. 2^{d} Reg^{t}:

Copy/ 27th March 1777 ~

Sir/

You will proceed wth the Detachm^{t} of the 1st Reg^{t} under ;your command to Fort Johnston & there deliver the men. The Military Stores, Provisions &c you will order to be deliver'd to the publick store Keeper & Commissary in Ch^{s}. town taking a receipt for the same__ \

To
Captn White
4th Reg^{t}

I am yours &c sign'd

Fran^{s} Marion
L^{t}. C^{o}. 2^{d} Regim^{t}.

Coppy of a letter to Captn Vanderhorst at Wappoo _

Sir/ omitted 26th March 1777 ~

Shou'd you not come up to us before we get in Town Harbour you will proceed directly to Fort Moultrie on Sullivants Island & land your men & their join their Regim^{t}: laving 1 Subaltern & a small Guard to the ammunition & Provisions ordering the Schooner to land all Military Ammunition & Provisions in Ch^{s}. Town & deliver y^{m} to the publick Store Keeper & Commissary taking a receipt for the Same__ you will make a return of all Provisions made use of on board your Vessell to & from Georgia giving orders to the officer who delivers the articles above mention'd to make a return of the Military Stores & Provisions wch may be deliver'd to the StoreKeeper & Commissary I am Sir &c Sign'd

To Cap^{n} Vanderhorst Fran^{s}. Marion

By L^{t}. Co. Marion Reg: Orders 27th March 1777 ~

1 Serjeant & 6 men to go on board the Schooner ar Ropers Wharf that brought some Troops_from Georgia & take Charge of the ammunition & Stores left there __

Parole Huger 28th March by Gen: Howe

For the day to morrow Captn Ashby Town Guard L^{t}. Mason
Magazine Guard L^{t}. Dunbar, Brickhouse L^{t}. P Gray

R.O By L^{t}. C^{o}. Marion

A Court Martial to sett to Morrow morning at 11 o'Clock to try such Prisoners as shall be brought before them Evidences to attend, Presed^{t}. Capt Ashby, members L^{ts} Burke & Shubrick

Gen Orders __

L[t] C[o]. Marion will soon as possible report to Gen: Howe the Military & other Stores receiv'd for the use of the Detachm[t] which went to Georgia & also what part of y[m]: were return'd to this State on his arrival. NB accord[g]. to last Court's Sentence, Serj[t] Marlow for misbehavour to Cap: Jervey was reprimanded __

By L[t] C[o]. Marion, March 29th: Parole Motte By Gen Howe

For the day to morrow Capt[n]. Oliphant, Town Guard L[t] Hall,
Magazine Guard L[t] Burke, Brickhouse L[t] Warley

30th March

Parole Pinckney __ by the Gen[l].

For the day tomorrow Cap[n] Potts, Town Guard L[t] 5th Reg[t].
Magazine D[o]. L[t] from 5th. Brickhouse L[t] from 5th.

R.O The Regim[t]. to parade at 4 o'Clock every afternoon (Sundays excepted) to exercise, all Officers to attend It is expected the officers will be particular in making their men turn out at that hour, any Soldier who absents himself without leave may expect to be punish'd

A R. Court Martial to sett to morrow at 10 o'Clock to try such Prisoners as shall be brought before them, Evid: to attend

President Capt[n]. Lessesne, Members L[ts]. Gray & Hart

March 31st

Parole Roberts by y[r]. General

For the Day to morrow Capt[n] Blake, Town Guard L[t] Hart
Magazine L[t]. Warley, Brickhouse L[t]. H Gray Pritchards guard L[t]. from 5th. ~ N.B. Serjeant Flannagan [552] was (according to the Sentence of last Court Martial) reduc'd to the Rank of a Private & was to receive 100 lashes for misbehaviour towards L[t]. Gordon [553] w[l]. on duty, but had the Lashes remitted, Jn[o] Apshear for absence without Leave receiv'd 100 Lashes. [554] Rob[t] Potts for D[o]. & loosing his Arms & accoutrements receiv'd 100 Lashes & put under Stoppages 'till the Gun, Pouch & Bayonet are paid for __ Sam: Kinney [555] & Fran[s]. Dupres remanded for want of Evidence John Clements & Tho[s] Welch to remain in Confinement

Gen: Orders By Gen Howe

A Garrison Court Martial consisting of a President & six members is to sett on Wednesday next within the hour appointed by the Con: Congress & at such place as the Presid[t]: shall appoint for the

[552] Patrick Flannagan enlisted in the 2nd South Carolina Regiment on 4 November 1775 as a sergeant. He was reduced to private on 31 March 1777 and he received 100 lashes for misbehavior towards Lieutenant Gordon while on duty., but the lashes were remitted.

[553] John Gordon had become a second lieutenant in the 5th South Carolina (Rifle) Regiment during 1777. He was promoted to first lieutenant on 21 January 1778. He was an adjutant of Colonel Henry Lee's Legion during 1780 but he was killed in the battle of Eutaw Springs on 8 September 1781.

[554] John Apsheare (Apshead) enlisted in the 2nd South Carolina Regiment on 4 November 1775. On 31 March 1777 he received 100 lashes for being absent without leave.

[555] Samuel Kinney enlisted in the 2nd South Carolina Regiment on 11 March 1776. He was court martialed for an unnamed offense on 31 March 1777, but the charges were dropped for want of evidence. In 1779 he was under Captain Charles Motte. He was in the siege of Savannah.

Tryal of John Eustas [556] a Soldier in Col°. Pinckney's Battallion for abusing & striking Jn° How, an Inhabitant in this Town, a Capt. from 5th Regiment to be Presidt of this Court, the other members to be taken from 2d & 5th Battallion by Detachmt.~ Adjutant D'Elliant will summon them according to Detail, Jn°. How is to furnish the Evidences & the Adjutant will warn him to attend __ Gen: Howe thinks proper to inform the several officers of Guards who have reported to him some matters worthy of notice, that it has not as yet been in his power to pay that attention to their recommendation wch they so justly deserve & wch otherwise he shou'd wth. Pleasure have done ___

R Orders by Lt. C°. Marion By a former Order no Officer is to give leave of absence to more than 2 men at a time, tis expected this order will be more particularly attended to for the future__ It has frequently been in Orders for one Comd: Officer of a Company to attend Roll Call, Commanding officers of Companies are desir'd to see this Order punctually complied wth. as they must answer in future for all neglect in their Company__

By Lt. C°. Marion April 1st 1777 ~

Parole Marion ___________ by Gen; Howe

For the day to morrow Capn Shubrick, Town Guard Lt. P Gray

Magazine Lt. Roux Brickhouse Lt. Hall

Officers for the Garrison Court Martial 1 Capn & 2 Sub: from 5th Regimt. Lt Mazyck, Shubrick & Burke 2d regt

Gen: Orders. . . . .By Gen Howe__

The General is much displeas'd wth a great number of Soldiers he observ'd drunk in the Street within this day or two, many of them wth the great risque of their health; and in danger of being run over by Carriages, laying in the most beastly Situation in the Publick Streets __ It is wth Concern he expresses a Doubt that officers of Companys are not sufficiently attentive to their Men, or they could not so frequently get drunk, or at least be confin'd to their Barracks while they were so __ It is the duty of every Soldier who is sober to be carefull of those who are in Liquor and to convey them to their Barracks whenever they meet them in the Street, whether they happen to belong to their particular Battallion and Company or not, & any Soldiers who in future shall neglect to perform this Office of humanity, will certainly be consider'd as guilty of a Breach of Duty, wch will make him liable to a Court Martial __

By Lt. C°. Marion April 2nd Parole Cattel By yr Genl.

For the Day to morrow Captn Lesesne For the Guards from 5th

NB. Lt. Perreneau was reliev'd by Lt Warley being sick at the Town Guard, Lt Hall was warn'd to take Lt. Masons Guard because he is sick also __

According to the Sentence of the Court Martial the 31st March last Thomas Welsh & Jn° Clements for absence from the Command was dismiss'd__ Frans. Dupres for abusing & striking Serjt Coffer receiv'd 30 Lashes __ [557]

---

[556] John Eustace (Eustas) enlisted in the 1st South Carolina Regiment on 4 November 1775. On 31 March 1777 he was court martialed for abusing and striking John How, a resident of Charleston.

[557] Lewis Coffer enlisted in the 2nd South Carolina Regiment on 4 November 1775 as a sergeant. He was appointed the Sergeant Major of the regiment on 20 October 1777. On 27 January 1778 he was charged with being drunk on duty and assaulting several men. He was acquitted of being drunk, but he was found guilty of striking Corporal Nehemiah Watt and was reprimanded. He was discharged on 13 July 1778 and was appointed the regimental sutler on 25 July 1778.

R.O. A Court Martial to sett to morrow 10 o'Clock to try such Prisoners as may be brought before them __ Evidences to attend_ President Capt Lesesne members L^{ts}. H Gray & Warley__

Commanding Officers of Companies are desir'd to make their men clean their arms & keep them so; any Soldier who shall appear on Parole, or Guard wth. their Arms dirty or rusted, to be confin'd for disobedience of Orders__ & suffer accordingly__ Maj^{r} Horry & Adjutant D'Ellient will be particular in confining such Soldiers who do not comply wth this Order __Commission'd & non Com^{d}. Off^{rs}: are desir'd to make their men dress properly whenever they turn out on Parade, at roll call, for Guard or Exercise & make them keep Silence while under arms, the comd^{g}: Officer is asham'd to see the Men in such Disorder on the parade__

By L^{t}. C^{o}. Marion **2nd April 1777 ~**

Gen: Orders By Gen: Howe

The officers of those Guards from wch James Kelly & Josh. Stone have escap'd are immediately to furnish the General in writing wth such reasons as they have to offer in Excuse for this Accident, and also to inform him what Centries were on duty when the men escap'd & why they have not been confin'd for such unpardonable neglect__

April 3rd

Parole Livingston By the Gen:

For the day to morrow Cap^{n}. Shakelsford, Town Guard Lt Burke

Magaz: Guard Lt Warley Brickhouse Lt Dunbar

R.O. Commanding Officers of Companies to make a return of what Bayonets & Pouches wanting for their Comp^{ys}. by to morrow 10 o'Clock in the Forenoon; tis expected they will make such men pay for what they may have lost & replace them as soon as possible, no Officer off Duty to be absent from Exercise in afternoon without Leave from the commanding Officer, such who may be sick are to acquaint him immediately __

Gen after Orders

The Escape of the Prisoners the other Night from the guard is a Circumstance exceedingly displeasing to the Gen: such accidents can hardly ever happen but from a Want of Vigilance, & by neglect of duty, an Escape of this sort is an injury to the Service & a reflection upon the conduct of the Guard when it happens, it therefore behooves every officers & Soldiers of the Guards to exert themselves to prevent its happening, No Prisoners Committed to the Guard is upon any Pretense whatsoever to go out, but in Charge of some person belonging to the Guard who will certainly be made answerable for the return of the Prisoner, any Centries who suffer any Persons Confin'd in the Guards to pass out without the order of the Commanding Officer or Serj^{t}. of the Guard, will be try'd & punish'd by a Court Martial __

Regimental After Orders

The regimental Court Martial which was order'd yesterday to sett today Cap: Oliphant Presid^{t}. L^{t}. Warley & Roux Members to sett to morrow 10 o'Clock in the forenoon to try William Henson of Captn: Blakes Comp^{y} & give a definitive opinion of said Henson as the Prisoner acknowledg'd the Crime laid to his Charge & the Court think him guilty of the Crime, but no Punishment is adjudged nor is he acquitted __[558]

[558] William Henson enlisted in the 2nd South Carolina Regiment in 1777 in Captain Blake's company. He deserted in June 1777. He was promoted corporal in the 1st Vacant Company on 20 July 1778. He was reduced

to private and confined to the Black Hole for 8 days on 27 December 1778 for striking and abusing Sergeant Simpson. He was in Captain Dunbar's light infantry at Savannah in 1779.

April 4th

Parole Thompson, . . By the Presidt:

For the day to morrow Captn Motte, Town Guard Lt H. Gray
Magaz: Do. Lt. Warley, Brickhouse Do Lieutenant Hart __

April 5th

Parole McIntosh By the General

For the day to morrow Captn. Ashby for the Guards from 5th Regt. N.B. The Guard at Hobkaw to be reliev'd to morrow morning __

Gen: Orders By Gen: Howe

The Officer of the Guard for this day will immediately inform the officer of the day what repair is necessary for the Guard Room to secure the Prisoners, who will after receiving this information wait on his Excellency the President & request the Favour of him to order the repair to be made wth all possible Expedition, he will also apply to his Excellency for such other matters as are requistd to the same purpose, Colo. Pinckneys Battallion will be receiv'd on Friday next, Colo: Robert's Regimt: the Monday following Colo Mottes & Colo Hugers' the Friday after that, The review will be between the Hours of ten & four on each day and on their respective posts

April 6th

Parole Roberts__ By the Presidt

For the day to morrow Captn. Cogdell of the 5th Regimt.
Town Guard Lt. Hall Magaz: Do. Lt. P. Gray
Brickhouse Lt Roux__

R.O. Lieut. P Gray is appointed a first Lieut in the Regt. and is to be obey'd as such, he is to join Captn Oliphants Company and do duty in the same as his first Lieutenant

A Regimental Court Martial to sett to morrow morning at 10 o'Clock to try all Prisoners brought before them Evid: to attend, Cap: Blake Presidt. Lieut. Mazyck & Hart members

April the 7th

Parol Beekman. By Gen: Howe___

For the day to morrow Captn Oliphant, for the Guards from 5th Regt. Pritchards Guard Lt Shubrick

By Colo. Motte **Charlestown April 7th 1776** [559]

R.O. The order of the 30th last month to be strictly complied wth: Commanding Officers of Companies to see that the arms & accoutrements of their Companies are immediately put in good order & that their men be fitted wth: Regim: Caps __ All the officers to provide themselves wth. white waistcoats & Breeches, as also a pair of blk half Gaters and to have Bayonets to their Fuzees, by the day the Regimt: is to be review'd

The Articles of War to be read at the head of the regimt to morrow afternoon.~

The Drummers & Fifers to be out at practice every forenoon from Ten to Twelve O'Clock (Sundays excepted)

[559] This was mistakenly written down as April 7th 1776 and not 1777.

April 8th

Parole Laurens By the Presidt.

For the day to morrow Captn. Potts, Town Guard Lt. Dunbar
Magazine Lt. Martin,[560] Brick house Lt. Mazyck

R. O. A Return of Arms wanting to each Compy: to be made immediately and deliver'd to the Adjutant, The Regim: Court Martial that satt Yesterday to meet again to day at 10 o'Clock to try all Prisoners brought before them –

Gen: Orders, By General Howe

The General desires to be inform'd if an order issued the 5th Inst respecting the officer of the day & the Officer of the Guard for that day has been Complied with Those Officers will therefore report accordingly to Morrow morning - The Officer of the day from this day is desir'd to make a particular Enquiry into the Conduct of Serjt. Curtis,[561] Jo. Whitfield [562] & Jo. Holmes [563] now under Confinemt: for the Escape of Js. Kelly of Colo Thompsons Battallion under Sentence of a Gen: Court Martial for desertion & report the Circumstances to the General to morrow Morning

April 9th

Parole Elliott__ By the Genl.

For the day to morrow Captn Blake, Town Guard Lt the 5th.
Magazine Lt. Burke, Brickhouse Lt. from 5th.

R.O. The Pay Bills to be deliver'd to the Paymaster immediately- The Co: is very much displeas'd that he is oblig'd to remind the officers of this part of their duty he hopes they will not give him reason to be so again- as he is determin'd to make an Example of the first that disobeys an order that has been so often repeated-

N.B. According to the Sentence of last Court-Martial, Thos. Walker for Misbehaviour & insolence to Lt. Eveleigh receiv'd 100 Lashes Wm. Baldwin for Neglect of duty rec'd 100 Lashes Corp: Fazier of Cap: Moultries Co: for Neglect of duty mulct 14 days Pay for the Benefit of ye Regt. Hospital-[564]
The other Prisoners remanded for want of Evidence

---

[560] John Martin was commissioned as a second lieutenant in the 2nd South Carolina Regiment on 16 February 1777 (the Officer's list says 3 March 1777). He was promoted to first lieutenant on 16 February 1778 in Captain Mazyck's company. When the British threatened Charleston in May 1779 he was assigned to Captain Mazyck's company. He fought in the siege of Savannah. Later he was promoted to Captain. He was taken prisoner at the fall of Charleston. He was exchanged during November 1780 and served until end of the war.

[561] Benjamin Curtis enlisted in the 5th South Carolina Regiment as a corporal on 20 May 1776 and was a sergeant in January 1777. On 8 April 777 he was charged with allowing James Kelly to escape from confinement, but the charges were dropped. He transferred to the 1st South Carolina Regiment after the regiments were consolidated in February 1780.

[562] Joshua Whitefield was in the 2nd South Carolina Regiment. On 8 April 777 he was charged and found guilty of allowing James Kelly to escape from confinement.

[563] John Holmes enlisted in the 5th South Carolina Rifles on 20 March 1776. On 8 April 777 he was charged with allowing James Kelly to escape from confinement, but the charges were dropped. He was in Captain Dunbar's light infantry at the siege of Savannah. He was in Marion's partisans in 1782. On 29 October 1782 he was found guilty of disobeying orders and sentenced to do double the length of time he was drafted for.

[564] Samuel Frazier was in the 2nd South Carolina Regiment under Captain William Moultrie. He received 35 lashes on 7 July 1777 for being in "liquor" on guard duty. He was with Sumter's partisans in 1781, under Captain Samuel Martin and Lieutenant Colonel William Polk.

Gen: Orders By Gen: Howe-

Serj^{t}. Curtis & Jno Holmes of the 5th Reg^{t} having been confin'd upon Suspition of their being accessary to the Escape of J^{s}. Kelly a deserter under Sentence: & whereas it appears they were not concern'd in it, they are therefore to be discharg'd from Confinem^{t}. & return to their duty

The Officer of the Guard is immediately to send a Serjeant & a file for Joshua Whitefield & to have him well iron'd & securely kept 'till further Orders, as he appears to be the Person who aided & abetted Kelly in his Escape-

By C^{o}: Motte/ **Charlestown 10th April 1777 -**

Parole Armstrong By the President

For the day to morrow Captn. Shubrick town Guard L^{t}. Hart, Magaz: D^{o}. L^{t}. Hall, Brickhouse L^{t}. Warley

R. O. A Reg^{t}. Court Martial to sett this Morning at 11 o'Clock to try such Prison^{rs}. as may be brought before y^{m}. Evidences to attend Presid^{t}. Cap: Ashby L^{ts}. H Gray & Martin Members -

April 11th

Parole McDonald, By the General

For the day to morrow Captn. Moultrie, Town Guard L^{t}. 5th

Magazine D^{o}. L^{t}. 5th. Brickhouse from 5th.

R. O. A Regimental Court Martial to set this morning at 10 o'Clock to try such Prisoners as may be brought before y^{m}, The Evidences to attend in time- President Captn Moultrie L^{ts}. Mason & Peter Gray Members-

N. B. According to the Sentence of the Court Martial the 10th Ins^{t}. Jno Hawkins for absence from Roll Call & neglect of Duty receiv'd 50 Lashes,[565] Owen Griffin on Suspicion of Selling B Delony's Reg^{l}. Gun for want of sufficient Proof was acquitted,[566] Robert Ivey for Misbehaviour & Insolence to M^{rs}: Mouatt [567] &c receiv'd 100 Lashes [568]

April 12th

Parole Lee, By the President

For the day tomorrow Captn Jarvey

Town Guard Lt Reaux

Magazine from the 5th Reg^{t}.

Brick House guard Lt Martin

Hobkaw guard new relief

NB According to Sentence of last court James Hooper for misbehaviour was repremanded, Richd Williamson for drunkenness mult 14 days pay to purchase necessary, by his Captn.

[565] John Hawkins enlisted in the 2nd South Carolina Regiment on 4 November 1775. On 11 April 1777 he received 50 lashes for being absent from roll call and neglect of duty. He was in the 2nd Vacant Company in 1778. When the 2nd Vacant Company was disbanded on 5 October 1778 he was assigned to Captain Charnock's company. In 1779 he was under Captain Adrian Proveaux.

[566] Jean Baptiste Delauney enlisted in the 2nd South Carolina Regiment on 4 November 1775. He was sentenced to receive 100 lashes for desertion on 24 June 1777, but he was pardoned. On 2 August 1777 he was reprimanded for being absent from guard duty. He was discharged on 1 July 1778.

[567] Possibly the wife of Lieutenant John Mouatt of the 1st South Carolina Regiment.

[568] Robert Ivey enlisted in the 2nd South Carolina Regiment on 4 November 1775. On 11 April 1777 he received 100 lashes for misbehavior and insolence to Mrs. Mouatt. He was discharged on 8 July 1778. He served 31 days as a lieutenant in Marion's partisans during 1782.

Orders 13th April by Genl Howe } Parole Horry [569]
For the day Tomorrow Captn Shakelford [570]
Town guard Lt. from 5th Regt
Magazine- Lt. Mazyck
Brick House Lt. P Gray.

Orders 14th April by G. Howe } . . . . Parole Baltimore . . . . by presidt.
For the day tomorrow Captn Spencer
Town guard Lt Warley
Magazine from 5th Regt
Brick house Lt. Burke
Pritchards from 5th

Regtl. Ordrs. All the Officers off duty to be on parrade this Afternoon Precisely at four O'Clock

A Regimental Court Martial to sett this morning at 10 o'C. to try such prisoners as shall be brought before them evid: to attend

Captn Lessesne Presidt.
Lts. T Hall & R Mason Members

Genl Orders by G. Howe

The Review of Colo Roberts Corps of Artillery is postponed till Thursday- Adjutant Delliant will therefore cause this order to be transmitted to the Colo.

N. B According to Sentence of the above court Petr Fagan [571] for breakg. his gun to receive 100 lashes 50 remitted & stoppage to be made to pay for the gun-
Adam Meek for Absents from roll call 39 lashes

By Col: Motte **Chs town 1777 ~**

Orders 15th April by G. Howe } . . . . . Parole Thompson . . . .by Gen
For the day tomorrow Captn Ashby
Town Guard Lt. from 5th Regt.
Magazine Lt. Hart
Brick House Lt Thos Hall

---

[569] The handwriting changes on this date, indicating a new orderly was chosen.

[570] Captain Shackelford served in the 5th South Carolina Regiment (1st South Carolina Rifles) until he resigned on 7 December 1777.

[571] Peter Fagan (Fagen) enlisted in the 2nd South Carolina Regiment on 4 November 1775. He was under Captain Blake during 1777 and 1778. On 14 April 1777 he was supposed to receive 100 lashes for breaking his gun, but it was remitted and his pay was stopped to pay for the gun instead. On 12 May 1777 he received 50 lashes for being drunk on parade. On 14 May 1777 he received 75 lashes for neglect of duty. On 24 June 1777 he received 25 lashes for letting a prisoner escape while guarding him. On 7 July 1777 he received 50 lashes for quitting his post. On 2 August 1777 he was reprimanded for being absent from guard. On 22 October 1777 he received 77 lashes for being drunk on guard. He received 100 lashes on 7 November 1777 for being absent for two days and nights. On 20 January 1779 he receieved 50 lashes on the bare back with switches for selling his regimental blanket.

Complaint are again made that the Hospital is in want of wood, Mr Will is therefore Immediately to furnish it, & is expressly Order'd to prevent that article being wanted for the feuter [572]

The Commanding Offrs. of Guards where any prisoners is taken ill is Immediately to Inform the Director of the Genl Hospital of it, who is to examine & report to the general if it tis requests to remove him

---

Orders 16th April by G Howe } .... Parole Mason .... by the Presidt
For the day tomorrow Captn Cogdell
Town Guard from the 5th Regt.
Magazine Lt Reaux
Brick House from 5th Regt.

~~Regtl. Ordrs.~~ A garrison court martial to sett for the tryal of Joshus Whitfield of the 5th Regt. on fryday next, Confined for having Aided & Abetted Jams Kelly a Soldier under Sentence of a General Court martial, in his escape, the Court is to consist of One Captn as presedt. & Six Subaltern Officers to be taken from the 2nd and 5th Regts. & to sett Within the hours presented by the Continental Congress at the Usual place of holding court martials, the Adjutant of the 5th Regt. will warn the Officers According to detail & Order Sergt. Curtis & Robt. Ivey as evidences, & all such other persons as may appear to know any thing of the fact- The Review of the 2nd Regt is postponed till Fryday the 25th Instant & the reviews of the 5th Battallion till Munday morning -

By Colo Motte **Chs town 1777**

Regimental Orders- No Officer to absent himself from exercise in the Afternoon without Leave from the Commanding Officer of the Regiment- who expects that the Offrs. Commdg. Compns. will immediately provide such men with arms & Accoutrements who want them & that they see them well cleaned & in good order

---

Orders 17th April by G Howe } .... Parole Moultrie .... by GH
For the day tomorrow Captn Oliphant
Town Guard Lt. Martin
Magazine Lt. P. Gray
Brick House from the 5th Regt.

Regtl. Ordrs. - No non Commissioned Officers or Soldiers to fire off his gun in or near the Barracks yard without Leave of the Commanding Officer of the Regt. Any person Disobeying this order may depend on being Severely punished, the men to be made Acquainted with this Order by the Offrs. of each Compy.

A Regimental Court Martial to sett this forenoon at 10 O'C: to try such prisoners as may be brought before it, Evid: to attend

Captn Richd Shubrick Presedt
Lts. Jacob Shubrick & Dunbar Members

For the Garrison Court Martial tomorrow Captn Lesesne Presedt. Lts. Hall, Warley, & Hart Membrs

---

572 Philip Will was in the German Fusiliers in Charleston in 1775. He served as a barrack master general from 25 March 1776 to 25 December 1779. He was a deputy quartermaster general from 17 March 1779 to 16 August 1780.

NB according to Sentence of the above court martial Pet[r] Deviney for neglect of duty received 50 lashes [573] W[m] M[c]Culloch for ditto rec[d]. 100

Orders 18th April } .... Parole Gadsden .... by G.H
By Gen Howe } For the day tomorrow Capt[n] Potts
Town Guard L[t] Mazyck
Magazine from 5th Reg[t].
Brick House L[t] Shubrick

By Col[o] Motte **Ch[s] town – 1777 -**

A Garrison court martiall to sett tomorrow consisting of one Capt[n] as presid[t] & 6 Subalterns to be taken from the 2nd & 5th Battallions According to detail & to be warned by Ad[j]. Barten this court is to try such prisoners as shall be brought before them, the Adjutant will inform himself who are the evidence & Order them to attend, the Court to sett within the hours prescribed by Cont[l]. Congress

The Sentence of the Garrison court martial ordered to try Joshua Whitefield is Approved off & ratified the Sentence to be executed tomorrow 10 O'C: in presence of the Battallion which Whitfield belongs

The Waggon master is to furnish the 2nd Continental Battallion of Georgia with a Waggon tomorrow morning as early as possible

---

Orders 19th April } ....Parole Howe .... by presed[t]
by G Howe } For the day tomorrow Capt[n] Shubrick
Town Guard from 5th Reg[t]
Magazine L[t] Shubrick
Brick House from 5th Reg[t].

The Garrison court martial Order'd to sett this day is postponed till further Orders

Regiment[l]. Ord[rs].

The Col[o]. desires all his Officers will merit him with their Favers at 12 OC: this day in the Barrack yard- All the Staff Offic[rs] to attend every afternoon when the Reg[t]. turns out to exercise

---

Orders 20th April } ....Parole Virginia .... by G.H
by G Howe } For the day tomorrow Capt[n] Charnock-
Town Guard L[t] Burke
Magazine L[t] Warley
Brick House from 5th Regem[t].

---

[573] Peter Deviney was in the 2nd South Carolina under Captain Thomas Moultrie. On 17 April 1777 he received 50 lashes for neglect of duty. He died on 18 October 1779 after the assault on the Spring Hill Redoubt in Savannah.

By Col° Motte **Chs town – 1777 -**

Orders 21st April
by Gen Howe

. . . . Parole Chs.Town . . . . Presedt

For the day tomorrow Captn
Town Guard Lt Hall
Magazine from 5th Regt.
Brick House from 5th d°
Pritchards Lt Hart

Regtl Ordrs _ The Col° Desires the Officers of the different Compys to Provide their men with such necessaries as they stand in need of for the Review & to Reimburse themselves by stoping the Mens Pay

The Pay master is not to pay the men till after review Day

Orders 22nd April
by G. Howe }

. . . . Parole Sumpter . . . . . . by GH

For the Day tomorrow Captn Harleston
Town Guard from the 5th Regt.
Magazine Lt Martin
Brick House Lt Reaux

The Genl has Accepted the resignation of Majr Adm McDonald & Lt Adams of Col° Mottes; they therefore no longer to be consider'd as Officers in the Continental Army, Kingham a Soldier in the 5th Batalion confin'd in the main guard is to be deliver'd by the Offr. of that guard when Col° Huger send for him that his conduct may be made the Subject of a Regemental Court Martial.[574]

Regemtl. Ordrs. All the Offrs. to Appear with their Fusees & fixt: Bayonets in Batallion every day at exercise the major only excepted- A Regtl Court Martial to sett this morning at 10 OC: to try such prisoners as shall be brought before them- Evid: to attend

Captn Charnock President
Lts. Burke & Baker members

By Col° Motte **Chs town 1777**

Lt David Adams having resigned his Commission wth leave is no longer to be respected as an Offr in the Regt.

Orders 23d April
By G Howe }

. . . . Parole Prince Town . . . . by presedt.

For the day tomorrow Captn Ashby
Town Guard from the 5th Regt.
Magazine from ___ 5th.
Brick House Lt Mason

Regtl: Ordrs.- The Majr. is desir'd to see that the Armourers are Constantly Imploy'd in Repairing the Arms of the Regt. & if he finds them neglect their duty to report them to the Commdg Offrs of the Regt.

---

574 Edward Kingham enlisted in the 5th South Carolina Rifles on 8 March 1777.

A Regimental court martial to sett this morning at 10 OC: to try such prisoners as shall be brought before them, all evidences to Attend

Capt[n] Harleston President
P Gray and Dunbar members

NB According to Sentence of last court martial John Thompson rec[d] 100 lashes- Tho[s] Bowen for neglect of duty 25 lashes, Corp[l] Gammel for do to be reduc't in the ranks as a private

---

Orders 24th April by G. Howe } ....Parole Williamsburgh.......... by G. H
For the day tomorrow Capt[n] Cogdell
For the town Guard, Magazine, Brick House } From the 5th Reg[t]

Reg[tl]. Ord[rs]. The men are to employ themselves all this day in cleaning & puting their Arms in good Order, any one Appearing with them otherways tomorrow may be Assured of being punished Severely- If the weather proves fine tomorrow the Cloathing is to be served out to each company, Command[g]. Off[rs] of Comp[ys]. are to Distribute them to their best men as far as they will go, & to be perticularly Careful that they be dressed in as neat & soldier like manner as possible, and to be ready to turn out at nine oC: in morning [575]

The Quarter master to Attend at 7 OC: to deliver the Cloathing

A Regimental court martial to sett at 11 OC: this morning to try such prisoners as may be brought before them, Evidences to attend

Capt[n] Ashby Presid[t].
L[ts] Hall and H Gray Members -

General Aft[r]. Orders by G. Howe

The Guards in town to be relieved tomorrow morning at 6 OC: precisely from Col[o]. Hugers Reg[t]. as Col[o]. Motte's is to be review'd they are after that to be mounted as usual till Munday Morn[g]: at 6 OC: at which time they are to be reliev'd by Col[o] Mottes that being the review day of Col[o] Hugers Batallion

The following promotions have taken place in Col[o]. Pinckneys Batalion, Capt[n] Scott made Major in room of Maj[r]: M[c]Donald resigned, L[t] Turner in Room of Capt. Scott promoted L[t] Jn[o] Williamson made first L[t]. in r. of L[t]. Turner promoted [576] & L[t] Glover made 2nd Lieut[t]. [577] They are to be Obey'd & respected accordingly - - - - - - - - - - - - - - -

---

[575] The blue coat "regimentals" were issued to the rank and file only when needed. The soldiers would wear the hunting frock for every day use. By 1779 each soldier would have their own regimental coat and would not have to turn it back in to the quartermaster. This was not done in Continental regiments in the other States, but since the South Carolina regiments were "garrison" regiments, they were able to issue and return clothing, as it was needed, without hurting the efficiency of the unit.

[576] John Williamson was promoted to 1st Lieutenant in the 1st South Carolina Regiment when Lieutenant Turner was promoted to captain. He was promoted to captain on 12 May 1779. He was taken prisoner at the fall of Charleston in May 1780. He was in the militia in 1782. He married Amelia Dixon and he died around 1790.

[577] Wilson Glover was born on 28 March 1758. He was commissioned a 2nd Lieutenant in the 1st South Carolina Regiment on 24 April 1777. He was promoted to 1st Lieutenant on 16 November 1778 when Lieutenant Gray was killed on the Frigate *Randolph.* He resigned on 24 November 1779. After the fall of Charleston he was in Marion's Brigade of Partisans. He died after 1790.

NB according to last court martial, Jerem: Mc Ginnes & B. Wilmon was reprimanded at the Halberts for neglect of duty[578]- & Rowland Thomas for the same crime rec^d. 100 Lashes

---

Orders 25th April } G Howe .... Parole Horry .... by presed^t
For the day tomorrow Captn Oliphant
Town Guard L^t from 5th Reg^t.
Magazine Lt Mazyck
Brick L^t. Pet^r. Grey

NB according to Sentence of Last Court Moses Bruce [579] for theft of 2 shirts & 1 p^r trowsers to be mult £10, & Reprimanded [580] -

By Col^o Motte **Ch^s town 1777**

Regimental Orders

The Regiment to be under Arms at 9 OC: this morn^g. & to be Drawn up in the Barrack Yard_ The men to be carefully Sized & their Arms & Accotrement Examined before they are marched off the parrade- The Col^o Expects that every man will Appear very clean his hair well Comb'd & powder'd & his shoes well Blacked [581]~ The Off^rs to be very particular in Appearing with White waistcoat & Breeches, & half gaters, Black Garters under the knee & a Black ribbon over their stocks- The Quart^r Master to deliver out 18 Cartridges p^r Man & as many flints as may be wanted by each Company

---

578 Benjamin Wilmot.

579 Moses Bruce enlisted in the 2nd South Carolina Regiment on 30 December 1776 in the Lieutenant Colonel's company. On 25 April 1777 he was reprimanded and mult £10 for stealing two shirts and a pair of trousers. On 3 November 1777 he had to run the gauntlet through the whole regiment for an unnamed offense. In 1778 he was under Captain John Blake. He received 50 lashes on the bare back with switches on 20 January 1778 for selling his blanket. In 1778 he was in the Lieutenant Colonel's company. On 4 November 1778 he was sentenced to run the gauntlet through the whole regiment for theft. On 20 January 1779 he received 50 lashes on the bare back with switches for selling his regimental blanket. He was in Captain Baker's company in 1779. On 11 February 1779 he had to run the gauntlet through the regiment three times for theft. He also had to do double duty for 14 days.

580 There are quite a few court martials that deal with the men selling their uniforms, and this is one of the reasons why the more expensive clothing, such as the blue coat regimental, was only issued out when needed. Prior to the invention of the cotton gin the production and procurement of clothing was time consuming and expensive for all social classes. If clothing became permanently stained or frayed around the edges it was not discarded, such as we see today. Today even the poorest of our citizens are able to purchase disposable garments with a frequency which would have astounded people in the Revolutioanry War. Americans of the 18th century owned far less clothes and wore them until they literally disintegrated. People patched and mended clothing repeatedly to extend its life span. When a garment could no longer be repaired it may have been turned into children's clothing or used for linings or quilts. In the British army the previous year's issue of a coat or waistcoat, became the next year's lining or waistcoat back. There was also a major business built around the used garment trade in England. A relatively unworn pair of pants, or waistcoat, could be easily sold for a nice sum of money, and this was a temptation that some soldiers could not resist.

581 Very few people wore wigs at this time, and it had gone out of fashion by the time of the Revolutionary War. Hair was worn so that it could be put up in a que or a tail. The hair of soldiers was powdered for full dress mode, with a white powder resembling flour.

Orders 26th April }
By G. Howe }

......Parole Maryland
For the day tomorrow Captn. Potts
Town Guard Lt Shubrick
Magazine Lt Burke
Brick House from the 5th Regt.
Barracks a Serjt:

Regtl. Orders

The men to put their Arms in good Order immedeatly & to Keep them so, the Regt. is not to turn out to exercise this After Noon, that the Arms may be well cleaned, anyone Appearing with them Dirty will be punish'd

A Regt. Court Martial to sett this Morning at 10 OC: to try such Prisoners as shall be brought before them Evidences to attend

Captn Oliphant president
Lts. Burke and Reaux

By Colo Motte **Chs town 1777 -**

Orders 27th Ap: }
by G. Howe }

...... Parole Dry ....
For the day tomorrow Captn. Blake
Town Guard Lt Dunbar
Magazine from 5th Regt.
Brick House 5th Regt.
Barracks a Sergeant

by presedt

The Review of Colo Hugers Regt is postponed till Wednesday next the Guards theirfore will be mounted as usual untill that day when they are to be relieved by Colo Mottes Regt at 6 OC: in the morning precisely

Regimtl. Orders

The Guards to be ready to march off tomorrow morning precisely at 6 OC:

Orders 28th April }
Genl Howe }

.... Parole Harnett .... by presedt
For the day tomorrow Captn Shubrick
Town Guard from 5th Regt.
Magazine Lt Hall
Brick House Lt Warley
Pritchard from 5th Regt - Barracks a Sergt.

A General Court Martial to be held on tursday next at Fort Moultrie within the hours prescribed by the Articles of war, for the tryall of Several persons belonging to the first regimt for desertion. The Adjutant of that Batallion will Summons the evidences to Attend of this Court Lt Colo Elliott is to be Presedt. the other members are to be taken ~~from the~~ accordg. to detail from the 1st, 2 & 5th Battalions, the Adjt: of Colo Hugers Regt. will transmit this Order to the several Corps

Genl Moultrie while Genl How remains Commander in Chief in this department (until further order'd to the contrary) is to take the Brigade that was alloted the Genl. & is to be Obey'd & respected accordingly

By Col° Motte **Chs town 1777 -**

The General is exceedingly displeased at the Escape from the prisoner at the Main Guard, the report in respect of the Bars of the windows is by no means an excuse, it was the Duty of the Offr of the guard to fix Centinels in that place if it was not secure, & he is highly culpible for not having done so, A Court Martial Aught & probable will be the Consequence of this Neglect; If however the Genl should in Lenity over look it he would have the Offrs. of the suceeding Guards to understand that a futer escape Either or through that place or by any other means will Certainly be made the Subject of a Court Martial

Regtl. Orders-

The regiment to turn out as usual to Exercise every After Noon

---

Orders 29th April } .... Parole Caswell ____ GH
By G Howe } For the day tomorrow Captn Charnock
Town Guard Lt Pimeneau
Magazine Lt Martin
Brick House Lt Reaux
Regtl Ordrs Barracks a Sergt.

For the Genl Court martial at fort Moultrie on thursdy. next Captns. Harleston & Charnock Lts. P Grey & Baker

A Monthly return to be given in to the adjutant by 10 OC: tomorrow morning, of each Compy A court martial to sett this morning at 11 OC: to try such prisoners as shall be brought before them

Captn Shubrick presedt
Lts. Shubrick & Burke members

Brigade Orders by General Moultrie 29th April

Mr Dellient will please to act as Brigade Major till one may be Appointed

The Orders of yesterday relating to my taking the Command of the Brigade (alloted Genl Howe by Genl Moore) to be sent Col° Roberts

Orderly Hours at 10OC: 6 men from the 2nd & 5th Regts to be Appointed as boat men to attend the Genl. & be exempted from all other Duties

---

Orders 30th Ap: } .... Parole Washington .... by presedt
by Genl Howe } For the day tomorrow Captn Moultrie
Town Guard Lt Mason
Magazine }
Brick house } From 5th Regt:
Barrack Guard a Sergeant

Orders 1st May } .... Parole Lee .... by GH
By G Howe } For the day tomorrow Captn Cogdell
Town guard Lt from 5th Regt.
Magazine Lt Shubrick
Brick House Lt Mazyck

Patrick O'Kelley

Barracks a Sergeant

NB According to Sentence of the last Court Martial Robert Masden for absence without Leave received 100 Lashes [582]

~ New Book ~[583]

[582] Robert Madsden was in the 2nd South Carolina Regiment. On 1 May 1777 he received 100 lashes for being absent without leave.
[583] This is the end of that orderly book and it has "New Book" written in the pages.

Officers in the S^{o}.Carolina Second Regiment [584]				
Officers Names	**Dates of Commiss^{n}**	**Names of the Lieut^{ts}**	**Dates of Commission**	
Col^{o}. Is: Motte	16th Sept: 1776	Thos. Dunbar	24th Feby. 1776	Prom - -
Lt Col^{o} Fran^{s}. Marion	. . . . .d^{o}	Thom^{s}. Hall	13th June 1776	
Maj^{r}. Pet^{r} Horry	. . . . .d^{o}.	~~Robt~~ H^{ry}. Gray	7th Augt 1776	
Captains-		Rich^{d}. Baker	16th Sep 1776	
Is: Harleston	4th Novr 1775	Adrian Proveaux	22nd Octr. 1776	
Charles Motte	. . . . . .d^{o}.	Richd Mason	27th Jany. 1777	
Anth^{y} Ashby resigned 16th Feb 1778	1st Decem^{r} 1775	Pet^{r} Gray	28th d^{o} 1777	
Wm Oliphant	24th Feby. 1776	Adanus Burke	6th May 1777	Resigned
Jno Blake	23^{d} Apr: 1776	John Hart	21st May 1777	
Richd Shubrick	23^{d} May 1776	Willm Galvan	21st Octr 1777	Resigned 25 Jly. 78
Wm Charnock	29th Augt. 1776	Second Lieut^{s}		
Thos Lesesne	16th Sept: 1776	~~Thos Dunbar~~	~~29th Feby 1776~~	2^{d} Regim: on assign
Thos Moultrie	22nd Oct:1776	~~Thos Hall~~	~~13th June d^{o}~~	2^{d} Regim: on assign
Danl Mazyck	6th May 1777	Henry Perreneau	12th Novr 1776	promoted 25 Novem^{r}.
Jacob Shubrick	21st Octr 1777	Albert Roux	26th Jany 1777	promoted 5 Decemb^{r}
Resignations	**Date when**	Jno Martin	3^{d} March d^{o}	prom 16 Mar 1778 to 1 L^{t}
Captn Wm Charnock	20th Oct: 1777	Wm Capers [585]	4th Augt d^{o}	prom 13 Mar 1778 to 1 L^{t}
Richd Shubrick } Dead - - - - - -	21st Novr 1777	Paul Warley	8th d^{o} - d^{o}	promo 18 July 1778
Henry Gray } made pay master	15th Decemb^{r} 16th Decemb^{r} 1778	Jno Bush [586]	26th Sept d^{o} __	resigned
Capt ^{n} Anth^{y}. Ashby	16th Feby 1778	Saml Guerrey [587]	28th Octr d^{o}	~~Resigned~~

[584] This is the officer's list that is referred to in other endnotes.

[585] William Capers was born on 13 October 1758. On 4 August 1777 he was appointed a second lieutenant in the 2nd South Carolina Regiment in Captain Motte's company. He became a first lieutenant on 24 February 1778 in Captain Shubrick's company. He was transferred to Captain Dunbar's light infantry company on 2 October 1779. He resigned on 20 January 1780. After the fall of Charleston he joined Marion's Brigade of Partisans. He became a captain under Colonel Richard Richardson. He died on 7 January 1813.

[586] John Bush enlisted in the 2nd South Carolina Regiment on 18 May 1776. He was commissioned a 2nd Lieutenant on 26 September 1777 in Captain Harleston's company. He resigned on 7 July 1778. He came back to the regiment as a second lieutenant on 10 May 1779 after the British took Savannah and threatened Charleston. He was assigned to Captain Moultrie's company. He was killed while carrying the 2nd South Carolina colors during the assault on the Spring Hill Redoubt at Savannah on 9 October 1779.

[587] Samuel Guerry was appointed a second lieutenant in the 2nd South Carolina Regiment on 28 October 1777 in Captain Shubrick's company. When Captain Shubrick died his company became known as the 2nd Vacant Company. When the 2nd Vacant Company was disbanded on 5 October 1778 he was assigned to the Lieutenant

L[t] Burke	22[d] d[o].	Pet[r] Foisson [588]	6th Decem[r] d[o]	
Capt[n] Jacob Shubrick Dead - - - - - - - - }	28th April 78	Henry Purcell Chaplain }	8th May 1776	
Henry Perreneau	15th July 78	John Hall Q Mast[r]	1 July d[o].	
W[m] Galvan . .		Jerim: Theus Surgeon	2nd August 1777	[589]
Capt[n] Blake _____	resigned 25th Ap[r]	Jn[o] Henry Rusche Surg[n]. Mate [590]	11 June 1778	resigned
Adjut. Downes [591]	d[o] 6th Octob[r]. 78	W[m] Downes Adj[t]	12th	Resigned 6th Octob[r]
		Joseph Kolb Ens[n]	14th July d[o]	
		Sylvester Springer Surg[n]. Mate [592]	27th d[o] - d[o]	

Officers in the S[o]. Carolina 2nd Regem[t] Nov[r] 16th 1778		
Captains	**Dates of Commission**	
Isaac Harleston	4th Nov[r] 1775	Promoted to Maj[r] in 6th Reg[t]
Charles Motte	Ditto	
William Charnock	29th August 1776	Resigned
Tho[s] Lesesne	16th Septem[r] 1776	
Tho[s] Moultrie	22nd Octob[r] 1776	
Daniel Mazyck	6th May 1777	
Tho[s] Dunbar	9th Novem[r] 1777	
Tho[s] Hall	16th Feb[y] 1777	
Rich[d] B Baker	25 Apr 1778	
Adrian Proveaux	27 April 1778	

Colonel's Company under the command of Lieutenant Richard Bohun Baker. He was promoted to 1st Lieutenant on 6 November 1778 (the Officer's list says 27 April 1778). He was transferred to Captain Hall's company in March 1779. When the British threatened Charleston in May 1779 he was assigned to Captain Motte's company. He died on 12 July 1779.

[588] Peter Foissin was appointed a second lieutenant in the 2nd South Carolina Regiment on 6 December 1777 in Captain Charnock's company. He was promoted to first lieutenant on 18 July 1778. He was taken prisoner at Charleston in 1780. Upon his exchange he joined Marion's partisans. He resigned his commission on 23 March 1782.

[589] Jeremiah Theus was appointed the chief surgeon in the 2nd South Carolina on 2 August 1777 and was taken prisoner at the fall of Charleston.

[590] John Henry Rasche (Rusche) was appointed surgeon mate to the 2nd South Carolina Regiment on 11 June 1778. He deserted on 26 October 1778 (Officer list says he resigned).

[591] William Downes was appointed the adjutant of the 2nd South Carolina Regiment on 12 March 1778. He resigned on 6 October 1778. He is also listed as John Downes.

[592] Sylvester Springer was appointed as the surgeon's mate to the 2nd South Carolina Regiment on 27 June 1778. He was taken prisoner at the fall of Charleston. After he was exchanged in October 1780 he served until the close of the war.

1st Lieutenants		
Richd Mason	27th January 1777	promoted to Captn 25th Novr 1778
Peter Gray	28th do. . . . . . do.	promoted 30 Decemr 1778
John Hart	21st May . . . . do.	
Albert Roux	5th Decemr. . . 1777	
Jno Martin	16th Feby ____ 78	
Wm Capers	24th do. . . . . . do.	
Paul Warley	25th April . . . do	
Samuel Guerry	27th do . . . . . .do	Dead
Peter Foissin	13th July . . . . do	
Ensigns		
Josiah Kolb [593]	14th July . . . 1778	promoted 1st Lt 15 July 1778
John Wickom	6th Novr 78	prom: to 1st Lt. 25th Novr 1778
Christopher Rogers Junr [594]	2d Lieutenant 3d March . . . 79	
George Ogier [595] . . . 2nd Lieut.		
Alexd Petrie [596] 1st Lt.	23 Feby 1779 ___	
Alexd. Humes [597] 2d Lieut . . .	promoted 11 July 79	
Cornelius Van Humstead Vieland	[598]	

593 Josiah Kolb enlisted on 12 December 1776. He was a sergeant under Captain John Blake in early 1778. He was appointed an ensign in the 2nd South Carolina Regiment 9 July 1778 in Captain Charnock's company. He was assigned to Captain Proveaux's company on 17 November 1778. He was promoted to first lieutenant on 20 November 1778 when Lieutenant Galvan resigned. He was taken prisoner at the fall of Charleston. Once he was exchanged, he served to the end of the war with General Francis Marion.

594 Christopher Rodgers (Rogers) was commissioned as a 2nd Lieutenant in the 2nd South Carolina Regiment on 28 February 1779 under Captain Charles Motte.

595 George Ogier was appointed a second lieutenant in the 2nd South Carolina Regiment on 3 April 1779. On 14 August 1779 he was promoted to first lieutenant under Captain Daniel Mazyck. He was taken prisoner at the fall of Charleston.

596 Alexander Petrie was commissioned as a 2nd Lieutenant in the 2nd South Carolina Regiment under Captain Charles Motte. On 23 February 1779 he was promoted to 1st Lieutenant and transferred to Dunbar's Grenadiers. When the British threatened Charleston in May 1779 he was assigned to Captain Motte's company. He was wounded in the siege of Savannah on 9 October 1779. He resigned in January 1780 and served as adjutant to the Charleston Militia from 1 November to 16 December 1782.

597 Alexander Hume was an ensign in the 2nd South Carolina Regiment under Captain Richard Bohun Baker. He was commissioned on 23 February 1779 as a 2nd Lieutenant and transferred to Captain Lesesne's light infantry company. He was promoted to 1st Lieutenant on 11 July 1779. He was transferred back to Captain Baker's company in August 1779 when Captain Lesesne resigned. He was killed on the assault on Spring Hill Redoubt in Savannah on 9 October 1779.

598 Cornelius Van Hempstead Vlieland was appointed a 2nd Lieutenant in the 2nd South Carolina Regiment on 17 July 1779 in Captain Baker's company. He was transferred to Captain Dunbar's company in August 1779 when it became the light infantry company. He was wounded at Savannah on 9 October 1779 and had to have his leg amputated, but he died a few days later.

James Gray [599]		
James Lagre [600]	20th Augt 1779	

By Col° Motte **Chs. town 17 -**

General Orders 2d May by G howe } .... Parole Georgia
For the Day tomorrow Captn McDonald
Town Guard Lt. Warley; Magazine from the 5th Regt - Brick House
Lt Hart- Barracks a Sergt.

Regimental Orders

Lt Burke to take charge of Captn Ashbys Compy till Order'd otherwise___The Regemt to turn out to exercise at 5 OClock in afternoon for the futer

---

G. Orders 3d May by G. Howe } ...... Parole Meade ...... GH
For the Day tomorrow Captn Oliphant
Town Guard Sub: from 5th Regt.
Magazine guard from 5th Regt. _____ Brick house Lt Hall __ Barrack a sergt
Hobcaw Command to be deliver'd by 5th Regt.

Regimental Orders

Devine Service will be performed by the Chaplain tomorrow Afternoon at half past 4 OC: in the new Barrack yard, the Col°. Expects all the Officers to Attend & that they will acquaint the men of their respective Compy. to be present, All such as are not present to be confined & reported _____ A Regimental Court martial to sett this morning at 11 OC: to try such Prisoners as may be brought before them, Evidences to be ordered to attend in time

Captn McDonald presedt _____ Lt Hall & Perreneau members

---

[599] James Gray was commissioned a 2nd Lieutenant in the 2nd South Carolina Regiment on 29 July 1779 in Captain Gray's company. He was mortally wounded carrying the regiment's colors in the assault on the Spring Hill Redoubt in Savannah on 9 October 1779.

[600] James Legare was commissioned a 2nd Lieutenant in the 2nd South Carolina Regiment on 20 August 1779 in Captain Hall's company. He was promoted to first lieutenant on 9 October 1779 after the assault on the Spring Hill Redoubt in Savannah decimated the officer corps. He transferred to Captain Gray's company on 4 December 1779. He was taken prisoner at the fall of Charleston, but was paroled on 26 June 1781. He served until the end of the war under General Francis Marion. He died on 14 January 1831.

NB Wm Collens Wm Francis,[601] Jno Batchelor[602] & Danl Pipken[603] were brought as Deserters by Jno Sergenor[604] & Ths Kidwell _____________

Orders 4th May by G Howe __ } .... Parol Beaufort ..... by Presedt
For the Day tomorrow Captn Potts
Town Guard from the 5th Regt __Magazine Lt Perrineau
Brick house Lt Reaux__Barrack a Sergeant

Genl. Howe Approves of the proceedings of the Genl Court martial lately held at Fort Moultrie which he ratifies Accordingly, with this only Exceptions that the punishment to which Robt Cunningham was sentenced is remitted, in Respect to the Court as they recommend him to mercy-[605] The Judgment of the Court as to the other prisoners may be carried into execution at such time & in such a manner as the Commanding Offr. for the time being shall direct, or as impowered if he thinks proper to pardon the Criminals, provided they agree to Inlist for the War ___Adjutant Delliant is to transmit this order to fort Moultrie

---

G. Orders 5th May by G Howe } ...... Parole Rutledge ...... by GH
For the Day tomorrow Captn Shubrick
Town Guard from 5th Regt
Magazine Lt Martin__ Brick House from 5th Regt.
Pritchards Guard Lt Warley___ Barrack a Sergt

Captn Jervey of Captn Hugers Battallion having informed the General that the Offrs of his Corps have from Some Misunderstanding refused to do duty with him upon which Occasion he petetioned for a Court of Enquiry

A court of Enquiry is therefore to be held on wednesday morning next at some Convenient place in Chs.town to take under consideration the conduct of Captn Jervey & to report to the general werther he has committed any Action which makes him worthy of the neglect which he has been treated at this court, those officers ~~those officers~~ who has refused Doing duty with the Captn. will appear & Charge him with the fact of which they suppose him Guilty __ the Court is to consist of one field Offr. as presedt & six other members to be taken According to detail from the first, Second & fourth Regimts. ie: one Captn & 1 Subaltern each, Adjt. Dilliant will take care that this order will be sent immediately to the several Corps ___

NB Colo Cattle was president of this court

---

601 William Francis deserted from the 2nd South Carolina Regiment in early 1777. He was captured and returned on 3 May 1777.

602 John Batchelor enlisted in the 2nd South Carolina Regiment on 18 December 1775. He deserted from the 2nd South Carolina Regiment in early 1777, but was captured and returned on 3 May 1777. He was court martialed for the desertion and received 100 lashes. He deserted again one month later on 10 June 1777.

603 Daniel Pipkin enlisted in the 2nd South Carolina Regiment on 4 November 1775. He deserted from the 2nd South Carolina Regiment in early 1777. He was captured and returned on 3 May 1777. He was court martialed and received 100 lashes, but deserted one month later in June 1777. He was a lieutenant in the militia in 1782.

604 Benjamin Sergenor was in the 2nd South Carolina Regiment under Captain Thomas Hall. He evidently was the sergeant they used to track down deserters. He was in the siege of Savannah.

605 Robert Cunningham was in the 2nd South Carolina Regiment. He was court martialed on 3 May 1777 for some unnamed offense, but the sentence was remitted by General Howe.

Regiment^{l} Ord^{rs}.

a Regem^{tl}: court martial to sett this morning at 10 OC to try such prisoners as shall be brought before them, evidences to Attend

Captn Shubrick president

L^{ts}. Hart and Martin members

Orders 6th May by G Howe } . . . . Parole Laurence
For the day tomorrow Captn Charnock
Town Guard L^{t} from 5th Reg^{t}
Magazine Lt Mason___ Brick house L^{t} P Grey___
Barrack Guard a Serg^{t}.

Reg^{tl} Orders

For the Court of Enquiry, Capt Moultrie & L^{t} H Grey

Corporal John Gammel of Captn Shubricks Comp^{y} is appointed a Serg^{t}. in the same & is to be Obey'd as such [606]

NB accord^{g} to sentence of last court Jeremiah McGinnis & Conrad Meyer was Acquitted [607]

By Col^{o} Motte **Ch^{s}. town 1777**
by presed^{t}

Orders 7th May by G. Howe } . . . . Parole Council . . . . .
For the day tomorrow Captn Harleston
Town Guard Lt Shubrick___ Magazine Lt Mazyck

Brick House from 5th Reg^{t} ____ Barracks a serg^{t}.

NB accord^{g} to last court W^{m}. Arnold,[608] Jno Batchelor, Danl Pipkin for Absents without leave rec^{d} 100 lashes each, W^{m}. Collins for d^{o}. 99 lashes & to run the gauntelope Jno ~~Thompson~~ Ja^{s} Thompson [609] & Robt Potts for quiting out post 99 lashes & run the guantelope [610]

---

[606] John Gammel enlisted in the 2nd South Carolina Regiment on 18 July 1775 under Captain Barnard Elliot. On 6 May 1777 he was promoted to sergeant in Captain Shubrick's company. He received 50 lashes on 16 April 1778 for being drunk.

[607] Conrad Meyers enlisted in the 2nd South Carolina Regiment on 4 November 1775. He was court martialed on 3 May 1777 for some unnamed offense, but the sentence was remitted by General Howe. On 12 May 1777 he received 100 lashes for being drunk on guard duty and losing his gun. He also was put on stoppages for the loss of the gun. On 12 June 1779 he received 50 lashes for killing a hog of Mr. Seats, and he had to pay for the hog. He served 172 days in the militia from 1780 to 1782 under Colonel Taylor.

[608] William Arnold enlisted in the Grenadier Company of the 2nd South Carolina Regiment on 8 July 1775, and again on 6 November 1775. He was captured and returned on 3 May 1777. He was court martialed and received 100 lashes. He became the drum major of the regiment on 28 July 1777. In 1782 he served 32 days with Marion's partisans.

[609] James Thompson was in the 2nd South Carolina Regiment in Captain Charnock's company. On 7 May 1777 he received 99 lashes and had to run the gauntlet because he quit his post. On 15 May 1777 he received 50 lashes on his "bare posterior" because he sold his waistcoat and breeches. He also was put under stoppages until they were paid off.

[610] When a soldier was given a punishment to "run the gauntlet" it was usually to give justice to the other men in the company for some offense such as stealing or leaving a post (and then one of the other men would have to take their place). The company, or regiment would have all the men line up in two lines. The convicted soldier would then have to run, or walk, through the middle of the line as the men hit them with their bare hands, switches, or the ramrods to the musket. If the soldier was to walk the gauntlet, then another soldier

---

Orders 8th May by G. Howe } .... Parole Armstrong ..... by presed[t]
For the day tomorrow Capt[n] Cogdell
Town Guard from 5th Reg[t] ___ Magazine from 5th Reg[t]
Brick House from L[t] Hart ____ Barracks a Sergeant

The court of Enquiry sett on Capt[n] Jervey have reported as follows,

The Court having taken in Consideration the Charge Alledged against Capt[n]. Jervey & the Evidences for and against him, are of Opinion that he is not worthy of the neglect with which he has been treated & Honorable Acquit him of any misconduct__ Signed W[m] Cattle L[t] Col[o] 1st Reg[t].

The Gen[l]. with pleasure Approve the Report & order Capt[n] Jervey to do duty as usual, & dissolves the court

---

Orders 9th May by G. Howe } .... Parole Eveleigh .... by GH
For the day tomorrow Capt[n] Potts
Town Guard L[t]. Hall___ Magazine L[t] Perrineau
Brick House 5th Reg[t] ___ Barracks a serj[t].

Brigade Orders by G Moultrie } One Subaltern one Serg[t]. & 15 Rank & file men to hold themselves in readyness to march in the country and take post about Peedee
the party to be Served with 22 round p[r] man & 1 spare flint, the men are to be picked out from the 2nd & 5th Reg[t]. Such as are acquainted with that part of the Country & those can be best Depended upon the off[r] of this party must Apply to Col[os]. Pinckney, Motte Roberts & Huger for a list of deserters, from their Respective Reg[ts]. The Commisary to be informed that this party will march off in a few days he is to supply them with their Rations while out on this command, The off[r] of this party will receive his Orders from the Commanding Off[r] of the Brigade L[t] Mason 2nd Reg[t] for this Duty with 10 Rank & file from the 5th Reg[t] 5 Rank & file & 1 Sergeant [611]

By Maj: Horry **Charles town ____ 1777**

Orders 10th May by G Howe } ....Parole Elbert .... by presed[t]
For the Day tomorrow Capt[n] Shubrick
Town guard L[t] Reaux ___ Magazine from 5th Reg[t].
Brick house L[t] Martin ___ Barracks a Sergeant
The Hobcaw guard to be relieved tomorrow

Regiment[l]: Orders by Major Horry } A court martial to sett this morning at 11 oC: to try such prisoners as may be brought before them__ evidences to attend
Capt[n] Moultrie, president L[ts]. Gray & Roux members

---

would walk in front of him with a bayonet pointed at the convicts chest, and another soldier would walk behind with a bayonet pointed at his back, so he would not walk too fast through the punishment or try to run away.

611 After the news that the British had captured Fort McIntosh in Georgia many men of the South Carolina line had deserted. So many had left that this expedition mentioned in the orders, to track down those deserters, was organized to bring them back.

The Commanding Offrs of each Compy. to give a report of all deserters with the place where they were inlisted & when; as soon as possible

Devine service will be performed by the Chaplain tomorrow Aftr noon at half past 4 OC, the Commandg Officer expects all the Offrs will attend & acquaint their men to be present all such as do not are to be confin'd & reported

---

Orders 11th May by G Howe } ....Parole Habersham .... by GH
To the Day tomorrow Captn. Charnock
Town guard Lt from 5th Regt ___ Magazine 5th Regt.
Brick house Lt Petr Grey ___ Barracks a sergt.

Orders 12th May by G Howe .... Parole Lynch .... by presedt
For the day tomorrow Captn Moultrie
Town Guard Lt Shubrick ____ Magazine from 5th Regt.
Brick house Lt Mazyck _____ Barracks a sergeant

Regimentl Orders

A regimtl. Court martial to sett this morning at Eleven OC: to try such prisoners as shall be brought before them, evidences to attend

Captn Charnock presedt Lts. Mazyck & Perreneau members

The Majr is truly sorry to find so little attention paid to orders more than once Issued, for the purpose of confining the soldiers to Barracks after tatoo beating, that very few Sergeants remain after that hour, by which neglect of duty the men are left to Commit many irregularities for which they suffer frequently the sentence of courts martial, _ It tis order'd that the Sergeant Major do once or oftener every night after tatoo beating Call over the roll of Sergeants in their respective Barracks & report next morning to the Major of the Regt. all such sergts as may be Absent

That a Sergeant of a company do once or oftener every night after tatoo Beat call over the roll of their respective Compy. & report to their Offrs. next morning such as may be absent after such roll-calling

By Maj: Horry **Charles town ____ 1777 ___**

And the Commandg Offr of the Regt Hopes the officers will be perticularly active to confine all such men reported absent in order that they may be brought before a Court martial & suffering Accordingly, the neglect of such Confinement are & will be ever prejudetial to the Regt. & the service in which it is engaged___

NB according to sentence of last court martial 10th Inst: Petr Fagan for drunkeness on parrade recd 50 lashes, ~ Benjm Willmot for repeated Absences from roll call recd 50 lashes ~ Conrade Meyers for being drunk on guard & loseing his gun recd. 100 lashes & to be under stoppages for the gun ~ Willm Simpson for drunkeness when for guard recd. 100 lashes ~ Emanuel Lopez for theft of 1 shirt 1 pr stockings, 1 handkirchf recd. 50 lashes & to pay £9 to him that lost them, who is May of the 5th Regt [612]

---

[612] This may be Thomas or Joseph May, both enlisted in the 5th South Carolina Regiment in April 1776.

Orders 13th May
By G Howe

............Parole...........
For the day tomorrow Captn Jervey
Town guard Lt Hart ___ Magazine Lt Hall
Brick house from 5th Regt ___ Barracks a sergt

Regimentl Orders by Colo Marion ___

A sergt. & 3 men to go to Colo Huger who will give them his orders to be their by 6 OC: in morning

A regimental court martial to sett tomorrow 10 OC: in the morning to try all such prisoners as may be brought before them Evidences to Attend

Captn Blake presedt.___ Lts. Grey & Warley members

NB According to last court Jno Clements for drunkenness & neglect of duty recd 50 lashes Edd Fry for do recd 30 lashes__ Chs Hatton was to receive 100 lashes but remitted

---

Orders 14th May
by G. Howe

.... Parole Page ....
For the day tomorrow Captn Harleston
Town guard from 5th Regt __ Magazine from 5th Regt
Brick house Lt Perrineau___ Barracks a sergeant

Regemtl: Orders by Lt Colo Marion — Commanding Offrs of Compys. to make their sergts. give the Surgeons of the regemt a List of their sick & where they are every day at roll call

Lt Reaux to hold himself in readyness to go a recruiting

NB according to sentence of last court martial Jas Thompson for selling his Regimental waist coat & Breeches recd 50 lashes on his bare Posterior & to be put under stoppages by Captn Charnock to replace the Jacket & Breeches [613] Petr Fagan for neglect of duty recd. 75 lashes

By Lt Colo Marion **Chs. town ___ 1777** by presedt

Orders 15th May
by G Howe

.... Parole Putnam ....
For the day tomorrow Captn Cogdell
Town guard Lt Mazyck__ for the Magazine Lt P. Grey
Brick house from the 5th Regt. ___ Barracks a sergt

General Moultrie is requested to Appoint some Offrs. of his Brigade to take an exact list of the prisoners of War in town & to make a report thereof at Head Quarters___The four following French Vessels having come to this State under perticular Surcumstances Vizt: The Union Captn Laroach; the marquies de la Chaletac Captn Poliguy; The Thunder Captn Anderson; The Andrea Captn Corronant; the men of these vessells are not upon any Account whatsoever to be inlisted in the Continental Batallions & if any of their men have been inlisted, the Commanding Offrs; of those Batalions into which they have enter'd are immediately to discharge them from the Service & have them safely convey'd to the Offrs. of the main guard in Chs Town who is to have them taken care of & deliverd when demanded to the master of the vessel to which they belong; the Adjutant is to transmit this order emmediately to the out posts___

---

613 In the 18th century the term jacket usually meant a shorter coat. In this orderly book the term jacket is referring to the waistcoat. This was the vest that soldiers wore under their regimental coat. It is not know whether this waistcoat was sleeved, or had no sleeves. It most likely did not have any sleeves, since this was the fashion.

The Honble the Continental Congress having entered into the following resolutions relative to some of Articles of War, Commandg. Offrs. of Brigad & of Batalions not formed into Brigades are emmediately to have them published to the Army that none may plead Ignorance thereof this Ordr. also to be transmitted to the out posts__

In Congress April 14th: 1777 ___

Resolved that from and after the publication hereof of the 2d Article of the 8 Section, the 1st Art: of the 11th Section, the 8th Article of the 14th Section & the 2d Art: of the 18th Section of the Rules and Articles for the Better Government of the troops, raised or to be raised & kept in pay by & at the Expense of the United States of America passed in Congress the 28th September 1776 shall be & they are hereby repealed & that the four following Articles be substituted in the place & stead thereof-

Art: 1st All Offrs. & Soldiers shall have full Liberty to bring in to any the forts or garrison of the United American States any quantity eatable provision Except where any Contracts ~~have~~ are/or shall be entered into by Congress or by their Orders for furnishing such contributed provisions

Article 2d If any Offr: shall find himself to be wronged by his Col°. or Commanding Offr. of the Regiment, and shall upon due application made to him be refused to be redressed, he may Complain to the Continental General Commanding in the state where such Regimt. shall be stationed. in order to obtain Justice; who is hereby required to examine into the Said Complaint & take proper measures for redresing the wrongs complained of & transmit as soon as possible to Congress a true state of such complaint with the proceedings had thereon.

Article 3d No Sentence of a General court martial shall be put in execution till after a report shall be made of the whole proceeding to Congress, the Commander in Chief or the Continental general Commanding in the State where such Genl. Court martial shall be held, & their on his Order being Issued for Carrying such Sentence into execution

Article 4th The Continental General Commanding in either of the American States, for the time being shall have full power of Appointing General Court martial to be held, & of pardoning or mitgating any of the punishment ordered to be inflicted for any of the offences mentioned in the above mentioned Rules & Articles for the better government of the troops, Except the punishment of offenders under the sentence of Death, by a General court martial, which he may Order, to be Suspended, untill the pleasure of Congress can be known which Suspension with the proceedings of the Court Martial, the said General immediately transmit to Congress for their Determination And every offender convicted by any regimental Court martial may be pardoned or have his punishment mitigated by the Col°, or Officer Commanding the Regt.

Orders 16th May by G Howe } .... Parole George Town ..... GH
For the day tomorrow Capn Potts
Town Guard Lt Mazyck ___ Magazine from 5th Regt.
Brick house from 5th Regt.____ Barracks a Serjeant

Brigade Orders by G Moultrie } The Genl desires the Captn of this day to give him a report of the prisoners of war now in town, Their name, Country & vocation, when taken & by whom, what place & when brought to this town

By Lt Col° Marion **Chs. town ___ 1777 __**

Regimental Orders 16th May __The Captains of the Regiment to take it by turns to Exercise the Batalion every afternoon beginning with the Oldest in Rank:

A Regtl court martial to sett tomorrow morning at 10 OC: to try such prisoners as shall be brought before them all evidences to attend

Captn Oliphant presedt. Lts Hall & Hart members

Orders 17th May by G Howe } ....Parole Congress.......... by presedt
For the day tomorrow Captn Oliphant
Town guard from 5th Regt ____ Magazine Lt. Shubrick
Brick House Lt. Warley ____ Barracks a Sergt.

Orders 18th May by G Howe } ....Parole Elbert........... By G H
For the day tomorrow Captn Blake
Town guard from 5th Regt. ___ Magazine Lt Hart
Brick House from 5th Regt ___ Barracks a Sergt.

Brigade Orders by G Moultrie } The Quartr Master of the 2nd & 5th Regements are Order'd to Draw Rations for the Boat Crew in the Service in proportion to the numbers of men

Regimtl. Orders __ A court martial to sett tomorrow morning at 10 OC: to try such prisoners as_may be brought before them all evidences to attend

---

Orders 19th May by G Howe } ....Parole Baltimore........ by G H
For the day tomorrow Captn ~~Shubrick~~ Charnock
Town guard Lt Perreneau ___ Magazine Lt from 5th Regt
Brick House from 5th Regt ___ Barracks a Sergeant

Orders 20th May by G Howe } ....Parole Nelson .... GH
For the Day tomorrow Captn Moultrie
Town guard from 5th Regt. ___ Magazine Lt Shubrick
Brick House Lt P. Grey ____ Barracks a Sergt.

A General court martial to be held in Chs.town at some common place for the Tryal of Lt Raphel of Col°. Roberts Corps of Artillery,[614] Charged by Captn P DeTrevelle with having Urgently & falsely

---

[614] John Raphael had enlisted during August 1775 in the Charleston Volunteer Militia under Captain Charles Drayton. He served as a lieutenant in the 4th South Carolina Regiment (Artillery) from 1776 to 1777. In May 1777 he was charged by Captain John Francis De Treville with "falsely aspersing his character in a manner unbecoming of an officer and a gentleman." Raphael was found not guilty, but he was warned by General Howe to not speak negatively about other officers. On 14 June 1777 he was again brought up on charges of conduct unbecoming an officer and a gentleman, but he was once again found not guilty. On 15 July 1777 he was once again brought up on charges, but this time for threatening a member of the court martial, and of threatening witnesses to his last court martial. He was once again found not guilty. He brought charges against Captain De Treville in October 1777 but the charges were found not to be founded upon the truth and Raphel had been lying. Shortly afterwards Lieutenant Raphel is listed as being dead. I do not know if this is from accident, suicide or a duel.

Aspersed his charracter in a manner Unbecoming an Officer & a Gentleman__Capt^{n} De Trevelle will produce Evidences to support this Charge
This Court to Consist of one field Officer as presed^{t} & twelve other members to be taken by Detachment from 1st. 2nd & 5th Batalion Each Batalion to furnish two Captains & two Subalterns, This Court is to try a number of prisoners of Different Batalions for Desertion & other Crimes, Command^{g}. Officers of those Corps who have prisoners Amenable to a Genl. Court Martial will take care to have them in town early on Munday Morning & Savely deliverd to the Off^{r}. of the main Guard, & will ~~offer~~ order the Evidences to appear and attend the Court

The Guard at Hobcaw to be relieved on Thursday morning from the 1st Batalion & the magazine Guard at Dorchester to be relieved by Detachment from the 2nd & 5th Consisting of One Subaltern one Sergeant & 16 privates This Relief to March to Dorchester on fryday next __the Off^{r} will attend at head Quarters for Orders before he setts off __ Adjutant Delliant will Transmit this Order Emmediately __ Lt Col^{o}. McIntosh to be presed^{t} of the General Court Martial

Regim^{tl}: Orders by Lt Col^{o} Marion } The Col^{o} Desires the presed^{ts} of court martial to seal the proceedings when ever it is Left at his house; the proceedings will be received for the feuter without its wrought fair & a sufficient room left between the Sentence of one & the begining of the next tryal to underwright & Sign his name

NB accord^{g}. to Sentence of Last court Fran^{s}. Fountain [615] for stealing Cloath from Jean Caddet [616] rec^{d} 50 lashes & is to pay for the Cloath Jno Clements for neglect of duty rec^{d} 100 lashes__ Jno Lyons for do rec^{d} 100 lashes [617] Woodrop Kelley for do acquitted[618]

Orders 21st May by G Howe } ....Parole Georgia......... by presed^{t}
For the day tomorrow Captn Jervis
Town Guard Lt Warley___ Magazine from 5th Reg^{t}
Brick house 5th Reg^{t} ___ Barrack a Serjeant

Regim^{tl}. Orders by Lt Col^{o} Marion } All the Officers off duty are desired to attend the funeral of Lt George Eveleigh at Mr Smiths house in Tradd Street this afternoon at 5 OC: One Subaltern one Sergeant & 24 Rank & file men to attend the funeral with Sufficient Number Blank Cartridges; the men are to receive their Coats out of the store & are to be as Clean as possible ___ The Officer who Command this party is to take an account of the men who receive their coats & see they are return'd before they are discharged

---

[615] Francis Fountaine enlisted in the 2nd South Carolina Regiment on 18 November 1776. On 20 May 1777 he received 50 lashes for stealing cloth from another soldier. He also had to pay for the cloth.
[616] John Caddy.
[617] John Lyons enlisted in the 2nd South Carolina Regiment on 4 November 1775. He was under Captain Blake. On 20 May 1777 he received 100 lashes for neglect of duty. On 2 July 1777 he received 50 lashes and had to pay Joseph Williamson £13.
[618] Woodford Kelly enlisted in the 2nd South Carolina Regiment on 19 November 1776. H was charged with neglect of duty on 20 May 1777, but he was acquitted. He was a sergeant in 1778.

By L[t] Col[o] Marion **Ch[s]. town 1777 __**

Orders 22nd May by G Howe } .... Parole Williamsburg .... by GH
For the Day tomorrow Capt[n] Moultrie
Town guard from 5th Reg[t] ___Magazine L[t]. H. Grey
Brick House L[t]. Hart ___ Barracks a Sergeant ___

Reg[tl]. Orders

For the Command tomorrow for Dorchester L[t] Pemmineau

A Regimental Court Martial to sett tomorrow 10 OC: in forenoon to try such prisoners as shall be brought before them Evidences to attend

Capt[n] Moultrie Presed[t]. ___L[ts] Mazyck & Shubrick members

NB Josiah Stone returned from Desertion & was pardon'd [619] by Intersession of Capt[n] Fogartee,[620] & Jn[o] Huger Esq[r] [621] __ Sergeant Marlow, John Serginer & Jessey Barefield [622] was sent on Command after Deserters

---

Orders 23[d] May by G Howe } ....Parole Virginia .... by presed[t]
For the Day tomorrow Capt[n] Charnock
Town Guard from the 5th Reg[t] __ Magazine L[t] Mazyck
Brick House from the 5th Reg[t] __ Barracks a sergeant __

Reg[tl]. Orders

The Court martial Order'd to sett yesterd[y]. is to sett today Capt[n] Charnock president L[ts] H Grey & Warley members

NB Lieut[t] Perreneau march'd off for Dorchester with his Detachment yesterd[y]. about 11 OC. morning ~

Orders 24th May by G Howe . . } .... Parole Congress .... by GH
For the day tomorrow Capt[n] Conyers [623]
Town Guard from the 5th Reg[t].__ Magazine ~~Lt Grey~~ from 5th Reg[t]
Brick House L[t] Shubrick ___ Barracks a Sergeant

Brigade Orders by G Moultrie } To L[t] Rich[d]. Mason
Sir. You will march with your party tomorrow morning to peedee there take post at some convenient place & form your head quarters sending out two or three small parties

---

619 Pardons were promised to the deserters so that it would be an incentive to bring back those that left after the British invaded Georgia.

620 James Fogartie served as a lieutenant under Captain Benjamin Marion in the Berkeley County Militia Regiment during 1776.

621 John Huger was the captain of the Charles Town Volunteers, in the Charles Town Militia. He was a lieutenant colonel of the Charles Town Militia during the siege of Charleston in May 1780.

622 Jesse Barefield enlisted in the 2nd South Carolina Regiment on 4 November 1775. After the siege of Charleston he switched sides and formed a mounted Loyalist militia unit. He fought against his old commander, Francis Marion, at Blue Savannah. He was wounded on 15 November 1780 at White's Plantation when he was able to capture Francis Marion's nephew, Gabriel Marion. Jesse's brother, Miles, was killed in this fight. Gabriel Marion was executed after he was captured by a mulatto named Sweat. In 1782 many of the Loyalists returned back to the Patriot cause, since the end was near and they knew the British would be leaving. Jesse Barefield switched sides again and served 72 days in the militia in 1782.

623 Clement Conyers.

to inquire after deserters from the Continental Troops & have them Apprehended & brought to Chs Town, Till you have a sufficient number to come down with, you will confine those you apprehend in the Gaol of the District you may be in [624] __You may promis a pardon to all those men who will deliver themselves to you & may give them a permit to Chs town, You move your party to Different parts of the Country as you may think most likely to take up Deserters, Your are to continue in this service three or four weeks unless Counter Order'd, Or unless you shoud have certain intelligence of this state being invaded, then you are to march your party to Chs town with all expedition You are to take notice of such persons who harbour or entertain Deserters & have them prosecuted as the law directs.

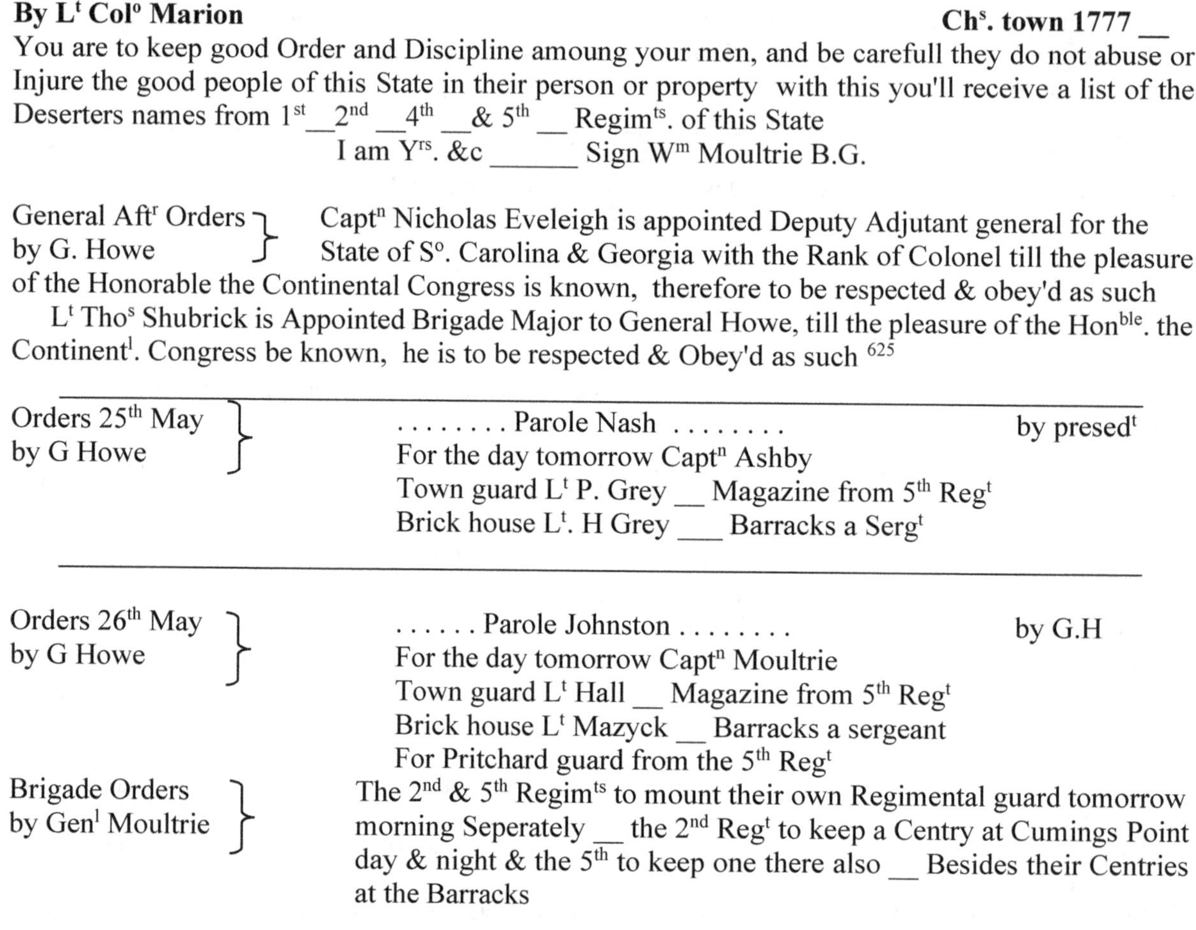

By Lt Colo Marion **Chs. town 1777 __**

You are to keep good Order and Discipline amoung your men, and be carefull they do not abuse or Injure the good people of this State in their person or property with this you'll receive a list of the Deserters names from 1st__2nd __4th __& 5th __ Regimts. of this State

I am Yrs. &c ______ Sign Wm Moultrie B.G.

General Aftr Orders by G. Howe } Captn Nicholas Eveleigh is appointed Deputy Adjutant general for the State of So. Carolina & Georgia with the Rank of Colonel till the pleasure of the Honorable the Continental Congress is known, therefore to be respected & obey'd as such

Lt Thos Shubrick is Appointed Brigade Major to General Howe, till the pleasure of the Honble. the Continentl. Congress be known, he is to be respected & Obey'd as such [625]

Orders 25th May by G Howe }
.... Parole Nash .... by presedt
For the day tomorrow Captn Ashby
Town guard Lt P. Grey __ Magazine from 5th Regt
Brick house Lt. H Grey ___ Barracks a Sergt

Orders 26th May by G Howe }
...... Parole Johnston .... by G.H
For the day tomorrow Captn Moultrie
Town guard Lt Hall __ Magazine from 5th Regt
Brick house Lt Mazyck __ Barracks a sergeant
For Pritchard guard from the 5th Regt

Brigade Orders by Genl Moultrie } The 2nd & 5th Regimts to mount their own Regimental guard tomorrow morning Seperately __ the 2nd Regt to keep a Centry at Cumings Point day & night & the 5th to keep one there also __ Besides their Centries at the Barracks

---

[624] Gaol is the 18th century word for Jail.

[625] Thomas Shubrick, Jr. was born on 17 December 1756. In January 1777 he became a first lieutenant in the 2nd South Carolina Regiment and served as brigade major to General Howe from 24 May 1777 to September 1778. On 21 June 1778 he became a captain in the 5th South Carolina Regiment. He served as a captain and aide-de-camp to General Greene during 1781. He died in 1810 at the age of 54.

Regim[tl] Orders

A court martial to sett tomorrow morning at 10 OC. to try such prisoners as will be brought before them evidences to attend

Capt[n] Charnock presed[t]. L[ts] Shubrick & Proveaux members

NB according to sentence of last court John Crawford for beating Mary Burgess receiv'd 50 lashes [626] __Jn[o] Griffin remitted __ Ralph Ingram rec[d] 75 lashes [627] Adam Creighton for lett a prisoner escape Acquitted __ Nich[s]. Flinn for Selling a Cutteau of Jam[s] Semple, order'd to return the s[d] Cutteau & Semple to pay him what he owes him[628] __Hugh Darborough remanded for want of evidence [629] Arthur Crawford for absents without leave sentenced to receive 100 lash pardon'd on Acc[t] of his good Character __[630]

By L[t] Col[o] Marion **Ch[s]. town __1777 __**

Orders 27th May by G Howe } ....Parole Jersey ....
For the day tomorrow Capt[n] Moultrie
Town guard L[t] Shubrick __ Magazine from the 5th Reg[t]
Brick house a Sergeant ~

The Deputy Adjutant General is to attend at Head quarter every day at 12 OC. to receive orders, Which Brigade Majors & Adjutants of Batalions will attend to receive of him __ He is to transmit to the Commanding Officer in Georgia a copy of the Resolution of the Hon[ble] the Continental Congress dated at Philadelphia 14th April 1777, Relative to the Articles of War; and to inform him it is the Generals orders the S[d]. resolution is made public that no Soldiers may plead Ignorance thereof

Complaint having been made to the gen[l]. that one of the speres belonging to the magazine at Cuming Point has been taken away not withstanding a Guard is kept their __ Gen[l]. Moultrie will cause the strictest inquiry to be made who was Off[r]. of the guard when the accident happen'd; & to take every step to Detect & punish the person or persons for such neglect of duty and likewise to give the strictest order to prevent any feuter Accidents

---

[626] Mary Burgess was most likely the wife of John Burgess, the quartermaster sergeant.

[627] Ralph Ingram enlisted in the 2nd South Carolina Regiment on 26 November 1775 as a drummer. He was under Captain Charles Motte from 1777 to 1779. He received 75 lashes on 26 May 1777 for beating Mary Burgess. On 18 March 1779 he was reduced to private for being drunk on guard duty. He was in the siege of Savannah.

[628] A cutteau is a knife.

[629] Hugh Durborough (Derberry, Durbury) served in the 2nd South Carolina Regiment under Captain Shubrick during 1777. He received 50 lashes on 15 November 1777 for being absent for 6 months without leave. After Captain Shubrick's death he served in the 2nd Vacant Company until it was disbanded on 5 October 1778. He was then assigned to Captain Hall's Company. He was one of two drivers in the 2nd South Carolina on 28 July 1779. He had a wagon and four horses that had been contracted from Gerrard Nelson. The other driver was Thomas Cadwell. He was in Captain Mason's company at the siege of Savannah. After the fall of Charleston he served in the militia.

[630] Arthur Crawfod enlisted in the 2nd South Carolina Regiment on 4 November 1775. He was sentenced to receive 100 lashes for being absent without leave, but was pardoned on account of his good character. He was discharged on 13 November 1778. He went home to Union County and then enlisted in 1779 in the militia as a sergeant under Captain James Crawford, Major Samuel Otterson and Colonel Farr. He was later under Colonel Brandon at his blockhouse in Fair Forest District. In 1781 he was back under Captain James Crawford and Colonel Thomas Brandon.

NB according to sentence of last court, Josiah Stone for absents from roll-call rec[d] 100 lashes Regimental Barrack Guard a Sergeant 1 Corp & 9 privates

The expedition against British Florida started on 1 May 1777 when Colonel Samuel Elbert departed Sunbury and sailed to the St. Johns River with 400 Georgia Continentals. Colonel John Baker was to go overland with 200 militia and rendezvous with Elbert. Button Gwinnett, the new Georgia governor, was determined to take Florida from the British.

Howe was not as enthusiastic, but he did send Sumter with his 6th South Carolina Regiment to Fort Howe and the 3rd South Carolina Rangers to Sunbury. The Georgia continentals were commanded by Lachlan McIntosh, but things began to go bad when Gwinnett ordered the arrest of McIntosh's brother on the charges of treason. Though the charges were dropped, the two men began an intense hatred towards each other that would eventually lead to a duel in which Gwinnett would be killed.

Elbert's flotilla was delayed and Colonel Baker encountered the Loyalists of Colonel Thomas Brown instead. In the initial skirmish Baker's men fled from the Loyalists. Brown's Rangers and Indians pursued and were able to ambush the militamen at Thomas Creek on 17 May 1777. The Georgians had three men killed, nine wounded and 31 taken prisoner. All but 16 of the prisoners were put to death. The survivors rendezvoused with Elbert and his Continentals on Amelia Island.

Elbert's ships were stranded in the shallow narrows there and the combined force of Continentals and militia marched to the St. Mary's river and were determined to make a stand there. Due to internal squabbles between Thomas Brown and British Major Mark Prevost, there was no last stand.

Due to the scandal caused by McIntosh killing Gwinnet, a signer of the Declaration of Independence, McIntosh would be transferred up north to Washington's army. He would take command of the North Carolina Brigade during the winter at Valley Forge.

Orders 28th May by G Howe } .... Parole Livingston
For the day tomorrow Capt[n] Charnock
Town guard from the 5th Reg[t] __ Magazine L[t]. Dubose
Brick house a Sergeant

The General has accepted of the Resignation of L[t] Rolando of the 5th Reg & he is no longer to be consider'd as an Off[r]. if the Continental Army [631]

Regimental Orders

Capt[n] Ashby to sett as a member of the General court martial in room of Capt[n] Oliphant who is taken sick __ A monthly return to be given to the Adjutant from each Comp[y]. on fryday morning before 9 OC:

NB Lieut[t]: Mason with his party marched off this morning after Deserters and with recruiting orders__________ for Peedee___________

---

[631] Lieutenant Rolando's first name is lost to history. He was a lieutenant in the 5th South Carolina Rifles who resigned on 19 May 1777. He attempted to lure some of the rank and file to leave the unit too, which led to some desertions.

Orders 29th May by G Howe } ....Parole Elbertt .... by GH
for the day tomorrow Captn Moultrie
Town guard Lt H Grey __ Magazine Lt Hall
Brick house a sergeant

Colo Huger is to act as Colo. Commandant of Genl. Moultries Brigade untill the General returns to town & is to be respected & obey'd as such

By Lt Colo Marion **Chs. town 1777**

Regimental Orders 29th May } A court martial to sett tomorrow morning at 10 OC: to try such prisoners as shall be brought before them, evid: to attend
Captn Moultrie presedt. Lts. P Grey & Martin members-

Omitted in General Orders of this day

Lieut: Wm Ronson Davis of the 5th Batalion is promoted to the rank of Captain & is to be respected & obey'd as such [632]

Orders 30th May by G Howe } ....Parole Marion
For the day tomorrow Captn Lesesne
Town guard Lt P Gray __ Magazine from 5th Regt.
Brick house a Sergeant

The general has Accepted of the resignation of Lt Alland Belen of the 1st Batalion, who is theirfore no longer to be considered as an Offr. in the Continental Army ______[633]

Regimental Orders

For the feuter the pay Bill of each Compy. is to be made monthly & given in three days before pay day

Devine Service will be performed at the New Barracks at 5 OC: in Aftr. noon It is expected all the officers & soldiers will attend

---

Orders 1st June by G Howe } ....Parole Nelson .... by presedt
For the day tomorrow Captn Moultrie
Town guard Lt Mazyck __ Magazine from 5th Regt.
Brick House a Sergeant

Orders 2nd June by G Howe } ....Parole Bonamtown .... by GH
For the day tomorrow Captn Charnock
Town guard from 5th Regt. __ Magazine Lt Shubrick
Brick House Lt Dunbar
For the Guard at Pritchard Lt Proveaux

---

632 William Ransom Davis was appointed as a captain in the 5th South Carolina Regiment on 29 May 1777. He resigned on 12 August 1779 after the failed attack on Stono Ferry. He served with Marion's partisans in 1780-1782. He died on 19 December 1799.

633 Allard Belin was commissioned a lieutenant in the 1st South Carolina Regiment in 1776. He resigned on 31 May 1777. He was appointed the assistant deputy quartermaster general on 22 June 1779. He was later in Marion's partisans.

Orders 3[d] June
By G Howe } .... Parole
For the day tomorrow (Capt[n]. Lesesne sick) __ Capt[n]: Moultrie
Town guard L[t] Hall __ Magazine from 5th Reg[t]

Brick house guard L[t] Martin ________ Reg[tl] Orders a court martial to sett to morrow morning to try such prisoners as will be brought before them evidences to Attend ________
Capt[n] Moultrie president L[ts] H Grey and Burke members _____

By L[t] Col[o] Marion **Ch[s]. town 1777 __**

Orders 4th June
by G Howe } ....Parole Baltimore .... by presed[t]
For the day tomorrow Capt[n] Charnock
Town Guard from 5th Reg[t]. __ Magazine L[t] Mazyck
Brick house L[t] P. Grey

Brigade Orders
by G Moultrie } The 4th Resolve of the Continental Congress dated 14th Apr: last Issued by G[l]. Howe the 15th May past are to be read at the head of Each Batalion According to S[d] Orders __

Reg[tl] Orders The Major to read the 4th Resolve of the Hon[ble] the Continental Congress, agreable to Brigade Orders of this day, this Aft[r]:Noon at the parrade to the Regiment

Gen[l] Aft[r]. Orders
by G Howe } M[r] Richard Montcriefe is Appointed a Second L[t] in Col[o]. Hugers Batalion & is to be respected & Obeyed as such [634]

Orders 5th June
by G Howe } ....Parole Lee .... by G.H.
For the day tomorrow Capt[n] Moultrie
Town Guard from the 5th Reg[t].__ Magazine L[t] Dunbar
Brick house L[t]. Shubrick

Orders 6th June
by G Howe } ....Parole Franklin .... by presed[t]
For the day tomorrow Capt[n] Ashby
Town guard from 5th Reg[t]. __ Magazine L[t]. Warley
Brick house L[t]. Hart

The General has received the proceedings of the General Court martial which he will consider in due time __ The court martial is dissolved

---

Orders 7th June
by G. Howe } ....Parole Morris town .... by Pres:
For the day tomorrow Capt[n]. Blake
Town guard from 5th Reg[t]. __ Magazine L[t] Hall
Brick house from 5th Reg[t]. _______

---

[634] Richard Moncrief was commissioned a 2nd Lieutenant in the 5th South Carolina Rifles on 4 June 1777. He was promoted to 1st Lieutenant on 21 January 1778.

Reg[tl]. Orders

a court martial to sett this morning to try such prisoners as shall be brought before them evidences to attend

Capt[n]. Ashby presed[t]. L[ts]. Mazyck & Martin members

NB according to last court martial Benj[m]: Wilmott for entering on board a vessell received 100 lashes

Robert Pinhorn [635]
Jn[o] Whitely [636]
John M[c]Dowel } for refusing to do duty rec[d]. each 50 lashes

By L[t] Col[o] Marion **Ch[s]. town 1777**

Orders 8th June
Gen[l] Howe } . . . . Parole Hancock
For the day tomorrow Capt[n] Oliphant
Town Guard L[t] Jacob Shubrick __ Magazine 5th
Brick House L[t] Pet[r] Gray

Brigade Orders by Col[o] Is: Huger

One Capt[n]. two Subalterns 2 Serg[ts]. 2 Corporals & 23 privates from the 2nd & 5th Regiments do hold themselves in readyness immediately with 12 Rounds p[r] man, the Quarter master to provide this detachment immediately with three day provisions

Aft[r] Brigade Orders

One Sergeant 1 Corporal & ten privates with 12 rounds p[r] man to hold themselves in readyness to Join the above Detachment under the command of Capt[n]. Conyers

For the above command Capt[n]. Conyers L[t] Jones from the 5th [637]
L[t] Warley from the 2nd Regements.
NB The above detachment march't off at six OClock

Regimental Orders

Devine Service will be performed this Aft.Nn. at 5 OC: it is expected Officers & soldiers will attend____

Orders 9th June
Gen[l]. Howe } . . . . Parole Adams . . . .
For the day tomorrow Capt[n] Shubrick
Town Guard L[t] Hart __ Magazine from 5th Reg[t].
Brick house L[t]. Dunbar

---

635 Robert Pinhorn (Penhorn) enlisted in the 2nd South Carolina Regiment on 20 November 1775. He received 50 lashes on 7 June 1777 for refusing to do duty. He was reprimanded on 25 October 1777 for being drunk on guard and abusing the officer of the guard. He was in Captain Motte's company during the siege of Savannah in 1779.

636 John Whitley was in the 2nd South Carolina Regiment. He received 50 lashes on 7 June 1777 for refusing to do his duty. He was in Captain Thomas Dunbar's company in 1779. He received 25 lashes on 24 June 1777 for being absent from roll call.

637 John Jones is listed by Heitman as being in the 6th South Carolina Regiment during 1777. He later served as a lieutenant under General Marion till 1782.

Brigade Orders by Col° Huger } One Subaltern 2 Sergts. & 24 Rank & file men from the 2nd & 5th Regiment with 12 Rounds pr man to hold themselves in readyness immediately

Officer for this detachment Lt Gordon from the 5th & 6 men & a sergt. from the 2nd Regt. 1 Sergt. 1 Corp: 16 privates

NB this party marched to the Exchange at one OC:

Regiml. Orders by Lt Col. Marion } Lt. Martin to hold himself in readyness to go a recruiting

A Regimental Court martial to sett tomorrow morning at 10 to try such prisoners as may be brought before them, Evidences to Attend

By Lt Col° Marion **Chs. town 1777**

Regiml. Orders Continues

Commanding Officers of Compys. are desired to make their men gett their Arms in good order & keep them so, The Col°. is really ashamed to see the men on parrade with their arms so very dirty Which proves a neglect of duty & Disobedience of Orders many time repeated & hope the Officers for their own Credit will be perticular in this point of their duty for the feuter

---

Orders 10th June by G Howe } .... Parole Burke ....
For the day tomorrow Captn Conyers
Town Guard Subaltern from 5th Regt.
Magazine Lt Hart __ Brick House Lt Heny. Gray

The General having considered the proceedings of the Genl Court martial Lately held for the tryall of Lt Raphel of the Artillery, Charg'd by Captn De Treville for having Ungenteel & falsely Aspersed his Charractr in a manner unbecoming a Gentleman & an Officer, & for the tryall of a number of prisoners of different Corps for Desertion & other Crimes find that the sentence against Lt. Raphel thus, That he was no wise Criminal but rather Indelicate the court therefore find him not guilty, Lt Raphel in consequence of his Acquital will do duty as usual __ the General however think incumbent on him for the sake of Service to observe that Indelicacy in the conduct of one Offr. to an Other, in a profession so pure as that of a Soldier Aught upon every Occasion to be avoided as inconsistant with that nicety of honour which gives dignity to the Carracter __ the Genl thinks proper to suspend his determination till a further day when those Sentance which inflict Capital punishments, all those inflicting Corporal punishments he approves of & Ratifies, the Sentence respecting those Criminals belonging to the Corps in town will be carried in execution at such times & in such manner as the Commanding Offr. of those Corps shall think proper, those under Sentence belonging to outtposts are to receive their punishments at those posts in the manner & at the time the Commandg. Offr. there shall Direct Who are to Order proper persons to receive the Criminals of their several Corps from the Offr. of the main Guard & Convey them safely to the place of punishment

The Commanding Offr. of the main guard for the time being are in the most perticular manner Injoined to be carefull of Robt Potts of the Second Regement & John Coker [638] of the first, an escape

---

[638] John Coker enlisted in the 1st South Carolina Regiment on 4 November 1775. On 10 June 1777 he was imprisoned for an unnamed offense, and the officer of the guard was warned to make sure that he did not escape, because it would be an "unpardonable neglect of duty." He deserted on 18 January 1779.

of Either of those prisoners will be consider'd as an unpardonable neglect of Duty, __ the Adjutant General will transmit a Coppy of there Orders to the out posts, with a copy of the Sentence of the Gen[l]. Court Martial relative to the Criminals of each Corps

---

Orders 11th June
Gen[l] Howe } .... Parole Putnam
For the day tomorrow Capt[n] Charnock
Town Guard L[t] Proveaux __ Magazine from 5th reg
Brick house from 5th reg[t]

A detachment from the Corps of Artillery Consisting of one Sergeant & 20 privates under the Command of a Subaltern Off[r]. are to march to relieve the Malitia Guard at the magazine at Beaufort on Monday next Col[o] Roberts will give the necessary Orders to the officer & direct in what manner they are to go

NB this morning L[t]. Martin with pet[r] are: went a recruiting [639]

Serg[t]. Newton & Corp: Keel [640] went after the following deserters viz Pipkin, Odom [641] Henson & W[m] Davis [642] & Batchelor who went away last night

---

Orders 12th June
G Howe } .... Parole Davis .... by GH
For the day tomorrow Capt[n] Moultrie
Town guard L[t] Pet[r] Gray __ Magazine L[t] Mazyck
Brick house from 5th reg[t].

Orders 13th June
by G Howe } ....Parole Palmeto .... by presed[t].
For the day tomorrow Capt[n] Harleston
Town Guard from 5th reg[t]. __ Magazine L[t]. Shubrick
Brick house from 5th reg[t].

The Guard on Morris Island to be Called in by the Commanding Off[r] of fort Moultrie as soon as the powder is removed from the Island [643]

---

[639] In the original orderly book this name is not clear, and looks like "potts rpr". They did not write names in the method that the military does today, such as last name first, then first name last. So the closest name that resembles the original handwriting is Peter Area. Peter Area enlisted in the 2nd South Carolina Regiment on 8 April 1777 as a fifer in Captain Charles Motte's company. This would make sense because on many recruiting duties there would be musicians accompanying the officer. On 12 June 1779 Peter Area received 25 lashes for gaming.

[640] Isaac Keel enlisted in the 2nd South Carolina Regiment on 4 November 1775. He was reprimanded on 7 November 1777 for being absent from guard. He was promoted to sergeant on 18 December 1777 in Captain Mazyck's company. He was discharged on 30 June 1778.

[641] Archibald Odom (Odum) was in the 2nd South Carolina Regiment. He deserted on 10 June 1777. After the fall of Charleston he was a private, lieutenant and then a captain in Marion's partisans. He died on 1 April 1845.

[642] William Davis was in the 2nd South Carolina Regiment. He deserted on 10 June 1777.

[643] Morris Island is beside the present day Folly Beach on Folly Island. There has been a lighthouse on Morris Island since 1767. During the War Between the States the lighthouse was destroyed, and a new lighthouse was built there in 1876. The lighthouse became surrounded by water and was decommissioned. The light was bought by private funds in 1999 as is currently owned by the State of South Carolina.

Orders 14th June by G Howe } ....Parole Fort Johnson .... by GH
For the day tomorrow Captn Blake
Town Guard L^{t}. Hart __ Magazine from 5th reg^{t}:
Brick House Lt Henry Gray

Col^{o} Roberts having by Letters Acquainted the Genl that from some information given him Against Lt Raphel he was under the disagreabl nesessity of requesting a court of inquiry on his conduct, A General Court of Inquiry is therefore to sett in Ch^{s}.town at some Convenient place on tuesday morning to inquire in to the Concuct of Lt Raphel of the Artillery reg^{t}. Accused of having behaved in a manner unbecoming the Charracter of an Off^{r}. & Gentleman, this Court to Consist as a field Off^{r}. as Presed^{t}. & six other members, Lt Col^{o}, Marion to be president, the 1st 2nd & 5th Reg^{ts}. to furnish for this court one Captn & 1 Lt Each, Col^{o}. Roberts will furnish the Evidences against Lt Raphel the Adjutant Genl. is to transmit this Order to the Out Posts & to give the Lieut^{s}. immediate notice to prepare for his ~~tryal~~ Defense

Brigade Orders by Genl Moultrie } A Court of inquiry to sett on munday to examind in to the conduct of Captn Jervey of Col^{o}. Hugers regim^{t}. on a Charge of several of Col^{o}. Hugers Officers for Behavour unbecoming an officer & a Gentlemen, Presed^{t}. Maj^{r}. Horry 2 Capt^{ns}. & 1 Subaltern from Col^{o}. Motte's reg^{t}. 2 Capt^{ns}. & 1 Sub: from Col^{o}. Roberts reg^{t}.

Regimental Orders by Lt Col^{o} Marion } It is expected for the feuter that all the men reported fit for duty will be parraded in the After noons at Exercise & no Off^{r}. except the Commanding Off^{r}. or Maj^{r}: will Discharge them though there may be some not fit for Exercise any Soldier who shall go of the parade without Leave from the Commanding Off^{r}. or Major shall be try'd for Disobediance of Orders & suffer accordingly

The men mentioned Deserted in the morning reports, their names to be put On the Back of such report with the day of the Month he deserted & to continue it as long as he is reported

Any Sergeant of a Guard who permits any prisoners to go with or without an Orderly to town or out of the Barrack yard (except to the necessary) shall be try'd for Disobediance of Orders & suffer accordingly, nor is he to permit any liquor to be brought in the Guard house without leave from the Captn. of the day or a field Off^{r}. Any man who may be found carrying liquors in the Guard house on any pretence shall be try'd for Disobedance of Orders & be punished Accordingly

For the G Court of Inquiry Captn Ashby L^{t}. Shubrick For the Brigade Court of Inquiry Capt^{ns} Harleston & Shubrick & Lt Mazyck

By Lt Col^{o} Marion **Ch^{s}. town 1777 __**

15th Jun Gen Howe } ....Parole Laurence .... by Presed^{t}
For the day tomorrow Captn Lesesne
Town guard from the 5th Reg^{t}.
Magazine Lt Hall __ Brick house a sergeant

The General having Accepted the resignation of Lt Paul Warley of Col^{o} Motte's Batallion he is no longer to receive the respect and Obediance due to an Officer in the Continental service

---

Orders 16th June G Howe } .... Parole Rancolph .... by GH
For the day tomorrow Captn Motte
Town Guard from 5th Regt.
Magazine Lt Proveaux __ Brick house a Sergt.

Regtl. Ordrs G. M. } A court martial to sett this morning at 10 OC: to try such prisoners as shall be brought before them evidences to attend _ Captn Lesesne Presedt Lieuts. Dunbar & Proveaux Members

Brigade Orders by Genl. Moultrie } A Brigade court martial to sitt on Wednesdy morning to try Sergt. Burgess Confined by Genl Howe's Order __Genl Howe must have notice where and when the court setts as he will Attend & give his Charges, the prisoners who has absented themselves and returnd of their own Accord will also be try'd by the sd. court Majr. Beekman presedt. Captns. Oliphant & Lesesne from 2nd Regt. one Captn. 1 Lieutt from the Artillery & 5th Regts. Members

One subaltern 1 Sergeant & 25 privates with 3 rounds pr man to attend the funeral of Doctr Airs this afternoon at 4 OC: [644]

Orders 17th June G. Howe } ......Parole Middleton .... by presedt
For the day tomorrow Captn Blake
Town guard Lt. from 5th Reg __ Magazine Lt. Petr. Gray
Brick House a sergt

Brigade Ordr by G Moultrie } all the guards in town to be relieved for the feuter at six OClock in the morning

By Lt Colo Marion **Chs. town 1777 __**

Orders 18th June G Howe } ....Parole Cattell
For the day tomorrow Captn from 5th Regt
Town guard Lt Mazyck
Magazine from 5th Regt __ Brick house a sergt.

Brigade Orders G Moultrie } The Court of Inquiry Orderd to examine into the conduct of Captn Jervey of Colo. Hugers Regimt: for charges laid against him for behaviour unbecoming an Offr & a Gentlm. do unanimously Acquit Captn. Jervey & is of opinion that his conduct has not been inconsistent with the Character of an Offr. & a Gentlm. __ the Court of Inquiry of which Majr: Horry is president is dissolved & the sentence Approves of __ the Sergeants of the 2nd Regt. having petitioned Genl Howe in favour of Sergt. Burgess Genl Howe in consequence of this salutation Looks over his Offence & hope he will behave better for the feuter he is therefore released from his confinement & Order'd to return to his Duty

Regimtl: Orders Lt Colo Marion } Roll Call to be at 6 OClock ^in morning & 5 in Aftr: for the feuter
It is positively order'd that no Soldier for the feuter do mount guard without their Hairs combed & beard shaved clean & their Cloaths as clean as possible with their shoes, stockins or their Leggins on, their Arms clean & in good order, any solders who do not Obey this Order may depend on being severely punish't

---

[644] Doctor Air's first name has been lost to history. He was the doctor to the military. Upon his death on 3 July 1777, he was replaced by Doctor William Keigh.

Commanding Officers of Comp^ys^. to see their men for duty Obey the above orders & to answer for their neglect

The Adjutant is Possively orderd not to march a man of the parrade for Guard that do not comply with the above Orders

Lt Col^o^ Marion has had frequent complaints made by the Generals of the men mounting guard with long beard & without shoes and stockings & their arms not Cleaned

Command^g^. Off^rs^. of Companies are desired to stop one half of their mens pay & get such necessarys as will make their men Appear as Decent as possible

The Col^o^. Acquaints the Regiment that he is Oblige to take notice of all Disobediance of Orders & neglect of duty in a manner not don hitherto ______

By Lt Col^o^ Marion **Ch^s^. town 1777 __**

Orders 19th June } ......Parole Dicipline .... by GH

G Howe }

For the day tomorrow Captn Lesesne

Town Guard from the 5th Reg^t^.

Magazine Lt Shubrick __ Brick house Lt Burke

The General with surprise & displeasure has observed slovenly, indecent & dirty in which the Soldiers have of late upon almost every Occasion Appeard inconsistent witheir health disgracefull to the Army censurable at all times & when upon duty absolutely unpardonable he laments the inattention of Officers of Companies to their men to which this degeneracy must in great measure have been owing & which it was their absolute duty as much as possible to have prevented, They cannot surely suppose that their whole duty consists in apppearing at & in end in parrade or that their reputation is not concerned in the Appearance of their men, if however they do it behoves them to Adopt ideas more consistant with their own credit & the good of the service; the uncombed unshaved & dirty condition of many soldiers even upon duty, the rusty improper condition of their Arms & Degeneracy in other particulars of late to discernable denote past inattention & will if not corrected in feuter be deemed & treated as Disobedeance of Orders __ the relief of Centries sent from guard even to the president's door & Head Quarters come up with floped Hatt, bare legs, long beards [645] & uncombd hair in short in a manner so shamefully dirty & Indecent that Officers of Guards permitting it may with too much Appearance of Justice be accused of inattention & neglect __ the General hopes that reformation will follow reprehension in all persons & every department where requisite, the soldiers will therefore take care to appear at all times but particularly when for duty in a manner as decent as the Situation of things will permit __ and all Adjutants are warned against receiving them & all officers of relief against marching them off till that is the case.

---

[645] The fashion of the times in the 18th century was to be clean shaven. Soldiers in both the British and the American armies had to also be clean shaven. An exception to this was the Hessians, who were allowed to grow moustaches. Soldiers had to shave, at a minimum, before they went on guard duty. This averaged out to once every three days. So when a soldier is being reprimanded for having a long beard, it is one that is three days old. It is not a beard that would be considered "long" by modern standards.

the relief of Centries sent from guard even
to the president's door & Head Quarters
come up with floped Hatt, bare legs,
long beards & uncombd hair in short
in a manner so shamefully dirty &
Indecent that Officers of Guards
permitting it may with too much
Appearance of Justice be accused of
inattention & neglect

By L[t] Col[o] Marion **Ch[s]. town 1777 __**

It is painful to the General to have occasion to remonstrate against any impropriety in the conduct of officers & men he has the Honour of Commanding, he wishes them to be assured that he never has or never shall do it but where duty exacts it of him & that he has never served with any Officer or men in his Opinion more respectable than those he is now with

Regimental Orders L[t] Col. Marion } A return of the Arms & Accoutrements & number of good and bad Wanting in each Company to be made by Saturday at 10 OC: in forenoon to the Adjutant as also what caps are wanting to compleat _ Commanding Officers of Companies to be perticular in examine their mens arms that noon may be given in fit for service when they are not

The Quarter master to make a return of what good & bad arms & accoutrements in the regimental store & in the Armourers shop who is to have made as many caps as he may be wanting M[r] Delliant will give him the number

The Col[o]. saw several men mount guard this day contrary to Orders of yesterday he Assures Commanding Officers of Companies if they do not make their Subalterns, Sergeants & men comply with that order strictly, that it is not in his power to prevent Disagreeable Consequences

The pay masters is Order'd to stop the pay of those men which Commanding Off[rs]. of companies may desire him not exceeding one half of their pay at a time

---

Orders 20th June G Howe } .... Parole Clinton .... by presid[t]
For the day tomorrow Capt[n] Moultrie
Town Guard L[t] Hall
Magazine & Brick hous from the 5th Reg[t].

The court of inquiry orderd to sett upon the conduct of L[t] Raphel of the corps of Artillery has reported that that Gentleman has acted no way unbecoming an Officer & a Gentleman & that the charge against him has not been supported the report is Approved of he is therefor is to be receivd & respected as Usual, by his Corps & the army & to do duty Accordingly __ The court is dissolved

By L[t] Col[o] Marion **Ch[s]. town 1777 __**

Orders 21st June G Howe } .... Parole Washington .... by GH
For the day tomorrow Capt[n]. Jervey
Town Guard from 5th Reg[t].
Magazine L[t] Proveaux __ Brick house L[t] Pet[r] Gray

Reg[tl] Ord[rs] Col[o] Marion } A regimental court martial to sett this morning at 10 OC: to try such prisoners as shall be brought before them evidences to attend
Presed[t]. Capt[n]. Moultrie L[ts]. P Gray & Burke member

The L[t]. Col is surprised to find Gentlemen sign morning reports without seeing their men on parrade, he orders therefore that no morning reports shall be certify'd without knowing it is right which cannot be done without Attending roll call

---

Orders 22nd June
G Howe } . . . . Parole Midlebrook . . . . by presedt

1777

For the day tomorrow Captn Ashby
Town Guard Lt Thos Hall
Magazine Lt Burke __ Brick House Lt Hurt
relief at Dorchester Lt Petrie
Pritchard Lt Gordon ~

Regtl Orders
Col. Marion } On Saturday 10 OC. in the morning divine Service will be performed by the Chaplain in St Michaels Church,[646] all Offrs. and men are desired to Parrade with their side arms at the new barracks at 9 OC. in morng from which the regiment will be marched to the Church __ It is expected the men will be clean & neat as possible with their hairs powdered __ the men to receive their Coats from the quartermaster that day for which Commanding Officer of Companies to give a receipt

By Lt Colo Marion **Chs. town 1777**

Orders 24th June
G Howe } . . . . .Parole Washington
For the day to morrow Captn Blake
Town Guard Lt from 5th Regt
Magazine Lt H Gray __ Brick house Lt Proveaux

Brigade Orders
by Genl Moultrie } The court martial of which Major Buchanan was Presedt. is desolved__ Genl. Moultrie approves of the sentences of the several prisoners tryd by the sd Court but remit the sentences on those who absented themselves without leave & that was seduced by Rolando formerly an Officer in the fifth Regt. & also on a promis of Genl Howe's to pardon them on delivering themselves up they are to be repramanded & to join their respective Regiments

Generals Howe and Moultrie being at Fort Johnston last Saturday were very much surprised to find that important post left without any field Officer of the Artillery It is therefore Orderd that Colo. Roberts or one of his field Officers be constantly there & reside within the fort __ Genl Moultrie expects this order will be strictly comply'd with [647]

NB The following men was sentenced by the Brigade court 18th Inst.

---

[646] Much of 18th century Charleston still exists today. St. Michael's church was established in 1751, and the church that is mentioned here still exists today on Broad Street.

[647] On 13 June 1777 the Marquis de Lafayette arrived in America. He landed at Georgetown, which was safer than trying to get into Charleston due to the British patrols off the coast of South Carolina. He traveled to Charleston and he was entertained by Gasden, Rutledge, Howe and Moultrie. The officers escorted the Marquis to Fort Moultrie, to show him the site of the victory in 1776. When they arrived they discovered that there were no artillery officers there, causing some embarrassment. This embarrassment would lead to General Moultrie placing Colonel Owen Roberts under arrest and courtmartialling him. That night there was a five hour dinner in LaFayette's honor, drinking toasts and conversing in broken French and English. The Marquis stayed to watch the anniversary of the Battle of Fort Sullivan.

Jn° Baptis Delany 100 Lashes __ Andrew le Bland 99 & picketed [648]- Alex Simmon [649] 99 & d° Vincent Maroney [650] 70 __ Jn° Cadet 50 all pardoned

~~~~~~

Regimental Orders
Lt Col° Marion } A Reg^{tl}. Court martial to sett this morning at 11 OC to try such prisoners as shall be brought fefore them __ evidences to attend __ Captn Ashby Presed^t. L^{ts} H Gray & Mazyck members __

NB the following prisoners was tryd for absents without leave
Phillip Thomas [651] __ Joseph Hughes [652] __ Ben^j: Clements [653] each to receive 50 lash Pardon'd
Joseph Stone for neglect of duty selling his regimentals & stealing a gun was Sentenced 99 lashes in three days & to be drummed out of the regim^t. with a halter round his neck which he received

NB accord in to Sentence of last Regimental court martial

John Cook [654] for Breaking confinement & Speak Insolently of the field Officers was reprimanded by the Major

---

[648] Picketting was a punishment in which the offender was bound around one wrist with a cord and the other end of that cord was attached to whatever was handy. That might be a pole, gibbet, stout tree limb etc. The man was left suspended above the ground dangling by his wrist. A narrow post was driven in the ground, which would allow the man to support his weight with difficulty by balancing with one foot on the post or picket. So whether the man elected to hang by his wrist or balance himself on the picket it was going to be a painful experience. Spending an hour or more under those conditions would be an experience that most soldiers would not care to repeat on a regular basis. This form of punishment was used in the American service through the early 19th century.

[649] John Alexander Simmons.

[650] Vincent Maroni enlisted in the 2nd South Carolina on 20 November 1775. He was sentenced to receive 70 lashes for desertion on 24 June 1777, but he was pardoned. In 1779 he was under Captain Charles Motte. He was killed during the assault on the Spring Hill Redoubt in Savannah on 9 October 1779.

[651] Phillip Thomas was in the 2nd South Carolina Regiment. On 24 June 1777 he was sentenced to receive 50 lashes for being absent without leave, but he was pardoned. In 1779 he was under Captain Proveaux.

[652] Joseph Hughes enlisted in the 2nd South Carolina Regiment on 19 June 1776. He served in Captain Shubrick's company in 1777. On 24 June 1777 he was sentenced to receive 50 lashes for being absent without leave, but he was pardoned. After Captain Shubrick's death he served in the 2nd Vacant Company until it was disbanded on 5 October 1778. He was then assigned to Captain Dunbar's Company. He was killed at Savannah on 9 October 1779 during the assault on the Spring Hill Redoubt.

[653] Benjamin Clements was in the 2nd South Carolina Regiment. On 24 June 1777 he was sentenced to receive 50 lashes for being absent without leave, but he was pardoned. He reinlisted on 15 July 1777. On 26 August 1777 he was promoted to corporal. He deserted again on 7 January 1778.

[654] John Cook enlisted in the 2nd South Carolina Regiment from the Beaufort District. He was under Captain Shubrick. He was in the Battle of Fort Sullivan in June 1776. He was reprimanded on 24 June 1777 for breaking confinement and speaking insolently to field officers. He was discharged on 15 July 1778.
~~~~~~

Corp: Jones Jn° Crawford Jerem[h] Early Ja[s] Smith	for absents from roll call was reprimanded Jam[s] Hooper rec[d] rec[d] 25 lash	Nicholas Bilboa [655] Anth[y] Wthsph both for absents from roll call was reprimand	Serg[t] Burgess for letting a prisoner escape was Acquitted Jn° Whitley Pet[r] Fagan 25 lashes

By L[t] Col° Marion **Ch[s]. town 1777 ___**

Orders 25th June
Gen[l] Howe } ....Parole Liberty
For the day tomorrow Capt[n] Shubrick
Town Guard L[t] from 5th Reg[t].
Magazine L[t] Dunbar __ Brick house L[t] Mazyck

Regim[tl]. Orders
L[t] Col° Marion } The Sergeants of the Guards having mistook the orders of the 14th Inst: has frequently given liberty to prisoners to walk in the Barrack yard, it is therefore Orderd that no prisoner shall be permitted to be out of the guard room on any pretense whatsoever (Except to the necessary with guard) witout Leave from the Commanding Officer of the Regiment any sergeant or Corporal of Guards who do not Observe this Order strictly must expect to be punished for disbedeance of Orders

Officers servants may be excused from Exercise in afternoon when such Officer are on guard but not Otherways

A Reg[tl] court martial to be held tomorrow at 11 OC: In forenoon to try such prisoners as shall be brought before them all evidences to attend __Capt[n] Shubrick presed[t]. L[ts]. Hall & Hart members__

Where there is no officer to attend Companys at roll calls a sergeant is to make & sign morning reports

---

Orders 26th June
G Howe } ....Parole Drayton
For the day tomorrow Capt[n] Charnock
Town Guard L[t] Hart
Magazine from 5th Reg[t] __ Brick house L[t] Hall

NB according to sentence of last Brigade Court Martial Josiah Stone rec[d] 100 lashes in 3 several times & drumm'd out of the regim[t]. with a halter ab[t] his neck

---

Orders 27th June
G Howe } .... Parole Moultrie
For the day tomorrow Capt[n] Lessne
Town guard L[t] from 5th Reg[t]
Magazine L[t] Proveaux __ Brick house L[t] Perreneau

In Commeration of the 28th June last on which day the good conduct and Spirited Behavior of the Officers & men of this state deservedly Obtained honours for themselves & render'd Esential service to their Country & common cause of America the following firing are to take Place at fort Moultrie 13 pieces of Cannon, at fort Johnson 11, at Broughtons Battery 7, at Littletons 7, at Elliots at Gadsdens Warff 7

---

[655] Nicholas Bilboa enlisted in the 2nd South Carolina Regiment on 7 October 1776. He was reprimanded on 24 June 1777 for being absent from roll call. He was discharged on 11 October 1777.

By Lt Col^o Marion **Ch^s. town 1777 ___**

The firings to begin at fort moultrie and when finished then to commence at fort Johnson then at Broughtons then at Litteltons & to finish at Elliots Col°. Huger (as Genl Moultrie is sick) will order an Officer with a proper number of men from the 2nd & 5th Reg^ts to get the Guns in Order at Elliots & to direct the firings at that place

Capt^ns: Grimball [656] & Darell [657] will be so Obliging to order that at the Batterys where the command, the signal for Begining the firing which will be one piece of Cannon from Broughtons Battery which will probable be about 10 OC: __ The 2nd & 5th Regiments will parrade at some convenient place tomorrow morning precisely at 10 OC: when a feu de Joye is to be fired,[658] the Commanding Off^rs. at fort Moultrie will turn out the men of that fort at such time as the tide will premit & think proper and fire either a feu de Joye or in plattoons, Tho as the former will not probable be heard in town the latter will be most elagible this firing is to be answered by the corps at fort Johnson who are to take it up in the manner observed at fort Moultrie some signal should be agreed upon between the two forts the General thinks proper to add that he hopes the Common Soldiers will not disgrace the festivity of the day by any improper behaviour, the Adjutant General will immediately transmit this Order the Command^g Off^r at fort Moultrie & acquaint Col° Roberts, Capt^ns Grimball & Darrel there with

Regimental Orders by Lt Col° Marion } Commanding Off^rs of companies to apply to the Quart^r Master for their mens coats this after noon in proportion to the number of men in each Comp^y and tomorrow to supply their men with Leggens all who have had a pair for last year to give Col° Marion their names the Quarter Master to take a receipt from an Off^r of a Comp^y for what Cloathing he delivers

A Number of Ladies in this town have been so kind as to Order a genteel dinner to be given the Soldiers tomorrow in memory of their good Behaviour the 28th June last past at fort Moultrie & the officer of the Regiment presents them with a Hogshead of Claret & 42 barrels Beer

Col° Marion hopes the men will behave with Sobriety & Decency in honour to those Ladies who have been so kind as to give them so genteel a treat, for Soldier being seen in the Street Drunk or Riotous will be scandal to the Regiment & prevent any further notice being taken of them, he hopes they will keep the Barracks & not go in town that day & should any man be overtaken in liquor the Sergeants and Corporals will have them put Quitely in their Barracks for which reason the Col°. Insists that every Sergeant & Corp^l. will stay on the Barrack yard that they may take care of the men in their Company the Serg^t. Maj^r. in perticular is to stay in Barrack yard & keep Good Order amoungst the men [659]

General Moultrie will be on the parrade tomorrow morning & is expected the man will take care to be very clean in respect to him

---

656 Thomas Grimball was born in 1745 and was a Charlestown attorney. He served as a captain of the Charlestown Battalion of Artillery from 1775 to 1778. He was promoted to major and served until the fall of Charleston in May 1780. He was at the Battle of Port Royal in 1779 and was captured at the fall of Charleston. He was exiled to St. Augustine. He loaned South Carolina £80,000 during the war. He died on February 1783.

657 Joseph Darrel was a captain in the Charlestown Artillery and he was the captain of the Littleton Bastion in 1775. He was in the siege of Charleston in May 1780. He later was promoted to major.

658 A feu de joi is French for "fire of joy". The soldiers would line up and fire their muskets, from left to right, having a continuous running fire.

659 All of the officers were at their own party and were not with the men that night.

Orders 28th June by G Howe **1777** } .... Parole Moultrie
For the day tomorrow Captn Moultrie
Town Guard Lt Mazyck
Magazine from 5th Regt. __ Brick house a Sergeant

Genl. Howe thinks proper to suggest to the ~~nesessity~~ Army the nesessity their is for propriety & conduct upon this memorable day, and hopes the Soldiers will not suffer festivity & rejoycing to degenerate into riot & disorder he wishes the men to confine themselves as much as possible to their Barracks that their Excess / should any happen / may not meet with any mobs nor to have any least hand in any riotous proceedings whatsoever & forbids upon pains of his highest Displeasure, the least Offer Insult or Injury to the persons or property of the Inhabitants of this Capitol, A Soldier Should at times consider himself as ordained to protect & defend the persons & support & maintain the Rights & Priviledges of his fellow citizens & Constituted for this noble purpose desdain every thing which Counter Acts it the Genl. hopes the conduct of the soldiers on this occasion will demonstrate they act under the Influence of such Considerations [660]

Brigade Ordrs by Colo Huger } 1 Sergt & fifteen privates from the 2nd & 5th Regiments to take charge of Elliots Battery at half past 10 OC. this party to be commanded by a Subaltern of the 2nd Regt } R. O. for this day Lt Gray

Lt Colo Marion **Chs. town 1777 ___**

Orders 29th June G Howe } .... Parole ....
For the day tomorrow Captn Jervey
Town Guard Lt from 5th Regt
Magazine Lt Hart __ Brick house Lt Dunbar

Regtl. Orders Lt Colo Marion } A regtl court martial to set tomorrow morning at 10 OC: in morning to try such prisoners as may be brought before them evidences to Attend & to try Wm Edmunds Sergt. Majr: Captn. Charnock will furnish the Evidences against the Sergt. Majr:

Captn Harleston Presedt. Lts H Gray & Hall members

Orders 30th June G Howe } ....Parole Hall .... by presedt
For the day tomorrow Captn Harleston
Town guard Lt Proveaux
Magaxine Lt Hall __ Brick house from 5th Regt

The General highly approves the respectable conduct of the Soldiers on Saturday last he feels himself interested in every thing wch concerns them and consequently cannot help but take pleasure in what contribute to their conduct

Brigade Orders by Genl Moultrie } Captn James McDonald of the So.Carolina Second Regt. Commanded by Colo Isaac Motte having resigned his Commission the 5th of May last is therefore no longer to be Considerd or respected as a Captn in the Continental Service

---

660 The day was celebrated throughout the city. There was the firing of the cannons, a military parade, the ringing of bells and an evening of entertainment. This would all be repeated within a week when the city celebrated the first anniversary of declaring independence from Britain.

Orders 1st July
G Howe } .... Parole Freedom .... by the Gen H
For the day tomorrow Captn Motte
Town Guard from 5th Regt.
Magazine Lt Burke __ Brick house Lt P. Gray

Regtl Orders
by Lt Colo Marion } The court which sett yesterday for the tryall of Sergt Major Edmunds reports they are of Opinion that the Sergts Confinement & the abuse he received is more than adequate to the disrespect shown Captn Charnock & therefore do Acquit him the sentence is approved of & the sergt. is released from confinement asd Ordered to return to his duty.

A court martial to sett tomorrow morning at 10 OC: for the tryal of Sergt Rayboult for disobedience of Orders 27th June Last & for threatening the life of abusing Mary Singleton an inhabitant of the town 28th Inst: [661] also for the tryal of all other prisoners Mr Saml Glaton to be warned as Evidence Captn Motte presedt Lts Burk & Gray members [662]

Lt Colo Marion **Chs. town 1777**

Orders 2nd July
G Howe } .... Parole Mifflin .... by presedt
For the day tomorrow Captn Ashby
Town Guard Lt Dunbar
Magaxine 5th Regt __ Brick house Lt Hart__

The 2nd & 5th Regts to parrade on fryday morning upon the shortest notice their arms & accoutrements to be in the nicest order

Regtl: Orders:
Lt Col: Marion } A Regtl. Court martial to sett tomorrow morning at 10 OC: to try such prisoners as shall be brought before them. evidences to attend

Captn Ashby presedt: Lts Hall & Proveaux members

NB according to Sentence of last court Jno Lyons for stealing 2 shirts of Jo: Williamson to receive 50 lashes & to pay £13 __ Sergt. Raybout Acquitted __

Orders 3d July
G Howe } .... Parole Lee .... by GH
For the day tomorrow Captn Oliphant
Town Guard Lt Hall
Magazine Lt P. Gray __ Brick house from 5th Regt.

General Moultrie will order an Offr with a proper detachment to Laurens Battery near Genl. Gadsden's warff to conduct the firing which is to be tomorrow at that place perticular firing will be directed at fort Moultrie and fort Johnson & the Garrison of each is to turn out in honour to that day when the declaration of Independence was published in this state by which America was delivered from the thraldom of great Briton who by reiterated Insults and Injuries & by the most cruel & tryannical invasion of every darling rights & priviledge had renderd all further Union with her Absolutely impossible to minds not absolutely lost to every sense of freedom

---

661 This may be the wife of Matthew Singleton, captain in the Volunteer Horse Company of St. Mark's Parish.

662 This name is written very small and may be Glaton, Felton or Fubon.

The firings is to begin at fort Moultrie is to be taken up by fort Johnson & will be carried on by Broughtons, Lyttleton's, Cravens, Granviles, & Laurence's Batteries in Succession [663] __ fort Moultrie fires 21 guns, fort Johnson 17 Broughtons 14 Lytleton's 9 Craven's Granviles & Laurence's 5 each in all 76.__ The Signal for beginning the fire will be a Signal hoisted from the steeple of St. Michaels Church which will probable happen about 12 OClock the strictest attention is to be paid that no mistake may happen

The Regiments in town are to parade precisely at 8 OC: in the morn^g. and to go through the common firings finishing by a General Volley

The Garrison at fort Moultrie is to turn out at such time & in such a manner as the Commanding Off^r. there shall direct & it is to be follow'd by similar firings by the Garrisons at fort Johnson it may be therefore Proper that the Commanding Off^r there should be Acquainted with the manner & time of firing at fort Moultrie

Brigade Orders Genl Moultrie } One Subaltern 1 Serg^t. & 15 Privates from 2nd & 5th Regiment to march to Laurences Battery at about 11 OClock to answer the firings with 5 guns, the Subaltern from 5th Reg^t: with a Serg^t. & 5 men and 10 men from the 2nd Reg^t: __ Lt Dunbar for this duty

Regiment^l Orders by Lt Col^o Marion } Tomorrow the Regiment is to turn out on the Green & will appear as clean as possible with their hair well comb'd & powder'd his arms must be in the best Order. Any Soldier who do not comply with this order must expect to be Severely punished Sergeants are Orderd to see their men as above by 8 OC: in morning at which time the regiment will be parraded, The Articles of war to be read to the men this Afternoon by the Eldest Captn. present

---

Orders 4th July G Howe } .... Parole Hancock .... by presed^t
For the day tomorrow Captn Blake
Town Guard from 5th Reg^t:
Magazine Lt Perrineau __ Brick house Lt Proveaux

Orders 5th July G Howe } .... Parole Eveleigh .... by GH
For the day tomorrow Captn Shubrick
Town Guard L^t from 5th Reg^t.
Magazine Lt Dunbar __ Brick House Burk

Reg^tl Orders by Lt Col^o F. M. } A court martial to sett this morning at 10 OC: to try such prisoners as shall be brought before them evidences to attend Captn Blake presed^t Lieut^ts. Dunbar & Burk members

NB According to sentence of Last Court Jno Michael Jacks for absents without Leave rec^d 99 lashes & to make up the time of his absents [664]

---

663 Craven's and Granville's Batteries were the oldest in Charleston. They were located on the waterfront of the city and had been built in 1704 in order to protect the city from Indians, Pirates, French and Spanish raiders.

664 John Michael Jacks was in the 2nd South Carolina Regiment. He received 99 lashes on 5 July 1777. He also had to make up the time of his absence. On 2 August 1777 he received 50 lashes for being drunk.

Orders 6th July G Howe } .... Parole Independence .... by presedt
For the day tomorrow Captn. Charnock
Town Guard Lieutt. Hart
Magazine Lt Hall __ Brick house Lt. from 5th Regt.

Regtl. Orders by Lt C: M } Divine Service will be performed this afternoon at the new Barracks where it is expected Offrs. & Soldiers will attend

Lt Colo Marion **Chs. town 1777**

Orders 7th July G Howe } .... Parole Caswell .... by G.H.
For the Day tomorrow Captn. Lesesne
Town Guard Preveaux __ Brick house Lt from 5th Regt.
Pritchards guard Sub: from 5th Regt:__ Magazine from 5th Regt.

A general Court martial to sett on wednesday between the hours prescribed by the Honourable the continental congress at some convenient place in Charles town for the tryal of Colo Owen Roberts of the Corps of Artillery put under arrest by Brigadier Genl. Moultrie & charged of quiting his post & being to frequently in town without leave of the Commanding Offr. in Chief or the Commanding Offr. of the Brigade to which he belongs And also for having observed that an Order Issued by Genl: Moultrie was unpresedented & ungentleman like

Of this Court Genl. Gadsden is to be president the other members are to be taken from the 1st, 2nd, 4th, & 5th Regts. Each Corps to furnish a field Offr. & two Captns. who are to be taken according to detail _____

Brigadier Genl. Moultrie is to produce the evidences the Adjutant Genl. will transmit this order to the out posts & serve Colo. Roberts with a Copy of it.

The Adjutant Genl is as immediately as possible to be furnished by the Commanding Offr. of every Corps with an exact list of those Commissioned Offrs of their respective Corps & the manner in which they stand for General duty, that a proper Roster may be made

After Orders by G Howe

Colo Huger having reported that all the Captns of his Batalion were out upon recruiting service the 1st & 2nd Regts. are therefore to furnish one Captn. each for the Genl. court martial of next wednesday more than was expressed in the order of this morning

Brigade Orders Genl. Moultrie } A Subaltern Offr from each Compy of 2nd & 5th regts is to reside in the new Barracks when there is room to receive them. __________

Regemtl Ordrs Lt Col. Marion } The Soldiers having very frequently do their Occasion near the Barracks & on the green near the Chaplains house Contrary to Decency and good order which may Occasion some disorder prejudical to the men, It is therefore Order'd that all sergeants & Corporals who see any soldiers doing their Occasions near the barracks & on the green to confine them, and any soldier so confined may expect to be try'd for disobediance of orders & suffer accordingly [665]

A court martial to sett on Tuesday 10 O'Clock in Morng to try such prisoners as shall be brought before them evidences to attend

Captn. Lesesne presedt. Lieutts. Shubrick & Petr Gray members

---

[665] The soldiers doing their "Occasion" was urinating and defecating right beside the barracks, instead of walking to the latrines.

Aft[r]. Reg[tl]. Orders.

The officers who have attestations of men they have Inlisted are desired to bring them & receive what is due which they are to pay the men what money may be yett due to them

All the Soldiers In the Reg[t]. not yet Inlisted for the war & has been two years in the service; may now Enlist during the War, & will receive immediately thirty dollars p[r] man & one hundred Acres of Land at the conclusion of the war they may apply to the Officers of their Companies who are desired to call on the Commanding Officer for money when wanted

NB according to sentence of last court Tho[s] Bowen, Sam[l]. Frazer, Tho[s]. Lewis, Jn[o]. Brown & Timothy Green for being in liquor on guard rec[d] each 35 lashes. Sam[l]. Henderson, pardoned, Pet[r]. Fagan for quitting his post rec[d] 50 lashes, Jam[s]. Blenchfield rec[d]. 50 lashes W[m]. M[c]Dowel for d[o] rec[d]. 30 lashes.[666] Franc[s] Dupree for stealing a p[r] silk stockings from Serg[t]. Munroe rund the Gauntlet

For the gen[l]. court martial L[t]. Col: Marion Capt[ns]. Motte Oliphant & Lesesne _______

Orders 8th July
G Howe

. . . . Parole Cattel . . . . by GH
For the day tomorrow Capt[n] Moultrie
Town Guard L[t] Burke
Magazine from 5th Reg[t] __ Brick house L[t] P. Gray

Commanding Officers of Corps are for the feuter to report to the Commanding Off[r]. in Chief for the time being, the Names & Rank of such officers as they send out upon duty & the nature of the duty upon which they are sent, this report may be lodged with the Adjutant General

---

Orders 9th July
G Howe

. . . . Parole Virginia . . . . by presed[t]
For the day tomorrow Capt[n] Ashby
Town guard from 5th Reg[t]
Magazine Guard L[t] Hart __ Brick house a sergeant

Brigade Orders
by Gen[l]. Moultrie

A Brigade court martial to sett tomorrow to try Serg[t]. Munro confined by his Excellency the president Capt[n] Charnock presed[t] L[ts]. Henry Gray, & Rich[d] Moncriff memb[rs]

L[t] Col[o] Marion **Ch[s]. town 1777**

Orders 10th July
G Howe

. . . . Parole Chatham . . . . GH
For the day tomorrow Capt[n] Blake
Town Guard L[t] from 5th Reg[t].
Magazine L[t] Perrineau __ Brick house L[t] Hall __

Reg[tl] Orders
L[t]. Col: Marion

A Regimental court martial to sett this morning to try such prisoners as may be brought before them evidence to attend

Capt[n] Blake Presed[t] - Lieut[ts] Shubrick & Hart members

---

666 William McDowell enlisted in the 2nd South Carolina Regiment on 4 November 1775. He received 30 lashes on 7 July 1777 for quitting his post. He received 50 lashes on 2 January 1778 for being drunk on guard duty. He received 50 lashes on 16 April 1778 for being drunk on guard again, and then on a third occasion he received 50 lashes with a cat-o-nine tails for being drunk on guard.

NB according to Sentence of Last Brigade court Sergeant Munson tryd for steal some lead was acquitted

---

Orders 11th July G Howe } ....Parole America .... by GH
For the day tomorrow Captn Shubrick
Town Guard Lt P. Gray
Magazine Lt Preveaux __ Brick house a sergt of 5th

Regtl Orders by Lt. Col: Marion } a court martial to sett tomorrow morning at 10 OC: to try such prisoners as shall be brought before them all evidences to attend

Captn Shubrick presedt. Lieutts. Mazyck & Dunbar members

Orders 12th July by G Howe } .... Parole Hancock .... by presedt
For the day tomorrow Captn Charnock
Town Guard Lt. Burke
Magazine Lt Hart __ Brick house from 5th Regt

Regtl Orders Lt C. Marion } Israel Dyer of Captn Mottes Compy is appointed a sergeant in the same & is to be obey'd as such [667]

Orders 13th July G Howe } .... Parole ....
For the day tomorrow Captn Moultrie
Town Guard Lt Hall
Magazine Lt. Dunbar __ Brick house a sergeant

Regemtl Orders Lt Col: Marion } Lieutt. Daniel Mazyck is appointed a Captain in the 2nd Regiment & is to be respected & Obey'd as such Captn Mazyck is to take charge of the Compy late McDonalds who resign'd the 5th of May last

A Court martial to sett tomorrow morning 10 OC: to try such prisoners as shall be brought before them evidences to attend
Captn Moultrie presedt. Lieutts. Preveaux & Perreneau members
Divine service will be prerformed tomorrow at the Barracks 5 OC: all Offrs & soldiers to attend

Lt Colo Marion **Chs. town 1777**

Orders 14th July G Howe } ....Parole Washington .... by presedt
For the day tomorrow Captn Mazyck
Town Guard Lt. from 5th Regt.
Magazine Lt. Preveneau __ Brick house Lt Proveneaux
Pritchard Guard Lt Jacob Shubrick __

---

[667] Isaac (or Israel) Dyer had enlisted in the 2nd South Carolina Regiment on 4 November 1775. On 12 July 1777 he was promoted to sergeant in Captain Motte's company. On 16 August 1777 he was sentenced to be reduced in the ranks for being drunk on guard but he was pardoned and reprimanded. On 28 October 1777 he was demoted back to private and was fined 14 days pay for being drunk on guard. He was promoted to corporal on 10 February 1778. He was discharged on 29 June 1778.

The Gen[l]. has accepted the resignation of M[r] Oliver Hart Surgeon's mate in the Continental Second Batalion in this state [668]

Orders 15th July G Howe }	.... Parole Patriotism .... For the day tomorrow Capt[n] Harleston Town guard from 5th Reg[t]. Magazine L[t] Burke __ Brick house L[t] P. Gray	GH

The General court martial Orderd to sett for the tryall of Col[o]. Owen Roberts of the Corps of Artillery put under arrest by Brigadier Gen[l]. Moultrie have reported as follows

The Court taking the matter between Brigadier Gen[l] Moultrie & Col[o]. Roberts into their most serious deliberations are of Opinion that Col[o]. Roberts according to the strictest & literal sense of the 3rd Article of the 13th Section of Rules & Articles of War, has been frequently absent without leave, but it being a Custom & usageby no means unpresidented at fort Johnson & fort Moultrie for the Commanding Officers to repair to town when ever inclination or necessity should prompt him without petitioning that Indulgence from any Superior officer, & it Appearing that Col[o]. Roberts has used this Discreationary power with great prudence & moderation even when enjoying it that the better cloathing & disposition of men engaged his Attention & they likewise take their most Serious deliberation the charge exhibited against him agreeable to the 1st Article of the 7th Section & being of Opinion that the Col[o]. is by no means ameneable to this court for the same do the former charge Acquit him

The General Approve of the Sentence of the court martial releases Col[o]. Roberts of his arest who is to be respected and Obey'd as usual

The Court to continue setting for the purposes & with the alterations following ___

L[t] Col[o] Marion **Ch[s]. town 1777**

The necessity of public service requiring the presents of Brigadier Gen[l]: Gadsden at his command on Sullivants Island, he is discharged from his attendance on the General court martial, Col[o] Huger will be added to the court & is to act as president, as several Officers of Col[o]. Roberts Corps of Artillery are to be try'd & some the officers of their corps now members of the court martial are among the persons who Charge them they are also discharged from that Court & are to be replac't by one Capt[n]. from 1st & one from 2[d] Batalions, they are to try L[t]. Marzill [669]& L[t] Raphel charged by Captains Eph:

---

[668] Oliver Hart was born on 5 July 1723. He was married twice, the first time to Sarah Bress and the second time to Anna Marie Gimball. He became a surgeon's mate in the 2nd South Carolina Regiment in November 1776. He resigned on 14 July 1777 to become a minister. When the British invaded South Carolina he returned to the military and became the surgeon's mate for the 3rd South Carolina Rangers. He was captured at the fall of Charleston in May 1780. He was exchanged and continued to serve until the end of the war as a surgeon in Marion's partisans. He died on 31 December 1795 at the age of 72.

[669] La Marzelle served as a lieutenant in the 4th South Carolina Regiment (Artillery). He resigned on 13 September 1777 possibly from having charges brought against him by his fellow officers. He was found not guilty of those charges.

Mitchel[670] Jn°. Detreville & Brooke Roberts [671] & by Lieut[ts]: W[m] Mitchell, Jn° Wickly,[672] Ja[s]. Wilson,[673] Ja[s] Mitchell W[m]. Dornam,[674] Jn° Frazier [675] & Rob[t] Simpson,[676] in the following manner, L[t] Raphel with Behaving in a manner unbecoming an Officer & a Gentleman, in having threatning to take the Life of a person who gave evidence against him at a Court martial, & for having threatning the reveng himself on the members of a court of Inquiry who satt on his conduct

Lieut. Marzell for behaving in a manner unbecoming a Gentlen: and an officer, by endeavouring by false representation to disturb the peace and harmony of his Corps & for behaving in a Scandalous and ungentleman like manner in drawing upon an Off[r]. who was unarmed & Indeavouring to take his Life __ The Officers Charging L[ts] Raphel & la Marzelle are to furnish the evidences, the court is also to try several prisoners for desertion & will be furnisht with their names & the evidences against them in due time, The Court will proceed with all possible expedition; the Charge against L[t] Beaubien are to the Generals Oppinion not proper for the subject of a Gen[l]. Court martial, he shall therefore as soon as can be made Convenient Order a court of Inquiry on the Conduct of that Officer.

Orders 16th July by G Howe } .... Parole Pinckney ....
For the day tomorrow Capt[n] Ashby
Town Guard L[t]. from 5th reg[t].
Magazine L[t] Hart __ Brick house L[t] Dunbar

Regimental Surgeons are for the feuter when they find it necessary to remove the sick of their several Regiments to the General Hospital to obtain orders for that purpose from their Col°. or Commanding Officer of the Corps for the time being, & to transmit to the Director general of the Hospital a signed return of their sick in which shall be specify'd the names of the men, the Comp[y]. they belong to the nature of the disorder, the time they have been Ill & the manner in which they have been treated the

---

670 Ephraim Mitchell was commissioned a 2nd Lieutenant in the 1st South Carolina Regiment on 17 June 1775. He was promoted to 1st Lieutenant in November 1775. He became a captain in the 4th South Carolina Artillery in May 1776. He was promoted to major on 20 June 1779. He was captured at the fall of Charleston in 1780. In 1782 and 1783 he served as a major in the Continental Regiment of Artillery. He died on 16 March 1792.

671 Richard Brooke Roberts was born in 1758. He married Everarda Catherine Sophia Van Braam Houckgeest. He was a captain in the 4th South Carolina Artillery. He was in the siege of Savannah during 1779 and was in the siege of Charleston in 1780 as the aide de camp for General Lincoln. He died on 19 January 1797 at the age of 39.

672 John Wickly was commissioned in the 4th South Carolina Artillery on 26 November 1776 as a lieutenant. He was promoted to captain on 20 June 1779. He was taken prisoner at the fall of Charleston.

673 James Wilson was born in Ireland in 1745. During November 1776 he served as a first lieutenant in the 4th South Carolina Regiment (Artillery). He became a captain-lieutenant on 31 May 1778. On 13 November 1779 he was court martialed with some other officers for being absent from the camp after the failed assault on the Spring Hill Redoubt in Savannah. The court considered the "particular circumstances which induced them to over stay their leave of Absents" and recommended that he be discharged from arrest. After being taken prisoner at the fall of Charleston he was exchanged and served to the close of the war. He died in 1825.

674 William Donnom was commissioned a 1st Lieutenant in the 4th South Carolina Artillery on 16 April 1776. He was promoted to captain in 1778. He was killed at Savannah on 9 October 1779.

675 John Frazer was a lieutenant in the 4th South Carolina Regiment in 1777. He was killed at Beaufort on 9 February 1779. The battle for Beaufort happened on 3 February, so it is unclear whether or not he was wounded then, and died later, or died for other causes.

676 Robert Simpson was a lieutenant in the 4th South Carolina Artillery in 1777.

Director General will furnish the Surgeons of each ~~Company~~ Regiment with a proper form of the return

Regimtl Orders
by Colo. Motte }
a Regimental Court martial to set tomorrow morning at 9 OC: to try all Prisoners brought to it, Evidences to attend in time

Captn Ashby presedt. Lts. H. Gray & Hall members

Lt Shubrick to Join Captn. Lesesne's Compy. Lt. Petr. Gray to join Captn Charnocks & Lt. Burke Captn Oliphants & to do duty in the same till further orders

NB according to Sentence of Last court 14th Inst. Petr Bourderhose for absence without leave recd 100 lashes,[677] Joseph Grenada recd 50 do.[678]- John Whyley for quitting his Guard recd 50 lashes

---

Orders 17th July
G Howe }
. . . . Parole Motte . . . .
For the day tomorrow Captn. Blake
Town Guard Lt. Preveaux
Magazine Lt. Hall __ Brick house Lt. from 5th regt.

Orders 18th July
G Howe }
. . . . Parole Success . . . .
For the day tomorrow Captn Charnock
Town guard Lt. Preveaux
Magazine from 5th regt.__ Brick house Lt. P. Gray

Colo. Huger having Informed the General that Majr: Beekman from some point of delicacy was desirious of being Discharged from the Genl: court martial Appointed for the tryal of Lieutts Raphel & la Marzelle A Captn. from Colo. Mottes Regt. is immediately to be warned to attend & to sett in Genl. Court martial at 9 OC: tomorrow morning in the place of Majr. Beekman who is discharged from that duty

RO for the Genl Court martial tomorrow Captn. Moultrie

Orders 19th July
G Howe }
. . . . Parole Charleston
For the day tomorrow Captn Mazyck
Town Guard from 5th regt.
Magazine Lt. Burke __ Brick house Hart

Regtl Ordrs
Colo. Motte }
A court martial to sett on Munday morning 10 OC: to try such prisoners as shall be brought before them, evidences to attend
President Captn Harleston __ Lieutts P Gray & Hart members

---

[677] Peter Bourdeshaw (Bourdeshoe, Bourdeshore) enlisted in the 2nd South Carolina Regiment on 4 November 1775. He received 100 lashes on 16 July 1777 for being absent without leave. He deserted on 12 January 1778. He was a horseman in the militia under Captain Joseph Bouchillon in 1781 and 1782.

[678] Joseph Grenada (Grandau) enlisted in the 2nd South Carolina Regiment on 26 October 1776. On 16 July 1777 he received 50 lashes for being absent without leave.

Col° Motte **Ch^s. town 1777 __**

Orders 20th July } .... Parole Maxwell ......
G Howe For the day tomorrow Captn Harleston
Magazine guard Lt. from 5th regt.
Town guard Lt Perrineau __ Brick house Lt Hall

R.O.
NB. According to sentence of the court 18th Inst: Adam Meek for being drunk on guard & abusing the Inhabitants of the town recd 100 lashes __ Arthur Jackson pardon'd of 50 lashes for letting a prisoner escape [679]

Orders 21st July } ...... Parole Maryland
G Howe For the day tomorrow Captn Ashby
Town guard from 5th regt:
Magazine Lt Preveaux __ Brick house Lt Dunbar

Commanding Officers of Corps & Batalions are Immediately to have a strict survey of the Arms of their men and to report to the general the exact state of them, he is anxious to receive their reports for perticular reasons; Officers of Companies having long since Issued to be carefull that each man of their Company had a form exactly fitted to his gun, & as the Calibars may not be equal, Commanding Officers of Battalions were directed to have Boxes made prepared to deposit cartridges in separate Bundles which Bearing some mark to distinguish to which gun they belonged, that no mistake, confusion or delay might happen in serving them out; the Genl. wishes to be informed if those orders were complyed with and therefore desires Commanding Officers of Batalions will examine & report to him as immediately as possible & why it has not been comply'd with / if that can possible be/ that it may be immediately done the General is desirous of receiving as immediately as possible an exact return of the Ammunition, cannon & other military stores at fort moultrie & fort Johnson and also an Account of what stores may be wanted in each place, having lately had a return of this last from fort moultrie: he is sorry so soon to request an other: but he has unfortunately mislaid the former report

Genl. Moultrie being out of town Col° Huger takes the Comm: of the Brigad till his return__ Captn Mazyck of Col°. Mottes Batalion is added to the Genl Court martial now setting & to attend Immediately __

Regimtl. Ordrs. } The Commanding officer of each company to give in a return tomorrow
Col. Motte morning by 10 OC: to the Quarter Master of the number of coats waistcoats & breeches & spatterdashers received by them from him

Col° Motte **Ch^s. town 1777 __**

Orders 22nd July } .... Parole Gates
G. Howe For the day tomorrow Captn Blake
Town Guard Lieut. Hart
Magazine from 5th reg: __ Brick house Lieut: P Gray

The Adjutant General will apply to the Keeper of the Arsenal and obtain of them an exact return of the stores there is of every kind; he will be very perticular in the number and size of the shot, the

---

679 Arthur Jackson was in the 2nd South Carolina Regiment. He received 50 lashes for letting a prisoner escape on 20 July 1777.

Quantity and Quality of Cartridge paper Cannon d^{o}__ and have this report Compleated as expeditiously as possible

Regem^{tl}. Ord^{rs}.
Col^{o} Motte } An exact state of the Arms & accoutrements of each company to be made immediately & deliver'd to the Col^{o}. by the Commanding Officer of each Comp^{y}.

The Quart^{r} Master to deliver a return to the Commanding Off^{r}. of the Regim^{t}. by 11 OC: today of the quantity of Ammunition he has in store

NB according to sentence of last court W^{m}. Smily & Thos Lewis for quiting the Guard & Insult^{g}. D^{r}. page rec^{d}. 60 lashes each

Jacob Dunbar rec^{d} 100 lashes for stealing publicks lead to receive it on his posterior

Ja^{s}. Hooper for beating Francis Sipson's wife rec^{d} 50 lashes [680]

After Orders

Corporal James Anderson of Captn Ashbys Comp^{y} is Appointed Sergeant in the same & is to be obey'd as such __

Orders 23^{d} July
G Howe } .... Parole Maxwell ....
For the day tomorrow Captn Charnock
Town guard Lt Shubrick
Magazine 5th reg^{t}: __ Brick house Lt Perrineau
Dorchester Command Lieut. Hall

Regem^{tl}: Ord^{rs}
Col^{o} Motte } The Col^{o}. has had Complaints that the men of his Reg^{t}: frequently fire their arms off thro the windows of the Barracks it is therefore his Orders that no firing of muskets be allowed in or near the Barracks and he perticularly desires the Captain of the day will Apprehend any one that may disobey this order & report him, who may depend upon being severely punished __ he has had also complaints that many of the men Appear on the Green near the Barracks in a very indecent manner, such as runnin with one an other intirely naked and playing many lacivous tricks a Scandal to the reg^{t}: they belong to, he orders that nothing of the Kind may happen again on pain of his Displeasure & Assurance of his being punisht, The men to be made Acquainted with this order by having it read to them by an officer of each Company

Col^{o} Motte **Ch^{s}. town 1777** __

Orders 24th July
G Howe } ....Parole Hariette
For the day tomorrow Captn Harleston
Town guard L^{t}. Roux
Magazine Lt H^{y} Gray __ Brick house from 5th reg^{t}:

Regiment^{l}. Orders
Col: Motte } A reg^{tl}: court martial to set this morning at 10 OC: to try such prisoners as shall be brought before it evidence to attend

Captn. Charnock presed^{t}. Lieut^{ts}. Henry Gray & Roux members

NB this court did not sett for want of L^{t}. H Gray in attendance

---

[680] Francis Simpson served in the 2nd South Carolina Regiment under Captain Thomas Hall during 1777. On 29 October 1777 he received 100 lashes over two days for disobedience to orders and impertinent language to his sergeant. He was at the siege of Savannah.

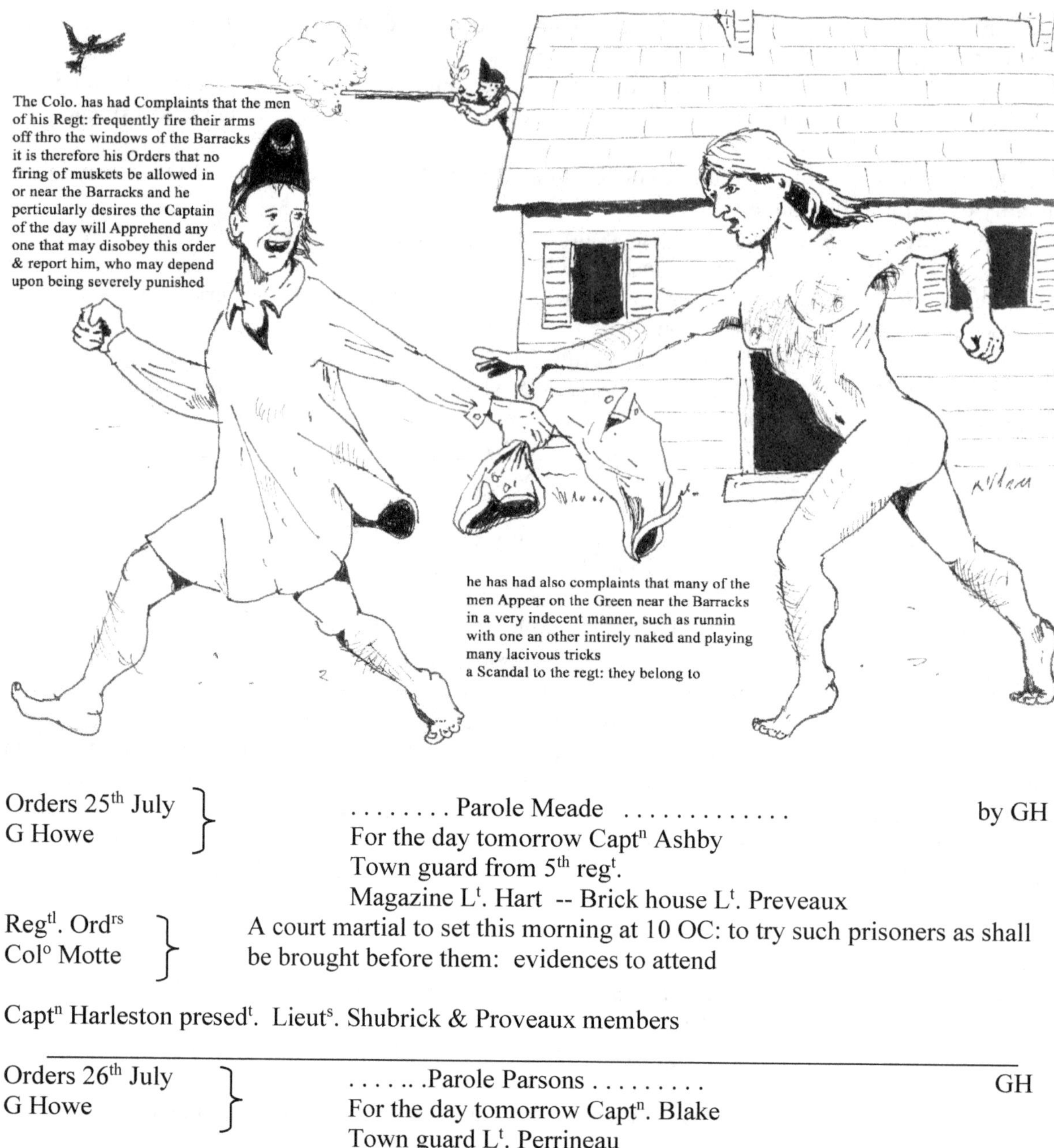

Orders 25th July
G Howe } .... Parole Meade .... by GH

For the day tomorrow Captn Ashby

Town guard from 5th regt.

Magazine Lt. Hart -- Brick house Lt. Preveaux

Regtl. Ordrs
Colo Motte } A court martial to set this morning at 10 OC: to try such prisoners as shall be brought before them: evidences to attend

Captn Harleston presedt. Lieuts. Shubrick & Proveaux members

Orders 26th July
G Howe } ...... .Parole Parsons .... GH

For the day tomorrow Captn. Blake

Town guard Lt. Perrineau

Magazine Lieut. Shubrick __ Brick house 5th regt:

Colo. Huger being renderd incaple of Attending the general court martial by sickness Colo. Motte is added to the court & is to act as president & is to attend the court Accordingly

Regim[tl]. Ord[rs] }
Col° Motte }
The Col°. disapprove of the unmilitary practice of the sergeants beating the privates & therefore orders that they will not take that liberty again, but if the men misbehave they are to confine ~~them~~ & have them report to the Command[g]. Off[r]. of the Reg[t]. who will take care to have them punished

Devine Service will be performed at the Barracks tomorrow after noon at 5 OC: by the Chaplain, Orderd that the regem[t]. attend perticularly at that hour

---

Orders 27th July }
G Howe }
...... Parole Parsons .... by presid[t]
For the day tomorrow Capt[n] Charnock
Town Guard from 5th reg[t]
Magazine L[t] Roux
Brick house Lieu[t]. H[y] Gray

Col° Motte **Ch[s]. town 1777 __**
Orders 28th July }
G Howe }
...... Parole Beaufort .... GH
For the day tomorrow Capt[n] Mazyck
Town Guard L[t] Preveaux
Magazine 5th reg[t]: __ Brick house Lieu[t]. Hart
Guard at pritchard from 5th reg[t]:

Orders 29th July }
G Howe }
.... Parole Arnold .... by presid[t]
For the day tomorrow Capt[n] Harleston
Town guard Lieut: Roux
Magazine 5th reg[t]: __ Brick house L[t]. Shubrick

Regim[tl]. Ord[rs] }
Col°. Motte }
John Proby of Capt[n] Ashby's company is appointed armorer to the reg[t]. and is immediately to make a return to the adjutant of the tools he finds in the Armourers shop & what are wanting to carry on the business

---

Orders 30th July }
G Howe }
.... Parole Brunswick .... GH
For the day tomorrow Capt[n] Motte
Town guard Lieut: H[y] Gray
Magazine Lieut: Burke __ Brick house 5th reg[t]:

General Howe has rec[d]. the report of the Gen[l]. Court martial which he will consider and determine upon in due time, the court is dissolved

Watch word Washington

Orders 31st July }
G Howe }
Watchword Ch[s] town }
...... Parole Brunswick .... by presid[t]
For the day tomorrow Capt[n] Ashby
Town Guard Lieut: Hart
Magazine Lieut: Shubrick __ Brick house L[t] Preveaux

Reg[tl]. Ord[rs]. }
Col: Motte }
Commanding Off[rs] of Comp[ys]. to make a monthly return of their respective comp[ys]. immediately and Deliver them to the adjutant

---

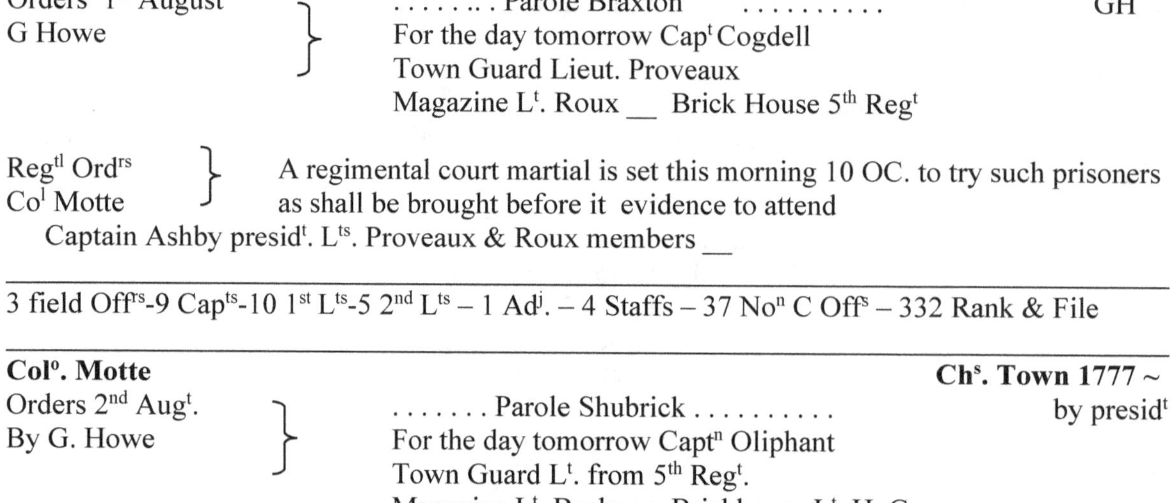

Orders 1st August
G Howe
. . . . Parole Braxton . . . . GH
For the day tomorrow Capt Cogdell
Town Guard Lieut. Proveaux
Magazine Lt. Roux __ Brick House 5th Regt

Regtl Ordrs
Col Motte
A regimental court martial is set this morning 10 OC. to try such prisoners as shall be brought before it evidence to attend
Captain Ashby presidt. Lts. Proveaux & Roux members __

---

3 field Offrs-9 Capts-10 1st Lts-5 2nd Lts – 1 Adj. – 4 Staffs – 37 Non C Offs – 332 Rank & File

---

Colo. Motte **Chs. Town 1777 ~**
Orders 2nd Augt.
By G. Howe
. . . . Parole Shubrick . . . . by presidt
For the day tomorrow Captn Oliphant
Town Guard Lt. from 5th Regt.
Magazine Lt. Burke __ Brickhouse Lt. H. Gray

The proceeding of the General court martial lately held upon the conduct of Lieuts. Raphel & La Murzel have been maturely considered by the General; The charges against them were various and of such a nature as must infallible render them unfit for services had they been supported to the Satisfaction of the court, The Accusations against these Gentlemen found their way to the General by a petition from a Body of Officers, & Claimed that Attention which it will be within his inclination to pay to the Address of persons is respected. He was happy in having an Opportunity of referring the matter immediately to the consideration of General court martial then sitting, which having been Ordered upon a particular Occasion acquiring officers of higher rank and greater experience than are usually requisite give this cause the best Opportunity of being fairly and Accurately examind, some progress had been made in the tryal, when the Gentlemen complaining petition to exhibit new charges respecting to them & a desire. The genl. had to search the bottom prevailed upon him to give an Additional Order that this new mater might be also considered. The Accusers have by their means had every Opportunity of supporting their Charges, & the Accused of Justefying them selves before a court whose Judgement & Integrity as not to be questioned & whose decision can result from no principle by Justice. This court have reported After due consideration & weighing the proofs brought in support of the Charges against Lieut. Raphael that they find him not guilty & do therefore Acquit him. Of Lt. Lamarzel they report that after having heard the Testimony of these witnesses brought in support of all the Charges exhibited against him, & also having heard his defense & taking the whole in to their various consideration Are of opinion he is not guilty & do there fore Acquit him with Honour Which sentences the General Approves of & release Leiuts. Raphael & LaMarzel from their Arrest & Orders them to return to Duty & enjoins their being received & respected Accordingly; he thinks proper to add that as it is the duty of every offr. to petition for a Court martial upon the conduct of any Officer who in his Opinion deserves it. The Accused in the present case should suppose that the Accusations exhibited against them arose from a belief that they were well founded & that information were made against them because Duty expected it. The Accusers by having brought them to that test which alone can establish their own innocence remain satisfy'd with the consciousness of having done their duty. Pay difference to the sentence of the court by receiving Lieut. Raphael & LaMarzel with that respect which is due to brothr. Offcrs. The General refers the Other Sentence of the court martial for further Considerations.

Regiment. Ord^rs^.
By Col^o^. Motte } A court martial court martial to set this morning at 10 OC: to try such prisoners as shall be brought before it evidence to attend Captn. Oliphant presed^nt^. L^ts^. H. Gray & Bush members

NB. By sentence of last court martial Thom^s^. Shores for absents from roll call & Drunkenness to receive 50 lashes pardoned – Jam^s^. Stanton for being drunk for Guard received 100 lashes [681]– Adam Meek for the same rec^d^. 100 lashes. Pet^r^. Fagan for Absent from Guard was reprimanded – W^m^. M^c^ullock for Absent from Guard rec^d^. 100 lashes – Jno. Babtis Deloney for the same reprimanded. Jno. Fenwick for drunkenness rec^d^. 50 lashes – Jno. Mich^l^. Jack rec^d^. 50 d^o^./ ^r^ Reg^t^

Orders 3^d^ Aug^t^
by G Howe } . . . Parole America . . . GH
For the day tomorrow Captn Blake
Town Guard Lt Proveaux
Magazine L^t^. Shubrick __ Brickhouse L^t^. Hart

R. Ord^s^
Motte } Lt Proveaux to Join and do duty in Captn. Ashby Comp^y^. Till further by Col^o^ Orders – The Reg^t^. To Attend divine Service the Aft^r^. Noon at 5 OC. The Col^o^. Expects all the Officers off duty will Attend

---

Orders 4th Aug^t^.
By G. Howe } . . . Parole Georgia __ Watchword Philadelphia
For the day tomorrow Captn. Charnock
~~Magazine~~ Town guard Lt Burke
Magazine L^t^ H Gray – Brick house Lt Roux
Guard at pritchards Lt Proveaux

The proceedings of the General court martial Ordered to try Simon Gibbons [682] & James Pritchard, Soldiers in Col^o^. Pinckneys Battalion for desertion have been considered by the general. He approves the Sentence but remits the punishment of boath. The pardon intended to Gibbons is out of respect to the court who recommended him to mercy – that to Pritchard begins – it appeared upon his tryal he returned Willingly. Might have escaped & did not, above he was very Active in taking up other Deserters.

Co^l^. Motte **Ch^s^. Town 1777**

Brigade Orders
Genl. Moultrie } A return to be made as usual the Beginning of every month by the Commanding Off^r^. of the Different Corps belonging to the Brigade, of their respective Corps

Reg^t^. Ord^rs^.
Col^o^. Motte } For the feuter the Maj^r^. is to Exercise Regim^t^ & in his Absents the Eldest Off^r^. on the parrade to do that duty

---

681 James Stanton served in the 2nd South Carolina Regiment during 1777. On 2 August 1777 he received 100 lashes for being drunk on guard. He was under Captain Charles Motte at the siege of Savannah in 1779.

682 Simon Gibbons enlisted in the 1st South Carolina Regiment on 8 April 1777. He was sentenced to be punished for desertion on 4 August 1777, but General Robert Howe pardoned him. He became a fifer on 23 September 1777. He reverted back to a private on 10 March 1778.

Second Lieut[s]. Burke and Hart are appointed first Lieutenants
M[r] W[m]. Capers is Appointed Second Lieut[t]. and M[r]. Jeremiah Theus is appointed Chief Surgeon to the Reg[t]. As such they are to be Obeyed and Respected – M[r] Capers is to Join & do Duty in Capt[n]. Motte Comp[y].

NB according to sentence of last court, Tho[s]. Welch for being drunk on guard [683] & Timothy Downing for d[o]. and Absent each to receive 100 lashes each; __ W[m]. Hide for the same Crime was reprimanded [684]__ W[m]. Ashford for neglect of duty, 50 lashes. Jam[s]. Smith for drunkenness and abusing Serg[t]. Anderson, received 100 lash –

Orders 5th Aug[t]. by G Howe } ... Parole George Town – Watchword Albany
For the day tomorrow Capt[n]. Th[o]. Moultrie
Town Guard L[t] Hart
Magazine L[t] from 5th reg[t]. – Brick house L[t] Proveaux

Orders 6th Aug[t]. by G Howe } ... Parole Dorchester – Watchword Boston
For the day tomorrow Capt[n]. Mazyck
Town Guard L[t] Roux
Magazine L[t] Shubrick – Brick House L[t] from 5th reg[t].

Orders 7th Aug[t] by G Howe } ... Parole Biddle – watchword Virginia
Town Guard Lieut[t]. Hart
Magazine L[t] Burke – Brick house L[t] H Gray

Brigade Orders By Gen[l]. Moultrie } A court martial from the second and fifth Reg[ts] to sit tomorrow morning at 10 OC: to try J[ms]. Clark of the Second Reg[t] [685]& Others for breaking & abusing a negroe the Property of M[r]. Hint of Ch[s]. town. They will inform M[r]. Hint when & Where they sett. – Capt[n]. Cogdelle of the 5th reg[t]. President – from the 2nd Reg[t]. L[ts]. Dunbar & Shubrick; from the 5th reg[t]. L[ts]. Gordon & Keith

Regiment. Ord[s] by Col[o]. Motte } A court martial to sett this morning at 10 OC: to try such prisoners as shall be brought before them evidence to Attend in time
Capt[n]. Mazyck presid[t]. – L[ts] H Gray & Burke members –

---

[683] Thomas Welsh

[684] William Hyde enlisted in the 2nd South Carolina Regiment on 17 December 1776. On 4 August 1777 he was reprimanded for being drunk and absent from guard. He was sentenced to do duty every other day for 14 days on 26 August 1778 for disobedience of orders and being absent from guard. When he was not on guard duty he was confined. He served under Captain Daniel Mazyck at the siege of Savannah. He deserted on 15 October 1779 after the assault on the Spring Hill Redoubt.

[685] James Clark enlisted in the 2nd South Carolina Regiment on 4 November 1775. On 7 August 1777 he was tried for breaking and abusing a "Negro". He received 100 lashes on 2 February 1778 for being absent without leave. Fifty of these were remitted since he returned on his own. On 24 September 1778 he was reduced in rank and ordered to run the gauntlet for theft. In 1779 he was in Captain Thomas Dunbar's light infantry at the siege of Savannah.

Col° Motte **Ch^s^. Town 1777 –**

Orders 8th Augt. By G Howe } . . . Parole Washington ~ watchword Brunswick
For the day tomorrow Captn. Motte
Town Guard Lt Roux
Magazine Lt Shubrick – Brick house from 5th regt.

Regtl Orders Col°. Motte } Mr. Paul Warley is appointed a Second Lieutt. In the Regt. He is to be Obeyed and respected as such

Lt Burke is to do duty in Capt Oliphants Compy. ~ Lt Hart in Cap Ashby's – Lt Proveaux in Captn. Moultries ~ Lt Worley in Captn. Blakes till further Orders

NB according to sentence of last court Sergt. Holladay, Corp Kidwell & McKann for gameing was reprimanded [686]

Adm. Meek on suspicion of theft was Acquitted

Orders 9th Augt. by G Howe } . . . Parole Burgh ~ watchword Dorchester
For the day tomorrow Captn. from 5th Regt.
Town guard Lt from 5th Regt.
Magazine Lt Proveaux ~ Brickhouse Lt Martin

All continental Officers of the Army Who are members of Assembly may Consider themselves as Exempted from duty (military) when ever they Choose or shall be acquired to Attend the Honbl. House of Assembly.

Regt. Ords Col°. Motte } A court martial to sett this morning at 9 OC: to try such prisoners as shall be by brought before us, evidence to Attend
Captn. Motte presdt. Lieuts. Proveaux & Hart members

Divine service will be performed tomorrow Afternoon at 5 OC: when tis expected all Offcrs. & Soldiers will Attend

---

Orders 10th Augt. by G Howe } . . Parole Arnold ~ Watchword Beaufort
For the day tomorrow Captn. Oliphant
Town Guard Lieutt. Henry Gray
Magazine Lt from 5th Regt. __ Brickhouse Lt Burke

Orders 11th Augt. By G Howe } . . . Parole Nanby ~ Watchword George Town
For the day tomorrow Captn. Blake
Town Guard Lieutt. Shubrick
Magazine Lt from 5th regt. ~ Brick house Lt Roux

His Excellency the president having informed Genl. Howe that by the Advice of the Honourable Council, by virtue of a Resolution of the Honourable the Continental Congress passed the 16th day of September 1776, by which the council conceived that each state was Invested with power to Appoint all Officer under the rank of General Officers he had Nominated and ~~appointed~~ filled up Commissions for the Office of Adjutt. General, Quarter Master General & Muster Master General,

---

686 William McKann was in the 2nd South Carolina. On 8 August 1777 he was reprimanded for gambling.

of the South Carolina Regiments in the Army of the United States & tho the General had hitherto never construed the resolution referred to in so Comprehensive a since, always till now Understanding they had reserved the Staff of the Army to themselves, Yett in Difference to the Abilities and Judgements of persons so respectable & in Order when even possible to view the utmost regard to the Civil Authority of the State which both as a Citizen & Soldier it will be his wish & inclination for each to do; He orders and directs that in consequence of the above maintained Appointment Nicholas Eveleigh, Es^q^., be rec^d^ respected and Obeyed by the Army as Adjutant General, Stephen Drayton, Es^q^. as Quartermaster General,[687] & W^m^. Massy, Es^q^. as muster master General of the S^o^. Carolina Regiments in the Army of the United States of America until the pleasure of congress can be Known.

Regimn^tl^. Orders by Col^o^. Motte } A reg^tl^ Court martial to sitt this morning at 10 OCC: to try all such prisoners as shall be brought before them. Evidence to attend.
Captain Oliphant Presed^t^.~ Lieut. Dunbar & Worley members

The Col^o^. Desires that the Off^rs^ Will be very exact in Obliging the sergeants & privates to wear their regimental caps agreeable to General & Regimental Orders ~ All such disregard this order may depend on being swiftly punished – Capt^ns^ or Commanding Officers of Comp^y^. are to furnish Caps for any of their men that may have lost theirs & to put them under stopages for the payment of them.

NB according to sentence of last court notes Moses Therrel for carrying a prisoner in a tavern was to receive 39 lashes but pardoned ~ Andrew Bland for disobedience of orders rec^d^. 50 lashes

Col^o^. Motte **Ch^s^. Town 1777**
Orders 12th Augt. by G Howe } . . . Parole Burke . . . Watchword Rutledge
For this day tomorrow Captn. Charnock
Town Guard Lt Proveaux
Magazine from 5th reg^t^. ~ Brick house Lt Martin –

The General thinks proper to publish the following resolution of the Honb^l^. The Continental Congress which was Officially sent him

In Congress 18 June 1777

Resolved; that a General Officer Commanding in a separate Department be Empower'd to Grant pardons to Order Execution of persons Condemned to Suffer death by General court martial, without being Obliged to report the matter to congress or the Commander in Chief.

Copy from the Journal of Congress Sign'd

W^m^. Houston Dep: Sec^t^: [688]

---

[687] Stephen Drayton became a major and aide-de-camp to General Howe on 29 November 1777. He served as aide-de-camp until he was appointed to acting deputy quartermaster general of the Southern Department on 11 August 1777. He was promoted to colonel on 12 November 1778 and appointed as the official deputy Quartermaster general of the Southern Department.

[688] William Churchill Houston was born in 1746 in South Carolina. In the 1750s he moved to North Carolina and attended Crowfield Academy. He went to college at Princeton, New Jersey, and graduated in 1768. He stayed at the college, becoming a professor and teaching mathematics and natural philosophy. Despite his political career, he stayed on as a professor in the college until 1783. He became friends with John Adams. He was elected an officer of the Somerset County Militia in February 1776, but he resigned to return to the college. He may have returned and seen action when the British forces moved onto Princeton after the Battle

NB according to sentence of Last Court martial Andrew Bland for escaping from confinement & riotous in town; Edward Pinrice [689] & Charles Lucas [690] for riotous Behaviour when Confined & for abusing Doct^r^. Theus. Each to receive 100 lashes at 4 different times ~ And^w^. Hendrickson [691] for Absent without Leave rec^d^. 100 lashes

---

Orders 13th Augt. by G. Howe }
... Parole Laurence ... Watchword Congress
For the day tomorrow Captn. Lisesne
Town Guard Lt Burke
Magazine from 5th reg^t^. ~ Brickhouse L^t^ H Gray

Reg^tl^ Ord^rs^ By Col^o^. Motte }
A reg^tl^ Court martial to sett this morning at 9 OC: to try all prisoners brought before it: evidences to attend
Captn. Charnock presid^t^. L^ts^. Shubrick & P. Gray members
The court to sett tomorrow 9 OC:

Orders 14th Augt. by G. Howe }
... Parole America,
For the day tomorrow Captn. Moultrie
Town Guard Lt P. Gray
Magazine Lt Shubrick ~ Brickhouse Lt Hart

NB according to last reg^t^. court martial Arch^d^. M^c^Donald for always being Dirty [692] ~ Pet^r^.Upthe Groves for taking peaches from a nigro ~ Sergeant Munroe for breaking & abusing an Inhabitant of the town. [693] Where all reprimanded.

---

of Trenton in 1776. In March 1777 he was elected the position of Deputy Secretary of the Continental Congress. In September 1777 he was appointed as a representative in the New Jersey legislature. In May 1779 he was elected as the New Jersey representative in the Continental Congress and he continued to serve there until 1785. He married Jane Smith and had five children with her. He died on 12 August 1788 of tuberculosis.

689 Edmund Penrice enlisted in the 2nd South Carolina Regiment on 2 July 1775 under Captain Barnard Elliot. On 12 August 1777 he received 100 lashes at four different times for abusing Doctor Theus, the regimental surgeon, and for "riotous behavior" while in jail.

690 Charles Lucas had enlisted in the 2nd South Carolina Regiment on 4 November 1775. On 12 August 1777 he received 100 lashes at four different times for abusing Doctor Theus and for "riotous behavior" while in jail. On 7 October 1777 he "broke confinement & deserted". He was supposed to receive 50 lashes on 10 November 1777 for being absent without leave, but it was reduced to a reprimand.

691 Andrew Hendrickson enlisted in the 2nd South Carolina Regiment on 22 January 1777. On 12 August 1777 he received 100 lashes being absent without leave. On 22 August 1777 he received 40 lashes for being absent without leave.

692 Archibald McDonald enlisted in the 2nd South Carolina Regiment on 4 November 1775 and re-enlisted on 15 August 1777. On 14 August he was reprimanded for always being dirty. He deserted and was apprehended in April 1778. On 15 June 1778 he was supposed to receive 100 lashes on the bare back with switches for the desertion, but it was remitted.

693 Hugh Monroe.

Col°. Motte **Ch^s. Town 1777 ~**

Orders 15th Augt. by G Howe } . . . Parole Galvan . . Watchword Thirteen
For the day tomorrow Captn. Mazyck
Town Guard Lieut^t. Dunbar
Magazine L^t from 5th reg^t. __ Brickhouse Lt Proveaux

Brigade Ord^s. By Genl. Moultrie } The prisoners of war in the main guard to be removed to the new Barracks ~ One Sergeant & twenty five men from the 2nd. 4th & 5th Reg^ts to hold themselves in readyness to go to Edisto. Each man to have Eighteen rounds & 2 spare flints. Lt Galvan will take the command of this party ~ from 2nd reg^t. 10 men – from 4th, one Serg^t. 4 men ~ from 5th 5 men

This detachment was sent to Edisto to defend against British row galleys and privateers out of St. Augustine. The privateers were known as "refugee boats" and carried 40 to 50 men armed with muskets and boarding pikes and usually two 6-pounders.

Reg^tl Ord^rs. by Col°. Motte } a reg^tl Court martial to sett this morning at 9 OC: to try such prisoners as shall be brought before it, evidences to Attend Captn. Moultrie presed^t. ~ L^ts. Dunbar & Proveaux members

NB according to sentence of last court Nehemiah Watt [694] & Thos. Welch for gaming rec^d. each 75 lashes ~ Serg^t. Gamble for behaving in a disorderly manner, do Smith & being out of Barracks; was reduced to the ranks ~ W^m. Smiley for quarreling & fighting was reprimanded ~ Corp^l. Kidwell & Owen Griffin for want of evidence remanded

---

Orders 16th Augt. by G Howe } . . . Pendlton . . . Watchword 26.
For the day tomorrow Captn. Motte
Town guard Lieut^t. Henry Gray
Magazine Lt Proveaux ~ Brickhouse from 5th reg^t.

Reg^tl Ord^r. Col°. Motte } The Col°. Desires all the Off^rs Will meet him today on the parrade, particularly those of the Genl. Court ~~martial~~ Assemble

A court martial to sett this morning at 11 OC: to try Corp^l. Kidwell. Robt Lucas & Rubin Minor.[695] All evidences to Attend

Captn. Mazyck presid^t. – Lieut^ns. H Gray & Proveaux members

NB. This court did not sett for want of the member tis sett a munday

---

694 Nehemiah Watt was in the 2nd South Carolina Regiment. He received 75 lashes on 15 August 1777 for gambling. He was promoted to corporal and then sergeant in 1778. In 1779 he was under Captain Thomas Hall and was at the siege of Savannah.

695 Reuben Minor enlisted in the 2nd South Carolina Regiment on 4 January 1777 in the Grenadier Company. He was promoted to sergeant in the Light Infantry Company under Captain Thomas Hall on 17 March 1778. He was in the siege of Savannah.

NB

according last court Serg[t]. Oneal for refusing to do duty & abusing Serg[t]. Maj[r] [696] & Serg[t]. Dyer for drunkenness when on guard, were sentenced to be reduced to the ranks but was pardoned & reprimanded

---

Orders 17th Aug[t]. by G Howe } ... Parole Moultrie ... Watchword 45.
For the day tomorrow Capt[n]. Ashby
Town Guard Lieut[t]: from 5th reg[t].
Magazine L[t] Burke ~ Brickhouse L[t] P. Gray

Reg[tl] Orders Col[o]. Motte } A court martial to sett this morning at 10 OC. to try such prisoners as shall be brought before it evidences to Attend
Capt[n]. Ashby president

Lieut[s]. H Gray Memb[r] L[t] Martin

Col[o]. Motte **Ch[s]. Town 1777 –**

Orders 18th Aug[t]. by G Howe . . } ... Parole Heyword ... watchword 10.
For the day tomorrow Capt[n]. Cogdell
Town guard Lieut[t]. Perrineau
Magazine L[t] Hart ~ Brickhouse from 5th reg[t].

Reg[tl] Orders Col[o]. Motte } M[r]. Daniel M[c]Neil is appointed Surgeon's Mate to the Reg[t]. Ordered that he be respected as such [697]

Orders 19th Aug[t]. by G. Howe } ... Parole Lee ... watchword 12...
For the day tomorrow Capt[n]. Oliphant
Town guard Lieut[t]. Martin
Magazine L[t] H Gray ~ Brickhouse L[t] from 5th reg[t].

Reg[tl] Orders Col[o]. Motte } The Regiment is to be mustered tomorrow After noon Order'd that the muster Rolls be made out immediately by the Command[g]. Off[rs] of each Comp[y]. who are desired to see that every man belonging to their respective Comp[ys] appear on the parrade except such as are sick & on duty, the muster roll must be made out agreeable to a form left in the hands of the Adjutant the Staff Off[rs] to Appear & be enter[d] in the muster roll of the Granadiers Comp[y]. As also the Drum maj[r]. Serg[t]. Maj[r]. Fyfe Maj[r]. & Quart[r]. Mast[r]. Sergeant ~ the Off[rs] that are members of the General Assemble to be present at the muster ~ the Battalion is not to turn out this evening that the men may have time to clean themselves, their Armes & put their Accoutrements in good Order, the Off[rs] to see this Order Swiftly comply'd with

---

696 Sergeant Major William Edmunds.

697 Daniel McNeil was appointed as the surgeon's mate to the 2nd South Carolina Regiment on 18 August 1777. He was replaced by John Henry Rasche on 11 June 1778.

Orders 20th Augt.
by G Howe . . } . . . Parole Cattel . . .
For the day tomorrow Captn. Blake
Town Guard Lieutt. From 5th regt.
Magazine Lt P Gray ~ Brickhouse Lt Capers

Brigade Orders
by G Moultrie } One Captn. one Subaltern, 1 sergt. & 40 rank & file with 3 rounds pr. Man to Attend the funeral of Baron Massenbourg at 5 OC: this Aftr. Noon from Mr. Hatfields six Captains (if not too many to be had) some Lieuts to be invited as Bearers. All the Offrs in town to be invited to his funeral ~

Regtl Ordrs.
Colo. Motte } The regiment to be under Arms in the Barrack yard at half past 5 OC. this After noon & the men to be drawn up two deep – the Colo. Expects that the Offrs will exert themselves in Obliging the men to Appear clean & neat

NB this order is postponed till tomorrow ½ past 5 OC: in Aftr. noon, & the Offrs to Attend the funeral of Baron Massenbourg

Colo. Motte **Chs. Town 1777**

Orders 21 Augt.
by G. Howe } . . . Parole Middleton
For the day tomorrow Captn. Charnock
Town guard Lieutt. Shubrick
Magazine Lt Martin ~ Brickhouse Lt Burke

Regtl Orders
Colo. Motte } Ordered that Sergeants of the Barrack guard do not Suffer any prisoner or prisoners that he may have in charge to walk in about the barrack yard. But to keep he or them Close confined, without permission to the company from the Commanding Officer of the Regiment.

A Court martial to sett this morning at 10 OC: to try such prisoner as shall be brought before it: evidences to Attend

Captn. Blake presdt. – Lieutns. Roux and Warley members

Orders 22nd Augt.
by G Howe } . . . Parole Congress
For the day tomorrow Captn. Lisesne
Town guard Lieutt. Dunbar
Magazine Lt Roux ~ Brickhouse Lt H Gray

The Honourable Brigadier Gadsden having resigned his commission is no longer considered as a Continental Officer.

Captn. Mouatt of Colo. Pinkney's Battalion having resigned his Commission is no longer to be considered as an Officer in the Continental service [698]

---

[698] William Moualt served in the 1st South Carolina Regiment until he resigned in August 1777.

Lieut[t]. Simion Theus is appointed a Capt[n]. in Col[o]. Pinckney Battalion in room of Capt[n]. Mouatt resign'd he is to be Obeyed & respected as such. [699]

Brigade Ord[rs] by G. Moultrie } One Subaltern 1 sergeant & 25 rank & file from 2nd & 5th regiments with 12 rounds & 2 spare flints p[r]. Man to march immediately to Eveleigh's warf, where they will find some Boats ready to take them on board to Carry them to M[r]. Galvan's Vessel Laying at Edisto they are to return with the vessel and to Defend them from any of the enemys vessels

Reg[tl] Orders Col[o]. Motte } A court martial to sett this morning at 10 OC: to try all such prisoners as shall be brought before it. Evidence to attend ~ Capt[n]. Charnock president Lieut[s]. P Gray & Capers

NB according to sentence last court James Oakes for absent without leave rec[d]. 100 lashes ~ And[w]. Hendrickson for d[o] Rec'd 40 lashes ~ Ab[m]. Berlin for d[o] was pardoned on amount of his age & sickness

Jam[s]. Hooper
Tho[s]. Windsor [700] } Acquitted for want of Evidence

Col[o]. Motte **Ch[s]. Town 1777**
Orders 23[d] Aug[st]. by G Howe }

..... Parole Thompson
For the day tomorrow Capt[n]. Moultrie
Town Guard Lieut[t]. Burke
Magazine L[t] from 5th reg[t]. ~ Brickhouse Lieut[t]. Capers
For the Command of Dorchester L[t] from 5th reg[t].

Lieut[t]. Jackson of Col[o]. Pinckneys Battalion is Appointed first Lieut[t]: of the same in the room of Lieut[t]. Theus promoted he is to be respected and Obey'd as such [701]

---

[699] Simeon Theus was born on 1750. He became a second lieutenant in the 1st South Carolina Regiment on 17 June 1775 and was promoted to first lieutenant during May 1776. He became a captain on 18 August 1777 when Captain Moualt resigned. He was paymaster during 1779 and 1780. He was captured at the fall of Charleston but was exchange and became a brevet major on 30 September 1783.

[700] Thomas Windsor served in the 2nd South Carolina Regiment under Captain Thomas Dunbar. On 22 August 1777 he was charged with being absent without leave, but he was acquitted due to lack of evidence. He received 100 lashes on 19 March 1778 for stealing a shirt. He received 50 lashes with a cat-o-nine tails on 2 May 1778 for selling liquor contrary to orders. He received 100 lashes on 6 July 1779 for theft. He was with Captain Dunbar's light infantry at the siege of Savannah.

[701] William Jackson became a second lieutenant in the 1st South Carolina Regiment during May 1776 and was promoted to 1st lieutenant on 18 August 1777 when Simeon Theus was promoted to captain. On 9 October 1779 he was promoted to captain under Colonel Pinckney. He was a major and aide-de-camp to General Lincoln from 1 September 1780 to 9 February 1781. He was captured at the fall of Charleston. He was a prisoner on parole until May 1783. He died on 17 December 1828.

Orders 24th Augt. by G Howe } .... Parole Maryland ... by GH
For the day tomorrow Capt Harleston
Town Guard Lieutt: from 5th Regt.
Magazine Lt P. Gray ~ Brickhouse Lt Perrenneau

Ruben Price a prisoner in the main guard having been reported to the General as Labouring under complaints which make his removal to the general Hospital necessary.[702] He is therefore to be removed to & received at the General Hospital, the Offr. of guard their must be particularly carefull that he does not Escape

Regtl Ordrs. Colo. Motte } Divine service will be performed by the chaplain at the usual time & place. The Offrs to Attend

---

Orders 25th Augt. by G. Howe } ... Parole Howe ... by presidt.
For the day tomorrow Captn. Ashby
Town guard Lieutt: Roux
Magazine Lt Shubrick ~ Brickhouse Hart
Pritchard Guard Lt H. Gray

Regtl Ordrs by Colo. Motte } A court martial to sitt this morning at 10OC: to try such prisoners as shall be brought before it: evidences to Attend
Captn. Harleston presidt. ~ Lieutt. Proveaux & martin members

---

Orders 26th Augt. by G Howe } ... Parole Spottswood ... by GH
For the day tomorrow Captn. Oliphant
Town Guard Lieutt: Martin
Magazine Lt Capers ~ Brick house Lt Proveaux

NB according to sentence of last regt. court Edwd. Murphy for drunkenness recd. 100 Lashes. George horn for disobedience of orders fined 7/6 for the regt: Hospite.[703]

Colo. Motte **Chs. Town 1777**

Orders 27th Augt. by G Howe } ... Parole Virginia..... by G H.
For the day tomorrow Captn. Cogdel
Townguard Lieutt: from 5th reg:
Magazine Lt Worley ~ Brickhouse Lt P Gray

Regt. Ordrs. Colo. Motte } A return of the number of regimental Caps wanted by each Compy. to be made out immediately and deliver'd to the Adjutant, such men as have lost their caps are to be supplied with others by their Captains, who are Orderd

---

[702] Reuben Price was in the 2nd South Carolina Regiment. He was under confinement, while in the hospital on 24 August 1777.

[703] George Horn enlisted in the 2nd South Carolina Regiment on 10 July 1775 under Captain Bernard Elliot. He was fined £7/6 on 26 August 1777 for disobedience of orders. He received 50 lashes on 18 September 1777 for being absent from roll call. He received 50 lashes on 29 October 1777 for stealing potatoes.

to stop their pay for the payment of them. A return of shirts and shoes wanted by each Comp^y^. to be out immediately

---

Orders 28th Augt. by G Howe } . . . Parole Williamsburg
For the day tomorrow Captn. Potts
Town Guard Lieut^t^: Hart
Magazine Lt Burke ~ Brickhouse L^t^ from 5th reg^t^.

A General ^court^ Martial to sett on Monday next at some convenient place in Charles Town within the hour prescribed by the articles of war of which Major Huger is to be president, the rest of the members is to be taken from the 1st, 2nd & 5th Battalions according to detail in the following manner each Corps is to furnish 2 Capt^ns^ & Subalterns, this court is to try Lieut^t^. Raphael of the corps of Artillery for disobedience of Orders & neglect of duty, Major Beekman will furnish the evidence ~ they are also to try Captn. De Trevelle of the same Corp, Charged by Lt Raphael in the following manner; 1st For having stole his horse, 2nd for attempting to defraud him of a sum of money, 3d for having purjured himself in the last court martial, 4th for having attempted to ruin his Charracter & destroy his reputation by many groundless charges, 5th that Captn. De Trevelle has been guilty of the theft, 6th that he has been guilty of Cowardice; 7th that he bound over Lt Raphael to the peace in an unofficer like and Cowardly manner, Lt Raphael is to furnish the evidences to support these charges ~ the court is also to try James Orange [704] of the 2nd regiment for Desertion. Captn. Harleston will furnish the evidence to support this fact he is also to be tryed for having without provocation when prisoner in confinement in the most outrageous manner, insulted and abused Lt Proveaux threatening with Knife in his hand & in the course of this abuse he repeatedly Damned the Continent & continental Congress, saying he was good English blood & would Support the Cause of Great Britain to his Last __ Lieut^t^. Proveaux will furnish the evidence

Col^o^. Motte **Ch^s^. Town 1777**
Orders 29th August by Genl Howe . } . . . Parole Washington . . . . by pres^dt^.
For the day tomorrow Captn. Charnock
Town Guard Lieut^t^. Perrineau
magazine Lt Dunbar ~ Brickhouse L^t^ from 5th reg:

Adjutant Delliant is Appointed Brigade Major to Brigadier Generl. Moultrie & is therefore to be respected and Obeyed Accordingly

Reg^tl^ Orders Col^o^. Motte } A R. court martial to set this morning at 10OC: to try such prisoners as shall be brought before it Evidence to be Order'd to Attend in ~ Captn. Blake presid^t^. L^ts^. Dunbar & Proveaux members

---

704 James Orange enlisted in the 2nd South Carolina Regiment on 22 September 1777. On 28 August 1777 he received 100 lashes on the bare back with a cat-o-nine tails because he insulted and abused Lieutenant Adrian Proveaux while being a prisoner and threatened him with a knife. He also repeatedly damned Congress, stating that "he was good English blood & would Support the Cause of Great Britain to his Last." He received 100 lashes for desertion on 18 October 1777, then he deserted again, one month later on 17 November 1777.

Orders 30th Augt. by G Howe }
... Parole Sinclair
For the day tomorrow Captn. Lesesne
Town Guard Lieutt. Preveaux
Magazine Lt Martin ~ Brick house Lt Capers

Regtl Ordrs. Colo. Motte }
For the General court martial Captn. Ashby, Blake & Charnock & Lieuts Hall and Roux –

NB according to sentence of last court Jno. Clements for drunkenness on guard recd. 50 lashes ~ Thomas Stafford for Absent without Leave 100 lashes, pardoned.[705] William Simpson for Abusing Ed. Fry, reprimanded at the Halberts.

Orders 31st Augt. by G. Howe }
... Parole Biddell
For the day tomorrow Captn. Moultrie
Town Guard Lieutt: P Gray
Magazine Lt Hart ~ Brickhouse Lt Worley

Regtl Ordrs Colo. Motte }
Divine Service this Evening as usual
Commanding Offrs of Compys to Give in Monthly return of their respective Companies tomorrow morning at 8 OC: to the Adjutant
A Court martial to set tomorrow morning at 9 OC: to try such prisoners as shall be brought before it: evidences to Attend
Captn. Moultrie presidt. ~ Lieuts Shubrick & Dunbar members.

The Monthly return 1st September
3 field Offrs 10 Captns~10. 1st Lieut:~5. 2nd Lts.~1 Adjs~1 Chapl: 1 paym:~1 q. Mast: 1 do Serg. 1 Surgeon~1 Surgeons mate ~ 27 Sergts ~ 17 drums & fifes 3 Armorers 338 rank & file

---

Orders 1st Sept. by G Howe }
... Parole
For the day tomorrow Captn. Motte
Town Guard Lieutt: Perreneau
Magazine Lt from the 5th regt.
Brickhouse Lt Capers ~

Lt Colo. Marion **Chs. Town 1777**

Orders 2nd Sept: by G Howe }
.... Parole Lee
For the day tomorrow Captn. Oliphant
Town guard Lieutt: Dunbar
Magazine Lt Galvan ~ Brickhouse Lt Shubrick

---

705 Thomas Stafford served in the 2nd South Carolina Regiment under Captain Thomas Dunbar. On 30 August 1777 he was supposed to receive 100 lashes for being absent without leave, but was pardoned. He was sentenced to receive 100 lashes on 2 October 1777 for desertion, but was pardoned by General Howe. He was supposed to receive 100 lashes for theft on 6 July 1779, but he only received 50. The others were remitted. He was with Dunbar's light infantry at the siege of Savannah.

Orders 3d Sept: by G Howe } . . . Parole
For the day tomorrow Captn. Lisesne
Town Guard Lieutt: H. Gray
Magazine Lt Proveaux ~ Brickhouse Lt Warley

The 1st. 3d & 6th Continental Battalions in this state to form one Brigade to be under the Command of Brigadier General Moultrie. The 2nd & 5th Battalion to form another Brigade of which Colo. Huger is to be Colo. Commandant till further Orders. ~ The Corps of Artillery to receive Orders from the Commander in Chief ~ The Orderly hours at head Quarters will be at 12 Oclock.

Regtl Orders by Lt Colo. Marion } Article the 4th of the 13th Section
No Officers non commissioned Officer or soldier shall fail of repairing at the time fixed at the place of parrade of Exercise or other rendezvous Appointed by his Commanding Officer, if not prevented by sickness or some other evident necessity, on the penalty of being punished according to the nature of his Offence by the sentence of a Court martial

A regimental court martial to be held tomorrow at 10 OC. in the fore noon to try such prisoners as shall be brought before them, evidences to Attend

Captn. Lesesne president ~ Lieuts. P. Gray & Perreneau members ~ ~

Orders 4th Sept: by G Howe } . . . Parole Roberts
For the day tomorrow Captn. Moultrie
Town guard Lieutt: Hart
Magazine Lt Burke ~ Brickhouse Lt P. Gray

Orders 5th Sept: by G Howe } . . . . Parole Thompson
For the day tomorrow Captn. Motte
Town Guard Lieutt: Shubrick
Magazine Lt Dunbar ~ Brickhouse Lt Galvan

Orders 6th Sept: by G Howe } . . . Parole Biddle
For the day tomorrow Captn. Lisesne
Town Guard Lieutt: Martin
Magazine Lt Perreneau ~ Brickhouse Lt H Gray

Regtl Ordrs. Lt Colo. Marion } a court martial to sett this morning at 9 OC: to try sergt. Marlow & all other prisoners that may be brought before it; evidences to attend ~
Captn. Motte presidt. ~ Lts. H Gray & Proveaux members

NB according to sentence of Court 4th Inst: Richd. Williamson for defrauding Hendrickson is to run the gauntlet 3 times & put under stoppages to repay Hendrickson. Jas. Hooper for drunkenness, reprimanded

Colo. Motte **Chs. Town 1777 –**

Orders 7th Sept: by G Howe } . . . Parole Woodford . . .
For the day tomorrow Captn. Moultrie
Town Guard Lieutt. P. Gray
Magazine Lt Proveaux ~ Brickhouse Lt Warley

Reg^{tl} Ord^{rs}.
Col^{o}. Motte } Order'd that the Regimental Orders of the 3^{d} Instant be strictly Complyed with, Such who disobey it may be assured of being treated as he deserves

---

Orders 8th Sept.
by G Howe }

.... Parole Randolph ...
For the day tomorrow Captn. Motte
Town Guard Lieut^{t}: Hart
Magazine Lt Shubrick ~ Brickhouse Lt Burke
Pritchards Guard Lt Martin

Reg^{tl} Ord^{rs}.
Col^{o}. Motte } The Col^{o}. Desires all the Off^{rs} of duty will meet him this After noon on the Parrade as also those of the General court martial.

All Officers who have been on the recruiting service are to make out their Accounts & Call on the Col^{o}. to have them settled

Six men from the regiment that are well Acquainted with the upper part of Great Peedee & Adjoining North Carolina to be in imedeat readyness to Join a Like party from the 1st and 5th Reg^{ts} to go after Deserters they are to receive 6 rounds, 2 flints & 2 days provisions p^{r} Man

---

Orders 9th Sept^{r}.
by G Howe }

.... Parole Annapolis
For the day tomorrow Captn. Ashby
Town Guard Lieut^{t}: Shubrick
Magazine Lieut^{t}: Dunbar ~ Brickhouse Lt Galvan

Reg^{tl} Ord^{rs}.
Col^{o}. Motte } a reg: court martial to set this forenoon to try all such prisoners as shall be brought before it, all evidence to Attend
Captn. Motte presid^{t}. ~ Lieut^{t}. Mason & Worley members

---

Orders 10th Sept:
by G Howe }

..... Parole Schuyler –
For the day tomorrow Captn. Lesesne
Town Guard Lieut^{t}: Warley
Magazine Lt Mason ~ Brick house H Gray

Lt Colonel Elliot is added to the Gen. Court martial now sitting and is to act as president in the room of Major Huger who has obtained leave of Absents upon a very particular and urgent Occasion

After Orders

Six men taken from the 2nd & 5th regim^{t}. According to detail to be Added to the main guard tomorrow, a Centry to be posted at Lyttleton & one to Craven Battery, who are to receive strict Orders to be carefull and vigilant that no Danger may happen to the Battery, Cannon or military stores, neglect of duty will be punished wth. Severity.

Nich: Eveleigh – DAG

Lt Col^{o}. Marion **Ch^{s} Town 1777 –**

Orders 11th Sept:
by G Howe }

.. Parole Weedon ....
For the day tomorrow Captn Motte
Town guard Lieut^{t}: Burke
Magazine from 5th reg: ~ Brick house Lt Jacob Shubrick

Reg^tl Orders
Lt Col^o. Marion

Lt Proveaux thinking himself Angered by a report of his Beheavour when on the Brick House Guard, some time ago request in court of Inquiry; a Regimental court of Inquiry is therefore Order'd to sett to today at 11 OClock to Inquire in the beheaviour of Lt Proveaux when on Guard at the brick house some time in August Last. Serg^t. Coffer will Summon the man who lives at Col^o Laurences house to Appear against Lt Proveaux, who will Summon the Sergeant and Corporal of the Guard at the time this report took rise & such other persons he may think proper – this court to consist of Major Horry as president, three Capt^ns & three Subalterns as members

For this Court Maj^r. Horry presid^t.
Captn. Motte, Lesesne & Moultrie – Lieut^nts H. Gray. Burke & Shubrick

After Orders by Lt Col^o. Marion

The Regimental court of Inquiry Order'd to Inquire in the Beheaviour of Lt Proveaux on the Brick House Guard; Report as follows –

The Court mett according to Orders and after due consideration on the testimony of the Evidence Produced to them, Are Unanimously & fully of Oppinion that Lt Proveaux conduct while on Guard the 26th July last has been Inconsistent with the Character of an Officer & a Gentleman and Deserves the Attention of a General court martial –

The Court of Inquiry of which Major Horry was president is dissolved –

Lt Proveaux
Sir

you are hereby Order'd under Arrest for behaving unbecoming an Officer and a Gentleman on the 26th of July Last when on the Brick house Guard – Dated July 11th: 1777 – Fr^n. Marion, L^t. Col^o.

NB the Lt Col^o Imediatly applied for Lt Proveaux tryal by a General Court Martial –

Lt Col^o. Marion **Ch^s town 1777 –**
Orders 12th Sept:
by G. Howe

. . . Parole Carolina
For the day tomorrow Captn. Ashby
Town Guard Lieut: Galvan
Magazine Lt Perrenneau ~ Brickhouse Lt Dunbar

Lt Col^o. McIntosh of Col^o. Hugers Battalion having Informed the Genl. that a regimental court of Inquiry Orderd to set on the Conduct of Lieu: Proveaux of the 2^d Reg^t. had reported that he had behaved in a manner So unbecoming an Officer and a Gentlemen, as to desire for their Opinion being try'd by a General court martial, The general therefore directs the General court martial now sitting to try Lieut^t: Proveaux for conduct unworthy an Officer and a Gentleman; Lt Col^o McIntosh will furnish the evidence

Captn. Sims White of the Corps of Artillery having resigned his Commission is no longer to be considered as a Continental Officer –

Reg[tl] Orders } Adjutant Delliant being sick Serg. Coffer is to act as Adjutant till further
L[t] Col[o]. Marion } Orders to be Obeyed as such

A reg[tl] Court martial was Orderd to sett day but was postponed –

After R.O.

The Order of this morning respecting Serg[t]. Coffer acting as Adjutant till M[r]. Dilliant was well, is now Suspended & Serg[t]. Coffer to Act only as Sergeant as formerly.

NB. According to Sentence of Last Court Jn[o]. Clements for riotous behaviour & cruelly beating his wife, rec[d]. 50 lashes ~ Serg. Maj[r]: Edmunds for absent without Leave was severely reprimanded ~ Edw[d]. Murphy for repeated absent from roll-call rec[d]. 45 lashes ~ Corporal Amos for Absenting from roll call & times, reduced to the ranks [706]

NB

about 9 OClock the alarm was beat at main guard & taken up at Barracks when the 2nd & 5th Regiments was under Arms & provided with 6 rounds p[r] Man & march't to the main Guard, which was composed of Capt[n]. Ashby L[t] Roux & 30 men, the Cartridges was by Order of L[t] Col[o]. M[c]Intosh taken from all the men & Left at the Guard House – after which the regim[ts] return'd – this Alarm was Occasion by riotous sailors of the randolph in Union Street 1 man was Dangerously wounded.

This was the Continental Navy Frigate *Randolph* of 36 guns, commanded by Captain Nicholas Biddle. The *Randolph* operated out of Charleston and captured British vessels. On 4 September 1777 the *Randolph* captured two transports, the *True Briton* and the *Charming Peggy*, 30 leagues southeast of Charlestown. The Royal Navy transports bound from Jamaica to New York offered a little resistance, but were easily taken by the American frigate. Both ships were laden with rum for the British army and navy. With them was the ship *Severn* that had been captured by an American privateer, then recaptured by the *True Briton*. There was also a French brig laden with salt that had been captured going from the West Indies to Charlestown. A small sloop with these vessels managed to escape capture and get away. The *Randolph* returned to Charlestown with her prizes on September 7th and the crew must have been celebrating in the town when they became "riotous". [707]

Orders 13th Sept[r]. } . . . Parole Congress
by G Howe } For the day tomorrow Cap[t] Lesesne
Town guard L[t] Dunbar
Magazine L[t] Worley ~ Brickhouse L[t] P Gray

The Name of L[t] Co[l]. M[c]Intosh was by mistake inserted in the Orders of yesterday Instead of L[t] Col[o]. Marion who is to Furnish the Evidence against Lieut[t]: Proveaux ~ L[t] LaMazell of Col[o]. Roberts Corps of Artillery having resigned his commission is no longer consider'd as a Continental Officer

---

[706] Samuel Amos enlisted in the 2nd South Carolina Regiment on 4 November 1775 as a corporal. He was reduced in ranks for being absent from roll call on 12 September 1777. He was promoted to corporal again in 1778, but he demoted to private on 26 August 1778. He died on 4 June 1779.

[707] O'Kelley, *NBBAS, Volume One* pp. 201-205.

Reg^tl^ Orders
Col^o^. Motte } A court martial to set this morning at 10 OC: to try such prisoners as shall be brought before them, evidences to attend
Captn. Lesesne presid^t^ ~ Lieut^s^ P. Gray & Hall members ~

Orders 14th Sept^r^.
by G Howe } ... Parole Moultrie .... by presid^t^.
For the day tomorrow Captn. Shakelford
Town Guard Lieut^t^: from 5th reg^t^.
Magazine Lt Burke ~ Brickhouse Lt Hart

Orders 15th Sept.
by G Howe } ... Parole Gates .... GH
For the day tomorrow Captn. Motte
Town Guard Lieut^t^: Shubrick
Magazine Lt Capers ~ Brickhouse Lt Galvan
Pritchards Lieut^t^: Warley

Reg^tl^ Orders
Lt Col^o^. Marion } The Quarter master to deliver out what cloathing is made in proportion to the number wanting in each Company and take a receipt from an Officer A court martial to set tomorrow morning at 10 OC: in the forenoon to try such prisoners as shall be brought before them. Evidences to Attend.
Captn. Ashby presid^t^. Lieut^s^. Dunbar and P. Gray members

Genl. Aft^r^. Order by G. Howe

Several ~~prisoners~~ soldiers of Col^o^. Pinckneys Battalion having been confined in the main guard for desertion, the general court martial now sitting are to try them, Lt Col^o^. Cattle will furnish the evidence.

The repeated Orders found against Adjutants receiving men for duty & against Officer of reliefs marching them off with Arms in improper Order, having been treatted with unpardonable neglect, the general think proper to give notice to the Army that he is determined to reform this Error if possible; Adjutants & Officers of relief will therefore be carefull that it doesn't happen in feuter or they certainly will be try'd for disobedience of Orders Officers of comp^y^ from whos Inattention & negligence the slovenly uncleanly condition of the Arms most Certainly Originate are frequently to inspect their Arms of their men, they may depend that every soldier who may be seen on duty with arms not properly Clean'd will have his name & the Company he belongs to taken down, the Officers of the Comp^y^. will be made Answerable for neglect of duty & disobedience of Orders & the soldier most Certainly be punished.

Lt Col^o^ Marion **Ch^s^. town 1777**
Orders 16th Sept^r^.
by G Howe } .... Parole Virginia .... by pres^dt^.
For the day tomorrow Captn. Potts
Town Guard Lieut^t^. P. Gray
Magazine Lt Hart ~ Brickhouse L^t^ from 5th reg^t^.

Orders 17th Sept^r^.
by G. Howe } ... Parole M^c^Intosh ....
For the day tomorrow Captn. Motte
Town Guard Lieut^t^: Galvan
Magazine Lt Shubrick ~ Brick house Lt Dunbar

Reg^{tl} Orders
Col^{o}. Motte } The Articles of war to be read at the head of the regiment this afternoon – no one to be absent that can possible Appear – The Staff Off^{rs} to have notice to Attend ~ Such of the Officers ~~who~~ as have not Copys of the Articles of War may be provided with them by Applying to the Colonel –

Orders 18th Sept^{r}
by G Howe } . . . Parole Motte by pres^{dt}.
For the day tomorrow Captn from 5th reg^{t}.
Town Guard Lieut: Martin
Magazine L^{t} from 5th reg: ~ Brick house Lt Capers

The General court martial now siting are to try Sergeant Major Edmunds & Serg^{t}. Oneal of Col^{o}. Mottes Battalion Confined by Lt Dunbar, of the same, for mutiny – Lt Dunbar will furnish the evidence

NB according to sentence of last reg^{tl} court martial John Fenwick for stealing public Rounds rec^{d}. 100 lashes ~ Sergeants Munrow, Mathews, [708] Laurence & Coleman [709] for Absent from exercise was reprimanded ~ Corp^{ls} Henderson, [710] Conyers & League[711] for the same crime was reprimanded ~ Jno. Steal for being drunk in the rank was rec^{d}. 50 lashes ~ Geo^{r}. Horn for Absents from roll call rec^{d}. 50 lashes ~ Jno. Whitsel [712] & Alex^{d}. Stewart for absents from Guard rec^{d}. a reprimand ~ Malcom W. McPharlan for the same pardoned [713]~ And^{w}. LaBland rec^{d}. 50 lashes for the same crime ~ Alexis Seymour pardoned [714]

Orders 19th Sept^{r}.
by G Howe } . . . Parole Manley . . . GH
For the day tomorrow Captn. Ashby
Town Guard Lieu: from 5th reg:
Magazine Lt Hart ~ Brickhouse L^{t} P Gray

---

[708] Robert Matthews enlisted in the 2nd South Carolina Regiment on 4 November 1775 as a sergeant. He was reprimanded on 18 September 1777 for being absent from exercises. He was demoted to private on 25 October 1777 for quitting his guard, but was promoted back to sergeant a month later on 28 November 1777 in Captain Shubrick's company. He was demoted back to private in 1778. He was promoted to corporal on 5 March 1779 in Captain Hall's company and to sergeant on 14 March 1779 under Captain Adrian Proveaux. He was killed at Savannah on 9 October 1779.

[709] Robert Coleman enlisted in the 2nd South Carolina Regiment on 4 November 1775. On 7 November 1776 he became a sergeant. He was reprimanded on 18 September 1777 for being absent from exercises. He was discharged on 8 July 1778.

[710] William Henderson served in the 2nd South Carolina Regiment from 4 November 1775 to 12 May 1780. He was reprimanded on 18 September 1777 for being absent from exercises. At the beginning of 1779 he served under Captain Peter Gray. On 3 April 1779 he was promoted to corporal in Captain Mason's company. He served as a sergeant under Captain Thomas Hall in 1779. He was demoted to private on 28 September 1779. He was at the siege of Savannah. He was promoted back to corporal on 29 December 1779.

[711] Absolom League enlisted in the 2nd South Carolina Regiment on 20 November 1775. He was reprimanded on 18 September 1777 for being absent from exercises. He died on 15 December 1777.

[712] John Whitsett.

[713] Malcom McPharlan (McFarlan) enlisted in the 2nd South Carolina Regiment on 4 November 1775. He received a reprimand on 18 September 1777 for being absent from guard but he was pardoned. He re-enlisted on 1 November 1779 under Captain Charles Motte.

[714] Alex Simmor.

Thos. Stafford of the Second Reg^t^. Confined by Order of Col^o^. Motte for desertion is to be tryd by the General court martial now sitting Lt Dunbar is to furnish the evidence

Regim^tl^ Orders by Col^o^. Motte } A court martial to set this forenoon at 10 OC: to try such prisoners as shall be brought before it: evidences to attend
Captn. Motte president ~ Lieut^s^: P. Gray & Hart members

Col^o^. Motte **Ch^s^. Town – 1777 –**

Orders 20th Sept. by G Howe } .... Parole Parsons
For the day tomorrow Captn. Lesesne
Town guard Lieut^t^: Dunbar
Magazine L^t^ from 5th reg^t^. ~ Brickhouse Lt Shubrick

Reg^tl^ Ord^rs^. Col^o^. Motte } Orderd that Commanding Off^rs^ of comp^ys^ to make a return to the Col^o^. the Col^o^. by monday morning 10 OC: of the men's names of their respective Comp^ys^ and Opposite to each name to set down the different Articles of Clothing they have rec^d^; such men as have not rec^d^. their Regimental coats are to Apply to the Quarter master who has orders to deliver them, but not before the above returns are given in[715]

A reg^tl^ Court martial to sett this morning at 11 OC: to try such prisoners as shall be brought before it all evidence to Attend

Captn. Ashby presid^t^. ~ L^ts^. Shubrick & Dunbar members

Orders 21st Sept. by G Howe .... Parole Arnold .... GH
For the day tomorrow Captn. Motte
Town guard Lieut^t^. Martin
Magazine Lt Proveneau ~ Brickhouse L^t^ H Gray

Reg^t^. Ord^rs^. Col^o^. Motte } Divine service this afternoon at the usual time, the regiment to Attend

---

Orders 22nd Sept. by G Howe } .... Parole Moultrie ... by presid^t^.
For the day tomorrow Captn. Ashby
Town guard Lieu: from 5th reg:
Magazine Lt P. Gray ~ Brickhouse Lt Capers
Pritchards L^t^ from 5th Reg^t^.

Reg^tl^ Orders Col^o^. Motte } A court martial to set this morning at 10 OC: to try such prisoners as shall be brought before it ~ evidences to be orderd to Attend in time –

NB the sentence of court martial 20th Inst. David Parsons for Absents without leave rec^d^. 100 lashes. W^m^. Simpson for Breaking confinement pardon'd of 50 lashes

---

715 Prior to this time the Regimental coats were kept by the Quartermaster and only issued out when needed. After France declared war on Britain the United States was able to get clothing, powder and weapons more freely from their French allies and clothing deficiencies were not as big a problem as in the regiment of 1776.

**2nd South Carolina Regiment
1777**

Orders 23[d] Sept. by G Howe – } .... Parole Starke .... GH

For the day tomorrow Capt[n]. Potts
Town guard Lieut: Hart
Magazine L[t] Galvan ~ Brickhouse L[t] from 5th reg:
Command to Dorchester Lieut. Shubrick

The General court martial now sitting are to try W[m]. Ficklen of Col[o]. Mottes Battalion for Desertion. Col[o]. Motte is to furnish the evidence.

The quarter master General will procure as immediately as possible a pump for the new Barracks. He will take his excellency the presid[t] Instruction upon this Occasion; he is also so Orderd a Centry Box to be Carried to the Brick House Guard

Col°. Motte Chs. Town 1777

Regtl Orders by Col°. Motte } Orderd that the Officers for Duty be more punctual in their attendance on the parrade at the hours Orderd for the Assembling of the Different Guards. Such who Disobey this Order will be taken proper notice of

NB Sentence of last court Jams. Hooper for riotous Behaviour recd 100 lashes

---

Orders 24th Sept. by G Howe } . . . Parole Washington . by presidt.
For the day tomorrow Captn. Harleston
Town guard Lieutt. Perrineau
Magazine Lt Martin ~ Brick house Lt Capers

Regtl Orders Col°. Motte } A regtl Court martial to set this morning at 10 OC: to try all such prisoners as shall be brought before it, evidences to Attend
Captn. Harleston presidt. . . Lieu:: Perreneau & martin members

---

Orders 25th Sept: by G Howe } . . . Parole Nelson
For the day tomorrow Captn. Motte
Town guard Lieutt: Warley
Magazine Lt from 5th reg: ~ Brickhouse Lt P. Gray

The General court martial orderd to try Lt Proveaux have reported as follows ~ the Court is of Opinion that Lieutt: Proveaux is Guilty of the Charge which he stands Accuses & comes under the later part of the 21st Articles of the 14th Section, of the articles of war and Sentence that he may be reprimanded by the Commanding Officer of his Battalion in the presents of the Officers of that Corps only – The General Approves & ratify the Sentence which Col°. Motte will execute in the manner he thinks proper, After which Lt Proveaux is to be Discharged from his Arrest, return to duty & be respected & Obey'd as usual

Regimtl Ordrs. Col°. Motte } A court martial to set this morning to try all such prisoners as shall be brought before them, all evidence to Attend ~ Captn. Harleston presidt
Lts. Hart & H Gray members

The Col°. desires all the Officers of the regiments to call at his house tomorrow forenoon at 10 OC: The Adjutant to Order Lt Proveaux to attend the same place & Hour

---

Orders 26th Sept. by G Howe } . . . . Parole Ferguson
For the day tomorrow Captn. Ashby
Town guard Lieutt: from 5th reg:
Magazine Lt Hart ~ Brick house Lt Perrenneau

NB according to sentence of Genl. Court martial Lt Proveaux was reprimanded ~ & the prisoners try'd by regtl Court was pardoned

Col°. Motte **Chs Town 1777**

Orders 27th Sept. by G Howe } Parole Walton
For the day tomorrow Captn. Potts
Town Guard Lieutt: from 5th reg:
Magazine Lt Martin ~ Brickhouse Lt. H Gray

Reg[tl] Ord[rs].
Col[o] Motte } A reg[tl]. court martial to set this forenoon at 11 OC: to try such prisoners as shall be brought before them. The Evidences to have notice to Attend in time ~ Capt[n]. Ashby presid[t]. L[ts]. Martin & Capers members

---

Orders 28th Sept[r]
by G. Howe }

Parole Stark
For the Day tomorrow Capt[n] Davis [716]
Town guard Lieut[t]. from 5th reg
Magazine L[t]. Proveaux ~ Brick house L[t]. Capers

Reg[tl]. Ord[rs]
Col[o] Motte } divine service this afternoon at the usual time and place when it is expected officers and soldiers will Attend

---

Orders 29th Sept[r]
by G. Howe }

Parole Lee GH
For the Day tomorrow Capt[n]. Harleston
Town guard Lieut[t]. Warley
Magazine L[t]. from 5th Reg: - - - Brick house L[t]. Hart
Pritchards L[t]. P. Gray

Reg[tl]. Orders
Co[l] Motte } The creditors of W[m]. Hossman lately deceased, are to apply to Capt[n]. Harleston for the Settlement of their Accounts

A court martial to sett this morning at 9 O.C: to try such prisoners as shall be brought before it, the Evidences to be Order'd to attend at that hour.

Capt[n]. Mazyck pres[dt]. . . . L[ts]. P Gray and Hart members

After Ord[rs]

Order'd that the Quarter master do visit the barrack rooms every morning to see the men do not injure them & if they do to report it immediately to the Commanding Officer of the regiment

A monthly return of each Company to be given in to Maj[r]. Dilliant tomorrow forenoon

NB Serg[t]. Bulleat for neglect of duty was reprimanded & mult 7 days pay [717]
Jn[o]. Thompson for drunkenness on guard rec[d] 50 Lashes ~ Alex[d] Stewart for stealing a blanket was pardoned by petition of the owner & was reprimanded to the court of this day.

Orders 30th Sep[t]
by G Howe }

Parole Middleton . . . . by presid[t]
For the day tomorrow Capt[n] Motte
Town Guard Lieut[t]: from 5th reg
Magazine L[t]. Perrineau
Brick house Martin

---

[716] Harmon Davis became a first lieutenant in the 4th South Carolina Regiment (Artillery) during May 1776. On 29 May 1777 he became a captain. He was wounded at Savannah on 9 October 1779 and was taken prisoner at the fall of Charleston. After being paroled he remained inactive for the rest of the war. He became a brevet-major on 30 September 1783.

[717] David Boilliet.

Col° Motte **Ch^s town 1777**

Reg^tl Orders Col° Motte 30th Sept } The Chaplain having made a Complaint of his being insulted by several men belonging to Mottes regim^t the commanding Off^r of the Reg^t orders that it may never happen again any one disregarding this Order, upon being found out may assured of being severely punished

Ordered that the Command^g Off^rs of the Comp^ys do immediately that the sick of their respective Comp^y be moved to the regimental Hospital & that they do not suffer any to be in town without a permit from the Col° or Commanding Officer

A reg^tl Court martial to set this morning at 10 OC. to try all Prisoners brought before it evidence to Attend ~ Captn Harleston president Lieut^t Galvan and Perreneau members
NB Serg^t Laurense for refusing to do duty was mult 14 days pay
John McDowl for being drunk on guard rec^d 50 lashes ~ Jn° Steel for d°. rec^d 100 lashes .
Jn° Smith for rioting & drunkenness was reprimanded at the halberds

---

A Monthly return of the reg^t 1st Oct:
8 Field Off^rs 10 Capt^ns 10, 1st Lieut^n 7, 2nd L^ts _ 1 Adj^t 1 Chapl: _ 1 q^rt mast^r: 1 Surgeon
1 S^r mate 20 Serg^ts _ 16 Drum & fifes 3 Arm^rs _ 320 Rank & file

---

Orders 1st Octob^r by G Howe } . . Parole Drayton . . . by presid^t
For the day tomorrow Captn Ashby
Town Guard Lieut: Capers
Magazine Guard L^t. Proveaux ~ Brick house L^t. from 5th reg.

Reg^tl Orders Col° Motte } Orderd that L^t. Martin join & do duty in the Granadier Comp^y. Mr John Bush is appointed a Second Lieut^t. in the Regim^t and to be respected & Obey'd as such; L^t. Bush to do duty in Captn Harlestons Company till further orders
~~ Captn Harleston is Added to the General Court martial
NB. by sentence of C Martial Aron Harris was mulct 1 day pay for stealing wood [718]

Orders 2nd Octob^r by G Howe } . . Parole
For the day tomorrow Captn. Potts
Town Guard Lieut: Warley
Magazine Hart ~~ Brick house L^t. from 5th reg:

Orders 3^d Octr. by G Howe } . . Parole Howe. . . . . by presid^t
For the day tomorrow Captn. Lisesne
Town Guard Lieut: from 5th reg^t.
Magazine L^t. Perrineau – Brick house Lt Galvan

---

[718] Aaron Harris served in the Grenadier Company of the 2nd South Carolina Regiment during 1777. He was fined one day's pay on 1 October 1777 for stealing wood. He received 100 lashes on 20 May 1778 for theft. He was in the Grenadier Company under Captain Thomas Dunbar in 1778 and 1779. He was at the siege of Savannah.

The General Court Martial ordered to try Serg[t]. Oneal of the 2nd Battalion in this state have reported that his Behavior to his Officer L[t]. Dunbar was Insolent & Criminal in a very high Degree Contrary to all good Order and Discipline & therefore they sentence him first to be reduced to the rank & then to receive 100 lashes on his bare back with a Cat of nine tails, this sentence the general Approves of & ratifys: Col[o] Motte will Order it to be carry'd in execution in the manner & time he thinks proper . . the s[d]. Court having Agreeable to Orders, also tryd Gabriel Scott of Col[o]. Pinkneys Regiment for desertion report that he is Guilty [719] & Sentence him to receive 99 lashes on his bare back with Switches & that he also be piquetted for one hour and half both part of his punishment to be Inflicted at Different times & at the head of the reg[t] to which he belongs ~ the General also Approves and ratifys this sentence, Directs the Officer of the main guard to deliver the Criminal to L[t]. Col[o] Cattle or any Off[r] Appointed by him to receive him, Col[o]. Cattle will Order the Execution in the manner and at the time he thinks proper but he is impowered by the General to shorten the time of piqueting if he thinks necessary ~~ Tho[s]. Stafford & W[m]. Ficklin soldiers in the 2nd Battalion being also tryd by the court for desertion was sentenced to receive 100 lashes but being recommended by the court as Objects of mercy, the General in respect to the Court extends pardon to the s[d]. Stafford & Ficklin who are pardoned Accordingly

Reg[tl] Ord[rs] Col[o] Motte 3[d] Oct } Ordered that no one under the degree of Second Lieutenant belonging to the Regiment do presume to lodge out of the Barracks without Wrighten permission from the Commanding Officer; Such as disobey this order may Assured being severely punished ~ the Major is desired to see this Order be strictly comply'd with

The Officers are desired to Attend the funeral of M[r] Press Smith from the Reverend M[r] Smith at ½ past 4 OC: this Afternoon

---

Orders of 4th Oct. by G. Howe } . . Parole Rutledge . . by presid[t]

For the day tomorrow Capt[n] Jervey

Town Guard Lieut: Martin

Magazine L[t] from the 5th reg: ~ Brickhouse L[t]. H. Gray

The General court martial Orderd to try Capt[n] De Treville of the Corps of Artillery for sundry charges exhibited against him by Lieut: Raphel of the Same Corps Report as follows ~ That they consider the first charge as very cruel & malitious founded upon no truth & therefore do honorably Acquit Capt[n] De Torville ~ The 2nd Charge they say that it does not appear to the court after the most minute inquiry that they was the last Intention of fraud and that honourably Acquit Capt[n] Detreville of this infamous charge

The 3[d] Charge they report to be Equally groundless & Malicious as the first and that it falls unsupported by any evidence, They also honourably acquit him of the 4th Charge ~ 5th Charge they report in the nature not Cognizable by the Court

The 6th & 7th charge they likewise consider as Groundless & honourable Acquit Capt[n] De Treville ~ These several Determinations of the court the General Approves of & ratifys Releases Capt[n] DeTreville from his arrest, directs him to return to Duty & Orders him to be respected & received by the army as usual.

---

[719] Gabriel Scott served in the 1st South Carolina Regiment under Captain Benjamin Cattell in 1775. He received 99 lashes on his bare back with switches for desertion on 2 October 1777. He also was ordered to be picketted for an hour and a half for the same crime. The punishments were delivered on two different times in front of his regiment.

Reg^{tl} Ord^{rs} } Ordered that Sergeant ONeal receive punishment at the head of the regiment this
Col^{o} Motte } Afternoon A reg^{tl} Court martial to sett this morning at 10 OC. to try all prisoners brought before it – the Prisoners to be Acquainted by the Adjutant that they are to be brought to their tryal and all evidences to be orderd to attend at that hour ~ Captn. Lesesne presid^{t}. _ L^{ts}.Martin & Capers Members

aft^{r} Orders

Ordered that the Sentence of the General court martial against Serg^{t} Oneal be Suspended till further orders.

Orders 5th Oct. } . . . Parole Elbert . . . by presid^{t}
by G Howe } For the day tomorrow Captn Motte
Town guard Lieut: Preveaux
Magazine L^{t} from the 5th reg^{t} ~ Brick house L^{t}. Capers

Regim^{t} Ord^{rs} } Divine service as usual this afternoon
Col^{o} Motte }

NB by sentence of Last court Thos Stafford & Wm Ficklin was to receive 100 lashes each, was pardoned

---

Orders 6th Oct^{r}. } . . Parole Ferguson. . . GH
by G Howe } For the day tomorrow Captn Ashby
Town guard Lieut: from 5th reg:
Magazine L^{t}. from 5th reg^{t}. ~ Brick house L^{t}. Warley
Pritchards L^{t}. Hart

Reg^{tl} Ord^{r} }
Col^{o} Motte }

A Reg: court martial to sett this morning at 10 OC: to try all prisoners brought before it, all evidence to attend

Captn Motte presid^{t} – L^{ts} Galvan & perreneau members

Aft^{r}. Ord^{rs}

The Officers of duty are desired to attend the funeral of Lt Raphel of the Artillery, this afternoon at 5 OC: at the house of Dr Labertas in Green & Tradd. The Battalion is not to turn out this After noon but the men are to be imployd in putting their Arms in the best Order

Brigade Orders } Orderd that a detachment from the 2nd & 5th reg^{ts}. of one Subaltern, 1 Serg^{t}.
Col^{o} Motte } 1 Corp^{l}. and 1 drum & fife and 24 ~~Rank and file~~ privates to attend the funeral of Raphel of Reg^{t} of Artillery from the house of Dr Labertas in green Street at 5 OC: this After noon

Col^{o} Motte **Ch^{s} Town 1777**

Orders 7th Oct. } . . Parole Liberty . .
by G Howe } For the day tomorrow Captn Lesesne
Town guard Lieut: Dunbar
Magazine Lt Perreneau ~ Brickhouse Lt from 5th reg:

NB Ch^{s}. Lucas Broke confinement & deserted –

Orders 8th Oct. by G Howe }

. . Parole Mohawk
For the day tomorrow Captn Potts
Town Guard Lieut: Hy Gray
Magazine Lt. from 5th reg: ~ Brickhouse Lt. Martin

Regtl Ordrs Colo Motte } A reg: court martial to set this morning at 10 OC: to try such prisoners as shall be brought before it, the prisoners to be acquainted that their tryal is to come on at that time & the evidences to be Ordered to Attend ~ Captn Lesesne presidt. Lts Proveaux & martin members

---

Orders 9th Oct. by G Howe }

. . Parole Mifflin . . .
For the day tomorrow Captn Davis
Town Guard Lieut: Warley
Magazine Lt. from 5th reg. ~ Brickhouse Lt from 5th reg

NB. by regtl Court the follow men was sentence, Sergt Munroe for neglect of duty was mult 1 Mo. Pay, one half remitted – Wm Robert for abusing a negroe when on Centry, reprimanded ~ Wm Clark for abusing & ill treating Ab: Moses, 35 lashes remitted [720]– Wm McCallister for calling Wm Ficklin a murderer recd. 25 lashes [721] ~ Samuel Henderson for Absents without leave recd 30 lashes –

Orders 10th Oct. by G. Howe }

. . Parole Huger . . . .
For the day tomorrow Captn Motte
Town guard Lt. Perreneau
Magazine Lt. from 5th reg: ~ Brickhouse Lt P. Gray

The commanding Officers of each Battalion & corps in Charles town, fort Moultrie and fort Johnson is to Order a report to be made by their respective Quartr Master to the Quarter master General of the Quantity of fire wood that has been recd. by each Corps from the 1st January 1777 to 1st Oct. Instant for the use of the Officers and men specifying the quantity deliverd to each Officer

Orders dated 7th

---

720 Unable to find any information on Abner Moses. He may have been the "Negro" that Private Roberts abused.

721 William McCallister had enlisted in the 2nd South Carolina Regiment on 4 November 1775. He received 25 lashes on 9 October 1777 for calling Private Ficklin a murderer. He was under Captain Thomas Moultrie at the siege of Savannah in 1779.

Capt[n] Kershaw [722] & Fran[s] Taylor [723] L[t] Jam[s] Garlie [724] of Col[o] Thompson's Battalion having Apply'd and Obtain Leave to resign their commission are no longer to be consider'd & respected as Continental Officers ~ L[t] Uriah Goodwin is appointed Capt[n] in the room of Cap[t] Kershaw:[725] L[t] Thom[s] Mairhall in room of Capt[n] Tayler resigned and are to be respected & Obey'd Accordingly [726]

Col[o] Motte **Ch[s] town 1777**

Reg[tl] Orders 10 Oct. by Col[o] Motte } A court martial to sett at 11 OC. This forenoon to try all prisoners brought to it, the prisoners to have notice given them that they are to be try'd & the Evidences to attend on time

Prese[d] Capt[n] Ashby L[ts] P Gray & Pereneau members

The Col[o] desires all the Officers off Duty & if convenient to them that are members of the general court martial now setting will meet him at his house this morning at 11 OC:

Orders 11 Oct. by G. Howe } . . Parole Woodford . . . GH

For the day tomorrow Capt[n] Ashby

Town guard Lieut: Martin

Magazine L[t] from 5th reg: - Brick house L[t]. H. Gray

All Officers Obtaining furlow are in wrighting to give in to the Adj[t]. General the time of Absence allowed them, where they may be found if wanted, & upon their return to duty, immediately to inform him of it & all Officers sent out upon Command are upon their return as immediately as possible to be reported by the Adjutant of their respective regiments to the Adjutant General

Reg[tl] Ord[rs] Col[o] Motte } The Quart[r]. Master to make a report in wrighting this day to the Quart[r] Master General of the quantity of wood he has rec[d]. for the use of the regim[t]. from the 1st of Jan[y]. last to the first Instant Agreeable to the Gen[l]. Orders yesterd[y]. and is at the same time to deliver in a Copy to the Commanding Officer of the Reg[t]. –

A court martial to set this morning at 10 OC: to try all prisoners brought it, all evidences to Attend ~ Capt[n] Motte presed[t] ~ L[ts] Galvan & Martin Memb

---

722 Eli Kershaw was born in 1745 in England. He became a captain in the 3rd South Carolina (Ranger) Regiment on 18 June 1775. On 7 October 1777 he resigned from the regiment. In 1779 he was a lieutenant colonel and he was taken prisoner at the fall of Charleston. He died while being transported to Honduras as a prisoner in December 1780.

723 Francis Taylor had been born in 1750 in Virginia. He was appointed as a second lieutenant in the 3rd South Carolina (Ranger) Regiment under Captain Charles Heatley and Colonel Thomson on 12 September 1775. On 6 May 1776 he became a first lieutenant. He became a captain during February 1777 and resigned on 7 October 1777. During 1781 he was in the militia.

724 John Garlie served as a lieutenant in the 3rd South Carolina (Ranger) Regiment until he resigned on 7 October 1777.

725 Uriah Goodwin had been born in 1750 in Virginia. He volunteered on 1 July 1775 as a cadet in the 3rd South Carolina (Ranger) Regiment. He assisted Lieutenant Thomas Charlton in escorting a wagon of powder to Saxe-Gotha. Patrick Cunningham captured them on 31 October 1775. During May 1776 he became a lieutenant in the 3rd South Carolina Regiment. He was promoted to captain on 9 October 1777 when Captain Kershaw resigned. He was wounded in the battle of Stono Ferry on 20 June 1779. He was taken prisoner at the fall of Charleston and exchanged in November 1780. He was killed at the battle of Eutaw Springs on 8 September 1781.

726 Thomas Marshall had been commissioned a first lieutenant in the 3rd South Carolina (Ranger) Regiment during April 1776. He was promoted to captain on 9 October 1777 when Captain Taylor resigned.

NB according to sentences of last court John McCaid rec[d] 100 lashes for neglect of duty [727] Jn[o] Thomson for stealing Beef re[d]. 50 lashes – Dempsey Thomas for Insolence to his Corp[l] was reprimanded at the Halberts [728] – Jam[s] Bladwell for theft rec[d] 100 lash [729] Jn[o] Davis for Disobedience of orders was publicly reprimanded & Ask Serg[t] Coffers pardon [730]

Orders 12th Oct. by G Howe } . . Parole Thompson . . . GH –
For the day tomorrow Capt[n] Potts
Town Guard Lieut[n]. from 5th regt.
Magazine L[t]. from 5th reg: ~ Brickhouse Lt. Proveaux

The Quarter master General will examine the main guard & report what reports are wanting also whatever else he shall find required to be done there to render it comfortable and Secure, he will also report what neccessarys are wanted for the prisoners confined there

---

Orders 13th Oct. by G Howe } . . Parole Pinckney. . by presid[t]
For the day tomorrow Capt[n] Lesesne
Town Guard Lieut: from 5th reg:
Magazine: L[t] Warley
Brick house L[t]. P. Gray – Pritchards Guard L[t] from 5th reg[t].

Col[o] Motte **Ch[s] Town – 1777 –**

Regimental Quarter masters are directed to be very exact in Reporting the Quantity of wood deliver'd to their several corps and what proportion each Officers receive of which they are regularly to transmit a report examined & Signed by the Commanding Officer of the Corps / for the time being / to which they belong, to the Quarter master General every month.

---

727 John McCade (McCaid) enlisted in the 2nd South Carolina Regiment on 4 November 1775. He received 100 lashes on 11 October 1777 for neglect of duty. He received 39 lashes on 24 October 1777 for slandering Sergeant Burgess. He was sentenced to do guard every other day for eight days due to being drunk on guard. When he wasn't on guard duty he would be confined. He received 50 lashes on 27 October 1777 for an unnamed offense. He received a reprimand on 4 November 1777 for being absent without leave. On 11 June 1778 he received 25 lashes on every other day for four days. This was for breaking into the regimental store and stealing all the clothing there. Between each punishment he was to be confined until 7 o'clock in the morning. He deserted on 10 November 1779.

728 Dempsey (or Demsey) Thomas enlisted in the 2nd South Carolina Regiment on 11 July 1775 under Captain Bernard Elliot. He was reprimanded at the halberts on 11 October 1777 for insolence to his corporal. He later served as a corporal in the 3rd South Carolina (Ranger) Regiment under Captain David Hopkins. He was in the 4th South Carolina Artillery under Captain James Mitchell from November 1779 to January 1780.

729 James Bladwell (or Blackwell) enlisted in the 2nd South Carolina Regiment on 2 May 1777. He received 100 lashes on 11 October 1777 for stealing.

730 John Davis enlisted in the 2nd South Carolina Regiment on 1 April 1777. On 11 October 1777 he was publicly reprimanded for disobeying Sergeant Lewis Coffer's orders and he had to publicly apologize to Sergeant Coffer. On 8 November 1777 he was put under stoppages for £8 for an unnamed offense. He was promoted to corporal on 8 October 1778. He served in Captain Dunbar's Grenadier Company, but when that company became a light infantry company he was transferred to Captain Hall's company. He was promoted to sergeant on 8 October 1778 in Captain Charnock's company. He was in Captain Mason's company at the siege of Savannah. He was wounded on 28 September 1780 at Black Mingo while serving with Marion's partisans.

Reg[tl] Ord[rs]. Col° Motte } a court martial to set this morning at 10 OC: to try such prisoners as may be brought to it; the prisoners to have notice given them that they are to be brought to their trial & the evidences to be Orderd to Attend in time
Capt[n] Mayzyck presid[t]. L[ts]. mason & P. Gray members

All the Officers off duty are desired to meet the Col° at his own house Tomorrow 10 Oclock in the forenoon

---

Orders 14th Oct. by G Howe } . . Parole Martin G.H
For the day tomorrow Capt[n] Jervey
Town Guard Lieut: Perreneau
Magazine L[t]. from 5th reg: ~ Brickhouse L[t]. H[y] Gray

The quarter master General is immediately to report to his Excellency the presd[t]. What alterations and repairs are necessary to the watch house when the main guard is kept & require the favour of him to give orders that they may be made, he is also to provide a sufficient quantity of Blankets for the use of the Continental prisoners when confind in Guard rooms, also potts for them to Cook with He is to order Ammunition Chests with Locks & Keys for each Guard in town & also for the Guards at Dorchester and pritchards & ship Yard, if they are not chests enough allready made he is immediately to employ workman to make them and to report his proceedings to the General

Reg[tl] Ord[rs] Col° Motte } The Officers are to keep an exact account of what wood they receive from this day and to give a Copy to the Quartermaster whenever he asks it –

Aft: Ord: The Bad weather preventing the Off[rs]. from meeting the Col°. this morn[g]: he desires they will be at his house precisely at 10 OC: tomorrow morning

A Regtl. Court martial to set tomorrow morning at 10 OC: to try such prisoners as shall be brought before it, all evidences to be warned in time

Capt[n] Ashby president Lieut[nts] H[y]. Gray & Proveaux members

NB sentence of Last court. Corp[l] Kidwell for abusing and striking Serg[t]. Coffer was reduced to the ranks ~ Jn° Fenwick for neglect of duty rec[d]. 50 lashes; & Liome Husbands for striking W[m]. Jones to ask his pardon & was reprimanded

---

Orders 15th Oct. by G Howe } . . Parole Drayton . . by presid[t]
For the day tomorrow Capt[n] Motte
Town Guard Lieut: from 5th reg:
Magazine L[t]. Proveaux
Brick house L[t]. P. Gray

Col° Motte **Ch[s] town 1777 –**

Gen[l] Orders 15th Oct:

All Officers of guards when prisoners of war are committed to their charge are immediately to furnish the Adjutant General with an exact list of their names, the rank they bear in the Army or navy of the Enemy, Regim[t]. or ship they belong to, by whom and when taken, by whose Order Sent to the Guard & from what person received; they are also to give unto him a Similar list of all prisoners committed to them belonging to the Enemy, private ships of war, armed vessels & merchantmen, whit the Adjutant General is directed regularly to enter into a book kept particularly for that purpose

Reg^{tl}. Ord^{rs} } Orderd, that two more Subalterns be added to the court martial that is
Col^{o} Motte } to set today at 11 OC: in forenoon
L^{ts}. Martin & warley added members of the above court –

Orders 16th Oct: } . . Parole Middleton . . GH
by G Howe } For the day tomorrow Captn Cogdell
Town Guard Lieut: Hart
Magazine Lt Hall ~ Brickhouse from 5th reg:

The General court martial order'd to try Sergeant maj^{r}. Edmunds have reported that he is not guilty of mutiny but comes under the last Article of the last Sect: of Articles of war, for which they sentence him to be reduced to the ranks for behaving Contrary to good Order and military Discipline – The General Approves of ratifys the Sentence ~ the Corporal punishment which Serg^{t}. Oneil was Sentence to receive / is in consequence of the request of L^{t}. Col^{o} McIntosh / remitted the rest of the sentence to be executed ~ the Sentence against James Orange of Col^{o} Mottes Battalion are as follows; with regard to the first Crime that of desertion the Court are of Opinion that he is Guilty of it; but taking into Consideration the many favourable Circumstance in the evidence, they only Sentence him to receive 100 lashes on the bare back with a Cat of nine tails, Respecting the 2nd Offence for which he was Orderd to by try'd, the Court are of Opinion he is guilty of the crime under which he was Arraigned and not withstanding he threw himself on their mercy, Yett reflecting on the Atrociousness of the Crime they Cannot but think, they extend it to him by sentencing him to receive on hundred lashes on the bare back with a Cat of nine tails; they further humbly recommend this Sentence may be executed a few days after the former The General Approves of ratifys both Sentences, which are to Carry'd into execution in time and manner as Col^{o} Motte shall direct, the Court Martial is dissolved

NB according to Sentence of Last Regim^{tl}. Court martial Peter Ficklin for neglect of duty rec^{d}. 50 lashes

Col^{o} Motte **Ch^{s} town – 1777 –**

Orders 17th Oct. } . Parole Motte . . . by presid^{t}
by G Howe } For the day tomorrow Captn Potts
Town Guard Lieut: from 5th reg:
Magazine L^{t}. Roux ~ Brick house L^{t}. H^{y}. Gray

The Barrack master is immediately to report to the Quartermaster General what Quantity of wood he has engaged for the Army, with what persons he has Contracted for it & what price

Reg^{tl} Orders } Order'd that the general Orders of Yesterday be read at the head of the reg^{t}
Col^{o} Motte } this Afternoon by the Lieut: Col^{o}. and that he see the Sentences against Serg^{t}. Maj^{r}. Edmunds & Serg^{t} Oneil put into execution agreeable to said Orders and that the first part of the Sentence against James Orange be Inflicted at the same time, After which he is to be confined in the Regimental Barrack Guard

NB sentence of reg. court Serg[t]. Smith for beating an inhabitant of the town was reprimanded before the Serg[ts] of the regim[t] [731]~ & the Sentence of Gen[l]. Court. Serg[t]. maj[r]. Edmunds & Serg[t]. Oneil was reduced to the ranks & Jn[o]. Orange rec[d]. 100 lashes & reprimanded In Guard –

Orders 18th Oct. by G Howe } . . Parole Weedon . . . GH
For the day tomorrow Capt[n]. Blake
Town Guard Lieut[n]. Martin
Magazine L[t]. from 5th reg: - Brick house L[t]. Capers

Reg[tl]. Ord[rs] Col[o] Motte } All Officers off duty are punctually to Attend Exercise ever after noon (Sunday only excepted) agreeable to Orders repeatedly given for that purpose, such as disobey this Order may be Assured being taken notice of ~ the regiment is to turn out to Exercise for the feuter at 4 OC: in after noon

A reg[tl]. Court martial to set this forenoon at 10 OC: to try such prisoners as may be brought to it the prisoners to be Acquainted that their trial is to come in the evidences to be Order to Attend in time.

Capt[n] Blake president L[ts]. Dunbar and mason members

The late Serg[t]. Maj[r]. Edmunds is to join & do duty in Capt[n]. Lesesne comp[y].

NB sentence of last court Adam Meek for sleeping on his post rec[d]. 50 lashes & 50 more next mundy

---

Orders 19th Oct. by G Howe } . Parole Miflin . . . by presidt.
For the day tomorrow Capt[n] Charnock
Town guard Lieu[t]. Preveaux
Magazine L[t]. from 5th reg: - Brickhouse L[t]. Warley

Exact lists of Necessarys furnished by the Quart[r]. Mast[r]. Gen[l]. for the use of the Guards and prisoners confined in them are to be made out Signed by him & Given to the Off[rs]. of those Guards, to whom the Articles were deliverd from them relieving Off[rs] are to receive these lists and if upon examination any deficientey may be found they are make report thereof, that the Officer in whose Guard the deficientcy happen may be made Accountable

Col[o] Motte **Ch[s] town 1777**

Reg[tl] Orders Col[o] Motte } Divine Service to be performed this Afternoon at 4 Oclock
To W[m] Dilliant /
It is my Orders that you wait on Capt[n] Oliphant & put him under Arrest _ It tis my further Orders that he remain at his Quarters and not be seen out and you may Acquaint him that he is put in Arrest for Absenting himself from the Regiment without my leave – J.M

---

[731] John Smith served as a sergeant in the 4th South Carolina (Artillery) Regiment under Captain Richard B. Roberts from 1 April 1777 until Roberts' death on 8 July 1780. He was reprimanded before the sergeants of the regiment on 18 October 1777 for beating an inhabitant of the town.

After Orders To Mr Dilliant

Captn Oliphant having made satisfactory Apology for Absenting himself form the reg^t^. without my leave, I do therefore Order that you take off his Arrest & Order him to his duty in the reg^t^. as formerly . . . .

JM.

Orders 20th Oct. by G Howe } . . Parole Elbert .
For the day tomorrow Captn Lesesne
Town Guard Lieut. From 5th reg:
Magazine Lt. Bush ~ Brick house Lt P Gray

The Quartr. Mastr. Genl. has reported that of the Blankets provided for the prison at the main guard so late as on Wednesday & Thursday last, five of them are already lost _ Guard rooms According to the disipline the General had been taught have all been consider'd and Aught always to be a place of Absolute Safety for whatever may be deposited in them & duty exacts of all Offrs Commanding them, their utmost exertion to make them so, How the Officer or Officers in whose tour of duty the above Articles were lost out of the Guard room, can reconcile to their feelings such want of Attention and negligence of duty as may with the Appearance of too much Justice be imputed to them, is to the General a Circumstance of as much surprise as Anxiety. _ he therefore direct the Commandant of the Brigade to inform himself wether the Q.M.General deliver'd the the Blankets reported the number to the offr. Who relieved them and Wether they made report to the subsequent relieving Offrs; & if not why they neglect doing what was absolutely their duty, & lay his proceeding before the General

Regtl Orders Colo Motte } A regimental court martial to set this morning at 10 OC: to try such prisoners as may be brought before ~~them~~ the prisoners to be informed that their trials is to come on at that hour & the Witnesses to be Orderd to attend –

Captn Charnock presidt – Lieuts P. Gray & Bush members

Afr. Ordr. Douglas Oneil of Captn Blakes Compy. is Appointed a Sergeant in Captn Charnocks Compy & is to be Obey'd as Such

---

Orders 21st Octr. by G Howe } . . Parole Vivat . . . . GH
For the day tomorrow Captn Mazyck
Town Guard Lieut. Hart
Brick house Lt. from 5th reg:

Colo Motte **Chs town 1777**

The General has accepted the resignation of Captn Wm Oliphant of Colo Mottes Battalion, he is no longer to be Considered and Obey'd as a Continental Officer ~ Lieut. Jacob Shubrick is Appointed Captn. In the room of Oliphant resigned & 2nd Lieut: Galvan is Advanct to the rank of 1st Lieut. In the Room of Lt. Shubrick promoted they are to be respected and Obeyd as Such

Regtl. Ordrs. Colo Motte } A reg: court martial to Set to day at 10 OC: to try such prisoners as shall be brought before it – the prisoners to be informed their trial comes on at that hour & the Evidences to be Orderd to Attend ~ Captn. Lesesne presidt. Lieuts. Hart and Roux members

---

Orders 22nd Oct. by G Howe } . . Parole Dry . . . by presid[t].
For the day tomorrow Capt[n] Harleston
Town Guard Lieut from 5th reg.
Magazine L[t]. Martin ~ Brick house L[t]. Capers

Col[o] Huger is Directed to wait upon his Excellency the president with the report of the Officer of the day, respecting the want of wood and to inform him of the great Sufferance of the Army for the want of that necessary Article for a Considerable time past Owing as tis suppord to the Conduct of M[r] Phil[p]. Wills who was and Aught to have furnished that Article ~ Officers of Guards are strictly forbid upon any Occasion Whatever to quit their Guards & as strictly injoined to exert themselves to prevent the men of their guard or any prisoners in their charge from giting drunk, they are to consider themselves, the Guard of every thing Committed to their charge, and will Certainly be made Answerable for any thing lost; It tis painfull to the general to be Obliged to point out Duty, so Obvious & so Absolutely Official, that not to know them is impossible and not to practice them unpardonable, the however laments that Some information he has Received give him to believe that it has been necessary ~ Any Soldier who gitts drunk upon Guard will punisht with the utmost severity & it will considerd as a breach of duty in the Off[r]. of the guard, if he does not confine & report him, Not more than two men at a time are upon any Occasion whatever (duty excepted) to be permitted to leave the guard and even this Liberty is to be admitted with the Caution & if abused will be withdrawn, These two latter Orders to be Copy'd and set up in every Guard room ~ Officers of the day are Directed to be particularly attentive that all Orders respecting Guards be strictly Comply'd with and punctually to report every neglect of duty ~ The General has accepted the resignation of Lieu[t] W[m] Hayward of Col[o] Pinckneys Battalion he is therefore no longer to be considered as a Continental Officer ~ [732]

Col[o] Motte **Ch[s] town 1777**

L[t]. Larvacher is Appointed 1st Lieu[t]. In the room of L[t]. Hayward resigned and to be respected and Obey'd as such [733]

Reg[tl] Orders Col[o] Motte } Lieut: Galvan is to Join & do duty in Capt[n] Lesesne's Comp[y]. till further Orders – the remainder of the Sentence inflicting 100 lashes is to be done this Aft[r] Noon

NB

By a Sentence of the Gen[l]. court James Orange recd. 100 lashes – by reg[tl]: Pet[r] Fagan for being drunk on guard rec[d] 77 lashes – W[m]. Baldwin for d[o]. rec[d] 50 lashes Jn[o]. Walvey for Absents without leave rec[d]. 39 lashes [734]

---

732 William Heyward served in the 1st South Carolina Regiment until he resigned on 22 October 1777.

733 St. Marie de Levashche was commissioned a second lieutenant in the 1st South Carolina Regiment during October 1776 and a first lieutenant on 12 October 1777 when Lieutenant Hayward resigned. On 9 October 1779 he became a captain. He was captured at the fall of Charleston. After being exchanged during June 1781 he served until the end of the war.

734 John Wieley was in the 2nd South Carolina Regiment. He received 39 lashes on 22 October 1777 for being absent without leave. He received 100 lashes on 7 February 1778 for sleeping on his guard post.

Orders 23d Oct. by G. Howe } . . Parole Middleton . . by presid[t]
For the day tomorrow Capt[n]. Cogdell
Town Guard Lieut[n]. Proveaux
Magazine L[t]. Warley ~ Brickhouse L[t]. from 5th reg:
Command at Dorchester L[t]. Jones of the 5th reg:

The Officers of the day is immediately to examine the different rooms in town and to be exact in noticing the Situation they are in & when he is relieved, he is to examine it again Accompanied with the Off[r]. who relieves him; if any Damage has been done he is to report it immediately that the Officer of the Guard in whose time it happened may be made Accountable for it; this method is to be continued by further Officers of the day till further Orders ~ Guards in town for the futer to be relieved punctually at 8 OC: in the morning; ~ Maj[r]. Thomas Shubrick is to Act as a Brigade Maj[r]. to the Brigade of which Col[o]. Huger is Commandant till further Orders and is to be Obeyd Accordly

Brigade Orders by Col[o] Huger } Col[o]. Huger injoins the Officers to take notice of the men on duty & if they see them negligent on their post are to reprimand them or report them to the Off[r]. of the Guard they belong to that they may be punished.

At Retreat beating all guards are to be under Arms, & see the numbers of prisoners Committed to their charge Secured. ~ three privates to Attend Capt[n]. Debraham to morrow morning at 7 OC: 2 from the 2nd & one from the 5th Battalions [735]

Reg[tl]. Ord[rs]. Col[o] Motte } Joseph Culp Corp[l]. in Capt[n] Blakes comp[y]. Is Appointed a Sergeant in the same and to be Obey'd as such – [736]

A court martial to set this morning at 10 OC: to try all prisoners that shall be brought before them all evid: to be warned to Attend

Capt[n]. Harleston presid[t] ~ L[ts]. Proveaux & Warley members

NB by sentence of Court Jn[o]. Clements for selling a public Blanket rec[d]. 75 lashes

Col[o] Motte **Ch[s] town – 1777 –**

Orders 24th Oct[r]. by G Howe } . . Parole Maxwell . . .
For the Day tomorrow Capt[n]. Blake
Town guard Lieu[t]. from 5th reg.
Magazine L[t]. Bush ~ Brick house Lieu[t]. P. Gray

Reg[tl]. Ord[s]. Col[o] Motte } A court martial to set this morning at 10 OC: to try such prisoners as shall be brought before it, All evidences to be Orderd to Attend in time.
Capt[n]. Blake presid[t]. Lieu[ts]. P Gray & Bush member

NB by sentence of Last court Jn[o] M[c]Caid for slandering & traducing Serg[t]. Burgess rec[d]. 39 lashes ~ W[m]. Simpson for assisting a prisoner at the main guard to escape rec[d]. 75 lashes

---

[735] J. Ferdinands Debraham served as a captain of the engineers. He was promoted to major of the engineers on 11 February 1778. He was captured at Charleston on 12 May 1780. After being exchanged on 22 April 1781 he continued to serve, and became a brevet lieutenant colonel on 6 February 1784.

[736] Josiah (or Joseph) Culp served in the 2nd South Carolina Regiment. He was promoted to sergeant on 23 October 1777.

Orders 25th Oct. by G Howe } . . Parole Burke . . . . GH
For the day tomorrow Captn. Charnock
Town Guard Lieut. Burke
Magazine Lt from 5th reg:
Brick house Lt Hart

Regtl. Ords. Colo Motte } Order'd that all the Offrs. Off duty do meet the Colo. at his house this fore noon at 10 Oclock ~ the Regt. is not to turn out to Exercise this Afternoon; the men to be employ'd in cleang: their Arms & putting their Accoutrements in good Order – the Sergeants to see this Order is complied with.

A reg: court martial to set this morning at 11 OC: to try all prisoners brought before it Evid: to be warned to Attend ~ Captn. Charnock presdt Lts Hart and Martin Members

NB by sentence of last court 24th Inst: Jams Greenwood for a neglect of duty recd 50 lashes ~ Tho. Welch for beating an inhabitant of the town recd 50 lash ~ Jacob Dunbar for falsely Accusing Sergt. Henderson of theft recd. 50 lashes ~ Robt Pinhorn for drunkenness on Guard & abusing the Offr recd. reprimand [737] ~ Robt Champneys for do. reprimanded [738] ~ Frederick Johnson [739] for Abusing Mr. Wm. Bull [740] red. 25 lashes, 25 remitted ~ Sergt. Mathew for quitting his guard reduced to the ranks ~ Alexr Ferguson for neglect of duty reprimanded [741] ~ Jeremiah McGinnes for Stealing a gun from Alexes Seymour ~~recd~~ to receive 100 lashes & have his pay stopt to make good the gun; the lashes remitted

---

Orders 26th Oct. by G. Howe } . . Parole Drayton . . .
For the day tomorrow Captn. Lesesne
Town guard Lieut. Perreneau
Magazine Lt. from 5th reg: ~ Brick house Lt. Hall

Amunition Chests having been Orderd long since for each Battalion, Commandg: Offrs. are to report to the Q.M.Genl. wither they have been recd. and what number they are also to report to him what number of Amunition Chests they are furnished with that more may be Orderd if Necessary.

Regtl. Ords Colo Motte } The Regiment to Attend Divine Service this After noon at 4 OClock –

---

737 Robert Penhorn was in the 2nd South Carolina Regiment under Captain Motte in 1779.

738 Robert (Roger) Champness (Champneys) enlisted in the 2nd South Carolina Regiment on 4 November 1775. He was reprimanded on 25 October 1777 for being drunk on guard and abusing the officer of the guard. He was promoted to corporal on 27 July 1778 in Captain Motte's company. He was reprimanded on 3 October 1778 for leaving his guard. He was supposed to have been reduced in rank on 20 November 1778 for unbecoming language to Sergeant Davis, but the charge was remitted. On 11 January 1779 he was court-martialed for an unnamed offense, however he was given mercy.

739 Frederick Johnson enlisted in the 2nd South Carolina Regiment on 4 November 1775. He was under Captain John Blake during 1777. He received 50 lashes on 25 October 1777 for abusing William Bull, but it was reduced to 25 lashes. He was discharged on 8 July 1778.

740 William Bull was an attorney in Charleston.

741 Alexander Ferguson served in the 2nd South Carolina Regiment from 1776 to 1779. He was reprimanded on 25 October 1777 for neglect of duty. He was under Captain Peter Gray at the siege of Savannah in 1779.

Col° Motte **Ch[s] town – 1777**

Orders 27th Oct. by G Howe } . . Parole Page . . . . GH
For the day tomorrow Capt[n] Jervey
Town Guard Lieu[t]. Marten
Magazine L[t]. from 5th reg: ~ Brick house L[t] Warley
Pritchards Lieu[t]. H Gray

The Q.M. General will as immediately as possible inform himself of the Continental Military stores in the State of S°. Carolina at what place Deposited and in whose charge and make particular report thereof at Head quarters

Reg[tl]. Ord[s]. Col° Motte } A reg: court martial to Set this fore noon at 11 OC: to try such prisoners as shall be brought before it, the prisoners to be made Acquainted their trial will come on at that time and the evidences to Ordered to Attend

Capt[n]. Lesesne presid[t]. L[ts]. H Gray & Proveaux Members

NB by sentence of last court James M[c]Caid & W[m]. Simpson rec[d]. 50 lashes each – Stephen Strickham,[742] Sam[l]. Horn [743] & Jn°. Castile [744] rec[d] 50 lash each for stealing potatoes

---

Orders 28th Oct. by G Howe } . . Parole Scott . . . . by GH
For the day tomorrow Capt[n]. Mazyck
Town guard Lieu[t]. Preveaux
Magazine L[t]. from 5th reg: _ Brick house L[t]. from 5th. reg:

Reg[tl]. Orders Col° Motte } Orderd, that all the Off[rs]. of duty do meet on the parrade this After noon at the usual time of the reg[ts]. turning out to Exercise the reg[t]. is not to turn out

NB by Sentence of last court Serg[t]. Dyer was reduced to the ranks & mult 14 days pay for drunkenness on Guard ~ Fran[s]. Simpson for disobedience of Ord[rs]. & impertinent Language to his Serg[t]. rec[d]. 50 lashes & to morrow to receive 50 more ~ Stephen Fillory for fighting with a bad Woman, reprimanded,[745] Conrad Fitner to receive 100 lash but suspended till Bernaby Bryan is examined. – [746]

Orders 29th Oct. by G Howe } . . Parole Starling . . . by presid[t]
For the day tomorrow Capt[n]. Harleston
Town Guard Lieu[t]. from 5th reg:
Magazine L[t]. Bush ~ Brick house L[t]. P Gray

---

[742] Stephen Strecham (or Streetham) served in the 2nd South Carolina Regiment under Captain John Blake. He received 50 lashes on 29 October 1777 for stealing potatoes. He received 100 lashes with a cat-o-nine tails on 4 December 1777 for being absent without leave.

[743] Samuel Horn was in the 2nd South Carolina Regiment. On 27 October 1777 he received 50 lashes for stealing potatoes. In 1778 he was under Captain Blake. In 1779 he was under Captain Richard Baker.

[744] John Castile enlisted in the 2nd South Carolina Regiment on 14 February 1777. He received 50 lashes on 29 October 1777 for stealing potatoes. He deserted on 1 June 1778.

[745] Stephen Fillery enlisted in the 2nd South Carolina Regiment on 20 November 1775. On 29 October 1777 he was reprimanded for fighting with a "bad" woman.

[746] Barnaby Bryan enlisted in the 2nd South Carolina Regiment on 4 November 1775.

Lt. Colo. Yates being Appointed deputy Muster master General in the Southern Department is to be respected and Obeyd Accordely [747]~ Willm. Massey Esqr is Appointed Deputy deputy muster master of the Continental troops in S^{o}. Carolina and is to be respected and Obey'd Accordingly

Brigade Ord^{s} } The Adj^{ts}. Of the 2nd & 5th reg^{ts}. are to report to the Brigade Maj^{r}. all off^{rs}.
Col^{o} Huger } who may Obtain leave of Absents or Sent out on Command also to inform him of their Return

Reg^{tl}. Ord^{rs}. } a reg: court martial to set this morning at 11 OC: to try such prisoners as shall be brought before them evidences to Attend in time
Captn Mazyck presid^{t}. – L^{ts}. Warley & Bush members

Col^{o} Motte **Ch^{s} town 1777**
Orders 30th Oct. } . . Parole Mifflin . . . . GH
by G. Howe } For the day tomorrow Captn Ashby
Town Guard Lieu^{t}. Galvan
Magazine L^{t}. Hart ~ Brick house L^{t}. from 5th reg:

Reg^{tl}. Ord^{rs}. } Order'd that the Commanding Officers of each compy. do give in a monthly
Col^{o} Motte } return of their respective companys to the Adjutants by tomorrow forenoon
~ the reg^{t}. to go through their firings with Cartrid^{gs}. Every Munday & Thursday till further Orders the Q. Master to deliver as many Cartridges as L^{t}. Col^{o}. Marion may Order [748]

After Orders

M^{r}. Samuel Guerrey is Appointed a Second Lieu^{t}. in the Regiment he is to be respected & obeyd as such ~ Lieu^{t}: Guerrey is to Join and do duty in Captn. Jacob Shubricks Company till further Orders

NB by sentence of last Court 24th Inst: Conrad Fetner rec^{d} 100 lashes; Barnaby Bryan was Acquitted for Suspicion of theft – – –

Orders 31st Oct. } . . Parole Stark . . . by presid^{t}
by G Howe } For the day tomorrow Captn. Cogdell
Town Guard Lieu^{t}. from 5th reg:
Magazine L^{t} ~ Brick house L^{t}.

Reg^{tl}. Ord^{rs} } Sergt. Coffer of Captn Blakes Comp^{y}. is Appointed Serg^{t}: Major to the Reg^{t}.
Col^{o} Motte } ~ Orderd to be respected & Obey'd as Such – – – –

---

747 William Yates served as deputy muster-master general of the Southern Department from 1777 to 1780. During 1782 he was a private in the militia. He died on 2 December 1789.

748 There are some historians who write that the soldiers of the time did not practice marksmanship, and that the soldiers did not aim at the enemy. Supposedly they merely pointed the weapons in the direction and fired. This is contrary to all the evidence in the manuals and in the orders, such as this one. Even though soldiers practiced, they did not kill or wound many of the enemy. This is because the military musket was a smoothbore weapon, and only had an effective range of around 100 yards. Major George Hanger wrote that "A soldier's musket, if not exceedingly ill bored…will strike the figure of a man at eighty yards; it may even at 100; but a soldier must be very fortunate indeed who shall be wounded…at 150 yards, provided his antagonist aims at him…I do maintain…that no man was ever killed at 200 yards, by a common soldier's musket, by the person who aimed at him."

A reg^tl: court martial court martial to set this morning at 10 OC: to try such prisoners as shall be brought before them; - the prisoners to be informed that their trial is to come on at that hour, & the Witness to be Orderd to Attend in time ~ Captn.

Orders 1st Novr by G Howe } . . Parole Habersham . . . GH
For the day tomorrow Captn Blake
Town Guard Lieu^t. Perreneau
Magazine L^t. Proveneux ~ Brick house L^t. from 5th reg:

The General having information that some wood for the use of the troops is Arrived in town & cannot be got to the Barracks for want of Waggons, the Q. M. Genl. is immediately to direct the Wagon Mast^r. Genl. to procure waggons to convey wood where necessary, and as the Army is Greatly Suffering for want of that Article, he is in this case of Necessity permitted to give more than the usual hire if they cannot be other ways procured

Reg^tl. Ord^rs. Col^o Motte } The reg^t. is not to turn out to Exercise this Saturday Aft^r. n^n. till further Ord^rs. the men to be employ'd on those days in puting their Arms in good Order, Mending their Cloaths & accoutrements & to have their Linin washed. Corporal Coleman of Captn. Blakes Comp^y. is Appointed a Serg^t. in the same & to be Obey'd as such – [749]

Col^o Motte **Ch^s town 1777**

Orders 2nd Nov. by G Howe } . . Parole Laurence. . by presid^t
For the day tomorrow Captn: Potts
Town Guard Lieu^t. from 5th reg:
Magazine L^t. Bush ~ Brick house L^t. Warley

NB By sentence of reg: court 31st Oct: Walter long [750] to receive 100 lashes for Absents wth out Leave pardoned ~ David Wiley for the same pardond ~ Theod^r. Brewer for neglect of duty Reprimanded [751]~ Jeremiah McGinnes for losing his pouch, to have his pay stopt to replace it [752]~ Thos. Welch for beating Phillips rec^d. 50 lashes [753]

---

749 Jacob Coleman enlisted in the 2nd South Carolina Regiment on 4 November 1775. He was promoted to sergeant on 1 November 1777 in Captain Blake's company. He was reprimanded on 1 December 1777 for neglect of duty. After the fall of Charleston he served 252 days as a sergeant under Colonel Hugh Horry in General Francis Marion's partisans.

750 Walter Long enlisted in the 2nd South Carolina Regiment on 4 November 1777. He was supposed to receive 100 lashes on 2 November 1777 for being absent without leave, but he was pardoned. On 13 July 1778 he was sentenced to receive 100 lashes for theft.

751 Theodosius (Thaddeus) Brewer enlisted in the 2nd South Carolina Regiment on 4 November 1775. He was reprimanded on 2 November 1777 for neglect of duty.

752 This is a cartridge box. In 18th century terms a cartridge "box" was the style that was worn around the waist on a belt. The cartridge "pouch" was the style worn over the shoulder. The 2nd South Carolina Regiment wore the style that went over the shoulder.

753 Richard Phillips enlisted in the 2nd South Carolina Regiment on 4 November 1775. He was in Captain Hill's company in 1779. On 6 March 1779 he was sentenced to receive 25 lashes for allowing an apron on one of the cannons to be stole while he was on guard. The apron was a piece of sheet lead used to cover the vent or touchhole of a cannon to protect against the elements. The soldiers would steal this to make shot, or possibly to sell at Haddrell's Point.

Orders Nov[r]. 3[d] by G Howe } . . Parole Walton . . . . GH
For the day tomorrow Capt[n]. Lesesne
Town Guard Lieu[t]. from the 5th reg:
Magazine L[t]. Hart ~ Brick house L[t]. P. Gray
Guard at Pritchards L[t]. Keith 5th reg:

Reg[tl]. Ord[rs] Col[o] Motte } A court martial to set this morning at 10 OC: to try such prisoner as may be brought to it, the prisoners to be Acquainted that their trial is to come on at that hour & the evidences ord[rd]. to Attend –
Capt[n]. Motte presid[t]. _ Lieu[ts]. Burke, Hart & Galvan members

---

Orders 4th Nov[r]. by G. Howe } . . Parole Putnam . . . .
For the day tomorrow Capt[n]. Jervey
Town Guard Lieu[t]. Galvan
Magazine L[t]. fr: 5th reg. ~ Brick house fr: 5th reg:

Reg[tl]. Ord[rs] Col[o] Motte } The Reg[t]. will be Musterd next payday After noon at 4 OC: Orderd that Command[g] Off[rs]. of Comp[ys]. have their muster Rolls ready agreeable to a form in the hands of Majr Dilliant who will furnish them with a Copy of the same

NB Sentence of last court Sam[l]. Henderson for Absents without leave rec[d]. 50 lash
Serg[t]. Henderson, Corp[l]. Roberts [754] & Jn[o] M.Cade was reprimanded

---

Orders 5th Nov[r]. by G Howe } . . Parole Congress
For the day tomorrow Capt[n]. Mazyck
Town Guard Lieu[t].
Brick house L[t]. ~ Magazine L[t].

Col[o] Drayton is Appointed to do duty of D.A. Gen[l]. during the Absents of Col[o]. Eveleigh who is absent upon leave

The General thinks proper to publish the follow Resolutions of ~~Cong~~ the Honourable the Continental Congress

In Congress Resolved that as several mitigating Circumstances appear in favor of Tho[s]. M[c]Kann Condemned by a General Court martial held in Charlestown, S[o] Carolina to be shot for desertion Brigadier Gen[l]. Howe Commanding the Continental forces in S[o]. Carolina be empowered to grant a free pardon to the Criminal if he shall thing such a step Conducive to the good of the Service & to the public wellfare, or order Execution if he think proper.[755]

---

754 Stephen Roberts enlisted in the 2nd South Carolina Regiment on 1 May 1777 and re-enlisted on 1 November 1779 under Captain Thomas Moultrie. He became a corporal on 6 May 1777 in the Grenadier company. He was reprimanded for being absent without leave on 4 November 1777. He was promoted to sergeant on 27 June 1778 in Captain Moultrie's company. He transferred to the 1st Vacant Company commanded by Lieutenant Richard Bohun Baker on 13 July 1778. He returned to Captain Moultrie's company and was at the siege of Savannah.

755 Soldiers were rarely given a sentence of death, and this is one of the few cases that are mentioned in the orderly book. In about a quarter of the cases the soldier did not actually get executed, but was pardoned. According to *Summer Soldiers a Survey & Index of Revolutionary War Courts-Martial* by James Neagles, there were 176 soldiers sentenced to death on the Patriot side during the war. Out of these 39 were pardoned.

Col° Motte **Ch[s] town 1777**

Resolved that a General Officer commanding in a separate department be empowered to grant pardons to or Order Execution of prisoners Condemned to suffer Death by Gen[l]. Court martials, without being Obliged to report the matter to Congress or the Commander in Chief – Copy of the Journals signed W[m]. C. Houston D. S.

In consequence of the first of these resolutions General Howe extends pardon to Tho[s] M[c]Kann, firmly persuaded that his reflections during his long Confinement have brought him to a proper sense of his Guilt, & that the Leniency showd him will Induce him by future good Conduct to Attone for it.

By the second resolve the General has it in his power to Order the immediate execution Tho[s]. Corker [756] & Rob[t]. Potts Sentenced by a General court martial to be shot for Desertion, and the Hainous crimes for which these criminals had been Condemned gives Justice the strongest claim to that life they have forfeited But in consideration that they have been the first Offenders Sentenced to death since he had been invested with power to grant pardons or order execution without reference to Congress of the Commander in chief, & Persuaded however profligate and they may have been, the dreadfull situation to which they reduced themselves must have awakened Sufficient honor and remorse in their minds not only to guard them against a repetition of that offence for which they stand condemned to death but also do induce them be every effort in their power to attone for past guilt, the General consents to pardon them upon Condition they Enlist for the war in those Battalions to which they belong, which will give them an Opportunity by faithfull futur service to show their gratitude for the undeserved mercy now Extended them, the Order of this day will Demonstrate to the Army how reluctant the General is to Punish he Owes it to service to Confess that in the undue Lenity he has shown he has perhaps has paid Attention to mercy at the expence of Justice, it will however vindicate the unalterable resolution he has fixed of Carrying into execution for the futer & that in the most rigid manner every sentence of a Court martial for Desertion, And he directs Officers of every rank in the Army to take the most effectual methods of making it known to the men and enjoins them to believe that no futer Deserter is to expect pardon ~ Commanding Officers of Brigades to which the Condemned prisoners belong will Appoint proper persons to make known to them the pardon extended them & to see that those to whom it is Conditionally granted to Comply with the Conditions after which they are to be deliver'd to the Order of the Commanding Officer of their several Battalions

---

Orders 6th Nov[r].
by G. Howe }

. . Parole Midleton . . .
For the day tomorrow Capt[n]. Harleston
Town Guard Lieu[t]. Warley
Magazine L[t]. Proveaux ~ Brick house from 5th reg:

Thomas McCann, Thomas Corker, and Robert Potts were all supposed to be shot, but were pardoned by General Howe.

[756] Thomas Corker was in the 2nd South Carolina Regiment. He was He was sentenced to be shot for desertion on 5 November 1777, but was pardoned by General Howe.

Reg^{tl} Ord^{rs} Col^{o} Motte } Orderd that in the Absence of the field Officers the regiments is to be Exercised by the eldest Off^{r}. in Rank on the 343arade Agreeable to a former Order and no one to take upon him to discharge the men without permission of the Commanding Off^{r}. till that duty is done unless the weather is bad

Col^{o} Motte **Ch^{s} town 1777**

The Officers are desired to give an Account to the Q. Master of wood they have received from the Barrack master and that to be done today.

Such of the Officers as have been on the recruiting Service to call on the Col^{o}. this forenoon to Settle their Accounts

A regimental Court martial to set this morning at 10 OC: to try such prisoners as shall be brought before them _ the prisoners to be Acquainted that their trial is to come on at that hour, & the Witnesses to be Ordered to Attend in time

Captn. Mazyck president Lieu^{ts}. Mason & Dunbar members –

Aft^{r}. Ord^{rs}.

The muster of the regimt^{s}. is put off till next munday at 4 OC: in the after noon, to which time the muster roll is to be made up

NB. Robt. Potts Sentenced to death by a G.C. Martial was discharged on Condition he Enlist for the war, to which he has agreed –

Orders 7th Novr. by G Howe } . . Parole Gates. . .
For the day tomorrow Captn. Ashby
Town Guard Lieu^{t}. Warley
Magazine from 5th reg: ~ Brickhouse L^{t}. Bush

Reg^{tl} Ord^{rs} Col^{o} Motte } A court martial to set this morning at 10 OC: to try such prisoners as shall be brought to it the Evidences to attend & the prisoners to be made acquainted their Tryal will be on that hour

Captn Harliston presid^{t}. Lieu^{ts} Hart & Galvan members

NB by sentence of the Court 6th Inst: Pet^{r} Fagan for Absents 2 day & nights rec^{d} 100 lash Jno Alwell for neglect of duty rec^{d}. 25 lashes ~ Corp^{l}. Keels & W^{m}. Hasman for Absents from Guard was reprimanded[757]

Orders 8th Novr by G Howe } . . Parole Drayton . . . .
For the day tomorrow Captn Blake
Town Guard Lieu^{t}.
Magazine L^{t}. ~ Brick house L^{t}.

Reg^{tl} Ord^{rs}. Col^{o} Motte } A Reg^{tl}. Court martial to set this morning at 10 OC: to try such prisoners as shall be brought before them, the prisoners to be Informed that their trial will come on at that hour & the Evidences to be Orderd to attend in time

---

757 William Hasman (Hasemon) enlisted in the 2nd South Carolina Regiment in 1777. He was reprimanded on 7 November 1777 for being absent from guard. On 20 July 1778 he was appointed a corporal in Captain Harleston's company. He was promoted to sergeant on 21 January 1779 and transferred to Captain Proveaux's company. He was in Captain Hall's company at the siege of Savannah.

Captn. Ashby presidt. Lieu^{ts}. H Gray & Martin members
NB by sentence of last court Jno Davis to be und^{r} Stoppage for 8 pounds.

Orders 9th Nov^{r}.
by G Howe }

. . Parole America . . . .
For the day tomorrow Captn Conyers
Town Guard Lieu^{t}. Perreneau
Magazine L^{t}. from 5th reg: _ Brick house L^{t}. H. Gray

All the Continental Officers in town off duty & as many at Forts Moultrie & Johnston that can be spared are requested to attend the funeral of Captn Rich^{d}. Shubrick tomorrow morning at 8 OC: from the house of Thomas Shubrick Esqr.[758]

Col^{o} Motte **Ch^{s} town 1777**

Reg^{tl} Ord^{rs}
Col^{o} Motte }

A party of one Captn. 2 Subalterns 2 Serg^{ts}. 2 drums 2 fifes and 40 rank & file to parrade at 8 OC: tomorrow morning to attend the funeral of the late Captn Richard Shubrick from the house of M^{r}. Thos. Shubrick, this party is to consist of as many Granadiers as are fit for duty & to be Completed from the Other Companies, all the Officers are desired to follow the Corps as mourners

The Commd^{g}. Off^{rs}. of Comp^{ys}. are to be very carefull that their must^{r}: rolls are ready in time & to have made out five Copies ~ the Regim^{t} to be under Arms in the Barrack Gard by three OC: tomorrow After noon to be mustered and the Off^{rs}. are desired to see that every man Appear that is able except such as are on Duty

---

Orders 10th Nov^{r}
by G Howe }

. . Parole Walton . . .
For the day tomorrow Captn Lesesne
Town Guard Lieu^{t}. from 5th reg:
Magazine L^{t}. Proveneaux ~ Brick house L^{t}. from 5th reg^{t}.
Pritchards Guard L^{t}. Martin

RO.

NB by sentence of reg^{tl} court martial the 9th Inst: Jno Baptist of Captn Jacob Shubrick Comp^{y}. for losing his coat to have his pay stopt to replace it; [759] Jeremiah Hill [760] & Serg^{t}.Rybould for neglect of duty mult 14 days pay each ~ Cor^{p}: Kersley for d^{o}. reducd to the ranks & mult

---

[758] Thomas Shubrick was the father of Richard, Jacob and Thomas, Jr. He was born in 1710. He settled in Charleston in 1739 and was a member of the merchant firm of Nickleson, Shubrick & Company. He commanded a merchantman until the 1740s, when he became a "wealthy and eiminent merchant". He mainly dealt with agriculture and forest products. He owned four trading vessels and was the co-owner of eight others. He had plantations on the Cooper River. When he died in August 1779 he owned 333 slaves.

[759] John Baptist enlisted in the 2nd South Carolina Regiment on 22 September 1777 under Captain Thomas Shubrick. On 10 November 1777 he had his pay stopped until he replaced his regimental coat. He deserted on 17 November 1777.

[760] Jeremiah Hill enlisted in the 2nd South Carolina Regiment on 4 November 1775 as a corporal. On 10 November 1777 he was fined 14 days pay for neglect of duty. He was demoted to private on 11 December 1777 for "scandalous and infamous behavior." He received 50 lashes on 20 February 1778 for selling rum contrary to orders.

14 days pay [761]~ Ch[s]. Lucas for Absents without leave was to receive 50 lashes but pardoned with a reprimand

---

Orders 11th Nov[r]. by G Howe } . Parole Gates . . . . GH
For the day tomorrow Capt[n]. Mazyck
Town Guard Lieu[t]. Warley
Magazine L[t]. from 5th reg[t]. ~ Brick house L[t] Mason

Reg[tl]. Ord[rs]. Col[o]. Motte } Orderd that all the muster Rolls of the different Comp[ys]. be sign'd Sworn to and Deliver'd to the must[r]. Mast[r].Gen[l]. today, and that the Command[g]:Off[rs]. of Comp[ys] enter an Exact copy of their Companies M. Roll into a book to be kept for that purpose

The Articles of war to be read this After noon at the head of the regim[t]. by Lieu[t]. Col[o]. Marion

A reg[tl] court to set this morning at 10 OC: to try such prisoners as shall be brought to it. The prisoners to be informed that their trial is to come on at that hour & the evidences to be Orderd to Attend in time

Capt[n]. Motte presidt ~ Lieu[ts] Mason & Bush members

Aft[r]. Ord[rs]. Lieu[t]. Dunbar is appointed a Capt[n]. & 2[d] Lieu[t]. Perreneau a 1st. Lieu[t]. in the regiment they are to be Obey'd & respected as such Capt[n]. Dunbar is to take the Command of the Granidiers Company & Lieu[t]. Perreneau to do duty in it till further Orders

Col[o] Motte **Ch[s] town 1777 –**

Orders 12th Nov[r]. by G Howe . . } . . Parole Philadelphia . . . by presid[t].
For the day tomorrow Capt[n]. Harleston
Town Guard Lieu[t]. from 5th reg[t].
Magazine L[t]. Bush ~ Brickhouse L[t]. P. Gray

The Q[tr] M[tr]. Gen[l]. will purchase a sufficient number of boxes for the use of the Guards, Complaints having been made to the General that some of the Soldiers have been pulling down the Inclosures of the new personage house in this town and also they have done Great Damage to the fencing & Inclosures of private persons, The General in the strictest manner forbids a repetition of such Injuries upon pain of the Severest punishment, he Calls upon the Officers of the line not only to exert themselves to the utmost to prevent futer enormities of this kind which disgrace an Army & render it hatefull but also to detest the persons if possible whose Transgressions have Occasion this order ~ Commanding Officers of Companies are required to have all Orders relative to the conduct of the Soldiers read at the head of their respective Companies & to be particularly Attentive that they may be made Acquainted with the Orders of this day.

Reg[tl]. Ord[rs]. Col[o]. Motte } Capt[n] Charnocks company to be under Arms in the barrack yard half after 10 OC: this morning

---

[761] Abraham Kearslick (Kearsley) served in Captain Peter Gray's company in the 2nd South Carolina Regiment. He was demoted to private on 10 November 1777 for neglect of duty, and fined 14 days pay. He was court martialed for drunkenness on 26 March 1779 and placed on probation. He was promoted to sergeant on 2 October 1779 and served under Captain Hall. After the war he moved to Alabama.

Orders 13th Novr by G Howe – } . . Parole Middlton
For the day tomorrow Captn Motte
Town Guard Lieut. from 5th reg:
Magazine Lt. Perreneau ~ Brickhouse Lt. H. Gray

Regtl. Ordrs Colo Motte } The Genl. Orders of yesterday to be read this & tomorrow After noon to the men by the Commanding Officer of each Compy.

A regtl. Court martial to set this morning at 11 OC: to try such prisoners as shall be brought to it, the prisoners to be informed that their trial will come on at that hour and the evidences to be Orderd to Attend in time

Captn. Harleston president Lieuts. Burke & Perreneau members _ _ _

Orders 14th Novr by G Howe } . . Parole Dicipline . . . GH.
For the day tomorrow Captn. Ashby
Town Guard Lieut. Capers
Magazine Lt. Proveaux ~ Brick house Lt from 5th reg:

The General Observing that not withstanding the repeated Orders to the contrary, Soldiers are sent upon duty with long beards, uncombed hairs & flopped hats, he therefore cannot forbear expressing his surprise that Offrs of Companies are not more Attentive to their men, if they did not suffer the men to appear in this unsoldierly manner upon parrade this Cause of Complaint Could not happen, and surely boath duty & a regard to Decentsy exact of them to prevent it, he has been led by the Appearance of the men, to Inquire and by that Inquiry has learned that Officers of Compy are not allways punctual at roll calls in the morning which he is concerned to hear, from whose this neglect of the men may in some measure proved, he therefore thinks proper to reinforce those Regimental Orders, which he has reason to think have been Ignored – enjoining one Commission Officer of each Company to Attend parrade every morning at Roll call, by Declaring he will not dispence with a neglect of this duty, and directs them to be particularly carefull that the men Appear as decent as possible Especially those for duty.

NB by Sentence of last reg: court Geo: Hughes for disobedience of Orders recd. 50 lashes [762]
Hugh Darbary for Absents 6 Mo. without leave 100 lashes 50 remitted[763]

Orders 15th Novr by G Howe . . } . . Parole Adams . . . by presidt.
For the day tomorrow Captn. Blake
Town Guard Lieut. Bush
Magazine Lt. from 5th reg: - Brick house Lt. P. Gray

---

[762] George Hughes enlisted in the 2nd South Carolina Regiment on 9 November 1776. He received 50 lashes on 15 November 1777 for disobedience of orders. In 1779 he was in Captain Proveaux's company. In 1781 he was a private and a sergeant in the partisans in the light dragoons under Captain William Smith, Lieutenant Colonel John Thomas, Jr. and General Sumter.

[763] Hugh Durborough.

Reg^{tl}. Ord^{rs}. Col^{o}. Motte } A court martial to set this morning at 10 OC: to try such prisoners as shall be brought before it, the prisoners to be informed that their trial will come on at that hour and the Evidences to be ordered to Attend

Captn. Ashby president ~ Lieu^{ts}. H^{y}. Gray & Galvan members

---

Orders 16th Novr. by G. Howe } . . Parole Arnold . . . . GH

For the day tomorrow Captn. Conyers

Town Guard Lieu^{t}. Burke

Magazine L^{t}. Hart ~ Brick house L^{t}. from 5th reg:

Reg^{tl}. Ord^{rs}. Col^{o}. Motte } Divine service as usual this After noon

Orderd that the Quarter master see the wood Sent for the regim^{t}. by the Barrack master be properly Distributed Among the Different Companies & that no no man take any away till that distribution is made & this Order to be made known to the men by an Off^{r}. of each Comp^{y}.

Orders 17th Novr by G. Howe } . . Parole Young . . . by presid^{t}.

For the day tomorrow Captn. Lisesne

Town Guard Lieu^{t}. Perreneau

Magazine L^{t}. from 5th reg^{t}. ~ Brick house L^{t}. H^{y}. Gray

Pritchards Lieu^{t}. Proveaux

Reg^{t}. Ord^{r} Col^{o}. Motte } all the Officers off duty to be on the Parrade this After noon at 4 OC: A reg^{t}. Court Martial to set this morning at 10 OC: to try such prisoners as shall be brought befor it, the prisoners to be informed their trial will come in at that hour & the Evidences to be Orderd to Attend

---

Orders 18th Novr. by G. Howe } . . Parole Hazelwood

For the day tomorrow Captn.

Town Guard Lieu^{t}.

Magazine L^{t}. ~ Brick house L^{t}

Col^{o} Motte **Ch^{s} town 1777 –**

G.O. Once commissioned Officer of a Comp^{y}. is for the future to sleep in the Barracks & is to be Attentive as possible and keeping the men of his Company at night in the Barracks, No Soldier is to be out of the Barracks after tattoo beating without a Wrighten leave from the Command^{g} Off^{r}. of his Comp^{y}. A transgression of this Order will certainly meet with punishment – Commanding Off^{rs} of Comp^{ys} are directed to be carefull in having this Order made known to their men It has allready been Orderd & particular Compliance with it expected that one Commissioned Off^{r}. is Allways to be present at morning Roll-call – the General cannot suppose that Any neglect of this Order will ever happen, if however Contrary to probability & his expectations it Should Occur, Adjutants of Battalions are to report to head Quarters the Comp^{y}. whose Off^{r}. are deficient, the General Laments the decline of Dicipline but too visible in that Diversion of the Army in town, he is Obliged to those Officers who he has Observe courting themselves to prevent it And Calls upon Officers of every degree to Second him in those Efforts which he is determine to make, to restore and improve dicipline

---

Orders 19th Novr by G Howe } . . Parole Lee . . . by GH.
For the day tomorrow Captn. Dunbar
Town Guard Lieut. Mason
Magazine Lt. Bush ~ Brick house Lt. from 5th Regt

Regtl Orders Col° Motte } A court martial to set this ~~Afternoon~~ Morning at 10 OC: to try such prisoners as shall be brought before it, the prisoners to have notice given them that their trial will Come on at that hour. Evidences to be Orderd to Attend
President Captn. Jacob Shubrick ~ Lieuts Mason & Bush members ~

The major is desired to fix upon five rooms in the barracks (or more if they can be spared) for the reception of an Officer of a Company Agreeable to General Orders of Yesterday & report the same as soon as he has done it, to the Col°.
Order'd that Lt. Martin take Charge of Captn. Moultries Compy. till Lt. Proveaux returns to town

---

Orders 20th Novr by G Howe } . Parole Weedon . . .
For the day tomorrow Captn Harleston
Town Guard Lt. from 5th reg:
Magazine Lt. from d°. – Brick house Lt. P. Gray

Regt. Ordr. Col° Motte } a court martial to sit this fore noon at 10 OC: to try such prisoners as shall be brought to it, the prisoners to be made Acquainted their tryal will Come on at that hour & all evidences to attend

Captn. Dunbar presidt. Lieuts. P. Gray & Burke members
Mr. John Perreneau is Appointed a Second Lieut. in the Regt. he is to be Obey'd & respected as such
[764]– Lieut John Perreneau is to Join Captn. Charnock Compy. & do duty in the same till further Orders

After Genl. Ordrs. by G Howe } Necessary returns are to be made by Commanding Officers of Battalions & Corps to head Quarters as immediately as possible specifying the necessaries which the men are furnished & also those that are wanting

Col° Motte **Chs town 1777 –**

Brigade Ordrs by Col° Huger } one Sargent & 10 privates from the 2nd Regt. with one Corporal & 4 privates from the 5th reg: to hold themselves in readyness by 9 Oclock to morrow morn: with 6 rounds pr. Man & 8 days provisions ~ the Sergt. for this Command will apply to Col° Is: Huger for his Orders

---

Orders 21st Novr by G Howe . . } . . Parole Gates
For the day tomorrow Captn Motte
Town Guard Lieut. from 5th reg:
Magazine Lt. Hart ~ Brick house Lt H Perreneau

Lieut. Dunbar is Appointed Captain in the room of Captn. R. Shubrick deceased

---

[764] John Perronneau was commissioned a second lieutenant in the 2nd South Carolina Regiment on 20 November 1777 in Captain Charnock's company. He resigned on 12 December 1777. He served in the Charlestown Militia Regiment under Captain James Bentham in 1778.

2nd Lieut. H Perreneau a first Lieu^{t}. in room of L^{t}. Dunbar prom: & M^{r}. John Perreneau is appointed a Second Lieu^{t}. in Col^{o}. Mottes Battalion they are to be respected and Obey'd Accordingly

Commanding Officers of Comp^{ys}. to Give a Copy of their Last muster roll to the Adjutant by tomorrow morning, also to make a return as soon as possible of what necessaries they have received & what are wanting Such as caps, shirts, Coats, waistcoats, Breeches, Spatterdashers,[765] Shoes, Blankets, Axes, Knapsacks,[766] havresacks,[767] Camp Kettles,[768] Arms & Accoutrements

A reg^{tl} Court martial to set this morning 10 OC: to try all prisoners as shall be brought to it, Evidences to be Order to Attend ~ Captn. Harleston presid^{t}. Lieu^{ts}. Galvan & Hart members

---

Orders 22nd Novr by G Howe } . . Parole Arnold . . .
For the day tomorrow Captn Ashby
Town Guard Lieu^{t}. Warley
Magazine L^{t}. from 5th reg: - Brick house L^{t}. Capers

Reg^{t}. Ord^{r}. Col^{o} Motte } Orderd that the Q. Master Supply the Officers belonging to the Regim^{t}. (and to none others) with the Blue Cloath rec^{d}. out of the publick store at the rate of £8.10p per yard

An officer of a Company to move into the regimental Barracks today Agreeable to the General Order of the 18th Nov^{r}.

Orders 23^{d} Novr by G. Howe } . . Parole Thompson
For the day tomorrow Captn Blake
Town Guard L^{t}. from 5th reg:
Magazine L^{t}. from d^{o}. – Brick house L^{t}. Mason
The Command at Dorchester L^{t}. Martin

Reg^{tl} Ord^{rs} Col^{o} Motte } – Divine service this Afternoon as usual Officers & Soldiers to Attend

---

Orders 24th Novr. by G Howe . . } . . Parole Lincoln . . . .
For the day tomorrow Captn Lesesne
Town Guard Lieu^{t} Bush
Magazine L^{t}. 5th reg: ~ Brick house L^{t} ~~from 5th reg:~~ Pet^{r} Gray
Pritchards L^{t}. from 5th reg:

Reg^{tl} Ord^{rs} L^{t}. Col^{o} Marion } a court martial to sett this morning at 10 OC: to try such prisoners as shall be brought before it, the prisoners to be informed by the Serg^{t}. Maj^{r}. that their trial is to come on at that hour, the Evidences to be orderd to Attend in time

---

765 Spatterdashers were gaiters that only came up to the mid-calf, covering the shoe.

766 Knapsacks were the 18th century equivalent of a rucksack. They were made of canvas and were plain or painted for waterproofing or covered in goat fur. The 2nd South Carolina most likely had knapsacks without the goat fur.

767 Haversacks were bags made of linen that were carried over the shoulder and contained the rations. Men would not normally carry these, but would be issued the haversack for campaigns or extended duty, already filled with rations per man.

768 The iron kettles that were carried by the regiment were not cast iron kettles, but were sheet iron and were tinned to prevent rusting. These kettles resembled a bucket, more than the three legged iron kettle that most people imagine to be carried.

L[t]. Col° Marion **Ch[s] town 1777**

For the Court martial Capt[n]. Charnock president Lieu[t]. Roux and P. Gray members.
NB. by sentence of last court W[m]. Simpson for disobedience of Orders rec[d]. 75 lashes, Conrad Fitner for misbehavour rec[d]. 50 lashes Jn°. Chaves for misbehavour to L[t]. Galvan is pardon [769] – Anth[y]. Hinds for disobedience of Orders reprimanded[770]

Orders 25th Nov. by G Howe } . . Parole Caswell . . . .
For the day tomorrow Capt[n]. Jervey
Town guard L[t]. from 5th reg:
Magazine L[t]. Hart ~~ Brick house L[t]. H[y]. Gray

M[r]. James M[c]Kenney is Appointed Second Lieu[t]. in Col° Hugers Battalion and is to be Obeyd & respected Accordingly [771]

Reg[tl]. Ord[rs] L[t]. Col°. Marion } Notwithstanding repeated Orders to the contrary, the Soldiers still Sleep out of Barracks, many of which are reported sick & are in town where the Surgion of the Reg[t]. cannot find them ~ Command[g]. Off[rs]. Of Comp[ys] are Desired to be particular in Obliging their men to sleep in the Barracks, & to send their Sergeants for all their sick now in town

It is hop'd the Officers of the regim[t]. will pay Attention to the Gen[l]. Orders of the 18th Inst: & Regimental the 22nd. –

The L[t]. Col°. is extremely sorry to see many Gentlemen neglect attending Exercise in the After noon, Contrary to Orders and the good of their Country which they have undertaken, which if perform'd with punctually & forthfullness will Rebound their honour and deserve the thanks of their Country

---

Orders 26th Nov[r]. by G Howe } . . Parole Saratoga . . .
For the day tomorrow Capt[n]. Shakelford
Town Guard Lieu[t]. from 5th reg:
Magazine L[t]. Capers ~ Brick house L[t]. from 5th reg:

Reg[tl]. Orders L[t]. Col°. Marion } As the L[t]. Col°. is not Acquainted what Officers of the regiment have Leave of Absents from Col°. Motte, all such are to give the Major the time had leave of Absents (such only who reside in Town) this to be done by Thursday 10 OC: in the fore noon the 28th Inst: _ the Maj[r]. will give a return of these to the Lieu[t]. Col°.

---

[769] John Chavis enlisted in the 2nd South Carolina Regiment on 7 October 1777. On 24 November 1777 He was supposed to receive 50 lashes for misbehavior to Lieutenant Galvan, but he was pardoned. He was reprimanded on 1 December 1777 for being absent from roll call. He received 50 lashes on 2 January 1778 for sleeping on his guard post. He was under Captain Lesesne in 1778. On 30 November 1778 he received 30 lashes on the bare back with a cat of nine tails for selling his regimental breeches. In 1779 he was in Captain Dunbar's Light Infantry at the siege of Savannah.

[770] Anthony Hinds served in the 2nd South Carolina Regiment after 4 November 1775. In 1778 he was under Captain Blake. After Blake resigned he was under Captain Richard Mason. He was discharged on 10 July 1778. He returned to Captain Mason's company and was at the siege of Savannah.

[771] James McKinney was commissioned as a 2nd lieutenant in the 5th South Carolina on 24 November 1777.

Orders 27th Novr. by G. Howe } . . Parole Buncombe .
For the day tomorrow Captn. Spencer
Town Guard L^t^. Proveaux
Magazine L^t^. Warley ~ Brick house L^t^. from 5th reg:

The Honourable the Continental Congress by Commission Bearing date the 27th. of Septemb^r^. Last, Appointed Francis Huger Esqr. Deputy Quarter master General for the State of S^o^. Carolina in the Army of the united States he is therefore to be Obey'd & respected Accordingly –

The General Consents that Sergeant Burgess be Issued in regimental Orders, as Quart^r^. Mast^r^. Serg^t^. in the quart^r^. Mast^r^. Genl. Department

Reg^tl^. Ord^rs^. L^t^. Col^o^. Marion } A court martial is to set tomorrow morning at 10 OC: to try Sergeant Coleman & several other prisoners now Confind, this Court to Consist of 1 Captn. as presid^t^. & 4 Subaltern Officers as members, Sergeant Coffin to warn the members today & Acquaint the prisoners that their trial will come on at the above hour, Also to Order all evidences to Attend

The above Court is to inquire in to the reason of Sergeant Marlow's over staying his time of furlough, which is from the 23^d^. of Septemb^r^. last to this day – Serg^t^. Marlow is Orderd to Attend the Court –

Captn. Mazyck president – Lieu^ts^. Burke, Hart, Galvan & Guerry members

---

Orders 28th Novr. by Gin^l^ Howe } . . Parole Congress
For the day tomorrow Captn Mazyck
Town guard Lieu^t^ from 5th reg:
Magazine L^t^. Bush ~ Brick house L^t^. P. Gray

Reg^t^. Ord^r^. L^t^. Col^o^. Marion } William Edmunds of Captn. Lesesne Comp^y^. & Robt. Mathews of Captn Shubricks Comp^y^. are Appointed Sergeants in the Regim^t^. they are to be Obey'd as such ~ Serg^t^. Edmunds is to Join and do duty in Captn. Moultries Company in the room of Sergeant Burgess who is Appointed by Genl. Howe, Quart. Master Sergeant to the D. Q. Mast^r^. Genl. Serg^t^. Mathews to do duty in Captn. J. Shubricks Comp^y^. till furth^r^. Orders

---

Orders 29th Novr. by G. Howe } . . Parole Congress. .
For the day tomorrow Captn Dunbar
Town Guard L^t^. Guerry
Magazine L^t^. from 5th reg: ~ Brick house Lt Mason

M^r^. Michal Keltigsen is Appointed Continental Waggon Master Genl. in the state of S^o^. Carolina & is to be respected Accordingly [772]

Commanding Officers of Battalions & Corps when vacancies happen by the Resignation or Otherwise, are for the future previous to any Appointments or Commission, to report to the General Commanding in Chief or in his Absents to the Officer Commanding the division of the Army to which they belong, by what means the vacancy happen & what Officer to Succeed to it in their respective Corps ~ The Deputy Adj: Genl. will Apply to his Excellency the president & request he

---

[772] Michael Kalteissen.

will be pleased to Order a report to be made of the pick Axes, spades & other Entrenching tools, which when Obtain'd the Q. M. Gen[l]. will bring to head Quarters

The Honourable the Continental Congress have Promoted Brigad[r]. General Howe to the Rank of Major General in the Army of the United States. Col[o]. Stephen Drayton & Capt[n]. John Fisheruard Grimkie are Appointed Aid de Camps to General Howe & are to be respected & Obey'd Accordingly[773]

Reg[tl] Ord[rs]. Col[o]. Marion } Orders has been Issued several times that the Officers orderly books will be Called for & Examined, It tis now Orderd that every Officers in the reg[t]. do Leave their Orderly books with the Major on the thirty first day of next December to be by him Examind, the Maj[r]. to report the state he finds them in ~ the Lieut. Col[o]. has put it to a distant largess that every Officer may have time to bring up their books, shoud any be Backward ~ As Keeping Orderly books is Absolutely necessary & is a part of an Officers duty the L[t]. Col[o]. cannot suppose that any Officer has not kept one, but shoud any be so negligent as not to have one they must expect to be taken notice of as a Disobeyer of Orders & Neglect of military Discipline

All Officers who have Attestation of the men they have Enlisted, to given them in to the L[t]. Col[o]. by munday next by 10 OC: in the forenoon

---

Orders 30th Nov[r]. by G. Howe } . . Parole Laurens .
For the tomorrow Capt[n]. Motte
Town guard Lieu[t]. Burke
Magazine L[t]. Guerry 2nd reg: _ Brick house L[t]. from 5th reg:

The 1st, 2nd, 4th and 5th Regiments are to be furnished with Camp Colours the grownd of the 1st is to be blew, the Insertion white, of the 2nd White the Insertion blew, the 4th red the Insertion yellow, the 5th yellow the Insertion Black, they are to be made in the Usual forms, the D. Q. Gen[l]. will provide the matireal and have them made as soon as possible, two to each Comp[y]. are requisite [774]

Reg[tl]. Ord[rs]. L[t]. Col[o]. Marion } The winter being very cold their will be no Service this Aft:noon at the Barracks

A monthly return to be made tomorrow morning & Deliverd Maj[r]. Horry by 11 Oclock precisely

For the futer no Soldier to be Allowed to stay on pritchards guard or at Dorchester longer than the usual time of relieving those Guards without a permission in wrighting from the Commanding Off[r]. of the regim[t].

---

[773] John Faucheraud Grimkè was born on 16 December 1752. On 29 November 1777 he became a major and aide-de-camp to General Howe and served until August 1778. When Colonel Eveleigh resigned, Major Grimkè was appointed Deputy Adjutant General for South Carolina and Georgia with the rank of Colonel. In 1779 he became a major in the 4th South Carolina Regiment (Artillery). On 20 June 1779 he became a lieutenant colonel. He was in the battles of Briar Creek, Savannah, Stono Ferry and the siege of Charleston. He was taken prisoner at the fall of Charleston. He died on 19 August 1819 at the age of 67.

[774] The colors and standards of a regiment averaged 6'6" x 6' and the poles for colors, standards and guidons was supposed to be 9'10". Camp colors were 18" square and the poles for camp colors are supposed to be 7'6" for those on the parade and 9' for those of the quarter and rear guards.

Monthly return for 1st Decem r 1777 –
1 Col o. 1 L t. Col o _ 1 Maj r. _ 10 Capt ns. _ 17 L ts. _ 1 QM _ 1 Chaplain _ 1 Surg n. 29 Serg t _ 17 Drum 2 Arm rs. _ 255 rank & file fit for duty _ 50 sick _ 15. on Com d _ 11 on furl: Total 881

Orders 1st Decem by G Howe } . . Parole Elliot . .
For the day tomorrow Capt n. Ashby
Town guard Lieu t. Perreneau
Magazine L t. from 5th reg: ~ Brickhouse L t. Hart
Pritchards L t. Warley

Reg t. Ord r L t. Col o. Marion } A reg: court martial to set tomorrow 10 OC. in the forenoon to try such prisoners as shall be brought to it, this Court to Consist of one Capt n. as president & 4 Subalterns as members, the Serg t. Major to Warn the members this day & give notice to the prisoners that their trial will come on at the above hour, also to Summon the evidences to Attend

One Subaltern of a Comp y. to Examine their mens rooms every Morning at roll-call & see what Damages are done to the floors &tc. And to bring all such men who may be found cutting the floor or otherway Damaging their Barracks (if the particular person who may do the damage cannot be found out all the men who are Quarterd in that room to be made Answerable for such Damages) and to be punish'd by a court martial

The Officers of the Company will be made Answerable for all Damages in their mens Quarters, that neglect bringing them to a court martial for such Damages

L t. Col o. Marion **Ch s. town 1777 –**

The Serg t. Maj r. to Examine the Regiment guard room every morning at relieving the guards & to Confine the Serg t. of that guard for all Damages done to the guard room while such Serg t. was on guard

NB Sentence of Last court Serg t. Coleman for neglect of duty reprimanded _ Bunker Tring for Absents from Guard rec d. 50 lashes & to be confine 3 days on bread & water [775]~ Jn o. Chavis for Absents from roll-call reprimanded ~ Nicholas Flinn for neglect of duty had 50 lashes remitted but Confind 2 days on bread & water ~

For the Court Martial Capt n. Ashby presid t ~ Lieu ts _Hall, Roux, H. Gray & Capers Members

Af t. Gen l. Ord rs G. Howe } as an Encampment will be formed & the 1st, 2nd, 4th, & 5th Regim ts. Assembled in Order to their being train'd together the field Officers of those Corps (as the D.A.Gen l. is out of town) are to meet at some convenient place as soon as possible & Establish a set of Manuevers (to be practiced by the whole, previous to their being called together, When done to be laid before the General for his Inspection

---

775 Bunker Thring.

Orders 2nd Decr by G. Howe } . . Parole Pinckney
For the day tomorrow Captn. Conyers
Town guard Lieut. Hall
Magazine Lt. from 5th reg: ~ Brickhouse Lt. from 5th regt.

Orders 3d Decr. by G. Howe } . . Parole Washington
For the day tomorrow Captn. from 5th reg:
Town guard Lt. H Gray
Magazine Lt. from 5th reg: ~ Brick house Lt. Capers

The following Order of his Excellency Genl. Washington has not till lately been Officially received, Genl. Howe Expects & is Determind to Exact their strictest Obedience to it from persons of every rank in that Division of the Army he has the Honour to command & he hopes the Salutary end it is Intended to answer will Induce all persons to Obey it without reluctance _ _

Head Quarters Morris town 8th May, Genl. Ordrs. by Genl. Washington

As few vice is attended with more pernicious consequence in Civil life so there an non more fatal in Military one than that of Gaming which often brings Disgrace & ruin upon Officers & Injury & punishment upon Soldiers, And reports prevailing which it is so to be feared are too well founded that this Destructive vice has spread its banefull influence in the Army and in a peculiar manner to the prejudice of the recruiting service The Commander in Chief in the most pointed & Explicit terms forbids all Officers & Soldiers playing at Cards, dice or at any Game except those of Exercise for Diversion, it being impossible if the practice be Allowed at all to Discriminate between Innocent play for Amusement and Criminal Gaming for pecuniary and sordid purposes, Officers Active to their duty will find Abundant Employment in Training & Diciplining their men providing for them & seeing that they ~~go~~ appear clean neat & Soldier like nor will any thing Redound more to their honour, afford them more solid Amusement or better answer the end of their Appointment than to Devote the Vacant moments they may to the study of military Authors, the Commanding Officers of every Corps is Strictly Enjoined to have their Orders frequently read & strongly Empressed upon the mind of the those under his Command, Any Office or Soldier or Other persons belonging to or following the Army Either in Camp or Quarters or the recruiting Service or elsewhere presuming under any pretense to Disobey their Order shall by try'd by a General Court Martial. The General Officers in each division of the Army to pay the strictest Attention to the Execution thereof, the Adjutant General is to transmit Copies of this Order to the Different Department of the Army also & Cause the same to be immediately published in the Gazette of Each state for the Information of Officers Dispersed in the Recruiting Service

By his Excellencys Command – Morgan Conners. A. G. [776] –

Regt. Ordrs. Lt. Colo. Marion } All the Officers off duty to be on the parrade tomorrow after noon at usual time of exercise to ballot for a Gentln. to fill up one of the Vacancies of the Regiment

---

[776] Morgan Connor was the acting Adjutant General of Washington's Grand Army in the north. He had been a major in the 1st Continental Regiment and then he was the lieutenant colonel in Hartley's Regiment on 9 April 1777. He was appointed the Adjutant General of the Continental on 19 April 1777 and continued to hold that position until Pickering was appointed in June 1777. He was the lieutenant colonel of the 7th Pennsylvania Regiment on 12 May 1779. He was granted a leave of absence on 2 December 1779, but he was lost at sea the next month in January 1780

Orders 4th Decem^r^ by G Howe – } . . Parole Heyward . . presid^t^.
For the day tomorrow Captn. Shakelford
Town guard Lieu^t^. from 5th reg:
Magazine L^t^. Mason ~ Brick house L^t^. Proveaux

As new Arms will shortly be served out to the Continental Regim^ts^. in the state, the Commanding Officers of each Corps are upon the receipt of them, to report to head Quarters the number they receive the number Distributed to each Company and the Names of the Officers to whom Deliverd ~ Officers of Company, are to make known to their men, that any Arms lost or Injured Otherways then in the Course of Service by Enevitable Accidents will be replaced or Repaired by Stoppages from their pay and that besides this they will most Certainly be punished, Off^rs^ of Comp^ys^ will be Attentive to Carry into execution their Order whenever necessary or they may depend upon being Themselves made Answerable for the loss & repair of Arms, when Commanding Off^rs^. of Companies are by promotion or otherwise removed from their Comp^y^. they are immediately to report to the Commanding Off^r^. of the Reg^t^. for the time being the Exact state of of the Arms when they were Appointed to when they left the company that Either they ~~or~~ the Off^r^. ~~Left~~ in charge of the Comp^y^. or the Off^r^. Succeeding to it may be made answerable shoud Occasion require it, Commanding Off^rs^ of regim^t^. for the time being are to take Certificates from the Off^rs^. of Companys for the Arms ~~for~~ which they have been served which with the reports to be made by those Orders are to be Carefully filed that they may be referred to Occasionally or if Enterd in a book kept for that purpose would be Better [777]

R.O NB

by sentence of Last court Joseph Reeves for Absents wth out Leave rec^d^ 100 lashes with Switches at 3 Different Days – Danl Spiller for d^o^. to rec^d^. 100 lashes wth Switches and to Enlist for the war without Bounty [778]– Stephen Streacham for the same rec^d^. 100 lashes wth a Catt – Thos. Hogatier for the same remitted 50 lashes[779]– Thos. Welsh for taking wood was mult 7 days pay

---

[777] After the Battles of Trenton and Saratoga France secretly allied themselves with the United States and began shipping what was needed most to the colonies in rebellion. The thing that was most needed were weapons and gunpowder. These items would be sent through "front" companies, such as Rodrigue Hortalex et Cie. The weapons and powder would come in crates or barrels marked as less lethal items. Normally a stand of muskets would consist of a musket, a bayonet, a cartridge box and a bayonet carriage. The weapon that was sent in the most quantities was the 1763 model French musket made at the Charleville or St. Etienne armory. Earlier models, such as the 1724 musket were also sent. The French muskets were .69 caliber, but the soldiers would fire a smaller .62 caliber ball. After the Battle of Brandywine Washington ordered that all cartridges be a mix of buck and ball. Three to five buckshot of .30 caliber would be placed in each round, so that when a soldier fired there would be multiple projectiles going towards the enemy.

[778] Daniel Spiller enlisted in the 2nd South Carolina Regiment on 10 July 1775 under Captain Bernard Elliot. He received 100 lashes with switches on 4 December 1777, and he had to re-enlist under Captain Blake for the duration of the war without receiving any bounty money. He deserted the next month on 1 January 1778.

[779] Thomas Hagartey served in the 2nd South Carolina Regiment after 17 November 1777. He received 100 lashes on 4 December 1777 for being absent without leave, but 50 were remitted. He was under Captain Richard Baker during 1779.

Orders 5th Dec
by G. Howe }

. . Parole Moultrie
For the day tomorrow Captn Spencer
Town guard Lieut. from 5th reg:
Magazine Lt. Bush ~ Brickhouse Lt. from 5th reg:

When the new arms are served out Care to be taken that as many of Equal Calibars be Chosen as possible it is supposed that Each Regt. may furnish itself with such as have bores alike; Bullet moulds is to be then provided to fit them & if the bores are Equal; to Each Regt. the mould Carrying four or five Bullets on each side will be sufficient, Commanding Offrs. of Regt. will have them made as soon as Possible, the Regt. they belong to is to be Marked on the moulds [780]

Regt. Ord.
Lt. Colo. Marion }

A court martial to sett tomorrow morning 10 OC: to try such prisoners as shall be brought to it, the Sergt. Majr. to Acquaint the prisoners that their trial will come on at that hour & to warn all evidences to Attend; this Court to Consist of one Captn. as presidt. & 4 Subalt: as members but if so many cannot be conveniently had then 2 membrs. – they are to be warned this evening & Desired to be punctual in their Attendance

Captn. Blake presidt. – Lieuts. Baker, Proveneaux, Burk & Guerry members

---

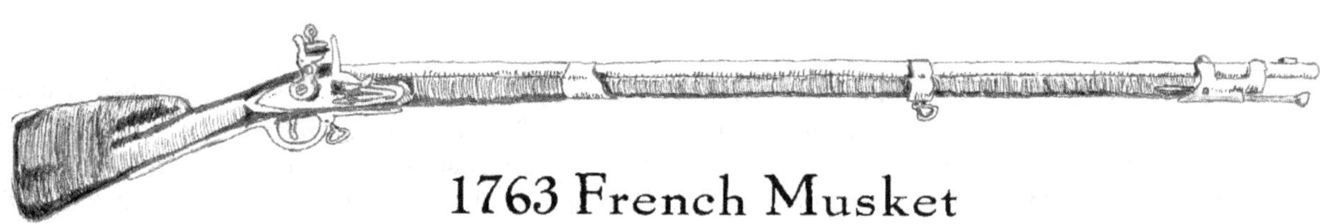

1763 French Musket

Orders 6th Dec
by G. Howe }

. . Parole Fayette . . . .
For the day tomorrow Captn. Mazyck
Town Guard Lieut P. Gray
Magazine Lt. from 5th reg:
Brick house Lt. Saml. Guerry

The Deputy Quartr. Mastr Genl. of this state have the Rank of Colo. in the army of the united States and is to be Obey'd & respected Accordingly,

the Honbl: the Continental Congress have Appointed Wm. Massey Esqr. D. Quarter master Genl. of the Continental troops of this State he is therefore to be Obey'd & Respected as such

---

780 Each musket was contracted out to different gunsmiths, who would make them according to a pattern. However due to the methods of the time each weapon may have a slightly different caliber, and it was a logistical problem to match each unit up with weapons of the same caliber. Muskets at the time would vary from .60 to .80 caliber

Peter Foissin Gentl^{m}. is Appointed a Second Lieu^{t}. in Col^{o}. Mottes Battalion and is to be Obey'd & Respected as such

---

Ord^{rs}. 7th Dec by G. Howe } . . Parole Huger
For the day tomorrow Captn
Town Guard Lieu^{t}.
Magazine L^{t}.

L^{t}. Col^{o}. Marion **Ch^{s}. Town – 1777**

G.O.

Captn. Shakelford of Col^{o}. Hugers Battalion having resigned his Commission is no longer to be Consider'd as a Continental Officer

Reg^{t} Ord^{r} L^{t}. Col^{o}. Marion } Divine Service this Aft^{r}. Noon as usual
The General Orders Issued the 3^{d} Inst: to be read to the regim^{t}. on Tuesday Afternoon 4 OC:

Orders 8th Dec by G Howe } . . Parole Biddle
For the day tomorrow Captn. Harleston
Town Guard from 5th reg:
Magazine from 5th reg: - Brick house L^{t}. Perreneau

Reg^{t}. Ord^{r} L^{t}. Col^{o}. Marion } A reg: court martial to set tomorrow at 10 OC: in the morning to try all prisoners brought before them this court to Consist of one Captn. as president & 4 Subalterns as members, the Serg^{t}. Maj^{r}. to warn the members and Acquaint the prisoners that their trial will come on at the above hour and Order all evidences to Attend; this to be done today

Captn. Harleston presid^{t}. – L^{ts}. Galvan, Hall, Roux & H Gray members

---

Orders 9th Decr. by G Howe . . } . . Parole Rutledge
For the day tomorrow Captn. Motte
Town Guard L^{t}. Hall
Magazine from 5th reg: - Brickhouse L^{t}. H Gray

Reg^{tl}. Ord^{rs}. Col^{o}. Motte } A regimental court martial to set tomorrow morning at 10 OC: to try such prisoners as shall be brought to it – the prisoners to be inform'd that their trial will Come on at that hour and the Evidences to be ordered to Attend in time

Captn. Motte president ~ Lieu^{ts}. Capers & Worley members

---

Orders 10th Decr. by G Howe . . } Parole Gates
For the day tomorrow Captn.
Town Guard Lieu^{t}.
Magazine L^{t}. ~ Brickhouse L^{t}.

Captn. Spencer of Col^o^. Hugers Battalion having resigned his Commission is no longer to be considerd as an Officer in that Corps ~ he is Appointed Assistant D. Q. M. Genl. for this State with the rank of Captn. & is to be respected & Obey'd accordingly

Reg^tl^. Ord^rs^ Col^o^. Motte } a regim^tl^. court martial to set Tomorrow morning at 10 OC: to try such prisoners as shall be brought to it – the Serg^t^. Major to inform the prisoners now in Confinement that their trial will come on at that hour and the Evidences to be Order'd to Attend in time –

Captn. Ashby presid: L^ts^. Baker & Bush members

Col^o^. Motte **Ch^s^. town 1777**

Orders 11th Decr by G. Howe } . . Parole White . . . .
For the day tomorrow Captn. Conyers
Town Guard Lieu^t^. from 5th reg:
Magazine L^t^. Bush ~ Brick house from 5th reg:

Reg^tl^. Ord^rs^. Col: Motte } NB R. 6th Inst:
by sentence of ~~last~~ court ^ Captn Dunbar presid^t^. Serg^t^. Anderson & Corp^l^. Hill for scandalous & Infamous Beheavour was reduced to the ranks – Serg^t^. Wickham for neglect of duty reprimanded – Samuel Henderson for neglect of duty to do Double duty for the war – David Vaughn for selling his Shoes L^t^. Burke to stop his pay to replace the shoes [781] – Emanuel Lopez for receiving a Stolen pocket book rec^d^. 35 Lashes wth Switches by a Court 9th Inst: Captn Dunbar also presid^t^ ~ Alex^d^ Stewart for Absenting himself from guard to be confind 8 days to mount guard every other day ~ By a Court held 10 Inst: Captn. Motte presid^t^.

Joseph Reeves for taking £27.22.6 Out of Capt. Charnocks portmanteau to be put under stoppages till that sum is paid, to receive 99 lashes the latter remitted, also for detouring & Losing a horse of his Captn. also a Saddle and Bridle & a p^r^. Plated spurs the whole amount to £138 ~ Sentence to receive 90 lashes with switches & be under Stoppages one half of his pay till that sum is paid – the Lashes remitted provided he Enlist for the war, wth held – Jno. Proby for selling a hand vice value £20: to be reduced from being Armorer & be under Stoppages to make good the vice – Serg^t^. Henderson for disobedience of Ord^rs^. reduced to ranks

---

Orders 12th Decr. by G Howe . . } . Parole Georgia . . . . .
For the day tomorrow Captn Jervey
Town Guard Lieu^t^. P. Gray
Magazine L^t^. S. Guerry ~ Brick house L^t^. from 5th reg:

Lieu^t^. John Perreneau of Col^o^. Mottes Battalion having resigned his Commission is no longer Considered as a Continental Officer

---

[781] David Vaughn served in the 2nd South Carolina Regiment in Captain Shubrick's company in 1777. His pay was stopped on 11 December 1777 to pay for the shoes that he sold illegally. On 13 July 1778 he was put under stoppages for losing his cap. After Captain Shubrick's death he served in the 2nd Vacant Company until it was disbanded on 5 October 1778. He was then assigned to Captain Hall's company. He was under Captain Adrian Proveaux during 1779. He received 60 lashes on the bare back with a cat of nine tails on 20 January 1779 for neglect of duty. On 3 February 1779 he was put under stoppages to pay for a musket that he lost.

John Sandford Dart Esq[r].[782] is Appointed Deputy Clothier General to the Continental Troops in the State of S[o]. Carolina untill the pleasure of the Hon[bl]. The Continental Congress be known there upon – the General thinks proper to publish in Order the following resolves of ~~Congress~~ this State

Saturday August 23[d]. 1777

Resolved

that every Soldier who hath or shall Enlist in any regiment of this state in the Continental Service shall receive Annually one Blanket one Coat one waist coat, one p[r]. Breeches one Hatt or Cap two shirts one Black Stock or Cravat two p[r] Stockings or Leggins & two p[r]. of Shoes

Resolved, that this house will make provision to Defray any Expenses Exeeding the twenty Dollars which may be incurred in Consequence of the foregoing Resolution

This Generous Donation Lays the Army under a high Obligation to the Hon[bl]. House of Assemble, the D. Clothier Gen[l]. in Consequence of it will immediately as possible, furnish each regim[t]. with the Articles mentioned in the resolve which he will take care to purchase with as much Oeconomy as possible he will take his Direction as to the Uniform of each regim[t]. from the Commanding Officer thereof & have Cloaths made to fit the men

Reg[t]. Ord Col[o]. Motte } Orderd that the Officers give in to the Quart[r]. Mast[r]. at the beginning of every Month an Account of wood rec[d]. by them the preceeding Month

Col[o] Motte **Ch[s]. Town – 1777**

Orderd that Serg[t]. monroe be reduced to the Ranks & be no longer respected as a Serg[t]. in the Regim[t].

NB. Sentence of last court Robert Potts for Stealing a hatt rc[d]. 50 lashes & to put under Stoppages of £10. to repay Tho[s]. Neate [783] – Ch[s]. Turbelle for drunkeness & Swearing in parrade rc[d]. 50 lashes [784]

Orders 13th Dec[r]. by G Howe }

. . Parole Laurens . . . .

For the day tomorrow Capt[n]. Lesesne

Town guard Lieu[t]. Burke

Magazine L[t]. Hart ~ Brick house L[t]. from 5th reg:

A Council of war to be held as immediately as possible at some Convenient Hall in Charles town to take in Consideration the matters which will be Laid before them, of this Council Brigadier General Moultrie is to be president the field Officers of the 1st, 2nd, 4th, & 5th Regiments to be members, if any field Off[rs]. should be absent on Duty or upon leave remote from the town, they need not be waited for, those in town or in its Neighborhood are punctually to Attend, Should any field Officers of 3[d] & 6th Regim[t]. be here they also are to be of this Council, further Orders at their meeting will be Communicated to them by the president [785] . . .

---

[782] John (Joseph) Sanford Dart was born in 1741. He served as deputy clothier-general from 12 December 1777 and was paymaster in the 1st South Carolina Regiment during 1780.

[783] Thomas Neat was in the 2nd South Carolina Regiment.

[784] Charles Turbelle was in the 2nd South Carolina Regiment. He received 50 lashes on 12 December 1777 for drunkenness and swearing in parade.

[785] This Council of War was to determine the ability of the regiments to support General Robert Howe's expedition into Georgia and Florida. Full details of the expedition are described in "*Nothing but Blood and Slaughter, Volume One*" by Patrick O'Kelley.

Orders 14th Decr. by G Moultrie } . . Parole Howe .
For the day tomorrow Captn. Mazyck
Town Guard Lieut. from 5th reg:
Magazine Lt. from 5th reg: - Brickhouse Lt. Perreneau

Regtl. Orders Colo. Motte } there will be no Divine Service performed at the Barrack this afternoon by the Chaplain

Order'd that the Regimental Barrack Guard be reinforced tomorrow morrow morning with three men and that a Centry be posted at the Colo. ~~Door~~ Door from that Guard, the Corporal to wait on the Colo: for his Orders to the Centry – this Centry to be Continued till Order'd otherwise by the Colo.

A party of one Subaltern 1 Sergt. 1 Corpl. 1 Drumr, 1 fifer 24 privates to Attend the funeral this After noon of Thomas Evance Esqr. Late pay master to the regt. from the House of the Deceased [786]– the party to parrade between 3 and 4 OC. ~~for~~ in the Afternoon, when the Officer Commanding the party is to see that the men are Clean & neat & properly
provided the officers are desired to Attend the funeral
Lt. Henry Gray to command the party

Orders 15th decemr . by G Moultrie } . Parole Parsons . . .
For the day tomorrow Captn. Dunbar
Town Guard Lieut. Proveaux
Magazine Lt. Hall
Brick house Lt. from the 5th reg:

Lt. Colo. Marion **Chs. town – 1777**

GO Lt. Henry Gray of Colo. Mottes Battalion having resigned his Commission is no longer to be Considered as an officer in the Continental Service –

Mr. Henry Gray is Appointed paymaster to the Second Continental Battalion of the State of So. Carolina & is to be respected Accordingly

Brigade Ordrs. by Colo. Huger } Lt. Martin of Colo. Mottes Battalion at Dorchester Guard to be relieved this morning. – Lt. Capers to relieve Lt. Martin

Regtl. Ordrs by Colo. Motte } A court martial to set this morning at 10 OC: to try all prisoners brought before it evidences for & against the prisoners to be warn'd in time
Captn. Mazyck presidt. Lieuts. Galvan & Saml. Guerry members . . .
NB. this court did not sett –

Orders 16th decemr by G Moultrie } . . . Parole Rutledge
For the day tomorrow Captn. Ashby
Town Guard Lieut. Warley
Magazine Lt. P. Gray – Brick house Lt. from 5th reg:
to hold themselves ~~immediate~~

---

786 Thomas Evance (Evans) paymaster to the 2nd South Carolina Regiment.

Sergeant & 10 privates ~~from~~ the 2nd & 5th regimts ∧ in readyness immediately with 6 rounds per man and 3 days provisions to Escort some waggons as far as Eutaw Springs where they will be relieved by another party from Colo. Sumpters Battalion, they must be men that can be depended on [787]

Regtl. Ordrs.
Lt. Colo. Marion } Second Lieut. Albert Roux is promoted to a first Lieut. in room of Lt. Henry Gray resigned, he is to be Respected & Obey'd as such Lt. Roux to Join & do duty in Captn. Charnocks Compy. in room of Lt. P. Gray transferred to Captn Motts Compy. Lt. Gray to Join & do duty in Captn. Mottes Compy. till further Orders

Lt. Petr. Foissin to Join & do duty in Captn. Charnocks Compy. till further Ordrs. The new Arms Latly recd. for the regimt. to be proved as soon as possible, Lt. Bush is to have and seen them proved and to take the Armourers and such other men in the regiment to perform this Service Lt. Bush will Apply to the Quartr. Master for powder & Ball to prove the Arms, who is orderd to Deliver what may be wanting a report to be made by Lt. Bush of what muskets may not stand proof & the number that is proof

Major Horry is requested to give the names of those Officers who do not appear at Exercise in the Afternoon, to the Commanding Offr. every Day – the pay bill to be given in tomorrow by 9 OC: in morning to Mr. Henry Gray pay master to the reg:

Orders 17th Decr.
by G. Moultrie }
. . Parole . . .
For the day tomorrow Captn. Conyers
Town guard Lieut. Hart
Magazine Lt. Saml. Guerry ~ Brick house Lt. from 5th reg:

Colo. Huger & Lt. Colo. Marion are requested to go over to haddrels point some day this week, which will be most convenient for them, & Examine the barracks there and make their report to the General wither they are fit to quarter the Soldiers in & what number of rooms are filled and how many men they can contain and also wither the barracks in town can take in more men then they have at present

Chs town 1777 –

Regtl. Ordrs.
Lt. Colo. Marion } A court martial to set tomorrow morning at 10 OC: to try such prisoners as shall be brought before them, this court to consist of on Captn as presidt and 4 subalterns as members the Sergt. Majr. to warn the members this day and Acquaint the prisoners their trial will come on at the above hour & over all evidences to Attend

Captn. Mott president Lieuts. Perreneau, Hall, Roux & Hart members

Aftr Reg: Ordrs.

Divine Service to be preformed by the Chaplain at the barracks tomorrow Afternoon at 4 OC. All Officers and Soldiers to Attend punctually – the Sergt. Major to Acquaint the Chaplain & Officers of this Order & Commanding Officers of Compys. to be particularly carefull to have their men clean & decent & oblige every man to attend that is off duty & not in the hospitles

---

Orders 18th Decr.
by Genl Moultrie }
. . . Parole Plumbard
For the day tomorrow Captn. Lesesne
Town Guard Lt. Perreneau
Magazine Lt. Hall ~ Brickhouse Lt. from 5th reg:

---

[787] 6th South Carolina Regiment.

Reg^{t}. Ord^{r} . } the court martial Orderd to Set today is postponed till tomorrow at
L^{t}. Col^{o} } which time they are to set
say Mazyck
Captn. <u>Lesesne</u> ^ presid^{t}. Lieu^{ts}. Roux, Baker, Warley & Mason members –

Orders 19th Dec^{r}. } . Parole Biddle . . .
by G Moultrie } For the day tomorrow Captn. Mazyck
Town Guard Lieu^{t}. Warley
Magazine L^{t}. Mason ~ Brick house from 5th reg:

Reg^{tl}. Ord^{rs}. } The Quarter master to Deliver to Commanding Officers of Comp^{y}. the Blankets
L^{t}. Col^{o} } now in the regimental store in proportion to the number of Men in each Comp^{y}. & to take a receipt ~ Commanding Officers of Comp^{ys}. will be particular in Keeping an Account to whom Given

Corporal Isaac Kiels of Captn. Mazycks Comp^{y}. is Appointed a Serg^{t}. in the Comp^{y}. & to be Obey'd as such –

Orders 20th Dec^{r}. } Parole Mifflin
by G Moultrie } For the day tomorrow Captn. Dunbar
Town Guard Lieu^{t}. Saml. Guerry
Magazine L^{t}. Bush – Brickhouse L^{t}. P. Gray

Col^{o}. Sumpter's Reg^{t}. is Orderd to march to Ch^{s}. town the munday following Christmas day next, where rooms are prepared at the new Barracks for their reception – the Commanding Officers of the Several Continental Reg^{ts}: in this State are Orderd to Apply to his excellency the presid^{t}. for what they want to Cloath their respective Battalions & to Deposit the same into the hands of the Deputy Cloathier General in Order to purchase Cloathing ~~&~~ who is to Account for the same

Lt Col^{o}. Marion **Ch^{s}. town – 1777**

Orders 21st Dec^{r}. } . . Parole Fayette
by G. Moultrie } For the day tomorrow Captn.
Town Guard Lieu^{t}.
Magazine L^{t}. - Brickhouse L^{t}.

A Corporal & 6 men from the 2nd & 5th regim^{ts} are Order'd to get in immediate readyness to go as a Guard to Wappo Cut to stop ~~to stop~~ all vessels from going out of the Harbour through that place without an Order from the Commanding Officer, each man to have three rounds there will be no service today at the Barracks

Reg^{t}. Ord^{rs}. }
L^{t}. Col^{o}. Marion } A reg: court martial to set tomorrow 10 OC: in morning to try such prisoners as may be brought before it Evidences to Attend the Serg^{t}. Maj^{r} to warn the members today & Acquaint the prisoners their trial will come on at the above hour
Captn Ashby: presid^{t}. Lieu^{ts}. Burke & Galvan members –

Orders 22nd Decr
G Moultrie } . . Parole Port Royal . . .
For the day tomorrow Captn. Lesesne
Town Guard Lieut. from 5th reg:
Magazine Lt. Mason ~ Brick house Lt. Mason
Pritchards Lieut. Bush

Lt. Tate of Colo. Hugers Battalion having resigned his Commission is no longer to be considered as a Continental Officer [788]

The Guard at Wappo Cut of a Corporal & 6 privates from the 2nd & 5th regts. are to be relieved by the same number of men from the Corps of Artillery next Wednesday

Orders 23d. Decr
By Gel Moultrie } . . Parole Hazelwood . . . .
For the day tomorrow Captn. Mazyck
Town Guard Lieut. Burke
Magazine Lt. S. Guerry ~ Brick house Lt. from 5th reg:
For Dorchester Command Lieut. from 5th reg:

Regtl. Orders
Lt. Col: Marion } A regtl. court martial to set tomorrow at 110 OC: in the fore noon to try such prisoners as shall be brought to it, the prisoners to be made Acquainted with the time of trial & all Evidences Warned to Attend in time ~ Captn. Mazyck presidt. – Lts Roux & Warley Members

---

Orders 24th Decr.
by G. Moultrie } . Parole Christmas . . .
For the day tomorrow Lt. Colo. Marion
Town Guard Captn. Dunbar Lieuts Warley & Lt from 5th reg:
Magazine Lt. Perreneau ~ Brickhouse Lt. Hart

One Captn. 2 Subalterns & 40 men for the town Guard tomorrow
The Captn. of the town Guard to send out a Subaltern & 4 men to patrole the Streets every two hours during the night, the field Offr of the day to visit the different guards & Centrys once in the Night; Orderd that above be provided by the D. Q. M. Genl. for Each Guard to hold 6 rounds pr. man to be Deliverd to the Different Commanding Officer of the Guards, Who are to be Accountable for them; one Drum & fife for the main guard one drum for the magazine & one for the Brick house, to remain with their guards till relieved

Brigade Ordrs.
Colo Huger } Detail for Guards tomorrow from 2nd Reg: 1 Captn. 2 Sub: 2 Sergt:
3 Corpl. 3 Drums & fifes & 46 privates – from 5th reg: 2 Sub: 2 Sergt:
2 Corpl: 1 drum 18 privates

Regt Orders
Lt. Colo Marion } All Commissioned & nonCommissioned Officers are to be particular at this time of festivity to have all such Soldiers whom may be found Drunk in the streets or Otherways sent to the Guard house there to remain till further Order; All soldiers are forbid going into town or out of Barracks After retreat Beating, Any Soldiers who may be found out of the Barrack Yard after that time to be Confind where they will remain for a week only to be taken

---

788 William Tate was in the 5th South Carolina Regiment. He resigned his commission on 22 December 1777. He was appointed a lieutenant in the 4th South Carolina Regiment (Artillery) on 8 October 1779. He was captured in Charleston in 1780 and was exchanged in October 1780.

out to mount guard every other day & After relieved Immediately put in Confinement; & those who may be found Drunk will be further punished – this Order to be made Known to the men this Evening

---

Orders 25th Decr. G Moultrie }
. Parole Franklin . . .
For the day tomorrow Majr. Huger
Town Guard Capt Ashby Lts. Mason & S Guerry
Magazine Lt. P Gray ~ Brick house from 5th reg:

Regtl. Ordrs. Lt. Colo Marion } A reg: court martial to set tomorrow between the hours of 8 in the morning and 3 in the After noon for the trial of all prisoners brought before them, the Sergt. Majr. to warn the members today and Acquaint the prisoners that their tryal will come on at the above time also to Order all Evidences to Attend – the president will fix the Hours of Meeting – Captn. Lesesne presidt. Lts Burke & Galvan members

NB only one man was tried Lt. Galvan refusing to try some prisoners for Neglect of duty because the evidence proved they were found sleeping on their post – which was reported to Genl. Moultrie

---

Orders 26th Decr G Moultrie }
. . Parole Boston . . .
For the day tomorrow Majr. Horry
Town Guard Captn. Lesesne Lts. Perreneau & Lt. from 5th reg:
Magazine Lt. Burke _ Brick house Lt. Hart _

Orders 27th Decr. G Moultrie }
. . Parole Whitemarscher . .
For the day tomorrow Lt. Colo. Marion
Town Guard Captn. Mazyck Lts Mason & Worley
Magazine Lt. petr. Gray ~ Brick house Lt. from 5th reg:

NB The 24th Instant Lt. Roux was put under an Arrest for Disobedience of Orders ~

Lt. Colo. Marion **Chs town 1777 –**

Orders 28th Decemr by G. Moultrie }
. . Parole Putnam . . . .
For the day tomorrow Captn. Ashby
Town Guard Captn. Dunbar Lts Hart & Lt. from 5th reg.
Magazine Lt. S. Guerry – Brick house Lt. from 5th reg:

Order'd that when any man of War are off the Barr, that a guard boat be kept at night going from fort Johnson to fort Moultrie & south from each fort to the other that they may be no Communication between the town & the Enemy that way as also to take up any suspected persons going out

No field Officer for duty till further Orders –

Regtl. Orders by Lt. Colo. Marion } A court martial to set tomorrow at 10 OC: in the morning for the trial of all prisoners brought before it, the Sergt. Majr. to warn the members today & Acquaint the prisoners that their trial will come one at that hour, also to Order all Evidences to Attend, as there is but few Subalterns in town fit for duty the Court is to Consist of one Captⁿ. the eldest in rank for Duty as prisid[t] & one Captⁿ & 3 Sub: as members; the members is to attend punctually & the Court is not to Adjourn before 3 OC: in Aft[r]. noon, without all the prisoners are try'd by that that time

Captⁿ. Ashby presid[t]. Captⁿ. Blake & L[ts] Burke, Galvan & perreneau

NB. this court did not do business for want of member L[t]. Perreneau

---

Orders 29th Dec[r].
by G Moultrie

. Parole Sullivant
For the day tomorrow Captⁿ. from 5th reg:
Town Guard Captⁿ. Blake L[ts] Warley & Perreneau
Magazine Martin ~ Brick house L[t]. from 5th reg:
Pritchards L[t]. Pet[r]. Gray

The Second Regim[t]. is Order'd to get in readyness to go to fort moultrie the 6th January to relieve the 1st Reg[t]. which is orderd to town – no hutts or Buildings about the fort are to be hurted or Demolished on any Account Whatsoever ~ the first reg[t]. may begin to remove their Bagage & as Soon as convenient, the Budgings that are private property the Gen[l]. will Indeavour to get them payed by the State

---

Orders 30th Dec[r].
by G. Moultrie

. . Parole Sunbury . . .
For the Town Guard Captⁿ. Dunbar L[t]. Hart & Guerry
Magazine L[t]. from 5th reg: - Brick house L[t]. Burke

Orderd that the Captⁿ. of the main Guard do visit at night all the Garrison Guards and make his report to Col[o]. Huger when relieved

L[t]. Col[o]. Marion **Ch[s] town 1777 –**

Orders 31st Dec[r]
by G Moultrie

. . Parole Virginia
For Guard tomorrow Captⁿ. Mazyck & 2 L[ts]. From 5th reg:
Magazine L[t]. Foissin – Brickhouse L[t]. Perreneau

The Deputy Q M Gen[l]. is Order'd to git some Vessels ready at Gadsden on munday next to provide Waggons to Cart the Baggage &c. belonging to the 2nd Reg[t]. to be Carried to fort moultrie

Reg[tl] Ord[rs].
L[t]. Col[o]. Marion

The Officers & Soldiers of the reg[t]. is to prepare their Baggage ready to be put on board the vessels by the 5th Jan[y]. that nothing may prevent going to fort Moultrie the day after – the Q. Must[r]. to pack up & git all the regimental Stores ready to be transported to Sullivants Island; he is also to send for the Old Arms now in the Store to the Arsenal & take a receipt from M[r]. John Calvert, this to be done immediately as possible – he is also to Hurry the Armourers in proving & marking the new muskets that they may be ready the 5th Jan[y]. – the Iron potts that have been given out to the men must be Called for, & the Q Must[r]. take a particular Account & make a report of them & all other Stores to the Commanding Officer –

After Orders

A Monthly return to be made and Given to the Maj[r]. by tomorrow 4 OC: in the After noon

A regim[tl] Court martial to Set tomorrow 10 OC: in forenoon to try all such prisoners as shall be Brought to it – the Serg[t]. Maj[r]: to warn the members today & Acquaint the prisoners that their trial will Come on at the above hour – this Court to Consist of one Captn. as presid^{t}. & 4 other Off^{rs}. as members, if they Shoud not be Subalterns a nough for members & the Court to be made up by as many Capt^{ns} as may be wanted, the Court not to Adjourn till 3 OC: in Afternoon without all the prisoners are try'd that may be brought to it

by Consent of the Commanding Officer

Captn. Ashby president Lieu^{ts}. Perreneau & Warley Members

NB no other Off^{rs} where to be had

1778

Monthly Return 1st Jany. 1778 –

1 Colo_ 1 Lt Colo_ 1 Majr._ 10 Captns._ 10 first Lieuts._ 6 Second Lieuts._ 1 Chaplain _ 1 Q Mastr _ 1 pay Mastr_ 1 Surgeon _ 24 Seargeants 1 do Majr._ 16 Drums & fifes; 8 Armrs. Mate _ 389 Rank & file men

NB the night after this return was made 2 men Deserted Charles Turbeville of Captn. Lesesnes Compy & Spiller of Blakes –

Lt. Colo. Marion — Charles town – 1778 –

Orders 1st Jany. by G Moultrie } . . . Parole Green . . .
For the town Guard tomorrow Captn. Harleston Lts. Perrenneau & Lt. from 5th reg:
Magazine Lt. Mason _ Brick house Lt. Warley

Regimtl. Orders Lt. Colo. Marion } The Q. Mastr. to make his fatigue men clean all round the the rooms that the men do not Inhabit particularly, the Guard, Barracks, & all Labinator,[789] Store, Doctrs. & the Offrs room – Commanding Offrs. of Compys. to make their men Scour & Clean their rooms, that no Slur of the Second reg: may be found on them by the troops which are to take place in them, this is to be done by Saturday Evening; the Lt. Colo. & Majr. will Examin them and see they are made as Clean as possible

---

Orders 2nd Jany. by G. Moultrie } . . Parole Germantown . . .
For Town Guard tomorrow Captn. Ashby Lt. Burke & 1 from 5th
Magazine Bush – Brickhouse Lt. S. Guerry

NB by Sentence of last reg: court Wm. McDowel for drunkeness on Guard recd. 50 lashes Wm. Bryan for neglect of duty re: sleeping on his post recd. 50 _ Danl Gordon for the Same recd. 50 lashes [790] _ Jno. Chavis for the same recd. 50 lashes

---

Orders 3d Jany by G. Moultrie } . . Parole Nantz. . .
For the town Guard Captn. Cogdel Lts. Hart & Foussin
Magazine from 5th reg: ~ Brick house Lt. from 5th reg:

---

[789] I am unable to determine what a Labinator was. It may have been the latrine, or it may have been the laboratory where they made ammunition for the artillery.

[790] Daniel Gordon enlisted in the 2nd South Carolina Regiment on 25 September 1777. He received 50 lashes on 2 January 1778 for sleeping on his guard post. He received 100 lashes on 26 March 1778 for being drunk on guard. He received 100 lashes on 26 August 1778 for disobedience of orders.

Orders 4th Jany by G. Moultrie } . . Parole Georgia . . .
For the ~~Town~~ Day Guard tomorrow Captn. Blake L^{ts}.
Town guard L^{t}. Perreneau
Magazine released by the 5th reg: - Brick rel^{d}. by 5th reg:
a Serg^{t}. and 9 privates from the 2nd Reg^{t}. are orderd to release the magazine guard at Hobcaw tomorrow in the room of the same number of men from the 1st Reg^{t}. who are to Join there Quarters the 5th reg: at the same time to relieve the Guard at pritchards Ship Yard, the main guard to be releavd by a Subaltern 1 Serg^{t}. 2 Corp^{ls}. & 29 privates from 2nd Reg^{t}. _ the magazine Guard by 1 Serg^{t}. 1 Corp^{l}. & 6 privates from the 5th reg^{t}. the Brick house Guard by a Serg^{t}. a Corp^{l}. & 9 privates form the 5th reg^{t}. – the Captn. of the day to be very particular in Examining all houses, store houses, Carriges & everything else give in Charge to the Different Centrys to make a report of the same to the Commanding Off^{r}. of the Brigade when relieved

Reg^{tl}. Ord^{rs}. by L^{t}. Col^{o}. Marion } a regimental court martial to set tomorrow 9 OC: in the morning to try all prisoners brought before them, the Serg^{t}. Maj^{r}. to warn the members today & Acquaint the prisoners that their trial will Come on at that hour & to Summon all evidences to Attend – this court to be composed of one Captn. as president & other officers as members the court not to adjourn till 3 OC: in the After noon without all the prisoners all try'd

The Q. Mast^{r}. to deliver out to Commanding Off^{rs}. of Comp^{y}. new Arms tomorrow afternoon taking a receipt for the same

Commanding Off^{rs} of Companies to see their men Deliver up their old Arms to the Q. Mast^{r} taking a receipt for them

The new arms to be Deliverd out from N. to as many as there is in the Eldest Comp^{y}. and from the last member in that to the next Eldest Comp^{y}. so on to the youngest

the Officer who receive the Arms for their Comp^{y}. to Enter the mans name in a book and the number on their muskets to be put opposite there names, that they may easily see at all times what number belong to each man

Commanding Officers of Companies are Charged with the Arms they receive & is to be made Answerable for all Loses; except where it may be ~~by~~ unavoidable

Captn. Dunbar will Choose the best Arms from the Old musket Deliverd up, to the number 75 if so many are Good – which will be numberd as soon as possible – he may have them Branded immediately with 2^{d} Reg^{t}. [791]

Orders 5th Jany. by G. Moultrie } . . Parole Windsor
Town Guard tomorrow L^{t}. Burke

Reg^{tl} Orders L^{t}. Col^{o}. Marion } One Captn. 2 Sub: 2 Sergts. 2 Corp^{l}. & 40 privates must be warn'd to hold themselves in readyness to mount guard when they Arrive at Sullivant's Island to be all Embarkt in one vessel, that they may

---

[791] By the time of this order the rest of the regiment would be carrying the new muskets, which most likely was the Model 1763 type French musket, while Captain Dunbar's Grenadier Company still carried the Long Land pattern British musket.

disembark regularly & march when Order'd, the men to be as clean as possible – Captn. Mazyck Lieuts Mason & Guerry for this duty

After Orders

one Subaltern and Eight men to Join the Guard on board the vessel at Gadsden Warff this After noon at 3 OC. when all the Baggage of the regimt is to be put on board, the Officer of that Guard will receive his Orders. – The General to beat tomorrow at 9 OC: in the morning Assemble & march at 10 – The Officers who do not put all their Baggage on board today must have them on board before the General Beats as the regimt. will Embark at 11 OC: precisely – All Officers & Soldiers to be on the parrade punctually at the hour Orderd for the beating the Assemble

For the Baggage Guard at Gadsden Warff Lt. Baker

NB. agreeable to Sentence of Last court Fredrick Simmons [792] recd. 50 lashes for Selling a Gun to David Williams [793] – Corpl. Kelley mult a weeks pay, for neglect of duty [794] – Thos. Fox for the same mult 5 days pay [795]– Thos Niel do. 5 day pay [796]

Lt. Colo. Marion **Chs. town – 1778 & Fort Moultrie [797]**

Orders 6th Jany.
by Lt. Col: Marion }

Parole
For the Quarter Guard at Gadsden Warff Lt. Capers

One Subaltern 1 Sergt. 1 Corpl. & 18 privates to mount a guard immediately on Gadsden Warff, to post ten Centrys so as to prevent any Soldier from going off the Warff without leave from the Commanding Officer

---

792 Frederick Simmons served in the 2nd South Carolina Regiment. He received 50 lashes on 5 January 1778 for selling his musket to David Williams. He was confined for 20 days on 4 May 1778 for being absent without leave. He also had to do guard duty every other day while he was confined. On 12 September 1778 he received 75 lashes with a cat of nine tails for being absent without leave. He was in Captain Dunbar's Grenadiers in 1778. He received 79 lashes on the bare back with a cat of nine tails on 19 January 1779 for neglect of duty and quitting guard. He was promoted to corporal in the Light Infantry Company on 20 January 1779. He was at the siege of Savannah.

793 Unable to find any information on David Williams. He may have been a civilian at Haddrell's Point. Every military post since the beginning of time has had a following of merchants who were on the sleazy side of society. Haddrell's Point was such a place, and it was located in between Charleston and Fort Moultrie. Soldiers could buy liquor, prostitutes, gamble, and sell items for money. Since a soldier did not have much, these items most likely were stolen from other soldiers or from the regimental stores.

794 John Kelly enlisted in the 2nd South Carolina Regiment on 4 November 1775 and was a corporal on 8 December 1777. He was fined a week's pay on 5 January 1778 for neglect of duty. On 13 July 1778 he was reprimanded for neglect of duty.

795 Thomas Fox was born in 1725. He enlisted in the 2nd South Carolina Regiment on 27 November 1775. He was fined five days pay on 5 January 1778 for neglect of duty. He was discharged on 28 November 1778. He died on 16 November 1822 at the age of 97.

796 Thomas Niel was in the 2nd South Carolina Regiment. On 5 January 1778 he was mult 5 days pay for neglect of duty.

797 The 2nd South Carolina Regiment was in transition, moving from the barracks in Charleston to Fort Moultrie. Regiments would rotate doing duty in between Fort Moultrie and Charleston. The last time the 2nd South Carolina Regiment had been in the barracks at Fort Moultrie was in August 1776.

NB. the regim[t]. marched to Gadsden Warff at 11 OC: this day, found only two boats ready to transport them to Sullivant Island, Capt[n]. Mazyck L[t] Mason & Guerry with the party Orderd yesterd[y]. Embarked with the Baggage on board one schooner – L[t]. Baker with 25 men on board the other boat proceeded to Sullivants Island there reembarked & Capt[n]. Mazycks party took possession of Fort moultrie & relieved the 1st reg[t]. Capt[n]. Ashby who was sent the day before to take an Acc[t]. of all the stores remained Commanding Officer. – the remainder of the regim[t]. & the Officers Campt on Gadsden warff waiting for Boats

---

Orders 7th Jan[y]. L[t]. Col[o]. Marion } . . Parole Gen[l]. Moultrie . . Count[r]. S[n]. 7

For the Fort Guard tomorrow Capt[n]. Dunbar L[ts]. Hart & Foissen rear Guard a Serg[t]. & 12 men

The Guards to Observe all Orders given them Yesterd[y]. with the Greatest Strickness to Exactness, – the Guards to be relieved at 8 OC: in the mornings roll-call ~~call~~ the same hour, retreat & tattoo Beat as usual

The Q Master to number the Officers & mens rooms agreeable to rank beginning with the Officers from the Center room & go from right to left

one Capt[n]. 2 Sub: 2 Serg[t]. 2 Corp: & 39 privates for the fort Guard tomorrow, One Serg[t]: a Corp: & 12 privates for the rear Guard NB the command[r]. of the reg[t]. Arrived at fort Moultrie 3 OC: in Afternoon this day

---

Orders 8th Jan[y]. Col[o]. Marion } . . Parole Dicipline - - Count[r]. S[n]. Exact

For the Fort Guard tomorrow Capt[n]. Ashby & L[ts]. Galvan & Capers rear Guard a Serg[t].

No person what ever to do their Occasion within the fort or within 20 yd[s]. of the walls on the outside, no bones or other filth or Litter whatever to be thrown in the fort all persons who may Disobey this Order may expect to be Severly punished – Commanding Officers of Companies to order two men & a Serg[t]. or Corp. as fatigue men to their Comp[ys]. who are to clean Dayly all filth which may be about their Barracks and to do other Company Dutys – It is Expect that an Officer will visit their mens quarters daily & See that Orders are Comply'd with

The Officer who see roll-call morning & Evening is not only to Call the mens names over, but to see they have their Arms & Accoutrements & in what Order when ever they find a man without any part of their Arms & Accoutrements they are immediately to Confine him & ring him to a Court martial, otherways they will be Liable for all Losses & will Certainly be Called upon for payment

That every Officer may have it in time to go to town or be Absent from Garrison the L[t]. Col[o]. desires the Genl[m]. to Observe, that no more than three Capt[n] to be Absent at a time from Garrison, that only one Subaltern of a Company to begin from the eldest in rank, if he shoud not Chuse to go the next may have the right, & two men of a Comp[y]. to have Leave of Absents at one time this is not meant to Include those men who Obtain furlowe but those only who may go to town

L[t]. Col[o]. Marion **Fort Moultrie – 1778 –**

As the regiment by being in town too long have lost a Great part of the Dicipline and tis necessary to reform all Abuses & Neglect of Dicipline the L[t]. Col[o]. Calls upon every Gentleman in the regim[t]. to Aid & Assist him to bring the regiment to true & Exact Dicipline that they may regain their former

Credit, & be an honour to themselves and their Country, he promises on his part that he will exert his utmost to so good a purpose and shall think no pain or trouble too great to Effect it, But must sink under no Further without the Assistance of the rest of the Officers – A Little perseverance with Attention ~~will under~~ to all parts of Duty will soon bring them to what we Could wish and make them Equal to the best troops in this State or in any of the United States of America – he begs leave to Observe a few regulations necessary for each Compy. – that besides the Orderly book for each Officer, on Aught to be provided for the Compy. which the Orderly Sergeant for the day Should Enter the Orders as soon as it Comes out & to Carry the book to all the Officers of his Compy. and not to have it on a Scrap of paper, which through negligence or Laziness may be lost) by which means the Sergeant will know all Orders as they may have full Access to it – all the men for duty or parrade to draw up before their own Barracks their to be Examind by the Seargts & when ready to be Examind by their Officer – though men may not be compleated with Cloaths yett such as they have shoud be put on to the best Advantage their Hairs Comb'd their face & hands made clean, the Orderly Sergt. may be the one who is Order'd for fatigue & Should see the men receive their provisions & property Distributed to each mess, all the men to be in messes of 6 & not Less than 5, to visit the men at meal time & see if their Victuals are well Cookt, to visit the Sick & report every thing which may happen During the day to the Commanding Offr. of the company – when ever any part of Duty is neglected or done in a Slovenly manner though ever so minute it finds to destroy Dicipline interely, that so necessary to never to over look any part whatever; many Small Crimes may be Committed , which would be best punished ~~by~~ in the Compy. by various ways much better & with Greater Effect than bringing them to a Court martial

One Corporal & 6 men with 12 rounds pr. Man for the Advance Guard tomorrow who will receive Orders as soon as they are ready to March this Guard to be relieved weekly

An Officer of the Guard ~~to visit~~ in garrison to visit the Centrys at night Once between each relief & to Send a Subaltern to visit the rear guard; and a Sergeant to patrole within the fort every half hour during the night – The Sergt. of the rear guard to visit his Centrys between each relief during the night, when he goes his rounds to Leave the Corporal the Charge of his Guard until he returns –

– Orders to the Sergts. of the rear Guard –

Sir, you are to Stop & ring too all vessels, boats or Cannoe which may attempt passing the Bridge Either up or Down & send the principle person with your Corporal to the Captn. of the fort guard & you are to examine all Such vessels or boat, & Give an Account what she may have on board particularly all such who may Come from town & detain them till you have Orders to the Contrary – You are not to Lett any boats Land near your guard without Examining them, without their shoud be an Officer belonging to the Continental forces of the United States of America or the president of this State or any of his Council – you are to make a report of any thing which may happen to the Captn. of the fort Guard – you are to give Orders to the Centry on the bridge not to Lett any Soldier go over the Bridge in the day time without a permit from Some Officer, nor Suffer any person pass after retreat beating without such a pass – this Order to be given to the relieving Sergt. & to be Continued till further Orders – – – – – –

Orders 9th Jany. }
Lt. Colo. Marion }

. . Parole Genl Washington . . . . Countr. Sn. 8

For the fort Guard tomorrow Captn. Blake & Lts. Bush & Gray

Rear Guard Sergt. Newman

A court martial to be held tomorrow 10 OC: in the morning in the Center room or the presidents – to try such prisoners as may be brought before them all Evid: to Attend – Captn. Ashby president Lieu^{ts}. Baker, Hart, Perreneau and Roux memb:

No Court martial to Consist of Less than five Officers for the feuter unless that number cannot be had, then 3 will be sufficient – more than one Captn. must Set as members when Subalterns are not to be had

1778

Orders 10th Jan^{y}: L^{t}. Col^{o}. Marion } . . Parole Paris . . Count^{r}. S^{n}. Morgan
For the Fort Guard tomorrow Captn. Mazyck L^{ts}. Perreneau & Hall
Rear guard Serg^{t}. Williamson [798]

The Battalion to exercise with small Arms every after noon (except Wednesday, Saturday & Sunday) when the tide suits, at beating the long roll all Officers & Soldiers off duty to Attend – the Maj^{r}. will order when the roll is to beat – Every Wednesday After noon the Battalion to exercise at the Cannon, & Saterday they are to show all their Arms & Accoutrements Blankets & regimental Clothing, an Officer of a Company to Attend and Examin them; on munday the Exercise will begin

At retreat Beating the Searg^{t}. Major to Attend the Commanding Off^{r} for the Counter Sign who is to give it to all the guards

1778

Orders 11th Jan^{y}: L^{t}. Col^{o}. Marion } . . Parole Amsterdam . . Count^{r}. S^{n}. 9 . . . .
For the Fort Guard tomorrow Captn. Dunbar Lieu^{ts}. Baker & Capers
rear guard Serg^{t}. Kolb [799] – Hobcaw Serg^{t}. Gammell

The Hobcaw guard to be relieved tomorrow by the same number as is now there, they are to be furnished with a weeks provisions, the Serg^{t}. will wait on the Commanding Officer when ready to march – the Q M Serg^{t}. to deliver this After noon the Blankets now in the store in proportion to the number of men in each Comp^{y}: taking a receipt for the Same – he is also to make his fatigue men sweep the Chimneys tomorrow by fixing a board or Broom to a long pole & run up and down the Chimney till they are made clean –

L^{t}. Col^{o}. Marion **Fort Moultrie – 1778 –**

Orders 12th Jan^{y}: L^{t}. Col^{o}. Marion } . . Parole Madrid . . Count^{r}. S^{n}. Blake
For the Fort guard tomorrow Captn. Ashby L^{ts}. Mason & Guerry
Rear Guard Serg^{t}. Jasper

Genl. Orders 11th Jany. by Genl. Moultrie rec^{d}. this day.

Orderd that one Captn. two Sub: 2 Serg^{ts}. 48 rank & file from 1st reg.
one Captn. two Subs. 2 Sergeants 48 rank & file form the 2nd reg:
one Captn. 1 Sub: 30 rank and file from the 4th reg^{t}.
one Subaltern & Serg^{t}. 19 rank & file from the 5th reg: - The above Detachment to be in readyness tomorrow morning to go on board the vessels drawn for each regim^{t}. to provide their men with 18 rounds per man and 50 rounds each man to be put in a military Chest on board the vessel they go in, the Captains and Subalterns who are to Command the different parties are to meet tomorrow morning at the new Barracks to draw for the vessels they are to go on board – the Officers

---

[798] Isaac Williamson served in the 2nd South Carolina Regiment under Captain Blake.

[799] Josiah Kolb served a a sergeant in the 2nd South Carolina Regiment under Captain John Blake in 1778.

Commanding parties are to take care to keep good Order & Dicipline amongst their men, and prevent them from Getting any Disputes with the Sailors and Assist the Capt[n]. of the vessel to the utmost of his power in Attacking and Opposing the Enemy – any Officer who Chuse to change his tour of Duty may have Leave by Acquainting first ~~Acquaint~~ the Command[g]. Off[r]. of their respective reg[t].

The names of the Officers going on this Command to be Given on to the General

Reg[tl]. Ord[rs] L[t]. Col[o]. } Commanding Officers of Companies to warn the members of men in their respective Comp[ys]. agreeable to the List the Serg[t]. Maj[r]. will show them, to hold themselves in readyness immediately to go on board the vessels which will be Appointed for them, as they are to Embark this After noon or early tomorrow morning – they are to parrade at 10 OC: today

the Q Mast[r]. to git the Ammunition ready immediately agreeable to the General Orders above

Capt[n]. Blake Lieu[ts]. Proveneux & Bush for the Command also Serg[t]. Marlow and Lawrence

After Order

the fort guard tomorrow to be reduct to one Serg[t]. 2 Corp[l]. & 33 privates with same number of Commissioned Off[rs] – the rear guard to 1 Serg[t]. 1 Corp: 9 privates one Centry form the fort guard to be plact before the Commanding Off[r]. door & prevent any persons taking the boards, which is there

The British had maintained a blockade of Charleston harbor since the Battle of Fort Sullivan. This blockade was not a continous one, but was done whenever the British had the time and ships to conduct the mission. By the middle of 1777 there was always a British warship outside the Bar. They were the four Royal Navy Frigates *Carysfort, Perseus, Lizard* and *Hinchenbroke.* The ships patrolled from Port Royal to Cape Romaine. These warships effectively bottled up the merchants in Charleston harbor.

In the fall of 1777 John Rutledge requested 150 Continental troops to serve as marines aboard a small fleet of ships that would be commanded by Captain Nicholas Biddle. Rutledge suggested that the Continental Navy frigate *Randolph,* in port having her hull scraped, and a number of South Carolina Navy ships go out and break the blockade. Biddle accepted the command of the naval force. Major General Howe called for a council of war and asked them to reconsider. If the soldiers went off to fight the Royal Navy he would be severely low on men to guard the city. The council of war agreed that the soldiers were needed more in South Carolina to stop any British raids from Georgia or Florida, however the council also agreed to allow a few men to become marines on board the South Carolina privateer fleet consisting of the frigate *General Moultrie* (twelve short and six long 6-pounders), and the brigs *Notre Dame* (sixteen 6-pounders), *Polly* (16 guns) and the Snow *Fair American* (20 guns).[800]

Orders 13th Jan[y]: L[t]. Col[o]. Marion } . . Parole Brandenbourg. . Count[r]. S[n]. L[t]. Bush

For the fort guard tomorrow Capt[n]. Charnock L[ts] ~~Roux~~ Gray & ~~Perreneau~~ Hart

rear Guard Serg[t]. Newman

[800] O'Kelley, *NBBAS, Volume One* pp. 201-205.

L^{t}. Col^{o}. Marion **Fort Moultrie – 1778**

Orders 14th Jan^{y}: ~~by L^{t}. Col^{o}. Marion~~ by Maj^{r}. Horry } . . Parole Marion . . Count^{r}. S^{n}. Ch^{s}. town –
for the fort guard tomorrow Captn. Mazyck L^{ts}. Baker & Perreneau
Rear Guard Sergeant Keils

That the non Commissioned Officers or privates order for Command do Join their respective companies and do duty in the same till further Orders

---

Orders 15th Jan^{y}: Maj^{r}. Horry } . . Parole Charles Town . . Count^{r}. S^{n}. Five
for Fort guard tomorrow Captn. Dunbar L^{ts}. Capers & Mason
rear guard Serg^{t}. Williamson – Advance Serg^{t}. O'Neil

that at the beating of the Long roll the following Guards be reinforced – The fort Guard with 6 privates – the rear with 8 privates – the Advance 1 Serg^{t}. & 6 privates – the major Strictly Enjoins the Commanding Officers of Guards to be very vigilant and Watchfull on their guards to visit the Centrys and see they are Allert on their post and to report to the Commanding Off^{r}. every neglect of duty – No Leave of Absents to be Granted to non Commissioned Off^{r}. or privates till further Orders

After Orders

all non-commissioned officers & soldiers to be in Garrison after tattoo beat^{g}. & their to remain till Revallese beating till further Orders

---

Orders 16th Jan^{y}: L^{t}. Col^{o}. Marion } . . Parole Vianna . . Count^{r}. S^{n}. Dresden
For the fort guard tomorrow Captn. Ashby L^{ts}. Gray & Guerry
For the Guard boat tonight L^{t}. Hart – rear Serg^{t}. Newton

The Officers who was orderd to hold themselves in readyness to go on Command in the Armed vessels are to Join & do Duty in the reg^{t}: till further orders –

All Officer who Obtain Leave of Absents to town are to return the day before its there tour for duty – those who do not Obey this Order punctually must expect to be taken notice of

no Soldier to have Leave of Absents to go to town after he has once stay'd longer than such Leave of Absents, without a good excuse & this the Command^{g}. Officer is to be a Judge – a report to be made to him of all who may out Stay their time of Absents –

Those Soldiers who was for Command to give the Ammunition they rec^{d} to Sergeant Fletcher immediately [801]

A Court martial to sett tomorrow at 10 OC: in the morning to try such prisoners as shall be brought befor it – all evidences to Attend

Captn. Charnock presid^{t}. – ~~Lieu^{ts}~~ Captn. Moultrie Lieu^{ts}. Hall Baker & Roux members

NB the men of war not being in sight the Guard boat did not go out –

Early on the morning of January 15th a fire swept through Charlestown, destroying 250 houses and 500 businesses and storage buildings to include the public library. This became known as the Great Fire of 1778. Arsonists from British ships were suspected and the idea to send out a fleet with marines to protect the coast became a reality.

The fleet consisted of the frigate *General Moultrie*, the snow *Fair American* and the brigs *Notre Dame* and *Polly*. With the South Carolina State Navy was the Continental Navy

[801] Quartermaster Sergeant William Fletcher.

frigate *Randolph.* The mission of the ships was to intercept privateers and the English ships coming from the West Indies loaded with supplies.

a Sergeant and twelve men from the 2d. Regt: to board the Genl. moultrie as soon as possible, they are to take their Instructions from Captn. Sullivant Commander of the Said Armed Vessel

On board the ships were men from the South Carolina Continental Regiments, consisting of Captain Joseph Joor's company from the 1st South Carolina Regiment,[802] Captain John Blake's company of the 2nd South Carolina Regiment, a half company from the 4th South Carolina (Artillery) Regiment, and a platoon from the 5th South Carolina Regiment under

[802] Joseph Ioor (Joor) became a first lieutenant on 17 June 1775 in the 1st South Carolina Regiment. He was promoted to captain during May 1776. He was killed when the *Randolph* was blown up fighting the Royal Navy Man-of-War *Yarmouth.*

the command of Lieutenant William Blameyer. Captain Joor's company drew the coveted assignment on the *Randolph.* Captain Blake was assigned to the *Fair American* and part of his company, under the command of Lieutenant Adrian Proveaux, was assigned to the *General Moultrie*. Lieutenant Blameyer was assigned to the *Notre Dame*. The 4th South Carolina was assigned to the *Polly*.[803] On January 27th, 1778 South Carolina launched its fleet.[804]

Lt. Colo. Marion **Fort Moultrie – 1778 –**

Orders 17th Jany: L t. Colo. Marion } . . Parole Constantinople . . Countr. Sn. 13 . . .
For the fort Guard tomorrow Captn. Charnock Lts. Harts & Perreneau
Rear guard Sergt. Holliday

Head Quarters Chs. town Jany: 17th: 1778

Please to dispatch a Sergeant and twelve men from the 2d. Regt: to board the Genl. moultrie as soon as possible, they are to take their Instructions from Captn. Sullivant Commander of the Said Armed Vessel __ __ __ __ __ __ __ [805] yours &c.

William Moultrie B. G.

Regtl. Ordrs. A sergt. and 12 men with 18 rounds per man to go on board the General moultrie Immediately, the Sergt. will receive his Orders from Captn. Sullivant Commander of Sd. Ship – the men for Command to go

For this Command Sergt. Lawrence

NB this Command went of at 3 OC: PM –

Orders 18th Jany: Lt. Colo. Marion } . . Parole Venice . . Countr. Sn. punctuality
the fort guard tomorrow Captn. Mazyck Lts. Baker & Capers
rear guard Sergt. Newman – Hobcaw Sergt. Smith –

the names of the Sergts. on fatigue and orderly to Each Compy. to be put dayly on the Back of each morning report; that the Majr. ~~Kn~~ may Know what Sergt. to call upon when he wants a man for duty out of the Compy. & what Sergt. do not attend the Sergts. call to receive orders – the Majr. will confine all such Sergts. who do not Attend who are orderly for the Company –

A court martial to set tomorrow 10 OC. in the morning to try all such prisoners as shall be brought before them, all evidences to Attend

Captn. Ashby president. Captn. Dunbar, Lieuts. Mason Guerry & Baker members

---

Orders 19th Jany: Lt. Colo. Marion } . . Parole Rome . . Countr. Sn. Trojan
the fort guard tomorrow Captn. Dunbar Lts. Mason & Guerry
Rear guard Sergt. Keils

---

803 William Blameyer served in the 5th South Carolina Regiment. He was promoted to captain in 1778. He resigned during November 1778.

804 O'Kelley, *NBBAS, Volume One* pp. 201-205.

805 Philip Sullivan of the frigate *General Moultrie*. He had also served as captain of the brigantine *Richard.*

NB by sentence of Court 17th Inst: Captn Charnock presidt. Jams. Anderson & Jno. ~~Maple~~ Muney for Absents wto Leave in town,[806] to be Kept in the Black hole for 8 days and put on Guard every other day, that is 4 days out of 8 Hugh munrow for Absents without Leave to receive 50 Lashes with Cat 9

for Misdemors in forging Geor Shed [807] hand to an Affidavit & Captn Ashbys to a pass to town to receive 50 more two day after & picketted 5 minute this Last remitted

Lt. Colo. Marion **Fort Moultrie – 1778**

Orders 20th Jany.
Lt. Colo. Marion } . . Parole Petersbourg . . C. Sn.
Fort guard tomorrow Captn. Ashby Lts. Gray & Burke
Rear Sergt.

Commanding Offrs. of Companies to Apply to Q Master for as many Blankets as will Compleat their Companies giving a receipt for the same

NB by sentence of Last court James Costelo [808] and Jacob Williams [809] for neglect of R.O – – duty was Stript & reprimanded at the Halberts –

Any soldier who is found Siting on his post when Centry to be relived and Confind, who may depend on being punished –

All orders respecting the men to be read to them at roll-call by an Officer of the Company

A court martial to set tomorrow 10 OC. in the morning for the trial of all prisoners brought before it – all evidences to Attend

Notwithstanding orders to the Countrary, the soldiers still do their Occasions under the platforms and Close to the walls on the outside; the Sergeants or Corporals of Fatigue in each Compy. must expect to answer for all filth and Nastyness about their mens barracks and will suffer Accordingly – the Centrys on the parapet is Order'd to Lett no man do these Occasions Close the walls on the outside – the Steps near the flagg staff to be taken up every night at retreat beating by the Guard

This filthy custom of the Soldiers doing their Occasions in and near the fort has already made a Disagreeable smell in Garrison which must bring Disorders on every individual if not prevented – all Officers to be particular in bringing every man to a Court martial who are guilty of such vile practices

those men who may be under the nesessity of doing their Occasions at night must Apply to the Sergeant who are to Let him out & see he returns

---

806 Benjamin Money enlisted in the 2nd South Carolina Regiment on 4 November 1775. On 19 January 1778 he was sentenced to spend eight days in the "Black Hole" for being absent without leave.

807 George Sheed was the commissary general until 30 April 1780. After the fall of Charleston he served again as the commissary general from March to October 1783.

808 James Castello was born in 1745. He enlisted in the 2nd South Carolina Regiment on 22 September 1777. He was reprimanded at the halberts on 20 January 1778 for neglect of orders. In 1779 he was under Captain Baker's company. He served in the regiment until 1783. He died on 10 August 1785 at the age of 40.

809 Jacob Williams served in the 2nd South Carolina Regiment. He was reprimanded at the halberts on 20 January 1778 for neglect of orders. He also served ten months in the light dragoons under Lieutenant Colonel Samuel Hammond.

Orders 21st Jany ~~Lt Colo. Marion~~ by Majr. Horry } . . Parole Schoulkil . Countr Sn. Attackt
Fort guard tomorrow Captn. Lesesne & Lts. Hall & Baker
Rear Sergt. Carter – Advance Corpl. Murphy

A court martial to sit this forenoon at 10 OC: to try all prisoners brought before them, Evid: to be orderd to Attend in time – Captn. Lesesne presidt. Lieuts. Hall, Baker Mason & Roux members

the Advance guard to be reduct tomorrow to a Corpl. and 6 men the out Centry usually posted from that guard to be discontinued till further Orders

After Orders by Majr Horry } NB by sentence of Last court Sergt. Majr. Coffer for Breaking Sergt. Marlows Barracks and for other abuses, Sentence to make good the dammage and to apologize for his conduct – David Manly of Captn. Moultries Compy. for Losing his Bayonet to be under stoppages to replace the Bayonet [810]

Orders 22nd Jany. By Majr. Horry } . . Parole Virginia . . Countr. Sn. 33 by FM
For the fort guard tomorrow Captn. Mazyck & Lieuts Mason & Gray – Rear Guard Sergt. Wickham

that on such afternoons as the Battalion turns out to exercise the Commanding Offrs. Of each Company present at Roll Call are to turn out from their respective Compy. all Such men as are new recruits or Such as are imperfect of the Manual Exercise to have them marched to and formed in the rear of the regimental 378arade, When the Officer for for Exercising the Battalion will appoint a Sergt. to take charge of them as an Aukward Squad with Orders to march them on the beach and to form at a proper distance from the Battalion, there to Instruct them in the Manual Exercise to March & Carry their Arms well and not to return or Discharge them but in Garrison & after the Battalion is discharged – – –

Orders 23d Jany. by Lt. Colo Marion } . . Parole Egypt. . Countr Sn Elbert
For the fort guard tomorrow Captn. Dunbar Lts. Burke & Guerry
Rear guard Sergt. Keels

As Long Hairs Gather much filth and take a Good Deal of time & trouble to Comb & keep it Clean & good Order – the Lt. Colo. recommends to every Soldier to have their Cut short to reach no further down than the top of the shirt Collar & trimmed upwards to the Crown of the head the fore top short without Toppee & short at the sides – those who do not have their Hairs in this mode must have them platted & tyed up, as they will not be Allowed to Appear with their hair down there Backs & over their forehead & down their Chins at the Sides which make them Appear more like wild Savages than Soldiers [811]

---

[810] David Manly enlisted in the 2nd South Carolina Regiment on 24 December 1776. On 21 January 1778 his pay was stopped until he paid for a bayonet he lost. He became a corporal under Captain Thomas Moultrie on 13 July 1778. He was in the siege of Savannah.

[811] The stereotype of the Revolutionary War soldier was the the men all had long hair or wore a wig. Wigs had gone out of fashion by the Revolution and only those who were older may have worn them. The description of the hair cut that Marion wanted the men to have would resemble a modern military haircut. In the 1st South Carolina Orderly book they go so far as to say that the men with long hair did not look manly: "The Coll. was in hopes that the Noncommissioned officers & privates would have Followed the Example of the officers in having their hair Cut Short, & is in Expectation that they will of their own accord follow so usefull a fashion,

The Major will please pick out three men to be regimental Barbers who are to be excused from mounting Guards or do fatigue duty they are dayly to Dress the mens head & Shave them before they mount guard; the men to pay them half a Crown a week each man – Any Soldier who comes on the 379arade with Beards or hair uncomb'd shall be dry Shaved immediately & have his hair dresst on the 379arade – the Orderly Serg[ts]. Or Corporal of Companies are to Call on & see the Barber dress & Shave their men that are for duty & See that they are Clean & their Cloaths are put on decently or must Expect to answer for the neglect

The Commissioned Officers are deserved to pay attention to their mens dress at all times particularly when for duty

No Off[r]. to take Charge or march off a Guard without the men have Complyd with the above Orders and are as Clean and decent as possible as the Circumstances of Clothing will permit –

No person to sell any Spirituous Liquors or Beer without leave from the Commanding Officer of the regim[t]. But may sell Candles, Soap ~~Cakes~~ or all Kind of Eatables

The Sutler M[r]. Young have Leave to sell one Gill rum & one Quart Beer p[r]. man a day & no more without a wrighten permit from an Off[r]. of the Comp[y]. the man belongs [812] he is not to Lett have or sell any liquors before the Guard is relieved or After retreat beating

Any persons who Cut Top or Bark any trees on this Island must expect to suffer Agreeable to a former Order

It tis Expected the Guards will pay the usual Compliments to all Field Off[r]. of the reg[t]. as well as the Commanding Off[r].

No Off[r]. to go to Haddrels point or off this Island without Leave from the Commanding Off[r]. – and all Off[rs]. To see the Orders of the day before they go any distance from the fort – Serg[t]. Newton of Capt[n]. Harleston Comp[y]. to do duty in Capt[n]. Mottes till Serg[t]. Laurence returns from Command

A court Martial to set this morning at 11 OC: to try all prisoners brought before them all evidences to Attend – Capt[n]. Moultrie presid[t]. Capt[n]. Dunbar L[ts]. Guerry Burke and Hart members

L[t]. Col[o]. Marion **Fort Moultrie – 1778**

Aft[r]. Orders one Serg[t]. & 3 privats to be Added to the Advance Guard immediately & to follow such Orders as he will have the Corporal analyze will relieve he is to post one Centry by day & 3 at night. James Campbell of Capt[n]. Ashby Comp[y]. Timothy Green of Capt[n]. Charnocks & David Stewart of Ashby are Appointed Barbers to the regim[t]. –

Orders 24th Jan[y]. } . . Parole Lisbon . . . Count[r]. S[n]. Portugal
L[t]. Col[o]. Marion } Fort the fort guard tomorrow Capt[n]. Ashby L[ts] Hart & Perreneau
Rear guard Serg[t]. Coleman

Lieu[t]. Baker is Appointed to Act as Adjutant till further Orders he is to be Obeyd as Such –

The Orders of the 25th June last to be strictly Complyd with

---

without Laying him under the Necessity of Essuing an Order for that purpose, However some of the Men may Prize & Effaminate Length of hair, Short hair is Certainly better for actual Service"

812 This is the John Young who had served as a sergeant in the 2nd South Carolina Regiment but had his leg shot away during the battle of Fort Sullivan on 28 June 1776.

NB this night L^{t}. Capers, one Serg^{t}. 1 Corp^{l}. & 18 privates went on board the guard Boat, & was during the night from Fort Moultrie to fort Johnson back & forwards the Watch word – Work –

Orders 25th Jan^{y}.
L^{t}. Col^{o}. Marion } . Parole Sparta . . $^{in\ town}$ Count^{r}. S^{n}. Malta

for the fort Guard tomorrow Captn. Charnock & L^{ts}. Hall & Mason & Guerry

Rear Guard Serg^{t}. Smith – Hobcaw Serg^{t}:

Guard boat tonight L^{t}. Pet^{r}. Gray

Commanding Officers of Companies will take care not to permit any Orderly Serg^{t}. or Corporal of their Comp^{y}. to be absent from Garrison on any Account whatever as they are Liable to be called on at Different hours & minutes and Should be allways in the way when the Adjutant calls on them

No Sergeant to march his men for guard on the Grand parrade before they are examin'd by a Commissioned Officer of the Company without their shoud be no Such Officer in Garrison

All Regimental Barbers who do not attend to shave & dress the men in time to mount guard may expect to be Severely punished

Commanding Officers of Companies are desired to see their mens having Cut and dress agreeable to orders of the 23^{d} Instant.

the Adjutant not to receive any men for duty that has not complied With orders as to dress and Arms – Orderly hours to be at 10 OC: in the morning precisely –

The Quarter master Serg^{t}. to go round the fort every morning boath in and out of it and report every filth that shoud be removed agreeable to orders and this to be done by orderly hours

all report of Guards to be made (except out post) at half after 9 OC: in the morning All Officers who obtain Leave of Absents to wait on the Commanding Officer at his return and Acquaint him therewith agreeable to former Orders – 30

The L^{t}. Col^{o}. is sorry to find that Orders to produce the Orderly books of each officer the first of december last has not been Completed with but by two which also he returns his thanks – he now Orders that the Orderly books of every Officer and their Company orderly books be produced to the Commanding Officer of the Reg^{t}. for Examination the first day of March next & Expects by that day every orderly book will be up to the day

A court martial to set tomorrow at 11 OC: in the fore noon to try all prisoners brought before it – all evidences to Attend

Captn. Lesesne president, Captn. Shubrick L^{ts}. Capers Gray & Roux members –

NB the guard Boat returnd watchword Johnston

L^{t}. Col^{o}. Marion **Fort Moultrie – 1778 –**

Orders 26th Jan^{y}.
by L^{t}. Col^{o}. Marion } . . Parole Lyeurgus . . Count^{r} S^{n}. 46 –

For the fort Guard Captn. Lesesne L^{ts}. Burke & Hart

rear Guard Serg^{t}. Carter – G Boat L^{t}. Perreneau & Serg^{t}. J. Coleman

---

Orders 27th Jan^{y}.
L^{t}. Col^{o}. Marion } . . Parole Solon. . . C S^{n}. Athens

For the fort Guard Captn. Moultrie L^{ts}. Marten & Capers

Guard Boat L^{t}. Mason & Serg^{t}. Burtell

Rear Guard Serg^{t}. Kolb

The Last court martial of which Capt[n]. Lesesne was presid[t]. gave in the following Accts Certify'd and the Sum forty Seven pounds 10 p in the hands of the L[t]. Col[o]: for the use of the Heirs who Died and was Killed in the service, any persons having Lawfull Claim must make it known before the first day of March next after which it will be delivered up to the treasury

Name	Amount	Remarks
Patrick Morin [813]........	£7.3	Died at fort Moultrie 1775
John Bonall..............	4.2.6	Killed 28th June in fort Moultrie
John Heukie	14.2.6	
Isaac Edwards.......	1.15.	
Rich[d]. Rogers	3.12	Diserted
Duncan M[c]Farlen [814].......	5	
Richard Bennet	8	
John Humphries [815]......	6.2.6	
Will[m]. Wheeler.........	7.7.6	Died in Gen[l]. Hospitle
	£ 47 – 10. 6	

Fort Moultrie 26th Jan[y]. 1778.

We, the president & members of the Court Martial of the above date do Certify that Capt[n]. Ashby has deliverd up to us the Sum of forty seven pounds 10 p. which he says is due to the Estate of the deceased & deserters of the Company made out as above –

Thom[s]. Lesesne presid[t].
Jacob Shubrick } members
Pet[r]. Foissin

Corp[l]. Robert Richie of Capt[n]. Shubricks Company is Appointed as a Sergeant in the same & is to be Obey'd as such [816]

Commanding Off[rs]. of Companies must be Carefull not to give Leave of Absents to any of their Sergeants & Corporals without first knowing from the Adjutants if they are for duty

NB according to Sentence of last court Serg[t]. Maj[r]. Coffin for being drunk on duty & for wantonly striking several men & Corp: Watt in particular & for partaking in order the Serg[t]. for duty [817]– is Acquitted for the Last Crime & reprimanded for the former – James Sample

[813] Patrick Moran enlisted in the 2nd South Carolina on 4 November 1775. He died on 16 June 1775.

[814] Duncan McFarlin enlisted in the 2nd South Carolina Regiment on 4 November 1775 and deserted on 2 December 1776. He was in the militia at the fall of Charleston in 1780.

[815] John Humphreys enlisted in the 2nd South Carolina Regiment on 4 November 1775 and was discharged on 2 July 1778.

[816] Robert Richey (Ritchie) enlisted in the 2nd South Carolina Regiment on 8 July 1775 under Captain Barnard Elliot. He reinlisted on 4 November 1775. He was promoted to sergeant on 27 January 1778 in Captain Shubrick's company. He was reprimanded for neglect of duty on 2 February 1778. He was discharged on 8 July 1778. He was a sergeant in the 3rd South Carolian Rangers under Captain John Hennington from 1 August 1778 to 31 May 1779. From March to July 1779 he was under Captain Lyell in the Rangers. After the fall of Charleston he served in the militia.

[817] Nehemiah Watt.

for siting down on his post remitted 60 lashes [818]– Rowland Walker for the Same pardond [819]– often being Lyed & slept – Thomas Smith for Absents without Leave rec^d^. 100 lashes

NB the guard boat did not go this night Occasion by the detatchment go away & not man Sufficient to man the boat –

L^t^. Col^o^. Marion **Fort Moultrie – 1778 –**

Orders 28th Jany. L^t^. Col^o^. Marion – } . . Parole Columbus . Count^r^. S^n^. Captn. Blake
Fort guard tomorrow Captn. L^ts^. Warley & Mason
Advance Serg^t^. Bartell – rear Serg^t^. Halladay

Genl. Orders 27th. Jany. rec^d^. this day – By Genl. Moultrie

The detachment that were Orderd to hold themselves in readyness to go on board the Armed vessels are to Embark immediately Rank

Captn. Blake 1 Subaltern 1 Serg^t^. 1 drum 1 fyffe & 34 ~~privates~~ & file of the 2nd Reg^t^. to Embark on board the Genl. Moultrie – 1 Subaltern, 1 Serg^t^. & 24 rank & file men to go on board the fair American Commanded by Captn. Morgan one Subaltern & a Sergeant and 15 rank & file to go on board the notre dame Commanded by Capt. Hall for this duty L^t^. Proveaux & Blameyer who are to draw Lotts for the Choice of ye two Barggs – NB Captn. Blake march'd his Command off at 2 OC: A. M.

Orders 29th Jan^y^: L^t^. Col^o^. Marion } . . Parole Methridates . . Count^r^. S^n^. 5
Fort Guard tomorrow Captn. Motte L^ts^. Gray & Hart
Guard Boat tonight L^t^. Hall – rear Guard Serg^t^. Kiels

A court martial to set this forenoon at 11 OC. to try all prisoners brought to it all evidences to Attend presid^t^. Captn. Motte L^ts^. Hart, Roux, Hall, & Gray

one Subaltern 1 Serg^t^. 1 Corp^l^. & 15 privates for the Guard boat tonight amunition as usial – The Off^r^. of the guard boats to see the men Deliver all their amunition to Serg^t^. Fletcher before he Discharge them also the Amunition for the Swivels and to make a report thereof – and to deliver the Bargs oars & saills and every thing belonging to her to the Serg^t^. of the rear guard who is to make a report of it and take charge of every thing which may be Landed near his Guard & Acquaint the Commanding Off^r^. of it Immediately

---

Orders 30th Jan^y^. ~~L^t^. Col^o^ Marion~~ Maj^r^. Horry – } . . Parole Randolph . Count^r^. S^n^ Success
Fort guard tomorrow Captn. Ashby L^ts^. Martin & Hall
Rear guard Serg^t^. O'Neil

that for the feuter all Soldiers when on duty and particularly when on Centry to support their Arms unless at Such times when Other duties exact a Contrary position, any Soldiers disobeying this

---

818 James Sample served in the 2nd South Carolina Regiment during 1777 and 1778. He was supposed to receive 60 lashes on 27 January 1778 for sitting down on his post, but it was remitted. He later served in the militia.

819 Rowland Walker served in the 2nd South Carolina Regiment under Captain Daniel Mazyck during 1778 and 1779. He was supposed to receive 60 lashes on 27 January 1778 for sitting down on his post, but he was pardoned. He was in the siege of Savannah.

Order may expect to be punished for it – Officer are desired to have this Order made Known to their men and to take notice of every neglect thereof

---

Orders 31st Jan^{y}.
Maj^{r}. Horry }

. . Parole Squadron . Count^{r}. S^{n}. 25
Fort Guard tomorrow Captn. Charnock L^{ts}. Capers & Warley
rear Guard Serg^{t}.

Monthly return 1st Feb^{y}.

one Col^{o} _ 1 L^{t}. Col^{o} _ 1 Major _ 10 Capt^{ns} _ 10 first Lieu^{t}. _ 6 second Lieu^{ts}.
1 Chaplain _ 1 Quart^{r}. Master _ 1 pay master _ 1 Surgeon _ 25 Sergeants
15 drums and fifes _ 2 Armourers mate _ 1 Serg^{t}. Maj^{r}. _ 1 Drum 1 fife majors
Rank & file 214 for duty _ 40 Sick _ 66 on Command _ 10 on furlow _ Total 330 _

L^{t}. Col^{o}. Marion **Fort Moultrie – 1778 –**

Orders 1st Feb^{y}.
L^{t}. Col^{o}. Marion }

. . Parole Lancaster . Count^{r}. S^{n}. Alfred
Fort Guard tomorrow Captn. Lesesne L^{ts}. Masion & Hall
Rear Guard Sergeant ~~J. Coleman~~ Newton
Hobcaw Serg^{t}. J. Coleman –

It tis orderd that no Soldier be permitted to go to town without their drest Clean and neat and for the future only one man out of a Comp^{y}. at one time to have Leave of Absents & to return in 24 hours punctually or be restrained for the future from going to town –

It tis Orderd that no Guards shall march off without all the Officers are in their proper places – all Officers for Guard to be on the parrade before the men and drest – those who may not Attend in time the Adjutant will Acquaint the Commanding Officer immediately who will order the next for duty for that Service – In Consulting Col^{o}. Motte I find that 2 p p^{r}. Man a week to the regimental Barbers is too much they are to have but one half of that sum p^{r}. man a week as they are Excused mounting Guards.

A Monthly return to be given in this day to the Adjutant

Orders 2^{d} Feb^{y}.
L^{t}. Col^{o}. Marion }

. . Parole Huperlee. . Count^{r}. S^{n}. Turk
Fort Guard tomorrow Captn. Moultrie L^{ts}. Martin & Capers
Rear guard Serg^{t}. Gammel

A court martial to set this morning at 11 OC: to try such prisoners as shall be brought before them all evidences to Attend – presid^{t}. Captn. Ashby, Capt^{ns}. Moultrie & Mazyck Lieu^{ts}. Martin & Foissin

NB by the sentence of this court Serg^{t}. Richie Reprimanded for neglect of duty – James Clark for Absents without Leave rec^{d}. 50 lashes 50 remitted on account his returning vollantarily the fourth time from desertion

---

Orders 3^{d} Feby.
L^{t}. Col^{o}. Marion }

. . Parole Morisinia . . Count^{r} S^{n}. Carolina
Fort guard tomorrow Captn. Mazyck L^{ts}. Warley & Mason &
Serg^{t}. Smith
rear Serg^{t}. Newman

Orders 4th Feby. L t. Col o. Marion } . . Parole Egbert . . . Countr. Sn. Long Island
For the fort guard tomorrow Captn. Dunbar Lts. Gray & Hart
Rear guard Sergt. ONeal – Advance Sergt. Richey

A court martial to be held today at 10 OC: in the forenoon to try all prisoners brought before them evidences to Attend – Captn. Charnock presidt. Lts. Hart, Martin, Hall & Roux members

---

Orders 5th Feby. . Lt. Colo. Marion } . Parole Quebeck . . . Countr. Sn.
Fort Guard tomorrow Captn. Motte Lts. Hall & Roux
Rear Sergt. Robert Coleman

Orders 6th Feby. . Majr. Horry } . Parole Crown point . . C Sn. 13.
Fort Guard tomorrow Captn. Ashby & Lieuts. Martin & Mason
Rear Guard Sergt. Jacob Coleman

Lt. Colo. Marion **Fort Moultrie – – 1778**

Orders 7th Feby. Majr. Horry } . . Parole Fort Johnston . . C Sn. Huger
Fort guard tomorrow Captn. Charnock Lts Gray & Hart
Rear Sergeant Burtelle

NB sentence of last court Jno. White of shubricks Compy. recd. 100 lashes for sleeping on his post –

Orders 8th Feby. Maj: Horry } . . . Parole Salem . . . C Sn. Epswich
Fort guard tomorrow Captn. Lesesne Lts Roux & Hall
Rear Guard Sergt. Sergt. Smith – Hobcaw Sergt. Fowler

Divine service will be performed in Garrison this forenoon by the Chaplain of the regt. As expected all Officers and Soldiers off duty will attend the same

---

Orders 9th Feby. Lt. Colo. Marion } . . Parole North Carolina . . C Sn. Georgia
Fort guard tomorrow Captn. Mazyck Lts. Martin & Capers
Rear Sergt. Bond [820]

Orders 10th Feby. Lt. Colo. Marion } . Parole Corsica . . C Sn. Paoli
Fort Guard tomorrow Captn. Motte. ~~Countr~~ Lieuts Warley & Mason
Rear guard Sergeant ONeil

The general Orders of 3d December Last to be read to the men every Saturdy. by an Officer of a Compy. – Agreeable to General Orders of the 4th decembr. last the quartr. Master to make a return of the Commanding Offr. of the Regt. by tomorrow 10 OC: in the morning of the number of new Arms received from the State and the number deliverd each Company and by whom received, also a return of wood received by each Offr. & Each Company from the 7th Jany. to 1st Feby. that it may be transmitted to the quartr. M. Genl. agreeable the Resolve of the Honourable the Continental Congress – all Offr. to render an Acct. of wood recd: by them to the Q. Mastr. agreeable to former Orders – the

---

820 William Bond enlisted in the 2nd South Carolina Regiment on 4 November 1775. He was promoted to sergeant and later was discharged on 29 June 1778.

Q. Mast^r^. to give in the same time and the Beginning of every Month the Quantity of forrage Rec^d^. for the reg^t^. and by whom rec^d^.

Those Off^rs^. who have rec^d^. Moneys from the L^t^. Col^o^. for the Recruiting Service are Desired as immediately as possible to settle with him as he is Oblige to Settle all his Accounts in the reg^t^. without Delay He hopes every Gentleman will Comply with this Order in fourteen days, from this date

Those Soldiers who will work at the regimental Cloathing will be excused all duty & receive three pounds for a Jacket & p^r^. breeches & five pound for each Coat – thirty Shillings per day will be Allowed for a man to cut out & act as foreman who must keep an Exact Account of all work done – – –

Orders 11th Feb^y^. } . Parole Wilmington . . Count^r^. S^n^. 29
L^t^. Col^o^. Marion } Fort Guard tomorrow Captn. Ashby L^ts^. Hart & Gray
Rear Serg^t^. Bartelle
Advance Serg^t^. Gammel.

L^t^. Col^o^. Marion **Fort Moultrie – 1778 –**

L^t^. Col^o^. Marion . . . .
Orders 12th Feb^y^. } . . Parole Biddle . . Count^r^. S^n^. Fleet
L^t^. Col^o^. Marion } Fort Guard tomorrow Captn. Charnock L^ts^. Roux & Hart
Rear Guard Serg^t^. Smith

A court martial to set this morning at 11 OC: to try all prisoners brought before them all evidences to Attend, Captn. Lesesne presid^t^. Captn. Dunbar & Lieu^ts^. Capers, Hall, & Martin Members

---

Orders 13th Feb^y^. } . . Parole Delaware . . Count^r^. S^n^. 78
L^t^. Col^o^. Marion } Fort guard tomorrow Captn. Lesesne L^ts^. Martin & Capers
Rear guard Serg^t^. O'Neil

Orders 14th Feb^y^. } . Parole Brandywine . . C. S^n^. Desist
L^t^. Col^o^. Marion } Fort guard tomorrow Captn. Moultrie L^ts^. Warley & Mason
Rear guard Serg^t^. Jacob Coleman

NB agreeable to sentence of last court Pet^r^. [821] & Arch^d^. Upthegrove [822] for desertion rec^d^.
99 lashes each – the piquetting remitted

After Orders Col^o^. Motte desires all Officers who have received Orders from him for the recruiting Service to get their Accounts ready to Settle with him next week –

---

[821] Peter Upthegrove (Upgrove, Updegroof) served as a drummer in Captain Blake's Company of the 2nd South Carolina Regiment. On 15 August 1777 he was reprimanded for taking peaches from a "Negro". He received 99 lashes on 14 February 1778 for deserting with his brother Archibald. He was also ordered to be picketted, but that punishment was remitted.

[822] Archibald Upthegrove (Upgrove, Updegroof) served in Captain Blake's Company of the 2nd South Carolina Regiment with his brothers Peter and Francis. He received 99 lashes on 14 February 1778 for deserting with his brother Peter. He was also ordered to be picketted, but that punishment was remitted. After this punishment he transferred to the 5th South Carolina Regiment.

Orders 15th Feb^y^. } . Parole Elk River . Count^r^. S^n^ Retreat
L^t^. Col^o^. Marion } Fort guard tomorrow Captn. Mazyck & L^ts^. Hart & Roux
Hobcaw Serg^t^. Bartell – rear Serg^t^. Hallowday

Orders 16th Feb^y^. } . . Parole Captn. Lee . Count^r^. S^n^. Light horse
~~Maj^r^. Horry~~ } Fort guard tomorrow Captn. Dunbar & L^ts^. Hall & Martin
Rear Serg^t^. Riche

Genl. Orders by
G Moultrie } a general court martial to set on Wednesday next in Charles town at such place as the president of the court Shall Appoint to try
Henry Martin of Captn Harlestons Company [823] & James Bleven [824] of Captn. Blakes of 2nd Contin^l^ regiment of foot & all other prisoners that may be brought before them, all evidences for and against the prisoners to be warned in time to Attend

from 1st reg^t^. 1 Captn. 2 Subalterns – 2nd reg^t^. 2 Capt^ns^. 1 Subaltern
3rd 1 Captn. 1 Sub: – 5th one Captn. 1 Sub: – 6th one Captn. 1 Sub:

After Ord^rs^.
by Maj^r^. Horry } Capt^ns^ Motte & Lesesne &n Lt Galvan is Appointed Members to Set on the Genl. Court martial Orderd to Set in Chas town on Wednesd^y^ next, they will Embark the first Opportunity going over to town in and to Attend that duty

L^t^. Col^o^. Marion **Fort Moultrie 1778 –**

Orders 17th Feb^y^. } . Parole Independence . Count^r^. S^n^. America
Maj^r^. Horry – } Fort guard tomorrow Captn. Charnock & L^ts^ Gray & Mason
rear guard Serg^t^. Coleman

A court martial to set this forenoon at 10 OC: to try all prisoners brought before them – Evidences to Attend – Captn. Charnock presid^t^. L^ts^. Hart & Foissin members –

Orders 18th Feb^y^. } . Parole Potomack . Count^r^. S^n^. Virginia
L^t^. Col^o^. Marion } Fort guard tomorrow Captn. Moultrie L^ts^ Foissin & Hart
advance Serg^t^. Fowler – rear Serg^t^. Smith

Captn. Ashby having having resigned his Commission the 16th Inst with Leave, is no longer to be respected or Obeyd as an Officer in the regiment –

L^t^. Thomas Hall is Appointed a Captn. in the room of Captn. Ashby resigned he is to be respected and Obeyd Accordingly

Captn. Hall is to take the Command of of the Company Late Captn Ashbys – Captn. Hall will take a particular Account of all Arms, Accoutrements & regimental Cloathing potts &^c^. &^c^. of his Comp^y^. & Enter them in his Company Book that he may give an Exact Account thereof when called on

Any soldier who shall be found firing their guns on the Island on any pretence whatever may expect to be punished – all Centries seeing any Soldier disobey this Order is to Call to the Serg^ts^. of

---

[823] Henry Martin enlisted in the 2nd South Carolina Regiment on 20 September 1777. After being court-martialed for desertion on 18 February 1778 he was sentenced to receive ninety-nine lashes with the cat-o-nine tails on the bare back and to be picketted for 15 minutes. He received 50 lashes on 30 March 1778 for being drunk and fighting in garrison. He served under Captain Peter Gray during 1779 at the siege of Savannah.

[824] Possibly James Bevins, who later served in the militia with Colonel Brandon Martin.

the guard & acquaint him there with, & Such Serg[ts]. shall immediately Apprehend such soldiers & Confine him for the Same – Any Cut Shott or Lead of any kind being found in possession of the Soldier he or they will be deemed the person who have stole the Aprons from the Cannon & may expect to Suffer for it

---

Orders 19th Feb . } . . Parole Williamsburg . . . C S[n] 29
L[t]. Col[o]. Marion } Fort guard tomorrow Capt[n]. Mazyck L[ts] Roux & Capers
Rear Serg[t]. Keels

Orders 20th Feb } . . Parole Chester . . . Count[r] S[n] 29
L[t]. Col[o]. Marion } Fort guard tomorrow Capt[n]. Dunbar L[ts] Marion & Warley
Rear Serg[t] Williamson

Gen[l]. Orders G. Moultrie 18th Feb[y]. } L[t]. Tho[s]. Hall of the 2nd Reg[t]. Commanded by Col[o]. Motte is appointed a Capt[n] in the Same in place of Capt[n]. Ashby who has resigned 16th Feb[y]. Inst:
A Dilliant B.M.

After Orders L[t]. Col[o]. Marion } Commanding Officers of Companies are to make out their muster rolls of their Comp[ys] by munday 10 OC: in forenoon when the reg[t]. will be musterd, all Officers & Soldiers are particularly desired to Attend

NB by sentence of Last Court Adam Meek for neglect of duty rec[d]. 100 lashes
Jaremiah Hill for selling rum Contrary to Orders rec[d]. 50 lashes –

L[t]. Col[o]. Marion **Fort Moultrie 1778**

~~C[n] S[n].~~

L[t]. Col[o] Marion

Orders 21st Feb } . . Parole Hamstead . C S[n]. Norwich by G H
L[t]. Col[o]. Marion } Fort guard tomorrow Capt[n]. Hall L[ts] Gray & Foissin
rear Serg[t]. Jasper –

Any Serg[ts]. of the rear guard who permits rum or other Spirituous liquor being landed near his Guard may expect to be punished & all Centries on the Bridge is to examine every man who comes from Haddrells point if he finds they have Rum or Spirituous Liquors they are to take it away Deliver it to the Serg[t]. of the guard who is with what he may find in boats to deliver it when releaved to the Command[g]: Off[r]. with the Names of those he may take ~~out~~ it from[825] – Any Woman Whom may be found bringing in or selling Liquors Contrary to Orders will be Whipt & drumm'd out of the regiment & those men who permit it will also Suffer as if they had sold such Liquors – this Order do not extend to Such Liquors which may be brought for the use of the Officers, or Such Gentlemen or Such Gentl[m]. ~~which~~ not belonging to the Garrison of what the Sutler may bring to this Island

Any Centry who Suffer himself to be relieved from his post without a corporal (or Lance Corp[l]. when necessity require) may entrust to be punished as if he had quitted his post without Leave – and all Sergeants or Commanding Off[rs]. of Guards who permits such unmilitary practice will be ~~punished~~ tried for disobedience of Orders & Suffer Accordingly

The regiment is to be as clean as possible when they turn out to be musterd munday –

---

[825] This was a bridge that led from a canal dug up to the back of Fort Moultrie and led to Haddrell's Point. The bridge was a floating bridge built upon boats.

Haddrell's Point

Orders 22nd Feb. L^t^. Col^o^. Marion } . . Parole Burke . . . C S^n^. 12 –
Fort Guard tomorrow Captn. Charnock L^ts^. Hart & Roux
Hobcaw Serg^t^. Kiels – Rear Serg^t^. Riche

Orders 23rd Feby L^t^. Col^o^. Marion } . . Parole Canada . . . C. S^n^. 14 –
Fort guard tomorrow Captn. Moultrie L^ts^. Capers & Warley
Rear guard Serg^t^. Carter

Orders 24th Feby. L^t^. Col^o^. Marion } . Parole Newbern . . C. S^n^.
Fort guard tomorrow Captn. Mazyck L^ts^ Baker & Foissin
Rear guard Serg^t^.

Richard Clark Armourers mate is Appointed Chief Armourer to the Reg^t^.[826] – and Nathaniel Rogers Armourers mate they are to be considerd Accordingly

Commanding Officers of Companies are to give the Armourer Such muskets as may want repair in their respective companies

L^t^. Col^o^. Marion **Fort Moultrie – 1778 –**

Orders 25th Feby. Captn. Charnock } . . Parole Genl. Moultrie . . C S^n^. 92 –
Fort guard tomorrow L^ts^. Hart & Capers
Rear Serg^t^. Smith – Advance Serg^t^. Jasper

A Court martial to set at 10 OC: this morning to try such prisoners as may be brought to it, the Evidences to Attend & the prisoners to warned that their trial comes on at that time

The Gentlemen who have reinlisted men to serve during the war are requested to give in to the Adjutant an Alphabetical List of all their names & he is desired to have a list of the whole made out & put up in the Guard room to prevent Impositions – for the Court martial Captn. Hall presid^t^. L^ts^. Roux & Foissin members

---

Orders 26th Feby. Maj^r^. Horry } . . Parole Gillon . . CS^n^. Commodore
Fort guard tomorrow Captn. __ L^ts^. Gray & Warley
rear Serg^t^. Carter

Orders 27th Feby. Maj^r^. Horry } . Parole Genl. Howe . C S^n^. Georgia
Fort guard tomorrow L^ts^ Baker & Foissin
rear Serg^t^. Kolbb

a court martial to sit at 10 OC: this forenoon to try such prisoners as shall be brought before them Evidences to Attend – Captn. Moultrie presid^t^. L^ts^. Hart & Capers members

Complaint having been made to the Commanding Off^r^. that the soldiers having leave to go over the Haddrels point & with their guns Doth greatly Determint the Inhabitants thereof & that they are considerable sufferers thereby – It is therefore Orderd that no Soldiers for the feuter to be allowed to go over to Haddrels point (Except Officers servants with a permit & without a gun) the Sergeants of the Rear guard will give his Centries that posted on the Bridge Orders Accordingly

---

Orders 28th Feb L^t^. Col^o^. Marion } . . Parole Captn. Joyner . . C S^n^. first
Fort guard tomorrow Lieu^ts^. Capers & Hart
Rear Serg^t^. Fowler

Monthly returns of Each Comp^y^. to be made & Given to the Adjutant by Munday next at 10 OC: in Morn^g^. – the Articles of war to be read to the men this After noon by an Off^r^. of their respective Companies

When Officers of the Reg^t^. Obtain Leave of Absents, they are immediately to call on the Major and Acquaint him theirwith & time of Absents Should the Maj^r^. be absent they are to Acquaint the Adjutant who are to Enter it in a book kept for that purpose when Such Officers return they are also

[826] Richard Clark served in the 2nd South Carolina Regiment under Captain Peter Gray during 1778-1779. He was appointed the chief armorer to the regiment on 24 February 1778. He was in the siege of Savannah

to Acquaint the Major or Adjutant, who are to note it down in their Book – No Officer to Leave the regiment without complying with this order

All Officers who march a guard or party which they may command are not to beat of play any march but the two which is appointed for the Battalion, – all Soldiers who go on furlow or return, to acquaint the major or Adjut who are Keep an Account of it & also deserter return

L^{t}. Col^{o}. Marion **Fort Moultrie – 1778 –**

Orders 1st March } . . Parole S^{t}. Maloe . C S^{n}. 19 .
L^{t}. Col^{o}. Marion } Fort guard tomorrow ~~C~~Lieu^{t}. Warley & Gray
Hobcaw Serg^{t}. Smith – Rear Serg^{t}. Ritchie

A court of Enquiry to be held tomorrow 11 OC: in forenoon to Enquire in to the conduct of M^{r}. Young the Sutler for Selling Liquors Contrary to Order and other Misdemeanors, on Complaint of Serg^{t}. Gammel, Thom^{s}. Thadwell & others – the Sergeant Maj^{r}. to warn M^{r}. Young and all evidences to Attend

Several Soldiers in the regim^{t}. having petitioned to the L^{t}. Col^{o}. to grant leave to Chas. Howell Simmons to teach them to read, Wright & arithmitick – the L^{t}. Col^{o}. have thought proper to grant their request & Exempt Ch^{s}. Simmons from all duties in the reg^{t}. provided he is willing to under take it

For the Court Martial L^{t}. Hart presid^{t}. L^{ts}. Capers & Foissin members

NB the court of Enquiry Acquit M^{r}. Young of the Crime had to his charge

---

Orders 2^{d} March } . . Parole Albany . . . C. S^{n}. Unite
Maj^{r}. Horry } Fort guard tomorrow L^{ts}. ~~Capers & Hart~~ Baker and Foissin
Rear Serg^{t}. Williamson

Orders 3^{d} March } . . Parole Canady . . . C S^{n}. Liberty
Maj^{r}. Horry } For Guard tomorrow L^{ts}. Capers & Hart
Rear Serg^{t} Kolbb . . . .

A court martial to set today at 12 OC: to try all prisoners brought before them – all evidences to be warned to Attend in time

Captn. Dunbar presid^{t}. L^{ts}. Gray & Warley members

Aft^{r}. Ord^{r} . Those Off^{rs}. who have not yet Attested their muster Rolls are to do it immediately & transmit them to the the muster master General without further Notice being given them

---

Orders 21st March } . . Parole New Hamshire . . C S^{n}. Casco Bay
Maj^{r} Horry } Fort guard tomorrow Captn. Charnock L^{ts}. Mason & Warley
Advance Serg^{t}. Kiels – Rear Serg^{t}. Bond

No Centries in future to go in Centries Boxes without it rains & then to keep a lookout through the holes on the Back sides of the Boxes made for that purpose – Officers of Guards are particularly desired to pay strickt attention to this order & to confine or otherways punish those Centries found negligent in so Essential a duty –

Aft^{r}. Ord^{rs}. That Officers who do sign Comp^{y}. morning reports, Presidents of Court martials Commanding Off^{rs}. of Guards & Commanding Off^{rs} of parties Sent from the reg^{t}. on Command, do in future wait on the Commanding Officer with their respective reports at his Quarters & if absent therefrom that they then leave such reports on the table for his Inspection also that all Officers returning to & Join the regiment do wait on the Commanding Officer & Acquaint him of such return & if the Command^{g}. Off^{r}. be Absent from his Quarters to leave on the table a note signifying his return to the reg^{t}.

Orders 5th March } . . Parole Quebeck . . C S^{n}. 14 . by L^{t}. Col^{o}
Maj^{r}. Horry } Fort Guard tomorrow Captn Moultrie L^{ts}. Gray & Foissin
Rear Sergeant Williamson

Orders 6th March } . . Parole St. Laurance . . C S^{n}. 45.
L^{t}. Col^{o}. Marion } Fort Guard tomorrow Captn. Mazyck L^{ts}. Hart & Roux

No Serg^{t}. or Corp^{l}. to have Leave of Absents from garrison but from the Maj^{r} or Commanding Off^{r}.

The General having Accepted the resignation of L^{t}. Adanus Burke he is no longer to be respected and Obeyed as an Off^{r}. in the regim^{t}.

A Court martial to set to day at 11 OC: to try such prisoners as shall be brought before them, evidences to attend – Captn Mazyck presid^{t}. L^{ts}. Capers & Martin mem^{r}

NB Michael Caps tried by this court for drawing provisions for his wife contrary to Orders, was found guilty & Acquitted by the Court – [827]

Orders 7th March } . . Parole Jno. Rutledge . . C. S^{n}. Resign'd
L^{t}. Col^{o} Marion } Fort guard tomorrow Captn. Dunbar L^{ts}. Martin & Capers
Rear guard Serg^{t}. Gammel

Monthly return 1st March

1 Col^{o}. _ 1 L^{t}. Col^{o}. _ 1 Major _ 10 Capt^{ns}. _ 8 first L^{ts}. _ 6 second L^{ts}. _ 1 Chaplain
1 Q Mast^{r} _ 1d^{o} Serg^{t}. _ 1 paymast^{r}. _ 1 Surgeon _ 1 Serg^{t}. Maj^{r}. 1 fife d^{o} _ 24 Serg^{ts}.
15 drums & fifes _ 3 Armourers _ 329 Rank & file

Orders 8th March } . . Parole Charles town. C. S^{n}. Lowndes
L^{t}. Col^{o}. Marion } Fort guard tomorrow Captn. Captn. Charnock L^{ts}. Warley & mason
Rear guard Serg^{t}. Bond – Hobcaw Serg^{t}. Smith

any Articles whatever which may be put in Charge of Centries & are lost the Centries when such things are found missing must expect to be punishd for it and the Corporal of that guard who placed such Centry will also be punished as if he had made away what was lost As it is his particular Business to see the Old Centrys Deliver & these & show every thing to the new who had in Charge

---

[827] Michael Capes enlisted in the 2nd South Carolina Regiment on 4 November 1775. On 6 November 1777 he was an armorer's mate. On 6 March 1778 he was found guilty of drawing rations for his wife, but no punishment was given to him. He was demoted to a private on 12 April 1778 and was no longer an armorer's mate. On 19 September 1778 he was reprimanded for abusing Sergeant Teague. He was discharged on 7 February 1779.

Commanding Officers of guards should Examin before they are releaved whatever his Centries have in charge & if any thing are lost lost to Confine the Centry and Corporal or Such Offr. of Guards are Liable to be called to ~~Acq~~ Acquaint for the Same

N.B. this day Lt. Colo. Marion went into the Country

Majr. Horry **Fort Moultrie – 1778 –**

Orders 9th March Majr. Horry } . . Parole Middleton . C Sn. Resolution
Fort guard tomorrow Captn. Moultrie Lts. Gray & Hart –
rear Sergt. Carter

A court martial to set this forenoon at 10 OC: to try such prisoners as shall be brought before it all evidences to Attend

Captn. Moultrie presidt. Lts. Roux & Foissen members

Orders 10th March Majr. Horry } . . Parole Drayton . . . C Sn. Delegate
Fort guard tomorrow Captn. Mazyck Lts Roux & Foissen
rear Sergt. Colb –

A Court martial to set this forenoon at 11 OC: to try all prisoners brought before it all evidences to Attend Captn. Mazyck presidt. Lts Martin & Capers memb:

NB by sentence of last court, Owen Griffin for theft recd. 100 lashes

---

Orders 11th March Majr. Horry } . . Parole Captn. Blake. . C Sn. Command
Fort guard tomorrow Captn. Dunbar Lts. Martin & Capers
rear Sergt. Holliday – Advance Sergt. Jasper

a court martial to set this forenoon at 10 OC: to try all prisoners brought to it evidences to Attend Captn. Dunbar presidt. Lts. Mason & Martin members

---

Orders 12th March Majr. Horry } . . Parole Second Regimt. . . C Sn. Unanimity
Fort guard tomorrow Captn. Hall Lts. Warley & Mason
Rear Sergt. Coleman

Such Officers who have not yet given in a list of the mens names Inlisted for the war or reinlisted for the war or three years Agreeable to Orders of the 28th Feby. last are desired to deliver the same to the Commanding Officer as soon as possible which the Adjutant or Majr. ~~will~~ of the regt. will enter in a book that will be kept for the inspection of Officers & to present future impositions – also whenever a man is reenlisted for the war or three years the Offr that Enlist him do immediately Acquaint the Major therewith that his or their names may also be entered

Genl. Orders by B.G. Moultrie dated 11th Inst: } all the troops in garrison to parrade tomorrow morning at 10 OC: (except those on magazine Guards) they are to be Supplied with 9 rounds of Blank Cartridges per man they will be ordered to March to Broad Street where they will be drawed up to Compliment the new president to fire such Salutes as will be Orderd by the Adj: Genl. – Broughtons Battery will be ready at 12 OC: tomorrow to fire 15 Cannons – Fort Johnston will follow firing 18 Cannons Fort Moultrie will then take up the fire & Conclude with discharging the Same number of Cannon – this Order to transmited to fort Johnston & fort Moultrie

---

Orders 13th March
Maj: Horry

. . Parole S°. Carolina . . C Sn. Affluent
For the fort Guard tomorrow Captn. Motte Lts Gray & Roux
Rear guard Sergt_ Coleman _

Genl. Orders
by B.Genl Moultrie

The Generall Orders the following resolution to be read at the head of every Corps in this State that every member may be 12th Inst: Acquainted with the same

Signed A. Dilliant B. Majr

In the General Assemble, 2d March 1778 –

Resolved, that instead of the cloathing hitherto allowed to the regiments of this State on the Continental establishment, Each nonCommissioned Offr. Drummers fifes and private shall in futer be Annually found with a coat waistcoat & breeches of Woollen Cloth, one Cap or Hat one Blanket four Shirts four pr. Stockings & four pr. Shoes two pr Breeches of Wool or Coars linen two waistcoat of the same, 2 Leathern stocks & 2 leathern garters And that five Watch Coats be allowed to a Company of men and so in proportion, but that this Allowance of watchcoats be not annually but to last till they are worn out – that each Offr. & Soldier be allowed their full Continental Rations, besides the half pound of Beef allowed by this State – and that if any person do not Chuse to receive his rations in time he may receive the same in money at 5/currency pr Rations

Resolved, that the futer daily pay of non commissioned officers of the several regiment of Infantry of this State be as follows Vizt: that of sergeant majr. 20 s. Quartr. Mastr. Sergt. 17/6. Drum Major 17/6, fife major and other Sergeants 15/~~Each~~ Armourers mate 15/each pr Diem

Resolved, that the dayly pay of the Subaltern Offrs. in the troops of the State be encouraged as follows Vizt: 1st Lieuts. 45/. 2nd Lieuts. 40/ Ensigns 87/6 & of the Quartr. Mastr. 40/, and that agreeable to the Spirit of the resolution of the Continental Congress the Adjutants be Allowed full Captains pay from the date of the resolutions of the Continental Congress respecting adjutants – that the Corporals drummers & fifes in the regiment of Artillery be allowed 12/6 pr. Diem, And the Subalterns Offrs. Adjutants & Sergts. the same pay respectively as those of the like Rank in the regiment above mentioned, and that in futer their shall be only one Captn. & a first & Second Lieut: to each Company in the regt. of Artillery – And the Col° in the regt. of Rangers be allowed per Diem to Commence from the date of his Commission as Colonel, the 1st Lieuts, 50/ the 2d, 50/ Adjutant 60/ agreeable to resolve of Congress, and all non Commissioned Officer in the same regt. in proportion to the pay of the regiments allowed the Offrs. respectively in the regt. of Infantry ~~~ And where as the Continental Congress the resolve of the 22nd day of November last Resolved that it be Earnestly recommended to the Severall States from time to time to exert these utmost Endeavours to procure to the addition of Cloathing made heretofore by Congress of Blankets, Shoes, Stockings Shirts & Other Cloathing for the Comfortable Subsistance of the Offrs. & Soldiers of their Respective Battalions and to Appoint one or more Persons to dispose Of such Articles to the Officers and Soldiers in Such proportion to the General Offr. from the respective states commanding in such army shall direct and at such reasonable prices as shall by the Clothier General in his deputy & being in Just proportion into the wages of the Officers and Soldiers and all Cloathing and all Cloathing hear after to be Supplied to the Offrs & the Soldiers of the Continental Army out of the public stores of the United States, Beyond the Bounty already Granted shall be Charged all at the Like price the Surplus to be paid by the United States provided that Effected measures be Adopted by Each State for preventing any Competition between their purchasing Agent & the Cloathier General or his Agents who is Severly directed to

Observe to Instructions of the respective States relative to the price of Cloathing purchased within such State therefore – Resolved that the above recievd 71/ resolution of the Continental Congress be Adopted by the States & carried in to ~~instruction~~ effect – Assented to Signed

Jn° Rutledge
5th March 1778 –

Regim[tl]. Orders } that at the turning out of the Battalion this after noon the above
Maj[r]. Horry } Resolution be read to the men at the head of the Battalion by the Adjutant
The new Commanding Off[r]. Cannot but observe the Liberal manner in which the late General Assemble has provided for better and more Compatible Sustenance of the troops of this State that truly merits their ~~Attention~~ Acknowledgment, that he Cannot doubt the same Ideas being Difused through all the ranks in the regiment that their perseverance in and Strict Attention to Every part of duty they only Can render Service Adequet to Such Generous bounty, that the Consideration of the trust reposed in them by their Country shoud Stimulate to Actions becoming their profession and the noble cause they are Engaged in & that at Length they may be the means of restoring to their Country peace and plenty abounding in their Country and Smiling on the Countenance of ever Individual therein

---

Orders 14th March } . . Parole General Assemble C S[n]. Honourable
Maj[r]. Horry } Fort guard tomorrow Capt[n]. Charnock L[ts]. Martin & Capers
Rear guard Serg[t].

The Commanding Off[r]. think proper to Inform Officers & Soldiers of the reg[t]. that whenever Leave of Absents are granted them from their post (which is Confined to Sullivants Island tho for ever so short a Distance that on their return they do Either verbally by themselves in Wrighting of by a message ~~Verbally by Themselves~~ Left at his Apartments – in order that he may at all times know those Officers and Soldiers present for duty if wanted - - - -

Maj[r]. Horry **Fort Moultrie 1778 –**

He Like wise has directions from Gen[l]. Commanding in the State and from the Col°. of the regiment to restrain and confine boath Officers and Soldiers to their post and to grant Leave of Absents to none but on the most pressing Occasions and then for as short a time as such Occasion will permit and recommend that when they do out stay the time Limited to their return (unless for every Just occasions made to the Command[g]. Off[r]) not to suffer or go at time of Absents to such defaulters for a very Considerable time after such defaulters the Command[g]: Off[r]. is determined & in duty bound to pay strict attention to Such recommendation and he hopes all Off[rs]. & Soldiers on their part will Conform to & pay Strict attention to the same which will Afford him much pleasure in his duty
After Orders ~ Second Lieu[ts]. Martin & Capers are Appointed first Lieutenants in the regim[t]. the former in the room of L[t]. Hall promoted the latter in room L[t]. Burke resigned, they are to be respected and Obey'd Accordingly – L[t]. Martin is to Join & do duty in Capt[n]. Mazycks Comp[y]. L[t]. Capers in Capt[n]. Shubricks Comp[y]. till further Orders

---

Orders 15th March Maj[r]. Horry } . . Parole Desertion . . C. S[n]. Death
Fort guard tomorrow Capt[n]. Mazyck L[ts]. Baker & Warley
Rear Serg[t]. Riche _ Hobcaw Serg[t]. Raybout

Gen[l]. Orders G. Moultrie 12th Inst } The general has received the proceedings of the General court martial of which L[t]. Col[o]. Henderson was president & will determine upon them in due time [828]– the Court is dissolved and the Officers to Join their respective corps immediately

Signed A Delliant B. M.

G. Ord[rs]. 13th G Moultrie } W[m]. Valentine Esq[r]. is appointed Deputy Commissary General for the troops in this State and is to be Obey'd as such [829]

L[ts]. Martin & Capers 2nd Lieu[ts]. in Col[o]. Mottes Battalion are promoted first Lieu[ts]. in the Same & are to be respected & obeyed as such

M[r]. John Downs is appointed Adjutant in Col[o]. Mottes reg: & is to be respected and Obey'd as such[830]

His Excellency the president returns thanks to the troops for the compliments paid him yesterday – the General approves of the Sentences passed by the Gen[l]. Court Martial of which L[t]. Col[o]. Henderson was president upon Henry Martin, Burrell Hill,[831] James Thomas,[832] Thom[s] Smith,[833] & Bartholemy M[c]Donald [834] & orders that the S[d]. Sentences be put in execution immediately

---

828 William Henderson was born on 5 March 1748. He became a major in the 6th South Carolina Regiment on 29 February 1775 and was promoted to lieutenant colonel on 16 September 1776. On 11 February 1780 he was transferred to the 3rd South Carolina Rangers after the consolidation of the regiments due to losses suffered at Savannah. He was taken prisoner at the fall of Charleston and was exchanged in November 1780. He was transferred to the 1st South Carolina Regiment on 1 January 1781. He was wounded in the battle of Eutaw Springs on 8 September 1781. On 30 September 1781 he was promoted to colonel and served to the close of the war. In 1782 he was promoted to brigadier general of state troops. He died on 29 January 1788 at the age of 40.

829 William Valentine (Valantine) had enlisted on 30 October 1775 in the Volunteer Company of the Colleton County Regiment of Foot under Captain Andrew Cummins. He later became a lieutenant in the 1st South Carolina. He was appointed the deputy commissary general in the Southern Department on 13 March 1778.

830 John Downs became an adjutant in the 2nd South Carolina Regiment on 13 March 1778. He resigned on 6 November 1778.

831 Burrell Hill served in the 1st South Carolina Regiment under Captain William Cattell from 4 November 1775. After being court-martialed for desertion on 15 March 1778 he was sentenced to enlistment for the duration of the war.

832 James Thomas served in the 1st South Carolina Regiment. On 15 March 1778 he was court-martialed for desertion, receiving 100 lashes on the bare back with a cat-o-nine tails. During 1780 he was under Captain William Jackson and Colonel Pinckney.

833 Thomas Smith served in the 3rd South Carolina (Ranger) Regiment under Captain George Lidell. On 15 March 1778 he was court-martialed for desertion and sentenced to enlist for the duration of the war.

834 Bartholomew McDonald enlisted in the 1st South Carolina Regiment on 4 November 1775 under Captain Edmund Hyrne. He was court-martialed for desertion and was sentenced to receive 100 lashes on his bare back with a cat-o-nine tails on 15 March 1778. On 1 February 1780 he re-enlisted under Captain Joseph Elliott.

The Gen[l]. also Approves the Sentence of S[d]. Court as awarded against James Oliver [835] Jn[o] M[c]Namara [836] & Jam[s]. Hurlock [837] for desertion, which Sentences are that they Suffer Death by being shot – the prisoners Sentenced to Death are to be removed to the guard house at Barracks where they are to have a room to themselves in Order to prepare for Death as the Sentence will be executed upon them on Wednesday the 25th of this month [838]

In Order that all Deserters may escape the fate which these unhappy criminals are to Suffer, the General takes this Opportunity of Giving public notice to all such ~~who~~ as have deserted from any of the regiments in this state on Continental Establishment that if they will Join their respective Corps on or before the 15th July next they will be pardoned, Such as Continue out at that time may be Assured that no method will be left unpracticed in order to Apprehend them & that when Apprehended they shall be tried by General court martial & have the sentence against them immediately put in execution. The Gen[l]. recommends it to such deserters as are willing to take Advantage of this public notice by returning to their Duty to deliver themselves up to a Magistrate who will give them a pass which will prevent them being taken up or interrupted in their way to their respective regim[ts] . By Sentence of the Gen[l]. Court Martial Henry Martin of the 2[d] Reg[t]. for desertion is Sentenced to receive 99 lashes on the bare back with Catt of nine tails & to be piqketted 15 Minutes, Burrel Hill of the 1st reg[t]. also for desertion to inlist during the war – Thomas Smith of the same reg[t]. for the same Crime to receive 100 lashes & to Enlist during the War – Barth[l] M[c]Donald of the same reg: & same crime to receive 100 lashes

The Quarter Mast[r]. of the 1st. 2nd. 4th & 5th reg[ts]. are orderd to Call on the Deputy Q. Mast[r]. Gen[l]. for the Camp Colours of their respective reg[ts].

---

[835] James Oliver was born in 1752 in Darlington County, South Carolina. He enlisted in the 2nd South Carolina Regiment on 4 November 1775 under Colonel Moultrie. He served under Captains Blake and then after Blake resigned, under Captain Mason. On 21 February 1778 he was tried for desertion and was condemned to die, but was pardoned if he reinlisted. He became a sergeant under Captain Thomas Dunbar in 1779. He was in the battle of Stono Ferry and the siege of Savannah. He was later demoted to private and served under Captain Turner. He transferred to the 3rd South Carolina (Ranger) Regiment and was taken prisoner in the fall of Charleston. Afterwards he was sent to Virginia, where he served in the militia.

[836] John McNamara enlisted in the 1st South Carolina Regiment under Captain Adam McDonald in 1775. He enlisted in the 6th South Carolina Regiment on 23 April 1776. He was sentenced to die by firing squad for desertion on 13 March 1778. Reverend Oliver Hart watched the execution of a Sergeant Malcolm in 1778. He wrote that "Sergeant Malcolm was blindfolded and made to drop to his knees. He was then shot, which was thrice repeated before he was quite dead."

[837] James Harlock enlisted in the 1st South Carolina Regiment on 4 November 1775. He was sentenced to die by firing squad for desertion on 13 March 1778.

[838] On March 25th the prisoners were marched to Hempstead Hill in Charleston and executed at 9:00. This is one of the few executions to occur, since most death penalties were pardoned or transmuted. The 1st South Carolina Orderly book explains the method the firing squad was selected, "The Deputy Quarter master Genl. is to provide 3 Coffens for the Crimonals to be Carried with the prisoners – one Sergt. & 2 men from the Granadier Company and 2 men from the Light Infentry Company & one from the other Companies of the first Regiment are to be appointed by Lots to be Drawn to Execute the prisoners.

G. Ord[rs] 14th Inst G. Moultrie } The Sentence of the General Court Martial respecting L[t]. Henry Perreneau is as follows, The Court having Maturely weighed the whole matter are of opinion that L[t]. Perreneau is not Guilty And do therefore Acquit him with honour – [839]

The General Cannot altogether Agree in Opinion with the Court, he however Confirms the Sentence & Discharges L[t]. Perreneau from his Arrest & Orders him to Join his regiment –

Signed A. Dilliant. B. M.

R. Ord[rs]. by Maj Horry } M[r]. John Down being Appointed Adjutant to the regim[t]. as by Gen[l]. Ord[rs]. dated 13th Inst:_ Lieu[t] Baker to be Acting Adj[t]. To the Same is therefore to return to & Join Capt[n]. Blakes Company & to do duty in the same as usual – the Commanding Officer returns his thanks to Lieu[t]. Baker for his Vigilance & Strict Attention to duty while acting as Adjutant to the regiment –

Orders March Maj[r] Horry } . . Parole Gen[l]. Moultrie . . C S[n]. Humanity
Fort guard tomorrow Capt[n]. Dunbar L[ts]. Mason & Gray
Rear Serg[t]. Oneal

The Quart[r]. Master is Orderd to keep a book for each mens cloathing to have the same ruled in Column for each Article & on delivery of any article to Enter the same with the mens names & also the time of delivery the Off[r]. to whom deliverd – A Strict Off[r]. also to keep a book & enter every Article in the Like manner – this book to be Kept Alphitically in order that should any Soldier in futer Apply to the Commanding Off[r]. for Cloathing Due him may have it in his (tho the Officers of the Comp[y]. be Attend) by recource to the Quart[r]. Mast[rs]. books have Justice done the Complanant –

Maj[r]. Horry **Fort Moultrie – 1778 –**

The Q. Mast[r]. to deliver this day to the Commanding Off[rs]. of Companies for each man in their respective companies, one shirt & one p[r]. Shoes & the Off[r]. receiving the same is to Sign his name in the Q. Mast[r]. book under the hand or title to whom deliverd – Ch[r]. Howel Simmons being Exempted from duty for the purpose of Teaching, Reading, Wrighting & Arithmatic to such Soldiers of the reg[t]. as are Willing to learn It is therefore Orderd that he keep a book in which every soldier that Enters do wright the time of his Entrance and if cannot wright to make his Mark with an Evidence Mark and that no more than 5 / p[r]. be paid by each Soldier

the Off[rs]. of the regiment having made a Liberal Submission for Educating the Soldiers Children of the reg[t]. will Deposite the same in the hand of L[t]. Pet[r]. Gray who is please to under take the same & who will Act as trustee for such Children – All Soldiers of the reg[t]. who have Children fit to School are desired to produce them to the trustee who will Enter them & will Attend in a more particular manner to see that Justice be done in their Education [840]

---

[839] Lieutenant Perrenneau was arrested by Lieutenant Colonel Marion on February 19th for "disobedience of orders and Neglect of duty". No details are known about this incident.

[840] In the 1st South Carolina Orderly book it states "tomorrow being St. Patricks day such Non Commissioned officers & soldiers as are Natives to the kingdom of Ireland are to be Excused Duty & the paymaster will pay them tomorrow the pay Due to them". Since the 2nd South Carolina Regiment was at Fort Moultrie, they did not get this day off.

Orders 17th March Majr. Horry . . . } . . Parole Gen. Howe . C Sn. St. Augusteen
Fort guard tomorrow Capt. Hall Lts. Galvan & Perreneau
rear guard Sergt. Williamson

Ruben Minor of the Granidr. Compy. is Appointed a Sergt. in the regt. & is to be Obey'd as such he is to Join & do duty in the Light Infantry Compy. till further Orders

---

Orders 18th March Majr. Horry } . . Parole Col° Williamson . . C Sn. 96 –
Fort Guard tomorrow Captn. Motte Lts. Roux & Warley
rear Sergt. Advance Sergt.

A court martial to set this forenoon to try such prisoners as shall be brought before it – all Evidences to Attend – Captn. Charnock presidt. Lieuts. Roux & Foissin members

---

Orders 19th March Majr. Horry } . . Parole Boston . . . C Sn. Navey
Fort Guard tomorrow Captn. Charnock Lts Capers & Baker
rear Sergt. Jaspier

NB by Sentence of Last court Thos. Windsor from stealg. A shirt recd. 100 lashes
Wm. Laws for drunkenness to do double duty for 8 days [841] – Wm Norman for Neglect of duty, was to receive 100 lashes but remitted

Aftr. Ordrs. M. Horry } The Hobcaw guard in futer to March through the road of Mr. Miller Leading over the damm called the Salt water damm belonging to Mr. Dart which will Save a mile & half to the Magazine but they are not to March or go through Mr. Darts plantn. & the Offr of that guard to strictly to Confine his men to his Guard & not to suffer them to Leave it without permission & to report when relieved all who Disobey this Order

Majr Horry **Fort Moultrie – 1778 –**

The Commanding Offr. thanks proper to Emit the following Rations as allowed to the Offrs. & Soldiers as by a Late Resolve of the Honbl. General Assemble, in Order that every individual may know what they have a right to receive – Offrs. that do not Chuse to have their Allowance in kind will inform the Q. Master the part they Chuse to receive

See Continental rations at the end of this book, with ye Adition of State

---

Orders 20th March . Captn. Motte } . Parole Horry . . Countr Sn. Huger
Fort guard tomorrow Captn. Mayzyck Lts. Warley & Mason
rear Sergt. Keels

A court martial to Set this morning at 10 OC: to try such prisoners as may be brought before them all evidences to Attend Presidt. Captn. Dunbar Lts. Gray & Warley

---

[841] William Laws was born on 17 December 1764. He enlisted in the 2nd South Carolina Regiment on 4 November 1775. He was sentenced to do double duty for 8 days on 19 March 1778 for being drunk. After the fall of Charlestown he served 63 days in the partisans under General Francis Marion. He died on 18 October 1812 at the age of 48.

The Q. M. Sergt. to have the Chimneys Swept in the Garrison & that the Shrowdes be immediately put in Order –

NB by Sentence of Last court this day. John Robertson of Mazycks Compy for Losing his Blanket to be put under Stoppages to replace it [842]

Lt. Mason taken sick was relieved by Lt. Gray

Orders 21st March
Majr. Horry } . . Parole Colo. Motte . . C Sn. McIntosh
Fort Guard tomorrow Captn. Dunbar Lts. Perrenau & Martin
Rear Sergt. ONeil

A court martial to Set this forenoon at 10 OC: to try such prisoners as shall be brought before it, All evid: to attend Presidt. Captn. ~~Mazyck~~ Hall Lts. Perreneau and Martin members

Commandg Offrs. of Compys to pay Attention to Orders Issued 24th last Month

Genl. Ordrs.
by Genl. Moultrie } The deputy Commissary Genl. in futer to Issue rations to the Officers & Men on Continental Establishment belonging to this State in the following Manner NB see at the end of this book –

Each rations to Consist of one pound & half of Beef or 18 Oz. pork

the quarter master or other persons drawing provisions for any regimt in futer, on the last day of every Month to make out an Abstract of the numbers of rations due each Offr. & Soldier Respectively & to deliver the same to the Deputy Commissary Genl. who is to Compare it with his books & finding it right shall Certify there on that the several Abstract are Just & that such a Sum as he shall find to be due should be paid to the respective pay masters of the regt. Corps or Detatchment – if a Commissary is Orderd to leave a post before the end of the Month the rations Abstracts are to be made out to the day of his or their leaving the post & Certified by the Commissary as afore Said

The deputy Commissary General in futer to Issue to the respective Offrs. of the Genl Hospitles the number of rations Allowed by the Continental Congress with the Addition of the half pound of Beef Allowed by this State

Major Horry **Fort Moultrie – 1778 –**

Orders 22d March
Majr. Horry } . . Parole Cadiz. . Countr Sn. Spain
For Fort guard tomorrow Captn. Hall Lts Baker & Capers
Habcaw Sergt. Carter – Rear Sergt. Newton

Orders 23d March
Maj: Horry } . . Parole Dartmouth . C Sn. New England
Fort guard tomorrow Captn. Motte Lts Mason & Warley
rear Sergt. Burtel

No leave of Absents to be granted to non Commissioned Officers but by Commanding Offr. or Major of the regimt.

A court martiall to set this forenoon at 10 OC: to try such prisoners as shall be brought before it. Evid: to Attend Captn. Charnock presidt. Lts. Galvan and Foissin members

---

842 Drummer John Robinson enlisted in the 2nd South Carolina Regiment on 19 January 1778 as a drummer under Captain Daniel Mazyck. On 20 March 1778 his pay was stopped until he paid for a blanket he lost. He was in the siege of Savannah.

NB by sentence of last court David Whyly for Insolent Beheavour Rec^{d}. 100 lashes Corp^{l} Conyers for absents without leave reduced to a private – Owen Griffin for Insolents & disobedience of ord^{rs}. rec^{d}. 100 lashes – Jno. McCaid for drunkenness on Guard to be put on Guard every other day for 8 days, & to be Confind the day he is not on guard – NB by the court this day James Smith for being drunk and Offering to sell his Shirt was reprimanded at the Halberts to be put on guard every other day for 8 days

---

Orders 24th March Maj^{r} Horry } . . Parole Bedford . . C. S^{n}. Salem
Fort guard tomorrow Captn. Charnock L^{ts}. Gray & Foissin
rear Serg^{t}. Wickam

Orders 25th March Maj^{r} Horry } . . New London . . C S^{n}. Connetticut
Fort guard tomorrow Captn. Mazyck L^{ts} Perreneau & Hart
rear Serg^{t}. Ritche – Advance Serg^{t}. Meyers

that in futer no Soldier do attempt taking out the breech pin of his gun as by such attempt many of them are broke but when nessesary to be taken out he is to carry it to the Armourer, who is Orderd to do it and to report to the Commanding Off^{r}. all such who disobey this Order & he or they may depend on being punished by a Court martial for disobedience of Orders, - This Order to be made known to them this after noon at roll-call by an Off^{r} of each Company [843]

A court martial to set this forenoon at 10 OC: to try such prisoners as shall be brought before it, Evidences to Attend in time

Captn. Motte presid^{t}. Lieu^{ts}. Baker & Capers members

---

Orders 26th March Maj^{r}. Horry } . . Parole Chatham . . C S^{n}. Liberty
Fort guard tomorrow Captn. Hall L^{ts}. Roux Martin
rear guard Serg^{t} Williamson

NB by sentence of Court yesterd^{y}. Jno. Tayler for neglect of duty on Guard received 50 lashes [844] – Danl. Gordon for drunkenness on Guard rec^{d}. 100 lashes

---

843 Breech plug.

844 John Barnet Taylor enlisted in the 2nd South Carolina Regiment on 1 April 1777. He was promoted to corporal on 10 February 1778. He received 50 lashes on 26 March 1778 for neglect of duty on guard. He received 99 lashes on 30 March 1778 for being absent without leave. He also had to mount guard every other day for eight days. On 11 June 1778 he received 25 lashes on every other day for four days. This was for breaking into the regimental store and stealing all the clothing there. Between each punishment he was to be confined until 7 o'clock in the morning. He was promoted to sergeant on 13 July 1778 in Captain Charnock's company. He was found guilty of desertion on 2 October 1778 and he was sentenced to receive 100 lashes on the bare back with switches. He actually received 70 lashes over a period of two days. Thirty lashes were not given "on Account of his Book." On 11 November 1778 he had to run the gauntlet for theft. On 20 November 1778 he was confined for losing his waistcoat and breeches, and he was put under stoppages to replace them. In 1779 he was under Captain Charles Motte. On 19 January 1779 he received 79 lashes on the bare back with a cat of nine tails for neglect of duty. On 29 March 1779 he received 100 lashes on the bare back with switches for losing his regimental clothing. He was also put under stoppages until he paid for the clothes. He served in the regiment until 15 November 1783.

Major Horry **Fort Moultrie – 1778 –**

After Orders
Maj^{r} Horry
26 March –

Notwithstanding orders given for the purpose of Keeping the Garrison inGeneral & mens rooms in Particular Clean and free from filth Little Attention are paid thereto – The Commd^{g}. Off^{r}. remind those whos particular business it is to see this order Complied with of such Neglect The Q Mast^{r}. Serg^{t}. to see his fatigue men do their General duty, and Orderly Serg^{ts}. Or Capt. Of Comp^{y}. to see their Comp^{y}. duty perform'd – the Command^{g}. Off^{r}. cannot think that the Orders for an Officer from Each Comp^{y} to visit their mens room dayly is complyed with, Otherwise so much filth would not remain in the mens rooms – He Assures them such a duty so Nessesary for the preservation of Health to every individual Cannot possible be over Lookt – the Serg^{t}. Maj^{r}. or Adjutant is desired once very day to Inspect the Garrison & wherever he finds any filth remaining (after the Orders of the day are given out) to report the same to the Commanding Off^{r}. that proper notice may be taken of those whose duty it tis to see it remov'd

---

Orders 27th March
Maj^{r}. Horry

. . Parole Winyaw . C S^{n}. Geo^{r}. Town by Lt. Col –
Fort Guard tomorrow Captn. Motte L^{ts}. Capers & Baker
rear Sergeant Kolb

Orders 28th March
Lt. Col^{o}. Marion

. . Parole Lowndes. . C S^{n}. president
Fort guard tomorrow Captn. Charnock L^{ts}. Foissen & Warly
rear Serg^{t}. Jasper

It has been Issued sevierly in Orders that every Off^{r}. to produce their Orderly books on a particular day, very few Complying with those Orders; It is not Orderd for the Last time as Standing Orders that every Off^{r}. in the regiment do keep an Orderly book and Insert in it, All Orders which have & may be Issued, And to produce their Orderly books at the end of every month to the Major of the regim^{t}. who is to make a report of Such books to the Commanding Off^{r}_ any Off^{r}. who neglect complying with this Ord^{r}. must expect to be taken notice of for disobedience of Orders

A court martial to set in the presid^{t}. room today at 11 OC: in the forenoon to try such prisoners as may be brought before Them all evidences is Orderd to Attend – The Serg^{t}. Maj^{r}. to acquaint the prisoners the hour their trial will come on & warn the Evidences

Captn. Mazyck presid^{t}. Lieu^{ts}. Capers & Hash members

The Adjutant to report to the Commanding Off^{r}. all Off^{rs}. off Duty who do not attend the Battalion Exercise – all Off^{rs} who may be sick or Ailing & not Capable to Attend the Battalion Exercise to make it known to the Commanding Off^{r}. this Ord^{r}. it is expected will be Strictly Complied with

Lt. Col^{o}. Marion **Fort Moultrie – 1778 –**

The Officers is general may expect to be called on to exercise the Battalion and go through such Manouvres as the Command^{g}. Off^{r}. will point out to them – he gives this Notice that they may not be Called on unexpectedly

When an Off^{r}. Obtains Leave of Absents to go to Haddrells Point or from Garrison without naming the time of his or their return they are to be at exercise in the Afternoon that day – No Off^{rs}. go from Garrison untill he see the Orders of the day without particular permission – this Order do not effect such who have Leave to go to town

At sunset on March 7th the Continental frigate *Randolph* spotted the Man of War *Yarmouth.* With the *Randolph* were several of the South Carolina Navy ships, including the *General Moultrie.* On board the *General Moultie* was Lieutenant Adrian Proveaux and half of Captain Blake's company of the 2nd South Carolina Regiment. The *General Moultrie* was able to stay with the *Randolph,* but the wind carried away the rest of the South Carolina fleet.

An hour after the sun went down the *Yarmouth* came alongside the *General Moultrie* and fired a warning shot, demanding to know who they were. The *Randolph* raised the United States colors and fired a broadside into the British ship. The *General Moultrie* quickly retreated from the fight, but the *Randolph* continued to fire, firing five broadsides for every one that the *Yarmouth* fired. The ships were so close that hand grenades were thrown from the tops of the ships onto each other's decks. The 1st South Carolina Regiment, acting as marines, fired their muskets at men on the deck of the *Yarmouth.* The *Randolph* also fired coehorn mortar rounds, which exploded on the deck of the enemy ship.

The *Yarmouth* fared the worst and had her topmast, mizzen, bowsprit and rigging all shot away. Captain Nicholas Biddle, commander of the *Randolph,* was wounded in the thigh and placed upon a chair on the deck so he could direct the fight. When it looked like the British would surrender the *Yarmouth* was able to fire a lucky shot into the Randolph's powder magazine. The explosion was huge, ripping the ship apart and raining debris onto the deck of the *Yarmouth.* One of the *Randolph*'s ensigns were blown into the forecastle of the *Yarmouth*, unharmed.

Only four sailors of the *Randolph* survived, out of a crew of 330. These four were found floating on wreckage of the ship, five days after the fight. Captain Joor's entire company of the 1st South Carolina was lost. The South Carolina fleet continued on their mission, and by 1779 they had captured 35 ships.[845]

Orders 29th March
Lt. Col°. Marion } . . Parole Unfortunate Randolph C. S^n^. blown up
Fort guard tomorrow Captn. Mazyck L^ts^ Mason & Gray
Rear guard Serg^t^. Smith – Hobcaw Serg^t^. Ritche

Orders 30th March
Lt. Col°. Marion } . . Parole Marshal Villars . . . . C S^n^. 61
Fort guard tomorrow Captn. Hall L^ts^. Hart & Perreneau
rear guard Serg^t^. Newton

A monthly return to be made and given to the Adjutant by Wednesday 10 OC: in forenoon ~~~ Serg^t^. Fletcher to give out to Commanding Off^rs^. of Comp^ys^. the Shoes now in the store in proportion to the number of men in each Comp^y^. taking a receipt for the Same, Command^g^. Off^rs^. of Comp^ys^. to give the Shoes to such of their men as want most & keep an Acct. to Whom Given

NB by sentence of Last Court M. Geo^r^. M^c^.Cormack for beinging drunk on Guard rec^d^. 75 lashes [846]– Henry Martin for drunkenness & fighting in Garrison rec^d^. 50 lashes – Drury

---

845 O'Kelley, *NBBAS, Volume One* pp. 201-205.

846 George McCormack enlisted in the 2nd South Carolina Regiment on 15 December 1777. He received 75 lashes on 30 March 1778 for being drunk on guard. He was under Captain Charles Motte in 1779. He was at the siege of Savannah.

Smith for the same & selling Liquors to be reprimanded [847]– Jn°. Barnard Tayler for Absents without Leave rec[d]. 99 lashes to Mount guard every other day for Eight days

---

Orders 31st March } . . Parole Luxembourg . . C S[n]. Duke
Lt. Col°. Marion } Fort guard tomorrow Capt[n]. Motte L[ts] Roux & Martin
rear Guard Burtell

A court m. to set this forenoon at 11 OC: to try all prisoners brought before it; evid: to Attend – – Lt. Roux presid[t]. L[ts]. Martin & Warley Memb:

After Orders – one Corp[l]. and 3 privates to parade at retreat beating to be a guard over the Materials for Building a Hospitle with orders not to Suffer any persons (except Carpenters under M[r]. Baldon) to take or Destroy any part thereof, this guard to Continue dayly till further Ord[rs]. & to be relived when the other Guards are

NB by sentence of the Court M. this day John Campbel for theft, to recived 50 lashes [848]& to be under stoppages to pay for a shirt of Corp[l]. Fellengharte [849] – Serg[t]. ONeil for beating Jacob Johnson was reprimanded

Major Horry **Fort Moultrie– – 1778 –**

Lt. Col°. Marion
Orders 1st April } . . Parole D[r]. Franklin . . C S[n]. 13.
Lt. Col°. Marion } Fort guard tomorrow Capt[n]. Lesesne L[ts] Capers & Warley
Rear Serg[t]. Rybbout – Advance Serg[t]. P. Smith [850]

Orders 2nd April } . . Parole Silus Dean . C S[n]. Agent by Capt M.
Lt. Col°. Marion } Fort guard tomorrow Capt[n]. Mazyck L[ts]. Mason & Gray
Rear Serg[t]. Carter

A court m: to set this forenoon at half after 10 OC: to try all prisoners brought to it: evid: to Attend: Capt[n]. Dunbar presid[t]. L[ts]. Mason & Gray members

NB Col° . Marion went to town

Orders by Capt[n]. Motte } Orderd that the Q. M. Serg[t]. do immediately have Col°. Mottes apartment put in Order

---

[847] Drury Smith served in the 2nd South Carolina Regiment under Captain Daniel Mazyck. He received 50 lashes on 30 March 1778 for being drunk and fighting in garrison, and additionally he was reprimanded for selling liquor. He was under Captain Mazyck at the siege of Savannah.

[848] This John Campbell is a different person than fife major John Campbell. This John Campbell enlisted in the 2nd South Carolina Regiment on 4 November 1775. He deserted on 12 February 1776, but returned shortly thereafter. He received 50 lashes on 31 March 1778 for stealing Corporal Tellinghart's shirt. His pay was stopped until he paid for that shirt. He was accused of theft on 13 July 1778, but he was acquitted. He was in Captain Harleston's company in 1778. On 5 January 1779 he received 100 lashes on the bare back with a cat of nine tails for stealing another shirt. In 1779 he was under Captain Peter Gray at the siege of Savannah.

[849] Corporal Fellinghart (Tellinghart) served as a corporal in the 2nd South Carolina Regiment in 1778. Unable to determine the first name of Corporal Tellinghart.

[850] Park Smith served in the 1st South Carolina Regiment under Captain William Hext.

NB by Sentence of Court M. this day Serg[t]. Holladay for neglect of duty was reprimanded – Jam[s] M[c]Donald for theft Acquitted – James Leghton for Suspicion of theft was Acquited & released [851]

Orders 3[d] April
Capt[n]. Motte } . . Parole Marion . C S[n]. Capt[n]. Motte by Lt. Col[o]. M
Fort guard tomorrow Capt[n]. Dunbar L[ts]. Hart & Perreneau
Rear Sergeant Kolb –

NB Lt. Col. Marion return 11 OC:

Orders 4th April
Lt. Col[o] Marion } . . Parole Marshal Villiroy . . C S[n]. 21.
Fort guard tomorrow Capt[n]. Hall L[ts]. Roux & Capers
rear Serg[t]. Holladay

Gen[l]. Orders 3[d] Ap:
by Gen[l]. Moultrie
Head Quarters – } Nicholas Eveleigh Esq[r]. Is Appointed by the Hon[bl]. the Continental Congress Deputy Adjutant General with the rank of Col[o]. for the State of S[o]. Carolina & Georgia he is therefore to be Obey'd & respected as Such

Henry Purcell Esq[r]. is also Appointed Deputy Judge Advocate for the States of S[o]. & Georgia & is to be respected & Obeyd as Such

Ferdinand Debram Esq[r]. is also Appointed Engineer with the rank of Major & is to be respected & obeyd as such

Resolve of the Hon[bl]. The Continental Congress at Yorktown Resolved, that the Commander in Chief or Commander of Department shall have full power & Authority to Suspend or Limit the power of Granting furlows or Leave of Absents & to reserve it wholly to himself ~~as he may think fit~~ or impart it to such Officers under him as he May think fit According as he shall Judge the good of the Service requires & that no Off[r]. under Colour or pretence of Authority to him Granted by the 2nd Article of the 4th Section or any other Article in the Rule & Articles of War presume to grant any furloughs or leave of Absents Contrary to the Orders of the Commander in Chief Or Commander of a Department on pain of being punished for Disobedience –

Transmited by
A Dilliant BM – Signed
Ch[s]. Thompson

Lt. Col[o]. Marion **Fort Moultrie – 1778**

Orders 5th April
Lt. Col[o]. Marion } . . Parole Patomack . . C S[n]. 44 –
Fort guard tomorrow Capt[n]. Motte L[ts]. Mason
Hobcaw guard Lt Warley & Serg[t]. Bond
Rear Serg[t]. Reybout

The fort guard to consist of one Capt[n]. 9 Sub: 2 Serjts. 2 Corp[l]. and 30 privates, the Centry before the Commanding Off[rs]. door to be taken off – The Hobcaw guard to be relieved tomorrow by one Subaltern, 1 Serg[t]. 1 Corp[l]. 1 drum & 12 privates –

Divine Service will be performed today by the Chaplain in the flagg Bastion, all Off[rs]. & Soldiers off duty to Attend the Soldier to be as clean as possible

---

[851] James Leaton enlisted in the 2nd South Carolina Regiment on 28 July 1777. In 1779 he was under Captain Peter Gray. He was with Captain Dunbar's light infantry at the siege of Savannah.

Orders 6th April . . Parole Dartmouth . . C S^{n}. Boston
Lt. Col^{o}. Marion } Fort guard tomorrow Captn. Charnock & L^{t}. Gray
Rear Serg^{t}. Coleman

Orders to L^{t}. Warley

S^{r}. you are to march with 1 Serg^{t}. 1 Corp^{l}. 1 drum & 12 privates to Hobcaw to relieve the Magazine guard there, You are to be carefull that no persons Approach too near the Magazine so as it may be any means by liable to be sett on fire by Evil & Designing Enemys of the States – You are to defend it (should they Happen on Occasion) to the Last Extremity – Should this powder Receiver or any person with Orders from the president want to go in the Magazine you are to permit him or them – As the Inhabattants have complained that the Soldiers rob & pillage the Plant^{ns}. round about that port in the night, You are to be very Exact to have the Roll-called as often in the night as you will think nessesary to restrain them from going to pillage or from Leaving the guard you are not to give leave to any Soldiers to go at any time to any plant^{n}. or house from the guard, Except those you send to Garrison for provisions & then they must keep the Directions and not call any plant^{ns}. –

Orders 7th April . . Parole Plymouth . . C S^{n}. 15
L^{t}. Col^{o}. Marion } Fort guard tomorrow Captn Lesesne Lt. Hart
rear Sergeant

No Off^{rs}. for the futer to grant leave of Absence from Exercise in the After noon to any Soldier Agreeable to former Orders & all Off^{rs}. Servants to exercise in Battalion agreeable to repeated Orders those who do not comply wth this order may expect to be punished

L^{t}. Col^{o}. Marion Fort Moultrie – 1778 –

A Court Martial to sett immediately to try Serg^{t}. Gammell for Disobedience of Orders of the 19th of march last

Capt Dunbar presid^{t}. L^{ts}. Roux & Capers members

NB this court Acquited Serg^{t}. Gammel

Orders 8th April . . Parole Moultrie . . Count S^{n}. 2 ...
Captn. Motte } Fort guard tomorrow Captn. Mazyck & L^{ts}. Perreneau
Rear Serg^{t}. Bybold – Advance Serg^{t}. Ritchie

A court martial to set this morning at 10 OC for the tryal of such prisoners as shall be brought before them Captn. Hall presid^{t}. L^{ts}. Mason & Gray members

The Commanding Off^{r}. is Extremely sorry to find the Off^{rs}. pay such little regard to Orders Particularly one Issued as late as the 28th March last which he hope they will pay Attention too

NB by Sentence of last court ~~Edward Murphy~~ Jno. McCaid rec^{d}. 50 lashes for beating Ed. Murphy

---

Orders 9th April
Captn. Motte } . . Parole Pinckney. . Countr. Sn. 87 –
Fort guard tomorrow Captn. Dunbar & Lt. Roux
reare Sergt. Coleman

a Court martial to set this morning at 11 OC: to try such prisoners as may be brought to them – Evidences to Attend

Captn. Charnock presidt. Captn. Mazyck & Lt. Roux members

NB by sentence of the court Sergt. Rybold for beating Lucy Dunns was reprimanded[852]– Thos. Guerring for neglect of duty recd. 75 Lashes [853]

Orders 10th April
Lt. Colo. Marion } . . Parole Bollinbroke . Countr. Sn. Prepare
For Fort Guard tomorrow Captn. Hall & Lt. Capers
rear Sergt. Rybold

Genl. Ordrs. by
Genl. Moultrie } The Detatchment now in town from Colo. Thompson's regimt. Consisting of a field Offr. 3 Captns. & 6 Sub: 6 Sergts. Drums & fifes & 150 Rank & file from Colo. Sumpters Regt. are to hold themselves in readyness to march at a moments Notice this Detatchment to be provided with one hundred rounds pr. Man the D Q M Genl. to furnish 3 Waggons for the Detatchment with an Amunition Chest for the Cartridges All Offrs now out on the recruiting Service to be Called in and no more to be sent on that duty till further Orders all Offrs non commissioned Offrs. & Soldiers that are out upon furlow are immediately to be called to Join their respective Corps & no more furloughs to be given for a longer time than 24 hours till further Orders the Commanding Offrs. of the Different corps of this State are to take particular care to have all their Arms & Accoutrements in good Order & that they have one hundred rounds of Cartridges pr. Man ready and to Apply to the Deputy Q: M. Genl. to furnish them with proper Ammunition Chests and the Number each Corps have Occasions for

Transmited by A Dilliant B. M –

In February 1778 the Congress debated on whether or not there should be an expedition to East Florida. Howe was aware of this and began to prepare for a possible expedition. On 12 March Lieutenant Colonel Thomas Brown's Rangers attacked Fort Howe and captured it with the loss of only one man. They killed two, wounded four and captured 23 prisoners.

Georgia Governor John Houstoun thought there were only five to six hundred men guarding St. Augustine, and that it could taken.[854] In April a group of 500 Loyalists, known as Scopholites, began to march from the backcountry of Georgia and South Carolina to rendezvous with Brown in East Florida. These Loyalists crossed the Savannah River south of Augusta, plundering homes along the way. Howe knew that if these Loyalists rendezvoused with Brown's Indians and Rangers, it would be a formidable force. Howe

---

852 Possibly the wife of John Dunn.

853 Thomas Guerin enlisted on 15 March 1777 in the 2nd South Carolina Regiment. He received 75 lashes for neglect of duty on 9 April 1778.

854 John Houstoun was born on 31 August 1744 in St. George's Parish, Georgia.. He was picked as a representative to the Continental Congress on 4 July 1775. On 8 January 1778 he was elected governor, serving from 1778 to 1779. Seven days after the fall of Savannah in 1778 Houstoun's term ended. He was elected governor again in 1784. He died on 20 July 1796 at the age of 52.

ordered Colonel Samuel Elbert to march with his Georgia Continentals on April 6th, to stop the Scopholites, but the Loyalists were mounted and Elbert's men were not. The Loyalists outdistanced the Continentals.[855]

Orders 11th April } . . Parole Digby . . C Sn. 7.
Lt. Colo. Marion } For the Fort Guard tomorrow Captn. Motte & Lt. Gray
reare Sergt. Jasper

The Honble. Council & Assemble have pas'd the following Law wch all Soldiers are to take notice of (this order to be read to the men today)
Be it Enacted by the Authory a fore sd. that all Deserters who heretofore deserted from Either of the Sd. regts. and not Capitally punished by Martial Law shall be compelled to serve in the regimt. from which he deserted for the full time for which he Originally Inlisted & that all deserters who may hereafter Desert from either of the Regimts. and not Capitallly punish'd by Martial Law shall be Compelld to serve in the regt. from which they Deserted double the time of their Absences from duty – And be it Enacted further by the Authority afore Said that two hundred Acres of Land Including the one hundred Acres Allowed by Congress, be reserved for & granted free of Expense & in fee Simple to every Soldier who hath already Inlisted or shall hereafter Inlist to serve in either of the regts. during the present war provided he doth faithfully compleat his time of Service & in case it shall so happen that any Soldier shall be slain in or Depart this life during this Contest his Heirs shall be Inteluded to the Said two hundred Acres of Land, & no Conveyance or Agreement to transfer Convey or sell the sd. two hundred Acres of Land or any part thereof before the Expiration of the war & the actual grant & Location thereof shall be Null and Void to all manner of purpose whatsoever and all the Lands in the forks between Tugaloo & Keowe river up to the new Charokee boundary line shall be & they are here by Reserved for such purpose, and no grants of Lands within the Sd. District or possession within the same shall be deemed legal & valid, till after the Expiration of the present war & till the Said Soldiers shall have their respective proportion of the Sd. Lands alloted & granted to them

---

Monthly return April } 1 Colo._ 1 Lt. Colo _ 1 Majr _ 10 Captns_ 10 first Lts._ 4 2nd Lts_ 1 Adj: 1 Chap. _ 1 QM_1 Surg: 1 Sergt. M_1 QM Sergt._27 Sergts._1 fife Majr. 13 drums & fifes, 2 Armourers ~~230~~ rank & file 338 ______

Lt. Colo. Marion **Fort Moultrie 1778 –**

Orders 12th April } . . Parole Cambden . . C Sn. 11
Lt. Colo. Marion } Fort guard tomorrow Captn. Charnock & Hart
Hobcaw Lt. Mason – rear Sergt. ~~Carter~~ Williamson

Michael Caps is reduced in the ranks and is no longer to be considered as Armourer's mate – Divine Service will be performed today all Offrs & Soldiers off duty to Attend

---

[855] O'Kelley, *NBBAS, Volume One* pg 208.

Orders 13th April }　. . Parole Consiea . . .　C Sn. 27
Lt. Colo. Marion　For Fort guard tomorrow Captn. Lesesne & Perreneau
rear Sergt. Colb.

A Court martial to Set this forenoon at 11 OC: to try all prisoners brought be it – evidences to Attend, the prisoners to be made Acquainted of the time of trial – this Court to Consist of prisidt. & Ten members if so many Can be had, but not less than five

Captn. Mazyck presidt. Lts. Perreneau, Roux, Gray & Capers

~~NB~~ by sentence of this court martial Sergent John Newton of Captn. Harlestons Compy. is reduced to do duty as a private in the rank this part of the Sentence is Approved of he is therefore no longer to be Obey'd as a Sergt. in the Regimt. the Othr. part of the Sentence will be Considered

---

Orders 14th April }　. . Parole Genl. Moultrie . .　C Sn. 19.
Lt. Colo. Marion　Fort guard tomorrow Captn. Mazyck & Lt. Roux
rear Sergt. Smith

A Court martial to set today At 11 OC. in forenoon to try such prisoners as may be brought to it – evidences to Attend, Captn. Dunbar presidt. Lts Waley & Hart members

Corpl. Alexandr McDonald is Appointed a Sergt. in the regt. & is to be Obey'd as such Sergt. McDonald is to Join and do duty in Captn. Halls Compy. till further Orders [856]

NB: by sentence of Court this day Sergt. Holladay for neglect of duty is Acquited – Jno. Griffin for quiting his guard to receive 75 lashes but remited

---

Orders 15th April }　. . Parole Tunens . . .　C Sn. 32
Lt. Colo. Marion　Fort guard tomorrow Captn. Dunbar & Lt. Capers
reare Guard Sergt. Bond

The guards to be relieved at 7 OC: in the Morning & tattoo at 9 OC: at night till further Orders

A court martial to sett today at 11 OC: to try such prisoners as shall be brought to it – evidences to Attend

Captn. Mazyck presidt. Lts. Hart & Capers members

Lt. Colo. Marion　**Fort Moultrie 1778**

Orders 16th Apr }　. . Parole Orleans . .　C Sn. New York
Lt. Colo. Marion　For Fort guard tomorrow Captn. Hall & Lt. Warley
reare Sergt. Coleman

NB by Sentence of last court Robt. Gamble to double duty for a fortnight neglect of duty Thoms. Nute for neglect of duty to double duty one week [857]– Jno. Gamble for Drunkenness Recd. 50 lashes [858]– Wm. McDowel for drunkenness on Guard recd. 50

---

[856] Alexander McDonald was born in 1750. He enlisted in the 2nd South Carolina Regiment on 4 November 1775 as a sergeant. He re-inlisted and was appointed as a sergeant on 14 April 1778 in Captain Hall's company. He was promoted to sergeant major on 16 November 1778. On 19 January 1779 he was reprimanded for neglect of duty. He was in Captain Dunbar's company at Savannah in 1779. He died in 1844 at the age of 94.

[857] Thomas Nutt enlisted in the 2nd South Carolina Regiment on 16 July 1777. On 16 April 1778 he was sentenced to do double duty for one week for neglect of duty. He was promoted to corporal on 28 July 1778. He was demoted to private on 17 November 1778. In 1779 he was in Captain Dunbar's light infantry at the siege of Savannah.

[858] John Gammel.

lashes – Ad[m] Creighton for the same reason rec[d]. 50 lashes – Phil[p] fry for fraud to be under stoppages one half of his pay till L[t]. Galvan be payed [859]– Tho[s]. Bowen for drunkenness on Guard – Ad[m] Meek for Seting on his post & drunkenness rec[d]. 75 lashes – Jn[o]. Sullivan for drunkenness on Guard rec[d]. 50 lashes

---

Orders 17th Apr } . . Parole Jersey . . C S[n]. Philadelphia
L[t]. Col[o]. Marion } For Fort guard tomorrow Capt[n]. Harleston L[t]. Baker
rear Serg[t]. Gammel

A court martial to set today at 11 OC: in fore noon to try such prisoners as shall be brought before it – evidences to Attend Capt[n]. dunbar president L[ts]. Gray & Perreneau members

NB by sentence of this court Edward Murphy for drunkenness & neglect of duty is to rec[d]. 100 lashes 50 remited – James Anderson for Absence without Leave was reprimanded – Sam[l] Murray for Losing his Gun, the court recommend him to be released to be alowed one Week to Look for it – approved provided he finds it otherways to be subjected to another trial - - - - - [860]

Orders 18th Apr } . Parole ~~Capt[n]. Blake~~ Peterborough C S[n]. ~~the Gantlet~~ Col[o]. Motte
L[t]. Col[o]. Marion } For the Fort Guard tomorrow Capt[n]. Motte L[t]. Hart
rear Serg[t]. Wickom

Commanding Officers of Companies to make a return of their mens Arms and Accoutrements that have been on Command on board the ship Gen[l]. Moultrie to the Adjutant on Munday at 11 OC: in the forenoon who is to make a gen[l]. Return thereof to the Commanding Off[r]. of the reg[t]. by one O Clock [861]

Orders 19th Apr } . . Parole Capt[n]. Blake. . C S[n]. 45
L[t]. Col[o]. Marion } For the Fort guard tomorrow Capt[n]. Charnock Lt Perreneau
Hobcaw L[t]. Gray – rear Serg[t]. Bond – Hospittle a Corp[l].

L[t]. Perreneau is to Join & do duty in Capt[n]. Charnocks Comp[y]. in room of L[t]. Roux who is to Join & do duty in Capt[n]. Dunbars Company till further Orders - - - - - -

---

Orders 20th Apr } . . Parole Lowndes. . . C S[n]. Presid[t].
Major Horry } For the fort guard tomorrow Capt[n]. Lesesne & L[t]. Roux
rear Serg[t]. Coleman – Hospitle a Corporal –

---

[859] Phillip Fry served as a drummer in the 2nd South Carolina Regiment during 1775. One half of his pay was stopped when he was found guilty of fraud against Lieutenant Galvan. He was under Captain Thomas Hall during 1779 at the siege of Savannah.

[860] Samuel Murray enlisted in the 2nd South Carolina Regiment on 4 November 1775. On 4 May 1778 he was confined for 14 days for being absent without leave. He also had to do guard duty every other day while he was confined. He served as a corporal under Captain Thomas Moultrie during 1779 in the siege of Savannah. He served in the partisans as a wagon master from 20 June to 6 September 1780 under Colonel Casey and General Sumter.

[861] This was the return of Lieutenant Adrian Proveaux's detachment that had been on board the *General Moultrie* acting as Marines. They were the first to tell eyewitness accounts as to what happened to the Frigate *Randolph*..

Certain Accounts are now recd. that the Randolph was Unfortunately blown up in an Engagement with the Yarmouth of 64 Guns

L^t. Col^o. Marion **Fort Moultrie – 1778 –**

Orders 21st Apr
L^t. Col^o. Marion } . . Parole Col^o. Pinckney . . C S^n. 19
For the fort guard tomorrow Captn. Dunbar L^t. Capers
reare Serg^t. McDonald – Hospitle Corp^l.

On Fryday next the 24th Inst: the Muster master Genl. Intends mustering the reg^t. it is Orderd that three Muster rolls for Each Company be prepared by 10 OC: in the forenoon that day at which time all Off^rs. & Soldiers off duty must be on the Parrade

---

Orders 22nd Apr
L^t. Col^o. Marion } . . Parole Conde . . C S^n. The Great
Fort Guard tomorrow Captn. Hall L^t. Baker
reare Serg^t. Ritche – Advance Rybold – Hosp: a Corp:

A court martial to set today at 11 OC: to try all prisoners brought before it evidences to Attend – Captn. Harleston presid^t. L^ts. Mason & Hart memb

---

Orders 23d Apr } . . Parole America . . C Sn. 18.
Lt. Colo. Marion } Fort Guard tomorrow Captn. Harleston Lt. Warley
Reare Sergt. Carter – Hospitle a Corpl.

A Court martial to set this forenoon at 11 OC: to try such prisoners as will be brought to it, evidences to Attend ~

Captn. Motte President Lts. Perreneau & Roux members

Genl. Ordrs. 21st Ap: } When a muster is made of any of the Corps of this State the Offr. (that
Genl. Moultrie } is to say the Captn & another officer of Each Company if not on furlow or detatchment) are to Sign & Swear to their respective Muster rolls before the Offr. there Commanding or some Civil Magistrate there present

As Certain Accounts are now recd. that the Randolph was Unfortunately blown up in an Engagement with the Yarmouth of 64 Guns by which Accident the whole Crew and Captn. Joor Lts. Grey & Simons with 2 Sergeants & 50 rank & file of the first ~~Regt~~ Continental Regt. of this State was destroy'd [862]

In Order that the Military of this State may testify their regard & Affection for their Brave but Unfortunate fellow Officers & Soldiers who thus Gallently perished in an heroic Attempt to humble the pride of our Inveterate Enemies – the Genl. Request the Offr. of the several Continental Corps in this State to wear for one month a Crape a round their Left Arms as mourning for them –

A Dilliant B M –

Orders 24th Apr } . . Parole Denmark . . C Sn. 4.
Lt. Colo. Marion } Fort guard tomorrow Captn Motte Lt. Mason
rear Sergt. Burtell – Hospitle Corp: Teague [863]

Corpl. Wm. Murphy of Captn. Mottes Company is Appointed a Sergt. in the Same he is to be Obey'd as Such

Lt. Colo. Marion **Fort Moultrie – 1778**

Orders 25th Apr } . . Parole Norway . . Countr. Sign 3.
Lt. Colo. Marion } For the fort guard tomorrow Captn. Charnock Lt. Hart
Rear Sergt. Smith – Hospitle a Corpl.

A Court martial to set this fore noon at 11 OC: to try all prisoners brought to it, evidences to Attend, this Court to be Composed of one Captn. as presidt. & four other Officers as members – those sett was Captn. Lesesne & Lts. Martin & Capers

After Orders

The Court martial Orderd to set ~~to~~ this day Captn. Lesesne presidt. having Sett with two members Less then was Order'd by the Lt. Colo . & the court having tried two prisoners which was not order'd for trial The proceeding of Said Court is made Void for want of regularity & having Sett not agreeable to Orders - - - - - - - -

---

862 Unfortunately any information on Lieutenants Gray or Simon of the 1st South Carolina Regiment are lost to history.

863 John (Josiah) Teague was born in 1759. He was promoted to sergeant in the 2nd South Carolina Regiment under Captain Mazyck on 30 June 1778. On 31 January 1779 he was mult a weeks pay for neglect of duty and disobedience of orders. He was in the siege of Savannah. He served in the regiment until 1 July 1783. He died in 1839 at the age of 80.

Orders 26th Apr
Majr. Horry } . . Parole Holland . . C Sn. States
For fort guard tomorrow Captn. Lesesne Lts. Perreneau
rear Sergt. Kiels – Hobcaw Lt. Foissin – Hospitle Corpl. Roberts

---

Orders 27th Apr
Majr. Horry } . . Parole Colo. Elbert. . CSn
For Fort guard tomorrow Captn. Moultrie Lt. Roux
reare Sergt. Coleman

Orders 28th Apr
Majr. Horry } . . Parole First regiment . . C Sn.
For fort guard tomorrow Captn. Dunbar & Lt. Martin
Rear Sergt. __ Hospitle Corpl.

A court martial to set this forenoon to try such prisoners as are brought to it all Evidences to Attend, Captn. Harleston presidt. Lts. Baker, Gray: ~~Galvan~~ Galvan & Perreneau members

Genl. Orders
Genl. Moultrie } One Captn. 2 Subaltern 2 Sergts. & 40 Rank & file from the Second regt. to be sent to town tomorrow morning to Attend the funeral of Captn. Jacob Shubrick of the Same regt.

After Orders
Lt. Colo. Marion } The men Order'd by the General to go to town tomorrow to the funeral of the Late Captn. Jacob Shubrick to be picked out immediately & get ready to embark on board the barge & Genl. Moultrie boat tomorrow by 7 OC: in the morning, they are to receive new Cloaths out of the Store and are to put themselves neat & Clean with their hairs Powder'd, When they return the Offr. Commanding the party is to see that the Cloathing recd. is returnd in the Store – Six men from the Compy. Late Captn. Shubricks are as Carriers & Sen. Offr. to Attend as bearers – the party to receive three rounds Blank Cartridges pr. Man & their Arms must be as Clean & as Bright as possible [864] – the Adjutant will see these Orders Complied with – The party & those Officers who to town on this Occasion is positively Orderd to return the same evening with the Boat

For the party	as Bearers		Offrs who may attend ye. funeral	
Captn. Charnock	Captn. Harleston	Captns. Dunbar	Lts. Baker	& Roux
Lt. Mason	Lesesne	Hall	Bush	For the fort guard tomorrow
Lt. Warley	Moultrie		Gray	
	Mazyck		Galvan	Lts Martin & Capers

---

[864] To keep the muskets bright Captain Barnard Elliot, of the 4th South Carolina Artillery, told his men "To polish the barrel of the fusee and keep it bright after being cleaned, every soldier must carry in his pouch a thick piece of buck skin, with which he is to rub the barrel will, as soon as he is relieved from his post as sentry or comes off guard; by the frequent repetition of this the polish becomes so long lasting as at length not to be spotted even by rain. Each man must also have in his pouch a worm and a wire pricker and 2 spare flints. It is recommended tat the stock of the gun be rubbed over with oil and wax, which will give it a gloss and prevent the wet from damaging it. The quartermaster will furnish the wax and oil and worm." Military muskets were not browned at this time.

L^t. Col^o. Marion **Fort Moultrie 1778**

Orders 29th Apr
L^t. Col^o. Marion } . . Parole Martenet C S^n. 72
For Fort guard tomorrow Captn. Dunbar & L^t. Baker
Advance Serg^t. Wickam – rear Kiels – Hosp^t. a Corp^l.
NB the funeral party orderd yesterd marches off 8 OC: AM - & return 8 PM

Orders 30th Apr
L^t. Col^o. Marion } . . Parole Frederica . CS^n. 45 by Col^o. Motte
For Fort guard tomorrow Captn. Hall L^ts. Gray & Hart
rear Serg^t. Bond
A monthly return of each Company to be made out & Given in to the Adj^t tomorrow 10 OC: in the forenoon

A Court martial to set today at 11 OC: in the forenoon to try such prisoners as shall be brought before it – all evidences to Attend
Captn. Motte presid^t. L^ts. Mason, Gray, Hart & Warley members

---

Orders 1st May
by Col^o. Motte } . . Parole Amity . . CS^n. 8.
Fort guard tomorrow Captn. Harleston L^ts. Perreneau & Roux
rear Serg^t. Williamson – Hospitle Corp: Raynes [865]
A reg^tl. Court martial to set this forenoon at 11 OC: to try such prisoners as shall be brought to it – the Adjutant to order all evidences to Attend and to inform the prisoners that their Trial is to come on at that Hour
Captn. Moultrie presid^t. L^ts. Mason & Martin Members

NB by the Sentence of Court 30th Apr: Jam^s. Hooper for Drunkenness & misbehavour rec^d. 100 lashes

---

Orders 2nd May
Col^o . Motte } . . Parole Union . . C S^n. 5.
Fort guard tomorrow Captn. Motte L^ts. Martin & Capers
Rear Guard Serg^t. Colb – Hospitle Corp^l. Tayler
The Order of the 28th Feby. Last to be strictly Complid with

The Col^o. recommends to any of the men that may think their time of Service is expired, not to refuse any duty that they may be orderd upon as they are Subject to be punish'd for such refusal (even if their time was out) till they are regularly Discharged – the Col^o. Assured all such as have served the time they inlisted for, that he will Discharge them as soon as that time comes about unless they chose to enter again in the Service of their country ~ the Off^rs. of the Different Companies are read this this Order to the men of their respective Companies this evening & tomorrow Morning at Roll call –

A court of Enquiry to set this forenoon at 10 OC: to enquire when Samuel Joiner of Captn Charnocks Comp^y. was Inlisted by whom Inlisted & for what time Inlisted and make their report

---

865 Robert Raine (Rain, Rains) enlisted in the 2nd South Carolina Regiment on 4 November 1775 as a corporal. He was in Captain Hall's company in 1779 at the siege of Savannah.

thereupon as soon as possible[866] – this Court to Consist of two field Officers & 3 Capt[ns]. L[t]. Col[o]. Marion to set as president, Members Major Horry Capt[ns] Motte, Dunbar & Hall –
After Ord[rs]. A reg[tl]. Court martial to set immediately to try such prisoners as shall be brought to it Evidences to Attend Capt[n]. Charnock president Lieu[ts]. Baker & Capers members

Col[o]. Motte **Fort Moultrie . . 1778**

Evening Orders, The court of Enquiry whereof L[t]. Col[o]. Marion was presid[t]. and Order'd to set this forenoon Report that the Court are Unanimous of Opinion that Samuel Joiner was Inlisted to serve three Years . . . .

Capt[n]. Blake having resigned his Commission is no longer to be Obey'd an Officer in the regiment [867]

NB by sentence of Reg[tl]. court this day Tho[s]. Windsor for Selling Liquors contrary to Orders rec[d]. 50 lashes wth. Cat 9 tails

---

Orders 3[d] May
L[t]. Col[o]. Marion } . . Parole Radnor . . Count[r]. S[n]. Georgia
Fort guard tomorrow Capt[n]. Charnock L[ts]. Baker & Warley
Hobcaw L[t]. Hart – Rear Serg[t]. Jasper – Hospitle a Corp[l].

Orders 4th May
Maj[r]. Horry . . } . . Parole Scott . . Count[r]. S[n]. first reg[t].
Fort guard tomorrow Capt[n]. Lesesne L[ts]. Mason & Gray
rear Serg[t]. Raybout [868] ~ Hosp[tl]. a Corp[l].

A court martial to set this morning at 10 OC: to try all prisoners brought to it
Evidences to attend Capt[n]. Dunbar presid[t]. L[ts]. Roux & Galvan members

NB by sentence of the court this day Sam[l]. Murray for Absence without Leave to be confined 14 days & to mount guard every other day ~ Frederick Simmons for the same Crime to be confined 20 Days & mount guard every other day – David Stewart for do. Confind 14 days & d[o] duty every other day – Jam[s]. Smith for d[o]. The same punishment – Allan M[c]Donald for the same crime the same punishment [869]

---

[866] Samuel Joiner enlisted in the 2nd South Carolina Regiment on 6 May 1776.

[867] John Blake resigned after half of his company was killed on the Randolph.

[868] Thomas Rybold.

[869] Allen McDonald served in the 2nd South Carolina Regiment in Captain Hall's company. He was confined for 14 days on 4 May 1778 for being absent without leave. He also had do guard duty every other day while he was confined. He was promoted to sergeant in the 2nd Vacant company commanded by Lieutenant Adrian Proveaux on 13 July 1778. He was confined on 20 November 1778 for drunkiness and he being too drunk to do his duty. He was remanded because he was too drunk to stand trial. On 15 February 1779 he was sentenced to be reduced to private and receive 30 lashes for being drunk on guard and quitting his post without leave. He was reduced but the lashes were remitted. He served 44 months in the Continental line from 1779 to 1783.

Orders 5th May }
Majr. Horry
. . Parole Thompson . . . Counter Sn. third Regement
For Fort guard tomorrow Captn. Moultrie Lts. Perrenneau & Foissin
Reare Sergt. Lawrence ~ Hospitle Corpl. Butler[870]

Orders 6th May }
Majr. Horry
. . Parole Carteel . Countr. Sn. Two by Lt. Colo.
Fort guard tomorrow Captn. Hall Lts. Roux & Marten
rear Sergt. Gammel ~ Advance Sergt. _ Hosptl. Corpl.

That one Sergt. 1 Corpl. & 6 privates be parraded immediately to go on board the the Carteel vessel Captn. Steel from St. Augustine & to Convey her up to town
Lt. Gray is to take charge thereof and to wait on his Excellency the Presidt. on Arrival in town & to give Orders to the Sergt. not to suffer any good to be taken out of Sd. vessel till further Orders

Aftr. Ordrs. }
Maj:
1 Sergt. 1 Corpl. & 6 privates to parrade immediately to go onboard the Cartel [871] Schooner Captn. Hatter from St. Augustine to convey her to town & not to Suffer any goods to be taken out of Said vessel till further Orders Lt. Foissin to Command this party

The Long Roll to beat at 5 OC: in After noon till further Orders

---

Orders 7th May }
Lt. Colo. Marion
. . Parole Prince Eugene . . Countr. Sn. 17.
Fort guard tomorrow Captn. Harleston Lts. Baker & Capers
Rear guard Sergt. Newman – Hosp: Corpl. McDonald

R. return – 1 Col. 1 Lt. Colo. 1 majr. 61 Captns. – 10 1st Lts. 4 Second do – 1 Adj: 1 Chap: 1 p. mastr. 1 Serg: 1 S. M. 1 FM. 1 DM. 1 Q M Sergt. 25 Sergts. – 15 D& Fifes 2 Arms – 327 Rank & file total of all ranks 444

Lt. Colo. Marion **Fort Moultrie 1778**

Orders 8th May }
Lt. Colo. Marion
. . Parole Madera . . Countr. Sn. 63.
For Fort guard tomorrow Captn. Motte Lts. Warley & Mason
Rear Sergt. Mynor [872] . . . . . Hosptl. Corpl.

A court Martial to set this forenoon at 11 OC. to try such prisoners as may be Order'd to it All evidences to Attend this court to consist of a Captn. as presidt. & 4 other Offrs. as members – Captn. Charnock presidt. Captn. Hall Lts. Gray Perreneau & Galvan memb:

Orders 9th May }
Lt. Colo. Marion
. . Parole Naples . . . Countr. Sn. Italy
Fort guard tomorrow Captn. Charnock Lts. Gray & Foissin
reare Sergt. Coleman ~ Hospitle Corp:

---

870 Samuel Butler enlisted in the 2nd South Carolina Regiment on 4 November 1775. He was promoted to corporal on 5 May 1777. He was discharged on 7 July 1778. He reenlisted on 1 July 1779 when the British threatened South Carolina.

871 A Cartel was a commissioned ship sailing under a flag of truce in time of war to exchange prisoners or to carry a proposal to the enemy.

872 Sergeant Rueben Minor.

A court martial to set this forenoon at 11 OC: to try all prisoners as may be brought to it, the evidences to be warned by the Sergt. Major to Attend

Captn. Lesesne presidt. Lts. Baker – Capers – Roux & Martin members

NB by sentence of the above court Sergt. Oneel for fighting & abusing Jno. Conner was reduced in the ranks as a private [873]– Wm. McDowell for drunkenness on Guard received 50 lash with a Cat

---

Orders 10th May } . . Parole Rome. . . Countr. Sn. 28.
Lt. Colo. Marion } Fort guard tomorrow Captn. Lesesne Lts. Roux & Martin
Hobcaw Lt. Perreneau ~ Rear Sergt. Gammel – Hosp: a Corpl.

Divine service will be performed this forenoon at the usual place by the Chaplain at the beating of the Long roll all Offrs. & Soldiers to Attend

---

Orders 11th May } . . . Parole General Lee. . . Countr. Sn. 99 .
Lt. Colo. Marion } Fort guard tomorrow Captn. Moultrie Lts. Baker & Capers
Rear Sergt. Jasper ~ Hospitle Corpl. Martin [874]

The Battalion is to have two field days in the week on Mundays & Thursdays and every other Wednesday to Exercise the Cannon & go through the parrapet firing the Other days the Commanding Offrs. of Companys are to exercise their men in the afternoons – No Offr off duty to be Absent on the Field days on any pretence whatever Except sickness & all Offrs. to Attend the Exercise of the Cannon & Command a Gun & a Plattoon in the Parrapet firings

A court martial to set this forenoon at 11 OC: to try all prisoners as may be Order'd to it – all Evidences to Attend – Presidt. Captn. Moultrie Lieuts. Warley, Mason – Gray and Foissin members

---

Orders 12th May } . . Parole Houston . . . Countr. Sn. Georgia
Major Horry . } Fort guard tomorrow Captn. Hall Lts. Warley & Mason
reare Sergt. Ritchie – Hospl Corp. McDonald

Orders 13th May } . . . Parole Huger . . Countr. Sn. Major . . .
Majr. Horry . . } Fort guard tomorrow Captn. Harleston Lts. Gray & Foissin
rear Sergt. Coleman – Advance Sergt. Laurance
Hospitle Corporal

Lt. Colo. Marion **Fort Moultrie . . . 1778 –**

Orders 14th May } . . Parole Fort Howe Countr. Sn. Altamahaw –
Lt. Colo. Marion } Fort guard tomorrow Captn. Motte Lts. Hart & Roux
rear Sergt. Coleman . . . . Hospitle Corp:

---

[873] John Conner enlisted in the 2nd South Carolina Regiment on 21 November 1775. He was promoted to corporal on 20 July 1778 in Captain Charnock's company. He was demoted to private again on 8 February 1779 in Captain Richard Mason's company.

[874] Joseph Martin enlisted in the 2nd South Carolina Regiment on 6 January 1778. He was promoted to corporal on 5 March 1778 in Captain Motte's company. He was promoted to sergeant on 27 July 1778. He was discharged on 8 February 1779.

A return of the Arms accoutrements & Blankets of those men who returned from Command on board the General Moultrie to be made and given to the Adjut: tomorrow 11 OC: in the forenoon, with the Numbers on each musket

A court martial to set this forenoon to try such prisoners as may be Orderd to it – evidences to attend, Captn. Hall presid^t^. L^ts^. Roux, Baker, Capers & Warley members

NB by sentence the above court Samuel Henry for absence from roll call several times rec^d^. 50 lashes – Nick^l^ Flinn to receive 50 lashes for striking Corporal Stone but is sick & the punishment postponed [875]

Orders 15th May } . . Parole S^t^. Marys . . Count^r^. S^n^. 15.
L^t^. Col^o^. Marion } Fort guard tomorrow Captn. Charnock L^ts^. Martin & Capers
rear Guard Serg^t^. Burtel – Hosp: Corp^l^. Reins . . .

NB Allwell of Captn. Charnocks Comp^y^. shott himself when on Centry, on the Habcaw Guard – + + + + + + + + + + + +

Return of Arms & Accoutrements & Blankets from on board the Ship General Moultrie
Command by Capt: Blake

Companies	Men	Muskets	Bayonets	Scabard	Belts	Frogs	Pouches	Blankets	Number of firelocks
Harleston . .	3	3	1	1	1	1	"	2	19 – 13 – 22
Mottes . .	4	4	3	"	"	"	2	"	50 – 44 – 53 – 57
Charnocks	2	2	2	2	2	2	2	1	159 – 165.
Lesesnes . .	1	1	1	1	1	1	1	1	188.
Moultries . .	3	3	2	"	"	"	2	4	250 – 230 – 222
Mazyck .	6	5	5	4	3	1	5	6	269 – 268 – 256 – 271 – 87
Dunbar . .	5	4	4	4	4	"	2	4	33 – 31 – 30 – 32
Halls . .	1	1	1	1	1	1	1	1	92.
1st Vacant [876]	7	7	7	7	7	7	7	7	136 –118.134.117.113.139.111
2nd Vacant [877]	2	2	1	1	1	"	"	2	293 – 296.
Total	34	32	27	21	20	13	22	25	

---

875 Benjamin Stone served in the 2nd South Carolina Regiment under Captain Daniel Mazyck in 1778. He was promoted to corporal. On 13 July 1778 he was reprimanded for persuading men not to reenlist. He was promoted to sergeant on 21 April 1779 in Captain Lesesne's company. He was in Captain Mazyck's company during the siege of Savannah. He died prior to January 1786.

876 This was Captain John Blake's company. Captain Blake had resigned after he returned from the sea voyage that took the lives of the men on the *Randolph*. When a commander resigned or died his unit was known as "Vacant" until that position was filled. This company would become the Lieutenant Colonel's company and be commanded by Lieutenant Richard Bohun Baker on 13 June 1778.

877 Captain Jacob Shubrick's company. Shubrick died on 27 April 1778. This company would become the Major's company and be commanded by Lieutenant Adrian Proveaux's on 13 June 1778. The 2nd Vacant Company was disbanded on 5 October 1778 and the men placed in other companies. Lieutenant Proveaux would be promoted to captain and assume command of a company in November 1778.

Orders 16th May
L^{t}. Col^{o}. Marion }

. . Parole Settelles . C S^{n}. Haulted . . by Maj Horry
Fort guard tomorrow Captn. Lesesne L^{t}. Baker & Warley
Hobcaw Serg^{t}. Murphy – Hosp: Corp^{l}. Smith

Orders 17th May
Maj^{r}. Horry }

. . Parole Altimaha. . C S^{n}. Advancing
Fort guard tomorrow Captn. Moultrie L^{ts}. Mason & Gray
Hobcaw L^{t}. Roux – rear Serg^{t}. . . . . Hospl. Corporal

Aft^{r}. Ord^{r}
Maj^{r}. }

Divine Service will be performed this Aft^{r}. Noon by the Chaplain. The Commanding Off^{r}. expects that all Off^{rs}. & Soldiers off duty will attend the Same at the Beating the Long roll

L^{t}. Col^{o}. Marion **Fort Moultrie 1778**

Orders 18th May
~~Maj^{r}. Horry~~ }

. . Parole Elliot . . Count^{r}. S^{n}. Artillery by L^{t}. Col^{o}.
Fort guard tomorrow Captn. Hall L^{t}. Hart & Foissin
rear guard Serg^{t}. Hospitle Corp^{l}.

A Court martial to set this forenoon at 11 OC. to try such prisoners as shall be Orderd to it, Evidences to Attend

Captn. Harleston presid^{t}. Lieu^{ts}. Baker & Hart members

---

Orders 19th May
L^{t}. Col^{o}. Marion }

. . Parole S^{t}. Johns . . Count^{r}. S^{n}. 33
Fort guard tomorrow Captn.
Rear Serg^{t}. Hospitle Corp^{l}.

Genl. Orders
by Genl Moultrie
13th May rec^{d}. this
day - - - - }

Head Quarters Ch^{s}. town
All Officers and other persons having Demands against the Public for recruiting service or Waggon hire, are to send in Certificates from the Col^{o}. of the Respective Regiments, Or Commanding Officer of the Detatchment, in which the Expenses were incurred to gether with an Attestation before a magistrate that the account is Just and such Officers when they send in their Accounts of recruiting Service are to Insert the Names of the Recruits in the Certificate

Transmitted by

A Dilliant B: M –

NB by sentence of the court 18th Inst: Timothy Downing for Drunkenness rec^{d}. 30 lashes

---

Orders 20th May
L^{t}. Col^{o}. Marion }

. . Parole Augustine . Count^{r}. S^{n}. Success
Fort Guard tomorrow Captn. Motte L^{ts}. Capers & Baker
Advance Serg^{t}. Gammel, Rear McDonald, Hosp^{tl}. Corp^{l}. Smith

A reg^{tl}. Court martial to set to day at 11 OC: to try all prisoners that may be Orderd to it; all evidences to Attend Captn. Charnock presid^{t}. L^{ts}. Foissin & Hart members

NB by sentence of the above court Aron Harris of the Granid^{r}. Comp^{y}. rec^{d}. 100 Lashes for theft – Conrad Fitner for abuse to ~~the these~~ fife Maj^{r}. to receive 20 lashes remitted Nick^{l}. Flinn tried 14th Inst: to receive 50 lashes 25 remitted

---

Orders 21st May
Lt. Colo. Marion } . . Parole Colo. Motte . . Countr. Sn. 44.
Fort Guard tomorrow Captn. Charnock Lts. Warley & Mason
Reare Sergt. Rybout – Hospite a Corpl.

Orders 22nd May
Lt. Colo. Marion } . . Parole Colo. Huger . . C Sn. 69.
Fort guard tomorrow Captn. Lesesne Lts. Gray & Foissin
Rear Sergt. Bond – Hosptl. Corpl.

Orders 23d May
Lt. Colo. Marion } . . Parole Elbert . . C Sn. 61
Fort Guard tomorrow Captn. Moultrie Lts. Hart & Perreneau
Rear Guard Sergt. Coleman – Hosptl. Corpl.

Orders 24th May
Lt. Colo. Marion } . . Parole Long Island . C Sn. Colo. Motte
Fort Guard tomorrow Captn. Hall Lts. Warley & Capers
Hobcaw Lt. Martin –
Rear Sergt. Fowler – Hospitle Corpl. Roberts

Lt Colo. Marion **Fort Moultrie – 1778**

Aftr. Orders 24th May
Lt. Colo. Marion } Three men from each company to warn'd this & every evening this week to parrade at troop Beating to make a booth for Divine Service on Sundays, also 2 Sergeants and 2 Corporals from the regiment to see to the work, the Q M Sergt. to give out Axes spades & hoes or any other tools which may be nessessary to carry the work in execution, & to take a receipt from the Sergt. of fatigue who is to deliver them at night to the Q M sergt. this work to be under the direction of the Majr. & Adjutant **NO** exercise in the aftr Noon till this work is completed

---

Orders 25th May
Lt. Colo. Marion } . . Parole Boston . . . C Sn. 32.
Fort guard tomorrow Captn. Harleston Lts. Mason & Gray
Rear Sergt. Raybout . . Hosptl. Corpl.

Orders 26th May
Lt. Colo. Marion } . . Parole Independence . . C Sn. America
Fort guard tomorrow Captn. ~~Charnock~~ Motte Lts. Hart & Foissin
rear Sergt. Minor . . . Hosptl. Corpl. Teague

Orders 27th May
Lt. Colo. Marion } . . Parole Treaty . . Countr. Sn. with France
Fort guard tomorrow Captn. Charnock Lts. Perreneau & Roux
Advance Sergt. Rear Sergt. Hosptl. Corp:

A court martial to set this forenoon at 11 OC: to try all prisoners as may be Orderd to it – all evidences to Attend – Captn. Moultrie presidt. Lieuts. Warley, Mason, Gray & Capers members

---

Orders 28th May }
L^{t}. Col^{o}. Marion } . . Parole Rockingham . C S^{n}. Notre dame
Fort guard tomorrow Captn. Lesesne L^{ts}. Baker & Capers
rear Serg^{t}. Williamson – Hosp^{tl}. Corp^{l}.

The barge will go to town tomorrow for the Soldiers of the 2nd reg^{t}. now on board the Notre dame, under Command of L^{t}. Proveaux, as soon as they arrive in Garrison. the Commanding Off^{rs}. of Comp^{ys}. who have men on board are to take a particular Rec^{t}. of their Arms, Accoutrements Blankets & regimental Cloathing & make a return thereof to the Adjutant as soon as possible; the Q M Serg^{t}. will make a return of the Amunition & flints returned

---

Orders 29th May }
Maj^{r}. Horry } . . Parole Genl. Lee . . C S^{n}. Exchanged
Fort guard tomorrow Captn. Moultrie L^{ts}. Mason & Warley
Rear Serg^{t}. Bartel – Hosp^{tl}. Corp^{l}. Teague

Orders 30th May }
Maj^{r}. Horry } . . Parole Capt^{ns} Parker & Smiley . . . C S^{n}. Boston frigate
Fort guard tomorrow Captn. Charnock L^{ts}. Gray & Foissin
Rear Serg^{t}. Smith – Hosp^{tl}. Corp^{l}. McCowen [878]

L^{t}. Col^{o}. Marion Fort Moultrie – 1778 –
Orders 31st May }
Maj^{r}. Horry } . Parole Brigg Washington C S^{n}. North Carolina by L^{t}. Col^{o}.
Fort guard tomorrow Captn. Lesesne L^{ts}. Hart & Roux
Hobcaw L^{t}. Capers – rear Serg^{t}. Raybold – Hosp^{tl}. Corp^{l}. Dyer

After Orders }
Lt Col^{o}. Marion } A monthly return to be given in to the Adjutant by tomorrow 10 OC: in fore noon by the Commanding Off^{rs}. of each Company without Fail
The Q M Serg^{t}. to have cleaned out the whole garrison by 10 OC: in forenoon he is to have one man p^{r}. Comp^{y}. Added to those he has Already for that purpose

Those Off^{rs}. who have neglected Complying with the orders of the 28th Inst: are to Comply with it immediately

---

Monthly return 1st June
1 Col^{o}. 1 L^{t}. Col^{o}. 1 Maj^{r}. 8 Capt^{ns}. _ 10 first L^{ts}. _ 4 Second L^{ts}. _ 1 Adj^{t}: _ 1 Chapl: _ 1 QM
1 pay m – 1 Surgeon – 1 Serg^{t}. Maj: _ 1 D Maj^{r}. _ 1 Fiffe M _ 1 Q M Serg^{t}.
20 Serg^{ts}. _ 15 Drums & Fiffes _ 1 Ch. Armourer _ 1 d^{o} Mate _ 329 rank & file men

---

Orders 1st June }
L^{t}. Col^{o}. Marion } . . Parole President . . C S^{n}. 57 –
Fort guard tomorrow Captn. Moultrie & L^{ts}. Baker & Martin
rear Serg^{t}. Coleman – Hosp^{t}. Fellinghart

---

[878] David McGowan enlisted in the 2nd South Carolina Regiment on 23 November 1775 as a corporal. He was promoted to sergeant on 12 July 1778 in Captain Charnock's company. He was discharged on 29 November 1778.

Gen[l]. Orders } Head Quarters Ch[s]. Town 29th May 1778
Gen[l]. Moultrie } A general court martial to set on Thursday 4th June next between the hours prescribed by the articles of war at such place in Ch[s]. town as the president shall Appoint to try all prisoners that shall be brought before it, All evidences to Attend

president L[t]. Col[o]. Mason – 2 Capt[ns]. 3 Subalterns from 2nd reg[t]. 2 Capt[ns]. 3 Subalterns from 3[d] reg[t]. - Capt[n]. 1 Sub[l]. From 5th reg[t].

A Dilliant B.M:

Orders 2nd June } . . Parole Granby . . C S[n]. 75.
L[t]. Col[o]. Marion } Fort Guard tomorrow Capt[n]. Charnock & L[t]. Warley & Mason
Rear G. Serg[t]. Burtell – Hosp[tl]. Henderson

A reg[tl] court martial to set this forenoon to try all prisoners as may be Orderd to it – evidences to Attend

Presid[t]. Capt[n]. Lesesne – members L[ts]. Gray Foissin, Hart, & Roux

NB by sentence of the above court Jn[o]. Thompson to be Confined to Day & do double duty – remitted

---

Orders 3[d] June } . . Parole Shelburn. . . C S[n].
L[t]. Col[o]. Marion } Fort guard tomorrow Capt[n]. Lesesne & L[ts] Foissin & Hart
rear Serg[t]. G – - Hospitle G Corp[l].

L[t]. Col[o]. Marion **Fort Moultrie – 1778 –**

Gen[l]. Orders } . . Head Quart[rs]. Ch[s]. town 30 May 1778
Gen[l]. Moultrie } Capt[n] Robert Goodwin [879] & L[t] W[m] Goodwin [880] of Col[o] Thompsons Reg[t]. Capt[n]. Edward Walsh of Col[o]. Hugers Reg[t]. having resigned their Commissions are no longer to be considerd as Continental Officers [881]

The Officers of the Different Corps are to draw rations for those servants only, who do not belong to the Army

Reg[tl]. Orders } The Off[rs] who are Appointed as members of the General court martial will
L[t]. Col[o]. Marion } Attend in Ch[s]. town tomorrow morning, they may go when the prisoners & ~~members~~ evidences are sent down in the Barge

---

879 Robert Goodwin was born in 1741. He became a captain in the 3rd South Carolina (Ranger) Regiment in 1775. He resigned in June 1778, and served in the militia under General Richardson. He was promoted to Colonel and from 7 February to 29 June 1779 he was under General Williamson. He was wounded at the fall of Charleston in 1780. He died sometime after 1785.

880 William Goodwin enlisted in the 3rd South Carolina (Ranger) Regiment in 1776 under Captain Robert Goodwin, his father. In 1778 he was commissioned as a lieutenant. He resigned in June 1778, and served in the militia under his father. He was promoted to captain and served under Colonel Thomas Taylor in 1780, 1781 and 1782. He died in 1801.

881 Edward Walsh (Welch) became a first lieutenant in the 5th South Carolina Regiment in 1777. On 20 January 1778 he was promoted to captain. He resigned on 30 May 1778.

Sergt. Park smith Solomon Long [882] & Jno. Jackson [883] to Attend as evidences against archibald McDonald – Sergt. Burtell, Solomn Long & J. Jackson against Wilkinson [884] – Members for the sd. Court martial

Captns. Thos. Moultrie & Dunbar – Lts. Proveaux Bush & Perreneau

---

Orders 4th June
Lt. Colo. Marion
. . Parole Middleton . . C Sn. Lynch
Fort guard tomorrow Captn. Harleston Lts. Roux & Martin
Rear do. Sergt. R. Coleman. Hospitle Corpl. Tayler

Orders 5th June
Majr. Horry
. . Parole Hayward. . . C Sn. Trapier
Fort guard tomorrow Captn. Charnock Lts. Baker & Worley
Rear do. Jacob Coleman

Orders 6th June
Majr. Horry
. . Parole McDougall . C Sn. Philadelphia
Fort Guard tomorrow Captn. Lesesne Lts. Mason & Gray
Rear do. Sergt. McDonald . . Hospitle Corp: Tellinghart

Orders 7th June
Lt. Colo. Marion
. . Parole Brigg Polley . C Sn. Captn. Anthony
For Fort Guard tomorrow Captn. Harleston & Lt. Roux
Habcaw Lt. Baker – Rear Sergt. Smith

Genl. Ordrs.
G. Moultrie
Copy of the resolution of Congress May 15th: 1778
Resolved unanimously, that all military Officers commissioned by congress who now are or may hereafter may be in the service of the united States of America and shall continue there in during the war and shall not hold any Office of profit under these States or any of them shall after the conclusion of the war, be intitled to receive annually for the time of seven years (if they live so long) one half of the pay of such Offrs. – Provided that no Genl. Offr. of the Cavalry, Artillery or Infantry shall be intitled to receive more than half of the pay of a Colo. of such Corps respectively, and provided that this resolution shall not attend to any Offr in the service of the united States, unless he shall have taken an Oath of Allegiance to & shall actually reside with in someone of the United States

Lt. Colo. Marion **Fort Moultrie 1778 –**

Resolved Unanimously, that every non commissioned Offr. and Soldier who hath or shall inlest in the service of these states for and during the war and shall continue there in to the end thereof shall be intitled to receive the further reward of Eighty Dollars at expiration of the war
U.S. Ordrs. Dated 7th Inst: Certified Henry Laurens presidt.

---

[882] Solomon Long was born in 1744. He enlisted in the 2nd South Carolina Regiment on 4 November 1775. He was promoted to sergeant but was reduced in rank later. He was promoted to corporal on 12 April 1779 and served under Captain Adrian Proveaux.

[883] John Jackson was born in 1737 in South Carolina. He enlisted in the 3rd South Carolina (Ranger) Regiment on 25 June 1775 under Captain Robert Goodwyn. During 1779 he was under Captain Felix Warley. He re-enlisted on 1 February 1780. His enlistment expired on 1 July 1781.

[884] William Wilkinson.

All Officers of the continental troops who hath not taken the Oath of Alligiance to the United States of America & Abjuration of George the third King of Great Britton are orderd to Attend to Head Quarters from 9 to 11 OClock in the morning to take the same

Transmitted by A Dilliant BM

A Return of mens Arms Accoutrements & Blankets from on Board the Notre dame a Detatchment Commanded by L[t]. Proveaux 29th May 1778

Companies	Men	Muskets	Bayonets	Belts	froggs	Pouches	Blankets	Number on each **Musket** – –
Capt Harleston	2	2	2	2	2	2	2	
Motte . .	2	2	2	2	2	2	2	
Lesesne . .	3	3	3	3	3	3	3	
Moultrie	2	2	2	2	2	2	2	229 – 241
Dunbar	1	1	1	1	1	1	1	
Hall . .	1	1	1	1	1	1	1	
2nd Vacant .	2	2	2	2	2	2	1	
Totall –	13	13	13	13	13	13	12	

Orders 8th June
L[t]. Col[o]. Marion }

. . Parole Laurense. . C S[n]. Congress
Fort guard tomorrow Capt[n]. Charnock L[ts]. Martin & Capers
Rear Serg[t]. Laurence – Hospitle Corp[l]. Webb[885]

Orders 9th June
L[t]. Col[o]. Marion }

. . Parole Burlington . C S[n]. Philadelphia
Fort guard tomorrow Capt[n]. Lesesne L[ts]. Warley &Mason
Rear Serg[t]. Coleman

A court martial to set tomorrow at 11 OC. in forenoon to try all prisoners Orderd before them, all evidences to Attend
Capt[n]. Harliston presid[t]. L[ts]. Gray & Foissin members

Orders 10th June
L[t]. Col[o]. Marion }

. . Parole Bristol . . C S[n]. 30
Fort Guard tomorrow Capt[n]. Harleston L[ts]. Gray & Foissin
Rear Serg[t]. Gemmel
Advance Guard Serg[t]. Burtell

---

885 Henry Webb served enlisted in the 2nd South Carolina Regiment on 9 April 1777. He was promoted to corporal on 9 March 1778. He was promoted to sergeant on 12 July 1778 and served until 1 July 1781. He was in Dunbar's Light Infantry company at the siege of Savannah. He was captured at the fall of Charleston. In 1782 he was in the militia.

Col^{o}. Motte **Fort Moultrie 1778**

Orders 11th June
Col^{o}. Motte } . . Parole France . . C S^{n}. 92.
Fort guard tomorrow Captn. Lesesne L^{ts}. Roux & martin
Rear Serg^{t}. Wickham

The Col^{o}. informs the Sergeants, Corporals, Privates, Drums, Fifes & privates of the Regiment that he had some time ago put in his hands £300, as a present from M^{r}. Samuel Wainwright to be applied for them as he and the other two fold Off^{rs}. shoud think proper for their Bravery for defence of this fort – they are of Opinion it will be but imploy'd towards an Entertainment for them on the 29th instant which is now orderd to be provided on that day, And the Officers have generously agree to make up the deficiency in the Expence of the same . . .

M^{r}. John Henry Rasche is appointed Surgeons mate to the Reg^{t}. – Orderd that he be respected as such .

The Reg^{t}. to be under arms tomorrow afternoon at 6 OC: to try Such prisoners as shall be Brought to it Evidences to be order'd to Attend in time ~ ~ ~ ~ ~ Captn. Lesesne presid^{t}. L^{ts}. Capers & Galvan Members

NB at a Court held 9th instant Presid^{t}. Captn. Harleston

Wm Ashford for neglect of duty rec^{d}. 39 lashes – Wm Baldwin for d^{o}. 39 Jno. McCade for ~~Stealing~~ Breaking open the reg^{tl}. Store & stealing all the Cloathing there to receive 25 lashes four different times with one day Between each punishment & to be confined in the Block from 7 OC: at night till 7 next morning. till he has rec^{d}. the whole punishment – Jno. Barnet Tayler for the same crime to receive the same punishment –

Orders 12th June
Col^{o}. Motte } . . Parole Spain . . C S^{n}. 77 . . b L^{t}. Col^{o}.
Fort guard tomorrow L^{ts} Capers & Warley
rear guard Serg^{t}. Raybold . Hosp^{tl}: Corp^{l}. Tellinghart

Order'd that the Officer do immediately provide themselves with Leather caps agreeable to a pattern fixed on & Left with M^{r}. Callipon Sadler in King Street Charles town and that they ware no Other kind of Caps

---

Orders 13th June
L^{t}. Col^{o}. Marion } . . Parole Prusia . . C S^{n}. 84.
Fort guard tomorrow Captn. Harleston L^{ts}. Mason & Gray
rear Serg^{t}. Laurence – Hospitle Corp^{l}. Tellinghart

L^{t}. Rich^{d}. Baker is to take the Command of the Company Late Capt. Blakes
L^{t}. Adrian Proveaux that of the Late Capt Jacob Shubrick agreeable to Genl. Orders

Genl. Ord^{rs}. 11 June
by Genl. Moultrie } The first Lieu^{ts}. Intitled to the Commission of Capt^{ns}. In the Reg^{t}. in the State of S^{o}. Carolina are Order'd to take the Command of the Companies which they are Intitled to till Commission can be had and they are to be Obey'd and respected Accordingly

Transmitted by A Dilliant –

L^{t}. Col^{o}. Marion **Fort Moultrie – 1778**

Orders 14th June
L^{t}. Col^{o}. Marion }
. . Parole Holland . . C S^{n}. 89.
Fort guard tomorrow Captn. Lesesne L^{ts}. Roux & Foissin
Hobcaw L^{t}. Warley – rear Serg^{t}.

The Maj^{r}. will chose out of the Regim^{t}. as many men as will exercise five field pieces with an Officer to train them every day till the 29th Inst: the field piece on the East Bastion is to be taken down for that purpose L^{t}. P. Gray will command this party with 2 Serg^{t}. 1 Corp^{l}. & 10 privates

---

Orders 15th June
L^{t}. Col^{o}. Marion }
. . Parole Portugal . . C S^{n}. France
Fort guard tomorrow L^{ts}. Martin & Capers
rear Serg^{t}. Gimmel – Hosp^{tl}. Corp^{l}. Rains

Orders 16th June
L^{t}. Col^{o}. Marion }
. . Parole Germany. . C S^{n}. 89.
Fort guard tomorrow Captn. Harleston L^{ts}. Mason & Foissin
rear Serg^{t}. Wickam – Hosp^{l}. Corp^{l} Watt

A regimental court martial to set this forenoon at 11 OC: to try all prisoners Ordered to it – evidences to Attend – presid^{t}. Captn. Harleston – L^{ts}. Galvan & Foissin members

Genl. Orders 15th June
Genl. Moultrie }
Archibald McDonald of 2^{d} reg^{t}. for Desertion sentenced by Genl. Court martial to receive 100 lashes on the bare back with Switches, remitted, Thos. Bowen of 2nd reg^{t}. for Desertion sentenced to receive 100 lashes but being recommended by the court is pardoned – James Smith of the 2nd reg^{t}. for Desertion to receive 100 lashes on the bare Back with a Cat onine tails – Willm Wilkinson of 2nd reg^{t}. for ~~Deser~~ Desertion is Sentenced to receive 100 lashes on the bare back with Cat. ONine tails well laid on – the General Approves the Above Sentences & Dissolve the Court martial of which L^{t}. Col^{o}. Marion was presid^{t}. & Order them to be put in Execution

– Transmitted – A Dilliant B M –

Orders 17 June
L^{t}. Col^{o}. Marion }
. . Parole ~~Savanna~~ Vienna C S^{n}. ~~Cumming point~~ 99
Fort Guard tomorrow Captn. ~~Dunbar~~ Lesesne L^{t}. ~~Foissin~~ Mason
rear Serg^{t}. Raybold – Hosp^{l}. Corp^{l}. McDonald

A reg^{tl} court martial to set this forenoon at 11 OC: for the tryal of all prisoners Orderd to it evidences to Attend

Captn. Moultrie presid^{t}. L^{ts}. Guerry & Martin members

---

by L^{t}. Col^{o}. M.

Orders 18th June
Maj^{r}. Horry. }
. . . Parole Poland . . C S^{n}. Cummins point
Fort guard tomorrow Captn. Dunbar Lt Foissin
rear Serg^{t}. Wickham – Hosp: Corp: Rains

A court martial to set this forenoon 11 OC: to try all prisoners that may be orderd before it evidences to Attend

Presid^{t}. Captn. Dunbar L^{ts}. Hart & Capers members

Aft^{r}. Orders
L^{t}. Col^{o}. Marion }
1 Serg^{t}. 1 Corp^{l}. & 12 privates with 18 rounds & 1 spare flint p^{r}. man to hold themselves in readyness to go as a guard to Maurice Island over

prisoners of war which will be delivered them at 7 OC: tomorrow Morning Opposite the rear guard on the Breach, the Serg[t]. & men must be Chose from those who may have had the Small pox, this Command to be relieved weekly – the Q M Serg[t]. to draw provisions for the prisoners dayly and to furnish the Guard with 3 tin Kittles & an ax, the Serg[t] will receive his orders when ready to march

L[t]. Col[o]. Marron **Fort Moultrie**

Gen[l]. Orders 17th June
Gen[l]. Moultrie } The troops at fort moultrie are to fire their Vollies and to conclude with a feu de joye, which is to be kept up by Col[o]. Hugers regim[t]. at fort Johnson in the same manner, this fire to begin at the before Sunsett, Col[o]. Thompson's reg[t]. is to fire after Col[o]. Huger they must draw up on the parrade behind the Barracks; This firing is in Commemoration of the Alliance between the court of France & the United States of America [886]

A General court martial to set on Fryday next at fort Johnson for the trial of Cap[t] George Cogdel of the 5th Reg[t]. on a Charge of Col[o]. Isaac Huger for Neglect of Duty & Disobedience of Orders: Evidences to Attend – President for the Court L[t]. Col[o]. Marion 3 Capt[ns]. & 2 Sub: from the 2nd – 3 Capt[ns]. 2 Sub: from 3[d]. – 2 Sub: from 6th

Transmitted – A Dilliant B. M.

R.O. Capt[ns]. Harleston, Mazyck & Hall & Lieu[ts] Mason & Martin are appointed for the G. Court Martial

---

[886] France had given the United states aid since 1776 by sending arms, munitions and clothing to the playwright Caron De Beaumarchais and his "front" company of Rodrigue Hortalex and Company. Benjamin Franklin would then have these items sent overseas to the United States, the Dutch island of St. Eustatia, or to the French island of Martinique. From there they would be sent to the United States. It was not quite clear if the United States would be able to defeat Great Britain until 1777, when the combined battles of Brandywine and Germantown, and then the victory at Saratoga, showed the French that the Americans had the will to fight and the ability to win. France declared their alliance with the United States on 6 February 1778. On 16 June 1779 Spain, an ally of France, declared war on Britain, but did not make an alliance with the United States. In 1776 the Dutch fired a salute when the United States ship *Andrea Doria* sailed into the harbor at St. Eustatia. This was the first time a country recognized the United States as a separate nation. In 1780, Catherine the Great of Russia formed the League of Armed Neutrality with Denmark and Sweden. This was to insure that their vessels would not be searched or seized during the current war between France, England and Spain. Britain knew that if the Netherlands joined this League of Armed Neutrality they may find themselves at war with most of Europe. Unlike popular opinion, Britain was not "the greatest superpower in the world" in the 18th century. That title could arguably go to Russia. Britain needed to declare war on the Netherlands, but they needed a reason. The reason that surfaced was the Scot's Brigade. This was a military unit that Britain "loaned" to the Netherlands in previous wars. Britain demanded that the unit be returned, but the Netherlands stated that they would not return the unit if it was to be used against the United States. Britain then declared war on the Netherlands for the reason of not returning the Scot's Brigade. With the United States, France, Spain and the Netherlands all at war with Britain it became a world war. Britain now had to determine where it needed to send its limited forces. Due to this "world war" Britain could no longer concentrate their full energy on the United States and instead focused on what they considered to be the real threat, that of France. In the end the British were not beaten by the United States militarily, but they ended the war in a way that was similar to how the United States ended the Vietnam and Afghanistan wars. Though Britain lost the campaign against the United States, they won the world war against the other nations and it created the British Empire of the 19th century.

Orders 19th June } . . Parole Defence . . . C S^n^. 16.
L^t^. Col^o^. Marion } Fort guard tomorrow Lieu^ts^ Capers & Hart
rear Serg^t^. Carter – Hosp^tl^. Corp^l^. Tayler

NB by sentence of a court 17th Jno. M^c^Bride rec^d^. 50 lashes for selling anothers property [887]

Orders 20th June } . . Parole Sloops . . . C S^n^. Pises
L^t^. Col^o^. Marion } Fort guard tomorrow Captn. Moultrie & L^t^. Baker
reare Serg^t^. M^c^Donald – Hosp: Corp: Tellinghart

Genl. Orders 19th } The General court martial orderd to set at fort Johnson this day is orderd
Genl. Moultrie } to set in Ch^s^. town tomorrow morn^g^. at such place as the presid^t^. of the Court shall Appoint

Transmitted A Dilliant B M

Orders 21st June } . . Parole Declaration of War . . C S^n^. France
Maj^r^. Horry } Fort guard tomorrow Captn. Dunbar & L^t^. Hart
rear Serg^t^. Kiels
For Hobcaw Lieu^t^ Foissin

Orders 22nd June } . . Parole United States . . C S^n^. America
Maj^r^. Horry } Fort guard tomorrow ~~Capt~~ Lieu^t^. Perreneau & Roux
Rear Serg^t^. Kiels – Hosp: Corp^l^. Roberts

Orders 23d June } . . Parole N^o^. Carolina . C S^n^. Georgia L^t^. Col:
Maj^r^. Horry } Fort guard tomorrow Captn. Moultrie & L^t^. Capers

A court martial to set tomorrow forenoon at 11 OC: to try all prisoners Orderd to it: Evidences to Attend

Captn. Mazyck presid^t^. L^t^. Galvan & Hart members

NB this Court did not sit

~~L^t^. Col^o^ Marion~~ Maj^r^. Horry **Fort Moultrie 1778 –**

The Regim^t^. will be mustered tomorrow by the Deputy muster Master at Beating the Long Roll all Off^rs^. & Soldiers off duty are to be on the parrade & these Muster Rolls for each Company are to be ready for Signing and Attested Immediately after the Reg^t^. is Mustered

---

887 John M^c^Bride was born on 6 October 1758. He was in the 2nd South Carolina Regiment under Captain Thomas Dunbar prior to 1778. In 1778 he was in Captain Moultrie's company. On 19 June 1778 he received 50 lashes for stealing and selling someone else's property. He received 100 lashes on 19 September 1778 for forging Captain Charnock's signature on a pass. He received 50 lashes on 12 July 1779 for being absent without leave. He was supposed to recieve 100 lashes on 9 August 1779 for selling his cloathing, along with Sergeant Newman and Simpson, however it was remitted when he turned evidence against the two sergeants. After the fall of Charleston he was a private in General Marion's Partisans. He was commissioned a lieutenant in August of 1780 when Captain Hamilton deserted to the British and James Witherspoon was promoted to captain. He was in the raid on Keithfield and was in the battles of Black Swamp, White Marsh, Black River, Georgetown, Santee and Eutaw Springs. He died on 9 January 1791 at the age of 33.

Orders 24th June } Parole Genl. Lee . C Sn. Philadelphia
Majr. Horry . } Fort guard tomorrow Captn. Mazyck & Lt. Baker
Maurice Island Guard Sergt. Burtell [888]
rear Sergt. Smith – Advance Sergt. McDonald

Orders 25th June } Parole Chas town . . C Sn. Savanna
Majr. Horry } Fort guard tomorrow Captn. Dunbar & Lt. Warley
reare Sergeant

A court Martial to set at 10 OC this forenoon to try such prisoners as may be Orderd to it. evidences to Attend

Presidt. Captn. Mazyck & Lieut. Baker & Capers members

Genl. Orders 19th June } The Deputy Clothier General is Order'd to deliver out of the Continental
by G. Moultrie } store, Cloathing to the Offrs. in such proportions as there may be Cloath and Linen to spare the Offrs paying ready Money to the D. Cloathier Genl.who will attend for that purpose at the store every Monday, Wednesday & Fryday from 7 OC: in morning to 12 OC. at noon –

Transmitted – A Dilliant B M

Orders 26th June } Parole Colo. Marion . . C Sn. 92 –
Majr. Horry } Fort guard tomorrow Captn. Charnock & Lt. Guerry
rear Sergt. Carter

The Q. M. Sergt. is Order'd to have the whole Garrison cleaned out Particularly under the Platforms –

The Captn. of the guard will give Orders to the Centries on the platform not to Suffer any Soldier to Meddle with the Cannon, Shot Carriages or other Utensils thereunto belonging, or with the shot that piled up at the foot of the Stairs leading to the platform (Gunner his mate & fatigueman only Excepted) who's duty is to keep the Artillery & Stores in proper Order – the Sergt. of the guard is therefore Order'd to Confine & report all Soldiers who is found Disobeying this Order & they may Depend on being punished by a Court Martial for such Disobediences

---

Orders 27th June } Parole Virginia . . C Sn. Wilkes
Colo. Motte } Fort guard tomorrow Captn Moultrie & Lt. Hart
rear Guard Sergt. Kolbb

The Colo. Acquaints his Officers that he had recd. a Card from his Excellency the president requesting the favor of their Company to Dinner at the State House in Chs. town on Saturday the 4th of next Month.

The Regiment to be on the parrade this Afternoon without Arms at 4 OC: - an Offr of each Company to Attend & Apply to Q Mastr. for Caps, Shirts, Shoes, Stockings, Shoe Buckles & Sleve Buttons who is hereby Orderd, to deliver & distribute them by companys beginning with the grenadiers & finishing with the Light Infantry the remainder to be Deliver'd tomorrow morning at 6 OClock –

---

[888] Morris Island.

The Officer are to take care that no part of the new cloathing be worn before next Munday, when he hopes they will exert themselves in making their men appear in a Soldier like manner, as they will be provided for every requisite for so doing ~

AS Silvister Springer is Appointed a Surgeons mate to this reg^{t}. Orderd that he be respected as such [889]

Corporal Stephen Roberts of Captn. Moultries company is Appointed a Sergeant in the same – Orderd that he be Obeyd as such

Hugh Davis of Captn. Mazycks Company is Appointed a Gunners mate to the ~~reg~~^{t}. Garrison

A regimental court martial to set today at 10 OC: to try such prisoners as may be brought to it, Evidences to Attend in time

Captn. Motte president Lieu^{ts}. Gray & Galvan members

Genl. Orders 26th June by Genl. Moultrie } In Commemoration of the 28th June 1776 the following fire is to take place & to begin at 12 OC: on Munday the 29th inst: June – Fort Moultrie begin by 13 Guns Fort Johnson by 13 Guns, Grimballs continue 13 Guns Dorrals also 13 guns, Laurens Battery finish by 13 guns – A pause of about 5 Minutes between each fort firing

The Deputy Commissary General is orderd for the futer to deliver no more than one ration per day to each Col^{o}. L^{t}. Col^{o}. Maj^{r}. Capt^{ns}. Subalt^{n} & Staff in the Continental Service and the Q Mast^{r}. in the Different Regim^{ts} are to make their returns Accordingly

Transmitted A Dilliant B. M:

NB by sentence of last court Serg^{t}. Marlow for Abusing M^{r}. Poyas was acquitted [890]

Orders 28th June Col^{o}. Motte } . . Parole Genl. Moutlrie . C S^{n}. Col^{o}. Motte by Maj^{r}
Fort guard tomorrow Captn. Mazyck L^{ts}. Perreneau & Roux
rear Wickham Serg^{t}.

The Reg^{t}. to be under arms tomorrow forenoon at 11 OC: & to be drawn up on any Grown the Major may think most convenient,[891] each man to be furnished with 3 blunt Cartridges, the Gunner of the Garrison to fire 13 Guns precisely at 12 OC: agreeable to Genl. Orders, he is also to be carefull that every thing on the platform be properly placed & to hoist the flagg at sunrise

The Q.M. Serg^{t}. is hereby Orderd to have the platform clean swept & to remove any dirt in or about the Garrison – this to be done early tomorrow morning as all to prevent any any Blankets or beding of any sort being Exposed to sight in the the fort – Command^{g} Off^{rs} of Comp^{y}. are desired to see that their Barrack rooms are will Cleaned out – the Col^{o} hopes the men will Act with such conduct as to give him Satisfaction & honour to themselves

L^{t}. Foissin to Join & do duty n Captn. Moultries comp^{y}. till further Orders – Divine Service will be preformed this afternoon at 5 OC: the Reg^{t}. to Attend Accordingly –

---

889 "AS" is capitalized in the original text, so it might be the abbreviation for the title of Assistant Surgeon.

890 John Poag was the owner of one of the two plantations on Haddrell's Point. He was a Charlestown merchant, slave trader and shipping owner. He served in the Royal Assembly in the 1760-70's. He was the commissioner for building the powder magazine in 1770. In 1773 he was appointed the Barrack Master and Storekeeper of Ordinance for British forces in the colony. He died on 15 December 1780.

891 Grown = Ground.

Major Horry **Fort Moultrie 1778**

Head Quarters Ch[s]. Town

Gen[l]. Orders 27 June by Gen[l]. Moultrie } The General Court Martial now Seting of which L[t]. Col[o]. Marion is president is to try Serg[t]. M[c]Colom;[892] Corporal W[m]. Johnston [893] & Phil[p]. McGuire [894] of the 1st Reg[t]. for Desertion & being found in Arms against the United States of America – All Evidences to Attend – all other prisoners that shall be brought before the Court are also to be tried And tis Orderd that their Evidences attend in time

After Gen[l]. Orders same date

All Off[rs]. who have Demands upon the Commissary for back rations are to give them immediately that they may be settled

Transmitted by A Dilliant B M

Aft[r]. Regiment[l]. Ord[r]. by Maj[r]. Horry } The troop to beat tomorrow morning at 5 OC: & the Guards for the day are as follows, for the fort guard 1 Capt[n]. 2 Sub 3 Serg[ts]. 2 Corporals. 1 drum & fiffe & 40 privates for the rear Guard 1 Serg[t]. 1 Corp[l]. & 6 privates, for Advance 1 Corp[l]. 3 privates – the Hospitle Guard to be taken off a Sergeant from Each Comp[y] to attend at the shaving & dressing the men of their Comp[y]. & to be very particular in seeing their Cloathes are well put on & their arms & Accoutrements are in the best order

---

Orders 29th June by L[t]. Col[o]. Marion } . . Parole Gen[l]. Moultrie . Count[r]. S[n]. 28th June
Fort Guard tomorrow Capt[n]. Dunbar & Lieu[t]. Capers
Habcaw Guard L[t]. Hart – rear Serg[t]. Smith

Orders 30th June Maj[r]. Horry } . . Parole first Regim[t]. . Count[r]. S[n]. Successfull
Fort guard tomorrow Capt[n]. Charnock & L[t]. Baker
Rear Serg[t]. Minor

Corporal Josiah Teague of Capt[n]. Mazyck Company is Appointed a Sergeant in the Same & is to be Obeyed as such

---

Orders 1st July Maj[r]. Horry } . . Parole Virginia . . Count[r]. S[n]. Maryland
Fort guard tomorrow Capt[n]. Moultrie & L[t]. Warley
rear guard Serg[t]. Teague

A Monthly return from Each company to be made out & Deliver to the Adj[t]. this day

---

[892] Charles McCollom enlisted in the 1st South Carolina Regiment on 30 March 1776 and re-enlisted in December 1776. On 28 June 1778 he was court martialed for desertion and being found in arms against the United States of America.

[893] William Johnston enlisted in the 1st South Carolina Regiment on 16 October 1776. He was promoted to corporal on 2 August 1777. He was demoted to private on 20 October 1777. On 28 June 1778 he was court martialed for desertion and being found in arms against the United States of America. He served in the 6th South Carolina Regiment under Captain George Warley in 1779. He re-enlisted in the 1st South Carolina Regiment on 1 February 1780.

[894] Phillip McGuire served in the 1st South Carolina Regiment. On 28 June 1778 he was court martialed for desertion and being found in arms against the United States of America.

Orders 2nd July
Majr Horry
. . Parole Jerseys . . Countr. Sn. Pinsilviania
Fort guard tomorrow Captn. Mazyck & Lt. Mason
rear Sergt. Coleman – reare Corpl. Webb

A court martial to set this fore noon at 11 OC: to try such prisoners as shall be Orderd to it – all evidences to Attend

Captn. Dunbar presidt. Lieuts. Roux & Capers members

Orders 3d July
Majr. Horry
. . Parole New York . Countr. Sign Connectecut
Fort guard tomorrow Captn. Dunbar & Lt. Guerry
rear Sergt. Jasper – Hospitle

Genl. Orders 3rd July
by Genl. Moultrie
In Commemoration of Independence of the united states of America proclaimed the 4th July 1776, the following firings is to take place on Saturday next about half an hour after 12 at noon Fort Moultrie begin by 13 Guns – fort Johnston next 13 Guns – the forts & Batterys in town are to follow Each Keeping 5 Minutes distance from Each firing – A Dilliant B M

~~Lt. Colo. Marion~~ **Fort Moultrie**
Major Horry

Orders 4th July
Majr. Horry
. . Parole Roade island Countr. Sn. Motte
Fort guard tomorrow Captn. Charnock & Lt. Gray
Rear Guard Sergt. Minor

~~Tomorrow the Garrison guard to be reduced to 2 Sub. 2 Sergts. 2 Corpl. 1 drum 1 fife & 24 privates, and the Officers thereof are strictly to Observe all former Orders respecting the Garrison a Captn. of the day to be Appointed who is to visit the garrison & rear guard once in the day & once during the nights to receive the reports of those guards, which with their own are dayly to be made to the Commanding Officer by 8 OC: in the forenoon~~

~~Orders 5th July~~
~~Majr. Horry~~
~~. . Parole Colo. Dry . Countr Sn. North Carolina~~
~~Fort ~~guard~~ the day tomorrow Captain Mazyck~~
~~Fort Guard Lieuts. Hart – Hobcaw guard Lt. Capers~~
~~rear Guard Sergt. McDonald~~

~~Corporal Alexd. Stewart of the 1st Vacant Compy. is Appointed a sergeant in the same Orderd that he be Obeyed as such~~

~~Genl. Orders 2nd~~
~~Genl. Moultrie~~
~~A General court martial to set on Tuesday next to try Lt. Perreneau of Colo. Mottes Battalion on a Charge of Colo. Isaac Motte for disobedience of Orders & neglect of duty also to try Lt. Jones of Colo. Thompson regt. on a Complaint of L. Taggart of the same regt. for sending him a Challenge to fight a Duel – All Evidences for & against the prisoners to be warned to Attend – President Major Horry, 6 Members from 2nd Regt. 4 Members from 3d . & 2 from 5th~~

~~Aftr. Orders~~
~~Majr. Horry~~
~~The members of the Genl. Court Martial to set on Tuesday next in Chs. Town are Captn. Moultrie & Charnock & Lieuts Baker, Warley & Foissin – the Barge will Attend the Members to town tomorrow~~

Orders 5th July } . . Parole Massachuset . C S^{n}. ~~North~~ New Hampshire
Maj^{r}. Horry } Fort guard tomorrow Captn. Mazyck & L^{t}. Foissin
Hobcaw Guard L^{t}. Perreneau – Rear Serg^{t}.

Genl. Orders } The Deputy Commissary Genl. in future to serve to serve out vinigar to
G. Moultrie } the troops in the continental service in this State According to the rations Allowed by the State, to begin by Fort Moultrie and Johnson, he is to allow rations to those Officers Servants only who do not belong to the army according to an Order Issued the 30th of May last, & he is to Issue them Accordingly

---

Orders 6th July } . . Parole North Carolina . . C S^{n}. S^{o}. Carolina
Maj^{r}. Horry } Fort guard tomorrow Captn. Dunbar & L^{t}. Roux
Rear Serg^{t}. Minor

A court martial to Sett this forenoon at 10 OC: to try all prisoners orderd to it all Evidences to Attend

Presid^{t}. Captn. Dunbar, Members L^{ts}. Gray & Guerry

Orders 7th July } . . Parole Georgia . . C S^{n}. 13
Maj^{r}. Horry } Fort guard tomorrow Captn. Charnock & L^{t}. Capers
Rear Guard Serg^{t}. Wickom

Major Horry **Fort Moultrie 1778**

Genl. Orders 7th July } Lieu^{t}. John Bush of Col^{o}. Motts regim^{t}. having resigned his Commission
by Genl. Moultrie } is no longer to be considered as a Continental Officer

The Genl. Court martial of which L^{t}. Col^{o}. Marion was president is dissolved the sentence respecting Capt Cogdell is as follows

The Court having Maturely weighed the matter are of Opinion that Captn. Cogdell is not guilty of neglect of duty or Disobedience of Orders and therefore Honourably Acquit him

The General Approves of the Sentence of the General Court Martial in which Captn. Cogdell is Honourably Acquitted, & orders Captn. Cogdell to Join his regiment

---

Orders 8th July } . . Parole Genl. Lee . . C S^{n}. 2.
Maj^{r}. Horry } Fort guard tomorrow Captn. Moultres & L^{t}. Baker
rear Serg^{t}. Williamson – Advance Serg^{t}. Raybout

A court martial to sett this forenoon at 11 OC to try all prisoners Ordered to it – all evidences to Attend – Presid^{t}. Capt Motte members L^{ts}. Hart & Warley

Lieu^{t}. Capers is orderd to Join & do duty in Captain Moultries Comp^{y}. till further orders

The Officers of the regim^{t}. are desired to reinlist no man who has more than 12 Months to serve, by Applying to the Major they will be Informed thereof

The Surgeon of the regim^{t}. will have what sick men are in garrison or Camp removed as soon as possible into the Hospitle

The orderly Serg^{t}. or Corporals of Companys for the day are every Morning after Roll-call, to Attend such men of their companys as complain of being sick & unfit for duty, to the Hospitle to be examin'd by the Surgeon or one of his mates & if found sick to deliver them to him Accordingly, otherwise to return them fit for duty

The Surgeon or in his Absence one of his mates, are Order'd whenever a main in the Hospitle is fit for duty to report the same immediately to the commanding Officer

---

Orders 9th July }
Majr. Horry }
. . Parole Genl. Schuyler . . . C Sn. 4
Fort Guard tomorrow Captn. Mazyck & Lt. Warley
rear Sergt. Fowler

Mr. Josiah Kolb is Appointed an Ensign in the regimt Orderd that he be Obeyed and respected as Such

Ensign Kolb is to Join & do duty in Captn. Charnocks Compy. till further Orders

Major Horry **Fort Moultrie 1778 -**

Orders 10th July }
Majr. Horry }
. . Parole Genl. Putnam . . C Sn. 6
Fort guard tomorrow Captn. Dunbar & Lt. Proveaux
rear Sergeant Murphy

Corporal Robert Watt of the Granidr. Company [895] & Wm. Brown late of Captn. Mayzycks Compy.[896] are Appointed Sergeants in the regimt. they are to be Obeyed as such & to do duty in the Granidier Company

---

Orders 11th July }
Majr. Horry }
. . Parole Genl. Gates . . C Sn. 8
Fort ~~Guard~~ the day tomorrow Captn Hall-Fort Guard Lts Mason & Guerry
Rear guard Sergent

Tomorrow the Garrison guard to be reduced to 2 Sub. 2 Sergts. 2 Corpl. 1 drum 1 fife & 24 privates – the Officers thereof are strictly to Observe all former Orders respecting the Garrison – a Captn. of the day to be Appointed who is to visit the garrison & rear guard once in the day & once during the nights to receive the reports of those guards, which with their own are daily to be made to the Commanding Officer by 8 OC: in the forenoon

---

Orders 12th July }
Majr. Horry }
. . Parole Colo. Dry . . C Sign No. Carolina
The day tomorrow Captn. Mazyck
Fort guard Lts. Hart – Rear Sergt. McDonald
Hobcaw Lt. Capers

Corporal Alexd. Stewart of the 1st Vacant Compy. is Appointed a sergeant in the Same Orderd that he be Obeyed as such

---

895 Robert Wade.

896 William Brown enlisted on 4 November 1775. He was in Captain Mazyck's company in 1778. On 10 July 1778 he was promoted to sergeant and transferred to the Grenadier Company. On 12 June 1779 he was demoted to private and transferred to Captain Mason's company. He was in the siege of Savannah. He served in the regiment until 1783.

Gen[l]. Orders 12th } A gen[l]. court martial to set on tuesday next to try L[t]. Perreneau of Col[o].
Gen[l]. Moultrie } Motts reg[t]. on a Charge of Col[o]. Isaac Motte for Disobedience of Orders & Neglect of duty, also to try L[t]. Jones of Col[o]. Thompson Reg[t]. [897] on a Complaint of L[t]. Taggart of the same reg[t]. for sending him a Challenge to fight a Duel [898] – All Evidences for & against the prisoners to be Orderd to Attend – President Major Horry, 6 Members from the 2nd Reg[t]. 4 Members from 3[d] reg[t]. & 2 from 5th.

Aft[r]. Orders } The members from the reg[t]. for the gen[l]. court martial Orderd to set in Charles
Maj[r]. Horry } town on Tuesday next Capt[n]. Moultrie & Charnock & Lieu[ts], Baker, Warley Guerry & Foissin – the Barge will Attend the members to town tomorrow

---

Orders 13th July } . . Parole Gen[l]. Sullevan . . C S[n]. Five
Maj[r]. Horry } the Day tomorrow Capt[n]. Dunbar
Fort guard L[t]. Mason – rear Serg[t]. M[c]Donald

The following men are Appointed Serg[ts]. In the regim[t]. & are to Join & do duty in the following Companies & are to be Obeyd as such
Corp[l]. John Tayler of Capt[n]. Charnocks Comp[y]. to be a serg[t]. in the same also
Tho[s]. Kidwell of the 2nd vacant Comp[y]. to be a Corpor[l] in Capt[n]. Charnocks
Douglas ONeal of Charnocks Comp[y]. to be a serg[t] in the 2nd vacant Comp[y]. [899]
Allen M[c]Donald of Capt[n] Halls to be a sergeant in the Same

Maj[r]. Horry **Fort Moultrie 1778**

Corp[l] David M[c]Gowan of Charnocks comp[y]. to be a Serg[t]. in the same
Corp[l]. Roberts of the Granid[r]. Comp[y]. to be a serg[t]. in the 1st Vacant d[o].
James Grover of Capt[n]. Moultries to be a serg[t] in Capt[n]. Lesesne comp[y]
Gen[l]. Orders } Lieu[t]. Capers is Orderd to take the Command of the Light Infantry
Maj[r]. Horry } Company till further Orders

---

[897] John Jones served in the 3rd South Carolina (Ranger) Regiment under Captain William Caldwell.
[898] William Taggard (Taggart) was appointed a 2nd Lieutenant in the 3rd South Carolina (Ranger) Regiment on 12 February 1778.
[899] Douglas O'Neal enlisted in the 2nd South Carolina Regiment on 9 September 1776. He was promoted to sergeant under Captain Blake in on 3 September 1776. He was sentenced to be reduced to private on 16 August 1777, but he was pardoned. He was sentenced to receive to receive 100 lashes on his bareback with a cat of nine tails, and be reduced to private for being insolent and "criminal in a very high degree" to Lieutenant Dunbar. His reduction to private was suspended on 3 October 1777. However he was reduced to a private just two weeks later on 17 October 1777 but Lieutenant Colonel McIntosh remitted the sentence two days after that. A day later he promoted back to sergeant on 20 October 1777 in Captain Charnock's company. He was reduced in ranks to private on 9 May 1778 fighting and abusing Private Conner. He was reprimanded on 31 October 1778 for beating Private Jacob Johnson. He was promoted to sergeant on 13 July 1778 in the 2nd Vacant company commanded by Lieutenant Adrian Proveaux. He was reduced in ranks again. He was promoted back to sergeant on 11 February 1779. He was in Captain Moultrie's company at the siege of Savannah in 1779.

Orders 14th July } . . Parole Cha[s] town . . . C S[n]. one
Capt[n]. Motte Day tomorrow Capt[n]. Mazyck – Fort guard L[t]. Gray
Rear Guard Serg[t]. Brown

A court martial to set this morning at 10 oClock to try such prisoners as may be brought before them – all evidences to Attend

President Capt[n]. Dunbar Members L[t]. Gray & Hart

NB by Sentence of this court Sam[l]. Henderson for loosing his regimental Coat to be under stoppages to replace it – David Vaughn for loosing his Cap to be under stoppages – Walter Long for theft to receive 100 Lashes, Drury Smith for theft Acquitted – Corp[l]. Stone for persuading men not to reenlist reprimanded – Corp[l]. Kelly for neglect of Duty reprimanded – John Thompson for Defrauding M[r]. Cross of a Sum of money, was remanded

---

Orders 15th July } . . Parole Heyward . . C S[n]. 10
Capt[n]. Motte For the day tomorrow Capt[n]. Dunbar
Fort guard L[t]. Capers – rear Serg[t]. ONeil
Advance guard Serg[t]. Webb

John M[c]Donald of the Granidiers Comp[y]. is Appointed a Serg[t]. in Capt[n]. Harlestons Comp[y]. he is to be Obeyd as such

The Hobcaw guard to be Commanded by a Serg[t]. till further Orders

---

Orders 16th July } . . Parole Gin Green . . C S[n]. 12.
Maj[r]. Horry for the day tomorrow Capt[n]. Motte
fort guard Ensign Kolb – Rear Serg[t]. Grover

Serg[t]. Wickom of Capt[n] Charnocks comp[y]. is Appointed Serg[t]. Major in the reg[t]. Orderd that he be Obey'd as such

The Capt[n]. of the day is to visit the Hobcaw guard once a day when Commanded by a Sergeant & report the Same Accordingly

A court martial to set this forenoon 10 OC: to try such prisoners as shall be Orderd to it, Evidences to Attend in time

President Capt[n]. Motte Members L[ts]. Rous & Mason

Maj[r]. Horry **Fort Moultrie 1778**

Gen[l]. Orders July 14 } The Deputy Q. M. Gel. Is to furnish major Debram with forrage for one
by G. Moultrie horse

John Francis of the 2nd Reg[t].[900] Francis Stevens of the 3[d] reg[t].[901] and John Butler of the same are to be tried by a general court martial now Siting for Desertion,[902] of wich court

---

[900] John Frances was in the 2nd South Carolina Regiment since 1776, under Captain Dunbar. He received 100 lashes with switches on 4 November 1778 for being absent without leave. On 14 July 1778 he was court martialed for desertion. He was in Captain Baker's company in 1778. On 13 December 1778 he received 25 lashes on the bare back with switches for neglect of duty.

[901] Francis Stephens was in the 3rd South Carolina Rangers. On 14 July 1778 he was court martialed for desertion.

[902] John Butler enlisted in the 3rd South Carolina (Ranger) Regiment on 26 March 1777. On 14 July 1778 he was court martialed for desertion.

Col°. Roberts is President & all evidences to Attend – all Officers that are going out upon Leave of Absence or Recruiting Services to Leave their names with the Brigade Major and all Officers that Arrives at head Quarters from Leave of Absence or recruiting are immediately to wait on the Commanding Officer – A Delliant B M

After Ord^rs^. Maj^r^. Horry } A court of Enquiry to set immediately to inquire into the Conduct of William Fletcher Q M Serg^t^. Respecting the Loss of the regimental Books – Presid^t^. Capt Mazyck members L^ts^. Gray & Hart

The Opinion of the above court are that Sergeant Fletcher is not guilty of any Misconduct or neglect of duty but that the Loss of the regimental Books was an Accident

NB by sentence of the court this day John Thompson for Defrauding M^r^. Cross was sentenced to be put under stopages one half of his pay till he Discharge the Note to M^r^. Cross – N. Flinn for drunkenness to double duty for 8 days

---

Orders 17th July Maj^r^. Horry } . . Parole Ch^s^. town . . C S^n^. 5
For the day tomorrow Captn. Mazyck
Fort guard L^t^. Proveaux – Rear Serg^t^. Rybout

All such men as have reinlisted on or after the 1st June last are to Join Such companys to whom such Officers belong at the time or reinlisting

G. Orders 15th July G. Moultrie } The General court martial now siting is not to proceed on the trial of L^t^. Perrineau as Col° Motte is willing to pass Over the Offence upon L^t^. Perreneau resigning his Commission the General therefore Accepts of L^t^. Perreneau resignation & Orders that he be no longer considerd & respected as a Continental Off^r^

A Delliant – B M –

Orders 18th July Maj^r^. Horry . } . . Parole Hancock . . C. S^n^. Fayete
For the day tomorrow Captn. Dunbar
Fort guard L^t^. Mason – Rear Serg^t^. Roberts

As the regiment is now much reduced by Discharges given to men whos time is Expired, Commanding Off^rs^. of Comp^ys^. are to give Leave of Absences but to one private Soldier at a time from their respective Comp^ys^. Commission Off^rs^. are to Apply to the Major for Leave of Absence after first Obtaining Obtaining Leave from their Commd^g^. Off^r^. of their Comp^y^. [903]

---

[903] The first enlistments that occurred in June and November 1775 were coming to an end. There had been no fighting in South Carolina since the Royal Navy was defeated at the Battle of Fort Sullivan in June 1776. Though the British had gained ground in Georgia, the accomplishments were mainly raids on forts and settlements. The British had not been able to capture any key cities, such as Savannah or Augusta, and the citizens of Charleston thought that it would almost impossible to take the city without the attacker losing most of their men. The only action there had been since Fort Sullivan was the failed expeditions to Georgia to fight the British in Florida. These expeditions were complete failures, and the British threat still did not seem credible. Due to this the soldiers did not reenlist in large numbers. There was much better money to be made by signing on as crew to the many privateers that were in the harbor.

2nd Lt. Warley is Appointed a first Lt. in the regt. Orderd that he be obey'd & respected Accordingly – Lt. Warley is to Join & do duty in Captn. Charnocks company till further Orders – [904]
Sergt. Sim

Majr. Horry **Fort Moultrie 1778**

Sergeant Simpson of Captn. Moultries company is Appointed QM Sergt. to the regt. Orderd that he be Obeyd as Such

Genl. Orders 17th July
G. Moultrie

Andrew Rutledge esqr. is Appointed Deputy Waggon master Genl. for this State & is to be respected & Obeyd As orderd

James Francis of the 2d. Regt. who was Orderd to be tried by a Genl Court Martial is to be deliverd to the Order of Colo. Motte [905]

Aftr R.O. A Dilliant B. M.

The provision returns from each Compy to be deliverd to the Q. M. Sergt. by 8 OC: every forenoon in Order that they may be ready to Send with the provision boat to town

---

Orders 19th July
Majr. Horr

. . Parole Lord Starling . . C Sn. 13.
For the Day tomorrow Captn. Motte
Fort guard Ensign Kolb – Hobcaw Sergt. Watt – rear Bond

Orders 20th July
Majr. Horry

. . Parole Genl. Mifflin . . C Sn. 20.
For the day tomorrow Captn. Mazyck
Fort guard Lt. Hart – Rear Sergt. ONeil –

The following Presenments takes place in the regimt. they are to be Obeyd Accordingly – Wm. Haseman of Captn. Harlestons compy. to be a Corpl. in ye same John Conner of Capt Charnocks to be a corpl. in the same – John Brook of Captn. Lesesne to be a Corporal in the same,[906] Wm. Jones & Wm. Rogers of Captn. Moultries Compy. to be Corpl. in the same [907]– Daniel Green of 1st Vacant Compy. a Corpl. in the same [908] Willm. Henson to be a Corporal in do.

---

904 Paul Warley.

905 James Francis was in the 2nd South Carolina Regiment. He was court martialed for an unnamed offense on 17 July 1778.

906 John Brooke enlisted in the 2nd South Carolina Regiment on 11 November 1776. He was promoted to corporal on 20 July 1778 in Captain Lesesne's company. He was discharged on 19 October 1778.

907 William Rodgers (Rogers) enlisted in the 2nd South Carolina Regiment during June 1775 under Captain Barnard Elliot. On 20 July 1778 he was appointed a corporal under Captain Thomas Moultrie. He was at the siege of Savannah. He was taken prisoner at the fall of Charleston. He also served in General Francis Marion's Brigade of Partisans in 1781 and 1782.

908 Daniel Green enlisted in the 2nd South Carolina Regiment on 22 July 1777 in Captain Blake's company. He was promoted to corporal on 20 July 1778 in the 1st Vacant Company. He was then under Captain Baker. He was promoted to sergeant on 19 November 1778 in Captain Lesesne's company. He was in Captain Moultrie's company at the siege of Savannah in 1779. He was taken prisoner at the fall of Charleston and exchanged. Afterwards he served in the militia under Captain Johnston.

A court of Inquiry to set Immediately on the conduct of ~~M^{r}. Young~~ M^{r}. Young Sutler on a Complaint of W^{m}. Manig of Captn. Mottes Comp^{y}. for beating & abusing him without provocation, all Evidences to attend [909]

Presid^{t}. Captn. Dunbar, Member L^{ts}. Roux & Proveaux

Orders 21st July
~~Genl~~. Maj^{r}. Horry } . . Parole Genl. Arnold . C S^{n}. Philadelphia
For the day tomorrow Captn. Dunbar
Fort guard L^{t}. Roux – rear Serg^{t}. Grover

The court of Inquiry orderd to set yesterday on the conduct of M^{r}. Young Sutler to the reg^{t}. report that he is guilty of the Charge alledged against him, but recommend that Maney d^{o} prosecute him thru Civil Law –

A court martial to set immediately to try all prisoners orderd to it Evidences to Attend – Presid^{t}. L^{t}. Mason Memb^{rs}. Lt Capers & Foissin

Genl. Orders 20th July
Gen. Moultrie } Col^{o}. Thompson reg^{t}. is to parrade on wednesday Morning 8 OC: at the new barracks to March to Cummins point, one Serg^{t}. & 12 rank & file to be draughted From Col^{o}. Thompsons Reg^{t}. to Execute the prisoners at that place at 10 OC. – One Subaltern 1 Serg^{t}. 20 rank & file from the 2nd Reg^{t}. 1 Subaltern 1 Serg^{t}. 20 rank & file from the 3^{d} Reg^{t}. are Orderd to be in there at 8 OC: on Wednesday next in the morning at the new Barracks to attend the Execution of the prisoners Sentenced to death, the D. Q. M. Genl. to provide a Cart with three Coffins, to be brought to the new barracks Early on Wednesday Morning their to Wait for further Orders –

Aft^{r}. Ord^{rs}.
Maj^{r}. Horry } L^{t}. Martin with 1 Serg^{t}. & 20 rank & file to parrade at beating the Long roll this Aft^{r}.Noon to hold themselves to Embark by 6 OC: tomorrow to go to Ctown & the New barracks in Order to Attend the Execution of prisoners sentenced to death agreeable to Genl. Orders – this party to have their Arms & Cloathing in the best Order & to have 3 rounds p^{r}. Man Deliverd them by the Q M Sergeant – NB this party salt off 3 OC: 21st July

---

Orders 22^{d} July
Maj. Horry } . . Parole Genl. Wayne . . C S^{n}. 6.
For the day tomorrow Captn. Charnock
Fort guard Lieu^{t}. Capers – Reare Serg^{t}. McDonald
Advance Serg^{t}. Minors –

No Soldiers or non commissioned Off^{r}. to have Leave of Absence from the regiment today –

---

[909] William Manning enlisted in the 3rd South Carolina (Ranger) Regiment on 1 June 1777. He was transferred to the 2nd South Carolina Regiment in 1778. He was promoted to corporal on 12 June 1779 in Captain Thomas Dunbar's company. He was with Dunbar's light infantry company at the siege of Savannah. William Manning was one of Charlestown's upper society and had written to Henry Laurens in 1775 that "it is very dangerous to put muskets and swords in the power of the vulgar, unless they are immediately employed." He noted that "idle soldiers, without strict control, generally rule their masters." The men of wealth in South Carolina and the rest of the colonies were always uneasy with the equality that the revolution created. It would have been interesting to note what William Manning thought now that he served in the ranks with the "vulgar" enlisted ranks.

Orders 23d July }
Majr. Horry

. . Parole Commodore Gillon C Sn. Captn. Parker
For the day tomorrow Captn. Moultrie
Fort guard Lieut. Warley – rear Sergt. Brown

Genl. Orders 21st July }
Genl. Moultrie

The Genl. Court Martial of which Colo. Roberts was president is Dissolved – the Genl. Approves of the Sentence of the Court passed on Lieut. Jones & the evidences against him, are of Opinion that Lt. Taggart did not bring Lt. Jones a general Court Martial with a Laudable Design of putting a Check to his pernicious custom of Duelling, otherways he woud have Orderd him Arrest on the receipt of that Anonimous note, for this & a variety of other reasons arising from the Surcomstances of the case, they return the prosecution frivolous & Dismiss it as such –

A Dilliant B. M.

Orders 24th July }
Colo. Motte

. . Parole Captn. Hall . . C Sn. 28
For the day tomorrow Captn. Dunbar
Fort Guard Lt. Guerry – Hobcaw Lt. Martin – rear Sergt. Webb

Genl. Orders 23d July }
Genl. Moultrie

a Genl. Court martial to set on munday next 29th Inst. at such place as the presidt. shall Appoint for the trial of Lt. Taggart of Colo. Thompsons regt. on a Complaint of Lt. Jones of the same Regt. for sending him a Challenge to fight a Duel – all Evidences to Attend

---

Orders 25th July }
Colo. Motte

. . Parole Richmond . . C Sn. 13. .
For the Day tomorrow Captn. Hall
Fort guard Lt. Proveaux – rear Sergt. Roberts

Mr. Lewis Coffer is Appointed Sutler to the regt. Orderd that no one else Sell any Beer of Spirituous Liquors on the Island[910]

Majr. Horry **Fort Moultrie 1778 –**

A regimental court martial to set at 10 OC: this forenoon to try such prisoners as may be brought to it – the evidences to be Orderd to attend in time & the prisoners made Acquainted that the trial will come on at that hour

Presidt. Captn. Charnock Members Lts. Capers Mason, Hart & Roux

---

Orders 26th July }
Majr. Horry

. . Parole Prize . . C Sn. Sloop
For the day tomorrow Captn. Motte
Fort Guard Lieut. Hart
Hobcaw Lt. Warley – Rear Sergt. Raybolt

No Soldier to go on board the Sloop now on shore in front of the Garrison, without leave granted them by the Commandg. Offr. this Order to be made known immediately to the men who may Depend on being punished if Disobeyed

---

910 Sergeant Coffer had been discharged a few days earlier and took the job of sutler after Mr. Young was found guilty of beating and abusing soldiers.

Orders 27th July } . . Parole Genl. Washington . C Sign America
Majr. Horry For the day tomorrow Captn. Charnock
Fort guard Lt. Roux – Rear Raybolt

Corpl. Joseph Martin of Captn. Mottes Compy is Appointed a Sergt. in the same also Willm. Clark & Roger Champneys of the same Compy. Appointed Corporals in Sd. Company, they are to be Obey'd Accordingly

A Court Martial to set at 10 OClock today to try all prisoners brought to it – evidences to Attend Presidt. Captn. Moultrie – Members Lt. Mason & Guerry

---

Orders 28th July } . . Parole Genl. Maxwell . . C Sn. Brigade
Majr. Horry For the day tomorrow Captn. Moultrie
Fort guard Lt. Capers – rear Sergt. McGowen

A court martial to set this forenoon at 10 OC: to try all prisoners as orderd to it – Evidences to attend President Captn. Hall members Lts. Warley & Hart

NB by sentence of this court 25th Inst: Wm. Hyde for Absence without Leave recd. 100 Lashes with Switches – Jno. Kelly for the same error recd. 50 lashes wth Cat – Timothy Downing for Drunkenness on guard recd. 50 do. wth Cat –

Orders 29th July } . . Parole Independance . . C Sn. 4.
Majr. Horry For the day tomorrow Captn. Dunbar
Fort guard Lt. Warley – Advance Corpl. Henson – rear Sergt. G

the advance guard to be relieved tomorrow by a Corporal & 3 privates to continue a Corpl. Guard till further Orders –

NB. agreeable to sentence of Last court Richard Williamson for losing his regimental Cloathing to be under stoppages to replace them

Major Horry **Fort Moultrie 1778 –**

Orders 30th July } . . Parole Pinckney. C Sn. one
Majr. Horry For the day tomorrow Captn. Moultrie
Fort guard Lt. Guerry – Reare sergt. Feast[911]

Captn. Dunbar & Hall & Lts. Mason, Roux & Capers are Appointed members for the Genl. Court martial Orderd to set in Chs. town they are to go up this morning to attend the president Colo. Huger at the State house by 9 OC:

Lieut. Hart & Foissin are to hold themselves in immediate readyness to go on the recruiting service – the rear guard to be reduced to a Corporal & 3 privates, & to post one Centry at the guard house

---

[911] James Feast enlisted in the 2nd South Carolina Regiment on 3 January 1776. He served as a private, matross, corporal and sergeant. On 29 August 1778 he promoted to sergeant under Captain Harleston. On 20 November 1778 he was confined for drunkenness and refusing to tell Captain Dunbar what crime a soldier had done, that Sergeant Feast had confined. He was remanded during the court martial for being too drunk. In 1779 he was under Captain Peter Gray at the siege of Savannah. On 1 March 1783 he was a private in the 1st South Carolina Regiment.

Orders 31st July } . . Parole Rutledge. . . C Sn. 12
Captn. Motte }
For the day tomorrow Captn. Motte
For guard Lt. Warley – Reare Corpl. Conyers

A monthly return to be given to the Adjutant tomorrow 10 OC: in forenoon from Each Company

A Court Martial to set today at 10 OC: to try such prisoners as shall be brought before them – Captn. Charnock presidt. Lts. Proveaux & Foissin members

A sergt. & 3 men to be added to the rear guard who is to take charge of it immediately & not to Suffer any non Commissioned Offr. or Soldier to pass the Bridge without a wrightin Order from the Commanding Officer till further Orders –

NB by sentence of the court M. Francis Dupree recd. 100 lashes wth switches

---

Monthly return 1st August –

Orders 1st Augt. } . . Parole Colo. Motte . . C Sn. Chs. town
Maj. Horry }
For the day tomorrow Captn. Carnock
Fort guard Lt. Baker – rear Sergt. Oneil

Orders 2nd Augt. } . . Parole Cheraws . . . C Sn. Pedee
Majr. Horry }
For the day tomorrow Capt Moultrie
Fort Guard Lt. Proveaux
Hobcaw Lt. Guerry – rear Sergt. Grover

Sergt. Minor with 2 Sergts. 2 Corporals 1 Drum 1 fife & 10 privates to be in immidiate readyness to march at 4 OClock this Afternoon on the recruiting Service & to receive his Orders when the party is ready from the Commanding Offr.

Orders 3d Augt. } . . Parole Majr. Horry . . C Sn. Success
Captn. Motte }
For the day tomorrow Captn. Charnock
Fort Guard Lt. Baker – Rear Sergt. McDonald

A court martial to set this morning at 10 OC: to try such prisoners as may be brought to it – all Evidences to Attend Presidt. Captn. Charnock members Captn. Moultrie & Lt. Baker –

Captn. Motte **Fort Moultrie 1778 –**

Orders 4th Augt. } . . Parole Gates . . C Sn. 13
Captn. Motte }
For the ~~Fort~~ Guard tomorrow Captn. Moultrie
Fort Guard Lt. Martin – Rear Sergt. Teague

~~Genl. Orders~~ the Officers desired to give in their own provision return at the same time the Company returns are made to the Q M Sergt.

Genl. Order } Commanding Officers of Batalions & Corps are to report to head Quarters the
G Howe } Officers sint out a recruiting & the money Given to Each for that Service they are also to report the Offrs. Absent on furlow the time of Leave Given to each

NB Sergt. Grover of Captn. Lesesne Company } A Dilliant
was reduced to the ranks for Beating a Corporal }

General Howe had recently returned from the failed Florida expedition of 1778. In April he had marched with 300 men of the 1st South Carolina Regiment to rendezvous with Colonel Elbert's Continentals at Frederica on St. Simon's Island. Elbert had captured the town on April 18th. Howe had also sent 1,100 militia under Colonel Williamson to rendezvous at Frederica. [912]

Howe and his army arrived at Fort Howe on the Altamaha River on May 9th, but Williamson's militia had not arrived. Unfortunately the men were ill equipped and were suffering from the tropical conditions. Howe crossed the Altamaha River on May 27th and camped at Reid's Bluff. He requested 300 slaves be sent from Georgia to cut a road through the jungle-like terrain, but only 56 slaves arrived. These slaves were divided into two companies and given to the two brigades.

Georgia Governor Houstoun did not offer any assistance in this expedition and was the cause of Howe's men not having enough equipment or supplies. The Georgia militia intercepted whatever supplies were being sent to Howe's force and took what they wanted. On June 14th Howe's army marched to Spring Branch and met with Houstoun. The Georgia governor promised Howe that his militia would rendezvous with the South Carolina army in three to four days.

On June 17th the army fought Thomas Brown's East Florida Rangers. The Rangers lost one man captured and eight horses. On June 19th Howe finally rendezvoused with Elbert's Georgians on the Great Satilla River, but he still had not seen Williamson's militia. Howe's army crossed the Satilla River on June 23rd by using rafts. The Georgia militia attempted to assist Howe's army, but Houstoun had them arrested for marching without orders. Houstoun refused to place his militia under the command of a Continental officer, and neither Howe or Houstoun was willing to yield any part of their command to the other. The two commanders did agree to attack Fort Tonyn, but as he departed the Georgia Governor ordered the guides of Howe's army to leave the camp and report to him.

Howe marched to Fort Tonyn on June 29th, but only found a burned and deserted post. Meanwhile, Houston's force found the British at Alligator River and had a bloody fight, where he had 13 of his men killed and several wounded. Brown's Loyalist Rangers only had one man killed and a few wounded.

On July 1st Houstoun asked Howe to help him fight the British and Howe agreed that he would cooperate. When Howe began to march to Houstoun's assistance, the Georgia Governor decided his militia could no longer fight due to lack of supplies. On July 8th Williamson and his militia finally linked up with Howe's force. Howe told Houstoun that he should command the whole force, but the Georgia Governor refused to relinquish control. To compound matters Williamson said that the South Carolina militia would only

---

[912] Andrew Williamson was born in 1730 in Scotland. He immigrated to the Ninety-Six district of South Carolina by 1758. He was unschooled and served as a cattle driver, then supplied provisions to the militia from 1760 to 1776. He was a lieutenant in the militia in 1760 and served in the Cherokee Indian War. By 1770 he was a major. In 1775 he ordered the arrest of Robert Cunningham, a known Loyalist. He lost the first battle of Ninety-Six in November 1775 to a superior Loyalist force. After the treaty of Ninety-Six he conducted the "Snow Campaign" which ended Loyalist forces in the South Carolina backcountry. He was the commander in chief against the Cherokee Indians in the expedition of 1776. He was promoted to colonel in 1776 and was made a brigadier general in 1778 of the Ninety-Six Regiment. He participated in the Florida expedition in May 1778. He was at the siege of Savannah in 1779. After the fall of Charleston in 1780 he was accused of being the "Benedict Arnold of the South" because he took British protection. He may have been a double agent, giving information to Colonel John Laurens in 1782. He died in 1786 at the age of 56.

take commands from him. To make matters even worse naval commander, Commodore Bowen, said that in all things dealing with naval matters his word must be supreme. The expedition came to a halt and disorder and chaos became the rule.

Due to the weather and the terrain Howe only had 400 Continentals who could fight. Houstoun's militia had deserted, leaving him only 550 men. The horses died in vast numbers during the march southwards and now they didn't have enough to scout or pull the wagons. The last straw was when Houstoun refused to meet with Howe during a council of war, so Howe finally had enough.

He withdrew his army on July 14th, and by July 30th most of the army had returned to Charleston. The British had learned from this and realized that there was no organization in the South. General Augustine Prévost advised Sir Henry Clinton that he should make an expedition into Georgia in the winter, when the heat and disease would not be as deadly.[913]

Orders 5th Augt. } . . Parole Howe . . C Sn. one
Captn. Motte }
For the day tomorrow Captn. Charnock
Fort Guard Lt. Warley – rear Sergt. Tayler – Advance Corpl. Connor

A court martial to set this morning at 10 OC: to try such prisoners as may be brought to it – Captn Charnock presidt. Lts. Gray & Martin Members

---

Orders 6th Augt. } . . Parole Spain . . C Sn. Friend
Captn. Motte }
For the day tomorrow Captn. Moultrie
Fort guard Lt. Gray – rear Sergt. Murphy

Orders 7th Augt. } . . Parole France . . . C Sn. 2
Captn. Motte }
For the day tomorrow Captn. Charnock
Fort Guard Lt. Martin – Rear Sergt. Teague

Corporal Cheatham of the light Infantry Company is Appointed a Sergt. in the same he is to be Obeyed as Such [914]

Orders 8th Augt. } . . Parole Genl. Moultrie . . C Sn. 13.
Captn. Motte }
For the Day tomorrow Captn. Moultrie
Fort Guard Lt. Warley – Rear Sergt.

A Court martial to set this morning at 9 OC: to try such prisoners as May be brought before it – Presidt. Captn. Moultrie Members Lt. Proveaux Gray

---

[913] O'Kelley, *NBBAS, Volume One* pp 209-211.

[914] Abia (Abial) Cheatham served in the 2nd South Carolina as a corporal on 25 August 1777 in the light infantry company. He was promoted to sergeant on 7 August 1778. He was discharged in November 1778

Orders 9th Augt. Captn. Motte } . . Parole Midleton. . C Sn. Congress
For the Day tomorrow Captn. Charnock
Fort guard Lt. Baker – Hobcaw Lt. Martin – rear Serg

The Officers who have recd. money from Colo. Motte for recruiting are to return their Attestation & What money they have remaing on their hands to the Colo. as soon as possible

NB by sentence of last court Sergt. Royboult was mult 14 days pay to be Given the Surgeon for the use of the Sick in hospitle –

Captn. Motte **Fort Moultrie – 1778 –**

Orders 10th Augt. Captn. Motte } . . Parole Drayton . . . C Sn. one
For the day tomorrow Captn. Moultrie
Fort Guard Lt. Proveaux – rear Sergt. McDonald

Orders 11th Augt Captn. Motte } . . Parole Mathews . . C Sn. 5
For the Day tomorrow Captn. Charnock
Fort guard Lt. Gray – rear Sergt. Watt

Orders 12th Augt. Captn. Motte } . . Parole Lawrence . . C Sn. 45
For the Day tomorrow Captn. Moultrie
Fort Guard Lt. Roux – Rear Sergt. Tayler

A court martial to set immediately to try such prisoners as may be brought to it – Captn. Hall presidt. Lts. Mason & Warley members

---

Orders 13th Augt. Captn. Motte } . . Parole Rockenham . . C Sn. 10
For the Day tomorrow Captn. Hall
Fort Guard Lt. Capers – Rear Sergt. McDonald

Genl. Orders 12th Augt. Genl. Howe } The whole of Colo. Sumpters regimt. now in town to take post at Haddrels point tomorrow morning the DQM Genl. or his Adjutant will therefore have vessels ready to Assist them Early in the Morning & inform Lt. Colo. Henderson where the vessels are

Major William Scott of Colo. Pinckneys ~~Company~~ Batalion is promoted to the rank of Lieut. Colo. in that regt. vice Lieut. Colo. Cattle deseased,[915] Captn. Thos. Pinckney to the Rank of Major vice Majr. Scott promoted they are to be Respected and Obey'd as Such

---

Orders 14th Augt. Lt. Colo. Marion } . . Parole Kings Bridge . . C Sn. 13
For the Day tomorrow Captn. Harleston
Fort guard Lt. Warley – Rear Sergt. Roybout

Order'd the Adjutant & Sergt. Major is Orderd not to receive any men for Guard without they are properly Drest & their Arms clean

the fort guard to be reduced tomorrow to 18 privates the Officers of the guard are to place only 6 Centrys & then in the most Advantageous corner

---

[915] The term "vice" means "in place of".

Orders 15th Augt. L t. Colo. Marion } . . Parole Lowndes . C Sn. Genl. Goodin
For the day tomorrow Captn. Motte
Fort guard Lt. Baker – rear Guard Sergt. Teague
A court martial to set this morning to try all prisoners Orderd to it all evidences to Attend Captn. Motte presidt. Lts. Proveaux & Baker members

Lt Colo. Marion **Fort Moultrie 1778**

Orders 16th Augt. Lt. Colo. Marion } . . Parole Colo. Pinckney . . C Sn. Haddrels
For the day tomorrow Captn. Charnock
Fort guard Lieut. Guerry – rear Sergt. Tayler
Hobcaw guard Lt. Capers

Brigade Orders 16th Augt. Genl. Moultrie } The 6th Regimt. to take charge of the Guard at Hobcaw Magazine & the 2nd Regt. to retire to do duty in the regt.

Orders 17th Augt. Lt. Colo. Marion } . . Parole Munmouth . . C Sn. Downes
For the Day tomorrow Captn. Dunbar
Fort Guard Lt. Proveaux & Mason – rear Sergt. McDonald
No Commissioned or non commissioned Officer or private to appear on parrade at roll-call or otherways, without being Drest in their Uniform, those Soldiers who are found without their shoes & stockings on parrade may depend on being confined & tried for Disobedience of this Order – the fort guard to be Augmented tomorrow with one Commissioned Offr & 21 privates & to plant one Centry more till further Ordrs.

Orders 18th Augt. Lt. Colo. Marion } . . Parole Richmond . . C Sn. Prize
For the Day tomorrow Captn. Harleston
Fort Guard Lts. Roux & Martin – Rear Sergt. Roybout

Orders 19th Augt. Lt. Colo. Marion } . . Parole Schooner Sally . . C Sn. 29.
For the day tomorrow Captn. Charnock
Fort guard Lts. Capers & Warley – rear Sergt. Stewart

Genl. Orders 17th Augt. G. Howe } The evidences which were neccesary to the inquiry of Lt. Taggart induced the Countermanding the Order of the 13th of Augt. having Arrived, a Court of enquiry is therefore to set on Wednesday next to inquire in to the Conduct of Lt. Taggart & to report wither he Deserves the Disrespect shewn him by those Offrs who have refused to do duty with him, the Officers who refuse to serve with him to Attend the court ~ the Court will Also inquire in ~~the~~ the conduct of Capt Coil [916] of the 6th Regt. with

---

[916] Thomas Coyle (Coit) served as a private, corporal, and a sergeant in the 5th South Carolina Regiment from 11 June 1776 until he was appointed a second lieutenant during 1777. On 22 January 1778 he became a first lieutenant. He was promoted to captain in the 6th South Carolina Regiment in 1778. He resigned on 3 September 1778 after being found guilty of conduct unbecoming an officer and a gentleman.

whom Captn. Doggat,[917] Warley,[918] Boyer[919] & Lieut. Lacey,[920] Buchannan,[921] Baker,[922] Pollard [923] & Daggat [924] have refused to do Duty & whom they have Charged with Conduct unbecoming of an Offr. & a Gentleman, the court is therefore to inquire wither he deserves the Disrespect shewn him, the Offrs. refusing to serve with him to Attend the Court – the court is to consist of a field Officer as president & 8 Offrs Members – Lt. Colo. McIntosh president of the court of inquiry, the Officers composing this court to be taken from the 1st. 3rd. & 5th regimts According detail 1 _____ from 1st. 2 Captns 3 Subs _____ 2nd two Captns. – 5th one Sub

N. Eveleigh D.A.G.
A Dilliant B. M.

Lt. Colo. Marion **Fort Moultrie 1778**

After Regtl. Orders Lt. Colo. Marion } Captns Motte & Hall for the general court of inquiry to sit in Ch. Town the ~~day~~ 21st Inst:

The regiment is to be musterd the 24th Inst: Commanding Offrs of Compy. will have muster rolls made out by that day

a Court Martial to set this forenoon to try all prisoners order to it Evidences to Attend Captn. Charnock president Lts. Capers & Warley Members - - - - NB this court did not sit

---

Orders 20th Augt Lt. Colo. Marion } . . Parole Sandy Hook. . C Sn. 17
For the day tomorrow Captn. Moultrie
Fort guard Lts. Proveaux & Guerry – rear Sergt. McDonald

---

917 Richard Doggatt was a captain in the 6th South Carolina Regiment under Lieutenant Colonel Henderson.

918 George Warley became a captain in the 6th South Carolina Regiment on 10 June 1777. He served in the 6th South Carolina Regiment until it was combined with the 2nd South Carolina Regiment in February 1780 due to the losses inflicted at Savannah. He was taken prisoner at the fall of Charleston but was exchanged in October 1780. On 30 September 1783 he became a brevet-major.

919 Alexander Boyer (Boyce) became a lieutenant in the 6th South Carolina Regiment during 1776. He was promoted to captain on 27 June 1778. He was mortally wounded at Savannah on 9 October 1779 and died the next month.

920 James Lacey served in the volunteer militia under Captain Benjamin Screven during 1775. He was promoted to captain in September but died a short time later on 20 September 1778.

921 John Buchanan was born in Northern Ireland. He served in the 2nd South Carolina Regiment then was commissioned as a lieutenant in the 6th South Carolina Regiment. He was assigned to the 2nd South Carolina on 1 March 1780, after the regiments were consolidated due to losses at Savannah. He was taken prisoner in Charleston on 12 May 1780. He was exchanged during June 1781 and served to the close of the war. He was also a captain in the 3rd South Carolina (Ranger) Regiment during 1782 and 1783 and became a brevet major on 30 September 1783.

922 Stephen Baker served in the 6th South Carolina Regiment.

923 Richard Pollard enlisted in the 6th South Carolina Regiment on 18 March 1776 and became a sergeant on 6 May 1776. He became a sergeant major on 2 June 1776. He was commissioned a lieutenant during February 1777 and promoted to captain in November with rank backdated to 20 June 1779. He was wounded in the battle of Stono Ferry and then was transferred to the 1st South Carolina Regiment in June 1779. He was taken prisoner at the fall of Charleston and confined at Haddrell's Point. He died in December 1803.

924 Joel Doggatt served in the 6th South Carolina Regiment under Captain Jesse Baker during 1779.

Orders 21st Aug[t]. } Col. Marion — . . Parole New York . . — C S[n]. 91
For the day tomorrow Capt[n]. Harleton
Fort Guard L[ts]. Mason & Roux – rear Serg[t]. Cheatham

Orders 22nd Aug[t]. } L[t]. Col[o]. Marion — . . Parole Bedford . . . — C S[n]. 27
For the Day tomorrow Capt[n]. Charnock
Fort Guard L[ts] Martin & Capers – rear Serg[t]. Murphy

A reg[t]. court martial to set this forenoon to try such prisoners as are Orderd to it – evidences to Attend – Capt[n]. Charnock president Capt[n]. Moultrie L[ts] Roux . Capers & Guerry Members

the rear guard to be Augmented 3 privates tomorrow & to place one Centry more near the Guard – Divine Service will be performed tomorrow by the Chaplain, all Soldiers are desired to Attend clean & decent

---

Orders 23[d] Aug[t]. } Capt[n]. Harleston — . . Parole – Aberquainny . . — C S[n]. 2.
For the day tomorrow Capt[n]. Moultrie
Fort guard L[t]. Capers & Warley – rear Serg[t]. Stewart

Gen[l]. Ord[rs]. 22[d] Aug[t]. } G Howe — the Court of Inquiry of which L[t]. Col[o]. M[c]Intosh was president Orderd to Inquire in the Conduct of L[t]. Taggart & to report wether he deserves the Disrespect Shewn him by those Officers of the 3[d] reg[t]. who have refused to do duty with him, the general Agrees in Opinion with the court & Orders L[t]. Taggart Returns to Duty & that he be Obey'd & Respected as a Continental Officer – the Opinion of the court respecting Capt[n]. Coyle the Gen[l]. has under consideration; the Court Dissolved ~ -

L[t]. William Fitzpatrick of the 3[d] reg[t]. having resigned his Commission is no Longer to be Considerd & obeyed as a Continental Officer [925]

Transmitted . . A Dilliant B. M.

L[t]. Col[o]. Marion **Fort Moultrie 1778**

Orders 24th Aug[t]. } Capt[n]. Harleston — . . Parole Shark. . — C S[n]. Capt[n]. Morgan
For the day tomorrow Capt[n]. Hall
Fort guard L[ts]. Baker & Capers – rear Serg[t]. M[c]Gowan

A regimental court martial to set this forenoon for the trial of such prisoners as shall be brought before it. Evidences to Attend

Presid[t]. Capt[n]. Charnock Members L[ts]. Mason & Foissin

NB: by sentence of court satt a Saturday Serg[t]. Roybout for drunkenness & Impertinence to L[t]. Mason was reduced to the ranks

Aft[r]. Orders } L[t]. Col[o]. Marion — as General Howe & Moultrie is to be in Garrison tomorrow the men for guard are to be Neatly Drest & powderd all other men in the regim[t]. is to appear neat & Decent whoever may Appear Otherways may expect to be taken notice of & Dealt with Accordingly

/ by G Howe

---

[925] William Fitzpatrick served as a lieutenant in the 3rd South Carolina Regiment in 1778. He resigned his commission on 23 August 1778 after the failed Georgia expedition.

Orders 25th Augt. L^{t}. Col^{o}. Marion } . . Parole Milford Haven . C S^{n} Col. Marion
For the day tomorrow Captn. Harleston
Fort Guard L^{t}. Proveaux & Mason – rear Serg^{t}. Brown

Genl. Orders 24th Augt. Genl. Howe } tomorrow being the Birth day of our great Ally the King of France Fort Moultrie is & fort Johnston are in Honour of the day to fire 21 Guns at fort Moultrie at one OC: precisely and five minutes after it finishes there it tis to be taken up by Fort Johnston

Col^{o}. Nicholas Eveleigh having resigned his commission as Dept^{y}. Adjut: Genl. for the State S^{o}. Carolina & Girgia he is no longer to be Considerd & Obeyed as a Continental Officer

Major John Foucheraud Grimkie is Appointed to Act as Dept^{y} Adjut: Genl. for the State of S^{o} Carolina & Georgia with the rank of Col^{o}. in room of Col^{o}. Nich^{l}. Eveleigh resigned, till the pleasure of Congress be known – Transmitted A Dilliant B M –

Orders 26th Augt. L^{t}. Col^{o}. Marion } . . Parole France . . . C S^{n}. 9.
For the day tomorrow Captn. Motte
Fort guard L^{ts}. Roux & Capers – rear Serg^{t}. Watt

NB by Sentence of Court 24th Inst: Serg^{t}. McDonald was reprimanded for Signing a false return – W^{m}. Hide for Disobedience of Orders & Absence from Guard to do duty every other day for 14 days & Confined the day he is not on Guard – Danl. Gordon rec^{d}. 100 lashes for Disobedience of Orders

L^{t}. Col^{o}. Marion **Fort Moultrie – 1778 –**

Orders 27th Augt. Captn. Harleston } . . Parole Abingdon . . C S^{n}. 28
For the Day tomorrow Captn. Charnock
Fort Guard Lieu^{ts}. Baker & Warley – rear Serg^{t}. Webb

A court martial to set this forenoon for the trial of such prisoners as shall be brought to it, Evidences to Attend

President Captn. Moultrie Members L^{t}. Mason & Guerry

Genl. Orders 26th Augt. Genl. Howe } The troops are to be divided in to two Brigades, the 1st_ 2nd_ & 6th Regim^{ts}.will form the first Brigade under the Command of Brigadier Genl. Moultrie – the 3^{d} & 5th Regim^{ts}. will be Commandd by Col^{o}. Isaac Huger Colonel Commandant & will form the second Brigade – the Artillery will receive their Orders from the Commander in Chief_ _

Transmitted . . A Dilliant BM

Orders 28th Augt. Captn. Harleston } . . Parole Harcourt. . . . C S^{n}. 7
For the Day tomorrow Captn. Moultrie
Fort Guard L^{ts}. Proveaux & Guerry – rear Serg^{t}. Cheatham

Genl. Ord^{rs}. 27th Augt Genl. Howe } The Hon^{bl}. the continental congress having passed Several resolution respecting the ~~Army~~ feuter government of the Army, the D. Adj: Genl. will transmit Copies to Commanding Off^{rs} of Brigade & Commanding Officers of Artillery who are to publish them to their respective Commands, that Commanding Officers of Regiments may govern themselves Accordingly

A return of the number of Officers in the Different corps with their rank & date of Commission or brevet is to be immediately made to the Adjutant General - - - -

Transmitted – A Dilliant BM~

Reg^tl. Ord^rs. L^t. Col^o. Marion } The Adjutant or in his Absence the Serg^t. Major to make out a return of the number of Off^rs in the reg^t. with their rank and date of Commission or brevets by tomorrow 9 OC: in fore noon & Given in to the Commanding Officers by that time to be sent out to the Officer Commanding the Brigade

---

In Congress 27th May 1778
Establishment of the American Army
1st Infantry

Resolved, that each Battalion of Infantry shall consist of 9 Comp^ys. one of which shall be of Light Infantry, the Light Infantry to be kept completed by Drafts from the Battalion & Organized During the the Campaign in to corps of Light Infantry that the Battalion of Infantry consist of

L^t. Col^o. Marion

	Pay p^r. Month
one Colonel and Captain	75 dollars
1 L^t. Colonel & Captain	60
1 Major	50
6 Captains Each	40
1 Captn. Lieutenant	26, 2-3^d
8 Lieutenants each	26, 2-3
9 Ensigns each	20
Paymaster } to be taken from the Line	20 } in addition to their pay as Officers in the Line
Adjutant	13
Q Master	13

Fort Moultrie – 1778 –

	Pay p^r. Month
1 Surgeon	60 Dollars
1 Surgeon mate	40
1 Sergeant Major	10
1 Quart^r. Mast^r. Serg^t.	10
27 Serg^ts. each	10
1 drum Major	9
1 fife Major	9
18 drums & fife each	7_1-3^d
27 Corporals each	7_1_3
447 privates each	6_2_3

Each of the field Off^rs to Command a Company
the L^t. of the Col^o. Company to have the rank of Captn. Lieu^tn .
Resolved, that the Adjutant & Quart^r. Master of a regim^t. be nominated by the field Officers out of the Subalterns & presented to the Commander in Chief, or the Commander of a Separate Department for Approbation; and that being Approved of they shall receive from him a warrant agreeable to such nomination

That the Paymaster of a regim^t. be Chosen by the Offers of the regim^t out of the Captains or Subalterns & Appointed by warrant as above: Other Officers are to risque their pay in his hands: the paymaster to have the Change of Cloathing & to distribute the same

Resolved, that the Brigade Majors be Appointed as heretofore by the Commander in Chief, or Commander in a separate Department, out of the Captains in the Brigade to which he shall be Appointed

That the Brigade Quart^r. Mast^r. be Appointed by the Quart^r. Mast^r. Genl. out of the Captains or Subalterns in the Brigade to which he shall be Appointable their Staff Appointment shall be Vacated: the present Aids-de-camp & Brigade Majors to receive their present pay & Rations:

Resolved, that Aids-de-Camp, Brigade Majors & Brigade Quart^{r}. Masters heretofore Appointed from the line, shall hold their present ranks & be Admissable in to the line again in the same rank they held when taken from the Line; provided that no Aids-de-camp, Brigade Majors or or B. Quart^{r}. Master shall have the command of any Officer who Commanded him while in the line

L^{t}. Col^{o}. Marion **Fort Moultrie – 1778 –**

Resolved, that wherever the Adjutant General shall be Appointed from the line he may continue to hold his rank & Commission in the line:

Resolved, That when supernumerary Lieutenants are continued under this Arrangement of the Battalions who are to do the duty of Ensigns they shall be hold their rank and to receive the pay such rank Intitle them to receive

Resolved, that no more Colonels be Appointed in the Infantry but to where any such Commission is or shall become vacant, the Battalion shall be commanded by a Lieu^{tn} . Colonel, who shall be Allowed the same pay as is not granted to a Col^{o}. of Infantry, And shall rise in promotion from that to the rank of Brigadier: and such Battalion shall have only two field Officers via: a Lieu^{tn} . Colonel and Major, but it shall have an Additional Captain

<u>May 29th 1778</u>

Resolved, that no persons hereafter Appointed upon the Civil staff of the Army shall hold or be Intitled to any rank in the Army by Virtue of such Staff Appointments

June 2nd 1778

Resolved that the Officers herein After mentioned be intitled to draw one Ration a day and no more; that where they shall not draw such Rations, they shall not be Allowed any Compensation in lieu thereof

And to the end that they may be Enabled to Line in a manner becoming their Stations.

Resolved, that the following sums be paid to them Monthly for their Subsistance Viz. to every Colonel 50 dollars, to every Lieu^{tn} . Colonel 40 Dollars, to every Major 30 Dollars to ever Captain 20 Dollars to every Lieutenant & Ensign 10 dollars, to every regimental Surgeon 30 Dollars to every Regimental Surgeons mate 10 dollars, to every Chaplain of Brigade 50 Dollars.

Resolved, that subsistance money be Allowed to Officers and Others on the Staff in lieu of Extra rations and that hence forward none of them be Allowed to draw more than one ration a day.

Order'd that the Committe of Arrangement be directed to report to congress as soon as possible such an Allowance as they shall think Addequate to the Station of the respective Off^{rs}. and persons employed in the Staff . . . . . . Extract of the Minutes Ch^{s}. Thompson Sect:

Orders 29th Augt. L^{t}. Col^{o}. Marion } . . Parole Chesterfield . . . C S^{n}. 7
For the Day tomorrow Captn. Hall
Fort Guard L^{ts}. Gray & Roux – rear Serg^{t}. Watt

A monthly return to be made of each comp^{y} & given in to the Adj^{t}. by Tuesday 10 OC in forenoon

Commanding Officers of Comp^{ys} who may have recruits are to make a Serg^{t}. of their company take them out every morning & evening & train them till they are fit for the Battalion

Corp^{l}. James Feast of Captn. Harlestons Comp^{y} is Appointed a Sergeant in the same, he is to be Obey'd as Such

Throughout 1777 there was not much success in recruiting for the Continental regiments. There was no real threat, and most of the fighting was happening in the north. Men who wanted adventure could sign on with a privateer and earn a profit. In March of 1778 the South Carolina Legislature passed the Vagrant Act to fill up the ranks. It ordered the immediate enlistment of "idle men, beggars, strolling or straggling persons" as privates in the six regiments.

Anyone who voluntarily enlisted received 100 acres of land in South Carolina in addition to the 100 acres of land that was promised by the Continental Congress. Bounties were also paid to recruits who were not over the age of forty five. This still did not produce the desired effect. A French official in Charles Town in 1778 observed that the garrison only amounted to 600 men and these were "recruited with some difficulty.

L^t^. Col^o^. Marion **Fort Moultrie - 1778**

Orders 30th Aug^t^. } . . Parole Richmond . . . C S^n^. 42
L^t^. Col^o^. Marion } For the Day tomorrow Captn. Charnock
Fort Guard L^ts^. Capers & Warley – rear Serg^t^. ONeill

A regimental court martial to set tomorrow 11 OC in forenoon to try all prisoners as will be Orderd to it, - Evidences to Attend
President Capt. Hall Members Lieu^ts^ Baker, Proveaux, Gray & Warley

---

Orders 31st Aug^t^. } . . Parole Scott . . . C S^n^. 7
Capt. Motte } For the Day tomorrow Captn. Moultrie
Fort Guard L^ts^. Baker & Warley – rear Serg^t^. Martin

Orders 1st Sept } . . Parole Horry. . . C S^n^. 25
Capt. Motte } For the Day tomorrow Captn. Hall
Fort Guard L^ts^. Proveaux & Guerry – rear Serg^t^. Feast

Orders 2nd Sept } . . Parole Light House. . . C S^n^. 99
L^t^. Col^o^. Marion } For the Day tomorrow Captn. Mott
Fort Guard L^ts^. Mason & Roux – rear Serg^t^. Stewart

A regimental court martial to be held today at 11 OC: in forenoon to try all Such Prisoners as shall be Orderd to it _ all evidence to Attend[926]
Presid^t^ Cap^t^_Mott Members L^ts^ Martin & Capers { by Captn Hall

---

Orders 3d Sept } . . Parole Denmark. . . C S^n^. Col^o^. Marion
L^t^. Col^o^. Marion } For the Day tomorrow Captn. Charnock
Fort Guard L^t^. Capers – rear Serg^t^. M^c^Gowen

The Battalion is to Exercise every After noon as formerly by all Officers & Soldiers to Attend, the Eldest Officer on parrade to Exercise the Battalion or Order the Adjutant to do it

---

926 There are more than one "company level" orderly books for this time period, and in one of them there is an additional entry of "a Serg^t^. & 3 privates to be added to the Advance Guard tomorrow".

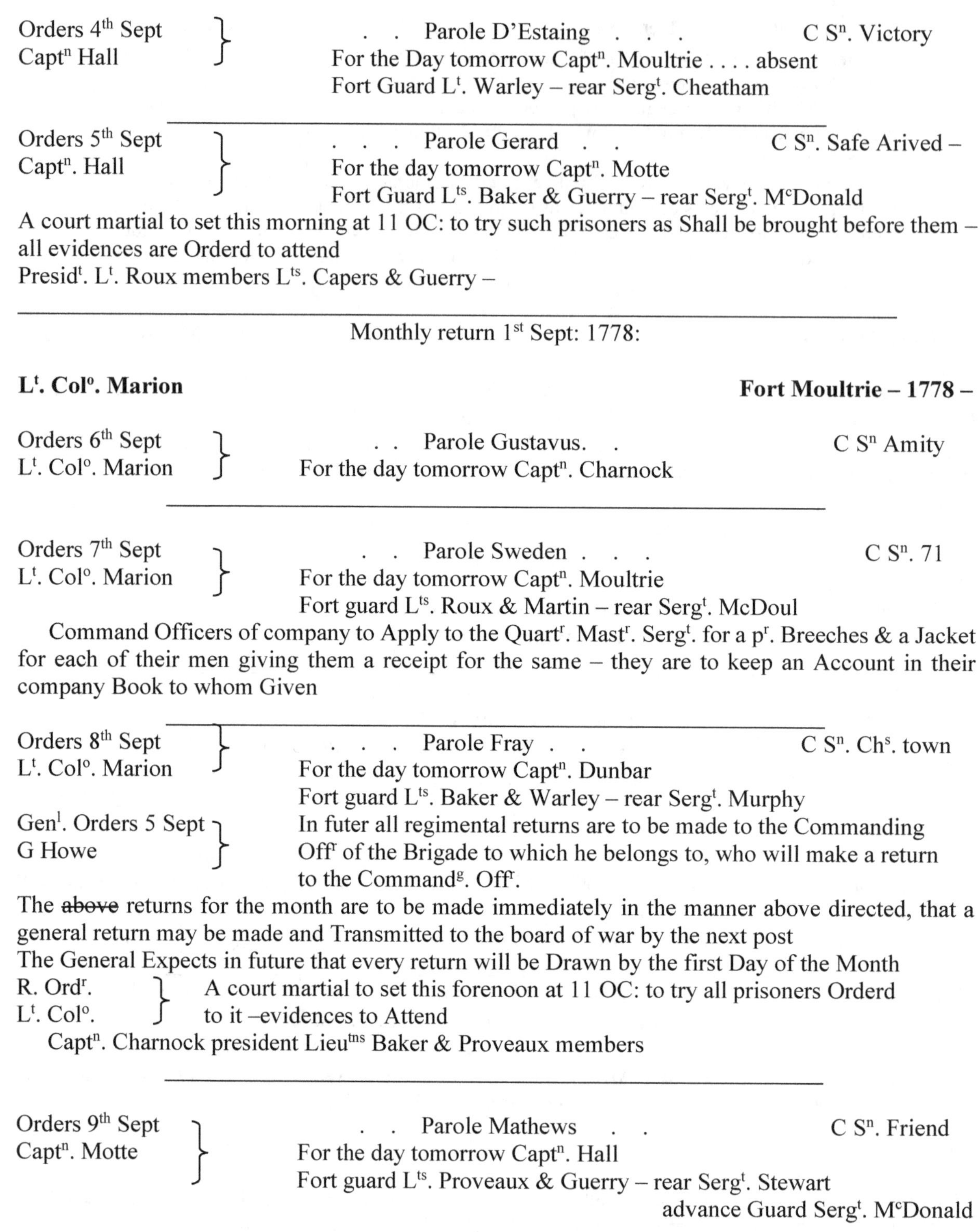

Orders 4th Sept
Captn Hall } . . Parole D'Estaing . . . C S^{n}. Victory
For the Day tomorrow Captn. Moultrie . . . . absent
Fort Guard L^{t}. Warley – rear Serg^{t}. Cheatham

Orders 5th Sept
Captn. Hall } . . . Parole Gerard . . C S^{n}. Safe Arived –
For the day tomorrow Captn. Motte
Fort Guard L^{ts}. Baker & Guerry – rear Serg^{t}. McDonald

A court martial to set this morning at 11 OC: to try such prisoners as Shall be brought before them – all evidences are Orderd to attend
Presid^{t}. L^{t}. Roux members L^{ts}. Capers & Guerry –

Monthly return 1st Sept: 1778:

L^{t}. Col^{o}. Marion **Fort Moultrie – 1778 –**

Orders 6th Sept
L^{t}. Col^{o}. Marion } . . Parole Gustavus. . C S^{n} Amity
For the day tomorrow Captn. Charnock

Orders 7th Sept
L^{t}. Col^{o}. Marion } . . Parole Sweden . . . C S^{n}. 71
For the day tomorrow Captn. Moultrie
Fort guard L^{ts}. Roux & Martin – rear Serg^{t}. McDoul

Command Officers of company to Apply to the Quart^{r}. Mast^{r}. Serg^{t}. for a p^{r}. Breeches & a Jacket for each of their men giving them a receipt for the same – they are to keep an Account in their company Book to whom Given

---

Orders 8th Sept
L^{t}. Col^{o}. Marion } . . . Parole Fray . . C S^{n}. Ch^{s}. town
For the day tomorrow Captn. Dunbar
Fort guard L^{ts}. Baker & Warley – rear Serg^{t}. Murphy

Genl. Orders 5 Sept
G Howe } In futer all regimental returns are to be made to the Commanding Off^{r} of the Brigade to which he belongs to, who will make a return to the Command^{g}. Off^{r}.

The ~~above~~ returns for the month are to be made immediately in the manner above directed, that a general return may be made and Transmitted to the board of war by the next post
The General Expects in future that every return will be Drawn by the first Day of the Month

R. Ord^{r}.
L^{t}. Col^{o}. } A court martial to set this forenoon at 11 OC: to try all prisoners Orderd to it –evidences to Attend
Captn. Charnock president Lieu^{tns} Baker & Proveaux members

---

Orders 9th Sept
Captn. Motte } . . Parole Mathews . . C S^{n}. Friend
For the day tomorrow Captn. Hall
Fort guard L^{ts}. Proveaux & Guerry – rear Serg^{t}. Stewart
advance Guard Serg^{t}. McDonald

Orders 10th Sept } . . Parole Drayton . . . C Sn. 2
Captn. Motte } For the day tomorrow Captn. Motte
Fort Guard Lts. Mason & Martin – rear Sergt. Stewart

Orders 11th Sept } . . Parole Hayward . . C Sn.
Lt. Colo. Marion } For the day tomorrow Captn. Charnock
Fort Guard Lieuts. Baker & Warley – rear Sergt. Watt

NB by sentence of Court M. 5th Inst: Frederick Simmons recd. 75 lashes with Cat for Absence without Leave

Lt. Colo. Marion **Fort Moultrie – 1778 –**

Orders 12th Septr. } . . Entrd in the 11th Septr. . . .
Lt. Colo. Marion } NB Sentance of a Court martial 5th Septr. Inst. . Frederick Simonds of Captn. Dunbars Compy. Recd. 75 Lashes with a Cat o nine tails for absent without Leave

---

Orders 12th Sept } . . Paroll Carteel . . C Sn. Marion
Lt. Colo. } For the day to Morrow Capt Moultrie
~~Lieut~~. Fort Guard Lieut. Provoux . . . . . .
Rear Guard Sargt. Webb . . . .

The fort Guard to be mounted to morrow with only one Subaltern

all offrs. have Leave to go to Haddrels Point after the Orders of the day is Issued in the morning provided they Return before Roll call & Battalion Exercise in the afternoon

---

orders 13th Septr. } . . Paroll Hamsted . . Cnt Sn. 7 - - - - - -
for the day to morrow Captn. Dunbar for the Fort Guard Lt. Mason
for the rear guard Sargt. Martin - -

orderd that all offrs who may Obtain Leave of Absence from the General or other Superior Offirs not Resident with the Regt. that Such Offrs. to Acquaint the then Commandg. offrs. present with Such Leave of Absence by Letter or Personally

---

orders 14th Septr. } . . Parole Marqs Fiatte. Ctr Sn 19 –
For the day to morrow Captn. Hall for the Fort Guard Lieutn . Martin
for the Rear Guard Sargt. Watt

A Regtl. Court martial to sit this fore noone at 11 OC: to try all Prisoners orderd to it Evidence to Attend Captn. Moultrie President members Lts. Martin & Warley

NB Michal Caps sentenced to be reprimanded for absence Sargt. Teague

Orders 15th Septr. ____
L^t^. Col^o^. Marion } Parole Rode Island Ct^r^ S^n^ Marrion
For the day to morrow Captn Charnock for the Fort Guard L^tn^ Warley
for the Rear Guard Sarg^t^. –

the Tattoo to beat for the Futer at 8 OClock at Night Guard Relieved at 7 OClock in Morning Exercise at 5 OC: in Afternoon no non Commissioned off^rs^. or Soldier to Sleep out of Garrison after this day without a permistion from the Com^dg^ Off^r^ of the Reg^t^.

Such Soldiers who now live out of the Garrison are orderd to to move in immedeately as all the Hutts will be pulled down to morrow Com^dg^. Off^rs^. of Comp^ys^ will see their men Comply with the above Order –

The Off^rs^ of the fort guard to send Sargts & men from his Guard After Retreat beating out of the Garison & take up & Confine all such men who may Contrary to orders remain out of Garrison to be continued till further Orders - - - - - - - - - - - - - - - - -

The Fatigue men to pull down all Hutts or houses to morrow (except such as the Com^dg^. Off^rs^ shall point out) that are out of Garrison they are to begin as soon as the Provision is Issued out the adj^nt^ & Sarg^t^ Maj^r^ will See they perform this Duty.

Orders 16th Sept^r^.____
by Captn. Charnock } Parole Hall – Ct^r^ S^n^. Success
for the day to Morrow Captn. Dunbar for the Fort Guard Lieu^tn^ . Proveaux for the Rear Guard Sarg^t^. Feast. for the Advance Guard Sarg^t^. Teauge –

Head Qu^ts^ Chas. Town 15th Septr. morn^g^. Ord^rs^. by G Howe

The Guard at Habcaw to Relieved very Early to morrow Morning by a Detachment from 2^d^. Reg^t^. Consist^g^. Of the usual Number.

Reg^tl^ orders by Capt Charnock

a Court martial to sett this Morn^g^. at 10 OClock to try Such Prisoners as may be Brought to it all Evidences to attend President Capt Dunbar members L^ts^. Mason and Proveaux

Orders 17th Sept^r^____
L^t^. Col^o^. Marion } Parole Statin IslandCt^r^. S^n^. New york
for the Day to Morrow Captn. Hall for the Fort Guard L^t^. Mason for the Rear Guard Sarg^t^. Taylor

A court martial to set to day at 11 O^clk^. in the forenoone to try all Prisiners Orderd to it –
all Evidences to Attend President Captn. Charnock *members* L^ts^. Warly & Mason

L^t^. Col^o^. Marion **Fort Moultrie 1778**

Orders 18th Sept^r^. } Paroll Kings Bridge
for the Day to Morrow Captn. Moultrie Fort Guard Lieu^t^. Martin
for Rear Guard Sarg^tn^ M^c^Gowin

Detail for the General Court martial to set on Friday 8 Septr. 1778. President Maj^r^. Beekman

5 Off^rs^. from the 2^d^. Reg^t^.
4 – from the 1st
3 – from the 6^th^_

Head Quarters Septr. 17th 1778 A Dilliant BM

Orders 19th Septr. L t. Colo. Marion } Agreeable to Brigade Orders the follow.g offrs is appointed to set on the Genl. Court Martial of which Majr. Beckman is President - - - -

Captn. Moultrie, Mazyck, Dunbar & Hall & Lt. Proveaux

Head Quarters Chas. Town 15th Septr. 1778
General Orders by Genl. How recd. this day after the above Brigade Orders }

A General court Martial to set at some Convenit. Place in Charles Town on Friday morning 10 OClock for the trial of Lt. Roux of Colo. Motte Batalian put in Arrist by Captn. Motte of the same Corps for Disrespect to his Comdg. Offrs. the court to Consist of one field officer as president & 12 other members taking them Accordg. to detail - - - - - - - - - - - - - -

from 1st Brigade 5 members
2st Brigade 7 members

Transmitted by A Dilliant BM

NB. By sentence of yesterday Court Jno. McBride of Captn. of Captn. Motts Compy. Recd. 100 Lashes for forgin Captn. Charnock hand to Pass the Bridge – By Sentence of a Court 14 Inst. Michal Capps was reprimanded for a busing Sargt. Teague - - - - - - - - - - -

Lt. Colo. Marion **Fort Moultrie**

Orders 19th June [927] } Parole White ~~Hall~~ plains — Ctr. Sn. Forty Five
For the day to morrow Captn. Hall
For Fort Guard Lieut. Warley
Rear guard Sergt.

Genl. Orders
17th June by Genl. How C town } Captn. Oliver Towles of the 3d Regd having Never been exchanged with the Enemy[928] is therefore to be considered as amenable to no Military duty untill he is Released from his parole by ~~Military~~ a proper & equal Exchange [929]

---

[927] This is a mistake, and it should read "19th September".

[928] Oliver Towles was born in 1734 in Virginia. He served as a sergeant in the 3rd South Carolina Regiment of Rangers under Captain John Caldwell and Colonel Thomson during 1775. When he was a lieutenant he was captured at Fort McIntosh on the Saltilla in March 1777. He was commissioned a captain in the 3rd South Carolina (Ranger) Regiment in 1778. During the fall of Charleston he was taken prisoner. He was killed by "Bloody Bill" Cunningham, during his "Bloody Scout" in November 1781.

[929] When a soldier was captured in the 18th century he would remain in captivity until he was paroled or exchanged. If he was paroled he would be allowed to go home. To get parole he might have to sign an oath stating he would no longer fight against his enemies, or that he wouldn't fight until he was exchanged. This was taken seriously in the 18th century, and if a soldier went back on his parole he could be executed if caught. If a soldier was exchanged they were "traded" for another soldier of equal rank that had been captured by the other side. If a soldier was paroled, and he received word that he had been exchanged, he would be allowed to fight. Until that time he could not fight or do military duty. In this situation Captain Towles had been captured at Fort McIntosh, and due to the surrender terms he was allowed to return to Charleston on parole. He could not command troops or do military duty until he had been exchanged. For officers this was a matter of honor, and other officers would make sure that he did not break his parole.

The Neccesity of Service required that the orders exemptin Officers who are members of asemble from all Duty during there attendance there upon Should be in some degree be Superceded, the Dep[y]. Adj[t]. Gen[l]. will therefore if recation Require warn some field officers as president of the Court Martial which is to set to morrow & of such off[rs] should be members of Assemble

---

orders 18th Sept[r]. } The General has thought proper to publish the following resalution of Congress

In Congress may 29th 1778.

Resolved that all military officers & soldiers in the service of the United States are of right & Aught to be Amenable to the laws of the States in which ~~in which~~ they reside in Common with other Citizens, but as the propriety of undertaking distant Expeditions and Enterprises or other Military Operations and the mode of Conducting them and Determine at his peril

Transmitted – A Dilliant B.M.

---

orders 20th Sept[r].
by Capt[n]. Hall }

Parole Independance Ct[r]. S[n]. Mason
For the Garrison Guard to morrow
Sargt[s]. M[c]Donald & Martin For the rear Guard Sarg[t] McGowing
Orders by L[t]. Mason

The Garrison & Habcaw Guards to be Reduced to morrow to a Sarg[t]. till further Orders –

L[t]. Col[o]. Marion **Fort Moultrie 1778**

orders 20th Sept[r].
by Capt[n]. Hall }

~~Parole Independance~~
Parole Charnock Ct[r]. S[n]_
For the garrison Guard to morrow Sarg[t]. Minor & Cheattem –
For the Rear Guard Sarg[t]. Feast

orders 22[d] Sept[r].
by L[t]. Col[o]. Marion }

Parole Voltair Count[r]. S[n]. 3
For the Garrison Guard to morrow Sarg[t]. Murphy and Watt
For the Rear Guard Sarg[t]. Taylor –

orders 23[d] Sept[r].

For Habcaw Guard Sarg[t]. Stewart –

---

by L[t]. Col[o]. Marion }

Parole Gen[l]. Sullivan Ct[r]. S[n]. Fayett
For the Day to morrow L[t]. Mason
For the fort Guard tomorrow Sarg[t]. M[c]Donald & M[c]Gowen
Rear Guard Sarg[t]. Chattam Advance Guard Sarg[t]. Martin –

Gen[l]. Orders by____
~~orders~~ Gen[l]. How
Cha[s] Town 19 Sep[r] } Col[o]. Motte of the 2[d] Batalion and Col[o]. Sumpter of the 6th having resigned their Commissions are no longer to be Considered as Contineltal Officers – [930]

A Detatchment of the 3[d] Regt. in proportion to the waggons now ready are to March for Orangeburg with all Possible Expedition [931]
Transmitted A Dilliant BM.

Reg[t]. orders____
by L[t]. Col[o]. Marion } L[t]. Col[o]. Marion having received a Very Polite Letter from Col[o]. Mot having thought it Nessesary to publish a paragraph Relative to the Regiment
The parragraph of Col[o]. Motte Lett[r] dated 19 Sep[t]. 1778
I most heartily and Sincearly wish You and the Corps every Blessing you possible can injoy & to assure you I shall ever have the Success & welfair of the Second Regimint greatly to Hart
Sined
Isaac Motte –

Gen[l]. Orders____
by Ge[l]. How } Head Quarters Cha[s]. Town Sept[r]. 21st 1778.
The Court martial now Seting is to try Charl[s] Troublefield a private in the 2[d] Contineltal Reg[t]. in this State for Desertion [932] L[t]. Col[o]. Marion will furnish the Evidences –

L[t]. Col[o]. West of 3[d] Reg[t]. having Resigned his Commission the 14th Inst. is no longer to be Respected & Obeyed a Contineltal Officer [933]

the Guard at Dorchester to be Relieved to morrow Mor[g]. by a Detachment from the 1st. Brigade Consist[g]. Of 1 L[t]. 1 Sarg[ts]. 2 Corporals & 13 Privates

Transmitted A Dilliant B.M.
Fort Moultrie 1778

---

930 Francis Marion became the commander of the 2nd South Carolina upon the resignation of Colonel Motte.

931 This may have been due to support Colonel Andrew Williamson and 500 South Carolina Militia that marched to the frontier to protect the settlements there. The month before, in August, Creek Indians attacked Wilkes County, Georgia and killed twenty settlers.

932 Charles Troublefield (Tumblefield) served in the 2nd South Carolina Regiment under Captain Lesesne in 1778 and 1779. On 22 September 1778 he was court martialed for desertion and was found guilty. He recieved 99 lashes on the bare back with a cat of nine tails over a period of three days, and he was picketed for seven minutes on one day and then eight minutes on another, for a total of 15 minutes. He received 100 lashes on the bare back with switches on 30 November 1778 for being absent without leave. On 13 December 1778 he received 50 lashes on the bare back with switches for "misdemeanor contrary to the articles of war".

933 Lieutenant Cato West was only a lieutenant. He had served in an independent ranger company under Captain Robert Ellison in 1775. In March 1777 he became a lieutenant in the 3rd South Carolina (Ranger) Regiment. He resigned on 14 September 1778 after the failed Georgia expedition.

orders by G[l]. How Sep[t]. 22[d] 1778 } Head Quarter Cha[s]. Town 23[d] Sept[r]
A sarg[t]. & 12 privates from Fort Johnston & a Sarg[t]. 1 Corp[l]. & 10 privates from Fort Moultrie to immediately Embark on board such Vessels and in Such A maner as his Excell[y]. the President shall Direct & to receive such ord[rs]. as he shall Issue - Sined R. How

To the Com[dn] Officer of Fort Moultrie Maj[r]. Gen[l].

The Party from Fort Moultrie to go on board the Pilate Boat Tryall & to put themselves under Command of the Master there of to Provide them selves with 24 Rounds p[r]. Man

Rawlins Lownds – [934]

Reg[t]. orders by L[t]. Col[o]. Marion } one Sarg[t]. 1 Corp[l]. 10 Privates to go on board the Pilot Boat immediately to be furnisht with 24 Rounds & 1 Spare Flint p[r]. Man

Orders to Sarg[t]. M[c]Donald –

Sir you are to Embark with y[e] . party on board the Pilot boat Tryall Comman[d] by Cap[t] Elisinore to follow [935] & Obey all Such Orders as you Shall receive from him or any of his Off[rs]. taking care to Keep your Men in good order [936]

Fort Moultrie 23[d] Sept[r]. 1778:
Fran[s]. Marion –
L[t]. Col[o]. 2[d] Reg[t].

orders 24th 1778 Sept[r] – } Parole Bulls Bay Ct[r]. S[n]. 17
For the Day to morrow Capt[n]. Motte
For the Day to morrow Capt[n]. Dunbar
for Fort Guard to Morrow L[t]. ~~Sarg[t]~~. Warly
For Rear Guard Sargent Feast

Com[dg]. Off[rs]. of Comp[ys]. to make a count Immediately as Possible of the Number of Arms & Accoutrements in their Respective Comp[ys]. with the Number on Each Muskit, they are also to make a Return of what Potts Kittles or Axes in their Comp[y] The Qut[r] Master Sarg[t]. to make an Exact

---

[934] Rawlins Lowndes was the President of South Carolina. He would later become Governor.

[935] Alexander Elsinore served as a harbor pilot under Captain Thomas Smith in Charleston in 1776. He commanded the pilot boat *Tryall* in 1778.

[936] This was in response to a British privateer operating near Charleston. The South Carolina Navy Board ordered Captain Stephen Duvall, of the *Eagle* pilot boat, "to proceed with all possible dispatch to Bulls Inlet in Quest of a small privateer Schooner which if you are so lucky to Meet there you are to Endeavour to take and bring to Charles Town, But should she not be there when you Arrive the Commissioners recommend that you Endeavour by every means in your power to Inform yourself which way they are gone If to the Northward that you pursue her as far as George Town Inlet and no Farther, Unless the Vessel is Actually in Sight If to the Southward that you look into every Inlet as far as Port Royal and after Scouring the Coast so far, if no Certain Intelligence where you may find her, You are to return to Charles Town. The Tryal pilot Boat Cap Elsinore is ordered to Sail with you, and to Obey all such Orders as you may give him during the Cruize the Commissioners recommend that you pay particular attention to, and keep a good harmony between your own Men and the Officers and Men of. the Detachment on board both Boats and that you be very particular in your directions to Cap Elsimore so as to prevent a separation of the two Boats"

Return of all the Arms Accurtiments Camp Kittles Havack Sack & all other Military Stores & Cloathn^{g} belonging to the Regim^{t}. this to be done as Soon a Possible –

Fort Moultrie 1778

The Q Mast^{r}. or his Sergeants is not at any Time to give out any Cloath^{g}. Camp Kittles haversacks or or any Arms or Accrutriments with out an Order from the Com^{dg}. Off^{r}. of the Reg^{t}. or field Off^{r}. he is Orderd not to give any forrage or Grain to any but a field Officer as no other have any Right to Forrage –
A Regimental court martial to Sett to Day at 11 OC^{lk}. to try all Prisinors orderd to it Evidences to Attend –
Captn. Hall pris^{d}.
NB L^{t}. Baker & Warley members
Jam^{s}. Clark Reduced & Rund the Gantlet one Jno. Griffin the same for theft

orders Septr. 25th

} Parole Cape Britton Ct^{r}. S^{n}. 19
For the Day to morrow Captn. Hall
For Fort Guard Lieu^{tn} Baker
For Rear Guard Sarg^{t}. Watt

Com^{dg}. Off^{rs}. of Comp^{ys}. to Apply to Qt^{r}. Mt^{r}. Sarg^{t}. for Garters for their men Giving a Rec^{t}. for the Same –

Orders Septr. 26 78

} Parole L^{t}. LauranceCt^{r}. S^{n}. 92
For the Day to morrow Captn. Motte
For Fort Guard L^{t}. Mason for Rear Guard Sarg^{t}. McDoal

Genl.
Orders by

Genl. How
24 Septr. 78 } Head Qut^{rs}. Chas. Town
The court martial ordred to set for the Tryal of L^{t}. Roux of the 2^{d} Contineltal Batalion in this State have reported as Follow

The Court is of opinion that L^{t}. Roux is not Guilty of the Charges under which he was Arraigned & do therefore Aquit him – the General therefore discharge L^{t}. Roux from Arrest & orders him to Return to Duty

Transmitted A Dilliant B. M.

Brig[dl] Orders____

L[t]. Col[o]. Marion – } Notwith standing Orders to the Contrary Some of the Garrison do their Occation under the Platform and in the Milons which are not fill up particularly those at S[o]. Ward I hope ever person will Remember that is Positively orderd that no Person shall do their Occations within the fort or within Thirty Yards of the walls[n] on the out Side, all those who disobey these Orders must Expect to be taken Notice of & Delt with Accordingly – [937]

---

orders 27th Sept[r]. } Parole Bay of Hunda Ct[r]. S[n]. 17
For the Day to morrow Capt[n] Dunbar
For Hobcaw Guard ~~Sarg[t]~~. Lieu[t]. Provaux
For the Fort Guard L[t]. Roux
For Rear Guard Sarg[t]. Teague

As the men are completed with Gaiters it is Orderd that they mount guard all ways with Gaiters & the Top over their Knees of their Breeches & not under, the Sarg[ts]. will be particularly carefull not to permit any man to mount guard, but as above –

The Hobcaw guard to be relieved to morrow by one Subaltern 1 Sarg[t] 1 Corp[l]. & 9 privates

Gen[l].____

orders 25th Sept[r]. by Gen[l]. How } an Officer to be warned for and immediately to attend the G[l]. C Marchal now Siting as a member in the Room of Capt[n]. Charnocks taken sick [938] Capt[n]. harleston of the 2[d] Contineltal Batalion in this State is appointed a member of the General Court Martial now sitting in the Room of Capt[n]. Charnock taken Sick

F Grimkie D Ad G

Reg[t]. Orders____

28th Sept[r]. Col[o]. Marion } Parole Greenland Ct[r]. S[n]. 7
For the Day to Morrow Capt[n]. Hall
For Guard L[t]. Grey
Rear Serg[t]. M[c]Doald

the Sarg[t]. Maj[r]. will take out all the New recruits morn[g]. & Evening & drill them & order a Sarg[t]. to attend him

---

orders by Capt[n] Motte } Parole Phillidelphia Ct[r]. S[n]. 19
For the day tomorrow Capt[n]. Motte
For the Fort Guard L[t]. Hart For Rear Guard Sarg[t]. Murphy

---

937 A merlon was the part of the fortification wall between two embrasures. The soldiers were defecating and urinating into the merlons that were still open on top and not filled in.

938 General Court Martial.

Cop^{l} W^{m}. Sims of the 2^{d} vacant Comp^{y} is appointed a Sargant in the Same he is to be Obeyed & Respected as Such – [939]
the Qt^{r}. Master Sarg^{t}. is to have the Shrowdes Immediately repaired & put in Order –

Fort Moultrie 1778

orders 30th Sept^{r}. } Parole Genl. Sullivan Ct^{r}. S^{n}. 39
For the Day tomorrow Captn. Dunbar
for the fort Guard L^{t}. Capers
for Rear Guard Sarg^{t}. McGowin
for advance Guard Sarg^{t}. Feast

Reg^{tl}. Orders by L^{t}. Col^{o}. Marion } No sargents to be appointed but by the Com^{dg}. off^{r}. of the Reg^{t}. as all such appointment will be made Voide –

L^{t}. Pet^{r}. Gray with a Sargeant & 4 Privates are to hold them Selves in readyness to go on the Recruting Service L^{t}. Gray will Chuse the Men for this party

Com^{dg}. Off^{rs}. of Comp^{ys} to give in a return to day what number of Linen waistcoats & Breeches they have Rec^{d}. & how many are wanting to Compleat their Respective Comp^{y}. Provision return to be given to the Qt^{r}. M^{str} Sargent before the Provision Boat Returns as all those who neglect so to do can not be Served with any Provisions the Next Day –

Orders 1 Octr. L^{t}. Col^{o}. Marion } Parole Notre Dame C^{t} S^{n} Hall
For the Day to morrow Captn Hall
Fort Guard L^{t}. Baker
Rear Sarg^{t}. –

A Monthly Return to be made and givin to the Adj^{t}. immediately

---

NB a monthly Return 1 L^{t}. Col^{o}. 1 Maj^{r}. 8 Capt^{ns}. 9 first L^{ts}. 2 Sec^{d} L^{ts}. 1 Ensine
1 Adj^{ts} 1 ~~Captn~~ Chap^{l}. 1 pay master 1 Qt^{r}. M 1 Surgion 2 mates 1 dr Maj^{r}.
1 Sarg^{t}. Maj^{r}. 2 M Serg^{t}. 18 Serg^{ts}. 13 drums & fifes 214 rank & file

---

Orders 2^{d} Octr. by Lt Col^{o}. Marion } Parole Bedford Ct^{r}. S^{n}. 17 –
For the Day to morrow Captn. Motte
For Fort guard L^{t}. Mason

orders by Captn Motte } Rear guard Sarg^{t}. Stewart
a Regimental Court martials to Set this morning to try Such Prisiners as may be brought before it Evidences to Attend President L^{t}. Mason
Memb^{rs} L^{t}. Warley & Hart –

---

939 William Sims (Simes) served in the 2nd South Carolina Regiment in the 2nd Vacant company. He was promoted to sergeant on 28 September 1778. When the 2nd Vacant Company was disbanded on 5 October 1778 he was assigned to Captain Moultrie's company.

Fort Moultrie 1778

orders 3d Octr. 1778 } Parole Laurina Cn Sn. Colo. Motte
For the Day tomorrow Captn. Dunbar
For the fort Guard Lt. Gray for the Rear Guard Sargt. McDowell

---

Genl. Orders by Genl. How } Head Qtr Octr. 2d 1778
The General Court martial ordered to try Chas. Troublefield and Jno. B Taylor Both of the 2d Regt. and Jno Pinker of the 5th for Desertion have Reported as follows [940] That Chs. Troublefield is Guilty and that they therefore Sentence being to Receive 99 Lashes on the Bare Back with a Cat of nine Tails and that he be Pitcked for a Quarter of an Hour – That Jno. B Taylor is also Gilty of Desertion that they therefore Sentence him to Receive 100 Lashes on the Bare Back with Switches the Genl. Approves and Ratifies the above sentences which the Comdg. officer of the Regt. to which they belong to will have Executed at Such times and in Such Manner as he shall think Proper Jno. Pinker of the 5th Regt. they report that as no Evidence Appearing Against him they Relieve him to the Guard, he is therefore to be Delivered to the Comdg. Officer of the Regt. to which he belongs to be tried of as he thinks Proper by a Regt. Court Martial – the Court Martial is Dissolved –

Transmitted by
A Dilient BM

The Sentence of the Court Martial that Sett yesterday Sentenced Corporal Champness Guilty of leaving his Guard and that he Reprimanded by the Adjutant

---

Orders 4th Octr } Parole Saratoga Cn. Sn. 84 –
For the Day to morrow Captn. Hall
For the Fort Guard Lt. Hart for the
Rear Guard Sargt. McDonald for Hobcaw Guard Lt. Mason –

Fort Moultrie 1778

orders 5th Octr. Lt. Colo. Marion } Parole Congress Ct Sn. 9
For the Day tomorrow Captn. Motte
Fort Guard Lt. Roux
Rear Guard Sargt. Staurt

The second vacant compy. to be Broke to day & the men to be Distributed in such Compys as the Colo. will point out – agreeable to Genl. Orders 27th Augt. & to resolve of Congress Lt. Proveaux is to Join & do duty in Captn. Moultries Compy. till Further Orders – Sargt. Sims to Join & do duty in Captn. ~~Captn~~. Moultries Compy. Lt RB Baker to take the command as Captn. Lieut. in the 1st vacant Compy which is to be the Lt. Colo. Compy. in future [941] Lt. Guerry to Join & do duty in the Lt. Colo. Compy till further Orders –

---

[940] John Pink enlisted in the 5th South Carolina Regiment on 17 June 1776 and re-inlisted on 1 January 1780. He was court martialed on 2 October 1778 for desertion, but he was released due to lack of evidence.

[941] A Captain-Lieutenant was a lieutenant in command of either the Colonel's Company or the Lieutenant Colonel's Company.

The men belon^{g} to 2^{d} vacant Comp^{y} Draw^{d} the Following men & are to Join & do duty in the Comp^{y}. as Follows

Thos Rawlins, W^{m}- Johnston, Jno. White, Jno. Hawkins — Capt Charnocks
Jno. Caddy in L^{t}. Col^{o} Marion

Corp^{l} Fred. Hughes [942], W^{m}- Wilkinson — in C^{t} Lesesne
Jno. Marrs [943], Tho Clark [944], Thos Jones [945] — in C^{t} Moultries

---

Jno. Jordan Combay [946], Joseph Hughes — in Captn. Dunbars
Hugh Dubarre, David Vaughn — in Capt Halls
Corp^{l} Jno. Mills [947], Wm Baldwin, Edwd. George, Jno. Perry [948] — in Captn Mazyck

Head Quarters the 14th Octr. 1778 by Genl. How the Returns of this month are expected at Head Quarters from Com^{dg}. Off^{rs}. of Brigade The Com^{dg}. Officer of the Corps of Artillery will have a Report made to the Genl. of the Fixed Amunition & the Artillery Stores that was returned after the Expedition of Georgia

Transmitted A Dilliant –

---

942 Frederick Hughes enlisted in the 2nd South Carolina Regiment on 12 May 1776 and was promoted to corporal on 22 September 1778 in the 2nd Vacant Company. When the 2nd Vacant Company was disbanded on 5 October 1778 he was assigned to Captain Lesesne's company. In 1779 he was in Captain Proveaux's company. He was demoted to private on 9 April 1779 for taking the aprons off the cannons when they were loaded.

943 John Mars enlisted in the 2nd South Carolina Regiment on 1 January 1776 and re-enlisted on 6 January 1778 in Captain Shubrick's company. After Captain Shubrick's death he served in the 2nd Vacant Company until it was disbanded on 5 October 1778. He was then assigned to Captain Moultrie's company. He was court martialed for desertion in April 1779 and sentenced to be executed by firing squad.

944 Thomas Clark enlisted in the 2nd South Carolina Regiment on 4 November 1775. He was in the 2nd Vacant Company in 1778. When the 2nd Vacant Company was disbanded on 5 October 1778 he was assigned to Captain Moultrie's company.

945 Thomas Jones enlisted in the 2nd South Carolina Regiment on 4 November 1775. He was in the 2nd Vacant Company in 1778. When the 2nd Vacant Company was disbanded on 5 October 1778 he was assigned to Captain Moultrie's company. He was in Captain Gray's company in 1779 during the siege of Savannah.

946 John Jordan Combay (Combe) was in the 2nd Vacant Company in 1778. When the 2nd Vacant Company was disbanded on 5 October 1778 he was assigned to Captain Dunbar's Grenadiers. He was promoted to corporal on 20 January 1779 and transferred to Captain Baker's company.

947 John Mills enlisted in the 2nd South Carolina Regiment on 31 March 1776 and he re-inlisted in Captain Shubrick's company on 1 April 1777. After Captain Shubrick's death he served in the 2nd Vacant Company. He was promoted to corporal on 19 July 1778. When the 2nd Vacant Company was disbanded on 5 October 1778 he was assigned to Captain Mazyck's Company. He was under Captain Adrian Proveaux at the siege of Savannah. His enlistment expired on 28 July 1781.

948 John Perry enlisted in the 2nd South Carolina Regiment on 4 November 1775. He served in Captain Shubrick's company in 1777. After Captain Shubrick's death he served in the 2nd Vacant Company until it was disbanded on 5 October 1778. He was then assigned to Captain Mazyck's Company. During the siege of Savannah he was assigned to Captain Adrian Proveaux.

Fort Moultrie 1778

orders 6th Octr. } Parole Hobcaw Ctr. Sn. 96 –
For the Day to morrow Captn. Moultrie fort Guard Lt. Warley
Rear Guard Sargt. Allen McDonald

A Regimentail court martial to set to day for the Trial of Such prisoners as will be Orderd. to it all evidences to attend

Presidt. Captn. Moultrie
Members Lt. Preveaux & Roux

Guards to be relieved at 8 oClock in Morning till Further Orders Genl. Orders Recd. this Day

Sargt. Simpson of the 2d Regt. is to be immediately sent for to attend the Grand Hospital as Orderly –

Majr. Dilliant BM – Jno. F Grimkie Det. G

NB. By sentance of a Court this day Corporal Kidwell is Suspended a Month to do duty as a Private for that time for Gaming Drury Smith for Gaming to Deliver the money he won of Kidwell up to Lt. Colo. for the use of the Hospital or Run the Gantlet 3 time through Regt.

Orders 7th Octr.
Lt. Colo. Marion } Parole Assemble Ct Sn. 69
For the Day to morrow Captn Mazyck
Fort Guard Lt. Baker
Rear Sargt. Martine advance Guard Sargt. Cheattem

Agreeable to ~~congress~~ Resolve of Congress dated 27th May 1778 the Pay Mstr of the Regiment is to take Charge & receive all the Cloathing beloně to the Regiment & Distribute them Accordingly to the Orders he shall from time to time receive from the Comdg. Officer of the Regt. he will receive from the Quarter Master a return of Cloathing given out for the Preasent Year & Enter them in his Books –

The Quarter Master is to deliver to the Pay master all the Cloathing now in the Regtl. Store an give him Exact return of such Cloaths he has Deliverd out to the Different Compys. for the Present Year The Fort Guard to be acquainted 3 privates & to Place a Cintry over Some plantns which will be brought in to morrow –

Fort Moultrie 1778

orders 8 Octr.
by Lt. Colo. Marion } Parole Majr. Horry Ctr. Sn 31
For the day to morrow Captn. Dunbar
For guard Lieut. Proveaux
Rear guard Sargt. Watt

NB agreeable to sentence of Genl. court martial Chas. Troublefield for Desertion recd. 99 Lashes with a cat 9 tails in 3 days & picketed 7 & 8 min after 2 different day – Jno. Barnet Tayler by the same court Recd. 70 Lashes with switches in two days for Desertion respected 30 lashes on Account of his Book.

Genl. Orders
by Genl. How } Head Quarters Chas Town 6 Octr. 1778.
Colo. Frances Huger having resigned his Commission of Deputy Quartr. Master Genl. to the State of So Carolina is no longer to be Respected or obeyed as a Contineltal Officer. Colo. Stephen Drayton is appointed to Act as Depty Quartr. Master General to the Contineltal

troops in the State of S° Carolina untill the pleasure of Congress can be had, with the rank of Colonel

\- - - - - - - - -

Captn. Roger Saunders of the 1st Batalion in Contineltal service of this State is no longer to be respected & obeyd he having resigned his Commission /at Supra / [949]
Captn. Alex^dr^ Petrie of the 5th Contineltal Batalion of this State having resigned his Commission is no longer to be respected & Obeyed as a Contineltal Officer

Transmitted by A Dilliant

Reg^tl^. Orders } Jn° Davis of Captn. Charnocks Comp^y^. is appointed Sarg^t^. in the same & is to be Obeyed as such

---

Orders 9th Oct^r^. } Parole Congress CS^n^ 13.
For the Day tomorrow Captn. Hall
for Fort Guard Lieu^t^. Roux
Rear Guard Sarg^t^. M^c^Donald

Reg^tl^. Orders by Captn Motte } A Court Martial to sett this morning to try Such Prisioners as may be brought before it President Capt Dunbar Members L^t^. Warley and Hart –

after Orders by Captn. Motte } A L^t^. and 1 Sargts. 1 Corp^ls^ and 6 Privates to go onboard the Cartell Imediately L^t^. Warly went –

Genl. Orders by Gibbow } a Sarg^t^. Corp^l^. and Eleven Privates to be in Readiness with 36 Rounds p^r^. Man to Recive their Orders from his Excellency the President – [950]

Orders 10 Oct^r^. } Parole Union Ct^r^ S^n^ one
For the Day to morrow Captn. Motte
for Fort Guard L^t^. Baker
For Rear Guard Sarg^t^. Sims

Parole Prepair Ct^r^. S^n^. 22 Fort Moultrie 1778

orders 11 Oct^r^. } For the Day to Morrow Captn. Moultrie
For Fort Guard Lieu^t^. Roux
For Hobcaw Guard Lieu^t^. Hart
For Rear Guard Sarg^t^. Stewart

---

949 Ut Supra is latin for "as above" and it means that the officer resigned on the date that is above, 6 October.

950 This is in response to the British privateers that had been in Bull's inlet had taken a ship of "Mr. Mottes". The South Carolina Navy board ordered "With the *Eagle* Pilot boat now under your Command you are to proceed with all possible Dispatch in Company with the Sloop *Missapotamia* Cap. Briggs in Quest of two Small privateer Boats said to be at Bulls, and by every means in your power Endeavour to take or distroy them The Commissioners of the Navy Recommend that you Endeavour all in your power to Cultivate Harmony and Friendship between your own people and the Continental Troops which go with you and that you always keep your Vessel Clear, and in good Order to Engage that you keep a Regular Watch of one half of the Troops, as well as boats Crew as well at anchor as under Sail, to prevent being Surprised by an Enemy and that you Continue to Cruize for Four days from the time you leave Chas Town Barr in such Station as to you appears most promising of Success unless the Sloop should return sooner in that Case you have Liberty to return also"

Commanding Offrs. of Companies will immediately as Possible make the Armours put all their Arms in the best of Order, & they are to begin by the Grenadiers & with the youngest Compy. – the Gunner of the fort will give in to morrow an exact return of Artillery stores & number of Different size shott the Quartr Mastr Sargt to give in a return of Muskets Catridge made up & number of flints also the Quantity of Lead and Loose Balls & Cartridge paper now in the Regimental store also Ammunition Chest...

When Howe returned from his failed expedition to Georgia his popularity with Congress was extremely low. He disagreed with Christopher Gasden as to who should control the troops in South Carolina, and this eventually led to a duel between the two men. Howe fired the first shot that clipped Gasden ear. When Gasden fired he aimed into the air. Howe declined a second shot. Later they both became friends.

Congress decided that Howe was no longer competent to command the Southern Department and on September 25th Congress issued orders that sent troops from North Carolina and Virginia to South Carolina. This was to prepare for a possible attack by the British. Howe was ordered to report to George Washington in the north and relinquish his command of the Southern Department to Major General Benjamin Lincoln.[951] Howe did not learn of his reassignment until October 9th. Lincoln did not arrive to take command until January 1779.

Genl. Orders 10th Octr. by Genl. How – } The Engineers are immediately to survey Fort Johnston & Moultrie & Report their Situation to the Genl. Comdg. Officers of Brigades are to report to the Genl. The Quantity of Cartridges with which their Brigades are furnished, the Amunition Chist and other military requisite they have & those wanting – and as the Present Aspect of Affairs requires that we should be in the best State of Defence as Possible they and all other Offrs are to Exert themselves to the utmoste to Effect a purpose in which the honour of the Army & good of the Common Cause are so esentially Concerned [952] The Commandg. offr of the Corps of artillery will immediatily report of

---

[951] General Lincoln arrived in Charleston on December 4th, and this was his first order. General Howe had already left to go to Savannah, so he was no longer commander of the Southern Department, but he did command the army in the field at Savannah. Benjamin Lincoln was born on 24 January 1733 in Hingham, Massachusetts. In May 1776 he was appointed a major general in the Continental army. He was in the battles of Bennington, Vermont and Saratoga, New York. In 1778 he was made commander of the Southern Department. He commanded the American forces both in the siege of Savannah and at the siege of Charleston, South Carolina. He allowed the British to box him in the city, which he surrendered with about 7,000 troops on 12 May 1780. He was exchanged and joined Washington's army in time to participate in Cornwallis' defeat at Yorktown in 1781. As Washington's second in command, he accepted the British surrender from Cornwallis' second in Command, Brigadier General Charles O'Hara. Due to the military protocol of the time officers of equal rank dealt with one another. Since Washington was higher in rank to O'Hara, protocol dictated the General Lincoln accept the surrender. He later served as Secretary of War of the Continental Congress from 1781-1783. After the war Lincoln led the militia against Shay's Rebellion in 1786. He was elected lieutenant governor of Massachusetts in 1788 and was the collector for the port of Boston from 1789 to 1809. He died in Boston on 9 May 1810 at the age of 77.

[952] After the failed expedition of Florida, Brigadier General Robert Howe expected the British to attack somewhere along the Georgia or Carolina coast. The attack never materialized in the month of October, but in November the British out of Florida raided up to Savannah and in December a British invasion fleet from New York landed in Savannah.

the Genl. State of the Militory Stores in his Department and also those Articles with which it is Nessary he should be furnished The Deputy Quartr Master Genl. to report the number & State of the Amunition Waggons that may be Ordered if nessesary – he is to have every thing in his Department put in the best order with all possible expedition Transmitted by A Dilliant BM

Orders 12 Octr. } Parole Deceptin Ctr. Sn. 88 –
For the Day to Morrow Captn. Mazyck
For Fort Guard Lt. Martin
For Rear Guard Sargt. McDonal

orderd. that the sutler or any other persons shall not at any time purchase Powder, Lead or Iron of any Sort nor Cloathing from any Soldier or any person belong to the Regt. in pain of being Tried by a Court Martial for Disobedience of Orders & Suffer Accordingly

Fort Moultrie 1778

Ordrd that soldiers or other persons belongg to the regt shall not sell any Powder Lead, Iron or Cloathg. of any Sort what ever to any person on pain of being tried by a Court martial for Disobedience of Orders & Suffer accordingly; any Powder Lead or Iron being found in the Possission of any Soldier he may depend on being Severely punished as if he had offerd. it for sail (except such powder or lead as may be given them for the use of the service) the Batalion to exercise the Cannon every wednesday as formerly, also the Parrapet firings

---

Orders 13th Octr. } Parole Camden Ctr. Sn. 6
For the Day to Morrow Captn. Dunbar
For Fort Guard Lt. Baker
For Rear Guard Sargt. Davis

Orders 14th Octr. } Parole Mexico Ctr. sn. 36
For the Day to Morrow Captn. Hall
For Fort Guard Lt. Mason
For Rear Guard Sargt. McDonald
for Advance Guard Sargt. McGowing

Ordered that a man from the fort guard is to be sent up the flagg staff every morning at sun rise & once an hour through the Day to look out & See what Vessels are off

---

Orders by Genl. How } Head Quarters Chas. town 13th Octr. 1778.
The Genl. Observed that many of the Soldiers appear without Bayonets Comdg. Offrs of Brigade will therefore take care that a Standing Order Issued some time Since Relative to the Loss or Ingury of Armes & Accrutriments otherwise then by Invocating Accidents in Actual Service is strictly Immediately carried into execution

Transmitted by A Dilliant BM

Orders 13th Octr. } Parole Chili Ctr. Sn. 63
For the Day to Morrow Captn. Motte
For ~~G Guard~~ the Cartel Lt. Martin
For the Fort Guard Lt. Roux
For Rear Guard Sargt. Feast

Orders 16th Octr. } Parole Brazil Ctr. Sn. 45.
For the Day to Morrow Capt Dunbar
For the Fort Guard Lt. Warly
For Rear Guard Sergt. Teague

Comdg. Offrs. of Compys. will make Mr. Gray PM. a return of the Number of Shoes wantg to Compt. their respective Compy. to 1 pr. Shoes pr. Man & Mr. Gray will be Down to day [953]

Fort Moultrie 1778

when an Offr falls in Batalion or mount guard they are all ways to have their Bayonets fixt & not at any time March without it Regimental orders Courtmartial to set to day 10 oClock for the trial of Such Prisiners as shall be Orderd before it Presidt. Captn. Dunbar Members Captn. Hall & Guerry – – NB Aquitte to the Sentence of the Court Martial Jno. Sheedy for Disobedience of Orders Sentenced to Receive 100 Lashes Remit upon Conditions of his Reinlisting for three Year or During the War [954] Captn. Motte was pleased to have the punishment remitted By sentence of the same Court Moses Newton for Loosing his fife Sentenced to be put under Stopgs to procure another – [955]

Orders 17th Octr. } Parole Monmouth Ct. Sn. 35
For the Day to morrow Lt. Hall
For the Fort Guard Lt. Guerry
For Rear Guard Sargt. Martine

Orders by Capt. Motte – } A court martial to set this morng. to try such Prisiners as may be brought to it President Capt. Charnock Members Lts. Mason & Roux Orderd that the Qtr. Master Sergt. Do have the Publick that is to near the Fort Gate removed as it is a Nuisance to the Garrison

B Orders by Gel. Moultrie } octr. 16th: 78 at Fort Moultrie when any thing appear in Sight to hoist the Flag for all, tho a Small one and when six or Eight Topsail Vessells to gether to hoist the Small Flagg under the Large one

Jno. Whitsett of the Colo. Compy. Confined by Captn. Dunbar For Disrespectfull Beheavour the Court is of opinion that he Prisiner is Guilty & do sentance him to receive 50 Lashes but Recomind him to to mercy Capt Motte aprove of the Sentence

---

Octr. 18th Orders } Parole Servant Ct Sn. 12
For the Day to morrow Captn. Motte
For Fort Guard Lieut Mason
For habcaw Guard Sargt. Davis
For Rear Guard Sargt. Sims

1778

---

953 Henry Gray, Paymaster.

954 John Sheedy served in the 2nd South Carolina Regiment in 1777 and 1778. On 17 October 1778 he was found guilty of disobedience of orders and he was sentenced to ro receive 100 lashes, or he could reinlist for three years or the duration of the war. He chose to reinlist.

955 Moses Newton served as a fifer in the 2nd South Carolina Regiment from 26 July 1777 to 15 November 1783. He was under Captain Richard Mason at the siege of Savannah.

Orders by Genl. How – } head Quarters Chas. Town 17 Octr. 78.
One Subaltern one Sarg^{t}. & Eight ~~Privates~~ Rank & file from Fort Moultrie one Sarg^{t}. & 12 Rank & file from fort Johnston to hold themselves in Readyness to go upon a Command they are to Act as a Covering to some Publick works upon Dewee's Island & are to be furnished with Tents twenty Rounds & a weeks provisions they will Rec^{d}. & obey such Orders as his Excellency the Presid^{t} Shall give and are to be Relieved weekly in the same proportion –

Morning Orders L^{t}. Col^{o}. Marion 18th Oct^{r}. – } the above Orders as far as it respects the 2^{d} reg^{t}. to be immediately Complied with L^{t}. Roux for this Com^{d}
a Sarg^{t} & 1 Corp^{l}. 9 Privates to relieve the Habcaw guard to morrow –

the Sarg^{t}. or Corporal who's Business it is to see their men for Guard will Certainly be punishd if they parrade them Dirty theyre to be particularly in making their men put on Clean Shirts Stockings or Garters & if they ware Lining Jackets & Breeches they must be Clean & their Coats or Wollen Jackets & Breeches will Brusht or they must answer for the neglect – –

Orders 19th Oct^{r} } Parole Raleigh Ct^{r}. S^{n}. 19
For the Day to morrow Captn. Moultrie
For Fort Guard L^{t}. Hart
For Rear Guard Serg^{t}. Martine

Orders 20th Oct^{r}. } Parole Alfred C^{t} S^{n}. Newyork
For the Day tomorrow Captn. Dunbar
For Fort Guard Lieu^{t}. Martine
For Rear Guard Sarg^{t}. Brown

Reg^{tl}. Orders 20th Octr. by L^{t}. Col^{o}. Marion } The Regiment~~al~~ is to be mustered by the D M Mast^{r}. on Thursday the 27th Instant Ordered that there muster roll for Each Comp^{y} be got Ready by that Day –

A court Martial to sit to day for the trial of such Prisiners as shall be Ordered before it – all evidences to attend Presi^{dt}. Capt Motte
Members Lieu^{t}. Guerry & Warley –

After Orders } one Subaltern one Sarg^{t}. 1 Corp^{l}. & 9 privates with 20 rounds p^{r}. man to hold them selves in Readiness to go on board the Cartel brigg now in the Road Lieu^{t}. Roux for this service who will Receive his orders as Soon as ready

NB L^{t}. Roux & party went on board immedately & Capt Motte reciev^{d}. a packet which he was ordered to carry to the Genl. & the President

---

Orders 21st Oct^{r}. } Parole Delaware Ct^{r}. S^{n}. 3
For the Day to morrow Captn. ~~Motte~~ Hall
For the Fort Guard Lieu^{t}. Warley
For Rear Guard Sarg^{t}. Teague
For advance Guard Serg^{t}. Taylor

When any boats come ashore from vessels inward bound the Sarg^{t}. of the guard is to make them as soon as Landed & inquire who or what they are & report it to the Off^{r}. of the Guard & not Suffer them to come to the fort or from the boat till he Receives his Orders for so doing; if it should be a flagg of France the Offrs is not on any Account to permit him or them to come a Shore without Leave first Obtain'd from the Commanding Offr. of the fort –

---

fort Moultrie 1778

orders 22d Octr. } Parole Hancock Ctr. Sn. Marion
For the Fort Guard tomorrow Captn. Motte & Lt. Baker
rear guard Sargt. Martin

Captn. to be added to the fort Guard tomorrow the party Orderd to hold themselves in Readiness to go to Dewee's Island is to Join & do duty in their Companies till further Orders, they are to deliver the Amunition tin Kittles &tc to Sargt. Major –

Orders 23d . Octr. } Parole Horry Ctr. Sn. Camelion
For Fort Guard to morrow Capt. Mason
For Rear Guard Sargt. Sims – – – – –

Orders 24th Octr } Parole Fair American Ctr. Sn. 9
For Fort Guard to morrow Captn. Hall & Lt. Hart
For Rear Guard Sargt. Webb

A Court Martial to sett to day at 11 OClk to try such Prisiners as shall be Ordered to it evidences to Attend Presidt. Captn. Moultrie Members Lieut. Hart & Roux –

Octr 25th. Orders } Parole Genl. Lincon Ctr. Sn. 39
For Fort Guard tomorrow Capt Charnock & Lt. Roux
For Rear Guard Sergt. Feast
For Habcaw Guard Lt. Martin

One Subaltern to be added to the Hobcaw guard to morrow Should the Command be orderd to Dewees Island tomorrow Lt. Roux is to be for that Service –

Orders 26th Octr. } Parole Genl. Green Ctr. Sn. 44
For Fort Guard to morrow Captn. Moultrie & Lt. Capers
For Rear Guard Sergt. Watt –

A Regtl. Court martial to set to day at 11 OClock to try such Prisiners as Shall be Ordered to it all evidences to attend this Court to Consist of a Captn. as Presidt. & four other members –

Jno. Rusch Surgions mate having left the Regiment without Leave is no longer to be considerd. as Surgions mate to the regt.

Presidt. Captn. Mazyck Membr. Lts. Mason & Hart Guerry & Kolb

---

Gen[l]. Orders 26th Oct[r]. by Gen[l]. How – } Brigadier Gen[l]. Moultrie & Col[o]. Comm[dt] Huger will give notice to the Off[rs]. at Fort Moultrie & Johnston & that they are invited to and as many as can be Spared from those fort may attend the funeral of L[t]. Col[o]. Elliott to morrow morning at Eight Oclock in Cha[s]. Town Gen[l]. Moultrie will order the Grenidier Comp[y]. of the 2[d] Reg[t]. to Attend upon this melancholy Occation & to Bring with them those Standard which were presented to the Reg[t]. by Mrs. Elliott [956] – –

Fort Moultrie 1778

Brigade Orders by Gen[l]. Moultrie } L[t]. Col[o]. Marion to attend the funeral of L[t]. Col[o]. Elliotts & the Grenadier Comp[y]. to be Completed to fifty man to Attend the same with 3 Rounds p[r]. man A Dilliant BM –

Reg[tl]. Orders by L[t]. Col[o]. Marion } The above orders respecting the 2 Reg[t]. to be immediately Complied with Capt[ns]. Dunbar will See his Comp[y] Completed to 50 men by Picking such men out of the Batalion Companys as may answer the intent - & to see they are Clean & in good order & to imbark to morrow at 6 OK in Morning L[t]. Roux is to be Relieved immediately that he may attend the Grenad[r]. Comp[y]. those Gentlemen who are want for Duty & tomorrow to attend the Funeral of L[t]. C[o] Elliott in Chastown agreeable to Gen[l]. Orders

---

Orders 27th Oct[r]. L[t]. Col. Marion } Parole Moultrie Ct[r]. S[n]. 28.
For the Fort Guard tomorrow Capt[n]. Dunbar & L[t]. warly
For Rear Guard tomorrow Sarg[t]. Brown

NB Greeable to the Court Martial that Sett yesterday Peter Area of Capt[n]. Mottes Comp[y]. Confined by Sarg[t]. Martin for Suspision of Theft but is A Quitted James Grover of Capt[n]. Moultries Comp[y]. Confined for Beating and Maiming Samuel Harry [957] The Court it is of oppinion the Prisinor is not Guilty of the Charge & therefore do A quit him Thimothy Downing of y[e]. Col[o]. Comp[y]. Confined by L[t]. Hart for Neglect of Duty the Court is of oppinion the Prisiner is Guilty of the Charge & do Sentence him to Rec[d]. one Hundred Lashes on the Bare Back with a Cat of Nine Tails Col[o]. Marion aProves of the Sentence But Remits the Punishment –

Orders 28th Oct[r]. L[t]. Col[o]. Marion } Parole PinckneyCt[r]. S[n]. 74
For fort Guard to morrow Capt[n]. Hall & Lut. Guerry
For Rear Guard Sarg[t]. Teague
For Advance Guard Serg[t]. Cheatten

---

[956] John Faucheraud Grimke wrote in his orderly book "The troops were drawn up before Lt. Colo: Elliots House, who, as soon as the Corpse was brought out & laid in the carriage, Rested their Firelocks. They then proceeded to Reverse their arms & wheeling by division to the left marched off the ground. When they approached the Church the Division filed off from their flanks in Indian File & formed a rank on each side of the Street: The whole Procession Halted. The Two ranks were then ordered to face inwards & to Rest on their arms reversed. the Procession went on thro the ranks & proceeded into Church: The Troops Shouldered their Firelocks, wheeled up from the Right & Left & formed again their Divisions. They then proceeded to the Funeral Ground, where being drawn up Two deep, They, after the Service was over, fired three volleys. The Troops were marched back to their Barracks, the Eldest officer in rank at their Head. All the officers trailed their Firelocks, carrying the butts foremost. The officers on Horse-back dropped the points of their Swords."

[957] Samuel Henry.

Orders 29th Oct^r^. } Parole Col^o^. Scott Ct^r^. S^n^. 88
For Fort Guard tomorrow Captn. Harleston & L^t^. Mason
For Rear Guard tomorrow Sarg^t^. McGowing
the Articles of War to be Read to the Batalion this afternoon by the ~~orders~~ Oldest off^rs^ hear & all off^rs^. & men to attend NB the Muster being on this Day the Articles of war was not Read – – –

Orders 30th Octr. L^t^. Col^o^. Marion } Parole Bertilley Ct^r^ S^n^. Count
For Fort Guard to morrow Captn. Motte & L^t^. Gray
For Rear Guard to morrow Sarg^t^. Feast
A Court Martial to set to Day at 11 OClock in the Morning to try all Prisiners ordered to it all Evidences to attend this Court Martial to Consist of a Presid^t^. & 2 members as the Articles of war was not Read Yesterday it is ordered to be Read to Day at 5 OC in the Afternoon to the Batalion all Off^rs^. & Soldiers to attend Pres^d^ Captn Hall Members L^t^. Gray Hart Roux & Capers

L^t^. Col^o^ . Marion **Fort Moultrie 1778**

Orders 31st Oct^r^. } Parole Hingist Ct^r^ S^n^.
For the Fort Guard tomorrow Captn. Charnock & L^t^. Hart
For hind Guard tomorrow Serg^t^. Taylor
A Regimental Court Martial to set Amediately for the trial of Such Prisoners as Shall be Orderd to it – Evidences to attend Presid^t^. Captn. Charnock Members L^t^. Mason & Ensign Polk [958]
A monthly return of Each comp^y^. to be given in to the Serg^t^. Wickham tomorrow morning by 9 OClock – - - - - -

---

Orders Novr. 1st L^t^. Col^o^. Marion } Parole Orangeought Ct^r^. S^n^. Virginia
For the Fort Guard tomorrow Capt Moultrie & L^t^. Roux
For Hobcaw Guard Lieu^t^. Capers
For Rear Guard Sarg^t^. M^c^Gowing - - - -

Orders 2 Novr. L^t^. Col^o^. Marion } Parole Gates Ct^r^. S^n^. 3
For Fort Guard to morrow Capt Mazyck & L^t^. Martin
For Rear Guard tomorrow Serg^t^. Sim
NB Agreeable to the Last Court John Martin Sharp Sentencd to Rec^d^. 50 lashes with a Cat of nine Tails on the Bare Back for Neglect of Duty the Com^dg^ Officer aprove of the Sentence But Remits the Punishment W^m^. Clarke of Captn. Mottes Comp^y^. Confined for Theft the Court are of oppinion that the Charge is not Sufficient Imported therefore they they do acquit the Prisoner –
NB by Sentence of the Court Martial that set Octr. 26th Rich^d^. Williamson of the Grenadier Comp^y^. Rec^d^. 50 Lashes on the bare back with a Cat o' Nine Tails for Drunkenness on Duty

---

958 Possibly Ensign Josiah Kolb. There was no ensign named Polk in the 2nd South Carolina Regiment.

Orders 3 Novr. by Capt Motte }
Parole Rutledge Ct^{r}. S^{n}. Five
For Fort Guard tomorrow Captn. Dunbar & L^{t}. Warley
For Rear Guard Serg^{t}. Murphy

A Court Martial to set this Morning 10 Oclock for the Trial of Such Prisoners as may be brought before it Presid^{t}. Capt Moultrie Members L^{ts}. Roux & Guerry – NB Agreeable to the Court Martial Mosey Bruce of the Lieu^{t}. Colonel's Comp^{y}. Run the Gantlet through the whole Regiment for Theft. Jno. Probey of Capt Hall Comp^{y} Confined by Lieu^{t}. Hart for absent without Leave the Court are of oppinion that the prisoner is Guilty of the Charge but that it was not willfully done they_therefore do recommend him to be Discharge – - -

---

Orders 4th Nov^{r}. }
Parole Gaurdaloup Ct^{r}. S^{n}. 17 –
For Fort Guard tomorrow Captn. Hall & L^{t}. Guerry
For Rear Guard Sarg^{t}. McGowen
For advance Guard Serg^{t}. –

NB Agreeable to the Court Martial that set Octr. 30th Jno. Francis of Capt Halls Comp^{y}. Rec^{d}. 100 Lashes with Switches for absence with out Leave Willford Smith L^{t}. Col^{o}. Comp^{y}. Confined & to Rec^{d}. 90 Lashes on the bare Back with Switches for absence without a Leave [959] Drury Smith Last & Jno. Smith Last Rec^{d}. 70 Lashes Each for absence without Leave Rich^{d}. Williamson Rec^{d}. 45 Lashes on the Bare Back with Switches for Quitting his Guard without Leave } Head Quarters in Chas. town 9 Novr. 1778. Genl. Orders } adjutant Robt. Simpson of the 5th Continetal Reg^{t}. of this State having resigned his Commission is no longer to be Obeyed as a Continental Officer [960] } The Hon^{bl}. The Assembly of this State have Resolved that all Off^{rs}. of the 1,2,3,4, 5, 6 Continental Reg^{ts}. to the Rank of Captain Should rise Regimentally & that all Off^{rs}. of and above the Rank of Capt Should Rise in the Line –

Transmitted Dilliant B. M.

Orders 5th Nov^{r}. }
Parole Dominico Ct^{r}. S^{n}. 71 –
For Fort Guard tomorrow Captn. Harleston & L^{t}. Mason
For Rear Guard tomorrow Sarg^{t}. Cheatem

NB Harleston Commanded _ _ _ _ _

Orders 6th Novr. L^{t}. Col^{o}. Marion }
Parole Martinico Ct^{r}. S^{n}. 92
Fort to Day Captn. Motte & Foissin
Fort Guard tomorrow Captn. Charnock & L^{t}. Hart
Rear Guard Sarg^{t}. Murphey.

The Long Roll to be at 4 Oclock in afternoon – when all Off^{rs}. & Soldier are to attend exercise except those on Duty the Battalion to be Served with Blank Cartridges twice a week as formally –

---

959 Willford Smith was in the Lieutenant Colonel's company in 1778. He received 90 lashes on the bare back with switches on 4 November 1778 for being absent without leave.

960 Robert Simpson served as the 5th South Carolina Regiment adjutant until he resigned in November 1778.

Orders 7th Novr. by Lt. Colo. Marion } Parole Lt. Christopher Ctr. Sn. 97.
Fort Guard tomorrow Captn. Hall & Lt. Roux
Rear Sargt. –

Divine Sarvice will be performed tomorrow by the Chaplin of the Regt. Itis expected all Offrs. & men will attend the men are all Ordered to be Clean or they will be taken notice of
A Court Martial to set immediately to try such Prisoners as Shall be Ordered to it Evidences to Attend – Presidt. Captn. Hall Members Lieut. Gray & Martin } NB this Court did not Sitt }

Orders by Captain Harleston } Orderd that a Court Martial do sit this Morning for the Trial of William Clark of Captn. Mottes Compy. Confined by Sargt. Murphy for Riot & Drunkinness

Prest Capt Dunbar Members Lts. Foissin & Hall

NB Willm Clark Sentence to Recd 50 Lashes but Remitted –

Fort Moultrie 1778

Orders 8th Novr. by Lt. Colo. Marion } Parole Cape Briton Ctr. Sn. 28
For fort Guard tomorrow Captn. Harleston & Lt. Martin
For Hobcaw Guard Lt. Warly
For Rear Guard Sergt. McGowing

Orders 9th Novr. by Majr. Horry } Parole Marion Ctr. Sn. 8
For Fort Guard tomorrow Captn. Motte & Lt. Capers
For Rear Guard Sargt. Feast

Orders 10th Novr. by Majr. Horry } Parole Baker Ctr. Sn. Gabriel
For Fort Guard tomorrow Captn. Charnock & Lt. Gurry
For Rear Guard tomorrow Sargt. Miner ___

Genl. Orders by Genl. Howe dated Octr. 18th Recd. this Day } adjutant Jno. Dowes of the 2d Continental Regiment in this State having Resigned his Commission the 6th Instant is not Longer to be Obeyed & Respected as a Continetal Offr. Transmitted Second time by Dilliant B.M.___

Genl. Orders by Genl. How dated Novr. 5th 1778 – } Captn. Harleston of the 6th Contintental Batalion in this State having Resigned his Comn the 9th ~~Instant~~ of August last & Captn. Coyl of the Same Regiment on the 30th Septr. Last 1 Lieut. Armstrong was Promoted to be Captn. in the Room of the first [961] & 1st Lieut. Lacy to be Captn. in the Room of the Second & 2d Lieut Brown was promoted to be first first Lieut Vice till Armstrong promoted [962] & 2d Lieut Redmond to be first Lt. Vice & Lt. Lacey promoted [963]
Capt Lacey having died on the 20th of Septr. & Captn. Armstrong on the 3d. October Last & Lieut. Hampton was promoted to be Captn. in the Room of the first [964] & 1 Lieut Buchanan to be Capt in the

---

[961] John Armstrong was born in 1749. He served as a captain in the 6th South Carolina Regiment in 1778 until his death on 3 October 1778.

[962] Benjamin Brown served as a first lieutenant in the 6th South Carolina Regiment under Colonel Henderson during 1778. He was promoted to captain after Captain Alexander Boyes died on 1 November 1779.

[963] Lieutenant Redmond served in the 6th South Carolina Regiment. His first name is not known.

[964] Henry Hampton was born in 1755. He served as a lieutenant in the 6th South Carolina in 1778. He was promoted to captain in November 1778. After the fall of Charleston he served as a lieutenant colonel in the light dragoons from April to September 1781 under General Sumter. He died sometime after 23 May 1826.

Room of the Second & 2 Lieu^t^. Milling was promoted to be first Lieu^t^. Vice[965] & Hampton promoted & Lieu^t^. Adair to be first [966]& L^t^. Vice Buchanan Promoted & Lieu^t^. Pollard take Rank as such on the 28th June 1778. 2 Lieu^t^. Doggart as Such on the 8th of May & 2^d^ Lieu^t^ Langford the 30th Octr 1778.[967] – Captn. W^m^. Blameyer of the 5th Cont^l^. Reg^t^. in this State having Resined his Com^n^. he is no Longer to be Respected or obeyed as a Cont^l^. Off^r^.

Transmitted by A Dilliant B.M.

Orders by Maj^r^. Horry } A Court Martial to Set this forenoon at 11 Oclock to try Such Prisoners as are ordered before it Evidences to attend Pres^d^. Capt Hall Members L^t^. Gray & Martin –

Fort Moultrie 1778

orders 11th Nov^r^. } Parole Genl. Starling Ct^r^. S^n^. 9
For fort Guard tomorrow Capt Moultrie & L^t^. Foissin
For Rear Guard Serg^t^. Tague
For advance Guard Sarg^t^. Stewart
NB Last Court Martial Jno. Taylor for Theft to run the Gantlet

---

Orders 12th Nov^r^. } Parole Turkes Island Ct^r^. S^n^. Burmudas
For fort Guard tomorrow Captn. Dunbar & L^t^. Mason
For Rear Guard tomorrow Sarg^t^. McGowings [968]
Command^g^. Off^rs^ of Comp^ys^. will have such Coates Changed in the store which have been given to their recruits & do not fit them & Apply for one Shirt p^r^. Man for them who have not Recev^d^. one for the Present Year

A Court Martial to set this forenoon at 11 OC. to try all prisoners as is ordered to it Evidences to attend Presid^t^ Capt Harleston for Members & Lt Capers & Guerry NB Simpson is to do Double Duty for a week for neglect of Duty

---

[965] Hugh Milling was born in Drumbo, County Down, Ireland in 1752. He was a volunteer in the battle of Fort Sullivan in June 1776. He enlisted in the 1st South Carolina Regiment on 4 November 1776 and served as a private under Captain Charles Pinckney. He served as a quartermaster, lieutenant and captain in the 6th South Carolina Regiment from 1777 to 1780. He was taken prisoner at the fall of Charleston and placed on parole until the end of the war. He died on 7 May 1837 at the age of 85.

[966] William Adair served as a lieutenant in the 6th South Carolina Regiment and was captured at the fall of Charleston. He later served as adjutant under Colonel Lacey. He died on 15 May 1808.

[967] Daniel Langford became a second lieutenant in the 2nd South Carolina Regiment during January 1777. He was promoted to first lieutenant in the 3rd South Carolina (Ranger) Regiment on 3 October 1777 under Captain Joseph Warley. In November 1778 he was a first lieutenant in the 6th South Carolina Regiment under Captain George Warley. After the consolidation of regiments in February 1780 he was assigned to the 2nd South Carolina Regiment. He was taken prisoner at Charleston on 12 May 1780 and was exchanged. On 30 September 1783 he became a brevet captain.

[968] Sergeant David McGowan.

Orders 13th Nov^{r}. } Parole Providence Ct^{r}. S^{n}. 99
For Fort Guard tomorrow Capt Hall & Ensign Cob
For Rear Guard tomorrow Sarg^{t}. Brown –

Information being given that many of the Soldiers take the Cloathing from their Brother Soldiers & Carry them over to Haddrells Point & then Sells them to Negroes & others it is therefore Ordered that the Serg^{ts}. of the Rear Guard give Orders to the Centry on the Bridge to Stop all Soldiers & Negroes which may being going Over to Hadrells & All their ~~Cloaths~~ sack baggs &^{tc} are Sarched & if any Shirts Shoes Stockings or any other Cloathing is found on them he is to Send them & the person or persons to the fort Guard – A Court Martial to set today 11 OC in forenoon to try all prisoners orderd to it Evidences to attend Pres^{dt} Capt. Motte Members Lt Kobb & Gray

---

Orders 14th Nov^{r}. } Parole ST Croix Ct^{r}. S^{n} 13
For Fort Guard tomorrow Capt Harleston & L^{t}. Gray
For Rear Guard tomorrow Sarg^{t}. Sims

after Orders (by Maj^{r}. Hory) No Vessell to Pass Fort Moultrie Either Going in or out without permission Granted by the Comd^{g}. Off^{r}. till Further Orders the Capt. of the Guard will Give his Centries the Necessary Orders for this purpose

---

Orders 15th Nov^{r}. } Parole Pee Dee Ct^{r} S^{n}_ Bluff
For Fort Guard tomorrow Capt Motte & L^{t}. Roux
For Hobcaw Guard L^{t}. Guerry
For Rear Guard Sarg^{t}. Stephen Roberts

NB Agreeable to the Last Court Martial Jno. Griffin of Capt Hall Comp^{y}.
Run the Gantlet through the Reg^{t}. for thift

Orders 16th Nov^{r} } Parole Camden Ct^{r}. S^{n}. Watteree –
For Fort Guard tomorrow Capt Moultrie & L^{t}. Martin
For Rear Guard tomorrow Sarg^{t}. Webb –

Fort Moultrie 1778

Orders 17th Nov^{r}. } Parole Romotions Ct^{r}. S^{n}. 2
For fort Guard tomorrow Capt Mazyck & L^{t}. Capers
For Rear Guard Sarg^{t}. McDoll

Head Quarters Charles town 14 Novr. 1778

Genl. Orders by MG. How } all men unprovided with Powder horn are as soon as possible to be furnished with them Comd^{g} Off^{rs} of Brigade will give Orders Accordingly they are to apply to the D Qtr Genl. for the Amunition Chist nessary to the men under command who is to Direct to furnish them with all Possible expedition The Genl. Calls upon the Off^{rs} of the Army of every degree to exert themselves to the Utmost to have the men under command & every thing their Several Department in the best Order as Possible for immediate Action

A Dilliant B. M.

Gen[l]. Orders
16 Nov[r]. by M G. How } The Hon[bl]. House of Assemble having Resolved that the Continental Reg[ts] Should Remain on the Usual Establishment except the Corps of Artillery, until the Pleasure of the Hon[bl]. Continintal Congress be Known thereupon the following promoted therefore takes Places

1st L[t]. Joseph Elliott is promoted to be a Capt in the first Reg[t]. Vice Capt[n]. Joor lost in the Randolph the 7th March 1778 2 L[t]. Benj[m] Postell [969] to be first L[t]. Vice Capt[n]. Elliot promoted 2 L[t]. Willison Glover to be 1 L[t]. Vice L[t]. Gray lost in the Randolph on the 7th March 1778 1st L[t]. W[m]. Hext to be Capt Vice Pinckney promoted to be Maj[r]. in the 1st May 1778.[970] 2 L[t]. W[m] Fishburn to be 1 L[t]. vice L[t]. Hext promoted [971] 1st L[t]. Ch[s]. Lining to be Cap[t] Vice Capt Cattle resigned on the 20th July 1778. / Gen[l]. Orders Continued / 2 L[t]. Cha[s]. Skirving [972] to be 1st L[t]. vice L[t]. Lining promoted 1 L[t]. Tho[s] Gadsden to be Capt Vice Capt Saunders resigned on the 6th Oct[r]. 1778 2 L[t]. Alexand[r]. Fraizer to be 1 L[t]. vice L[t]. Gadsden promoted[973] 1 L[t]. Rich[d]. B. Baker of the Second Reg[t]. to be Capt vice Capt Blake resigned 25th April 1778. 2 L[t]. Warley to be 1 L[t]. Vice L[t]. Baker Promoted 1 L[t]. Adrian Privaux to be Capt vice Capt Jacob Shubrick deceased 27 April 1778. 2 L[t]. Samuel Guerry to be 1 L[t]. vice L[t]. Provaux promoted 2 Lieu[t]. Peter Foissin to be 1 L[t]. vice L[t]. Perreneau resigned Galivant 15th July 1778 Jn[o]. Wickham Gentleman is appointed an Ensign in the 2[d] Continental Reg[t] this State this Com[n] to bare Date 6th Nov[r]. 1778 these Off[rs] are therefore to be respected & obeyed Accordingly & take Rank from the Day the Several vacancies happined

A Dilliant B. M.

Fort Moultrie 1778

Reg[tl] Orders by L[t]. Col[o]. Marion } Capt Rich[d]. B. Baker to take the Command of the Comp[y]. which is at present the L[t]. Col[o]

Orders that all those men which formerly belong to the 2[d] Vacant Comp[y]. Draughted from it to be given up & form the 10 Comp[y] in this Reg[t]. Capt[n]. Adrian Provaux to take the Com[d]. of the 10 Comp[y] Ensign Kolb to Join & do duty in Capt[n]. Provaux Comp[y] till Further Orders Ensign Jn[o]. Wickham to Join & do Duty in Cap[t] Mottes Comp[y] till Further Orders Sarg[t]. Alex[d]. M[c]Donald is appointed Sarg[t]. Maj[r]. to the Reg[t]. he is to be Obeyed Accordingly Rob[t]. Lance of Capt[n]. Moultries Comp[y]. is appointed fife Maj[r]. he is to be Obeyed Accordingly _ _ _ _ _

---

969 Benjamin Postell was born on 8 February 1759. He was promoted to 1st Lieutenant in November 1778 when Joseph Elliot was promoted to captain. He served under Captain Hezekiah Maham in the 1st South Carolina Regiment. He was taken prisoner at the fall of Charleston. He died on 22 December 1801.

970 Major Thomas Pinckney.

971 William Fishburne was born on 12 September 1760. He became a 2nd Lieutenant in the 1st South Carolina Regiment in 1778. He was promoted to first lieutenant on 6 November 1778. He was wounded in the battle of Stono Ferry on 20 June 1779. He was appointed a captain in Colonel Peter Horry's cavalry of Marion's Brigade of Partisans on 6 December 1781. He died on 3 November 1819 at the age of 59.

972 Charles Skirving became a 2nd Lieutenant in the 1st South Carolina Regiment on 20 December 1777. He was promoted to 1st Lieutenant on 6 November 1778. He became the regimental adjutant in 1779 and was promoted to captain on 1 December 1779. He was taken prisoner at Charleston on 12 May 1780.

973 Alexander Fraser was born in 1722. He became a 2nd Lieutenant in the 1st South Carolina Regiment under Captain Drayton on 31 January 1778. He was promoted to 1st Lieutenant on 6 October 1778. During 1780 he was under Captain Charles Lining and was taken prisoner at the fall of Charleston. Afterward he served as a lieutenant in Marion's Brigade of partisans. He died on 6 May 1791 at the age of 69.

Orders 18th Novr. } Parole Genl Moultrie Ctr. Sn. 12
For fort Guard tomorrow Capt Dunbar & Lieut Warley
For Advance Guard Sargt. Taylor
For Rear Guard Sargt. Sims

A Court Martial to set this Morning at 11 OC to try all prisoners Ordered to it evidences to attend Ensign Wickham will Act as Adjutant to the Regt. till Further Orders he is to be Obeyed accordingly

Prsdt Capt Hall Members Lieut. Hart & Roux Wm. McCullock Confined by Capt Moultrie for Loosing his Gun the Court Martial are of oppinion that the Prisoner is Guilty of the Charge & do Sentenced that one half of his pay to be taken every pay day till the Sum of £25 is made up –

Orders 19th Novr. } Parole Egg Harbour Ctr. Sn. 21.
For fort Guard tomorrow Capt Hall & Lt. Foisint
For Rear Guard Sargt. Miner _ _ _ _ _

Corporal James Cambell of Capt Halls Company is Appointed a Sargt. in the Same he is to be Obeyed Accordingly

Corporal Daniel Green of Captn. Baker Compy is appointed a Sargt. in Captn. Lesesne Compy. he is to be Obeyed as Such

When a Corporal is wanting in any Compy. the Comdg. Offr of Such Compy. may choose him & Recomend him to the Colo. for his Approbation & until he is Approved off by him he is not to be Considered a Corpl. no more than two Sargents & two Corpls will be Allowed to one Compy at Present those who have more they must be transfered to Such Compys as may wanting –

Fort Moultrie 1778

Regtl. Orders by Capt Dunbar Novr. 20th 1778 } Parole Marion Ctr. Sn. Horry
For fort Guard tomorrow Capt Lt. Mason & Kobb
For Rear Guard Sargt. Stewart

A Court Martial to set at 11 Oclock in the forenoon to try such prisiners as may be brought before it all Evidences to attend

Lt. Mason } President
Lt. Martin } Members { & Warley

Agreeable to the Sentence of the Court Martial Corporal Jones of Capt Moultries Compy is Confined for Neglect of Duty was Reprimanded

//Jno. Taylor of Captn. Mottes Compy. Confind for Loosing his Regiml waistcoat & Breeches put under Stopage to replace them _ _ _ _ _ _

//Jams Oakes of Captn. Baker Compy. Confined for Loosing his Regiml Waistcoat & Breeches to be put under Stoppages to replace them – _______________

//Robt. Dinnison Confind for Disobedience of Orders & unbecoming Language to Sargt. Davis to be Severly Reprimd [974]

---

[974] Robert Dennison enlisted in the 2nd South Carolina Regiment on 4 November 1775. On 20 November 1778 he was confined for disobedience of orders and unbecoming language to Sergeant Davis. He was "severly reprimanded".

//Thos Mills of Captn. Mottes Comp^{y}. Confined for Unbecoming Language to Sarg^{t}. Davis to Ask his Pardon _ _ _ _[975]
// Fred^{k}. Simmons & Jno. White Confind for Dissobediance of Orders Not Guilty - - - -
//Jno. Sullivan of Capt Charnocks Comp^{y} Confind by the Q^{tr} Master Sarg^{t}. on Suspition of Thift Not Guilty =
//Sarg^{t}. McDonald of Capt Halls Comp^{y}. Confind by the Sarg^{t}. Maj^{r}. for Drunkinness & being incapable of Duty remanded being too Drunk _ _ _ _ _ _
//Sarg^{t}. Feast of Captn. Harlestons Comp^{y}. Confined by Capt Dunbar for Drunkness & Refusing to give a Crime against a Man whom he Confined Remanded Being too Drunk
//Corp^{l} Champness of Capt Mottes Comp^{y}. Confined by Sarg^{t}. Davis for unbecoming Language to be Reduced But Remitted _ _ _ _

Orders 21st Nov^{r}. } Parole Nimrod Ct^{r}. S^{n}. 25
For fort Guard tomorrow Captn. Harleston & L^{t}. Gray
For Rear Guard Sarg^{t}. McDowell

Brigade Orders by Genl. Moultrie
Novr. 20th –

The Quarter Master of the First Second & Sixth Regiments in this State are to Apply immediately to the Deputy Quar^{tr} Master Genl. for the Ammunition Chest & Powder Horns each of them are wanting to Compleat their respective Reg^{ts}. According to a General Order issued for that purpose on the 14th Instant –

Signed A Dilliant B.M.
R.O. –

As the President & Generals Howe & Moultrie are to be at Fort Moultrie this Morning it is expected that every Soldier will appear Decent the Sergants of each Comp^{y}. are to See this Order Complied with immediatelly.

Orders 22^{d} Nov^{r}. } Parole Col^{o}. Elliott Ct^{r}. S^{n}. 2
Fort Guard tomorrow Capt Mazyk & Lieu^{t}. Hart
For Habcaw Guard Lieu^{t}. Fossine
For Rear Guard Serg^{t}. Sims

Genl. Orders 20 Novr by Genl. How – } Capt Ge^{o} Turner of the First Continental Battalion is appointed Aid de Camp to the Genl. with the rank of Maj^{r}. in room of Col^{o}. Stephen Drayton Promoted Ensign Josiah Kolb of the 2^{d} Continental Batallion is promoted to be a first Lieu^{t}. Vice L^{t}. Galvan resigned 15th July 1778 until the Pleasure of Congress is Known, these Off^{rs}. are therefore to be respected & obeyed Accordingly

A Delliant B. M.

---

[975] Thomas Mills enlisted in the 2nd South Carolina Regiment on 4 November 1775. On 20 November 1778 he was confined for unbecoming language to Sergeant Davis. He also had to apologize to Davis. He was under Captain Charles Motte in 1779. He was at the siege of Savannah. After the fall of Charleston he served in the cavalry under Lieutenant Colonel Maham from 17 August 1781 to 17 August 1782.

Orders 21st Novr. by Genl. How } The Honl Continental Congress having appointed Lt. Colo. Tumant Inspector of the Continental Troops in the State of So Carolina & Georga he is therefore to be Obeyed & respected Accordingly [976]

Congress having also ordered that untill a place of regulation for the Inspectures department now under consideration Shall be finally Arranged & Transmitted that he shall train, Exercise & Diciplin the Army in this Department in the manner Introducd & Practiced in the Genl. Army by the Inspectors Genl. [977] the Genl. therefore requires that Offrs of every Degree will chearfully and & assist the Inspector in a Matter so Consisting with the Good of Sarvice for which purpose Batalions & Corps will parade when when he shall Require it & adjutant of the Battalion in Town are by First to Give a Coppy of the General Orders of the Day at the Inspectors Quarters – A Dilliant – ....

Regt Orders by Lt. Colo. } a court martial to set tomorrow 11 OC in forenoon to try all Prisoners Orderd to it Evidences to attend President Capt Hall Member Martin & Kolb

---

Orders Nov. 23. 78 by Majr. Horry } - - - Parole Colo. Moultrie – Ctr. Sn. Georgia
For fort Guard tomorrow Capt Dunbar & Lt. Roux
For Rear Guard Sargt. Minor –

---

orders 24th Novr. by Capt Harleston } - - - Parole Manchester – Ctr. Sn. Baker
For fort Guard tomorrow Capt Hall & Lt. Martin
For Rear Guard Sargt. Teague _ _ _ _ _

Orders 25th Novr. by Lt. Colo. } – Parole Tinmouth - - Ctr. Sn. Charnock
For fort Guard tomorrow Captn. Baker & Lt. Capers
For Rear Guard Sargt. Stewart
For advance Guard Sargt. McDonald

A Court Martial to be held this forenoon at 11 OC. to try all prisoners ordered to it Evidences to Attend – Presd Capt Harleston, Members Lt. Gray & Mason

Q. M. Sargt. to give out Eight watch coates to the Offrs of the Fort Guard for the Sentries & two to the rear Guard the Officers of those Guards are to Deliver them to the Relieving Offr. & report the Same daily to the Comdg. Officer _ _ _ _ _

Orders 26th Nov. } _ _ _ Parole Genl. Moultrie – Ctr. Sn. 91
For fort Guard tomorrow Capt Harleston & Lieut Warley
For Rear Guard Sargt. Sims _ _ _ _ _

---

[976] Lieutenant Colonel Jean Baptiste Ternant was a French engineer serving as inspector general for the Southern Department. He was at the British capture of Savannah in December 1778 and he was at the siege of Charleston. He commanded militia at the skirmish at Gibbes Plantation in March 1780.

[977] This was the drill that was introduced by Baron Friedrich Wilhelm Baron von Steuben in 1778 at Valley Forge to Washington's Grand Army. This drill was a modified form of the drill that was used by the Prussian army. Prior to this the South Carolina Regiments used the British 1764 drill.

orders 27th Novr. } _ _ Parole Elbus _ _ _ Ctr. Sn. 10
For fort Guard Tomorrow Capt Motte & Lts. Guerry - -
For Rear Guard Sargt. Minor

(Orders by Capt Motte) A court martial to Sett this morning at 10 OClock to try all prisoners ordered to it president Capt Dunbar Members Lieut Hart & Roux. Agreeable to the sentance of the Court Martial James Grover of Capt Moultrie Compy. Recevd ~~30~~ 50 lashes on the Bare back with Switches for Defrauding the Sutler & to be put under Stoppages to Repay the Sutler the Same he defrauded him of. By Sentence of a Court Martial that Sett 25th Novr. 1778.
Wm. Clark of Capt Mottes Compy. Recid 30 Lashes on the Bare Back with Switches for abuse & threathining Corpl. Joneses Life _ _ _ _ _
//Joseph wright of Captn. Harleston Compy. Recid 30 Lashes on the Bare Back with Switches for Drunkenness & Neglect of Duty _ _ _ _ _ _ [978]
//James Gowin of Captn. Moultries Compy. was Sentenced to be Striped & tied to the halbirtts & there to be Reprimanded Sevearly – [979]

Fort Moultrie 1778

Orders 28th Novr. } _ _ Parole Colo. Huger – Ctr. Sn. ~~Mason~~
For fort Guard tomorrow Capt Moultrie & Lt. Mason
For Rear Guard Sargt. Stephen Roberts –

Orders by Genl. How dated Novr. 24th 1778 } The General being under the Necessity of Reparing to to Georgia; The Command of the Continental Troops in this State will be in his Absence be Vested in Brigadier Genl. Moultrie the Sentries at Head Quarters to be Continued as Usual & to be particularly diricted to be Careful & Vigilant by Orders of the Genl.

J..F.. Grimke Det Adj

In mid-October Sir Henry Clinton, with the British in New York, coordinated an invasion of Georgia that would use Loyalist forces out of Florida and British troops from New York. General Augustine Prévost was ordered to the St. Marys River to cooperate with the invasion fleet coming from New York under the command of Lieutenant Colonel Archibald Campbell. Prévost ordered his brother, Major James Mark Prévost, to collect cattle for the British army. As a diversion for this cattle raid Lieutenant Colonel Lewis V. Fuser landed near Sunbury, fifteen miles south of Savannah.

When Major Prévost and the 60th Regiment arrived at the Midway Meetinghouse he collected all the cattle he could and set fire to the buildings. When the British invaded

---

978 Joseph Wright served in the 2nd South Carolina Regiment under Captain Harleston. On 27 November 1778 he received 30 lashes on the bare back with switches for drunkenness and neglect of duty. On 5 January 1779 he received 100 lashes on the bare back with a cat of nine tails for quitting his post while on Sentry duty. On 25 March 1779 he received 20 lashes on the bare back with a cat of nine tails for neglect of duty. After the fall of Charleston he supplied beef for the Continental Line during 1781 and served in the militia during 1782.

979 James McGowen (McGowin) served in the 2nd South Carolina Regiment under Captain Thomas Moultrie for 34 ½ months. On 27 November 1778 he was stripped and tied to the halberds and reprimanded severely. He was wounded at Savannah during the assault on Spring Hill Redoubt in October 1779 and left a cripple.

Georgia there were only 200 Continentals in the four battalions in Georgia. Most of these men came from outside of Georgia. The four battalions were merged into one with the designation of the 2nd Georgia Battalion. Colonel John Baker and his Georgia Continentals were ordered to march to Midway Meeting House with "three days provisions of bacon & biscuit with 40 rounds ammunition each."

The British had been at the Midway Meetinghouse for several days collecting their looted cattle and provisions and were now heading south. At Bulltown Swamp, about four miles south of the Midway Meetinghouse, the Georgians encountered Prévost's army and fought a delaying action. Colonel Baker was wounded in the action but his army continued to skirmish, waiting for Brigadier General James Screven's reinforcements to arrive so that they could cut off Major Prévost's force. Brigadier General Screven arrived with his milita, but unfortunately it was only about twenty men.

Brigadier General Screven arrived with his Patriot militia, to cut off Major James Mark Prévost's invading force, but unfortunately it Screven only had about twenty men. Screven linked up with Colonel John White's force of about 100 men and marched towards the British. Hearing that Screven and White were moving on the road between Midway and Sunbury, Colonel Fuser ordered Brown and his Rangers to ambush the enemy. Brown picked thirty-two men and intercepted Screven's force. While Brown's force was concealed on the side of the road, Screven and White halted their forces and made a speech about the upcoming battle. At the end of the Patriot's speeches, when they were cheering about how they would defeat the British, Brown ordered his Loyalists to fire. Screven fell over, wounded in the first volley that also wounded Major James Jackson, Screven's second in command. Brown's men captured Screven, but one of Brown's Rangers shot him a second time after he was taken captive.

White moved his men back to prepared positions located at Midway Meetinghouse. Major James Mark Prévost arrived with the main force, and attacked. Most of the battle took place on the road, since the ground off the road was swamp or thick brush. Prévost's horse was killed by a cannon ball as it skipped down the road, but he was uninjured. Not knowing what else was coming their way the heavily outnumbered Georgia Patriot militia soon left the battlefield and withdrew towards the Midway Meetinghouse a few miles away.

Prévost discovered that his force on the road was unsupported and, not knowing what else was out there, he might be in danger of being cut off from General Augustine Prévost's main army, so he ordered a retreat out of Georgia. He took two thousand head of cattle and several slaves. Prévost sent the mortally wounded General Screven back to the Patriot lines under a flag of truce.[980]

Orders by Genl. Moultrie 25 Nov. 1778 } Capt William Charnock of the 2 South Carolina Battalion & Capt Thomas Jarvey Adjutant Deputy Muster Master having resigned their Commission are no Longer to be Considerd as Contl. Officer _ _ _ _ _ [981]

---

[980] A full account of these actions are described in "*Nothing but Blood and Slaughter, Volume One*".

[981] Thomas Jervey served as a lieutenant and captain in the 5th South Carolina Regiment. He was appointed adjutant deputy muster master general. He resigned on 25 November 1778. He died on 14 June 1796.

Orders 29th Novr. } _ _ _ Parole Colo. White _ _ Ctr. Sn. Midway
For fort Guard tomorrow Capt Mazyck & Lt. Gray.
For Rear Guard Sargt. Stewart
For Hobcaw Guard Lieut. Koll _ _ _ _

Genl. Orders November 28th 1778.} _ _ Parole Purysburgh.
Colonel Hendersons Regiment to hold them Selves in Readiness to march to Morrow with fifty Rounds per Man & the Comdg. Officer of Said Regt. to Report immediately what Arms are wanting to compleat this Corps _ _ _ _ _ _ [982]
//The Comdg. Officer of the different Corps of this State are immediately to Call in all Offrs & Soldiers now out upon furlough or Recruting Sarvice
//The Deputy Quarter Master Genl. to provide waggons To Transport five thousand pounds weight of Gun Powder to Georgia and also to apply to his Excellencey the Prisident for that Quantity from the Publick Magazine & also for five Thousand pounds weight of Lead _ _ _
//The Deputy Quarter Master Genl. or his Deputy to Attend on the Troops Going from this State to Georgia _ _ _ _ _ _ _
//The Director Genl. of the Hospital Send on Surgeon & one Surgeon Mate with a proper Medicine Chest for these Troops & those Ordered to Georgia _ _ _ _ _

A Dilliant B Major –

Headquarters Chas. Town Novr. 29. 1778
ordered that Colo. Pinckney Marion & Motte are to attend the Genl. to morrow at the State House at 10 OC in the Forenoon to meet his Excellency the President in Council at that house _ _ _ _

Fort Moultrie 1778

orders 30th Novr. by LtCol Marion } Parole Genl. Scriven _ _ Ctr. Sn. Savannah
For fort Guard tomorrow Capt Dunbar & Lieut Hart
For Rear Guard Sargt. Feast _ ___________

No Offrs. or Soldiers to have Leave of Absence from Garrison till further orders
10 Fatigue Men to be given to the Quarter M..Sargt. to go to Hadrells Point for moss for wading [983]
//Mr. McHail to Chose 6 men tomorrow to make Cannon Cartridges ~~the~~ _ _ _ _ _ _ _ _ _
//The Officers to be Particular in Examining & have their mens Arms in the Best Order _ _
orders by Majr. Horry - - } A monthly Return of Each Compy. to be made out and delivered to the Adjut by 9 OClock tomorrow Morng.

A Court Martial to set this forenoon at 10 OClock to try such Prisoners as are Ordered to it Evidences to Attend
– – Presdt. Capt Hall
– – Members Martin & Capers

---

982 Lieutenant Colonel William Henderson and the 6th South Carolina Regiment.

983 This was a common fatigue duty for the soldiers in Charleston. Charleston was surrounded by almost 400 artillery pieces, and these all required wadding in the construction of the artillery rounds. This wadding may have been a piece of wood or oakum in between the powder charge and the projectile, or in the case of Fort Moultrie, it would be the "Spanish Moss" that hangs from the trees in abundance.

Agreeable to the court martial Nicolas Flin of Capt Lesesnes Comp^{y}. to recv^{d} 30 Lashes on the Bare Back with a cat of Nine Tails for Selling his Regimital Blankete
Jno. Chavis of Capt Lesaines Comp^{y}. Reci^{d} 30 Lashes on the bare back with a cat of Nine tails for Selling his Regimental Lining Breeches [984]

Orders Decr. 1 } _ _ Parole New York – Ct^{r}. Sr. Rhodeisland [985]
Fort Guard tomorrow Captn. Hall & L^{t}. Roux –
For Rear Guard Sarg^{t}. Camble [986] –

Orders Decr. 2 } _ Parole Sundberry _ _ C^{n}. S^{n}. 29
For fort Guard to morrow Capt Baker & L^{t}. Martin
For Rear Guard Sarg^{t}. Stephen Roberts –
For Advance Guard Sarg^{t}. Martin
A Court martial to Set today at 11 OClock in forenoon to try Such Prisioners as Shall be Ordered to it all Evidences to Attend in time
Comd^{g}. Off^{rs}. of Companys to Apply to M^{r}. Simpson for caps for their Accounts –

Pres^{dt} Captn. Lesain Members L^{t}. Warley Guerry Mason & Gray agreeable to Sentance of Last Court Martial Chas Troublefield of Captn. Lesene Comp^{y}. Reciv^{d} 100 Lashes on the Bare back with Switches for absent without Leave –

Orders Decr. 3^{d}
L^{t}. Col^{o}. Marion } - - Parole Duke Charter Ct^{r}. S^{n}. 92
For fort Guard tomorrow Captn. Mason & L^{t}. Capers
Rear Guard Sarg^{t}. Tayler

Fort Moultrie 1778

Genl. Orders 1st Dec^{r}.
by Genl. Moultrie } first L^{t}. Richd Mason of the 2^{d} Battalion in the State S^{o} Carolina is apointed to be Capt in the Same Vice Capt Charnock resigned 25th Novr. Ensign Jno. Wickam of the Same regim^{t} promoted to first Lieu^{t} in Room of L^{t}. mason promoted to be Captn. _ _ _ _ _ _ _ _

Reg^{tl} Orders by
L^{t}. Col^{o}. M. } Capt Richd Mason to take the Command of the Comp^{y} State Capt Charnocks he is to take Acc^{t}. of the Comp^{ys} arms acoutriments & Cloathing & Keep an Acc^{t} thereof Lieu^{t} Jno Wickom to Join & do duty in Capt Harlestons Comp^{y} till Further Orders _ _

---

984 The person who was the orderly for this section of the orderly book was not able to spell as well as the other orderlies. Many of the words that he did not know are written out phonetically. So Captain Lesesne is spelled "Lesaine." The advantage of this is that the actual pronunciation of the 18th century names are known due to the phonetic spelling.

985 The monthly return of the regiment on 1 December 1778 lists the following companies, and the total of all ranks in the company – Captain Isaac Harleston – 28; Charles Motte – 31; Thomas Lesesne – 27; Thomas Moultrie – 23; Daniel Mazyck – 22; Thomas Dunbar – 32; Thomas Hall – 28; Richard Bohun Baker – 27; Adrian Proveaux – 22; 1st Vacant Company – 26; Total for the whole regiment – 277.

986 Sergeant James Campbell.

After orders Lt. Colo. } Those men who are Carpenters in the Regiment that will work at the repairs of Fort Moutlrie Shall be excused all Duty & receive a Doller pr. Day more than their pay soldiers they are to parade every Day at 6 OC in the morning to go to work Sargt. Jno. Roberts is to take an Account of all those who may go to work & preside over the G Repair under the Lt. Colo. Direction he is also to have a Dollar pr. Day all manner of tools nessessary will be given out those who Recid them must be Accountable for them ~~there~~ there Will be Negroes Labourers to Assist the Carpenters – the Names of those who will work to be given tomorrow Morning at 6 OClock to Sargt. Roberts Orders that Sargt. Jno. Roberts be immediately Relieved from the Hobcaw Guard with Directions to repair the Garrison

---

Orders 4th Decr. } __ Parole Colo. Bull Ctr. Sn. Savannah
For fort Guard tomorrow Capt Harleston & Lt. Warley
For Rear Guard Sargt. Brown

For fort Guard to be Reduced to day to 18 privates Sargt. & Corpl. & Drum as usual – the Centry at the Comd. Offrs Door ~~as usual~~ & one in the old Hogg Bastian to be taken off a Regt. Court Martial to set at 10 OClock this forenoon to try such Prisoners as shall be Ordered to it Evidences to Attend Presd Capt Moultrie

Members Captn Hall Lt. Hart & Roux & Foissin –

Orders 5th Decr. } . . . Parole Genl. Lincon Ctr. Sn. Genl. Moultrie
For fort Guard to morrow Capt Lesain & Lt. Guerry
For Rear Guard Sargt. Minor –

Orders 6th Decr. } - Parole Colo Pinckney Ctr. Sn. Drayton
For fort Guard tomorrow Capt Moultrie & Lt. Foissin
For Rear Guard Sargt. Teagu

ordered that one Commisiond offr & privates that Mount Guard their hair to be powdered Dailey the Adjutant only oblige the Regt Barbers to do this duty & no man to be Suffard to march on the parade without Complying with this order - - - -

Fort Moultrie 1778

Orders 7th Decr. } . . Parole Winyard Ctr Sn. Roberts
For fort Guard tomorrow Capt Mazyck & Kolb
For Rear Guard Sargt. Welsh –

A Regtl. Court martial to set this forenoon at 10 OClock to try all persons orderd to it Evidences to Attend

Captn. Dunbar Presdt members Capt Baker & Lt. Martin –

_ _ Agreeable to the Sentance of the Court Martial that ~~Orders 8th~~ Sett 4th December Jno. Fenwick of Capt Bakers Compy. receive 100 Lashes with ~~an~~ Switches for absent without Leave & to be put under Stoppages untill he makes good for the watch Coat –

---

orders 8th Decr. } . Parole Geo town Ctr. Sn. 5.
For fort Guard tomorrow Capt Dunbar & Lt. Gray
For Rear Guard tomorrow Sargt. Sims

A Regtl Court Martial to be held at 11 OClock this forenoon to try all Prisoners Ordered to it evidences to attend –

Orders 9th Decr.
Lt. Colo. Marion } Parole Cockspur Ctr. Sn. Tendar
For guard tomorrow Capt Baker & Lt. Hart
rear Sargt. Roberts

Genl. Orders 8th Decr. by
Genl. Lincolns – } The Genl. Earnestly recommends it to the Officers of all ranks to pay the Strictest attention to the Arms Amunition and Accoutrements of the men & see that they are in the Best order possible The Late movements of the Enemy so manifestly point to the prospect of this measure that the Genl. think it needed to him to urge any Regiments on the Subject A Return of the names & Ranks of all field Offrs of the Several Continental Batalions raised in this State to be made to head Quarters on Friday next at orderly time by the Major of Brigade Specifying the Batalion to which each Belongs

A Dilliant B. M.

Regtl. Orders
Colo. Marion ~ } the Colo. hopes that Due Attention will be paid to the above Genl. Orders so neccesary for the preservation of their & their mens arms & honour & the good of their Country at a time when their cannot be any Doubt of their Shortly be called into action all Soldiers of Duty to be sent to day to Haddrells with proper Sargt to get moss for wading the Sargt Majr. will go with this party & stop them from Stragglen about the Plantations –

Orders 10 Decr.
Colo. Marion } _ _ Parole Savannah Ctr. Sn. 51
For fort Guard tomorrow Capt Dunbar & Lt. Roux
Rear Guard Sargt. Taylor –

Genl. Orders 9 Decemr.
by Majr. Genl. Lincoln
Chas. Town } The Honbl Continental Congress have been pleased to pass the following resolution viz :
In Congress Septr 25th 1778,

Resolved that Majr. Genl. Lincoln take the Command in the Southern Department and Repair immediately to Chas. Town So. Carolina – As it is Absolutely Nessesary that the Genl. on his first Arrival shoud be Acquainted with the Strongest of the Army & the particular State of the Department of which he is appointed to take Command the Request the Offrs. at the Head of the Quarter Master General Commisary General Commisary of Adinence Store [987] & Medical Department to make a Return as soon as possible to Head Quarters of the Continental Stores & Situation of their Several Departments –

A Brigade Return of the army to be made on tuesday at orderly time which will be Duly at 12 OClock A Dilliant B M

---

[987] Should read "Ordinance"

Gen[l]. Orders by Gen[l]. Lincoln Dated 10 De[r]. 78. } A Gen[l]. Court Martial to Set on Saturday at Such time & place as the President shall order, for the Tryal of all Such Prisioners as may be brought before them all Evidences to be warned to attend in time

Pres[dt]. Lieu[t]. Col[o]. William Scott

Members from 1st Reg[t]. 2 Cap[t]. 2 Lieu[t] from 2 Reg[t].

3 Capt[n]. 2 Lieu[t]. from 6th Reg[t]. 1 Cap[t] & 2 Lieu[t]. –

Members for a Gen[l]. Court Martial Ordered to Sit on Saturday next from 2 Reg[t]. are Captain Lesesne & Mazyck & L[t]. Hart & Martin –

Orders 11 Dec[r]. } Parole Georgia Ct[r]. S[n]. 97

For fort Guard tomorrow Capt Baker & L[t]. Capers

For Rear Guard Sarg[t]. M[c]Dowell –

Orders 12 Dec[r]. } Parole Beauford – – Ct[r] S[n] Artillery

For fort Guard Capt Motte & L[t]. Warley

For Rear Guard Sarg[t]. Campble –

orders by Maj[r]. Horry } all Soldiers off Duty are Immediately to go over to Hadrills Point to get Moss for wading The Serg[t]. Maj[r]. & all Non Commissioned Off[rs] are to attend to men & to see that Each man bring as much Moss as he Conveniately Can Carry as ~~Shall be~~

A Court Martial to set Immediately to try Such Prisioners as Shall be Ordered to it Evidences to attend Pres[dt]. Lieu[t]. Guerry Members L[t]. Foissin & Kolb –

Fort Moultrie 1778

Orders 13 Dec[r]. . } Parole North Carolina . .C S[n]. 200

For G. Guard tomorrow Capt[n]. Dunbar & L[t]. Guerry

For Rear Guard Sarg[t]. P. Roberts

Agreeable to the Sentence of the Court Martial that Sett 7 Dec[r]. Cha[s]. Troublefield of Cap[t] Lesenes Comp[y] Rec[d]. 50 Lashes on the Bare back with Switches for Misdemeaner contrary to the articles of war –

//Jn[o]. Frenise of Capt[n]. Baker Comp[y] rec[d]. 25 Lashes on the Bare back with Switches for Neglect of Duty – [988]

//Thomas Crozier of Capt[n] Halls Comp[y] confined for Loosing his Reg[tl]. Stockings the Court Martial Sentenced him to be put under stoppages to replace them – [989]

//Samuel Henderson of Capt[n]. Lesaines Comp[y]. Sentenced to do Double duty for four Days for Drunkness on Guard – – – –

Agreeable to the Court Martial that Sett Decemb[r]. 12th Jn[o]. Griffen of Capt[n]. Halls Comp[y]. confined for Stealing a Regi[t]. Muskett the Court is of oppinion he is Guilty of the Charge & to Sentence him to Receive 50 Lashes on the Bare Back with switches & to be put under Stoppages untill the Gun is paid for –

---

[988] John Frances.

[989] Thomas Crozier enlisted in the 2nd South Carolina Regiment on 27 July 1778 under Captain Thomas Hall. On 13 December 1778 he was confined for losing his regimental stockings and was put under stoppages to replace them. On 21 January 1779 he received 50 lashes on his bare back with a cat of nine tails for neglect of duty. On 18 March 1779 he received 50 lashes on the bare back with a cat of nine tails for being absent without leave. On 22 July 1779 he was sentenced to receive 100 lashes for neglect of duty, but he was given mercy. He was killed on 9 October 1779 during the assault on the Spring Hill Redoubt in Savannah.

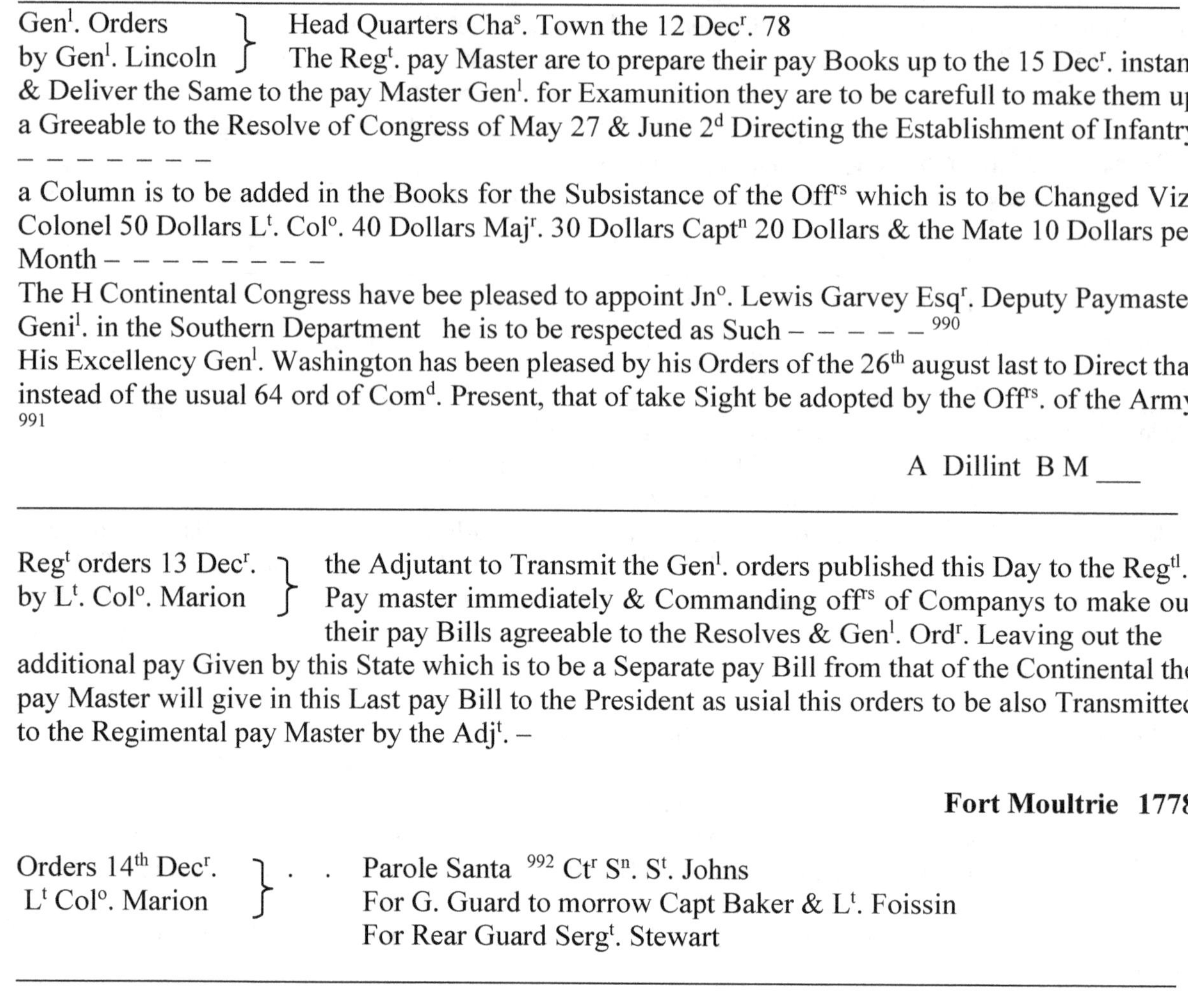

Genl. Orders by Genl. Lincoln } Head Quarters Chas. Town the 12 Decr. 78
The Regt. pay Master are to prepare their pay Books up to the 15 Decr. instant & Deliver the Same to the pay Master Genl. for Examunition they are to be carefull to make them up a Greeable to the Resolve of Congress of May 27 & June 2d Directing the Establishment of Infantry – – – – – – – –

a Column is to be added in the Books for the Subsistance of the Offrs which is to be Changed Vizt Colonel 50 Dollars Lt. Colo. 40 Dollars Majr. 30 Dollars Captn 20 Dollars & the Mate 10 Dollars per Month – – – – – – – – –

The H Continental Congress have bee pleased to appoint Jno. Lewis Garvey Esqr. Deputy Paymaster Genil. in the Southern Department he is to be respected as Such – – – – – [990]

His Excellency Genl. Washington has been pleased by his Orders of the 26th august last to Direct that instead of the usual 64 ord of Comd. Present, that of take Sight be adopted by the Offrs. of the Army [991]

A Dillint B M ___

Regt orders 13 Decr. by Lt. Colo. Marion } the Adjutant to Transmit the Genl. orders published this Day to the Regtl. Pay master immediately & Commanding offrs of Companys to make out their pay Bills agreeable to the Resolves & Genl. Ordr. Leaving out the additional pay Given by this State which is to be a Separate pay Bill from that of the Continental the pay Master will give in this Last pay Bill to the President as usial this orders to be also Transmitted to the Regimental pay Master by the Adjt. –

Fort Moultrie 1778

Orders 14th Decr. Lt Colo. Marion } . . Parole Santa [992] Ctr Sn. St. Johns
For G. Guard to morrow Capt Baker & Lt. Foissin
For Rear Guard Sergt. Stewart

[990] John Lewis Garvey served as the Deputy Paymaster General for the Southern Department from 1778 to 1780.

[991] In the British 1764 drill regulations the order for firing is, "Make Ready – Present – Fire". The new drill created by Baron Von Steuben replaced the commands with "Make Ready – Take Sight – Fire". This was replaced again within the year with the command "Take Aim" instead of "Take Sight". There is a misconception that soldiers in the Revolutionary War did not aim with their muskets. This is not correct and both the British and the Patriot armies trained their men to aim, often firing at targets to help their aim.

[992] The password "Santa" being so close to Christmas might be attributed to Santa Claus. On 26 December 1773 the name appears in the New York Gazette as "St. Nicholas, otherwise called St. A Claus". So the name Santa was known by the colonists prior to the Revolutionary War. There is a chance that the famous poem, "*A Night Before Christmas*" was not written by Henry Clement Moore, but was instead written by Major Henry Livingston, Jr. of the 3rd New York Regiment. The poem "*A Visit from Saint Nicholas/Night Before Christmas*" was first published by a Troy, New York newspaper in 1823. Livingston died in 1828, and Moore started to claim the poem in the late 1830s. There is also a chance that the password "Santa" might be the Spanish word for Saint, and it is mentioned in the context of St. Johns, the countersign.

Orders 15th Decr. L t. Col. } Parole Puritan Ctr. Sn. Generated
For G Guard to morrow Captn. Mason & Lieut. Kolb
For Rear Guard Sergt. Stewart

Genl. Orders 4th Decr by Majr. Genl. Lincoln } In Congress Octobr. 12 1778
whereas the Religion & good morals are the only Solid foundation of publick Liberty & Happyness Resolved that it be and it is hereby earnestly recommended to the Several States to take the most effectually measure for the Encouragment there of & for the Suppression of Theatrical Entertainment & horse Racing Gaming & Such Divertions as they are productive of Idleness & Depredation – [993]

Resolved that all offrs of the United States be & hereby are Strictly Enjoyned to See that the good & whole some Rule provided for the Preservation of moral among the Soldiers are daily & punctually observed

Extract of the Minutes Chs. Thompson Sectr.
Transmitted A Dilliant BM

Regtl. Orders L t. Colo – } The Regimental Taylors are ordered to Join & Do duty in their Respective Companies Except two the Colo. will give Leave to do Such work as the Officers & Soldiers may want done & are to be Exempted from Duty till further Orders –

Brigade Orders 15th Decr. Genl. Moultrie – } Sir as the Militia at the magazine Guard at Hobcaw is orderd away the Same number of men as usual under the command of an officer from 2d Regimt to Relieve them Signed William Moultrie
Brigd. Genl. –

Aftr R.. Ordrs L t. Colo. } one Subalter 1 Sergt 1 Corpl & 9 Privates to be gott ready ready immediately to march & to be charge of the Hobcaw Guard they are to be furnished with 2 tin Kittles 1 Ax & 12 rounds pr. man when they are ready to march the Offrs. Commanding the Party will Receive his Orders This party marched off immediately for Hobcaw Guard L t. Gray –

Orders 16th Decr. L t. Colo. Marion } Parole Chesterfield – Ctr. Sn. 74
For G. Guard tomorrow Capt Harleston & L t. Roux
For Rear Guard Sargt. Feast
For advance Guard Sargt. Green

---

993 This most likely did not go over very well. Benjamin Lincoln was from puritan Massachussetts, and most likely thought much of the Southern ways of doing things were outrageous. Charles Woodmason, an Anglican minister, wrote that South Carolinians, "delight in their present low, lazy, sluttish, heathenish, hellish Life, and seem not desirous of changing it." In the book "*Cracker Culture*", the author wrote "Gambling was widespread and accepted in a society that revered the racing of horses and other such sensual pleasures. Many Southerners spent as much of their time as they could "drinking, gambling, horse-racing, fox-hunting," and doing what a contemporary called "enjoying life." A Yankee described the activities and the arrangements at a Southern "gander pulling" as typical: "The whiskey kegs on the stumps, the gaming tables under the shades, the cock-fights in the pens, the horse-race out in the woods, will amuse the crowd to-morrow. And the fox-chase...will close the festivities. Horse racing was quite popular in the South. A Northerner who objected to racing wrote from South Carolina: "Curiosity induced me to go once, which will satisfy me for life." "In New England," boasted the president of Yale College, "horse racing is almost and cockfighting absolutely unknown."

Fort Moultrie 1778

Orders 17th Decr. } Parole Bolenbroke Ctr. Sn.
Lt. Colo. } For G Guard to morrow of Captn Mott & Lieut. Capers
For Rear Guard Sargt. Stewart

A Court Martial to set this fore noon at 11 OC to try all prisoners ordered to it Evidences to attend Presidt. Captn Dunbar Members Lt. Capers Guerry & Foissin & Kolb Genl. Orders by Genl. Lincoln dated 16 Decr. 1778 a Sargt a Corpl. & 12 Privates to Mount Guard in forenoon at the House Provided for Head Quarters Lieut Jno. Wickham is appointed adjutant to the 2 Regt. he is to_be Obeyed as Such A Dilliant BM

Orders 18th Decr. } Parole Harley Ctr. Sn. 12.
Lt. Colo. Marion } Fort Guard tomorrow Captn. Dunbar & Lieut. Warley
Rear Guard Sargt. McDowell

Morning Orders } 1 Sargt. 1 Corpl & 6 Privates to hold themselves in Readyness to march with 3 rounds pr man the Officer Commanding this party will Receive his Orders when ready which must march by 8 OClock this morning Sargt. Majr. took Command of the above Party. A Court Martial to set this forenoon at 11 OClock to try Such Prisoners as shall be Ordered to it Evidences are to Attend

Presd. Capt Baker Members Capt Mason Lt. Guerry Foissin & Kolb

---

Orders 19 Decr. } Parole Ormond – Ctr. Sn. Duke –
Lt. Colo. } For G Guard to morrow Captn. Baker & Lieut. Guerry
For Rear Guard Sargt. Feast –

The Capt of the Guard will give Sergt. Simpson a man from the fort Guard to be Centry where he Issues provisions with Orders not to admit any person in the house where Such provision is without Mr. Simpsons aprobation & to prevent all Riotts & Disorders this order to continue till orderd otherwise –

Genl. Orders by Genl. Lincoln 17 Decr. 1778 }
at a Genl. Court Martial where a Lt. Colo. William Scott was Presdt. Held the 14 Instant Terince McConnel a Soldier the first Regiment was tried for Desertion & found Guilty & Sentence to Receive 100 Lashes on the Bare Back [994] also Joseph Baiso a Soldr. in the Same Regt. Tried for Desertion was found Guilty & Sentence to Receive 100 Lashes on his Bare Back [995] the Genl. approves the Sentance & orders them to put in Execution tomorrow Morning at Guard Mounting the Genl. Court Martial of which Lt. Colo. Scott was Presdt. is dissolved

In future Genl. Court Martial not appointed for the Tryal of any Pulticular prisioner or Persons but of all Such as may be brought before them are to Continue to it or Adjourn from Day to Day as the Cause my Require, but are not to Consider them Selves disolved untill done by a General Order, when any Prisoners are Sent to the Main Guard his Crime must be Given to the Comdg. Offr. there of

---

[994] Terrence (Ferrence) McConnel enlisted in the 1st South Carolina Regiment on 29 December 1776 for the duration of the war. He was found guilty of desertion on 19 December 1778 and he received 100 Lashes on his bare back. He deserted again on 15 March 1779.

[995] Joseph Baiso was in the 1st South Carolina Regiment in 1778. He was found guilty of desertion on 19 December 1778 and he received 100 lashes on his bare back.

in writing & the names of the Witnesses to be Sent to the Main Guard for a Crime which may be tried by a Regimental Court Martial The Barrack Guard is to be Considered as the Main Guard till further Orders The off^{r} of the Main Guard to make his report to the Capt. of the Day, who it to report the Same at head Quarters Specifying the Prisoners Names the Crimes by whom & how Long Confined

A Dilliant BM

Orders 20th Decr. L^{t}. Col^{o}. Marion }
Parole Richmond C^{t}. S^{n} Huger –
For G. Guard to morrow Captn. Mason & Lieu^{t}. Foissin
For Hobcaw Guard Lieu^{t}. Roux / for Rear Sarg^{t}. Watt

Orders 21st Decr. L^{t}. Col^{o}. Marion }
Parole Step Moultrie Ct^{r} S^{n}. 7
For G. Guard tomorrow Capt. Harleston & L^{t}. Kolb
For Rear Guard Sarg^{t}. Webb –

Orders 22^{d} Decr. L^{t}. Col^{o}. Marion }
Parole EngagementCt^{r}. S^{n}. at Sea
For G. Guard tomorrow Capt. Lesesne & L^{t}. Gray
for Rear Guard Serg^{t}. Feast –

Orders 23^{d} Decr. L^{t}. Col^{o}. Marion }
Parole Fleet Ct^{r}. S^{n}. S^{t}. Eustatius –
For G. Guard to morrow Capt Moultrie & L^{t}. Hart
For Rear Guard Serg^{t}. Watt
advance Guard Serg^{t}. Teague

The Battalion to turn out this afternoon at 2 OClock without arms & to go to Haddrells for Moss for wading, the Serg^{t}. Maj^{r}. will take Command & See they do not Stragle & make them bring a turn p^{r}. Man, the other Serg^{ts}. will attend also –

Orders 24 Decr. Capt. Lesene }
Parole Crises – Ct^{r}. S^{n}. 9.
for G Guard to morrow Capt. Mazyck & L^{t}. Martin
for Rear Guard Sarg^{t}. McDowell

Orders 25th Dec^{r}. }
Parole Capt. Stone Ct^{r}. S^{n}. Battle
for G. Guard tomorrow Captn. Dunbar & L^{t}. Capers
for Rear Guard Sarg^{t}. Brown

3 Privates to be added to the advance guard immediately & 3 $^{doz.}$ Cartridges to be sent to them

Fort Moultrie 1778

Genl. Orders by Genl. Lincoln }
Head Quarters 23^{d} Decemb^{r}. 78
Parole Drayton

The Hon^{bl}. Continental Congress have been pleased to Consider The appointement of Colonel Drayton in the Office of Deputy Quarter Master General for the Southern Departments he is therefore to be Respected & Obeyd as Such Edmund Massenburg Hyrne DAG

Orders 26th Decr. L^{t}. Col^{o}. Marion }
Parole Frederica Ct^{r}. S^{n}. Georgia
For G. Guard tomorrow Captn. Mason & Lieu^{t}. Warly
For Rear Guard Sarg^{t}. Green

Orders 27th Decr. L^{t}. Col^{o}. Marion }
Parole Bloody Point Ct^{r}. S^{n}. Detatchment
For G. Guard to Morrow to Capt. Harleston & Lieu^{t}. Guerry
For Hobcaw Guard to Morrow Lieu^{t}. Hart
For Rear Guard Serg^{t}. Webb

---

after Orders by L^{t}. Col^{o}. Marion } one Subalton 1 Serg^{t}. 1 Corp^{l}. & 1 Drum 1 fife & 25 Privates to be in Readyness to Embark Immediately to town with 18 Round p^{r}. Man & 1 Spare Flint the Barge to be ordered to be in readyness to Transport the party over; orders will be given as Soon as the party is ready to March –
The fort Guard to be Reduced to 15 privates the Rear Guard to a Sarg^{t}. 1 Corp^{l}. & 3 privates the Rear guard to take off the Centry on the Guard & Keep the Bridge Drawn after Retreat Beating. 6 Men to be recalled from the advance Guard tomorrow Morning that Guard to be only a Serg^{t}. a Corporal & 3 privates the Fort Guard will Consist of the usual Commissioned & Non Commissioned Off^{rs}. and to Plant only ye Sentrys the Centry in the Middle of the Platform fronting / the Southeast to be taken off

NB Corporal Henson of Capt. Bakers Comp^{y}. is reduced to the Ranks & was Confined in Black hole for 8 Days a greeable to Sentence of Last Court Martial for Striking & Abusing Serg^{t}. Simpson
Brig^{d}. orders Decr. 27th .78 by Genl. Moultrie – } Sir youll order immediately a Lieu^{t}. a Sarg^{t}. a Corp^{l}. 25 privates 1 Drum fife to take Charge of the New Barracks & will Receive further orders from me about their duty you'll also order four boatmen to be in Readiness when demanded for the Barge the party for the Barracks is to be Reliev^{d}. every Week Lieu^{t}. Martin Took Command of the above party –

Orders 28th Decr. Col^{o}. Marion }
Parole Cockspur Ct^{r}. S^{n}. 110.
For G. Guard timorow Capt. Lisisne & L^{t}. Kolb
For Rear Guard Sarg^{t}. Feast

L^{t}. Col^{o}. Marion **Fort Moultrie 1778**

Orders 29th Dec^{r}. }
Parole Tibee Ct^{r}. S^{n}. Genl. Bull
For G. Guard to morrow Captn. Mazyck & L^{t}. Gray
For Rear Guard Sarg^{t} –
In the future all off^{rs} names are to be inserted in the Provision Returns mentioning so many White Servants & so many Black Servants; no return to be taken without this orders is Completed with the Quarter Master Sarg^{t}. is to bring the Provision Return as Soon as they are made to him to be Corrected by the Com^{dg}. Off^{r}. Dayly before they are sent to town –
The Pay bills are to be made immediately agreeable to the form the pay Master Serg^{t}. with Captn. Mason –

On the morning of December 28th the British invasion fleet landed near Savannah, Georgia. They were slowed for 24 hours at Brewton Hill by the 3rd South Carolina Rangers under the command of Captain John Carraway Smith.[996] General Howe had 800 men to defend the city against almost 4,000 British and Provincial troops under the command of Lieutenant Colonel Archibald Campbell.

Howe planned his defense the best he could, but a slave showed the British a route through the swamp that allowed the Light Infantry and the 71st Highlanders to get behind the Patriot defenses. When Campbell made a feint at the defenses, the militia fled or surrendered. The Patriot line collapsed and the retreat became a rout.

Colonel Owen Roberts and the 4th South Carolina Artillery were able to hold the British back long enough for the Georgia Continentals to regroup and counterattack. Unfortunately they were overwhelmed and were pushed into the swamps. Howe managed to escape with a few men and officers into South Carolina, but he lost his army and the artillery. The Patriots had 83 killed or drowned, 11 wounded and 453 captured.

Along with the city of Savannah, the British captured nine Brass field pieces, thirty-nine iron field pieces, twenty-three coehorn mortars, ninety-four barrels of gunpowder and several ships, including a Spanish vessel carrying twenty-two guns. Five days later Howe relinquished his command to Benjamin Lincoln and headed north to Washington's army.

After the capture of Savannah, Governor John Houstoun ceased being the governor. The executive council of Georgia was tired of his politics and elected Lyman Hall as governor of Georgia.[997]

Orders 30th Decr. } Parole Coopsaw Ctr. Sn. Fleet
For G. Guard tomorrow Captn. Dunbar & Lt. Hart
For Rear Guard Sargt. McDowell
For advance Guard Sargt. Watt

The Surgion or his mate is to Acquaint the Commanding Offr. of the Regt. & Compys when ever he thinks it nessesary to Send a man from the Regtl. Hospital to the Genl. Hospital –
No man to be received in the Regtl. Hospital without an order from his Commandg. Offr. of Compy. or Regt. –

Orders 31st Decr. } Parole Genl. Lincoln Ctr. Sn. Genl. Moultrie –
For G. Guard to morrow Capt. Baker & Lt. Roux
For Rear Guard Sargt. Feast –

Genl. Orders by Genl. Moultrie } Majr. Brown of the 6th Regiment having Resigned his Commission is no more to be Considered as a Continental Officer [998]

Signed Wm. Moultrie B Genl.

---

[996] John Carraway Smith was born in 1751 in North Carolina. He had been a lieutenant in the 2nd South Carolina Regiment in 1775. He was promoted to captain on 16 September 1776. He was wounded at the siege of Savannah on 9 October 1777 when he was in the 3rd South Carolina Rangers. He was taken prisoner at the fall of Charleston and exchanged in November 1780. He was a captain in Maham's Dragoons and he resigned on 26 February 1782. He was made a brevet major on 30 September 1783.

[997] Full details of the capture of Savannah are described in "*Nothing but Blood and Slaughter, Volume One*".

[998] William Brown served as a captain and major in the 6th South Carolina Regiment in 1778. He resigned in December 1778.

Reg[tl]. orders by Maj[r]. Horry } A monthly Return from Each Comp[y]. to be made out & delivered to the Adjutant by tomorrow Morning at 8 OClock at 10 OClock this forenoone all the Men off Duty are to be Parraded by the Sarg[t]. Maj[r]. who with the Sarg[t]. off Duty are to take Charge of the Men & go to Hadrells point for a Turn of Moss for Wading No Leave of Abscen Can be Granted to Off[rs] or Men till further Orders –

1779

Orders 1 Jan[y] 1779 – } Parole New Year Ct[r]. S[n]. 79
For G. Guard to Morrow Cap[t]. Marion & L[t]. Warley
For Rear Guard Sergeant Teague

Orders by Maj[r] Horry __ } A turn of Moss to be brought form Haddrels Point this morning by the Men off duty, the Serj[ts]. to attend their Men -

L[t]. Col[o]. Marion **Fort Moultrie**

Orders 2[d] Jan[y]. } Parole Colonel Marion Ct[r]. S[n]. Cha[s] . Town -
For G. Guard to morrow Ccap[t]. Lesesne & Lieu[t].. Kolb
For Town Guard to morrow Cap[t]. Motte & Lieu[t]. Capers
For Rear Guard Sarg[t]

Gen[l] Orders by Gen[l]. Moultrie Dec[r]. 30: 1778 } Lieut Col[o]. Marion
Sir you are to Remain in Charles Town to forward all dispatches to head Quarters & to Give all Intelligence which you may think Necessary – You are to supply the Deputy Quarter Master Gen[l] for money for any of the above Services as you are to be Commanding Officer in Town untill Some Senior Officer should come in – I would not Mention your Particular Duty as you are well Acquainted I am Sir your most hum[l] Serv[t]

W[m] Moultrie BG

Orders by L[t]. Col[o] Marion ~~Jan[y]~~: 79 } M[r]. Jn[o] Hall Quarter Master to the 2[d] Regiment will transmit all Orders to the different Departments & out posts till further Orders [999] – Orderly hours to be at 11 OClock in the Forenoon Transmitted by John Hall Q[r]. Master 2 R

Orders 3[d] Jan[y] } Parole Col[o]. Huger. Ct[r]. S[n]. M[c]Intosh
For G. Guard to Morrow Cap[t] Mazyck & L[t] Gray
For Rear Guard Sarg[t] Cample

Orders 4th Jan[y] } Parole Ebenezer Ct[r]. S[n]. Retreat
For G: Guards to Morrow Cap[t] Dunbar & L[t]. Hart
For Rear Guard Sarg[t] Stephen Roberts

Orders by Maj[r] Horry - } Comd[g] Officers of Comp[ys] may give Leave of Absence to one Private of Comp[y] at a time to Go to Town but must Return at Night three Men to be Given to the Gunner of the Fort to assist in making up wadding for the

---

[999] John Hall served as the deputy quartermaster general during 1779. He resigned as deputy Quarter Master on 15 August 1779. He is not the same Lieutenant John Hall who was the regimental quartermaster of the 2nd South Carolina Regiment on 1 July 1776. That John Hall resigned during December 1776 and joined the British. He was also captured and hanged for treason at Ninety Six on 17 April 1779.

Cannon – The Centry Posted on the Right Curtain of the Garrison to be Removed to the front & to Observe Strictly the Orders Given to Sentries Posted there – A Court Martial to set this forenoon at 10 OClock to try Such Prisoners as Shall be Ordered to it Evidences to attend Presid^t Capt Mason Members Lt Martin & Gray –

Orders 5th Jan^y } Parole Col^o Pinckney Ct^r. S^n. Maj^r Pinckney
For G: Guard to Morrow Capt. Baker & Lt Roux
For Rear Guard Sarg^t McDonald
Orders by Maj^r Horry } A court Martial to Set this forenoon at 11 OClock to try all Prisoners brought to it – Ev: to Attend
Presd^t Capt Lesesne Members L^t. Kolb & Gray
Agreeable to the above court Martial Jno Cammel of cap^t. Harlestons comp^y. receive 100 lashes on the bare Back with a Catanine Tails for Stealing a Shirt [1000] – Bartholomew Hine of the Same Comp^y. Rec^d. 50 lashes on the Bare back with a Catanine Tails for taking away a whole Messes Provisions [1001]

Fort Moultrie 1779

Agreeable to the court Martial that Set 4th. Jany. Samuel Henderson of Capt. Lesesne Comp^y. Rec^d Fifty lashes on the Bare Back with a Cataninetails for Loosing the lock of his Gun when on Guard in Town
//Joseph Wright of Capt. Harleston Comp^y. Rec^d. 100 lashes on the Bare back with a Catanintail for Quitting his post when on Sentry
//William Simpson fo the Same Comp^y. Rec^d. 100 lashes on the Bare back with a Cataninetails for Quitting his post when on Centry _

Orders 6th Jan^y } Parole Capt: Hall Ct^r. S^n. Notredame
For G: Guard to Morrow Capt. Mason & Lieu^t. Martin
For advance Guard to Morrow Serg^t
For Rear Guard _________ Sarg^t Sims
Orders by Maj^r Horry } A Court Martial to set Immediately to try Such Prisoners as will be brought to it
Evidences to Attend – Presid^t. Capt Mazyck Members Lt Hart & Warley
The Commanding Officer is Ashamed to See the Dirty & filthy condition of the Garrison & therefore orders that the Sarg^t Maj^r do Immediately Parrade & & take charge of all Non Commissioned Off^rs & Men off Duty, which with the Qt^r. Master Serg^t fatigue Party are to have the whole of the Garrison Including the Platforms thoroughly Cleaned & all filth & Dirt & Rubbish Removed to a Proper distance from the Garrison – that the Serg^t Major do daily at 12 OC Examine the Garrison & Report to the Comd^g Officer, the Comp^y Orderly Sarg^t or Corporal for the day in whose division he finds

---

[1000] Private John Campbell, not fife major John Campbell.

[1001] Bartholomew Hine enlisted in the 2nd South Carolina Regiment on 1 July 1778. In 1779 he was in Captain Harleston's company. On 5 January 1779 he received 50 lashes on the bare back with a cat of nine tails for taking away the provisions of a whole mess. A "mess" was a group of soldiers who tented together and ate together. Each tent would hold a "mess". It was usually between 5 to 8 men. Hine was in Captain Peter Gray's company during the siege of Savannah in 1779.

any filth, it being their Duty to See the Same removed, & Such Reported Orderly Sarg^{t} or Corporal may Depend on being brought to a Court Martial & Suffering the Sentence thereof for Disobedience of Orders – Those Men that do not Live in Comp^{y} Barracks but in their Barracks Respectively & do not Remove their own filth daily the Serj^{t}. Maj^{r} is also Ordered to Report to the Comd^{g} Officer for disobedience of Orders & they may Expect to Suffer Accordingly – the Sarg^{t} Maj^{r} is Ordered to have this Order fully made Known to those it Concern that none may plead Ignorant thereof –

Orders 7th Jan^{y} } Parole Col^{o}_ Elbert Ct^{r}. S^{n}. Roberts
For G: Guard to Morrow Capt. Lesene & L^{t}. Guerry
For Rear Guard Serg^{t} Steph Roberts

Fort Moultrie 1779

Charles Town Jany 6th
Orders by L^{t}. Col^{o}. Marion __ } Sir, you will immediately Send to Town a Subaltern & one Serj^{t}. and 4 privates to the Guard at the new Barracks in the Room of Lieu^{t}.. Capers & men sent to Purisburgh Sign^{d} Francis Marion _____
Orders by Maj^{r} Horry } the order of yesterday respecting the Garrison to be continued & finnished this day –

Orders 8th Jan^{y} } Parole Militia Ct^{r}. S^{n}. Haddrells
For G: Guard to morrow Capt. ~~Mazyck~~ Moultrie & L^{t}. Kolb
For Rear Guard to Morrow Serg^{t} McDonald

Orders 9th Jan^{y} } Parole Col^{o}. Horry – Ct^{r}. S^{n}. Maj^{r} Hagen
For G: Guard to Morrow Capt. Mazyck, & Lieu^{t} Hart
For Town Guard to Morrow Capt Lesene & Lieu^{t} Gray
For Rear Guard to Morrow Sarg^{t} Campbell
Orders by Maj^{r} Horry } The paybills of the Different Comp^{ys} to be made out this day – The Q^{r}. Master Serjeant will Delliver to one of the Boats crew Two Tin Kittles & one for the use of the Town Guard _____
A Turn of Moss to be brought from Haddrells point this morning by the men of duty the Serjeants to attend them –

Orders 10th Jan^{y} } Parole Lowndes Ct^{r}. S^{n}. Gaddsen
For G: Guard to Morrow Capt. Dunbar & L^{t}. Roux
For Rear Guard to Morrow Sarg^{t} Watt.
Orders by Maj^{r} Horry } The Q^{r}. Master Serj^{t}. to deliver to the Coxswain of the barge 3000 Bullets to be delivered to a fatigue Party of the Artillery at Work in Lightwoods Bastion in Town – The Centry from the Rear Guard Posted at the Draw bridge is to be Continued on that Post by Night as well as during the day – the Centry Posted at the Gate of the Garrison after Retreat beating to be removed to the front Platform & that Post Continued Guard untill after Reveille Beating when the Centry is to be moved to the Garrison Gate [1002]

[1002] On 10 January, Fort Morris at Sunbury, Georgia fell to the British after a five day siege. There were 289 men captured, including a company of the 3rd South Carolina Rangers.

Orders 11th Jany } Parole Plombard Ct^r. S^n. Consul
For G: Guard to Morrow Capt. Baker & Lieu^t Martin
For Rear Guard to Morrow Sarg^t Davis

R.O. by Captain Harleston **Fort Moultrie 1779**

A Court Martial to set this Forenoon to try such prisoners as shall be brought before it __ In future no Non Commissioned Officer or Soldier shall presume to go a Hunting or Fowling without leave first obtained in writing from the Command^g Officer on pain of Being depicted a disobeyer of Orders & Suffering Accordingly For the Court Captain Provaux Members Captain Mason & Lieut. Kolb

G.O. by Lieutenant Colonel Marion __

No Officer or Soldier to have leave of Absence From any Out Post but from Morning to Retreat beating

Transmitted January 11th 1779 by Jno: Hall Q.M. 2^d. Reg^t

Orders 12th Jany } Parole Plenipotentiary Ct^r. S^n. Girard
For G Guard to Morrow Captn. Proveaux & Lt Warly
For Rear Guard tomorrow Sarg^t Teague

Pursuant to Yesterdays Court Martial Benjn Sojourner & Corporal Champney being recommended to Mercy had their Sentence remitted & were discharged [1003]__ A Court Martial to sit this forenoon for the trial of such prisoners as shall be brought before it – Presid^t. Capt. Motte Members Lieutenant Hart & Roux

//Agreeable to the above court martial Daniel Crabb [1004] and Hardy Flowers [1005] both of Capt. Dunbars comp^y. confind. by Lieu^t. Rouse for loosing their Regimental Blankets the court Sentence the prisoners to be put under Stoppages for to make good their Blanketts __ __ __

//John Proby of capt^n. Halls comp^y. confind by Order of Lieu^t. Col^o. Marion for Disobedience of orders The court are of opinion the prisoners is Guilty of the charges & do Sentence him to Receive fifty lashes on the Bare Back with a cat onine Tails the Comd^g officer approves of the above Sentences –

Fort Moultrie 1779

Orders 13th Jany } Parole Capt. Bonneau Ct^r Georgetown
For G. Guard to Morrow Captn. Mason & L^t. Kolb
For Rear Guard to Morrow Serg^t Feast
For Advance Guard Serg^t McDowell

After Orders by Maj^r Horry } The Long Roll to beat every forenoon at 10 OClock & all Non Commis^d officers, Drummers, fifes & Privates off duty are to Attend & be formed on the Regimental Parade when the Sarg^t. Maj^r is to take Charge of them & Centry Daily

---

[1003] Benjamin Sojourner was in the 2nd South Carolina Regiment. On 11 January 1779 he was court-martialed for an unnamed offense, however he was given mercy.

[1004] Daniel Crabb enlisted in the 2nd South Carolina Regiment on 11 September 1778. In 1779 he was in Captain Dunbar's company. On 12 January 1779 he was confined and put under stoppages for losing his regimental blanket. He deserted on 24 February 1779.

[1005] Hardy Flowers enlisted in the 2nd South Carolina Regiment on 17 September 1778. In 1779 he was in Captain Dunbar's company. On 12 January 1779 he was confined and put under stoppages for losing his regimental blanket.

bringing a Turn of Moss from Haddrells for the use of the Garrisn till further orders – Such as Disobey this order the Sargt. Majr will Report to the Comdg offrs. who may depend on Suffering the Sentence of a Court Martial. No Leave of Absence to be Granted to any of the Men till further Orders –

Orders 14th Jany Parole Sunberry Ctr. Sn.Georgia
For G: Guard to Morrow Captn. Motte & Lt. Hart
For Rear Guard Sergt Murphey

Orders by Majr Horry __ } No Fires to be made on Hearths Laid or on fire tubs in the Upper the Soldiers Barrack on the Left wing – Such as disobey this order, the Orderly Compy Sergt or Corpl for the day is Ordered to Report to the Comdg Officer for Disobedience of Orders –

Orders 15th Jany } Parole Lincoln Ctr. Sn. Moultrie
For G: Guard to Morrow Capt. Moultrie & Lt. Roux
For Rear Guard Sergt Stewart –

Orders 16th Jany } Parole America Ctr. Sn. Beekman
For G: Guard to Morrow Captn Mazyck & Lt. Martin
For Rear Guard Sergt Green

Orders 17th Jany } Parole Craven County Ctr. Sn.Regt.
For G: Guard to Morrow Capt Dunbar & Lt Warley
For Town Guard to Morrow Capt Moultrie & Lt Kolb –
For Rear Guard to Morrow Sergt Teague

Orders 18th Jany } Parole Chas Town Ctr. Sn. Militia
For G: Guard to Morrow Capt Baker & Lt. Gray
For Rear Guard to Morrow Sergt Stewart

Orders by Majr Horry __ } All Stopages of Pay Ordered to be made by Sentence of Court Martial the Adjutant is Ordered to Receive Accordingly & to Account for the Same to the Comdg Officer of the Regt – A Court Martial to Set Immediately to try all Prisoners Ordered to it Evidences to Attend Presd. Captn. Proveaux & Memr. Lt. Kolb & Hart

Fort Moultrie 1779

Orders 19th Jany } Parole Colo Ash Ctr. Sn. North Carolina
For G: Guard to Morrow Capt Proveaux & Lt Hart
For Rear Guard Sergt Green

Orders by Majr Horry ___ } Commdg Officers of Compys will as Expeditiously as Possible make an Exact Inspection of the Arms & Accoutrements of their Respective Compys & make Returns of their Condition to the Commdg officer It is Positively Forbid that Centries hold any Conversation with any Person whatever while on their post as it takes of their Attention from their Duty and the Commdg Offr Calls on & hopes that Every Officer in the Regt will Join with him in detecting & Punishing all Such as are Guilty of so Unmilitary a Practice A Court Martial to Set at 10 OC this forenoon to try Such Prisoners as are brought to it Evids: to Attend Presidt. Capt Mason, Memr Lts. Roux & Warley __

//Agreeable to Court Martial that Sett 18th Jany Jno Taylor of Capt Mottes Compy Recd. 79 lashes on the bare back with a cat nine tails for Neglect of Duty –
//Fredk. Simmonds of Capt. Dunbars Compy. Recd. 79 lashes on the bare back with a cat Nine Tails for Neglect of Duty & Quitting the Guard
//Serjt. McDonald of Capt. Halls Compy Was Reprimanded for Neglect of Duty

---

Orders 20th Jany } Parole France Ctr. Sn.
For G: Guard to Morrow Capt Motte & Lt Roux
For Rear Guard Sergt. __
For Advance Guard Sergt Martin

Orders by Lt. Colo. Marion dated Chs. Town 19 Jany 1779 } The Pay Master to the 2d Regt is Ordered to Stop the Mens Pay agreeable to Sentences of Courts Martial, not Exceeding at a time one half of their Pay, & deliver it to the Adjutant of the Regt who is to apply it agreeable to the Sentence of Court Martial Signed Francis Marion Lt. Colo

---

Fort Moultrie 1779

Orders by Majr Horry Orders } No more Moss to be brought for the use of the Garrison, till further Horry __
Comdg Officers of Compys may Give Leave of Absence to one Private pr Compy to go to Town at a time & to Return by Retreat beatg to Garrison
A Court Martial to Set at 10 OClock this forenoon to try Such Prisoners as are Ordered to it Evidences to Attend –
Presidt Capt Motte Members Lts Kolb & Gray
The following Promotion takes place in the Regt & they are to be Obeyed Accordingly –
Corpl Wm Hairman of Capt. Harleston Compy to be a Serjeant Capt. Proveaux Company –[1006]
Wm. McCollough of Capt Moultries Compy to be a Corpl in the Same Fredk Simmons of the Grenadiers to be a Corpl in the Light Infantry Compy. –
Henry Savage, of Capt Masons Compy to be a Corpl. in Capt Baker Compy,[1007] Also Jordan Combe of the Grenadier Compy to be a Corpl. in Capt. Bakers Compy [1008]
//Agreeable to the Court Martial that Sett Yesterday James Quinn of Capt. Lesesne Compy. Recd. fifty lashes on the bare Back with a cat o nine Tails & Sentenced to be Put Under Stoppages for a Musket he lost on Town Guard
//Sergt. William Murphey of Capt. Motte Compy. Confind by his capt. for Disobedience of orders and Insolance of the court are oppinion the Prisoner is Guilty of the charge & do Sentence him to be severely reprimanded by the adjutant & Suspended for 15 days the Comdg. Officer approves of a reprimand

---

[1006] William Hasman.
[1007] Henry Savage of Captain Blake's company was promoted to corporal in the 2nd South Carolina Regiment on 20 January 1779 and transferred to Captain Richard Mason's company. He was reduced in rank on 9 February 1779 for neglect of duty. He was at the siege of Savannah.
[1008] John Jordan Combay.

// Serg^t^. M^c^Dowal of Capt. Harlestons Comp^y^. Confind by Capt. Lesesne for repeated Drunkiness when on Guard in Town – The court do Sentence the prisoner to be Reduced to the Ranks – The Comm^dg^ Off^r^ aproves of the Sentence But limits the punishment –

//Agreeable to the Last Court Marital Boses Bruce [1009] of cap^t^. Bakers comp^y^ . Rec^d^. 50 Lashes on the bare back with Switches for Selling his Regimental Blankett –

//Peter Fagan of the Same Comp^y^ Rec^d^. 50 Lashes on the bare back with Switches for Selling his Regimental Blankett–

David Vahaun of Capt. Proveaus Comp^y^ Rec^d^. Sixty lashes on the bare back with a Cat anine tails for Neglect of Duty __[1010]

Fort Moultrie 1779

Orders 20th Jan^y^ } Parole Col^o^. Motte Ct^r^. S^n^. Horry
For G: Guard to Morrow Capt. Moultrie & Lt Martin
For Rear Guard Serg^t^. Murphy __

orders by Maj^r^ Horry ___ } 3 Men to add to the Rear Guard & Sentry to be posted on the platform of the new battery with orders not to Suffer any lumber to be Taken from or near Said Battrey without orders from Capt. Harleston or the Com^dg^ Officer __ __ _

orders by Maj^r^ All Vessells outward Bound that are Suffered to Pass fort Johnston are also to pass fort Moultrie unless particularly Ordered to the Contrary by the Com^dg^ Officer of this Garison –

A Court Martial to sit this forenoon for the trial of such prisoners as shall be brought before it Presid^t^ Capt Baker Members L^t^. Hart & Warley

---

Orders 22nd Jan^y^ } Parole: Toujours Fidelle Ct^r^. S^n^. Intrepid
For G: Guard to Morrow Capt. Baker & L^t^. Warley
For Rear Guard Serg^t^. Stewart

A Court Martial to sit this forenoon for the trial of such prisoners as shall be brought before it – Presid^t^ Capt Proveaux Members Lieu^t^. Hart & Roux ____________

Agreeable to the above court Martial Thomas Crosier of Capt. Halls Comp^y^ . Rec^d^. 50 lashes on his bare back with a cat o nine tails for neglect of duty

Orders 23rd Jan^y^ } Parole Virtue Ct^r^. S^n^. Prevail
For G. Guard to Morrow Capt. Proveaux & L^t^. Gray
For Rear Guard Sarg^t^. Martin

Orders 24th Jan^y^ } Parole Pinckney Ct^r^. S^n^. M^c^Intosh
For G Guard to Morrow Capt. Mason & Lt Roux
For Town Guard to Morrow Capt. Mazyck & L^t^. Hart
For Rear Guard Serg^t^ Campbell

Orders 25th Jan^y^ } Parole Laurens Ct^r^. S^n^. Resigned
For G. Guard to Morrow Capt. Motte & L^t^. Martin
For Rear Guard Serg^t^ Stewart

1009 Moses Bruce.

1010 David Vaughn.

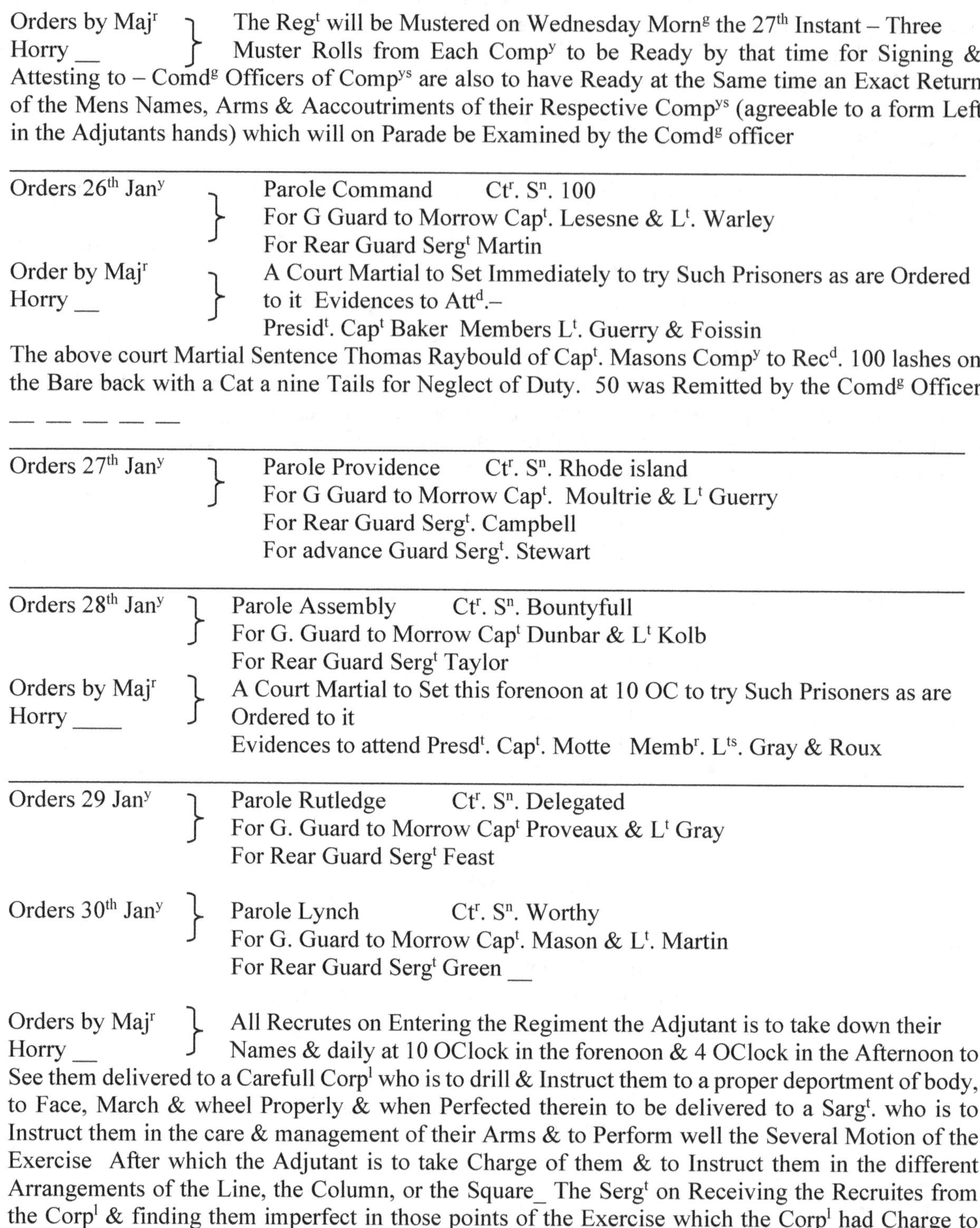

Orders by Maj^{r} Horry __ } The Reg^{t} will be Mustered on Wednesday Morn^{g} the 27th Instant – Three Muster Rolls from Each Comp^{y} to be Ready by that time for Signing & Attesting to – Comd^{g} Officers of Comp^{ys} are also to have Ready at the Same time an Exact Return of the Mens Names, Arms & Aaccoutriments of their Respective Comp^{ys} (agreeable to a form Left in the Adjutants hands) which will on Parade be Examined by the Comd^{g} officer

---

Orders 26th Jan^{y} }
Parole Command Ct^{r}. S^{n}. 100
For G Guard to Morrow Capt. Lesesne & L^{t}. Warley
For Rear Guard Serg^{t} Martin

Order by Maj^{r} Horry __ } A Court Martial to Set Immediately to try Such Prisoners as are Ordered to it Evidences to Att^{d}.–
Presid^{t}. Capt Baker Members L^{t}. Guerry & Foissin

The above court Martial Sentence Thomas Raybould of Capt. Masons Comp^{y} to Rec^{d}. 100 lashes on the Bare back with a Cat a nine Tails for Neglect of Duty. 50 was Remitted by the Comd^{g} Officer
— — — — —

Orders 27th Jan^{y} }
Parole Providence Ct^{r}. S^{n}. Rhode island
For G Guard to Morrow Capt. Moultrie & Lt Guerry
For Rear Guard Serg^{t}. Campbell
For advance Guard Serg^{t}. Stewart

Orders 28th Jan^{y} }
Parole Assembly Ct^{r}. S^{n}. Bountyfull
For G. Guard to Morrow Capt Dunbar & Lt Kolb
For Rear Guard Serg^{t} Taylor

Orders by Maj^{r} Horry ____ } A Court Martial to Set this forenoon at 10 OC to try Such Prisoners as are Ordered to it
Evidences to attend Presd^{t}. Capt. Motte Memb^{r}. L^{ts}. Gray & Roux

---

Orders 29 Jan^{y} }
Parole Rutledge Ct^{r}. S^{n}. Delegated
For G. Guard to Morrow Capt Proveaux & Lt Gray
For Rear Guard Serg^{t} Feast

Orders 30th Jan^{y} }
Parole Lynch Ct^{r}. S^{n}. Worthy
For G. Guard to Morrow Capt. Mason & L^{t}. Martin
For Rear Guard Serg^{t} Green __

Orders by Maj^{r} Horry __ } All Recrutes on Entering the Regiment the Adjutant is to take down their Names & daily at 10 OClock in the forenoon & 4 OClock in the Afternoon to See them delivered to a Carefull Corp^{l} who is to drill & Instruct them to a proper deportment of body, to Face, March & wheel Properly & when Perfected therein to be delivered to a Sarg^{t}. who is to Instruct them in the care & management of their Arms & to Perform well the Several Motion of the Exercise After which the Adjutant is to take Charge of them & to Instruct them in the different Arrangements of the Line, the Column, or the Square_ The Serg^{t} on Receiving the Recruites from the Corp^{l} & finding them imperfect in those points of the Exercise which the Corp^{l} had Charge to

Instruct them in, is to Remand Such Recrutes to the Corporal for better Information & the Adjutant is Likewise to Remand to the Serg^{t} for better Instruction all Such Recruites as he finds deficient under the Management of the Serg^{t} __ __ __ __ __ __ __ __ __ __ __ __ __ __ __ __

NB. Sentence of a Court Martial. Presid^{t}. Capt. Mason

Members L^{ts} Foissin & Guerry Serg^{t}. Teague was Sentenced to be mulct of a Weeks Pay for Neglect of Duty & Disobedience of Orders which was Approved by the Comd^{g}. Off^{r} – [1011]

Orders 31st Jan^{y} } Parole Bounty Ct^{r}. S^{n}. Success
for G. Guard to Morrow Capt Motte & Lt Warley
for Town Guard Capt Dunbar & L^{t}. Roux
for Rear Guard Serg^{t}. Martin ___

Orders by Maj^{r} Horry ___ } A Monthly Return from Each Comp^{y} to be made to ~~morrow~~ the Commd^{g} Officer by 10 OClock tomorrow forenoon The Q^{r}. Master Serg^{t}. to make at the Same a Return of Wood & Forrage delivered the Reg^{t} the Present Month.

Orders 1st Feb^{y} } Parole Col^{o}. McDaniel Ct^{r}. S^{n}. Camp
For G Guard to Morrow Capt. Mazyck & L^{t}. Guerry
For Rear Guard Serg^{t} Green

Orders 2 Feby Parole Genl Howe Ct^{r}. S^{n}. Fort Moultrie
For G. Guard to Morrow Capt Baker & Lt Gray
For Rear Guard Serg^{t} McDonald

Orders by Maj^{r} Horry ___ } A Court Martial to Set at 10 OClock this forenoon to try Such Prisoners as Shall be Ordered to it –
Evidences to attend Presid^{t} Capt Proveaux Members L^{ts} Martin & Gray

---

Orders 3 Feb^{y} } Parole Col^{o} Hicks Ct^{r}. S^{n}. Capt. Dewit
For G. Guard to Morrow Capt. Proveaux & Lt Hart
For Rear Serg^{t}__ advance Serg^{t}___

By Sentence of a Court Martial held Yesterday David Vaughn of Capt Proveaux Comp^{y} was Sentenced to be Put under Stoppages to make Good a Musket which he Lost __ __ __ __

Prévost ordered Major Valentine Gardiner to conduct a naval landing and occupy Port Royal Island. Prévost hoped that the landing would attract Lincoln's army about thirty miles away, and would draw off any Patriot troops that could stop Campbell from capturing Augusta. The 200 British troops landed on Hilton Head Island and occupied Beaufort, South Carolina on 3 February.

What Gardiner did not know was that Moultrie had arrived in Beaufort with General Bull and his 1,500 militia the day before the British landed at Port Royal. The only Continental troops in this battle were nine men manning the brass 2-pounder, commanded by Captain de Treville of the 4th South Carolina Artillery.

[1011] On 30 January Lieutenant Colonel Archibald Campbell captured Augusta, Georgia without incident. They were greeted by cheering Loyalists. On the opposite side of the Savannah River General Andrew Williamson's army faced the British, waiting for them to move.

Moultrie tried to deploy his forces in a wooded swamp but discovered that the British had beaten him to the wooded terrain. The South Carolina militia lined up across the road in the open, near the Halfway House, 200 yards from Gardiner's force. Moultrie placed two 6-pounders in the middle of the road and a 2-pounder on the right in some woods. Two signers of the Declaration of Independence, Thomas Heyward, Jr. and Edward Rutledge commanded the artillery in the road. In a twist the Patriots were in the open and the British troops took to the trees.

After Moultrie's force started taking casualties he ordered the men into the trees on the side of the road. The two sides attempted to flank each other, but to no avail. The British left wing charged Moultrie's riflemen. The riflemen, having no bayonets, fled the field. After 45 minutes ammunition began to run out for both sides. The British pressed both of the American flanks and Moultrie ordered his artillery back. The British stopped their advance due to having only 93 cartridges left for the entire line. Even after searching the dead of both sides they were only able to produce 300 more cartridges. The British soldiers gave the retreating Americans three cheers, which they did not return.

As the sun set the British withdrew from the town. The British lost 40 killed or wounded and seven captured, compared to the Patriots eight killed and 22 wounded. Due to the heavy loss of troops the British did not try to mount any operations into South Carolina until Prévost moved against Charlestown in May of 1779. [1012]

Fort Moultrie 1779

Orders 4th Feb^{y} } Parole Col^{o}. Giles Ct^{r}. S^{n}. Ten
for G Guard to Morrow Capt Mazyck & L^{t}. Warley
for Rear Guard to Morrow Serg^{t} Stuart
for Fort Johnson Guard Capt. Motte & Lt Martin

Orders 5th Feb^{y} } Parole Fort Johnston Ct^{r}. S^{n}.
For G Guard to Morrow Capt. Baker & Lt Guerry
For Rear Guard Corp^{l} Simmons [1013]

Orders by Maj^{r} Horry ___ } A Captain, a Subaltern, 2 Serg^{ts}, 2 Corp^{ls}, 1 Drum^{r}. 1 fifer & 24 Privates to Embark this Morn^{g}. for Fort Johnston as a Reinforcement to that Post – Capt. Motte will take the Command of the Party & on their Arrival At that Post & to Observe Strickly Such Orders as he shall Rec^{d} from L^{t}. Col^{o}. Marion

For the future the Recrutes are not to be mentioned as Such in the Morning Reports, but Reported fit for Duty or otherwise as the Case may be, as there is a Necessity for Putting them on Immediate Duty – Comd^{g} Officers of Comp^{ys} are to Morrow Morning to Apply to the Q^{r}. Master Serg^{t}. for a Pair of Shoes for Each Man of their Respective Comp^{ys} & for a Pair of Stockings for Each Serg^{t}, Giving a Receipt for the Same – The Recrutes as they join the Reg^{t} the Q^{r}. Master Serg^{t} are to Compleat in Cloathing taking a Receipt from the Command^{g} Officer of Such Companies as they are Ordered to do duty in – Also Commd^{g} Off^{rs} of Comp^{ys} are to Arm & Accoutre Immediately Every Recrute on joining their Respective Comp^{y} – The Order of the 30th Ult^{o}. to be Strictly Observed

---

[1012] Full details of the Battle of Beaufort are described in "*Nothing but Blood and Slaughter, Volume One*".

[1013] Jesse Simmons served in the 2nd South Carolina Regiment under Captain Peter Gray in 1779 during the siege of Savannah.

Orders 6th Feby } Parole Heyward Ctr. Sn. Artillery
For G. Guard to Morrow Capt Proveaux & Lt Gray
For Rear Guard Corpl Hughs –

Orders 7th Feby } Parole Discipline Ctr. Sn. Exact
For G. Guard to Morrow Capt Mazyck & Lt Hart
For Town Guard to Morrow Capt Baker & Lt Warley
For Rear Guard Corpl Savage _____

Fort Moultrie 1779

Orders by Majr Horry ___ } 4 doz flints & 20 doz Cartridges to be delivered to the Comdg Offr of the fort Guard who is at Retreat beating to See that Each Man of his Guard is Served in wt. a flint & 3 Cartridges & to receive them again before he is Releaved & to deliver them as well as Whatever Else he had in Charge to the Releaving officer & In his Report to the Comdg Officer to Insist each & Every Article so delivered as also Every other Circumstance, Matter or duty during his Guard – No Reports shall otherwise be Received from Tattoe beating to Ravallie Beating a Party of a Sergt. or Corpl. 4 Privates are Every hour to Pattrole as far as the Narrow defile to the Eastward of the Garrison & round about the Garrison & to take up & Confine all Persons found without the Same & who Cannot Give the Countersign are also to apprehend or Give timely notice to the Guard of any Sculking Parties of the Enemy being near or about the Garrison in order to 505urprise the Same. The Centries on the fort Guard are to Receive Every Quarter of an Hour a Visit from an Offr or Non Comd Offr of the fort Guard – The Rear Guard to be Visited at Least Once during the Night by an Officer from the fort Guard –The whole Guard is to be Turned out at Every Relief – One Offr to be always in the Guard Room during the day & all the officers of the Guard during the Night, unless at Such times as one of them is Visiting the Centries or the Rear Guard – After Tattoe Beating the Commandg Officer of the Guard is to be Particularly Carefull to Suffer non Noise or disturbance or Riot to be made in Garrison & to Confine all Such are Guilty of the Same – From the above Order it is ~~that~~ obvious that Some danger of the Enemy is apprehended at this Post & that in the Present Allarming Situation of our Country it Behoves our Utmost Exertion & the Commandg Offr Cannot Doubt of the Chearfull & Ready Obedience of Offrs & Men to Strict duty, which alone Can Render them a Credit to their Country & Honr to them Selves

Fort Moultrie 1779

After Orders by Majr Horry } Commandg officers of Companies are by 8 OClock tomorrow forenoon to make a Return of the Present State of their Respective Compys agreeable to a form Left in the Adjutants hands & to deliver the Same to the Comdg Officer by that time – One half of the Officer Servants now in Garrison are to be taken At a time to Augment the Guards at Night till the Regiments further Reinforced with Recrutes – The Gunner of the fort is to Morrow by 10 OC in the Morning to Prepare thirteen Pieces of the Heaviest Cannon in Garrison in Order to Proclaim Jno Rutledge Esqr. as Governour & Thomas Heyward Esqr. as Lieut Governor of this State [1014]– The time of Firing will be made Known by the Commdg officer tomorrow Morning

---

[1014] Thomas Heyward was born in 1749. He had studied law in England and came back to America in 1771. He served in the Second Continental Congress and was a signer of the Declaration of Independence. In 1776

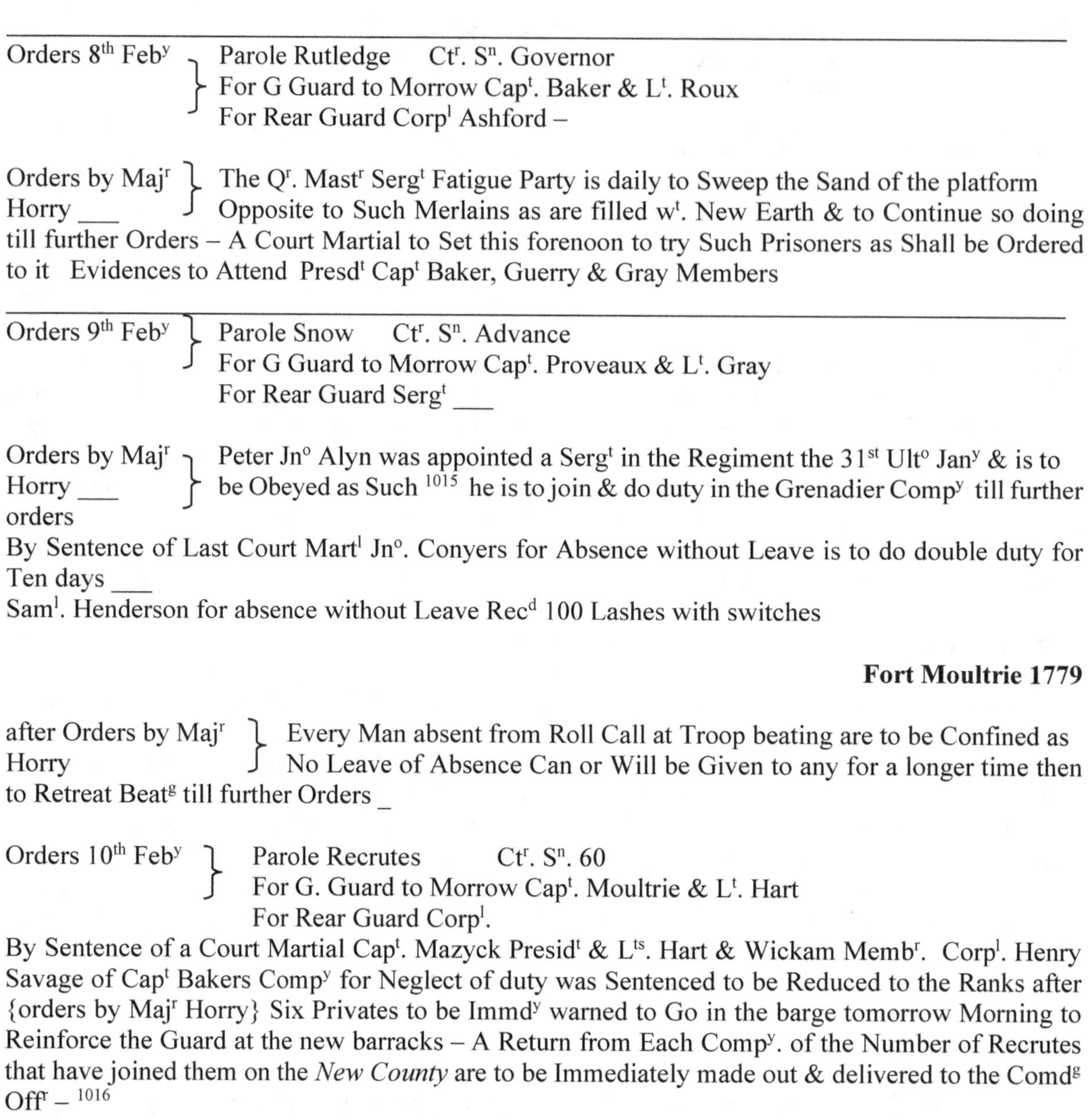

Orders 8th Feb^y } Parole Rutledge Ct^r. S^n. Governor
For G Guard to Morrow Capt. Baker & L^t. Roux
For Rear Guard Corp^l Ashford –

Orders by Maj^r Horry ___ } The Q^r. Mast^r Serg^t Fatigue Party is daily to Sweep the Sand of the platform Opposite to Such Merlains as are filled w^t. New Earth & to Continue so doing till further Orders – A Court Martial to Set this forenoon to try Such Prisoners as Shall be Ordered to it Evidences to Attend Presd^t Capt Baker, Guerry & Gray Members

Orders 9th Feb^y } Parole Snow Ct^r. S^n. Advance
For G Guard to Morrow Capt. Proveaux & L^t. Gray
For Rear Guard Serg^t ___

Orders by Maj^r Horry ___ } Peter Jno Alyn was appointed a Serg^t in the Regiment the 31st Ult^o Jan^y & is to be Obeyed as Such [1015] he is to join & do duty in the Grenadier Comp^y till further orders

By Sentence of Last Court Mart^l Jno. Conyers for Absence without Leave is to do double duty for Ten days ___

Saml. Henderson for absence without Leave Rec^d 100 Lashes with switches

Fort Moultrie 1779

after Orders by Maj^r Horry } Every Man absent from Roll Call at Troop beating are to be Confined as No Leave of Absence Can or Will be Given to any for a longer time then to Retreat Beat^g till further Orders _

Orders 10th Feb^y } Parole Recrutes Ct^r. S^n. 60
For G. Guard to Morrow Capt. Moultrie & L^t. Hart
For Rear Guard Corp^l.

By Sentence of a Court Martial Capt. Mazyck Presid^t & L^ts. Hart & Wickam Memb^r. Corp^l. Henry Savage of Capt Bakers Comp^y for Neglect of duty was Sentenced to be Reduced to the Ranks after {orders by Maj^r Horry} Six Privates to be Immd^y warned to Go in the barge tomorrow Morning to Reinforce the Guard at the new barracks – A Return from Each Comp^y. of the Number of Recrutes that have joined them on the *New County* are to be Immediately made out & delivered to the Comd^g Off^r – [1016]

---

he was elected a captain of the Charles Town Battalion of Artillery and was wounded at Beaufort on 9 February 1779. He was taken prisoner at the fall of Charleston on 12 May 1780 and sent to St. Augustine. He was exchanged during April 1781. He died on 6 March 1809 at the age of 60.

[1015] Peter John Allen (Allyn) enlisted in the 2nd South Carolina Regiment in February 1779. He was promoted to corporal on 1 February 1779 in Captain Peter Gray's company. He was promoted to sergeant on 9 February 1779 and transferred to the Grenadier Company. He was demoted to private on 11 June 1779 and was transferred back to Captain Gray's company. He was in the siege of Savannah.

[1016] The *New County* might have been the name of a ship.

Orders 11th Feby } Parole Col° Giles Ctr. Sn. Charles Town
For G Guard to Morrow Capt. Mazyck & Lt. Roux
For Rear Guard Sergt. Wood [1017]

Orders by Majr Horry } The following Recrutes are Appointed Serjeants in the Regiment & are to be Obeyed as Such.__ __ __ __ __ __ __ __ __ __ __ __
Joseph Turner to join Capt. Harlestons Compy & to do Duty ye Same[1018] __ Hugh Newman to join Capt. Motte Compy & to do duty in ye Same __ Noble Barnard to join Capt Lesesnes Compy & to do duty in ye Same [1019]__ __ __ __ __ __ __ __ __ __ __ __ __ __ __ __
Douglas ONiel to join Capt Moultrie Compy & to do duty in ye Same [1020]__ __ __ __ __
William Wood to join Capt Mazyck Compy & to do duty in ye Same [1021]
No Outward bound Boat or Vessell is to Pass Fort Moultrie till further Orders – The Offr Comdg the Rear Guard is to Suffer no boat with Lime to Pass the Bridge till further Orders[1022] A Court Martial to Set Immediately to try Such Prisoners as shall be Ordered to it Evidences to Attend
Presidt. Capt Baker Members Lt Roux & Guerry

Fort Moultrie 1779

NB} by Sentence of the above Court Martial Moses Bruce for Theft was Sentenced to & did Run the Gauntlet three times thro' the Regimt & is to do double duty for 14 days
James Fitzsimmons for Neglect of duty was Sentenced to Receive 100 Lashes, but his Good Character Recomd him for Mercy & Mercy was approved off by the Comdg. Officer – [1023]

Orders 12th Feby by Majr Horry } A Sergt to take Charge of the Rear Guard & 3 Privates to be Added thereto– the Sergt. will Receive his Orders from the Comdg Offr when Ready to March – fifty Cartridges & 1 doz flints to be Immediately Sent to the Rear Guard for the use of the Same which is to be deliverd to Each Relievg Officer & Reported Accordingly to the Comdg Offr by Each Officer so Relieved __ __ __ __

1017 William Wood served in the 2nd South Carolina Regiment under Captain Daniel Mazyck. He was promoted to sergeant on 11 February 1779. He was in the siege of Savannah.
1018 Joseph Turner was promoted to sergeant on 11 February 1779 in Captain Harleston's company of the 2nd South Carolina Regiment. He later served under Captain Peter Gray in 1779 at the siege of Savannah. He was also at the fall of Charleston. From 7 May to 28 August 1781 he served 113 days in the militia. He and his son also hunted and drove cattle for 62 days under Captain George Robinson.
1019 Noble Barnett was promoted to sergeant in the 2nd South Carolina Regiment on 11 February 1779 in Captain Lesesne's company. He was in Captain Adrian Proveaux's company on 1 July 1779.
1020 Douglas O'Neal had been in the 2nd South Carolina since 1776, but he had been reduced to private on 9 May 1778. He was promoted to sergeant on 11 February 1779 in Captain Moultrie's company.
1021 These men were not new recruits, but had been sergeants in the 2nd South Carolina Regiment during an earlier enlistment. The British had captured Savannah, Georgia in December 1778 and then had made attempts to attack South Carolina in February 1779. The old sergeants and corporals of the 2nd South Carolina decided to return to duty when an invading force was threatening their home.
1022 Lime is a key component in making mortar which is an essential part of any military architectural project.
1023 James Fitzsimmons enlisted in the 2nd South Carolina in 1777. He served under Captain Charles Motte in 1779. On 11 February 1779 he was sentenced to receive 100 lashes for neglect of duty, however mercy was granted him due to his good character. He was at the siege of Savannah.

All Provision Returns to be made out & Sent to the Qr. Master Sergt by 9 OClock in the forenoon in Order that the Same may be Sent to Town by the Provision Boat otherwise no Provision Can be Sent for them & the Qr. Master Sergt is to Receive no Returns after the Provision boat is Gone, unless by Particular Orders from the Comdg Offr – He is Also Ordered to Serve out no Rations to any Compy where no Sergts or Corpls attends to Receive the Same – Comdg Offrs of Compys Are Immediately to make out a Return in the form of a Monthly one of their Respective Compys to the Present time & to deliver the Same to the Comdg Officer

Parole General Lincoln Ctr. Sn. Genl. Moultrie
For G. Guard to Morrow Capt Baker & Lt. Guerry
For Rear Guard Sergt McDonald

Orders to the Sergt of the Rear Guard Sir
You are to take Charge of the Rear Guard, to Post two Centries, one at the Draw Bridge & the other a Little distance in front of & on the Left of your Guard House & towards the Sand Beach – After Tattoe beating you are to Suffer no Person to Pass the Bridge, who Cannot Give the Counter Sign or any boat to Pass the Bridge or to Land near your Guard without Examining the principal Person on board, who you are to Send to the Comdg. Officer of the fort Guard for his Examination & to detain the boat till Orders Received to Suffer the Same to Pass – Your are to Give your Centries Proper Orders for the above Purpose – At Retreat beating your are to Give a Flint & 3 Cartridges to Each Man of your Guard & to Order that they do not Load but when there is a Necessity; & at Troop beating you are to receive what is not Expended & to deliver them as also whatever Else you had in Charge to the Officer that Relieves you & to Report the Same with Every other matter or Circumstance Relative to & during your Guard to the Comdg. Offr of the G. Guard – You are also to Examine Every Person Passing the bridge or boat Landg near your Guard for Rum or other Spiritous Liquors (Officer Servants & the sutler or his Servants only Excepted) & to Send Such Person with Such Liquors found on them to the Comdg. Officer of the fort Guard – Should your Guard at any juncture of time Require a Reinforcement you are Immediately to Send to the fort Guard for the Same – am

Yours &c &c
P. Horry Majr. 2d Regt.

Orders 13 Feby } Parole Commissioners Ctr. Sn. Recruiting
For G: Guard to Morrow Capt. Moultrie & Lt Roux
For Town Guard Capt. Baker & Lt Guerry
For Rear Guard Sergt O'Neil

Orders by Majr Horry __ __ } The following Recrutes are Prefered in the Regiment & are to be a ~~Sergt in Captn Bakers Compy~~ Obeyed Accordingly –
William Jasper to be a Sergt in Captn. Bakers Compy – [1024]
Jno Marlow to be a Sergt in Captn. Halls Compy [1025]
Jno McDonald to be a Corporal in the Grenadier Compy

[1024] Like the other "recruits", William Jasper was not a recruit and was the hero of the Battle of Fort Sullivan, and had been a quartermaster sergeant under Captain Thomas Dunbar.
[1025] John Marlow had been a sergeant in the Regiment before. He was discharged and then re-enlisted on 13 February 1779.

Orders 15th Feby } Parole Head Quarters Ctr. Sn. Removed
For G. Guard to Morrow Capt. Mazyck & Lt Warley
For Rear Guard to Morrow Sergt Cammell [1026]

Orders by Majr Horry ___ } The Majr finds that Notwithstanding the Orders of the 8th of July Last for all Such Men as are Reported Sick in the Morning Reports to be Carried to the Surjeon or his Mate for Examination; the Same is Greatly Neglected, by which means many do Sulk duty with Impunity & Good Men of the Regt Suffer by doing their Duty – It is therefore Ordered that the Surgn or his Mate do daily at 12 OClock Report to the Comdg Offr all Such Men as on Examination he finds fit for Duty as also the Names & Compys of Such Men as are Reported Sick on the Compy Morning Reports but were not brought for Examination & as it is the duty of Every Orderly Sergt or Corpl for the day to Carry the Sick of their Respective Compys for Examination; Such as Neglect the Same may depend on Suffering the Sentence of a Court Martial

Harris Dewet of Capt Mottes Company is appointed a Corpl. in ye Same[1027] Daniel Andrews of Capt Proveaux Compy is appointed a Corpl. in the Same [1028] they are to be Obeyed as Such. –

A Court Martial to set Immediately to try all Prisoners Ordered to it Evidences to Attend __ Presdt. Capt Proveaux Members Lts Guerry & Wickham –

By Sentence of the above Court Martial, James Hooper for Rioting & drunkenness in Town was Sentenced to Receive 100 Lashes wt. Switch but was Pardoned by the Comdg. Officer _

Sergt. Allen McDonald for Drunkenness on Guard & Quitting the Same without Leave was Sentenced to be Reduced to the Ranks & to Receive 30 Lashes – The former was Confirmed but the Latter Punishment was Remitted by the Comdg Offr

John Gaught for Drunkenness on Guard was Sentenced to Receive 50 Lashes, but was Remitted by the Major[1029]– Corpl. Stewart for Drunkenness on Guard was Reduced to a Private – [1030]

Saml. Henderson for Loosing his Regtl Clothing is to be Put under Stoppages till the Same are Replaced –

Fort Moultrie 1779

Orders 16th Feby } Parole Militia Ctr. Sn. Relieved
For G: Guard to Morrow Capt. Proveaux & Lt Guerry
For Rear Guard to Morrow Sergt ONeil __

Orders by Majr Horry __ } The Qr. Master Sergt is to deliver no More Woolen Waistcoats & Breeches to Recrutes that join the Regt. as Summer is near they will Receive the Same of Linnen –

1026 James Gamble.

1027 Harris Dewitt was born in 1756. He was promoted to corporal in the 2nd South Carolina Regiment under Captain Motte on 15 February 1779. He was in Dunbar's Light Infantry company at the siege of Savannah.

1028 Daniel Andrews was promoted to corporal in the 2nd South Carolina Regiment on 15 February 1779. He was in Captain Proveaux's company. He was somehow captured by the British on 1 September 1779, but made his escape and rejoined the Regiment. On 7 October 1779 he was reduced in rank and he received 75 lashes on the bare back with a cat of nine tails for insolent behavior and abusive language to Doctor Theus.

1029 John Gaught (Gought) was in the 2nd South Carolina Regiment. On 15 February 1779 he was sentenced to receive 50 lashes for being drunk on guard, but it was remitted by Major Horry.

1030 David Stuart.

The Sergt Major is Every day after the Orders of the Same is Given out to Parade all New Appointed Sergts & Corpls & to Instruct them in their Exercise & every other Part of duty Incumbent on them to Know & to Continue So doing till further Orders –
A Court Martial to Set this forenoon to try Such Prisoners as Shall be Ordered to it Evidences to attend __ Presdt. Capt Moultrie, Members Lts Roux & Guerry __ __ __ __
By Sentence of the above court martial Jno Conner of Capt. Masons Compy. was Sentenced to be Reduced & to do duty as a private Soldier for Drunkenness on Guard By Sentence of the Same Court Martial Edward Murphy of Capt. Moultrie Compy was Sentenced to be stopped of one months pay to be Used for the Regtl Hospital –

Orders 17th Feby } Parole Capt Tomplat Ctr. Sn. Advance
For G. Guard to Morrow Capt. Moultrie & Lt Roux
For Rear Guard to Morrow Sergt Murphy
Orders by Majr Horry __ } Three Men to be this Morning to be added to the Rear Guard – Comdg Offrs of Compys are to Apply to the Qtr Master Sergt for a Pouch for Each Recrute of their Respective Compys & to Give a Receipt for the Same –
The Grenadier Compy is to Return their Present Arms into the Regtl Store & to Receive a New Set there from wc. the Comdg Officer of that Compy is to Give the Qtr. Master Sergt. a Receipt for – [1031]

Orders 18th Feby } Parole Genl. Moultrie Ctr. Sn. Chas Town
For G Guard to Morrow Capt. Mazyck & Lt. Martin
For Rear Guard Sergt Roberts
Orders by Majr Horry __ } The Qtr. Master Sergt. is to Serve to Each Recrute as they join the Regt a Linnen Waistcoat & Breeches instead of Wollen ones

Moultrie 1779

Orders 19th Feby } Parole Pinckney Ctr. Sn. Friend –
For G Guard to Morrow Capt. Proveaux & Lt. Gray
for Rear Guard Sergt Campble

Orders 20th Feby } Parole Command Ctr. Sn. 150 –
For G. Guard to Morrow Capt. Motte & Lt Roux
For Rear Guard to Morrow Sergt Marlow –
Orders by Majr Horry ___ } A Court Martial to set Immediately to try Such Prisoners as Shall be ordered to it Evidences to attend
Presdt. Capt Mazyck Members Lt Guerry & Wickham __
after orders by Majr Horry } Comdg Officers of Compys are as Expeditiously as Possible to make a Return of What Cloathing their Respective Compys have Received out of the Last made Cloathing for the Regt; also to Receive from their Recrutes the New Pouches & to deliver the Same to the Qr. Master Sergt. who is to Return them to Town as unfit for Service – All the Men of the 5th Regt now at this Post that are fit to March are to hold themselves in Immediate Readyness. –

[1031] This may mean that the Grenadier company was now equipped with the French musket, like the rest of the regiment.

Orders 21st Feb^{y} }	Parole Detachment Ct^{r}. S^{n}. 2^{d} Reg^{t} For G. Guard to Morrow Capt. Mazyck & Lt Warley For Rear Guard to Morrow Serg^{t} Stewart For Town Guard to Morrow Capt. Proveaux & L^{t}. Martin
Orders by Maj^{r} Horry __ }	No Men for the Grenadier or Light Infantry Companies to be taken for Guard tomorrow.

Orders 22 Feb^{y} }	Parole Col^{o}. Horry Ct^{r}. S^{n}. 4 For G Guard to Morrow Capt Baker & Lt Roux For Rear Guard Serg^{t} Martin

---

Orders 23^{d} Feb^{y} }	Parole Maj^{r} Horry Ct^{r}. S^{n}. 1 For the day to Morrow ~~Capt~~ L^{t}. Warley For G Guard to Morrow Serg^{t}. Stewart For Rear Guard Serg^{t} Feast .

Genl. Orders 22nd Feby by B. Genl. Moultrie } Sir you will order from your Reg^{t} one field Off^{r}. 2 Drums 2 fifes with one hundred & fifty rank & file to March to Purisburgh with all Expedition you will apply to Col^{o}. Drayton for flints, Kettles Waggons & all Nessesarys that may be wanted for their march

I am S^{ir}. Y^{r}. H^{bl}. Servt. Wm Moultrie B. G.

To Lt Col^{o}. Fr^{s} . Marion

General Moultrie had received information that the British had left 200 men in Savannah and were going to march upon Charlestown. Moultrie ordered the 2nd South Carolina Regiment to reinforce Lincoln's army at Purisburgh, Georgia to block any movement by the British.

There were two armies in the field to stop the British. Lincoln's army of 4,000 men at Purisburgh and Ashe's army of 2,300 men at Briar Creek. British Lieutenant Colonel James Mark Prévost had withdrawn his army of 5,000 out of Augusta and set a trap for Ashe at Briar Creek. Ashe followed and was totally defeated by Prévost suffering around 200 killed, 41 wounded, 173 captured and 106 missing. The British only lost 5 killed and 11 wounded.

Between the loss at Savannah and the defeat at Briar Creek the Georgia Continental line ceased to exist. One historian wrote that Briar Creek can be considered one of the bloodiest, one sided engagements in American military history. The British lost less than 1% of their force, while the Americans lost 30% of theirs. Moultrie wrote that losing this battle prolonged the war by another year, and let the British take South Carolina.[1032]

[1032] Full details of the Battle of Briar Creek are described in "*Nothing but Blood and Slaughter, Volume One*".

Fort Moultrie 1779

Reg[tl] Orders 23[d] Feb[y] by L[t]. Col. Marion – } For the above Com[d] Cap[ts] Lisesne, Moultrie, Dunbar & Baker, with their Officers & men belonging to their Comp[ys], to be Com[d] by Major Horry – The Grenad[r]. & Light Infantry Comp[ys] to Consist of 3 Serg[ts]. 1 Drum 1 Fife & 40 rank & file Cap[t] Moultries & Bakers to be 2 Serg[ts] & 35 Rank & file, they are to be taken from the Recruits belonging to the Remaining Battalion Comp[y] to Compleat that number in each Comp[y]. the men are to be completed with Cloathing, a p[r]. Gaiters & one p[r]. Stockings – This Detachment to be ready to March by Thursday morning Early, when they are to be furnished with a powder horn, ¼[w/] powder & 12 dozen Ball p[r]. Man, pouches will be given them as soon as ready – The Q[tr] M Serg[t]. to git ready 6 Ammunition Chests Containing 2500 Cartridges Each & 1000 flints to be carried with the Detatchment – [1033]

L[t]. Peter Gray is Appointed Cap[t] in the Reg[t]. in room of Cap[t]. Is: Harleston promoted to a Maj[r] in 6th Reg[t] he is to be Respected & Obeyed as such ____________

Capt[n]. Pet[r]. Gray to take the Com[d] of the Comp[y]. late Capt[n]. Harleston ____

M[r]. Alex[d]. Petrie is appointed a first Lieu[t]. in the Reg[t] he is to be respected & Obey'd as Such ____

L[t]. Petrie to Join & do duty in Capt[n]. Dunbars Comp[y]. till further orders & is to march with that Comp[y].

M[r]. Alex[d]. Hume is Appointed an Second Lieu[t]. in the 2 Reg[t]. & is to be respected & Obeyd as such __

Lieu[t]. Hume is to Join & do duty in Capt[n]. Lesesne Comp[y] till further Orders & to March with his Comp[y] ____

The Battalion to parrade tomorrow morn[g]. at 8 OC: - no Off[r]. or man that is order for Com[d] to be put on Duty tomorrow & if any of Such men is on Guard they must be Reliev[d]

[1033] This detachment under Major Peter Horry was to march to General Ashe at Briar Creek. Major Horry and his detachment arrived too late to take part in the battle. General Moultrie in his memoirs wrote, "General Ash's affair at Brier-Creek, was nothing more than a total rout; never was an army more completely surprised, and never were men more panic struck… most of them threw down their arms, and run through a deep swamp, 2 or 3 miles, to gain the banks of a wide and rapid river, and plunged themselves in, to escape from the bayonet; many of them endeavoring to reach the opposite shore, sunk down, and were buried in a watery grave; while those who had more strength, and skill in swimming, gained the other side, but were still so terrified, that they straggled through the woods in every direction; a large body of them were stopped early the next morning at Bee's-creek bridge, about 20 miles, by a detachment of the second regiment, under Captain Peter Horry, marching to camp, who told me he had just heard of the affair at Briar-Creek, and saw a large body (2 or 300) of the fugitives coming in a hasty and confused manner, most of them without their arms, and Gen. Ash and Bryant with them…drew up his men at the bridge: Gen. Ash rode up to him, and requested that he would stop those men; that they were running away: Gen. Bryant said they were not running away; Gen. Ash insisted they were; Capt. Horry then asked of the two generals who was commanding officer; it was answered Gen. Ash: then, sir, I will obey your orders: and presented fixed bayonets, and threatened to fire up on the fugitives, if they attempted to come forward, which stopped them: afterwards Capt. Horry proceeded to camp, with his detachment, and Gen. Ash and Bryant brought back the fugitives."

Fort Moultrie 1779

Orders 24th Feby ___ } Parole Detatchment Ctr. Sn. Ready
For the day to Morrow Captain Mazyck
for G Guard to Morrow Sergts Taylor & McDowell
For Rear Guard to Morrow Sergt. Stewart

Orders 24th The troops going to head Quarters to be Completed with havre Sacks, this Detatchment to be furnished with 27 tin Kittles; belts, pouches will be sent them & horns to Compleat each man & if there is a Number of Stockings & Shirts sufficient they are to have a pr. Stockings & 1 Shirt pr. Man, to begin by the Light Infantry Compy

Orders 25th Feby } Parole Williamson Ctr. Sn. Success
For the day to Morrow Lt. Warley
For G. Guard to Morrow Sergt Davis & Camble
For Rear Guard to Morrow Sergt. Feast

Orders by Capt. Motte } James Clackworthy [1034] of Capt. Masons Compy is Appointed a Corpl in the Same he is to be Obeyed as Such __ __ __ __

Thos. Galloway of Capt. Grays Compy is Appointed a Corpl in the Same he is to be Obeyed as Such — — — — — [1035]

A Court Martial to Sett Immediately for the Tryal of Sergt Murphey of my Compy for Gaming Contrary to the orders Issued by Genl Washington [1036] – Presidt Capt Hall Members Lts Warley and Guerry _______

NB This Court was Countermanded ____________

Orders 26th Feby by Capt Motte } Parole Augusta Ctr. Sn. 16
for the Day to Morrow Lt. Guerry
for G. Guard to Morrow Sergt Murphy
For Rear Guard Sergt Marlow

Orders 27th Feby Capt Motte } Parole Majr Horry Ctr. Sn. Honour
for the Day to Morrow Capt. Mazyck
for G. Guard to Morrow Sergts Feast & Davis
For Rear Guard to Morrow Sergt McDonald

Fort Moultrie 1779

1034 James Clatworthy was in the 2nd South Carolina Regiment under Captain Richard Mason. He was promoted to corporal on 25 February 1779. He was promoted to sergeant on 14 March 1779. He was in Captain Mason's company at the siege of Savannah.

1035 Thomas Galloway enlisted in the 3rd South Carolina (Ranger) Regiment on 20 November 1776. He enlisted in the 2nd South Carolina Regiment on 2 February 1779 and was promoted to corporal on 24 February 1779 under Captain Peter Gray. He was at the siege of Savannah.

1036 There are several orderly books from the different companies of the 2nd South Carolina Regiment. In one of the orderly books this passage is written as "Captain Gray's company", in this particular orderly book it is listed as "my" company, which means that Captain Gray was the person who kept this orderly book for his company.

Orders 28th Feb[y] by Cap[t] Motte } Parole Drayton Ct[r]. S[n]. 5
For the day to Morrow Cap[t]. Proveaux
For G. Guard to Morrow Serg[t]
For Town Guard to Morrow Cap[t]. Mason & L[t]. Warly
For Rear Guard to Morrow Serg[t] _

A Monthly return to be made tomorrow by 10 OC & Given in to the Adjutant – the Serg[t]. Maj[r] will make a return of what men are left from the Four Comp[ys] that are gone on Command and Give in to the Com[dg] Officer–

Likewise What Number of Men Women and Children of the 5th Reg[t]. likewise of Such Children as belong to Either Reg[t]. that is above Nine Years of Age –

Orders 1st March Cap[t] Motte } Parole Lawrence Ct[r]. S[n]. 4 [1037]
for the Day to Morrow Cap[t]. Gray
for fort Guard Serg[t]. M[c]Dowell & Feast
for Rear Guard Serg[t]. Taylor

Orders 2d March Cap[t] Motte } Parole Lynch Ct[r]. S[n]. Friend
for the Day to Morrow L[t]. Martin
for G. Guard Serg[t]. Wood
for Rear Guard Serg[t] Hausman

Orders 3 March L[t] Col. Marion } Parole Hutson Ct[r]. S[n]. 8
for the day to Morrow L[t] Warley
for G. Guard Serg[t] Campbell & Turner
for Rear Guard Serg[t] Davis ___

Orders 3d March L[t] Col Marion } No Officer to take a Soldier as a Servant that is a man who are able to do Duty nor no off[r] is to take any Soldier as a Servant ~~nor no Officer is to~~ who Draws Rations for a Black Servant No Officer to give Leave of Absents to any Soldier But the Comd[g] off[r] for the time being; it tis Expected the above Orders will be Punctually Observed till further Orders M[r]. Christopher Rogers Jun[r]. is appointed an 2 Lieu[t]. in the Reg[t]. & is to be respected & Obeyed as Such, his commission to be Dated 28th Feb[y]. 1779

Fort Moultrie 1779

2d Lieu[t] Rogers is to join & do duty in Capt[n]. Motte Comp[y] till further Orders –
L[t]. Guerry is to Join & do duty in Capt[n]. Halls Comp[y] till further Orders –
all the men & soldiers wives belonging to those Comp[ys] gone on Command are to do duty & Draw Rations in Capt[n]. Mottes Comp[y] till further Orders Including the men belonging to the fifth regiment now in Garrison
A Court Martial to Sett this Morning to try Such Prisoners as May be brought to it Evidences to Attend Pres[dt] Cap[t]. Mazyck Members Cap[t]. Gray and L[t] Warley
Orders by Cap[t] Motte

[1037] In the other company orderly books they have a countersign of "One" for this day. On several occasions the countersigns are different from each orderly book, but the rest of the information is the same, and even the parole is the same. Each company may have had their own countersign for each day, though on some days the countersigns are the same for all.

2nd South Carolina
Regiment
1779

Orders 4th March } Parole Col°. Marion. Ctr. Sn. Friend
For the day to Morrow Capt Mazyck
For G. Guard to Morrow Sergt Stewart & Wood
For Rear Guard Sergt Turner –

Orders 4th March Motte } A Court Martial to Sett this Morning to Try All Such Prisoners as by Capt. May be brought to it Evidences to Attend – President Capt. Mazyck Members Capt. Proveaux & Lt. Martin The Sergt. and Corpl will See their Men Clean and their Hair Well Combed, as no more Powder Can be had for them –

Orders 5th March } Parole Middleton Ctr. Sn. 4
for the day to Morrow ~~Capt~~. Lt Martin
for G. Guard Sergt Davis & Campl
for Rear Guard Sergt Feast

Orders 6th March } Parole Heyward Ctr. Sn. 3
for the Day tomorrow Lt Warley
for G. Guard to Morrow Sergt Taylor & McDowell
for Rear Guard Sergt Wood

Orders by Capt. } A Court Martial to Sett this Morning at 10 OClock to try Such Prisoners as may be Ordered to it Presidt Capt. Gray Members Lts. Warley and Wickam _ _ _ _ _

Robert Mathews of Capt. Proveaux Compy is Appointed a Corpl in Capt. Hall's Compy he is to be Obeyed Accordingly –

NB Agreeable to Sentence of Court Martial that set March ye. 4th Andrew De Bland of Capt Masons Compy for Desertion was sentenced to receive 100 Lashes with the Cat a Nine Tails - The Comdg Offr was pleased to Remit 40 – By sentance of the same Court Abraham Berlin of Capt Grays Compy for Desertion Received 50 Lashes - Also Richd Phillips of Capt. Hill Compy. for suffering an Apron to be stole off of the Gun While on Centinel was Sentenced to Receive 25 Lashes –

Orders 7th March Capt Motte } Parole Majr Harleston Ctr. Sn. 6
for the Day to Morrow Lt. Guerry
for G Guard to Morrow Sergt Campbell & Hausman
for Rear Guard Sergt Turner
for Town Guard to Morrow Capt. Mazyck & Lt. Warley

Agreeable to the Sentence of the Court Martial that Sett Yesterday Joseph Pain of Capt. Gray's Compy Was Sentenced to received 50 Lashes on the Bare Back with a Catt of Nine Tails and Recommended to Mercy [1038] – Sergt. Stewart By the Same Court was Sentenced to be Reduced to the ranks as unworthy of that office The Comdg Offr Approved of the Above Sentences & orders the Said Stewart do Duty as a Private in his Compy –

[1038] Joseph Pain enlisted in the 2nd South Carolina Regiment on 9 July 1775 under Captain Bernard Elliot. During 1779 he was under Captain Peter Gray. On 7 March 1779 he was sentenced to receive 50 lashes on the bare back with a cat of nine tails for some unnamed crime. He was recommended mercy. He deserted on 1 October 1779 during the siege of Savannah.

Orders 8th March — Parole Recruits Ct^{r}. S^{n}. 9
for the day to Morrow Capt. Mason
for G. Guard to Morrow Serg^{t} Murphy & Wood
for Rear Guard Serg^{t} Marlow -

Orders by Capt. M. — The Comd^{g} Officer is Much Surprised to See how Very Slovenly the Reports are Given in the Morning, He Expects that they will be more Circumspect – Likewise that the Officers and Men will Attend More to Standing Orders Issued the 17th Augt Last

Ordered that the Officer of the Day Constantly Stay in the Garrison Except when he Visits the Different Guard till relieved, when he will make out a Report in Writing of the Different Guard & What May be Particular During his time of Duty –

Orders 9th March
Capt Motte — Parole Expedition Ct^{r}. S^{n}. Liberty
For the Day to Morrow L^{t}. Martin
for G. Guard to Morrow Serg^{t} McDowell & Feast
for Rear Serg^{t} Marlow

Orders 10th March
Capt Motte — Parole Ashepoo Ct^{r}. S^{n}. 59
for the day to Morrow Capt. Hall
for G Guard Serg^{t}. Murphy & Wood
for Rear Guard Serg^{t} Marlow

Capt Motte **Fort Moultrie 1779**

Orders 11th March — Parole Williamsburgh Ct^{r}. S^{n}. 76
for the day to Morrow Capt. Mason
for G Guard to Morrow Serg^{t} Taylor & McDowell
for Rear Guard Serg^{t} Campbell

Orders 12th March — Parole Happiness Ct^{r}. S^{n}. 13
for the day to Morrow Lt Martin
for G Guard Serg^{t}. Murphy & Wood
for Rear Guard Serg^{t} Marlow

Orders by Capt Motte — A Court Martial to Sett this Morning to try Such Prisoners as May be brought before it President Capt Hall Members Lt Martin and Ensign Rogers ___

Orders 13th March — Parole Hampton Ct^{r}. S^{n}. Motte
For the day to Morrow Lt Guerry
For G. Guard to Morrow Serg^{t} Murphy & Taylor
For Rear Guard to Morrow Serg^{t} McDowell

Orders 14th March — Parole Maj^{r} Pinckney Ct^{r}. S^{n}. 13
for the day to Morrow Capt Proveaux
for Town Guard to Morrow Capt Hall & Lt Martin
for G. Guard to Morrow Serg^{t} Murphy & Clatworthy
for Rear Guard Serg^{t} Matthews

Orders by Capt. Motte } Corpl James Clatworthy is Appointed a Serjt. in Capt. Mason's Compy he is to be Obeyed as Such – George Brewton of Capt Mottes Compy is Appointed a Corpl in Capt Mason's Compy he is to be Obeyed as Such [1039] Corpl Robert Mathews of Capt Hall's Compy is Appointed a Sergt in Capt Proveaux Compy he is to be Obeyed as Such –

Orders 15th March } Parole Colo. Henderson Ctr. Sn. 49
for the Day to Morrow Capt Mason
for G. Guard to Morrow Sergt Taylor & McDowell
for Rear Guard Sergt Campble

Orders by Capt Motte } John Burtell of Capt Proveaux Compy is Appointed a Corpl in Capt Hall Compy he is to be Obeyed and Respected as Such –

Orders 16th March } Parole Harnett Ctr. Sn. 39
for the Day to Morrow Lt Warley
for G. Guard to Morrow Sergt Turner & Clatworthy
for Rear Guard to Morrow Sergt Haisemann

Orders by Capt Motte } The Articles of War to be Read this Afternoon to the Battalion and Soldiers are to Attend –

Lt Colo Marion **Fort Moultrie 1779**

Orders 17th March } Parole Liberty – Ctr. Sn. America
for the Day to Morrow Lt Guerry
for G: Guard Sergt Murphy & Taylor
for Rear Guard to Morrow Sergt McDowell

Orders by Capt Motte } A Court Martial to Sett this Morning to try Such Prisoners as may be brought before it Evidences to Attend – Presdt Capt Mazyck Members Capt Mason & Lt. Guerry

---

Orders 18th March by Lt Colo Marion } Parole Briar Creek Ctr. Sn. Genl. Elbert
for the Day to Morrow Ensign Rogers
for G. Guard to Morrow Sergt Campbell & Haisman
for Rear Guard Sergt Turner

The Orders of the 30th January to be strictly Observed – Corpl Jno Burtel of Capt Hall Compy is Appointed a Sergt in the Same he is to be Obey'd as Such & George Valley of Captn. Masons Compy. is appointed a Corpl in Captn. Halls Compy. he is to be Obey'd as Such –[1040]

[1039] George Brewton (Bruton) was born on 21 February 1745. He enlisted in the 2nd South Carolina Regiment under Captain Charles Motte on 24 February 1779. He was promoted to corporal on 14 March 1779 in Captain Mason's company. On April 3rd he was promoted to sergeant and served in Captain Motte's company again. He was in the siege of Savannah in 1779. He later served in the militia. He died on 28 August 1815 at the age of 70.

[1040] George Valley was in Captain Mason's company of the 2nd South Carolina Regiment in 1779. He was promoted to sergeant on 18 March 1779 and transferred to Captain Hall's company.

Marmaduke Average of Capt[n]. Mazycks Comp[y]. is Appointed a Corp[l] in the same he is to be Obey'd as Such __ __ __[1041]
William Oliver of Capt[n]. Mottes Comp[y] is Appointed a Corp[l] in the same he is to be Obey'd as such __ __ __[1042]
Agreeable to the Last Court Martial
Thomas Crosier of Cap[t] Halls Comp[y] Reci[d] 50 Lashes on the bare back with a Cat of nine Tails For absence with out Leave ~
Ralph Ingram of Cap[t] Masons Comp[y] was Reduced to the Rank of private Soldier For Drunkenness on Guard

---

Orders 19th March } Parole Command Ct[r]. S[n]. 2
for the Day to Morrow Cap[t] Mazyck
for G Guard Serg[t] Burtell & Mathews
for Rear Guard Serg[t] Clotworthy

Orders by Cap[t] Motte ___ } The Q[r]. Master Serg[t] will deliver out to the Different Comp[ys] a Coat and a p[r]. Breeches to Such Men as the Comd[g] Off[r] of the Different Comp[ys] think Most Proper – Such of the Grenadiers as are fit for Duty are Imediately to furnished and Completed with Arms Cloathing and Accrutrements A Cap[t] one L[t]. 1 Serg[t]. 2 Corp[l] and 18 Privates to be in Readiness to March tomorrow to Head Quarters, this Party to be Provided with 18 Rounds p[r]. Man, and Haversacks & 3 tin Kettles – The Grenadiers to be Included in the Number –
Cap[t] Proveaux and L[t]. Martin for this Command –

Orders 20th March } Parole Cap[t] Proveaux Ct[r]. S[n]. 51
for the Day to Morrow Cap[t] Mason
for G. Guard to Morrow Serg[t] Teague & M[c]Dowell
for Rear Guard Serg[t] Feast

Orders by Cap[t] M } The Q[tr] Master Serg[t]. is Immediately to have all the Rooms in the Square that are Not Locked up Cleaned out and the Keys to be Delivered to the Comd[g] Off[r] – He will Likewise have All the Chimneys Swept and the Garrison Cleaned Out

---

Orders 21st March } Parole Victory Ct[r]. S[n]. 45
For the Day to Morrow Ensign Rogers
For Town Guard Cap[t] Mason & L[t]. Guerry
For G: Guard to Morrow Serg[t] Murphy & Turner
for Rear Guard Serg[t] Burtell

1041 Marmaduke Ethridge served under Captain Daniel Mazyck in the 2nd South Carolina Regiment. He was promoted to corporal on 18 March 1779. He was at the siege of Savannah.

1042 William Oliver enlisted in the 2nd South Carolina Regiment on 9 February 1779 under Captain Charles Motte. He was promoted to corporal on 18 March 1779. He was killed in the siege of Savannah on 9 October 1779.

Orders 22nd March } Parole Congress Ctr. Sn. 92
for the Day to Morrow Lt. Warley
for the G Guard to Morrow Sergt Murphy & Taylor
for Rear Guard to Morrow Sergt Burtell

Orders by Capt Motte ___ } A Court Martials to Sett this Morning at 10 OC to Try all Such Prisoners as may be brought to it.
Presidt Capt Mazyck Members Capt Gray & Ensign Rogers
after orders by Capt Motte } The Rear Guard to be Reinforced to Morrow With 3 Privates – The Sergt. of that Guard Will Plant one Centry More –

Orders 22d March Capt Motte } Parole Union Cn. Sn. America
for the Day to Morrow Captn. Mazyck
for G Guard to Morrow Sergt Taylor & Feast
for Rear Guard to Morrow Sergt Davis

Orders 23d March Capt Motte } Parole Congress Cn. Sn. 92
for the Day to Morrow Capt Hall
for G. Guard to Morrow Sergt Haiseman & Turner
for Rear Guard to Morrow Sergt Wood __

Orders 24th March Capt Motte } Parole Orangeburgh Cn. Sn. Camp
for the Day to Morrow Lt Warley
for the G: Guard to Morrow Sergt Murphy & Taylor
for Rear Guard to Morrow Sergt Burtell

Orders by Capt Motte } A Court Martial to Sett this Morning at 10 OClock to try All Such Prisoners as May be brought to it President Captn Mazyck
Members Capt Gray and Ensign Rogers–
After Orders by Capt Motte} The Rear Guard to be Reinforced to Morrow With 3 Privates – The Sergt. of that Guard will Place one Centry More –

Orders 25th March } Parole Kinlock Cn. Sn. 37
for the day to Morrow Ensign Rogers –
for G. Guard to Morrow Sergt Feast & Davis
for Rear Guard to Morrow Sergt Cample

Orders by Capt M } The Battalion to Exercise every every Munday and Thursday at the Small Arms and Wednesdays at the Cannon as formerly – The Compys are to be Posted as follow Vizt Capt Mottes & Mazyck's in the two front Bastions from Right to Left – Capt Hall and Capt Proveaux on the Right and Left of the front Platform and Capt Masons & Gray in the Square and Magazine Bastions in the Same Order till Otherwise Countermanded – The Eldest Officer on Parade to Command till further Orders __ __ __ __ __ __ __ __ __ __ __

Agreeable to the last court Martial
John Hyrn Drum^{r} of Capt Halls comp^{y} confind by Captn. Mazyck for Disobedience of Orders the Court warr of opinion the prisoner was Guilty of the Charges & did Sentence him to Rec^{d} 25 lashes on the Bare back with a cat of nine Tails But on accounts of his former Good behaviour did Recommend him to mercy _______
Con^{d} Fitner of Capt Halls Comp^{y} Reciv^{d} 25 Lashes on the Bare back with a Cat a nine Tails for Disobedience of orders ______________
James Smith of Capt Halls Comp^{y} Reciv^{d} 25 lashes on the bare back with a Cat a nine tails for Willfully abusing his Musket ________________
Thomas Welch of Capt Halls Comp^{y} Rec^{d} 25 lashes on the Bare Back with a Cat a nine Tails for Wilfully Abusing his Musket __ Joseph Wright of Capt Proveaux Comp^{y} Rec^{d} 20 Lashes on the Bare back with Cat a nine Tails for Neglect of duty

---

Orders 26th March } Parole McKintosh C^{n}. S^{n}. 5
For the Day to Morrow Lt Warley
for G. Guard to Morrow Serg^{t} Murphy & Tayler
for Rear Guard Serg^{t} McDonald

Orders by Capt Motte } A Court Martial to Sett this Morning to try Such Prisoners as May be Ordered to it Presd^{t} Capt Gray Members L^{t}. Wickam Ensign Rogers
Evidences to Attend – Agreeable to the above court Martial Serg^{t} Campble of Capt Halls comp^{y} was Suspended For one month & to do duty as a private Soldier for Drunkenness:
Corp^{l} Kearslick for Drunkenness was Sentenced as above

Orders 27th March
Captn Mazyck } Parole Col^{o} Marion C^{n} S^{n} Merit
for the Day to Morrow Ensign Rogers
for G. Guard to Morrow Serg^{t} Feast & Davis
for Rear Guard Serg^{t} Hausman –

Orders by Capt Mazyck Jacob Davis of one of the Georgia Battalions is to do duty in the 2^{d} Reg^{t} Till further orders he is to join cap^{t} Mazycks Comp^{y} _______
What Reg^{tls} Coats are left in the Store The Q^{r}. M. Serg^{t}. must Issue them to the old Soldiers – Particularly to the Barge Men
A court martial to Set immediately to Try Such prisoners as are brought before it – Evidences to attend Presd^{t} Capt Hall Members Lieut^{t} Wickham & ~~Ensign~~ Lieutenant Rogers __ __ __ __ __ __ __ __ __ __

Orders 28th March
Captn Mazyck } Parole Govr. Rutledge C^{n} S^{n} 79
For the Day to Morrow Captn Hall
for G Guard to Morrow Serg^{t} McDowell & Taylor
for Town Guard to Morrow Capt Mayzyck & Rogers
for Rear Guard to Morrow Serg^{t} Wood

Orders 29th March } Parole Windsor C^n^ S^n^ Camden
for the day to Morrow Captn Mason
for G. Guard to Morrow Serg^t^ Feast & Davis
for Rear Guard to Morrow Serg^t^ Haseman

Orders by Capt M. } The Q^r^. Master Serg^t^ will See the Orders of the 17th October Last Strictly put into Execution ______

Agreeable to the last court Martial

Jno Taylor of Capt Mottes Comp^y^. Rec^d^ 100 lashes on the Bare back with Switches for loosing His Reg^t^ Cloathing and to be put under Stoppages until he payes for the Clothes

Captn Motte **Fort Moultrie 1779**

Orders 30th March } Parole Thompson C^n^ S^n^ 3
for the day to Morrow Lt Warley
for G. Guard Serg^t^ Turner & Wood
for Rear Guard Serg^t^ Clatworthy

Order by Capt Motte } A Court Martial to Sett this Morning at 10 OClock to try such Prisoners as May be brought to it Presd^t^ Capt Mazyck Members Capt Gray and L^t^. Guerry – Evidences to Attend – The Command^g^ Officer find that the Relieving of the Town Guard Interferes Much with the Battalion Exercise on Munday Therefore Thinks Proper to Alter the Same to Tuesday the other Day or before Ordered –

Orders 31st March } Parole Henderson C^n^ S^n^ 99
for the Day to Morrow Lt Guerry
for G. Guard to Morrow Serg^t^ M^c^Dowell & Feast

Orders by Capt M } for Rear Guard to Morrow Serg^t^ Turner
Jacob Kalckoffen is Appointed a Serg^t^ in Capt Mazyck's Comp^y^ he is to be Obeyed as Such __ __ __ __ __ __ __ __ __ __ __ __ [1043]

A Monthly return to be made of Each Company and Delivered to the Adjutant to Morrow by 10 OClock __ __ __ __ __ __ __ __ __ __ __ __ __ __ __

Agreeable to the last court martial Jacob See of Capt Masons Comp^y^. was Sentenced to Recive nintynine lashes on the back with a cat a nine Tails for absence without leave [1044] The Comd^g^ officer approved of the Sentence But Remited 60 lashes as the Doct^r^. having reported that the Prisoner Could Not Receive the Remainder Jacob See Sentenced to be put under Stoppages For the musquit he lost

— — — — — — — — — — —

[1043] Jacob Kalckoffen was in the 2nd South Carolina Regiment under Captain Daniel Mazyck. He was promoted to sergeant on 31 March 1779. He was at the siege of Savannah.

[1044] Jacob See was in the 2nd South Carolina Regiment under Captain Mason. On 31 March 1779 he was sentenced to 99 lashes on the bare back with a cat of nine tails for absence without leave. Sixty lashes were remitted because the doctor said that Jacob See could not survive. He also was put under stoppages for a musket he lost while being absent without leave. On 5 December 1779 he received 100 lashes on the bare back with a cat of nine tails for deserting and attempting to escape to Charleston after the failed assault on the Spring Hill Redoubt.

Orders 1 April } Parole Santee   Cn Sn 42
for the day to Morrow Capt Gray
for G. Guard to Morrow Sergt. Davis & Haisman
Orders by Capt Motte } For Rear Guard to Morrow Sergt. Turner
The Long Roll to beat at 5 OClock in the afternoon and the Troop at 7 in the Morning till further orders – A Court Martial to Sett this Morning to try Such Prisoners as may be brought before it Evidences to Attend – Presdt Capt. Mazyck Members Lt Warley & Guerry

---

Orders 2 April Lt Colo. Marion } Parole Winyaw   Cn Sn 34
for the Day tomorrow Capt Mason
for G. Guard Sergt. Matthews & Kalcoffen
Rear Guard Sergt Bartell ___
Order by Capt Motte } A Court Martial to Sett this Morning at 10 OClock at the Presidt appartment for the Tryall of Such Prisoners as May be Ordered to it Presidt Capt Mason Members Lts Warley and Guerry

---

Orders 3 April Lt Colo. Marion } Parole Orangebourg   Cn Sn 43
for the Day to Morrow Capt Gray
for G. Guard Sergt Mc.Dowell & Davis
for Rear Guard Sergt. Haisman
No Sergt. to sign a morning report when there is an Offr of the Compy with the regiment, nor no morning Reports to be brought to the Comdg Offr. without being Signed by an Officer Commissioned or non Commissioned Officer – this and former Orders respecting Sergts Roll Call to be punctually Observed __
No officer to Change their tour of Duty but on very Extraordinary Occasion, that cannot be well Avoided as it often Creates Confusion in Duty
Corpl Geo Brewton is appointed a Sergt in Captn. Mottes Compy, he is to be obey'd as such _
Samuel Brown of Captn. Proveaux Compy. is appointed a Corpl in Capt Grays Compy he is to be Obeyed as Such – [1045]
Wm Henderson of Captn. Grays Compy is Appointed a Corpl in Captn. Mason Compy he is to be Obey'd as such –

Orders 4th April Lt Colo. Marion } Parole Lincoln   Cn Sn 45
for the day to Morrow Lt Guerry –
for G. Guard to Morrow Sergt Wood & Clatworthy
for Rear Guard Sergt. Matthews –
for Town Guard Capt. Hall & Lt. Warley
Agreeable to the last Court Martial that Set Apr. 2d Andrew Deblong of Capt Masons Compy. Recd 35 lashes on the Bare Back with a Cat a nine Tails for absence Without leave [1046]

[1045] Samuel Brown had enlisted in the 1st South Carolina Regiment on 24 May 1776 and was discharged on 3 June 1778. He enlisted in the 2nd South Carolina Regiment and served under Captain Proveaux. He was promoted to corporal in Captain Peter Gray's company on 3 April 1779. He was at the siege of Savannah.
[1046] Andrew Bland.

Cap[t] ~~Motte~~ Mazyck **Fort Moultrie 1779**

Orders 5th April } Parole Maj[r] Horry C[n] S[n] 7
for the day to Morrow Ensign Rogers
for G Guard to Morrow Serg[t] Murphy & Taylor
for Rear Guard to Morrow Serg[t] Kolcoffon
R.O. by Cap[t] Mazyck} A Court Martial to Set to Morrow to try such Prisoners as Shall be brought before it Evidences to attend Presid[t]. Cap[t] Gray Members L[t]. Guerry & Wickam

---

Orders 6th April
Cap[t] Mazyck } Parole Gen[l]. Huger C[n] S[n] 18
for the day to Morrow Cap[t] Mason
for G Guard Serg[t] Davis & Haisemann
for Rear Guard Serg[t] Turner
A ^Serg[t]^ Corp[l]. and three Privates are to hold themselves in Readiness to go up to Town tomorrow Morning as a Guard to the Gen[l]. Hospital he will receive his order from L[t]. Col[o]. Marion

---

Orders 7th April
Cap[t] Mazyck } Parole Congress C[n] S[n] Cap[t] Mazyck
for the day to Morrow Cap[t] Gray
for G. Guard to Morrow Serg[t] Turner & Matthews
for Rear Guard Serg[t] Burtell

Orders 8th April
Cap[t] Mazyck } Parole Purisburgh C[n] S[n] 35
for the day to Morrow L[t] Guerry
for G. Guard Serg[t] M[c]Donall & Taylor
for Rear Guard Serg[t] Davis ___

Orders 9th April
Cap[t] Mazyck } Parole Col[o]. Taylor C[n] S[n] 57
for the day to Morrow Ensign Rogers
for G. Guard to Morrow Serg[t] Turner & Wood
for Rear Guard Serg[t]. Clatworthy
Agreeable to the last court Martial Corp[l] Hughes of Cap[t] Proveaux Comp[y] was Reduced to the Ranks for taking the Aprons of the cannon in the night when loaded __ __ __ __ __
Thomas Inling of Cap[t] Mazyck Comp[y] Recie[d] 80 lashes with Switches on the Bare Back at two Different times for Stealing a Blanket __ __ __ __ __ __ __ __ __ __ __[1047]
Jn[o] Cadday of Cap[t] Proveaux Comp[y]. Recieve 100 lashes on the Bare back with Switches at four Different Times for Absence without Leave ________

1047 Thomas Inling (Inkling) served in Captain Mazyck's company of the 2nd South Carolina Regiment. On 9 April 1779 he received 80 lashes with switches on the bare back for stealing a blanket. This punishment was spread out on two different occasions.

Capt[n] Motte **Fort Moultrie 1779**

Orders 10th April } Parole Cap[t] Mason C[n] S[n] 45
for the day to Morrow Cap[t] Gray
for G. Guard to Morrow Serg[t]. Taylor & M[c]Dowel
for Rear Guard to Morrow Serg[t]. Burtell
R Orders by Cap[t] Motte } A Court Martial to Sett this Morning at 10 OC at the Presidents Appartment to try Such Prisoners as may be Brought to it – Presd[t] Cap[t] Mason Members Cap[t] Gray and L[t] Guerry __ __ __ __ __ __ __ __ __ __ __ __ __
After Orders by Capt[n] Motte } Ordered that the Pay Bills of Each Comp[y] be Immediately Made Agreeable to a Resolve of Congress passed the 27th day of May 1778 and Delivered into the Pay Master, Cap[t] Mazyck and Ensign Rogers with Capt Mazyck's Comp[y] to hold themselves in Readiness to go over to Fort Johnson on Munday Next – The Regiment Will be Mustered on Munday Next When the Comd[g] Officer of Each Comp[y] will have his Muster Rolls Ready for the Deputy Muster Master __ __ __ __ __ __ __ __ __ __ __ __ __

Orders 11th April Capt[n] Motte } For Parole Virginia C[n] S[n] 51
for the day to Morrow Cap[t] Mason
for G. Guard to Morrow Serg[t] Turner & Clatworthy
for Rear Guard to Morrow Serg[t]. Matthews
For Town Guard to Morrow Cap[t].Mason & Lieu[t]. Guerry

Orders 12th April L[t] Col[o] Marion } Parole Komet C[n]. S[n]. Lost
For the day to Morrow L[t]. Warley
for G. Guard to Morrow Serg[t]. Taylor & M[c]Dowell
for Rear Guard Serg[t]. Burtell
The Rear Guard to Stop all Boats at the Bridge going to Town & to Acquaint the Comd[g] officers of the Garrison Immediately till further Orders
Solomon Long of Capt[n]. Mottes Comp[y] is appointed a Corp[l] in Cap[t] Proveaux Comp[y] he is to be Obey'd as Such
James Reid of Capt[n]. Halls Comp[y] is appointed Gunners Mate to fort Moultrie he is to be considered as Such _______

Cap[t] Motte **Fort Moultrie 1779**

Orders 13th April } Parole Eagle C[n] S[n] 97
for the day to Morrow Cap[t] Mason
for G, Guard to Morrow Serg[t] Sims & Clatworthy
for Rear Guard to Morrow Serg[t] Turner
Orders by Cap[t] Motte___ } Commanding Officer of Each Comp[y] Will Imediately make out a Return of what arms are good Bad and Wanting to Compleat them and Given to the Comd[g] Officer by 12 OClock – [1048]

[1048] Moultrie wrote in his memoirs, "About this time Capt. Morgan arrived from St. Eustatia, with a fresh supply of arms and ammunition which were much wanted; we could not have moved anywhere without them,

---

Orders 14th April } Parole Drayton Cn Sn 5
for the Day to Morrow Lt. Warley
For G. Guard to Morrow Sergt. Taylor & Bertell
For Rear Guard to Morrow Sergt. McDowell

Orders by Capt Motte } The Gunner is Imediately to make a Return of all the Ordinance Stores Now at Fort Moultrie & the Qr Master Sergt a Return of all the Arms Accoutrements Rum Provision of what Kind so ever in the Store ______

Orders 15th April Capt Motte } Parole Fairfield Cn Sn Liberty -
for the Day to Morrow Capt Hall
for G. Guard to Morrow Sergt Sims & Turner
for Rear Guard to Morrow Sergt. Feast

Agreeable to the last court martial Daniel Savage [1049] of Capt Halls Compy Recieved 50 lashes on the Bare back with a cat o nine Tails for Absence with out leave __ __ __ __ __

Orders 16th April } Parole Hampton Cn Sn 19
For the day to Morrow Capt Proveaux
for G Guard to Morrow Sergt Newman & Matthew
for Rear Guard to Morrow Sergt Clatworthy

Orders by Capt Motte } Ordered that the Surgeon or his Mate do Call at the Comdg Officer Appartment every Morning & take off the Names of Such Men as are Reported Sick, and that the Surgeon Do Make a Report Every Munday Morning of the State of the Hospital – A Court Martial to Sett this Morning at 10 OClock for the Tryal of Such Prisoners as may be Ordered to it Evidences to Attend __

Presidt. Capt Proveaux Members Capt Mason & Lt. Warley ___

After Orders by Capt Motte } A Sergt a Corpl and Six Privates to be furnished Six Rounds pr. Man to go on Board the Cartel Schooner in five Fathom Hole – The Sergt. will receive his Orders from the Commanding Offr

---

Orders 17th April Capt Motte } Parole Friendship Cn Sn 13
for the day to Morrow Capt Hall
for G Guard to Morrow Sergt Taylor & Brewton
for Rear Guard to Morrow Sergt McDowell

---

as we lost, at the affair at Brier-creek, upwards of one thousand stand of arms; and we were obliged almost always to arm all the reinforcements that came from North-Carolina."

[1049] Daniel Savage had initially been in the militia of the Bloody Legion of Hilton Head under Captain John Leacraft. He was in the 2nd South Carolina Regiment in Captain Hall's company in 1779. On 15 April 1779 he received 50 lashes on the bare back with a cat of nine tails for being absent without leave.

Fort Moultrie 1779

Orders 18th April } Parole Black Swamp C^n S^n 95
for the day to Morrow Capt Proveaux
for G. Guard to Morrow Serg^t Newman & Matthews
for Rear Guard to Morrow Serg^t Simes
for Town Guard to Morrow Capt Mason & L^t. Warley

Orders 19th April } Parole Ashepoo C^n S^n 51
for the day to Morrow Capt Gray
for G: Guard to Morrow Serg^t Taylor & Bartell
for Rear Guard Serg^t. Brewton

Agreeable to the last court martial Thomas Roybould of Capt Masons Comp^y. Recev^d 30 Lashes on the bare Back with Switches for absence without Leave

---

Orders 20th April } Parole Major Grimkie C^n S^n 71
for the day to Morrow Lieu^t Guerry
for G Guard to Morrow Serg^t Feast & Simes
for Rear Guard to Morrow Serg^t Davis

Orders by Maj^r G. Lincoln __ __ } Head Q^tr Black Swamp [1050]
April 15th 1779

Sir / as Congress have Appointed an Auditer through whose hands all pay Rolls and Abstracts must pass you'll please order the Pay Rolls of that part of the Second Reg^t now With you and the Other Continental troops to be Made up to the 1st Day of April, in future you Will Send your Rolls once a Month & Warrants will be forwarded to you for the Sums due therein

Signed – B. Lincoln M. G.

Orders by L^t. Col^o Marion } Ordered that the pay Rolls be made up agreeable to the above and According to the Resolves of Congress of 27th of May last Are to be transmitted to head Quarters as soon as Possible The Additional pay Given by this State to be in a Separate pay Roll which will be paid by this State as Usual –

[1050] Black swamp was a strategic point 25 miles from Purisburgh, which commanded the Yemassee Bluff opposite Abercorne. Sergeant William Jasper of the 2nd South Carolina Regiment had been detached to conduct special operations missions across the Savannah River. William Moultrie wrote, "He was a brave, active, stout, strong, enterprising man, and a very great partisan. I had such confidence in him, that when I was in the field, I gave him a roving commission, and liberty to pick out his men from my brigade, he seldom would take more than six: he went often out, and returned with prisoners before I knew he was gone. I have known of his catching a party that was looking for him. He has told me that he could have killed single men several times, but he would not, he would rather let them get off. He went into the British lines at Savannah, and delivered himself up as a deserter, complaining at the same time, of our ill usage to him, he was gladly received (they having heard of his character) and caressed by them. He stayed eight days, and after informing himself well of their strength, situation and intentions, he returned to us again: but that game could not be played a second time. With his little party he was always hovering about the enemy's camp, and was frequently bringing in prisoners." During this time period Jasper and Sergeant John Newton, from the 2nd South Carolina Regiment, crossed the Savannah River and captured Captains Scott and Young. They then returned to Patriot lines and brought their prisoners to General Lincoln at Black Swamp.

Orders 21 April } Parole Maj^r Hartiston C^n S^n 39
for the Day to Morrow Capt Hall
for G Guard to Morrow Serg^t Burtell & Brewton
for Rear Guard to Morrow Serg^t Matthews
Orders by L^t. Col^o. Marion ____ } M^r. Geo. Ogier is Appointed a Second Lieutenant in the Reg^t he is to be respected Obbeyed as Such to take rank the 3^d April _ _ _ _
L^t. Ogier is to Join & do duty in Captn. Mazycks Comp^y till further Orders _ _ _ _ _

Capt Motte **Fort Moultrie**

Benjn Stone of Captn. Lisesne Comp^y is appointed Sergeant in the same he is to be Obey'd Accordingly _ _ _ _ _ _ _ _
No Cloathing to be given out with out an Order from Col^o. Marion
The men to be trained to the Exercise now in use up at head Quarters which Captn Dunbar will Show the meathod while he stays [1051]
Orders by Capt Motte } A Court Martial is to set this Morning at 11 OC to try Such Prisoners as may be brought to it Evidences to Attend Presd^t
Captn. Hall Members Capt Proveaux & Gray _ _ _ _ _ _ _ _ _ _ _ _

Orders 22^d April Capt Motte } Parole Camden C^n S^n 31
for the day to Morrow Capt Proveaux
for G. Guard to Morrow Serg^t McDowell & Sims
for Rear Guard to Morrow Serg^t Feast

Orders 23^d April } Parole Portland C^n S^n 79
for the Day to Morrow Capt Gray
for G. Guard to Morrow Serg^ts
for Rear Guard to Morrow Serg^t
Orders by Capt Motte _ _ _ } A Serg^t. 1 Corp^l. and Six Privates to be in Readyness immediately With 12 Rounds P^r. Man and a camp Kettle to go to Morrises Island – the Serg^t and Party who has had the Small Pox to be in this Service, the Q^r. Master Serg^t to give this Party each with two days Rations p^r. Man – Whatever Woman Will go on this Command will be paid by the State –

Orders 24th April } Parole Scharborough C^n S^n 31
for the day to Morrow L^t. Kolb
for G. Guard to Morrow Serg^ts Taylor & McDowell
for Rear Guard to Morrow Serg^t Feast

[1051] This would be the new drill created by Major General Steuben.

Whatever Woman Will go on this Command will be paid by the State

Lincoln __ __ } G.Orders by G. Purisburgh 24th Feby. 1779
The Deputy Barrack Master is ordered to Supply the Offrs & Soldiers with the following Provisions of wood & Candles –

To a Brigadier Genl.
½ a Cord of wood 1lb of Candles pr. week } To a field offr. and Capt ¼ a Cord of wood & 1lb of Candles pr. week
To a Subaltern of Each Compy. ¼ a Cord of wood & 1lb of Candles pr. week
To non Commissioned Officer & not Less than Six or more than Ten in a Mess ¼ a Cord of Wood and ¼lb Candles pr. week

Transmitted by
Jno. Hale Q. M. Q. R..

Lt. Colo Marion **Fort Moultrie 1779**

Orders 25th April } Parole St. Assaph Cn Sn 31
for the Day to Morrow Capt Proveaux
for Town Guard to Morrow Capt Hall & Lt. Guerry

for G. Guard to Morrow Sergts Roberts & Stone
for Rear Guard to Morrow Sergt. Campbell

Orders 26th April
Captn M } Parole Abington Cn Sn 24
for the Day to Morrow Capt Mason
for G. Guard to Morrow Sergts Taylor & Feast
for Rear Guard to Morrow Sergt McDowell

Orders 27th April
Lt. Colo Marion } Parole Rockingham Cn Sn 42
for the Day to Morrow Lt Kolb
for G. Guard to Morrow Sergts Campbell & Stone
for Rear Guard to Morrow Sergt Davis

When the Officer of the day go his rounds he is to be considered as Grand rounds & to Receive the parole from the officer of the Guard, but he must first give the Counter Sign ~
A court martial to be held today at 10 OClock in the forenoon, to try Corpl Kidwell on a Charge of Vincent Marony Evidences to Attend __ __ __ __ __ __ __
Commanding Offr. of Companys may give Leave of Absence to one man at a time for 24 hours __
For the above Court Martial Capt Motte Presdt Capt. Proveaux & Lt Kolb Members, NB the Court acquitted Corpl Kidwell __
Agreeable to the court martial that Sett April 21st Amos Tubs of Capt Proveaux Compy Recievd. 100 lashes on the Bare back with Switches at four different Times For theft ___ [1052]

[1052] Amos Tubbs served in the 2nd South Carolina Regiment under Captain Adrian Proveaux during 1779. On 27 April 1779 he received 100 lashes on the bare back with switches for theft. This punishment was given out at four different times.

Orders 28th April } Parole Rubicon   Cn Sn Julius
Lt. Col° Marion } For the day to morrow Capt Proveaux
for G. Guard to morrow Sergts Feast & Simes
for Rear Guard to Morrow Sergt. Taylor

The Gunner will send the field Piece now at the rear Guard with 12 rounds of round Shott & Cartridges with Nessessarys for it to the Advance Guard today & Send a Spare Carrage for it to be mounted on the platform of the Battery – the fatigue men for this Duty –

Lt. Col° Marion **Fort Moultrie 1779**

Orders 29th April } Parole Embargo   Cn Sn 18
For the day to morrow Capt Mason
for G. Guard Sergt Davis &
for Rear Guard Sergt Campbell

Orders 30 April } Parole Augusta . . . . C Sn Williamson
Lt. Col° Marion } For the day tomorrow Lieut Warley
Fort Guard Sergeants Roberts & Taylor; Rear Sergt Feast

The monthly returns to be made out & given in by tomorrow at 10 OC: in forenoon to the Adjutant –

Orders 1st May } Parole Savanna . . . . Cn Sn Georgia
Lt. Col° Marion } For the day tomorrow Captn Mason
fort Guard Sergts Campbell & Chatworthy - Rear Sergt Davie

Ten pr. Shoes to be Deliver'd out to Each Compy, to be Given to those who want most & an Account kept to Whom given

On April 22nd 30 Loyalists attacked a small guard post at Black Swamp that was guarded by soldiers from the 6th South Carolina Regiment. The Loyalists painted themselves as Indians. The house there was burned and the 6th South Carolina withdrew. The next morning they returned and they were reinforced by 100 men of the 5th South Carolina Regiment under Lieutenant Colonel Alexander McIntosh. An additional 20 Catawba Indians were also placed there to counter anymore Loyalist raids, however on April 28th British General Prévost landed 300 men at Purrysburgh and due to being outnumbered, the post was abandoned.[1053]

Orders 2nd May } Parole Col°. McIntosh . . . . .   Cn Sn Success
Lt. Col° Marion } For fort guard tomorrow Captn
Fort Guard Sergts Sims & Stone

When any man Obtain leave to go to town & stay Longer than his Leave of Absents he is to be sent for immediately & confined who Shall be brought to a Court martial & Suffer Accordingly

---

[1053] O'Kelley, *NBBAS, Volume One* pp 270-271.

Major General Augustine Prévost made a diversionary attack towards Charleston so that Major General Benjamin Lincoln and his army would withdraw from Augusta. Lincoln recognized the move for what it was and refused to leave Georgia. What both commanders did not realize is that the city did not have a very good defense, and once Prévost discovered that there were no obstacles in his path, the diversionary attack towards Charleston became an actual attempt to capture the city. Lincoln sent an additional 1,000 men to Black Swamp to delay Prévost if he decided to cross the Savannah River. Prévost did cross the river on April 28th and Moultrie withdrew all his forces except the 100 men of the 5th South Carolina that had been at the Black Swamp post. These men were to delay the British while Moultrie could find a good place to stop Prévost's army. Moultrie requested artillery from Governor Rutledge to stop Prévost, but neither Rutledge nor Lincoln thought that Prévost would attempt to attack Charleston.

Moultrie ordered the 5th South Carolina Regiment to rendezvous with him at Coosawhatchie River. He decided to take a stand at Tullifinny Hill and placed guards on all the crossing points of the Tullifinny River. Lincoln did send 250 hand picked North Carolina light infantry under Colonel John Laurens to assist Moultrie.[1054]

Orders 3d May } Parole Genl. Moultrie . . . . Cn Sn Success
For the day tomorrow Capt Gray –
Fort Guard Sergts Davis & Campbell – Rear Guard Sergt Turner

Orders 4 May } Parole Beaufort . . . . . Cn. Sn. 79 by LtC M
For the Day to morrow Lt Warley
Fort Guard tomorrow Sergts Taylor & Stone –Rear Guard Sergt Matthews

On May 3rd Moultrie sent Laurens and his Light Infantry to bring in the rear guard of Captain Shubrick's company of the 5th South Carolina Regiment before it was cut off by Prévost. Laurens was supposed to bring the 5th South Carolina back to Tullifinny Hill, but instead he decided to make a stand with his 400 men on the west bank of the Coosawhatchie River.[1055]

Laurens wanted to win fame for his action, but his position was not a good one. The British decided to fire long range artillery at the Patriots. Laurens had no artillery and was powerless to do anything. Major F. Skelly, with Maitland's force, wrote that the Patriots "crossd the River, burnt the bridge, and made a trifling stand on the opposite side. They killed and wounded four of our men. We drove them from thence, waded the river, found a few of their dead, pursued them cross Tullifinny ferry which we forded, saw their rear at Poketallago River, gave them a few Cannon Shots".

Several of his men were killed and wounded, and Laurens himself was wounded in the arm by an artillery projectile that also killed his horse. Once Laurens left to seek medical aid, Captain Shubrick wisely had his men abandon the position.

[1054] O'Kelley, *NBBAS, Volume One* pp 272-274.

[1055] Tuliffinny Hill is located two miles north of Coosawhatchie, beside interstate 95.

Moultrie was angered by the young officer's stand and realized that his position was no longer defendable. He ordered the bridge over the Tullifinny River destroyed and he retreated towards Charleston. The militia with Moultrie's army became more demoralized with each bridge they burned, so they deserted in droves.[1056]

Orders 5th May } Parole Coosehatchee . . . . C. Sn. Retreat
For the day tomorrow Lieut Guerrey
Fort Guard tomorrow Sergts McDoual & Sims[1057] –Rear G. Sergt Feast
Sergeants Stephen Roberts & Stone with all the men belonging to the Light Infantry Compy. to hold themselves in readyness to March & Join their Compy – they must be ready by tomorrow morning

— — — — —

The Adjutant will see them well Armed & Accoutered, to be furnished with one pr. Shoes & a haversack, & 15 rounds, 1 spare flint pr. Man –
Commanding Officers of companys to make a return of the number of Bayonets wanting in their respective companies

---

Orders 6th May Parole Chulifency . . . . . . C. Sn. Ashepo
Lt. Colo Marion For the Day tomorrow Lt Ogier
Fort Guard Sergeants Bartell & Matthew
NB Rear Sergt
Sergt Stone with the men orderd yesterday Sett off this morning

Lt. Colo Marion **Fort Moultrie 1779**

Orderd that all Officers ware their side arms on parrade & platform at the time of Exercise
Captn. Motte with his Company to hold themselves in readyness to Relieve Captain Mazyck & his Compy. at fort Johnston Tomorrow morning[1058]

1056 O'Kelley, *NBBAS, Volume One,* pp 272-274.
1057 Sergeant John McDowell.
1058 In the Huntingdon Library there are three rolls of microfilm containing the Francis Marion Orderly Book. On the second microfilm there is one orderly book containing January to May 1779. In the third microfilm there is another orderly book containing January to December 1779. These are two different orderly books from two different companies. Both are identical except for mistakes and corrections. The following pages were at the end of the orderly book that went from January to May 1779.

Captain Dunbar' s Compliments to Mrs. Bennet, has
Sent his y^{ds} the Bennet Three y^{ds} of Linen, 2 p^{c} Containing 24 Yards each & one p^{c}. 24 y^{ds} –
the Colonel Allows ¾ Yards to each Shirt. The sleeves must be made long & the Tails
rather short – he has also sent 2 Oz: of Thread & 54 Buttons & 10 ½ y^{ds} Linen for three
Shirts for the Serjeants
Wednesday Morn^{g} –

Roster 16th August
Captn. Moultrie
Captn. N Carolina
Lieu^{t}. Capers
L^{t}. N Carolina
Foissin
N Carolina
Petrie
Hume
Rogers
Vlieland

August 15th – 1779 –
Sir / I have mad 86 Shirts for the
Second Ridgment wich I wold
be glad you will giv an arder
for and you will much oblidg
your Humble Sarv^{t}.

Mary Bennett

to Con^{l}. Marrion –
<u>of 2^{d} Rigment</u> –

Rules in relieving Guards

When the guard is deliverd to the Officer who is to Command it he is to Examine the mens Arms, see if they are in Order and ~~he~~ if the men are shaved & dresst & what Amunition, Havresacks tin Kittles &tc which they may have agreeable to orders and to Count the number of men when he finds all Right he is then to take Charge of them, and march with a double step till they Come about ten yards from the Old guard they are then to take up the Single Step Beating the Battalion march, ~~the old guard~~ when the front of the new is ready Opposite to the right of the Old they are to Beat also the same march if boath belongs to the same regiment if not their own Battalion march & rest their arms the new guard marched in, they are to form in a rank intire Opposite the old with his right facing their Left, then halt the drums & fifes Ceasing & the men Dresst, they are to come to rest ~~their firelocks~~, the two Commanding Officers will meet in the Center of the two guards, the Officer of the Old will then give him what Order he had, all Other Officer to keep their places, the Old guard will Shoulder their firelocks & the new do the same, When the sergeant will number the men & take their names down, the Corporal of the new with the Corporal of the Old will take off the necessary Centries & relieve them of the Old guard – while this is doing the Commanding Officer of the Old will shew the prisoners and everything he has in charge to the Commanding Officer of the new Leaving the men with supported arms & the other Officers in their proper places who is not to suffer a man to speak, or make the least noise, when the Old Centries are returnd & falls in their ranks, they are to be Counted to see if they are all present, unfix their Bayonets, the new to fix theirs, the Old will face to the right & troop of his Guard the new to rest till the Old is Clear from them, then shoulder & take up the Grown the Old stood on with arms Shoulderd, front, & recover their Arms face to the right & dismiss them with Bayonets fixt, their Arms must be Lodg'd Either in front of the guard or in the Guard room

The Old Guard will ~~come~~ March beating the troop to the parrade where they was received, the Commanding Offr. will have the men Examind if they have all their Arms & Accoutrements, Havresacks tin kittles & every thing else which he marcht them off with, before he Dismist them, & make a report of what may be Lost ~~as well~~ and of what he deliverd the relieving Officer

I am very Sorry that you Shold make so grate a mis Stake in Charging me with 14 pieces of Lining for I was very perticuilar in Charging my self-with the peecis that was sent wich Capt Dunbar Can-prove the first peecis was three and 10 yd for the Sargents wich was June 23th –

2 peecis –	one	– – –	25 yards
-----------	one	_ _ _	24 -------
-----------	one	_ _ _	24 -------
Sergents Lining –	one	_ _ _	10 ½ ---
			23 – ½

Capt provox Sent his July 1th – three peecis

one peece	25 – yards
one peece	24 ------
one peece	23 ------
	72 – yards

Capt Dunbar Sent five peecis July 11th as follows –

one peece	25 yards
one - -	24 yards
one - -	25
one - -	25 ------
one - -	23 ------
	122 yards

ps – the total yards that I had was 277 – yards & ½
and out of that I made 86 shirts wich I beg you will last up &
See what wast I have made in the Eleven peecis of lining
I beg you will show this to Capt Dunbar that he may setle the mis Stake
as it maks me ver unesy to be Charg^d. With what I Nevour had and
you will much oblidg your Humble Sarvent
August 25th – 1779 – Mary Bennett [1059]

[1059] Possibly the wife of Richard Bennett, who deserted in August 1776.

Pay List for the Southern Department

Rank	p^{r}. Month Calend^{r}	Rations p^{r}. diem	When resolved by Congress	Rations in State of So. Carolina with additional rations of + Continental	
	Dollars				
Brigid^{r}. Genl. in Chief for table included .	250	12	Ap^{r}. 15th 1777	12	
Brigadiers	125	. .			a Ration Consisting of
Brigade Major . . .	50	. . . .	11th d^{o}.		
Sicritary to a Brigd^{r} Commanding in Chief . .	50	. . .	June 17th d^{o}		
D Adj: General	75	. . .	Feby. 22nd d^{o}	6	
Col^{o}. of Reg^{t}. of Infantry .	75	6	Oct: 7th: 1776	2	~~rations of~~ ½^{oz} beef or fresh pork
L^{t}. Col^{o}. of d^{o}.	60	5		7½	3 ½^{oz} peas or Beans 1 pint Milk
Major	50	4	. . . . . .	6	p^{r}. Diem ½ pint rice or Corn meal
Captn.	40	3	. . . .	4 ½	p^{r}. Week-1 q^{rt}. Spruce bier p^{r} Diem
Lieutenant	27	2	. . . . . .	3	or Mollasses p^{r}. Comp^{y}. of 100 men 9
Ensign	20	2	. . . . . .	2	Gall^{s}
Adjutant	40	4	. . . . . .	3 ½	3 Candles p^{r}. Week to women
Quart^{r}. Master	27 ½	2	. . . .	2	2 Gall^{s} vinegar to 100 d^{o} p^{r}. Week
Surgeon	60	4	Ap: 8th: 1777	4	6 Ounces Butter p^{r}. Man a week
D^{o}. Mate	40	2	.	2	
Chaplain	40	3	. 11th d^{o}	3 ½	
Paymaster	40	3	March 29th d^{o}	3	
Serg^{t}. Major	9	1	. . .	1	
Q. M. Sergeant	9	1	. . .	1	
Drum & fyfe Major . .	8 ¾	1	. . .	1	
Sergeant	8	1	. . .	1	
Corporal	7 ¾	1	. . .	1	
Drums & fyfe – each .	7 ¾	1	. . .	1	
Privates	6 ¼	1	. . .	1	
D. Paymaster of y^{e}. Army	50 –	6			
Assistant d^{o}.	26 👌				
Soldiers acting as fatigue Or pioners wth. Q. mast^{r} More than privates	1/8 —				
L^{t}. Col^{o}. Commandant of Rangers	75	6			
Maj^{r}. & other Inf^{y}. Off^{rs}. in proportion					
Surgeons the Same as L^{t}.					
Paymast^{r}. D^{o}. –					
Privates in rangers, to Provide themselves horse Gun & provisions for them Selves & horse	12 ½				

Be Cool and Do Mischief

£. 5..10 for this Book

Fran^s. Marion L^t. Col: 2nd Reg^t.
May 16th 1777 –

Staff of the American Army to August 1777 –

His Excellency Geo^{r}. Washington Command^{r}. in Chief

Names	When made Brigad^{r}. Genl.	When made Major Generals	Names	When made Brigadier General	When made Maj^{r}. General
Charles Lee .	. . . .	June 15th 1775	Arthur Wayne .	Feby. 21st 1777	
Ph^{p}. Schuyler . .	. . . .	. d^{o} . . .	J^{o}. Mar^{sh} Vasuain	. d^{o} . . .	
Israel Putnam	. . . .	. d^{o} . . .	Geo^{r}. Weldon	. d^{o} . . .	
Horatio Gates	June 15th: 1775	May 16th: 1776	Pet^{r}. Mullenbourgh	. d^{o} . . .	
Willm. Heath . .	22nd 1775	~~Augt. 9th d^{o}~~	W^{m}. Woodford . .	. d^{o} . . .	
Joseph Spencer	. d^{o} . . .	. d^{o} . . .	Geo^{r}. Clinton .	March 25th 1777	
Jno. Sullivan .	. d^{o} . . .	. Ditto . .	Edward Hand .	Ap: 1st: - d^{o}	
Nathl. Green . .	. Ditto . .	. d^{o} . . .	Charles Scott . .	. d^{o} . . .	
Benedict Arnold	. d^{o} . . .	May 2nd 1777	Learned .	. d^{o} . . .	
W^{m}. Thompson	March 1st: 1776	P^{r}	Jedediah Huntington	May 12th. d^{o}	
W^{m}. Lord Starling	. Ditto . .	Feby. 9th 1777	Thos. Conway . .	May 18th. d^{o}	
Robt. Howe	. d^{o} . . .	Octr. 1777	Benj: Lineau	. . . .	Feby. 19th 1777
Thos. Mifflin . .	May 16th. 1776	Feby. 19th . d^{o}	Marq: Delafayette .	. . . .	July 31st d^{o}
Nixon . .	Augt. 9th d^{o}		Mons: De Coudre .	. . . .	Augt. d^{o}
Arthur S^{t}. Clair	. d^{o} . . .	Feby. 19th. d^{o}			
Alex^{d}. McDougall	. d^{o} . . .				
I. H. Parsons	. d^{o} . . .				
Saml. Clinton	. d^{o} . . .				
Adam Stevens	Septr. 4th. d^{o}	Feby. 19th d^{o}			
W^{m}. Moultrie .	Ditto				
Laughlin McIntosh	. d^{o} . . .				
W^{m}. Maxwell	Octr. 22nd: 1776				
W^{m}. Smallwood	. d^{o} . . .				
Roche DeTornay	Novr. 5th. d^{o}				
Chas: deBone	Decem^{r}. 1st: d^{o}	Resigned			
Henry Knox	27th d^{o}				
Fran^{s}. Nash	Feby. 5th 1777	Dead this d^{o}			
Enock Poor	27th d^{o}				
Jno. Glover	. d^{o} . . .				
C. Patterson	. d^{o} . . .				

May 7th ~

Parole Chs . Town ___ C Sn. 16

For the Day to Morrow Capt ~~Mason~~ Gray ~
For G. Guard to Morrow Serjeants Feast & Sims
Rear Guard to Morrow Serjeant McDowall

Morning Orders } The Regiment to hold themselves in readyness to march at a minutes warning, they are to carry nothing with them but their Blankets – all their Baggage must be put in one of the Regimental Stores – the Officers will put all their Baggage in two rooms ~~Store~~ & Leave their rooms to be Occupied by the troops Coming in – the Adjutant will immediately make room in the privates Barracks for one hundred men, ~~Present~~ & some rooms to be Given in the Offrs Barrack, for the Officers coming in

No man to be Sufferd to go off the Island & all Boats to be Detained, the Bridge to be Kept Drawn till further Orders

Commanding Offrs of Companies to see their mens Arms in good order – the Guns which are repaired & operating to be given out to Each Company in proportion to the wanting

Sergeant Simpson with the Gunners mates only to be Left in Garrison, every man that can stand on his leggs must March

The field piece at the Advance Guard must be sent for immediately

Moultrie slowed the march of the British army under Prévost by burning the bridges across the rivers and felling trees along the road. The British were slowed by themselves when they stopped to loot the plantation homes along their route. Prévost further weakened himself by detaching groups to escort flour and the loot to Georgia. Moultrie knew that the target was now Charelston and he urged Lincoln to return. South Carolina Governor Rutledge departed from Orangeburgh with 500 militia on May 4th, arriving in Charlestown on the 7th. Lincoln, who was always cautious, held a council of war and his officers agreed that he should return to Charleston. He departed Augusta on May 6th, but moved very slowly; not realizing the city was in danger of being seized by Prévost.

Moultrie ordered that the Charlestown militia occupy the left of line at Charlestown Neck. The Upcountry militia would occupy the right of the line. Colonel McIntosh and the 5th South Carolina Regiment would occupy the redoubt on the right side of the line, and Colonel Marion and 100 men of the 2nd South Carolina would occupy the redoubt on the left. Lieutenant Colonel Harris and the Camden militia would occupy the advance redoubt in front of the line. The remainder of the 2nd South Carolina, with Pulaski's infantry, would occupy the half-moon redoubt in the center of the line as a reserve force.[1060]

~ 8th ~

Parole Jacksonbourg . . C Sn. Maj Bull

For the Day to Morrow ~~Lieutenant Warley~~ Captn. ~~Motte~~ Akins [1061]
For G: Guard to ~~Day~~ Day ~~Serjeant~~ Lieutn . Warley
For G: Guard tomorrow Lt. Guerry
For Rear Guard to Morrow Serjeant ~~Matthew~~ Newman

[1060] O'Kelley, *NBBAS, Volume One* pp 274-276.
[1061] Unable to determine who Captain Akins was.

The fort guard to be Augmented to 1 Subaltern 2 Serg^{ts}. 2 Corporals & 21 Privates, the rear to 9 privates
Captn Akins will Order a ~~must~~ report to be made out Every morning, of the State of his men & given in to the Commanding Off^{r}. Directly after roll-call, he will order a Sergeant to Attend the Sergeants call to receive orders dayly – he will be carefull not to give Leave of Absents to more than two men at a time – he will Augment his Guard at the ~~Bridge~~ Advance Guard, to 1 Serg^{t}. 1 Corporal & 9 privates

~ May 9th-
Parole Honour C Sign – 2

For the Day to Morrow captain ~~Provaux~~ Mason –
For G: guard to Morrow Lieu^{t} Warley ~
For Town Guard to Morrow Capt Mason & L^{t}. Ogier
For Rear Guard to Morrow Serjeant ~ Burtell
Orders by Capt Motte } Captn. Proveaux and Captn Halls Companies to be Imediately Completed to fifty Men each from the Other Companies. The Quarter Master Serj^{t} Will furnish there Companies with One hundred Rounds p^{r}. Man and Six spare flints – L^{t}. Thomas of Capt Akins Company to relieve L^{t}. Guerry Imediately at the Garrison Guard [1062] – Such Men as belong to Capt Proveaux and Capt Hall's Companies to be Imediately relieved from Guard – Ch^{s} . Town 9 May 1779
Gen Orders by Gen Moultrie } – Col^{o} Marion with all the Men belonging to his Reg^{t}. to Come over to Chas Town Except one officer and twenty Men to Keep Possession of Fort Moultrie –
After Orders by Capt Motte } Capt Gray and ~~two~~ One Serg^{ts} ~~two~~ One Corporals and twenty Privates to remain at Fort Moultrie till Further Orders –

May 10 – Charles Town
Parole Paloski Counter S^{n}_ Spirit ~

orders by L^{t}. col^{o}. marion the officers & Soldiers are To Sleep at the Barracks & they are not to be Absent From it at any time without leave
From the commanding officer ~
After Orders
L^{t}. Petrie to Join & Do Duty in Capt Mottes Compan^{y} till further Orders
M^{r}. John Bush is Appointed a Second Lieu^{t}. he is to be respected Obey'd as Such
L^{t}. Bush is to Join & do duty in Captn. Moultries Company till further Orders
L^{t}. Hart will Join Captn. Halls Comp^{y}.
L^{t}. Guerry to Join Captn. Baker Comp^{y}.
The Non commissioned officers & Men to remain in the respective Companys they now are in – L^{t}. Martin to Join Captn. Mazycks Comp^{y}.
G Orders
General orders by Genl. Moultrie May 6th – 1779
a Return to be made immediately by the command^{g}. officer of the different corps of the number of officers with their Rank Rank & dates of commissions and also the number of ~~Men~~ Rank & file in their Respictive corps –
R.O.

[1062] Unable to find any information about Lieutenant Thomas.

The Barrack Guard to be relieved tomorrow morning by the Companys which came from forts Moultrie & Johnston & to Consist of a Captain Subaltern 2 Serg^{t}. 2 Corp: & 28 privates
For the Barrack Guard to morrow cap^{t}. Motte & L^{t}. Warley

The defenders of Charleston had quickly constructed defenses across the Charleston Neck, between the Ashley and Cooper Rivers. Fort Johnson was blown up so that the British would not be able to use it. Four galleys defended the Ashley River from any attempt to cross. The frigate *Bricole* and the brig *Notre Dame* defended the Cooper River. The Raccoon Company of Riflemen, consisting of Indian warriors, had been delaying Prévost's army and were the last to come through the gates.

When Prévost's army appeared before the Charlestown gate Moultrie ordered the cannons to fire upon them. The cannonade continued slowly until dark. During the night Major Benjamin Huger of the 1st South Carolina and twelve others were killed by friendly fire when they were setting fire to tar barrels in front of the lines to light up the night. The artillery also set fire to every building between the Cooper and Ashley rivers.

Due to the fratricide Moultrie was finally given command of the whole army, and not just the Continentals. Governor Rutledge did not think the city could stand against the British and wanted to surrender. General Moultrie wanted to continue to hold out, but he yielded to the civilian leaders.

On May 11th a white flag was raised and surrender terms were negotiated. Prévost's son, Lieutenant Colonel James Prévost, demanded unconditional surrender of the town. He also stated that all inhabitants who rejected Royal authority would become prisoners of war. Moultrie refused to accept those terms. He told Rutledge that the city could be held. Rutledge made a proposal that the city would surrender if Prévost could guarantee the neutrality of the harbor for the rest of the war.

Moultrie gave the message to Lieutenant Colonel Laurens, who had been recuperating in Charlestown for his wounded hand. Laurens was asked to deliver the message, but once he found out what it contained, he refused. Moultrie found two other officers to deliver the message to Prévost.

Prévost refused to negotiate with Rutledge, who he considered to be a civilian authority and not a military one. Prévost again demanded unconditional surrender. Moultrie did not want to surrender the city and told Prévost that the negotiations are over. If he wanted Charlestown he would have to take it.

The next day the defenders woke to find that the British had left. Prévost knew that if he tried to storm the trench lines his force would be decimated. The British were low on ammunition and supplies due to the speed of their march to Charlestown. Prévost also knew that Lincoln's army was approaching, and he did not want to be surrounded. Prévost withdrew to James and John's Island to await transportation to Georgia. On the island there was plenty of rice and cattle to feed his troops. To protect his rear Prévost constructed defenses at Stono Ferry.

Lincoln arrived in Charlestown with his army on March 14th. Had he moved quicker he could have captured the British army. After the British had departed the 2nd South Carolina Regiment returned to their garrison at Fort Moultrie.[1063]

[1063] Full details of the Siege of Charlestown Neck are described in "Nothing but Blood and Slaughter, Volume One".

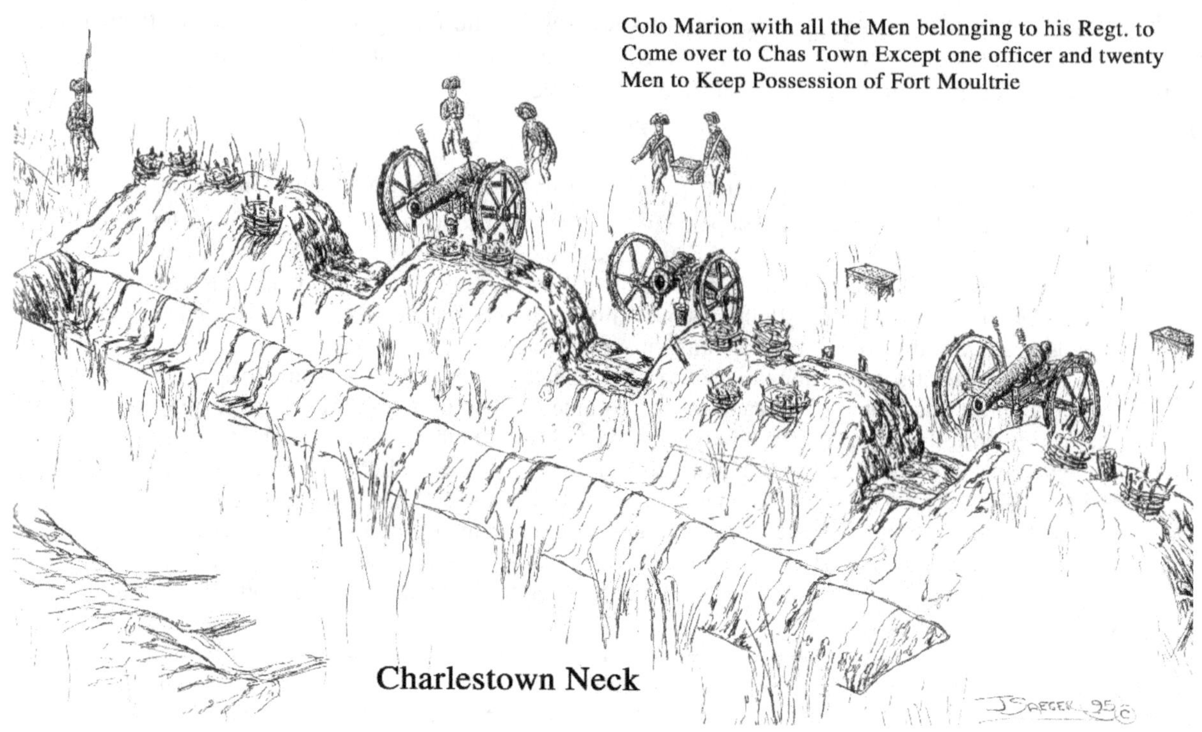

16 May Fort Moultrie

Parole Gen[l]. Moultrie C[n]. 33

one Subaltern 2 Serg[ts]. 2 Corp[l]. & 10 privates to march to Haddrells Barracks as a Guard to the public papers, to be furnisht from Col[o]. Maybank Detatchment of Malitia [1064]

The Malitia must take Barracks on the Left the 2nd Reg[t] on the Right

Capt[n]. Gray will Appoint the Appartments to the Officers agreeable to the rank

L[t]. Karwan of Col[o]. Maybanks ~~Detatch~~ Regim[t]. will act as Adjutant to the troops in Garrison till further Orders he is to be Obey'd & Respected as Such [1065]

Serg[t]. M[c]Doul to Act as Serg[t]. Maj[r]. to the 2nd Reg till Serg[t]. Maj[r]. M[c]Donald Joins the Reg[t].

~~1 Serg[t] 1 Corp[l]. & 6 privates to Augment the fort Guard on the 2nd Reg[t]. to releave as many of the Militia Now on that Guard~~

one Sergeant 1 Corporal & 6 privates to relieve the Same Number on the fort Guard & Join the remainder of the malitia on Guard ~

No Officer or Soldier to Leave the Island without Leave from the Commanding Officer

---

1064 Joseph Maybank was born on 19 December 1735. He was the commander of the Berkeley County Regiment of Cavalry on 2 April 1776. Colonel Maybank's regiment was sent to Fort Moultrie to free up troops for the assault on Major General Augustine Prevost's defenses at Stono Ferry. Colonel Maybank was wounded at Hickory Hill, Georgia on 28 June 1779. He died in 1783 at the age of 48.

1065 Thomas Karwon had been an ensign under Captain Benjamin Marion and Colonel Singleton in the Berkeley County Militia Regiment in 1775. In 1781 he supplied beef for Continental use and he also provided provisions and forage for the militia. During 1782 he served in Marion's partisans as General Marion's adjutant.

For the Day tomorrow Captn. Mazyck

For the fort Guard tomorrow Captn. Provaux & Lieu^{tn} from Col^{o}. Maybanks reg^{t}.

/17th May /

Parole Count Polaskie . . . C S^{n}. 24

For the Day tomorrow Captn. Gray

Fort Guard Captn. from Col. Maybanks reg^{t}. & L^{t}. Guerry

and one Subaltern from Col^{o}. Maybanks reg^{t}.

One Sergeant 1 Corporal & 6 privates to be posted this Afternoon at 5 OC: in the new Redoubt, to plant two Centries on the parapet - - - - - - -

One Sergeant 1 Corporal & 6 privates to be posted this Afternoon at 5 OC: between the cove & Sea, to plant 2 Centries where the Adjutant will Direct - - - - -

These two Guards will give immediat notice to the Commanding Off^{r}. of the Day of all vessels boats or men which may Approach their post - - - - - - - -

The Officer of the Day to visit ~~the~~ All Guards & Centries in and out of the Garrison twice During the night & report to the Commanding Guards to have their Centrys visited every half hour During the night

It is recommended to the Officers & Soldiers on Duty to be very watchfull & Diligent on their post, those who may be found Negligent may Expect to be taken ~~notes~~ notice of - - - - -

M^{r}. McNeil will as soon as possible fill all the Ammunition Chests with fixed Ammunition [1066]
It is positively Order'd that no Guns shall be fired in or out of the Garrison with out Leave from the Command^{r}.

Alarm post to be as follows

Captn Mazyck with his Company & one Captn 2 Subalterns & 20 Rank & file from Col^{o}. Maybanks reg^{t}. to ~~be~~ take post in the flagg Bastion

Captn. Provaux with his Comp^{y}. & one Captn. 2 Subalterns & 20 Rank & file of Col^{o}. Maybanks reg^{t}. to take post in the Old flagg Bastion on the Left

M^{r}. McNeil Gunner will Attend Captn. Mazyck & Proveaux

Captn. Gray with his Company & one Captn. 2 Subalterns & 20 Rank & file from Col^{o}. Maybanks reg^{t}. to take post in the Magazine Bastion

M^{r}. Davis Gunn^{rs} Mate Will Attend Captn. Gray [1067]

The Remainder of Col^{o}. Maybanks Regim^{t}. will take post in the N West Bastion

M^{r}. Reed Gunners mate will attend in the N W Bastion [1068]

Col^{o}. Maybank will take the Command from the Magazine ~~Bastion~~ to N West Bastions

Adjutant Kerwan with the Serg^{t}. Maj^{r}. will attend the Commanding Officer to carry Orders

Col^{o}. Maybank will Nominate the Officers & men for the Different posts as above Directed

---

[1066] Charles McNeil enlisted in the 1st South Carolina Regiment on 13 November 1775. He re-enlisted in December 1776. He served as the gunner of Fort Moultrie in 1779.

[1067] Hugh Davis served as a gunner's mate to the 2nd South Carolina Regiment under Captain Mazyck from 17 May 1778. He was appointed as the gunner's mate to Fort Sullivan on 27 June 1778. He was a private under Captain Daniel Mazyck in the siege of Savannah.

[1068] James Reid.

One Sergeant 1 Corporal & 6 privates to be ready immediately to march to Mr. Jonathan Scotts a guard to the powder there & Escort the waggons from their to the Hospitle here - [1069]

The Cart to be get ready immediately for the purpose [1070]

/ 18th May /

Parole General Lincoln . . C Sn. Caution

For the Day tomorrow Captn. From Colo Maybanks

Fort Guard a Captn. from Colo Maybank & 1 Lieut. & Lt ~~Warley~~ Kolb 2d Regt.

Lt. Warley to do duty in Captn. Mazycks Company till further Orders

A regimental court martial to set this forenoon to try Such prisoners as shall be brought before them Evidences to Attend Captn Mazyck presidt Lts Warley & Kolb members NB Jno. Conyers recd 25 lashes for quitting his rank without Leave

/ 19th May /

Parole General Williamson . . C Sn. Watchfull

For the day tomorrow Captn.~~ from Colo. Maybanks regt.~~ Mazyck

Fort Guard a Captn. 1 Subaltern from Colo. Maybank & Lt. Ogier 2nd Regt.

A regimental court martial to set today at 10 OC: in forenoon to try such prisoners as shall be brought to it Evidences to Attend - - -

Captn. Proveaux President, Lts. Guerry & Ogier members

NB Conrad Fitner recd 25 lashes for quitting his rank without leave

/ 20th May /

Parole Colo. Pinckney . . . C Sn. Brave

For the day tomorrow Captn. Proveaux

Fort Guard 1 Captn. 1 Sub: from Col. Maybank & Lt. Warley

/ 21st May /

Parole Stono . . . . C Sn. Steady

For the day tomorrow Captn. Gray

Fort guard 1 Captn. 1 Sub. From Colo Maybank & Lt. Guerry

/ 22 May /

Parole Johns Island. . . Countr Sign Rasalution

For the Day tomorrow Captn. Mazyck

Fort Guard 1 Captn. 1 Subaltern from Colo Maybanks & Lt. Kolb

---

[1069] Jonathan Scott owned one of the two plantations on Haddrell's Point. After the British captured Charlestown, Scott congratulated General Cornwallis when he defeated Gates at Camden. For this he was ordered out of Charlestown on 28 April 1782. He died before 3 April 1784.

[1070] After Prevost's army left Charlestown Neck they moved to Stono River. The 2nd South Carolina Regiment that was posted in the trenches at Charlestown Neck were sent back to Fort Moultrie. There was a belief that the British would try to take the fort.

/ 23 May /

Parole James Island C S^{n}. Spirit

For the day tomorrow Captn Proveaux

Fort Guard 1 Captn 1 Subaltern from Col^{o} Maybank & Lieu^{t}. Ogier ~

40 men to turn out at 2 OC: this even^{g}. As fatigue men

A Regimental court Martial to Set to day at 10 OC. to try all prisoners brought to it, evidences to Attend

Captn. ~~Gray~~ Mazyck President Lieu^{ts} Roux & Guerry memb^{rs}.

In future all Officers of the rounds Either Grand or ordinary rounds are to Give the parole to the Officers of the Guards

24th

Parole Wappoo . . . . . C S^{n}. Prize

For the Day tomorrow Captn. Gray

Fort Guard 1 Captn. from Col^{o} Maybank & 2 Subalts

~~Two Subalterns from 2^{d}. Reg^{t}. L^{t}. Guerry~~

/ 25th May /

Parole Edisto . . . . . Count^{r}. S^{n}. 97

For the day tomorrow Captn. Mazyck

Fort guard 1 Captn. 1 Subaltern from Col^{o} Maybank & Lieu^{t}. Roux

/ 26th May /

Parole General Huger . . C S^{n}. Attention

For the Day tomorrow Captn Proveaux

Fort Guard 1 Captn. 2 Subalterns from Col^{o} Maybank

It tis recommended to the Officers of Col^{o} Maybanks Detatchement not to give leave of Absents to More than two men at a time from each company & if the Companys are small not more than one & to return the same day, except some very Extraordinary Circumstances

/ 27th May /

Parole Paris . . . C S^{n}. 79

For the day tomorrow Captn Gray

Fort Guard 1 Captn 1 Subaltern from Col^{o} Maybank & Lieu^{t} Foissin 2nd Reg^{t}.

Regimental Orders; the Orders of the 3^{d} March last Respecting Officers Servants is to be strictly Observed –

In the provision returns it is Ordered that it be particularly mentioned if ration are for a White or Black Servants, if neglected the rations will not be Issued, & the Quart^{r} Mast^{r} or his Serg^{t} is Orderd to Issue such for Servants without tis strictly for White or Black Servant

A regimental court Martial to set today at 11 OC: in forenoon for the trial of such prisoners as may be Orderd to it, evidences to Attend –

Captn Gray presid^{t}. Lieu^{ts} Foissin & Kolb members

No Court

/ 28 May /

Morning Orders

The 2nd Regiment to hold themselves in readyness to March immediately, the Officers & Men will Give M^{r}. Simpson the care of all their Baggage, to be put in the Regimental stores – [1071]
The General to Beat at half after 7 OC. this Morning the Assemble & March at 8 OC.

The Guards will be relieved at 6 OC: Col^{o} Maybanks Detatchment will take all the Guards, & relieve those of the 2nd Reg^{t}.

M^{r}. Simpson will get ready all the Ammunition ~~ready~~ to move with 2nd Reg^{t}.

General Prévost retreated to John's Island when he received word that General Lincoln was advancing towards his position. Prévost had collected all the boats in the area and built a floating bridge across the Stono inlet to John's Island. Guarding that bridge was a redoubt at Stono Ferry.[1072] The 2nd South Carolina Regiment was ordered to move from Fort Moultrie to 13 Mile House near Stono Ferry to prepare for an assault on the redoubt. This order was later recalled and the regiment stayed at Fort Moultrie.

Ch^{s} . town

Genl. Orders 27 May by G Moultrie } Sir you will be ready to bring over all the men of the second Reg^{t} tomorrow morning with all their Arms and Accoutrements in proper Order and one hundred rounds p^{r} Man Each Barge to Send over some of the Ch^{s} town ~~Malitia~~ Artillery and Malitia to relieve you

L^{t}. Col^{o}. Marion fort Moultrie }
I am your humbl Servt
William Moultrie

28 May

Parole Artillery . . . C S^{n}. Disopointed

/ /

Charles town May 29th 1779

A Regimental Court Martial to set tomorrow Morning at 9 OClock for the tryal of such prisoners as shall be orderd to it – Evidence to Attend, the Major is orderd to send Evidences from Capt Masons Comp^{y}.

[1071] Sergeant William Simpson, the Quarter Master Sergeant.

[1072] The battle of Stono Ferry was located south of present day Rantowles and on the 12th fairway of "The Links at Stono Ferry" golf course. The two main square redoubts were destroyed during the construction of a railroad in the 19th century. The site is now in the gated golfing community, at the end of Boone Run Road. A third, round redoubt, was destroyed when the golf course was constructed in 1989. Karl McMillan, then vice president of the development company, was quoted in a *Post and Courier* news article that it was bulldozed because no one could assure him that it was a redoubt. In another article he explained that it was "right in the center of the fairway's landing zone and would have been damaged by golfers." The state interrupted the development pending an archeological investigation, but when Hurricane Hugo struck in September of 1989, that was forgotten. The golf course owners built a small replica of the redoubt 50 yards to the west, and placed a fake cannon on the top, aimed the wrong way to Johns Island.

Acquaint Matthew Scipper[1073] & Isaac Hyrnes [1074] of his Company for Absenting themselves without leave from s^{d}. Company - - - -

President Capt Dunbar

~ Members Capt Provaux Lieu^{ts} . Foissin, Kolb & Ogier

/ 30 May /

Parole Polaskee . . . . C S^{n}. Horse

For the day tomorrow Captn ~~Dunbar~~ Proveaux

Fort Guard Lieu^{tn} Ogier & a Sub: from Col^{o} Maybank

The Volanteer company Commanded by Captn Hall to Embark this Afternoon for Charles town

NB by sentence of last court Matthew Skipper & Isaack Herren are Sentenced to receive 100 lashes with Switches 4 Different times for being Absents without leave – approv'd of & to be inflicted in town

/ 31st May /

Parole Col^{o}. Bayler. . . . C S^{n}. 27

For the Day tomorrow a Captn. from Col^{o} Maybanks reg^{t}.

For Guard Lt Roux and 1 Sub. From Col. Maybank

/ 1st June /

Parole Beaufort . . .C S^{n}. 72

For the Day tomorrow Captn. Gray

Fort Guard 2 Subalterns from Col^{o} Maybanks reg^{t}.

R.O.

a regimental court martial to set at 11 OC: this morning for the trial of Such prisoners as shall be orderd to it evidences to attend – Captn Gray Presid^{t}. Lieu^{ts} Foissin & Kolb members –

2^{d} June

By Captn. Mazyck

to day Parole Lincoln – Countersign - Prudent

For ^ tomorrow a Capt from Colonel Maybank's Regiment For the Garrison guard Lieu^{t} Foissin and a Subaltern from Colonel Maybanks Reg^{t}.

3^{d} Jun

Parole Long Island . . C S^{n}. 9

For the Day tomorrow Captn Mazyck

Fort Guard L^{t}. Kolb & a Subaltern from Col^{o} Maybank –

1073 Matthew Skipper served in the 2nd South Carolina Regiment under Captain Daniel Mazyck. On 30 May 1779 he received 100 lashes with switches for being absent without leave. This punishment was done on four different occasions and it was done in the town. He was in Captain Mason's company during the siege of Savannah

1074 Isaac Herin (Herring) was born on 2 March 1761. He enlisted for a term of 16 months in the 2nd South Carolina during 1778 under Captain Richard Mason. On 30 May 1779 he received 100 lashes with switches for being absent without leave. This punishment was done on four different occasions and it was done in the town. He was wounded in the siege of Savannah and taken prisoner. He died on 23 October 1833

/ 4th June /

Parole Virginia . . . C Sn. 5.
For the day tomorrow a Capt from Col° Maybanks reg
Fort guard Lt. Ogier & 1 Sub. from Col° Maybank

A regimental court martial to set today at 11 OC: to try such prisoners as shall be orderd to it, evidences to attend

Captn. Mazyck president: Lieuts Roux & Ogier members –

/ 5th June /

Parole Baltimore . . C Sn. fleet
For the day tomorrow Captn ~~Proveaux~~ Dunbar
Fort Guard Lts ~~Rous~~ Foissin & 1 Sub: from Col. Maybank

/ 6th /

Parole Vigilant . . . . . C Sn. 24
For the Day tomorrow Captn Proveaux
Fort Guard Lt Kolb & 1 Sub: from Col° Maybank regt.

The Great number of dogs now in Garrison has become a Nuisance it is therefore Ordered that all Dogs be sent away by tomorrow evening those what remain after that time will be Orderd to be Killed

/ 7th May /[1075]

Parole Suffolk . . . . . C Sn. 42.
For the day tomorrow Captn from Col: Maybanks reg
Fort Guard Lt Ogier & 1 Sub: from Col° Maybank
One officer of the fort Guard to be constantly on the platform During the night till further Orders
The Officers of the Day to Visit the outguards at Least once in the day (besides what is already Orderd during the night) & See the Men are at their post, the Guard to turn out with Shoulderd Arms to the Officer of the Day As often as he visits there

8th May

Parole Williamsburg . . . . C Sn. McNeil
For the day tomorrow Capt Gray
Fort Guard Lt Roux & a Sub: from Col: Maybanks

NB this day Mr. McNiel Genl. D.A.

/ 9th May /

Parole General Scott . . . . . C Sn. 99
For the day tomorrow Captn from Col° Maybank
Fort Guard Lt Foissin & a Sub: from Col° Maybank
Col° Maybank regiment to furnish 10 men dayly Every evening & morning to draw the Bridge & push down – & they are to be warned every evening at roll call:

1075 This is a mistake and should read 7 June. The same mistakes were made for the June 8th – 10th entries.

/ 10 May /

Parole Carolina . . . C. Sn. 12

For the day tomorrow Captn Mazyck

Fort Guard a Captn. From Col° Maybank with one Sub. & Lieut Kolb –

The fort Guard to be Augmented this Evening at 5 OC: with 6 privates & plant 2 Centries more where Directed [1076]

The Quarter Guard to be Augmented at 5 OC: this Evening With, one Subaltern & one Sergt. 1 Corpl & 6 Privates, the Subaltern 1 Sergt 1 Corpl & 6 privates that augment the Picket Guard, to remain there from 5 OC: evening till next morning at sunrise, when they are to return to Garrison this mode to be continued until further Orders

The Long Roll to beat every morning at 8 OC when all the Officers in Garrison are at their Alarm post till Discharged at the same time all Guards will stand to their Arms till clear day light –

For the Piquit Guard this Evening Lt Ogier – this Offr will receive further orders when ready to march –

The Detatchement under Lt Warley when ordered to their alarm post is to take the first Curtain

The Sergeant Majr will take a number of men this evening when he goes to draw the Bridge & Collect every boat & Cannoe or the Vessels & put them in Charge of the Centry on the Bridges he will march the men sooner than usual for this purpose

When the out guards are Ordered in they are immediately to Join their respective companys –

/ 11 June /

Parole George town ________________ C Sn. 21

For the Day tomorrow Captn. ~~Proveaux~~ Dunbar

Fort Guard 1 Captn 1 Subaltern from Col°. Maybanks & Lt. ~~Hall Petrie~~ Roux

Picquit ~~a sub from Col° Maybank & Roux~~ Lt Petrie

Regimental court Martial to set this morning at 10 OC to try such prisoners as will be order to it evidences to Attend

Captn. Dunbar Presidt. Members Captn. ~~Proveaux~~ Gray Lts. Warley Foissin & Petrie

NB Sergeant Allyn & Brown [1077] was reduced to the ranks & Joshua Hall[1078] & Buchanan recd 50 lashes with a cat 9 tails for Desertion[1079]

12th Jun

Parole Santee . . . C Sn. St. Johns

For the Day tomorrow Captn. ~~Proveaux~~ Proveaux

Fort Guard ~~Captn Proveaux from Col° Maybanks~~ & Lieut Warley-2d Regt 1 Sub. Col. Maybanks

Piquet tonight a Subaltern from Col° Maybank

R. Orders

1076 A company of 40 men were sent to reinforce Fort Moultrie. General Moultrie was apprehensive about what the British were doing at Stono Ferry and reinforced the fort in case the British tried to assault it. The garrison of Fort Moultrie at this time was 300 men.

1077 William Brown.

1078 Joshua Hall served in the 2nd South Carolina Regiment after 17 September 1778 under Captain Peter Gray. On 11 June 1779 he received 50 lashes with a cat of nine tails for desertion. He was in the siege of Savannah.

1079 Unable to determine who Buchanan was.

Sergeants Allyn & Brown of the Granidier company reduced to the ranks agreeable to Sentence of a Court Martial – they are not longer to be consider as Such –

Allyn is to Join & do duty in Capt Grays Comp^y^. & Brown to Join & do duty in Captn. Masons Comp^y^. ~~Captn~~. Captn. Dunbar will Choose a man from those companys in their room –

Corporal Jno M^c^Donald and Reuben Dewitt of Captn Dunbars Company is Appointed Sergeants in the Same they are to be Obeyed as such – [1080]

W^m^. Manning of Captn. Dunbars comp^y^. is appointed Corporal in the Same he is to be Obeyed as such ~~~~~~~~~~~~~

A Garrison court Martial to set today At 11 OC to try such prisoners as shall be Order'd to it Evidences to Attend: Captn Mazyck presid^t^ Lieu^ts^ Foissin & Kolb of 2nd Regim^t^ & 2 Subalterns from Col^o^ Maybanks regim^t^.

NB By sentence of this court Conrad Myers was found guilty of Killing a hog of Mr Seats to receive 50 lashes & pay for it

Thos. Wats [1081] of Col^o^. Maybanks reg^t^. to be reprimanded for Abuse to Serg^t^. Kalchoffer – Peter Area to receive 25 lashes for gaming

____________________/ 13 June /____________________

~~Error – Parole Petersborg C S^n^.~~
~~For the day tomorrow Captn. Mazyck & L^t^. Foissin~~
~~Fort guard Captn. 1 Sub from Col^o^. Maybanks reg^t^.~~
~~Picquit tonight ~~Kolb~~ L^t^. Kolb~~
Error

Parole Petersbourg - C S^n^. Lauren
For the day tomorrow Captn from Col^o^ Maybanks
Fort Guard Lt Kolb & 1 Sub: from Col. Maybank
Picquet tonight Lt Foissin –
The Advance Guard to be relieved tomorrow morning by 1 Serg^t^, 1 Corporal & 9 privates from Col^o^ Maybanks regim^t^ & to be relieved weekly

____________________/ 14th June /____________________

Parole Ohio . . . . C S^n^. 33
For the Day tomorrow Captn Dunbar
Fort Guard Lt Petrie & 1 Sub: from Col^o^ Maybank
Picquet tonight a Sub: from Col^o^ Maybank
The Articles of war to be read this Evening at 5 OC to the 2nd Regiment by the Serg^t^ Maj^r^ and to Col^o^ Maybanks reg^t^ by Adjutant Kirwan at the same hour –

Col^o^. Marion hope the good harmony of subsisting between the Continental & State troops will continue, he Assures them he ~~will~~ does not nor will not consider them in any other light than friends

---

[1080] Reuben Dewitt served in the 2nd South Carolina Regiment Grenadier Company under Captain Thomas Dunbar. He was promoted to sergeant on 12 June 1779. He was at the siege of Savannah with Dunbar's light infantry.

[1081] Thomas Watts, Jr. served in the militia of Colonel Maybanks. On 12 June 1779 he was reprimanded for abuse to sergeant Kalckoffen. After the fall of Charleston he was in the militia under Colonel Myddleton.

to their country & therefore will be partial to Either, he hope the Officers in particular will be carefull that the present good harmony do Esential to their countrys will not be Interrupted

The Officer of the Day to Visit the Advance Guard once a day

15 June

Parole Ontario . . . . . C S^{n}. 51
For the day tomorrow Captn from Col^{o} Maybank
Fort Guard L^{t} ~~Ogier~~ Roux & 1 Sub. from Col. Maybank
Picquit tonight Lt Ogier
R. Ord^{rs} an Officer of a Company to train their men Every After noon when they do not Exercise the Cannon

After Order

1 Serg^{t}. & ~~Corp^{l}~~ and 12 Rank & file from 2 Reg^{t}. to be ready Immediately to go to Haddrels as Guard; to be relieved every 2 days

/ 16th June /

Parole Detract .. . C S^{n}. Canada
For the Day tomorrow Captn Proveaux
Fort Guard Lt Foissin & 1 Sub: from Col^{o} Maybank
Picquet to night L^{t}. ~~Kolb~~ Warley

/ 17th June /

Parole Louisiana – C. S^{n}. Missisippi
For the Day tomorrow Captn Gray
Fort Guard Lt Kolb & a Sub: from Col^{o} Maybank
Picquet tonight a Sub: from Col^{o} Maybank

/ 18th June /

Parole Orleans . . C S^{n}. 94
For the day tomorrow Captn from Col^{o} Maybank
Fort Guard Lt Ogier & a Sub: from Col^{o} Maybank
Picquet tonight Lt Petrie
A regimental Court martial to set this forenoon 10 OC: to try such prisoners Orderd to it evidences to attend Captn. Dunbar Presid^{t} & Lieu^{tn} Roux & Ogier members
Samuel McMillen of Captn. Mazyck company Appointed a Corporal of the same he is to be Respected as such[1082]

1082 Samuel McMillian was promoted to corporal in the 2nd South Carolina Regiment on 18 June 1779 under Captain Daniel Mazyck. He was in the siege of Savannah

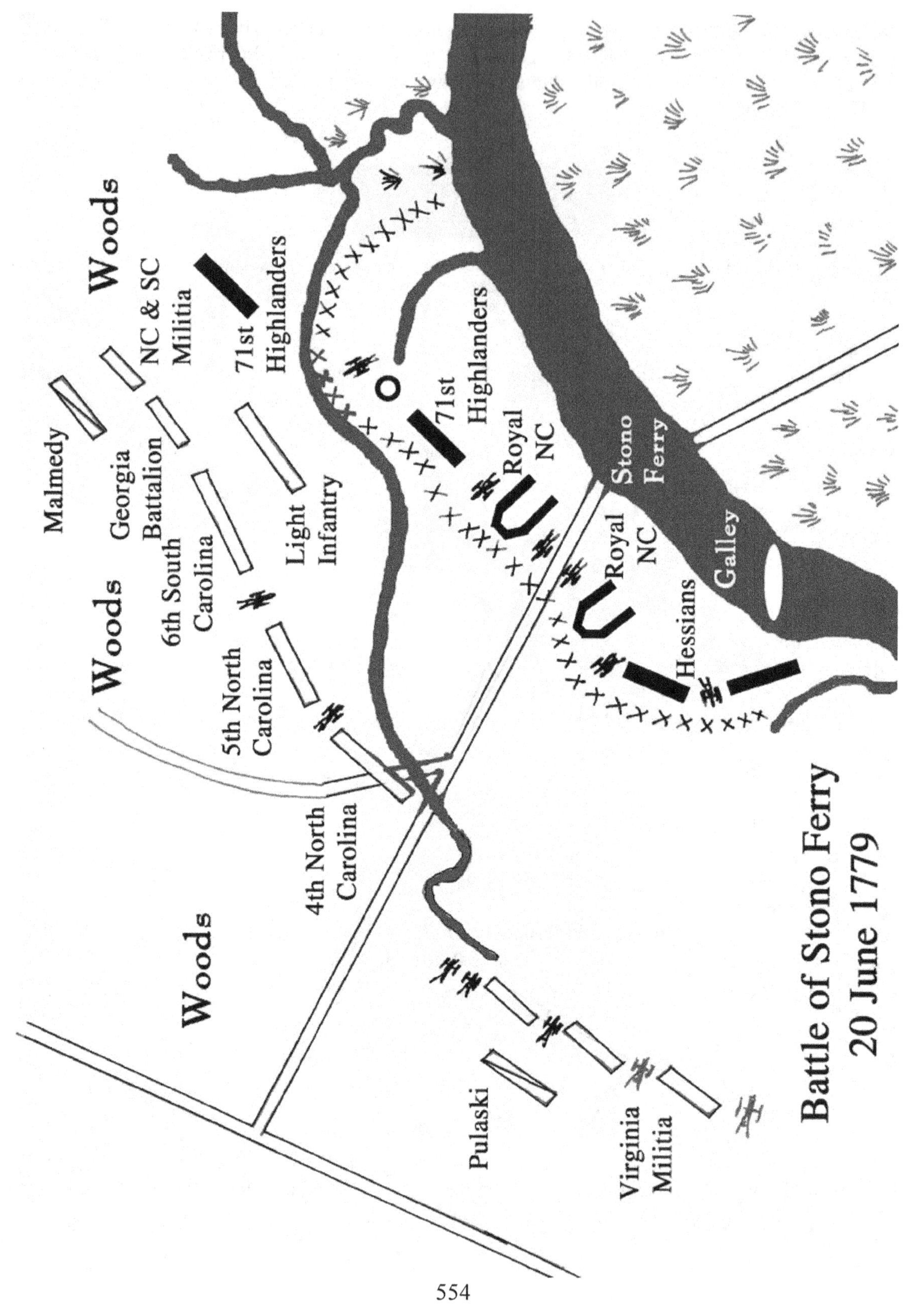
Woods
Woods
Malmedy
Georgia
Battalion
NC & SC
Militia
71st
Highlanders
71st
Highlanders
Royal
NC
Royal
NC
Stono
Ferry
Galley
Hessians
Woods
6th South
Carolina
Light
Infantry
5th North
Carolina
4th North
Carolina
Woods
Pulaski
Virginia
Militia
Battle of Stono Ferry
20 June 1779

On June 19th Colonel Moultrie received this order from Benjamin Lincoln, "immediately on the receipt of this… to throw over on James'-Island, all the troops which can be spared from the town; shew them to the enemy on John's-Island; carry your boats up Wappoo-cut, ready to throw your men on John's-Island in case an opportunity should offer without risking too much. If you should hear any firing in the morning at Stono-ferry and find the enemy on John's-Island moving from you, will endeavor to tread on their heels. I have written to Count Paulaski, to aid you in your movements." The 2nd South Carolina did not take part in the battle of Stono Ferry.

/ 19th June /

Parole Crosspoint . C S^n^. Lake George
For the Day tomorrow Captn Dunbar
Fort Guard L^t^. Warley & a Sub: from Col^o^ Maybank
Picquet to night Lt Roux

Prévost had been slowly evacuating his troops on any ships that arrived from Savannah. Prévost left the 71st Highlanders, the Hessian forces, and the Loyalists on the mainland at Stono Ferry to cover the withdrawal. Prévost then departed with the rest of the army for Savannah on June 16th. The front of the British defenses consisted of two square redoubts with a battery between them. They had three artillery pieces aiming down the road. Most of Lincoln's men were about to be discharged from service, so he knew that he had to act quickly to capture what British troops were left.

Moultrie was supposed to have led a detachment of 700 troops and one galley from Charlestown down to James Island. His mission was to fool Prévost into thinking that there might be an attack in his rear. Moultrie did not immediately execute these orders, due to a social event in Charlestown that many of his officers were attending.

Lincoln's army moved out to attack at midnight, and marched eight miles. Due to the confusion of his guides he formed his battle line three quarters of a mile in front of the British defenses. An hour after daylight Lincoln was finally able to get his line of battle in position for the attack. He tried to outflank the British, but the terrain was too broken for him to accomplish this.

The 6th South Carolina Regiment was able to drive back the Highlanders placed in front of the redoubts. In front of the redoubts was a deep creek that prevented the Continentals from storming the position. The British held their fire until the attackers were sixty yards away, then fired a devastating volley.

The North Carolina Continentals under Sumner were able to force the Hessians out of their redoubt, but an armed flatboat on the British side fired into the redoubts, keeping their heads down. The battle went on for fifty-six minutes, and only ended when the men began to run out of ammunition. Lincoln could see British reinforcements crossing the inlet, so he ordered a retreat. The mounted units and the Virginia militia covered the withdrawal of the army. Many of the Patriot militia who were listed as missing simply went home. Three days after the battle the British moved to Port Royal Island.[1083]

[1083] Full details of the Battle of Stono Ferry are described in "Nothing but Blood and Slaughter, Volume One".

/ 20th June /

Parole Onondaga . . C S^{n}. 49
For the Day tomorrow a Capt ~~from Col^{o}. Maybank~~ Proveaux
Fort guard Lt Foissin & a Sub: from Col^{o} Maybank
Picquit to night a Sub: from Col^{o} Maybank
G.O
It tis contrary to Orders that sentrys knowingly Set on their post it is hoped no sentrys will make use of that unmilitary practice as they cannot possible Do their duty properly

/ 21st June /

Parole Keywah . . . C S^{n}. Col^{o} Motte
For the day tomorrow Captn Gray
Fort Guard Lt Petrie & a Sub from Col^{o} Maybank
Picquit to night Lt Kolb

/ 22nd /

Parole Battle . . . . C S^{n}. Stono
For the day tomorrow Captn from Col^{o} Maybank
Fort Guard Lt Ogier & a Sub. from Col^{o} Maybank
Picquet to night a Sub. from Col^{o} Maybank –

23 June 1779

Parole Brave . . . C S^{n}. 27
For the Day tomorrow Captn Dunbar
Fort Guard Lt Roux & a Sub. from Col^{o} Maybank
Picquit to night Lt Warley

/ 24th June /

Parole ~~CS~~ Galleys – C S^{n}. 72
For the day tomorrow Captn from Col^{o} Maybank
Fort Guard tomorrow Lt Foissin & a Sub. from C. Maybank
Picquet to night a Sub: from Col^{o} Maybank
R. O A court martial to set at 11 OC: to try such prisoners as shall be Orderd to it, Evidences to Attend
President Captn Proveaux Members L^{ts} Warley & Petrie

/ 25th June /

Parole Telemachus – C: S^{n}. – Ulysses
For the Day tomorrow Captn Proveaux
Fort Guard Lt Petrie & a Sub: form Col^{o} Maybank
Picquet tonight Lt Kolb
A Court Martial to set at 10 OClock this Morning to try such prisoners as shall be Order'd to it, Evidences to attend Captn. Gray President
L^{t}. Petrie } Members { L^{t}. Kolb

/ 26th /

Parole Mentor . . . C Sn. Troy
For the Day tomorrow Captn Gray
Fort Guard Lt Ogier & a Sub from Col. Maybank
Picquet tonight a Sub: from Colo Maybank

June 27th 1779

Parole Utica . . . . C Sn. Ajax
For the Day tomorrow Captn Dunbar
Fort Guard Lt Warley & 1 Sub: from Colo Maybank
Picquit to night Lt Roux

/ 28 June /

Parole General Moultrie . . . . . . C Sn. Sir Peter
For the Day tomorrow a Captn from Colo Maybank
Fort Guard Lts Kolb & Foissin
Picquit tonight Lt from: Colo Maybank

/ 29 /

Parole Beaufort . . . C Sn. Polaskey
For the Day tomorrow Captn Proveaux
Fort Guard Lt Ogier & 1 Sub from Colo Maybank
Picquit tonight Lt Petrie

/ 30th June /

Parole MarbleHead. . C S. Boston
For the Day tomorrow Captn from Col. Maybank
Fort Guard Lt Roux & a Sub: from Colo Maybank
Picquit tonight a Sub. from Col. Maybank
the Cannon to be Exercised ~~to be~~ only every Wednesday Aftr. Noon, the Small Arms agreeable to Order of the 15th Instant

/ 1st July /

Parole General Scott . . C Sn. Brunswick
For the Day tomorrow Captn Gray
Fort Guard Lt Foissin & a Sub. from Col. Maybank
Picquit tonight Lt Warley

2d July

Parole Columbia . . C Sn. Burmuda
For the Day tomorrow Capt from Colo. Maybank
Fort Guard ~~tonight~~ Lt. Petrie & a Sub. from Colo. Maybank
Picquit tonight Lt. Kolb –

3rd July –

Parole Hector – C: S^{n}. Priam –

For the Day toMorrow Capt Gray

Fort Guard Lieu^{tn} ~~Petrie~~ Roux –

Picquit tonight a Sub. from Col^{o} Maybanks ~

His Excellency the Governor presents his Compliments to Col^{o} Marion & the other Officers of the 2^{d} S^{o} Carolina Continental Regiment & desires the pleasure of their Company to Dinner at the State House next Monday, at Three oClock ..

R. O: -

A Boat will be ready on Monday Morning to carry those Officers (off Duty) to Town that Accept his Excellency's Invitation. they are to Return to Fort Moultrie in the Evening –

The Orderly Sergeant or Corporal of each Company to turn out Two Fatigue Men immediately – they are to be under the direction of the Q^{r} Master Serg^{t}. –

4th. July –

Parole Andromeda – C: S^{n} Proveaux

For the Day to Morrow Captn from Col^{o}. Maybank

Fort Guard – Lieu^{tn} from D^{o}. –

Picquet toNight Lieu^{tn} Warley –

A Court Martial to sit immediately for the trial of such prisoners as shall be brought before it Evidences to Attend –

L^{t}. Warley – President

L^{t}. Foissin } Members { L^{t}. Kolb

July 5th –

Parole Vulcan – C S^{n} Venus

For the Day tomorrow Captn. from Col^{o} Maybank's

Fort Guard Lieu^{tn} Foissin

Picquet a Sub: from Col^{o} Maybank –

– July 6th –

Parole Jupiter – C: S^{n}: Leda

For the Day toMorrow Captn Gray –

Fort Guard Lieu^{tn}. From Col^{o} Maybank's –

Picquit tonight Lieu^{tn} Kolb –

Agreeable to the Sentences of this Court Martial the 4th Inst. –

Thos. Windsor & ~~Thomas Stafford~~ Reciev'd 100 Lashes & Thomas Stafford 50 Lashes – 40 being remitted. – (for Theft)

July 7th –

Parole Motezuma – C: S^{n}- Cortes –

For the Day toMorrow a Captn from Col^{o} Maybank's

Fort Guard Lieu^{tn} Ogier –

Picquet to Night a Sub: from Col° Maybank –
The Militia being Discharg'd After do this Detail as follows –
Captn. Gray for the day & L^t^. Ogier for Guards – this Picquet withdrawn

For the Day tomorrow Lieu^t^ Roux For Guard Lt Foissin
Captn. Vanderhorst produced the following Orders
(Copy)
Sir –
You are immediately to discharge the Militia under your Command at Fort Moultrie
July 6th 1779 (Sign'd) John Rutledge [1084]

The Militia being discharged the Guards were Reduc'd as follows –
Garrison Guard – 1 Sub: 2 Serg^ts^. 2 Corp^ls^. & 21 privates
Advance D°. – 1 Serg^t^. 1 Corp^l^. & 6 privates Rear d° as usual & the Picquit withdrawn –

July 8th –

Parole Cato ~ C: S^n^: Casar –

For the Day tomorrow Lieu^tn^ Kolb –
For the Guard Lieu^tn^ Ogier –
A Return of the strength of each Company to be immediately made & deliver'd to the Commanding Officer particularly specifying the different guards & commands which the Men are on . –

– 9th -

Parole – Semprorius – C: S^n^: Syphad

For the Day tomorrow Capt Provaux –
For Guard – Lieu^tn^ Roux –
R.O The Orders of the 7th ~~May~~ June relative to the Officers of the Day, to be strictly Complied with –

– July 10th –

Parole – Demosthenes – C: S^n^. Cicero

For the Day toMorrow Captn Gray –
For Guard Lieu^tn^ Foissin –

– 11th -

Parole Lucius – C S^n^. Jubal

For the Day toMorrow Lieu^tn^. Roux
For Guard Lieu^tn^ Kolb –

– 12th -

Parole Lincoln – C: S^n^. Brylis –

For the Day toMorrow Cap: Provaux
For Guard Lieu^tn^ Ogier –
R:O.

1084 After Stono Ferry the militia's enlistments were over and the North Carolina militia went home.

A Court Martial to set at 10 oClock this Morn^{g}: for the Trial of such prisoners as shall be brought before it – Evidences to Attend –

Capt: Gray – President
Lt Foissin } Members { Lt Ogier

General Orders by Major Genl Lincoln –
Rum to be Issued to Morrow but the Non Commissioned Officers, Drum & Fifes & privates present for Duty or such in Garrison & are not in the Regimental Hospital – Officers will draw once a Week
R:O.

The Surgion of the Regiment is immediately to furnish the Commanding Officer with a Return of all Necessaries wanted for the Hospital that he may be supplied with them –
No more than Two Women of each Company are to draw Rations – the Commanding Officer of each Company ~~of each Company are~~ to furnish the Q^{r} Master Sergeant with the Names of such as are Allowed to draw –
Agreeable to the Sentences of the Court Martial John McBride received Fifty lashes for Absence without Leave –

– July 13th: –
Parole – Proserpine – C: S^{n} – Pluto

For the Day toMorrow Captn Gray
For Guard – Lieu^{tn} Foissin
A Court Martial to set at 10 oClock this Morning for the trial of such Prisoners as shall be Order'd to it – all Evidences to Attend –

Captn Gray – President
Lt Roux } Members – { Lt Kolb

Sergeant Brewton was Tried by the above Court for Absence without leave & Acquitted

– July 14th -
Parole – Romeo – C: S^{n}. Juliet

For the Day toMorrow Lieu^{tn} Roux –
For Guard Lieu^{tn} Ogier –
R: O:
The Articles of War to be Read this Even^{g} at Roll Call by the Sergeant Major –

– 15th –
Parole – Paris – C: S^{n}. Helen –

For the Day toMorrow Captn Proveaux
For Guard Lieu^{tn} Warley

A Court Martial to sit at 10 oClock this Morning for the trial of such prisoners as shall be brought before it – all Evidences to Attend

Lt Roux – President
Lt Warley } Members { Lt Kolb

The Orders of the 15th February & 27 May to be particularly Attended to –

"No more than Two Women of each Company are to draw Rations"[1085]

July 16th 1779

Parole Mark Anthony . . . C S^{n}. Cleopatra

For the Day tomorrow Captn Dunbar

Fort Guard L^{t}.

Captn Mazyck, Proveaux & Grays ~~companys will~~ Company that are fit for duty to hold themselves in readyness to March at an hours warning, they will Apply to M^{r}. Simpson for one Shirt one Overhalls & 1 Lining Jacket p^{r}. Man What Arms & Accoutrements are wanting to Compleat them – it is expected that Captn. Mazycks Comp^{y}. will be ready to go to town this Afternoon –

Lt Warley will immediately Join Captn Masons Company in town – Lt Petrie to relieve L^{t}. Warley who is now on Guard

[1085] This is a drawing of my wife Alice O'Kelley.

Prévost had evacutaed his men to Beaufort but he did not have enough men to occupy the town. He constructed a defensive position on Edisto Island. Most of his troops were kept on board the ships and only a few men were allowed on shore to get water and wood. Colonel Daniel Horry kept an eye upon the British forces on Port Royal Island, but he did not have enough men to attack either.

On July 7th the 1st South Carolina Regiment under Colonel Pinckney was ordered to Port Royal ferry to reinforce Horry. Prévost's ships had begun shuttling the troops to St. Helena Island but Lincoln thought that there was still a chance that the British could try to take Beaufort or sail around to attack Fort Moultrie.

/ 17th July /

Parole Augustus Seasor . . C Sn. Emperor

For the day ~~tomorrow~~ today Lt. Petrie

Fort Guard a Sergt. –

Orders by Genl. Lincoln July 12. 1779

In the dayly returns for rum no Commissioned Offr to be included, company to draw but the non Commissioned Offrs. Drums and fifes & Ranks & file privates & those Sick in quarters

Offrs are to draw weekly for themselves & for their Servants who are not of the Army

All the provision returns & Return for rum are to be Signed by Commanding Offrs of Battalions or Corps who are desired to be exceedingly carefull they agree with the morning report which are to be made agreeable to the prentid form now in the hands of the Deputy Adjt General & are ready to be Delved out to the Several Battalions & Corps

July 14th

Commanding Offrs of Regts & Corps are Desired to be extreamly Attentive to the Orders of the 12th Instant particularly with respect to returns ~~for service~~ for Rations & to take the first Opportunity of Bringing to a Court Martial any Officer who may in future neglect to make proper returns of their respective Companys

No more than two Women in each company will be Allow'd to draw Rations & they are to be Appointed by the Commanding Officer of Regiments or Corps ~

July 16th

A General court martial to set at 10 OC: on Monday next in the new Barracks to try such prisoners as will be brought before them, Colo Bedaulph to be President [1086]~ the Detatchment of the 2nd Regt at fort Moultrie to furnish one Captn. & one Subaltern

Transmitted by

Thos. Hall
DAG

[1086] Charles Frederick de Bedaulx of Pulaski's Legion infantry. After the aborted ambush at the Old Race Tracks on 11 May 1779 Pulaski's infantry was no longer able to function. Pulaski was only able to field his cavalry after that skirmish.

Reg^{tl}. Ord^{rs}.

Captn. Moultrie & L^{t}. Hume are appointed members of the Genl. Court Martial to set on Munday next –

the names of such women Who Draw rations in the several Companys to be Given in today to the Commanding Officer

18th –

Parole Detatchment . . . . . C S^{n}. Beaufort

For the Day today ~~Captn. Dunbar~~ Lt Roux

For the Day tomorrow Lt Hart

Fort Guard ~~a serg^{t}~~ – Lt Petrie –

For the Future Officers & Servants are to be included in their Company returns for their rations of provisions & are not to draw separate ~~from the Q^{r} Mast^{r} Serg^{t}~~ as it Occasions Great Confusion in the Accounts, all the Staff are to draw together the returns for them to be made out by the Serg^{t} Major, the Gunner & the Mates to draw in the Company they properly belong to Also the Armourer & his Mate

---

July 19th –

Parole – Dejanira – C: S^{n}. Hercules

For the Day toMorrow Lt Roux –

For Guard – Lieu^{tn} Rogers

20th

Parole – Desdemona C: S^{n}. Othello

For the Day toMorrow, Lieu^{tn} Hart –

For Guard – Lieu^{tn} Foissin –

21st;.

Parole Cassius – C: S^{n} Brutus

For the Day toMorrow, Lieu^{tn} Roux –

For Guard Lieu^{tn} Petrie –

A Return of the Arms wanting in each Company to be immediately made & deliver'd to the Com^{dg} Officer [1087]

July 22^{d} -

Parole Charon – C: S^{n} Styx

For the Day toMorrow Lieu^{tn} Rogers –

Fort Guard Serjeants Roberts & McDonald

[1087] During this time Sergeant Jasper of the 2nd South Carolina was still detached from the regiment to conduct special operations behind the British lines in Georgia. Moultrie wrote to Lincoln, "Sergeant Jasper with a party of men wait upon you, desirous of something being given them to do. Your being immediately on the spot, will better enable you to judge of the most advantageous manner in which they may be disposed of. It is theirs and my wish that they may be employed at your discretion." Lincoln ordered "the Georgia troops, Captain Newman's company of horse" and Jasper's handpicked men to Georgia "to harass and perplex the enemy in that state." Full details of some of his missions are described in "*Nothing but Blood and Slaughter, Volume One*" by Patrick O'Kelley.

R: O.

A Court Martial to sit at 10 oClock this Morn^{g}. for the trial of all prisoners brought to it – all Evidences to Attend –

L^{t}: Hart President

Lt Roux } Members { Lt Rogers

As the Orders of the 20th June are not properly Attended to, the Orderly Serjeants or Corporals are desired to Acquaint every Man in Garrison with them, that none may plead Ignorance –

Geo. Carrick & Thomas Crozier were try'd by the above Court [1088]– the former for Neglect of Duty & Sentence'd to Receive One Hundred Lashes, but Recommended to Mercy – the Latter for Absence without Leave Sentenc'd Fifty Lashes but Remitted –

23^{d} -

Parole Hymen – C: S^{n}. Cupid

For the Day toMorrow Lieu^{tn} Hart

Fort Guard Serg^{ts} John Roberts & John McDonald

– 24th –

Parole Nonemia C: S^{n}. Polydon

For the Day toMorrow Lieu^{tn} Roux

Fort Guard Serg^{ts}: Bruton & Stone

M^{r}. Cornelius Van Hempstead Vlieland is appointed a Second Lieu^{tn}. in the regim^{t}. & is to be respected & Obeyd as such – to take rank from 17 July

L^{t}. Vlieland is to Join and do duty in Captn. Bakers Company

Joseph Wilkins of Captn Bakers Company is Appointed a a Corporal in the same & is to be Obey'd as Such ~[1089]

/ 25 July /

Parole Genl. McDougal . . C S^{n}. 5 and 37

For the day tomorrow Lt Petrie

Fort Guard Serg^{ts} Oniell, & Dewit

26th –

Parole Seneca – C: S^{n}. Nero

For the Day toMorrow Lt Rogers

Fort Guards Serg^{ts}: Green, & Bruten

1088 George Carrick enlisted in the 2nd South Carolina Regiment on 4 November 1775. During 1779 he was under Captain Thomas Dunbar. He was sentenced to receive 50 lashes on 22 June 1779 for absence without leave, but it was remitted. He was at the siege of Savannah.

1089 Joseph Wilkins served in the 2nd South Carolina Regiment under Captain Richard Bohun Baker. He was promoted to corporal on 24 July 1779.

27th

Parole Cambresis – C: Sn. Cyrus

For the Day to~~Morrow~~Day Lt ~~Roux~~ – Hart

Fort Guard Sergts: - Stone & McDonald

For the Day tomorrow Lt Roux

28th

Parole Norval C: Sn. Douglass –

For the Day toMorrow Lieutn.

Fort Guard Sergts: ~~McDonald & Stone~~ McDonald & Bruten

R: O.

A Court Martial to sit at 10 oClock this Morning for the trial of all prisoners brought before it – Evidences to Attend –

Lt. Hart President

Lt Roux } Members { Lt Rogers

The Commanding Officer is inform'd that several Soldiers go over daily to Haddrells without ~~a pass~~ permission

the Sergeant of the Rear Guard will be Answerable for any future Neglect

/ 29 July /

Parole Bedford . . C Sn. 97

For the Day to~~morrow~~ day Lt Roger

fort Guard Sergt ~~McDonald & Dewit~~ & Henderson

For the Day tomorrow Lt Hart

Mr. James Gray is Appointed a second Lieutn. in the Regimt he is to be respected & Obeyd as such To take rank from 23 July Inst: -

Lt. James Gray is to Join & Do Duty in Captn. Grays Compy. till further Orders

A Return of the Number men present in Each Compy to be Given in today to receive Cloathing – Commanding Offr of each Compy to keep an Exact Account In their Compy Book to whom Given

30 July

Parole Tinmouth C Sn. 79

For the day tomorrow Lt. Roux

Fort Guard Sergeants Bruten & Stone

A court martial to set today 11 OC: in forenoon to try all prisoners Orderd to it – evidences to Attend

Captn. Moultrie president

Captn. Dunbar Membr. Lt. Roux

Lt. Rogers Vlieland

John Hyrne Drumr is Appointed Drum Major in the regiment he is to be Obeyd as such.

The pay Rolls to be made out today & Delivered to the Commanding Offr ~~for his~~ it is hoped that it will not be delayed any Longer as it subjects the men to Great Inconvenientcy

/ 31st July /

Parole Somerset . . . C Sn. 35.
For the Day tomorrow Lt Rogers
Fort Guard Serjeants Oniell & Dewit –
After Orders
a monthly return to be given tomorrow 10 OC: in forenoon, to the Commanding Offr.

/ 1st August /

Parole ~~Galvan~~ New castle. C Sn. 45
For the Day tomorrow Lt Vlieland
Fort Guard Sergeants Stephen Roberts & Webb

2d

Parole Middlesex – C Sn. Wilkes
For the day tomorrow ~~Captn Dunbar~~ Lt Hart
Fort Guard Sergts McDonald, & Campbell
The Barrack Guard to be Relieved once a week
A court martial to sit at 11 OClock this fore noon to try all prisoners orderd to it, evidences to Attend
Lt Hart President
Lieuts Petrie & Humes members –
NB by Sentence of the above court moses Grumer was acquitted for Defrauding his mess of provisions,[1090] Humphrey Haines was Acquitted Sentenced 25 Lashes for sell his Ammunition to a Negroe but was remitted by – the Colo.[1091]

/ 3d August /

Parole Washington Cn. Sn. Virtue
For the day tomorrow Lt Roux
Fort Guard Serjts Brewton & Stone

4th August

Parole Rutledge C Sn. _ 20 _
For the day tomorrow Lt Hume
fort Guard Serjts Murphy & McDonald

5th

Parole Lincoln Cn Sn. 10
For the Day tomorrow Lieut Rogers
Fort Guard Serjts Webb & Henderson
Reare Guard Serjt Stephen Roberts

[1090] Moses Groom enlisted in the 2nd South Carolina Regiment on 1 August 1779 under Captain Thomas Moultrie. He was charged with defrauding his mess of provisions on 2 August 1779, but he was acquitted. He was given leniency on this charge since he had only been in the regiment one day. He was at the siege of Savannah.

[1091] Humphrey Haines served in the 2nd South Carolina Regiment after 2 November 1776 and was under Captain Charles Motte during 1779. On 2 August 1779 he was sentenced to receive 25 lashes for selling his ammunition to a "negro", but it was remitted by the Colonel. He was at the siege of Savannah.

6th August

Parole Genl. Huger C n. S n. 27.
For the Day tomorrow Lieutenant Vlieland
Fort Guard Serjts Green & Bernard [1092]
for Rear Guard Serjt. John Roberts

Captn. Lesesne

/ 7 August /

Parole Worcester . . C Sn. 27.
For the day tomorrow Captn Moultrie
Fort Guard Lt Roux

at the beating the Long roll at any time, the whole ~~regimt~~ Garrison are to turn out & imediatly repair to their alarm post (except the hours of Exercise), which are to be as follows Captn Moultrie's Company in the old Flagg Bastion, the Light Infantry in Magazine Bastion, Bakers, the East Curtain; Captn Moultrie will Command these 3 Compys.

Hall and ~~Baker~~ Mottes Company, in the front Curtain Granadiers in the flagg Bastion, which will be commanded in the Absents of Captn Dunbar by Lieut Hart ~

the Gunner & one of his mate to attend Captn Moultrie the other mate ~~to~~ With Lt Hart. – the Guard will Remain for the defences of the Gate ~ the Sergt Majr will carry all orders from the Commanding Officer ~

A court martial to set at 11 OC: this forenoon to try all prisoners Ordered to it, Evidences to Attend –

Captn. Moultrie President

Lts Roux, Petrie, Humes & Rogers members –

Commanding Officers of Companys will apply to Mr. Simpson for Shoes for such of their men are are present

8th August 1779

Parole Durham . . . C Sn. 105
For the day tomorrow Lieutm Petrie
Fort Guard Lt Hume

Genl. Ordrs 7th Augt. by Genl. Lincoln } Commanding Officers will immediately make a return of the Artificiers in their different Corps

Regt. Ordr. } Commanding Offrs of companys will give in to the Commandg. Offr. a return of the Different Artificiers in their companies today

The Cartridges given the men are to be Examin'd every Roll-call & the men made answerable for any Loss or Damage that is not avoidable

[1092] Sergeant Noble Barnett.

/ 9th August /

Parole Snowden. . C Sn. 37
For the day tomorrow Lt Rogers
Fort guard Lt. V.H. Vlieland
The court orderd to set on Saturday did not on account of Some of the members being Absent it tis therefore Orderd to sit today at 11 OC: in the forenoon with this alteration
Captn Dunbar & Lt Bush in the room of Lts Petrie & Humes
NB by sentence of the above court Chs. Caves recd. 100 lashes for Absents without Leave[1093] – Sergts Newman & Simpson was reduced to the ranks. Newman recd. 100 lashes & put under Stoppages for Embezelling Cloathing of the regimt. Simpson under Stoppages of the pay now due him & Discharged the service also for Embezelling the Cloathing belonging to the regt. _ McBride to receive 100 Lashes for selling his Cloathing, but remitted as he turned evidence against Newman & Simpson –

/ 10th August /

Parole Grannades . . C Sn. 113
For the Day tomorrow, Captn. ~~Dunbar~~ Capers
Fort Guard Lieut. ~~Roux~~ Petrie
Sergeants Simpson & Newman is reduced in the ranks for Embazelling the regimental Cloathing, they are no longer to be considerd Sergeants in the regiment

The Light Infantry Company Commanded by Captain Lesesne, is no longer to be considerd as Light Infantry but as a Battalion Company

Captn. Dunbars Company to be a Light Infantry Compy. Instead of Granidiers

Captn. Dunbar will Chose a 11 men from Capt Lesesne Compy to Augment his Light Infantry, & to Exchange those men in his Company that will not do for Light Infantry for other men in the remaining four Companys present – any of the men Left that belong to the Companys gone on Command, that may do for Light Infantry, are to be taken for that purpose, not Exceeding nine ~

As Congress has Order'd that a Light Infantry Compy shall be formed in each regimt. & Kept up by Draught from the Battalion Companys, tis hoped no Officer will be Displeased at the above regulations.

Sergt. Davis, Chs. Caves, McClean & McCoullock is to do Duty & Draw rations in Captn. Halls Compy. till further Orders[1094]

11th August 1779 **Capt Motte**

Parole America - C Sn. Independence
For the day tomorrow Captn Dunbar
For the Fort guard – Lt Roux –

[1093] Charles Caves served in the 2nd South Carolina Regiment under Captain Daniel Mazyck. When Captain Dunbar's company became light infantry he was transferred to Captain Hall's company. On 9 August 1779 he received 100 lashes for being absent without leave. He was in Captain Mazyck's company at the siege of Savannah.

[1094] James McClean (McClain) enlisted in the 2nd South Carolina Regiment on 5 August 1775 under Captain Bernard Elliot. He re-inlisted on 4 November 1775 and was discharged on 4 August 1778. He re-inlisted as a sergeant in Captain Dunbar's Grenadier company in 1779. When Captain Dunbar's company became a light infantry company he was transferred to Captain Hall's company.

12th Augt. **Lt Colo Marion**

Parole Richmond C Sign 88 –
For the day tomorrow Capt: Moultrie & Lt Capers
For fort guard Lt Humes ________

13th Augt.

Parole Windsor - C Sn. 107
For the Day tomorrow Lieut Capers. say Petrie
Fort Guard Lt Petrie say ~~Vlieland~~ Rogers
Any Persons found doing their Occasion, within the fort or within twenty yards of the walls on the outside may Depend on being tried by a Court Martial & Suffer Accordingly – the Guard and Centries are orderd to prevent any persons from Disobeying this order & if Detected to confine them immediately

the Alarm post for the North Carolina troops is the West Curtain & Bastion

______________________/ 14th Augt /______________________

Parole Light Infantry – C Sn. 204
For the day tomorrow Captn Dunbar
Fort Guard Lieutn Roux
Soap to be Issued to the troops in the proportion of ten pounds to a hundred men per week

______________________/ 15th Augt /______________________

Parole Harmony ________ C Sn. 13
For the Day tomorrow a Captn from N. Carolina
fort Guard Lt Vlieland
Head Quarters 12 Augt. 1779
Captn Lesesne of the 2d So Carolina Regt. having resigned his Commission is no longer to be considerd as a Continental Officer – Quartr Mastr Hall of the same Corps having resigned his Appointment is no longer to be Considerd as a Continental Offr.
Lt Saml Smith of the first So Carolina Regt having resigned his Commission is no longer to be Considerd as a Continental Officer[1095]
Captn.[1096] & Lieut. Davis[1097] of the 5th So. Carolina Reg: are having resigned their Commissions are no longer to be consider'd Continental Offrs. [1098]

Transmitted Thos Hall
D.A.G

In 1777 the ten North Carolina Continental regiments marched north to Washington's army and participated in the battles of Brandywine, Germantown, the winter at Valley Forge, and Monmouth. By Monmouth the ten regiments had been reduced dramatically, and they were reconsolidated into the 1st and 2nd North Carolina regiments. The 3rd North

1095 Samuel Smith served as a Lieutenant in the 1st South Carolina Regiment and resigned in August 1779 after the failed attack at Stono Ferry
1096 William Ransom Davis
1097 Unable to find any information on a Lieutenant Davis in the 5th South Carolina Regiment
1098 All these resignations may have been in part due to the failed attack at Stono Ferry.

Carolina regiment consisted mainly on paper and the officers were sent home to recruit men to fill the regiment.

In 1778 the North Carolina legislature had authorized a draft to fill the ranks of 2,648 nine month levies. Half of these men served in Washington's army in New York, and the other half went to South Carolina to oppose the British threat from that quarter. In March 1779 these men were commanded by Jethro Sumner and joined Lincoln's forces at Black Swamp. The 759 levies formed the newly organized 4th and 5th North Carolina Regiments. They were ill equipped and relied on South Carolina for much of their inventory. They participated in the Battle of Stono Ferry, and by July the levies had returned home. What remained were soldiers who were not levies and were Continentals on furlough, or men who had enlisted in the newly created 3rd North Carolina Regiment. These men were at Fort Moultrie with the 2nd South Carolina in August. [1099]

/ 16th August /

Parole Carolina C Sn. 12
For the Day tomorrow Lt Capers
Fort guard a Lt from N. Carolina troops

~~17th August~~

~~Parole Congress Cn. Sn. 50~~
~~For the day tomorrow Lt. Foissin~~
~~For Fort guard from No. Carolina troops~~

17th August

Parole Congress – Cn. Sn. 50
For the Day tomorrow Captain Dunbar
Fort Guard Lieutenant Roux –

18th August

Parole Genl. Waine Sn. Sn. 500
For the Day tomorrow Lt Foissin –
For Fort guard From No Carolina Troops
Gen orders by Genl. Lincoln } Capt Drayton of the first Regt of So Carolina having Resigned his Commission is no longer to be Considered to be a Continental Officer

19 August 1779

Parole Liberty Cn Sn. 45
For the day tomorrow Lieutn Petrie
For fort guard – Lt Rogers
The Bricklayers, Wheelwrights, House Carpenters & Gun Smiths (excepting the Armourers) in the Regiment are to get ready to go to Town –

[1099] Rees, John U. "'*The Pleasure of Their Number': 1778, Crisis, Conscription, and Revolutionary Soldiers' Recollections.*" ALHFAM Bulletin 33, no. 3 (Fall 2003): 23-34; 33, no. 4 (Winter 2004): 23-34; and 34, no.1 (Spring 2004): 19-28.

Head Quarters 14 Aug[t]. 79

Sir

The General requests you will please order the Bricklayers, Wheelwrights, House Carpenters and Gunsmiths, Artificiers in your Regiment, to attend L[t] Col[o] Grimkie of the Artillery – he may have occasion to employ them for a short time, when they will be orderd to the Island

L[t]. Col[o]. Marion — I am
Fort Moultrie — Sir your &tc
In his absence the Officer commanding the 2[d] Regiment — Signed Will Jackson ADC

20 August 1779

Parole Fairfield . . . . . . Count[r]. S[n]. 27.

For the day tomorrow Capt[n] ~~Moultrie~~ Dunbar

fort Guard L[t] ~~Hume Roux~~ Vlieland

Lieu[tn] Hume is to Join & do duty in Capt[n]. Bakers till further Orders –

Lieu[tn] Vlieland to Join & do duty in Capt[n]. Dunbars Comp[y]. till further Orders

/ 21st August /

Parole New haven — Count[r] S[n] 72.

For the day tomorrow Capt[n] Moultrie

Fort Guard Lieu[t]: Capers

/ 22 August /

Parole Norwarck — C. S[n]. 127.

For the Day to Day Capt[n]. Chapman [1100] in room of Moultrie sick

For the Day tomorrow L[t] Foissin

Fort Guard L[t] from North Carolina troops

23[d] August

Parole Gov[r]. Caswell — C[n]. S[n]. 11

For the day tomorrow ~~L[t] Petrie~~ L[t] Roux

For the Fort guard ~~L[t] Rogers~~ L[t] Vlieland

24 August

Parole Shain __________ C[n]. S[n]. Independance

For the day tomorrow Lieu[t] Petrie

For the Fort guard L[t] Rogers

[1100] Samuel Chapman had been commissioned in the 8th North Carolina Regiment on 28 November 1776. He was promoted to first lieutenant on 1 August 1777, and after the reorganization of the North Carolina regiments at Valley Forge he was transferred to the 4th North Carolina Regiment on 1 June 1778. He was promoted to captain on 5 April 1779 and served until the close of the war.

25 August 79 –

Parole – King of France C S^n. 16 –
For the day tomorrow – ~~Lt Hume~~ Capt Chapman
For the Fort guard Lieu^t Hume
General Ord^rs.

By M; G, Lincoln August the 10th 1779 } Captain Stephen Guerry having Resigned his Commission in the 5th South Carolina Regiment is No longer to be considered as A Continental Officer _ _ _ _ _ _[1101]

Genl. Orders August the 19th 1779 } Captain Lisle having Resigned his Commission in the third South Carolina Regiment is No longer to be Considered As a Continental Officer [1102]

Admiral D'Estaing was the 50 year old French Admiral who had been given command of a fleet in 1778. He had lost a chance for glory at Newport, Rhode Island due to a gale that made it impossible for him to land his troops. He had received orders to return to France, but he did not want to leave with the loss of Newport on his military career. D'Estaing sailed to the Grenadines, where he successfully captured the island of Grenada and St. Vincent.

General Lincoln, Governor Rutledge and Congress asked D'Estaing to join the Southern army in an attack on Savannah. Since it would take time to repair his fleet to sail back to France, D'Estaing stated that he would help. However, he gave them a warning that he would only stay off the American shores for two weeks, since the hurricane season was at hand. D'Estaing did not want a repeat of Rhode Island.

Moultrie wrote in his memoirs, "This information put us all in high spirits... we were sure of success; and no one doubted but that we had nothing more to do, than to march up to Savannah; and demand a surrender: the militia were draughted; and a great number of volunteers joined readily, to be present at the surrender; and in hopes to have the pleasure of seeing the British march out, and deliver up their arms; but alas! It turned out a bloody affair."

On the 16th of August D'Estaing sailed for the Georgia coast with thirty-seven ships, including twenty-two ships of the line, and 4,000 troops detached from duty in the West Indies.[1103]

1101 Stephen Guerry served as a captain in the 5th South Carolina Regiment. He resigned in August 1779

1102 John Lisle served as a lieutenant colonel in the militia during 1775. He joined the 3rd South Carolina (Ranger) Regiment as a lieutenant and was promoted to captain in 1779. He resigned in August 1779. In 1780 he joined the Loyalist forces as a lieutenant in the Rocky Mount Regiment. After the battle of Williamson's Plantation (Huck's Defeat) he brought his whole company back to Sumter's Brigade with all their arms and supplies that had recently been issued to them.

1103 O'Kelley, *NBBAS, Volume One,* pp 359-360.

26th August

L^{t}. Capers took the guard today in room Lt Hume Sick –
Parole Count D'Estaighn C^{n}. S^{n}. 24
For the day tomorrow ~~L^{t}. Foissin~~ Capt Dunbar
For the Fort guard Lieu^{t} Roux

L^{t}: Legare to take rank from 20 August 1779

August 27th 1779

Parole Sheldon C. S^{n}. 97.
For the day tomorrow Lt Foisson
Fort Guard L^{t} from the N Carolina ~~####~~Troops

M^{r}. James Legare is Appointed a Second Lieu^{t} in the 2nd Reg^{t} he is to be Obey'd & Respected as Such –

Lt James Legare to Join and do duty in Captn. Halls Company till further Orders

/ 28th Aug^{t} /

Parole Lt Govenor Count S^{n}. 79.
For the day tomorrow Lt Petrie
Town Guard L^{t} from North Carolina troops
R. O

a regimental court martial to be held today at 11 OC: to try all prisoners Orderd to it, Evidences to Attend

Captn Dunbar president Lt Roux & Rogers Members

NB Jno. Fenwick sentenced to receive 100 lashes with Switches for theft & selling his Shirt, his pay to be withheld for the Shirt

29 August

Parole Jay C S^{n}. 13 –
For the day tomorrow from Light Infantry [1104]
Fort Guard Light Infantry

The Troops now in Garrison to be exercis'd at the Cannon every Tuesday Afternoon, at the usual hours ~ . . . . .

Notwithstanding Orders to the contrary Soldiers are permitted to go over the Bridge to Haddrells Without a permit from the Commanding Officer

The Serg^{t}. Corporals & Centrys at the rear Guard will be punished in futer for all Disobedience of Orders

Those Soldiers (Officers, Servants excepted) who go over the Bridge without a pass Signed by the Command^{g} Officer may be Assured to Suffer Sevier for their Disobedience

[1104] This is a unit known as the Carolina Light Infantry that was under the command of Lieutenant Colonel John Laurens. This was a unit made up of other light infantry companies from various North and South Carolina regiments.

30 August 1779

Parole Stoney Point . . C Sn. Lt. Gibbon
For the Day tomorrow ~~Lt Petrie~~ Captn Chas motte
Fort Guard Lt Rogers [1105]
~~A month~~ A monthly Return to be made & given the Commanding Officer on Wednesday 1 Sept. at 10 OClock in forenoon ~
After Orders. a Sergeant 1 Corporal & 9 privates with a weeks provisions & 18 rounds & one Spare flint pr Man, to be immediately ready to go to Dewees Island, this party is to be from 2nd Regt & the Sergt will Call for orders as Soon as he is ready to proceed –
Orders Recd this day 3 OC. PM
Head Quarters Ch Town Augt 30th 1779
Sir The general requests that you will immediately on Receipt of this letter, order a Sergeants Guard from 2d Regt to Dewees Island, to be Stationed there as a covering party to the Workmen employ'd in cutting palmetto Loggs – and to be relieved, untill further order by You at Such time as may be most Convenient

I am Sr. Yr. M. O. Sert.
Will. Jackson
AD. Camp –

31st August 1779

Parole No Carolina Cn. Sn. Friend
For the day tomorrow Lieut Capers
For the Fort guard – from No Carolina Troops
The Gunners will attend the No Carolina troops, this afternoon to instruct them in exercising the Cannon –

/1st September /

Parole Colo Fleury C Sn. Lt. Knox
For the day tomorrow Light Infantry
fort Guard Light Infantry –
NB
~~A General court~~
A Garrison Court martial to set this forenoon at 11 OC: to try all prisoners Orderd to it – Evidences to Attend – Captn. Moultrie President
Lts Faisson & Hart from 2nd Regt.
2 Subalterns from the North Carolina troops

[1105] Christopher Rodgers served in the 2nd South Carolina Regiment under Captain Motte during 1779.

All the men able to March of the 2nd Regt & Captn Ramseys Detatchment to hold themselves in readyness to march at a Minutes warning

North Carolina Continental

/ 2nd Septr /

Parole Colo. Stewart C Sn. Brave
For the day tomorrow Lt ~~Petrie~~ Clark N:C [1106]
fort Guard ~~from N Carolina troops~~ Lt Petrie

The Garrison court martial orderd to set Yesterday are to set this day, Compose of the same

Tho Goodson Vacant Compy sentenced to receive 20 lashes with the cat 9 tails for sleeping on his post[1107]

September 3rd 1779

Parole Lincoln Count. Sn. 20
For the Day tomorrow Captn Moultrie
Fort Guard Lieut Rogers from N Carolina[1108]

/ Septr 4th

Parole, Count DeStaing C Sn. flute
For the day tomorrow Captn ~~eeeeeee~~ Dunbar
Fort Guard Lt from Light Infantry

/ 5th September /

Parole France Countr Sn. Spain
For the Day tomorrow Captn Chapman
Fort Guard Lieut Foissin from N. Carolina [1109]
4 OC PM

After Orders, All the men able to March of the 2nd Regt & Captn Ramseys Detatchment to hold themselves in readyness to march at a Minutes warning[1110]~ Captn Ramsey will give in a return immediately the number of men he has capable of Marching – the Officers & men are made Acquainted that no more baggage than what they can carry in their havresack, can be carried – the

[1106] Thomas Clark was an ensign in the 9th North Carolina Regiment on 28 November 1776. He was promoted to lieutenant on 1 February 1777. After the ten North Carolina regiments were reorganized at Valley Forge he transferred to 4th North Carolina Regiment on 1 July 1778. He was promoted to captain on 10 February 1779 and served to the close of the war.

[1107] Thomas Goodson was born in 1762. He enlisted in the 2nd South Carolina Regiment on 1 July 1779 under Captain Thomas Hall. On 2 September 1779 he received 20 lashes with a cat of nine tails for sleeping on his post. He was at the siege of Savannah. After the fall of Charleston he served with General Marion's partisans and was in the battle at Wadboo Bridge.

[1108] Christopher Rogers was not from North Carolina. In the Orderly Book there is a "fill the blank" space where a lieutenant from North Carolina was supposed to be assigned. The orderly wrote in Lieutenant Rogers name. Lieutenant Christopher Rogers (Rodgers) served in the 2nd South Carolina Regiment under Captain Charles Motte during 1779.

[1109] This is another "fill the blank" and Lieutenant Foissin was not from North Carolina.

[1110] Matthew Ramsey had been commissioned a captain in the 9th North Carolina Regiment on 28 November 1776. When the ten North Carolina Regiments were reduced at Valley Forge he transferred to the 4th North Carolina Regiment on 1 June 1778. He was in command of the North Carolina Light Infantry. Ramsey resigned in November 1781.

Quart^{r} Master Serg^{t} to give in an Acct of the Numb^{r} of Cartrige in Store, & to ball & fill all the Blank Cartridges immediately with all his time – he is to take as many men as will Compleat them this night ______________ [1111]

Genl. Orders by G. Lincoln Sept 5. 1779

All Officers now Absent from their Corps, will hold themselves in readiness to join them respectively All Soldiers on furlow will repair to their reg^{t}. immediately

Head Quarters at Ch^{s} town Sep^{t} 5th 1779

Sir

You will immediately order all your Officers & men capable of marching, to hold themselves in readiness to proceed at a Minutes warning to the S^{o}.Ward

Each man is to be Supplied with ~~40~~ 60 rounds of Cartridges ~~p^{r}. man~~ and to have his Arms & Accoutrements in compleat order

All the N^{o} Carolinians able to march are to proceed with you as it will be Exceedingly Difficult perhaps impossible to procure Waggons, you will intimate to your Officers & men, the Nessesity of taking with them as Little baggage as possible – little more than on Shift of Linen need be carried – please make a return by Whereabout what Number I may depend on.

L^{t}. Col^{o} Marion — I am S^{r}. Y^{r} most Obed^{t}. Servt.

Fort Moultrie — **B Lincoln**

___________________________________/ 6th Sept^{r} /___________________________________

Parole Motte — Count^{r}. S^{n} Major

For the Day tomorrow

Fort Guard L^{t}.

RO. { Commanding Off^{rs}. of Companys will immediately Examin'd their mens Arms & Accutrements & see they are in the best Order – all the Baggage of the Officers & men may be put in the regimental Store the care of Such will be Given to M^{r}. Harvey the Gunner; as we may expect to march every Minute,[1112] it is hoped the Off^{rs} & men will be ready as soon as Called upon to march instantly- no leave of Absents to be given to any man, on any Account whatsoever

The Col^{o}. Expects the men will have their clothes clean that they may make as Decent an Appearance in town as possible –

the Drum Major is Orderd to git all the Drums in the best Order immediately

---

[1111] General Lincoln ordered all officers and soldiers to join their respective regiments and march to Sheldon as soon as possible. The troops were all converging at Sheldon so as to march upon Savannah.

[1112] William Harvey enlisted in the 2nd South Carolina Regiment on 16 February 1778. On 1 August 1779 he was promoted to sergeant under Captain Thomas Hall. He was in the siege of Savannah.

Sir

As the N° Carolinians are unprovided You will please leave them on Island & move immediately with all Your own Regiments fir for a March to Savannah – bring your Camp Kettles – A Schooner is sent for you

No time it to be lost I am

Sir your hum Serv

Sign[d] B Lincoln

Col°. Marion

Sep[t]. 6th 1779.

After Order by Cap[t]. Moultrie

The Advance guard & the party at Dewees Island to be Relieved by the N° Carolinians immediately except the gunner & Mates

All the Men of the 2nd Regiment ^ able to march are to get themselves ready to embark immediately –

The Arms & accoutrements of the Workmen & others in Town, All the ammunition chests & Camp Kettles are to go with the Regiment

7th _ above Quart[r] House [1113]

Parole Marion – C: S[n]. Honour

For the day L[t] Roux

/ 8 /

Parole Aship - - C S[n] Dorchester

For the day today L[t] Foissin

The Gen[l] to Beat tomorrow Morning 12 OC: assemble & march at half after 2 – the waggon horses to be picquetted at retreat beating & to march with the troops – no Soldier to Leave their ranks without some very Extraordinary business and then must Obtain Leave from an Officer – It is Expected that the Soldiers will not Destroy Either rails or any thing belonging to the good people of the State – all Misdemeanor will be punished with the greatest Severity – the rear Guard to bring up all men who may Loiter or fall in the rear, & to bring them up

On the present Interesting behavour tis hoped that all Officers & Soldiers will Exert themselves in forwarding the march with the utmost Expedition [1114]

L[t] John Hart is appointed a Capt[n] in the Reg[t]. he is to be Respected and Obeyd as such

Capt[n]. Hart to take Charge of the comp[y] promoted Capt[n]. Lesesne

Capt[n] Motte is appointed a Major to the regiment and is to be respected & Obeyd as Such

L[t] Hume is promoted a first Lieu[t] he is to be respected & Obeyd as such, he is to do duty in Capt[n] Bakers Company

---

[1113] The Quarter House Tavern was located five miles from Charlestown, and was located where Success Street intersects with Meeting Street.

[1114] Lincoln began his march from Charleston to Savannah on September 8th after gathering an army of 1,500 men, mainly composed of militia.

/ 9 Sept /

For the Day today Lt. Petrie
Lt Vleland with such man as are Lame & Sick to form the Rear Guard & take charge of the waggons, & their contents, to follow the main body as soon as possible
2 Days provisions to be Issued to night
The Genl. to beat at 2 OC. in the morning assemble & march at half an hour After

10th Parole Tulefeneys C Sn. 4
For the day tomorrow Lt Legare
The Genl. to beat tomorrow at 3 OClock Assemble & march at 4 – –

September 11th

Parole Savannah - - Cn. Sn. _ 8

The French fleet arrived at Georgia on September 10th and moved to the Tybee Lighthouse at the mouth of the Savannah River. Fort Tybee had one 24-pounder and one 8½-inch howitzer, but it was no match for the fleet. After firing their guns a few times they abandoned the fort. That night a French detachment occupied the fort. On the stormy night of September 11th, 1,200 French troops landed unopposed at Beaulieu beach on Ossabaw Sound, south of Savannah. The French army established a camp three miles from the city.[1115]

Sept 12th

Parole Ebenezer- - Cn. Sn. - - 5

Sept 13th

Parole Pulaskie Cn. Sns Temple, & parker

Sept 14th

Parole Lawrence Cn. Sns. Barry. & Bark.[1116]

15th

Parole Savannah Cn. Sns. Arnold. & Wayne

16th

Parole Friendship Cn. Sns. France. & America

[1115] O'Kelley, *NBBAS, Volume One,* pp 360-361.

[1116] George Washington sent the Virginia Continental levies south to help stop the British invasion. On September 14th these Virginians, Parker's 1st Virginia Detachment and the 1st Continental Dragoons, joined Lincoln's army at Ebenezer 23 miles from Savannah.

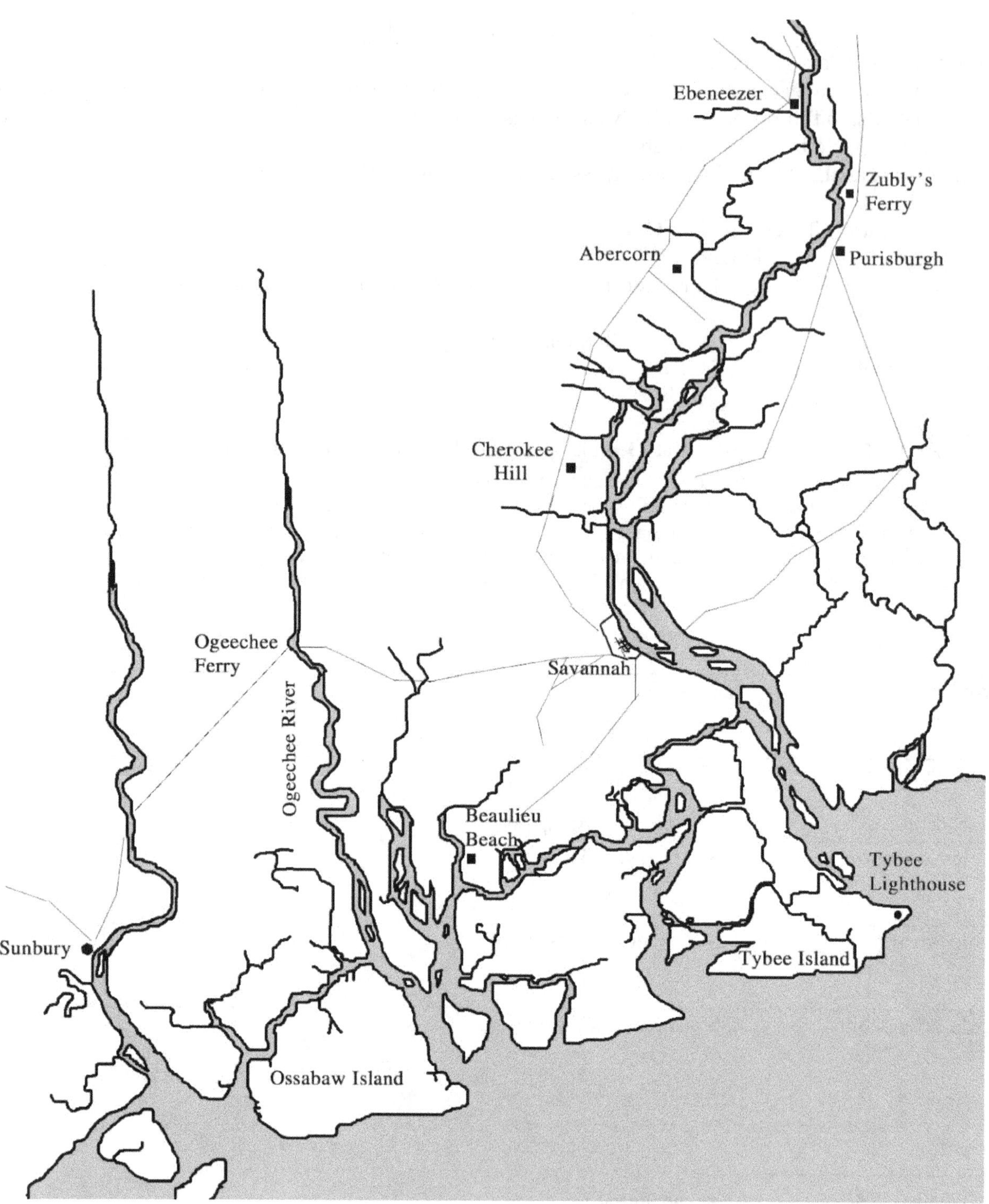

Campaign for Savannah
September – November 1779

Lincoln was able to cross some of his troops at Zubly's Ferry by using a single canoe they found. He told Pulaski to cross over with his troop using the single canoe. Pulaski sent one man across at a time with all his equipment, while his horse swim alongside him. Captain Bentalou was able to get thirty men across in this manner, then they were able to ride to D'Estaing, effectively linking up with the French.

On September 16th, D'Estaing sent a formal demand to Prévost that he surrender Savannah to the arms of his Majesty the King of France. Prévost asked for a 24-hour truce to allow him to confer with civil authorities in Savannah. D'Estaing agreed to his request, but it was not a wise decision. He could have captured Savannah at that moment by direct assault because the British garrison was unprepared for an attack.

Prévost used the delay to put soldiers, townspeople and several hundred black slaves to work around the clock, to finish the city's fortifications. Prévost withdrew the garrison at Sunbury, and Colonel Maitland was told to bring his troops down from Beaufort as fast as possible to reinforce Savannah. At the end of the truce Prévost's position was too formidable to attack. To make matters worse, it began to rain, and the siege lines could not be started until the weather cleared.[1117]

17 Sept:

Parole

RO

Lt. Roux of the Light Infantry compy is appointed a Captn in the Regimt he is to be Obey'd & respected as Such –

Capt. Roux is to take the Command of Captn Mottes Company promoted to a major

The Officers are requested to see that their mens arms are put in the best order, & to Examin their Cartridges & Supply them with Good Cartridges in the room of those which may be Dammaged,

A return of the number of Cartridges & flints Each man have in their pouches, ~~xxxxxxx xxxxxxx xxxxxxx~~, to be given in by 9 OC: this morning ~

It is Expected that, all Offrs. Servants will be Immediately furnish'd with Arms & Accoutrements [1118]

Cherochee hill, September 16th [1119]

Field Officer for this Evening Major Wise [1120]

Brigade Major Captain Linning

Weekly Returns of the Troops are to be Made to the Adjutant Genl Every Monday And Friday at Orderly time in the Morning, tis hoped this order will be punctually Complied with for Other wise it will be impossible to regulate the Detail of the Army forms May be had of the Adjutant Genl. – After orders / as the Militia cannot possibly be of any service on Horseback there being a Great proportion of Regular cavalry in the Army

The Genl. Orders that every regiment or Corps of Militia Now in camp or who may hereafter Arrive Except Captain Capt Elliots who were Commissioned as Horse be Dismounted And serve on

1117 O'Kelley, *NBBAS, Volume One,* pp 361-362.

1118 This was normal, as the servants, both black and white, had been ordered to train with the soldiers back at Fort Moultrie. Some of these men were indentured servants to the officers, but most were Black slaves who fought in the line with the 2nd South Carolina Regiment.

1119 Cherokee Hill is now the site Savannah/Hilton Head International Airport.

1120 Samuel Wise.

foot[1121] – The Militia belonging to the State of Georgia to be Anexced to the first those of South Carolina Now in Camp to the Second Brigade

The Q[r]. M[s]. Gen[l]. will provide a pasture in the Rear of the Camp where the Horses Must be sent Early tomorrow Morning with a Small Guard from each Regiment to prevent their Straying

M[r]. Bereford, [1122] M[r]. Simmons,[1123] M[r] John Izard [1124] And M[r]. Walter Izard [1125] Now Acting As Volunteers in the Army Are to be Respected as confidential Officers Attendant on Gen[l]. Huger – & All Orders from them are to be Obey'd And they Respected Accordingly

September 17th

Head Quarters Millings house 3 Miles from Savannah
Parole Bruglis. C[n]. S[n]. Courege & Conquer
Field Officer ~~for~~ the to Day Major Anderson [1126]
for tomorrow
B M for to day Major M[c]Intosh [1127]
for tomorrow Captain Lining
Four Orderly horsemen from Col[o]. Temples Dragoons are to Attend Constantly at Head Quarters & 2 from the same corps are to Attend the Field Officer of the Day as an Escort [1128] – An Orderly Serjt. From Each Brigade must also Attend at Head Quarters

The Guards to be Reliev'[d] Immediately

After Orders / Col[o]. Temples Dragoons are to join count Pulaskie tomorrow Morning[1129]
Col[o]. Horrys will Remain[1130] – Six Orderly Horsemen from the cavalry <u>under count Pulaskie</u> are to Attend at Head Quarters - & two to the Field Officer of the day – they will be paraded with the Guards Daily - - - -

---

[1121] This was Captain Barnard Elliot and his Mounted Independent Company. They were attached to Pulaski's cavalry during the siege. The militia of the South tended to ride everywhere they went. An observer noted, "Even the most indigent person has his saddlehorse, which he rides to every place, and on every occasion; for in this country nobody walks on foot the smallest distance, except when hunting; indeed a man will frequently go five miles to catch a horse, to ride only one mile upon afterwards." When the militia arrived the logistical problem of feeding both horse and man became predominant.

[1122] Richard Beresford was born in 1755. He served under General Huger in Georgia during 1778, and served as a captain and aide to General Moultrie. After he was captured at the fall of Charleston he was imprisoned in St. Augustine until 1781, when he was exchanged. He was a delegate to the Continental Congress from 30 May 1783 to 3 June 1784.

[1123] This may be Charles Howe Simmons.

[1124] John Izard served as an aide to General Huger. In 1775 he had been elected to the South Carolina House of Representatives from Goose Creek, St. James Parish.

[1125] Walter Izard was born in 1750. He was the son of Ralph Izard of Dorchester. He married Mary Fenwick on 7 November 1779. He died in 1788.

[1126] Richard Clough Anderson was in Colonel Richard Parker's 1st Virginia Detachment of Scott's Virginia Brigade. These are the Virginia troops that had marched south from Washington's army.

[1127] William McIntosh was a captain in the 1st Georgia Regiment.

[1128] Lieutenant Colonel Benjamin Temple's 1st Regiment of Continental Light Dragoons under Major John Jameson.

[1129] Count Kazimierz Pulaski and Pulaski's Legion of Horse and Foot (The American Legion).

[1130] Daniel Horry and the South Carolina Dragoons.

September 18th

Parole Languedock – Cn. Sns Ceres And Lively
Millings Genl. Orders by General Lincoln
Field Officer for tomorrow – Major Motte
Brigade Major – Major McIntosh
M, Orders / The Genl. Orders that No Soldier Either Regular or Militia presume on Any pretence Whatever to Farther than half Mile from camp without a written pass From the Commanding Officer of the Troop or Company to which he belongs, Any one found beyond the Limits prescribed May Expect to be punished for Breach of Orders –

No Horse whatever is to be pressed in the service without a Certificate being Delivered Acknowledging the Receipt of the Horse & Specyfying his Appraised Vallue

For Guard from 2d. Regt.	C	L	Serjts	R&F
	-	1-	-2-	-28

A Genl. Court Martial to set tomorrow Morning a seven OClock – Prest. Colo. Thompson,[1131] Members three Capts & 3 Subls from First three Captains & 3 Subls from the 2nd Brigade

The court is to try all prisoners brought before them – An Orderly Adjutant from Each Brigade, And All Evidences to Attend

The Genl. hopes the court will be as expeditiously as possible

Captain Handely,[1132] Captain Lucas,[1133] John Houstoun Esquire & John Jones Esquire[1134] Volunteers to the Army Appointed Confidentials Officers Attendant on Genl. McIntosh[1135] Are to be Obey & respected as ~~such~~ Accordingly

Brigade Orders by Genl. Huger

All Orders Respecting the Non Commissioned officers & privates are to be read at the head of each regt as soon as possible After being Issued – in order that Ignorance of them Might not be pleaded

A Return to be made Immediately of the Dates of Commissions of the Field Officer in the Brigade

---

[1131] William "Danger" Thomson.

[1132] George Handley was a captain in the 1st Georgia Continental Battalion. General McIntosh had been the commander of the Georgia Continentals at one time, so many of his officers were from Georgia. After the capture of Savannah and the battle of Briar Creek the Georgia Continentals were reduced and no longer considered an effective unit. Due to this there were many officers without an assignment.

[1133] Captain John Lucas, who had been in the the 4th Georgia Regiment.

[1134] John Jones had been a captain in the Georgia militia in 1776. He became General McIntosh's aide de camp. He was killed at Savannah on 9 October 1779 just feet from an enemy cannon.

[1135] Lachlan McIntosh was born in Scotland. He was selected as a delegate from the parish of St. Andrew for the Provincial Congress in 1775, that was held in Savannah. On January 7, 1776, he was appointed Colonel of Georgia troops and in September 1776, he was elected Brigadier General of the Continental troops of Georgia. In the fall of 1776 he challenged signer of the Declaration of Independence, Button Gwinnett, to a duel. McIntosh was wounded but Gwinnett died. Gwinnett's friends charged McIntosh with murder, but he was acquitted. He was transferred to General Washington's Headquarters so that the patriots in Georgia would remain united in the cause. At Valley Forge Washington appointed General McIntosh the command of the North Carolina Brigade. On 26 May 1776 McIntosh was appointed the commander of Fort Pitt. After building Fort McIntosh and Fort Laurens he was ordered south to participate in a campaign to recapture Savannah. He was captured after Charleston fell on 12 May 1780. He was exchanged on 9 February 1782. His confinement undermined his health and he was not in robust condition for the rest of his life. In 1783 he was made a Major-general. He was elected to Congress in 1784. He died on 20 February 1806.

19th { - Parole Washington Cn Sn cockspur & Tybe
Mr. Bards Plantation 2 Miles from Savannah
Parole Washington Cn Sns Cockspur & Tybee

Brigade Orders by Genl. Huger

The Genl. orders that the Soldiers will not burn or destroy the rails Fences or Any part of the Buildings on this plantation – that they will Not plunder On Any pretence Whatever – he hopes the Officers will Attend to the Execution of this Order And Report Any one who May be Guilty of a Breach of it

	Subl	St	R&File
Genl. Guard	1 =	2 -	33
Regtl Guard	-	1 =	10

RO. the Long roll to beat every morng at 7 OC. & ~~in the Eving~~ and Roll call at retreat beating, when every officers & soldiers are on the parade, the morning reports Every Morning as usual – the Ammunition Guns & Accoutrements to be Examind every morning at roll call –

No Officer or Soldier to Leave the camp Without Leave from the commanding Officer

Every Soldier who Disobey any Brigd or Regimental Order to be confined Immediately, that they may suffer for such Disobedience

Any Soldier who waist or otherways Destroy their Ammunition, unavoidably May Depend being punished severely –

It tis recommended to the Soldiers to Keep their Arms in good order as they may expect to be called in action every moment –

____________________/ 20 Septr. 1779 Bairds Plantn 2 Miles (from Savannah

Parole D:Estaing Countr Sign Frederick & Prussia

RO

A regimental court martial to be held Immediately to Consist of a presidt & 4 members for the trial of All prisoners orderd to it, Evidences to Attend

President Captn Moultrie

Members Captn Mazyck, Lts Martin, Faisson & Warley

Lt. Bush & Gray is Appointed to Carry the Colours & to take charge of them at all times, who are to do no other Duty Either in the Line or otherways, [1136]

Brigade Orders by Genl. Huger

Field Officer for the day Colo. Parker [1137]

[1136] One of the misconceptions of color bearers in the 18th century is that they were carried by the rank of Ensign. The colors were not necessarily carried by just one rank, but by whoever was thought capable to do the job. Color guards would be the strongest, most trustworthy men in the regiment. A unit assault may depend on whether the colors stay in place or not. If a unit lost their colors it was considered the ultimate insult and so it became the object of an enemy attack. Much of the firepower would be directed at the color guard and the color bearer. During the assault on the Spring Hill Redoubt on 9 October 1779, Lieutenants James Gray, John Bush and Sergeant Jasper were all shot down carrying the 2nd South Carolina's colors.

[1137] Richard Parker commanded the 1st Virginia Detachment of Scott's Virginia Brigade. He was killed on 25 April 1780 when he was shot in the forehead while he was looking over a parapet during the siege of Charleston.

	S	Serg[t]	
General Guard	2 -	2.	- 32 Rank & file
Regimental Guard		1 - 1	

Millings Gen[l]. Orders by General Lincoln

Field Officer for tomorrow Col[o]. Henderson

Commanding Officers of Brigade or Corps are Desired to send One Orderly for Every ten of their Men, sick in Hospital –

They are also Desired to give Directions that the Arms & Accoutrements belonging to the sick, be kept by the different Regts. or Brigades they being Frequently lost or Spoiled when sent to the Hospital

21st September

Bards B. O. by General Hugar

Parole La Fayette C[n] S[ns] Paris – Nantz

Field Officer for to day Col[o] Scriving [1138]

For tomorrow Lieut Col[o] Marion

	S -	Serj[t].	R
Gen[l]. Guard	1 -	2 -	22 Rank & file
		1 -	10

In future the Troop will beat at seven OClock in the Morning the Guards will be Paraded immediately After –

No Officer or Soldier will quit Camp without Leave from the Commanding Officer

The Gen[l] positively forbids the Discharging of Small arms in & About the Camps And hopes that every Officer will escort himself in detecting & bringing to punishment Any one who May be Guilty of this Shameful practice - -

The Picquets will Not suffer Any Person to pass towards the Enemies lines without a written order

Millings Gen[l] O; by Gen[l] Lincoln

Capt Gaddsden is Appointed Assistant to the Adjutant General of the Southern Department And is to be Obey'd & respected Accordingly

A party of 3 Officers & One Hundred Men for Fatigue immediately – they Are to take their Orders from Col[o] Laumoy [1139]

The Gen[l] Court Martial Held before Savannah by Order of Gen[l] Lincoln, Col[o] Thom[p] President have Reported, Michael Thomas W[m] Cuddoe and John Cole of the first Virg Battalion Charged with Desertion but Acquitted of that & Found Guilty of Absent without leave & sentenced to Receive 30 Lashes on the Bare Back – The Gen[l]. Approves the Sentence & orders it to be put in Execution to Morrow Morning At Guard Mounting

Williams Waily Serg[t] of the Sixth South Carolina Battalion Charged with Desertion pleads Guilty, & Sentenced to be Hanged, The Gen[l]. Approves of the Sentence & orders it to be put into the

[1138] Colonel James Skirving served as a captain in the Berkely County Regiment of Militia during 1775. He was later promoted to colonel of the regiment.

[1139] Jean Baptiste Joseph de Laumoy of the American Engineer Corps, which consisted of French Engineers in the Continental army.

execution the Day After tomorrow at Guard Mounting[1140]– John Findley A Soldier of the Fourth Battalion of South Carolina Charged with Desertion found guilty & Sentenced to receive one Hundred Lashes on his bare back with Switches at four several times [1141]
Gen[l] Approves of the Sentence & orders the Punishment to be Inflicted tomorrow & three Succeeding mornings at guard Mounting

Durham Ford a Serg[t] of the first South Carolina Battalion[1142] & Jeremiah Fleming a Waggon Master charged with Desertion Acquitted [1143]

On September 21st the French camp was set up within range of the British guns, 1,200 yards from Savannah. Due to the camp being in a heavily wooded section, no one was harmed when the British fired artillery at them. The trees were cut down for 3,200 yards to accommodate the camp. The land forces had spread across the southern side of the city, and American and French ships had cut off all the waterways leading to the city. By the end of that day the city was cut off. [1144]

September 22[d] 1779

Gen[l] O, by Gen[l] Lincoln – Millings
Parole C[n] S[ns] Allen
Lieu[t] Beverly Stubblefield of Col[o] Parkers Reg[t] is Appointed Brigade Quarter Master in Gen[l]. Hugers Brigade & is to be Obey[d] & respected Accordingly – [1145]

The court Martial of which Col[o]. Thompson is president is Disbanded –
Two Officers of the Artillery are to Superintend the Making of Fascines –

On September 22nd Lieutenant Guillaume and fifty Grenadiers were ordered to seize the British advance post. Unfortunately Guillaume had his men rush the post. The British easily repulsed this attack with their artillery, killing four of the French grenadiers and wounding three others. During the attack two soldiers of the French advance guard post were killed and two were wounded.

The next day Lincoln's army rendezvoused with D'Estaing. D'Estaing believed that Lincoln should have pinned Maitland at Beaufort, instead of coming directly to Savannah.

1140 William Wailey served in the 6th South Carolina Regiment. On 21 September 1779 he pled guilty to desertion and he was sentenced to be hanged.

1141 John Findley enlisted in the 4th South Carolina Regiment (Artillery) on 16 January 1776 as a matross under Captain Richard B. Roberts. On 21 September 1779 he was found guilty for desertion and was sentenced to receive 100 lashes on his bare back with switches on four different occasions. He later served in the militia.

1142 Denham (Durham) Ford served as a sergeant in the 1st South Carolina Regiment and was court-martialed on 22 September 1779 on the charge of desertion, but he was acquitted. During 1780 he served as a sergeant in the 1st South Carolina Regiment under Captain Charles Skirving.

1143 Jeremiah Fleming enlisted in the 6th South Carolina Regiment on 2 June 1777 as a sergeant. He was demoted to private on 12 January 1778. He served as a wagon master and was court-martialed on 22 September 1779 on the charge of desertion, but he was acquitted.

1144 O'Kelley, *NBBAS, Volume One,* pg. 364.

1145 Beverley Stubblefield of the 1st Virginia Detachment of Scott's Virginia Brigade. He was later promoted to captain and commanded a company in the 2nd Virginia Regiment.

D'Estaing believed Lincoln did this just to get military glory. Lincoln in turn blamed the French for Maitland's arrival in Savannah. He thought that D'Estaing should have blocked the Broad River.[1146]

September 23th 1779

Genl Orders by General Lincoln

Parole Charlestown – C^{n} S^{ns} Laumoy & Company

The Genl. is informed that Great Numbers of Men belonging to the Army in Stead of Attending their duty in Camp, are wandering About the country & spreading destruction wherever they go – in order to restrain this Licentiousness the Genl. Represents to the Officers, Not Only the Cruelty of such proceedings but the Dangerous Consequences which May ensue – Commanding Officers of Brigades, Regts & Companies of Militia as well as Continental Troops tis hoped will use their Utmost Influence & Authority to Restrain this Unwarrantable, Unmilitary practice – A condition which Not Only Reflects Disgrace on us as soldiers, but Immediately Endangours our Safety – The Roll is order:d to be Called Morning & ~~Evening~~ Noon And Night And Every One Absenting himself therefrom who May be found at Any Time

RO. A regimental ~~court~~ martial to set Immediately to try all ~~priso~~ners orderd to it Evidences to Attend

President Captn. Mason

L^{ts}. Foissin, Ogier, ~~Bush~~ & Gray Members

NB

Give Cornelius Constantine a Certificate of his Driving a Waggon for the 2^{d} Rg from the 15th Septr 1779 to 11^{t} April 1780 – [1147]

When the rain let up on the 24th the French begin their entrenchments. At 3:00 a.m. a trench was opened up 300 yards from the British lines. Six companies of French Chasseurs lay on their stomachs guarding the digging party that worked hard to complete the trench before sunrise. At 7:00 that morning a British attack was made by three companies of British Light Infantry. The French Grenadiers counterattacked. Captain Campbell and Lieutenant McPherson of the British lights were killed and the rest were chased back into their redoubts at bayonet point. The French had made the mistake of leaving their defensive works, and they were driven back by British artillery firing canister. The British had six killed and fifteen wounded. The French had 26 men killed and 84 wounded when they had left the trenches.[1148]

~~Parole Virginia~~ C^{n}_ S^{n}_

September 25th – 1779

Genl Orders by Genl Lincoln

Parole Thunderbolt – C^{n} S^{ns} Bailey & Abercorn

Field Officer for tomorrow

1146 O'Kelley, *NBBAS, Volume One,* pp 364-365.

1147 Cornelius Constantine served in the 2nd South Carolina Regiment under Captain Charles Motte in 1779.

1148 O'Kelley, *NBBAS, Volume One,* pg. 365.

The Troops to be Supplied with Provisions in the Evening for the Next Day & the Commissary is Ordered to have One Days Rations of Beef Always on Hand

The Escpress sets out for Charlestown Tomorrow Morning At Nine OClock

2 pieces of Artillery Are to be Annexxed to Each Brigade & the Remainder to be Encamped in the Rear of the Interval between the two Brigades –

Francis Kinlock Esquire is Appointed to Act as Aid De Camp to Genl Huger – He is therefore to be Respected & Obey^{d} Accordingly[1149]

The Genl Requests the favour of the Officer of the Day At Head Quarters

Septr 26 Parole Vigilance C S^{n} Felicity & Perseverance
Fatigue Duty

	S^{t}	Rank File	C	S	Sergt	R&F
Genl. Guard	1 -	10	1	1	1.	10

At 4:00 in the morning on September 27th, the 71st Highlanders and two cannon conducted a second sortie towards the French trenches. The Highlander's intent had been to lure the French into charging their lines, where 800 men were waiting, lying in prone on the ground. The Highlanders attacked and then quickly retreated. The Americans tried to gain the Highlander's right flank, while the French maneuvered on the left flank. Both sides fired upon each other by accident. Fifty French and American soldiers were killed or wounded by the friendly fire. The Highlanders only had three men wounded in this action.

Septr 27^{t} Parole Sagitare C Signs – Experiment

Field Officer for tomorrow Major Motte ~

R Orders A Court Martial to Sett Immediately to try All Such Persons as May be Ordered to it. Evidence to Attend – President Capt. Hall Members L^{ts}. ~~Petrie~~ Legare Hall ~~and~~ Hume and Warley.

On the night of the 28th the French sentinels thought that some of their own workmen in the batteries were another enemy assault force. For a half an hour the two sides fired at each other, killing two and wounding seven, in yet another friendly fire incident. The British added to the death and destruction by firing on the fleeing workmen trapped between the two lines. After that night the British placed out 100 marksmen each night to snipe at the work parties.[1150]

[1149] Francis Kinlock had been born on 7 March 1755 in Charleston, South Carolina. He served as a captain during 1776 and was in the battle of Beaufort in 1779. He was appointed an aide de camp to General Isaac Huger on 25 September 1779. In September 1779 he was wounded at Savannah. He was taken prisoner at the fall of Charleston on 12 May 1780. After he was paroled he was captured a second time in Virginia by his relative Captain David Kinlock, of the 17th Light Dragoons, attached to Colonel Banastre Tarleton. He died on 13 February 1826 at the age of 71.

[1150] O'Kelley, *NBBAS, Volume One,* pg. 366.

Septr 28 P. Philadelphia Cou^n^ S^n^_ GeorgeTown camden
Field officer for to Morrow
Gen orders / the whole army will turn out at nine OClock this Morning to continue the making fascines and the comm^d^ Officers will order those made by their Respective Regim^t^ To be carry^d^ to the common Repository on the right of the French camp by their Regimental waggons
Two captains four Subalterns Eight Serg^t^ and Sixty Rank & file form the 2^d^ Regiment are to leave Immediately they are to be paradded a half an hour before Retreat Beating with one days Provisions Cooked -

R O – 28 Sept^r
the Off^rs and men for Command to get their regiments Dressed immediately & to be on the parrade by 5 OC: this afternoon ~
Serg^t. Willm. Henderson is Reduced to the ranks & is no longer to be considerd as a serg^t in the regiment

~ Septr. 29 1779 ~

Parole C^n_ S^n_
Field officers for to morrow Major Lee [1151]
R.O. the monthly returns to be given in tomorrow by 10 OClock in the morning
B. Major Captain Lining

The Honourable the Continental Congress have been pleased on the 10th of August last to come to the following Resolutions Viz

Resolved that untill the further Order of Congress the Officers of the Army be intitled to Receive Monthly for their Subsistence Money the sums following to Viz –

Each Col^o & Brigade Chaplain five Hundred Dollars Every Lieut Colonel four hundred Dollars Every Major & Regimental Surgeon three Hundred Dollars Every Captain 200 Dollars Every Lieutenant Ensign & surgeon 100 Dollars

Resolved that untill the further Order of Congress the sum of ten Dollars be paid to Every Non Commissioned Officer Monthly for their Subsistence in Lieu of those Articles of Food Regimentally Intended for them & Not Furnished

An Excpress will set out from Head Quarters for Charlestown every Morning at Eight OClock.

The General orders that the Pay Roll of the Continental Troops be immediately made out to the first of October 1779

For Command, to be paraded at Retreat Beating two Captains, four Subalterns, six Serjeants, and fifty Rank and File

For Fatigue to be paraded half an Hour before Retreat Beating, two Captains, 4 Subalterns four Serjeants, and sixty Rank and File, they will take their Orders from Colonel Laumoy[1152]

[1151] Unable to determine who Major Lee is. He may be Captain William Lee of the True Blue Company of the Charles Town Militia. William Lee was captured by the British at the siege of Charleston in 1780, and he was still a prisoner of the British at St. Augustine in November 1780.

[1152] These men were detailed to work in the trenches. In a siege the approach trenches were dug to within 600 to 800 yards of the defenses, then a trench twelve to fifteen feet wide, and three feet deep, was dug at right angles to the approach trenches. Normally these would be dug in zigzag lines, so the enemy would not be able to fire down into the trenches. The dirt removed from the trenches was used to make a parapet four feet high in front of the trench. The first trench was called the first parallel. Between the parallels, short parallel trenches were dug on the flanks of the approach trenches. The infantry were gathered in all these to protect against possible sorties from the enemy aimed at destroying work parties, trenches, and artillery positions. At the head of the approach trenches a squad of soldiers would begin digging a narrow, shallow trench called a sap. The lead man would excavate a small trench only a foot and a half wide, and deep. He would push ahead of him a two-wheeled device called a mantalet, used to protect him from enemy fire. As he went along he would place gabions along the route. The men behind him would widen and deepen the sap and put fascines on the parapets.

	C	S	S	R&F
2d Regt. Command -	"	"	1.	5
Fatigue -	"	"	1.	6
Guard tomorrow	"	"	" 2	19
	"	"	4	30

G O September 30 1779.

For Fatigue to be paraded immediately two Captains , 4 Subalterns, 6 Sergts. and 80 Rank and File – The Party to be commanded by Lieut Col. Hopkin –[1153]

	C.	S.	S.	R&F
2d. Regt.	1.	1.	1.	8

As by An order of the 23 d. September Certain Limits were Appointed beyond which No Soldier was to pass without a Written Permit – The Genl Acquaints the Army that If Any one in Defiance of that Order should stray beyond the prescribed Limits and be taken Up & Confin'd in the french Camp, he Must Not Expect his Interposition to get him Liberated –

The Army will Draw One Gill of Rum pr. Man this Afternoon –

Octr 1st

Parole Rutlidge – Cn. Sn_ Bee & Farm

Field Officer for to Morrow

2d Regiment Guard

	C	S.	S.	Rank & file
Guard	1-	1.	1-	19
Fatigue	1.	1.	1.	8

Octr 2

Parole Conde - Cn_ Sn_ Ceasar & Scipio

Field officer for to Morrow Colo. Thompson

2d Regiment

	Ct	Sl	S	R
Commd	0	1	1	10
Fatigue	0	0	1	9
Guard	0	1	1	10
	0	2	3	29

before Savannah 2nd Oct: 1779

R.O, Commanding Officer of Companys will see their mens arms are in Good Order & Examine their Ammunition, those found dammaged to be Exchanged for good ones

one fatigue men Out of Each Company to be Appointed dayly & be Given to the Quartr M. Sergt to keep the parrade & the Streets in camp clean, who is orderd to Make Nessesary houses in the front, not Less than 200 yards from the first line of parrade –

Corporal Kerslick of Captn Grays Compy. is appointed a Sergt. & to Join & do duty in Captn Halls Company he is to be Obeyed accordingly

[1153] Samuel Hopkins of Colonel Richard Parker's 1st Virginia Detachment of Scott's Virginia Brigade.

John Dubose private in Captn Mazycks Comp^{y} is Appointed a Corporal in Captn Halls Comp^{y} he is to be Obey as such –[1154]
L^{t}. Capers to Join & do duty in Captn. Dunbars Comp^{y} till further Orders
The court of Inquiry ordered to Set this Morning Have Reported that after maturely considering the Evidence and Facts Relative to the conduct of Lieu^{t} Col^{o} Scott they are of oppinion that Col^{o} Scott acted impudently in leaving the Reg: without his commanding officers permission but Not think him Subject to a court Martial.
For fatigue to be warned to Night and to March to Morrow Morning at Reveille Beating under the command of cap^{t}. De Treville Two cap^{t}. four Sub^{s}. Four Serg^{t} – and Eighty Rank & file –

	Captn	S^{b}.	S.	R. file
Fatigue	0 -	1 -	1 -	8
Comm^{d}.	0 -	0 -	1 -	8
Guard	0 -	1 -	1 -	18

Octr 2 | Camp before Savannah G^{l} orders by G^{l} Lincoln
For command to be paraded at four oClock
Two cap^{t}. four Sub^{s}. Six Serg^{t}. and Eighty Ranks & file

They are to relieve the covering Party under Lieu^{t}. Col^{o}. Henderson Lieu^{t} Col^{o} Hoppkins is to relieve the above Command ________________ ____________ __________
The Court Martial of which Lieu^{t} colo^{l} Scott was President have Reported cap^{t} Mitchell of the artillery charged First of being absent from his alarm Post on the Night of 11th of august 1779[1155]
2^{dly}_ For beating and abusing a Matross on Gavins Erving in Violation of a Standing Regimental Order[1156]
3 ^{dly}_ for behaving with contempt and Disrespect to Colonel Beeckmen of that corps –
4^{thly}_ For Treating colonel Beeckman in the Character of commanding officer at that Time When the whole line was Turned out and the camp in an actual State of alarm with abusive and Scurrilous Language
5^{thly}_ For beheaving in a manner unbecoming an officer and a Gentleman –
On the first change the count was of oppinion that cap^{t} Mitchell was absent from his post on the Night of 11th of aug^{t}. 1779 – as it was evi^{d} that it was his duty to have been in the park at the Time of the alarm Not with standing Particular post or division had been asined him / but as he was absent a very Short Time and had Taken Every Precaution Nessesarry to Receive the Earliest intelgence of an alarm they think the charge Rather frivolous
on the 2dly _ that the court could not cognizant of it as punishment stood an evi^{d} to a Breach of that order
on the 3. 4 & 5th charges that they could not be Separately Determined but was of oppinion that Captn Mitchell did Treat Colonel Beeckman the commanding officer with abusing & Scurrilous Language at the Time when the hole line was Turned out and the camp in an actual State of alarm they therefore

---

[1154] John Dubose was born on 13 June 1738. He served in the 2nd South Carolina Regiment. He was promoted to corporal in Captain Hall's company on 2 October 1779. He was at the siege of Savannah. He died on 24 September 1800 at the age of 62.
[1155] William Mitchell.
[1156] Private Gavin Ervin served as a matross in the 4th South Carolina Regiment (Artillery).

Sentenced that cap[t]. Mitchell Should be Reprimanded in the presence of the officers of the artillery –

It Gives the Gen[l] Pain that he is under the Nessesity of disproving the Sentence of the court but is constrain to do it as he Think it Totally inadequate to the offence and that to prove it wold be Subversive of order & Discipline in the army -

~~Cap[t]. S[b] S.~~

a fatigue Party to be paraded Presently at Retreat beating to Relieve that now at work to consist of Four ~~Four Sub[t]~~ cap[t]. Four Sub[t]. four Serg[t]. Eighty Rank & file

	Cap[t].	S[b].	S.	R file
commd.	0 -	1	1 -	10
Fatigue	0 -	0 -	1 -	9
Guard	0 -	1 -	1 -	18

3 October 1779

Camp before Savannah G. O. by Gen[l] Lincoln
Parole S[t]. Peter C[n] S[ns] Diaden & Morris
F. O. for tomorrow Col[o] Parker
Brigade Major for tomorrow Capt[n]. Linning
The Brigade Arms are to be Discharged at Four OClock this Afternoon – four Captains Eight Subalterns Six Serj[ts] And two Hundred Rank & File – Col[o]. Pinckney Lieu[t] Col[o] Saunders And Major Anderson for Command[1157]

	S – R&F		Sb S R
for Guard	.2 – 20	Command	1 - 3 – 22

At midnight on October 3rd the French and American artillery began a bombardment from land and sea. An aide-de-camp to Prévost wrote, "The Town was torn to Pieces, and nothing but the Shrieks from Women and Children to be heard. Many poor Creatures were killed in trying to get in their Cellars, or hide themselves under the Bluff of the Savannah River." [1158]

A keg of beer was supposed to have been sent to a French ship, but instead a keg of rum was delivered. "The cannoneers being still under the influence of rum, their excitement did not allow them to direct their pieces with proper care." The drunken artillerymen fired on their own troops. The bombardment continued for two hours until it was ordered to stopped because the French were doing more harm to their own troops than the British artillery was. [1159]

[1157] John Sanders had served as a lieutenant in the Horse Shoe Company of the Colleton County Regiment of Foot in 1775. In 1779 he was a captain, then lieutenant colonel and colonel of the South Carolina Militia. He was captured at the fall of Charleston.

[1158] Gilbert Bodinier, Les Officiers De l'Armee Royale Combattants de la Guerre d'Indépeendance des Etats-Unis De Yorktown a l'An II, (Service Historique de L'Armée de Terre at Château de Vincennes, 1983), p. 470.

[1159] O'Kelley, *NBBAS, Volume One*, pp. 339 – 340.

4 Oct: Camp before Savannah Gen[l] O by Gen[l] Lincoln
Parole Iphigene C[n] S[ns] Amazon & Ariel
Field Officer for tomorrow ~~XXXXXXX~~
Brigade major Captain M[c]Intosh
The Continental Troops & Independante Troops are without Delay to be Completed to 40 Rounds for the Non Commissioned Officers & Soldiers fit for Duty – Those Regiments which have more than that Quantity will immediately Return the Superfluity into the Hands of the Brigade Quarter master
B. O. by Brigadier General Huger
Daniel Huger Esquire is Appointed as Aid De Camp to General Huger - & is to be Respected & Obey'd Accordingly[1160]

> At 4:00 in the morning on October 4th fifty-three heavy cannon, and fourteen mortars, began a bombardment of the town. The galleys in the Savannah River added their guns to the bombardment. One of the British defenders counted 187 shells thrown into the town by the mortars. The bombardment had little effect to the military, but killed many civilians. Unfortunately a French mortar threw shells into the French trenches and more men were harmed by friendly fire. The bombardment stopped four hours later when the batteries on the left of the line collapsed due to poor construction. A dense fog helped cover the workmen while they repaired the batteries and by 10:00 the bombardment began again.
>
> The firing continued off and on for the next five days. The casualties were mostly civilians. Throughout the bombardment it continued to rain. The soldiers in the trenches were living a miserable existence. Meyronnet de Saint-Marc wrote that the climate was "so extraordinary that during the day we were exposed to the most intense heat and night to bitter cold." [1161]

R. Orders 4th Octob[r]. } the Quarter Master Serg[t] to give or bring Immediately an Exact Account of all good Cartridges now in his possession –

	C	S	Ser	R&F
Command	1 -	0 -	2 -	10

For Command at four OClock this after noon Under Col: Skivers Lieu[t] Col Bougain[1162] and Major Harleston ______ 4 Captains, Eight Subalterns, 16 Serj[ts] & 200 Rank & File

2[d] Regiment	C.	Sb.	S.	Rank & file
	0 -	1 –	2 -	17 –

[1160] Daniel Huger was born on Limerick Plantation in St. John's Parish, Berkeley County on 20 February 1742. He was a member of the colonial assembly from 1773 to 1775, and he was a justice of the peace in 1775. He was a member of the State house of representatives from 1778 to 1780. He was a member of the Continental Congress from 1786 to 1788. He was also elected to the First and Second Congress from 1789 to 1793. He retired from Congress and managed his Wateree Plantation. He died in Charleston on 6 July 1799 at the age of 57, and is buried in St. Philip's Churchyard in Charleston.

[1161] O'Kelley, *NBBAS, Volume One*, pp. 340-341.

[1162] Lieutenant Colonel John Baptist Bourquin was an officer, but I do not know what unit he was in.

Oct^{r}_ 5th / Parole C^{n}. S^{ns} _
~~Field Officers for tomorrow~~
~~Brigade Major~~
Adjutant Maner for tomorrow
A fatigue party from Each Regiment to be immediately Employed in building Necessary houses One hundred yards in front of their Respective Regiments

The Vaults are to be Dug 2/2 feet Deep And to be Slightly Covered Over Every Morning – The Quarter Master will see this Order Complyed with

After Genl. Orders

Colonel White has Lost Near Camp a gold watch with a steal Chain & two Gold Rings hanging to it – if Any person will Deliver it to the Owner Or to the Adjutant General he shall Receive a Reward of 2 hundred Dollars [1163]

Octr 5th { Camp before Savannah G. O. by Genl. Lincoln
Parole Virginia C^{n} S^{ns} Boston & Georgia
Field Officer for tomorrow Colonel Williams[1164]
Brigade Major Captain Lining
A Court of Enquiry is to set this Afternoon at three OClock to Enquire into the Conduct of Captain Espey [1165] & Capt Wyche [1166] of the Georgia ~~Battalion~~ Militia they being Charged by Genl Prevost

---

1163 Colonel John White commander of the 2nd Georgia Battalion. On the night of 1 October 1779 with only two other officers, a sergeant and three privates, he tricked Lieutenant John French of DeLancey's Brigade into thinking that a larger force surrounded the camp. White had lit fires in the woods around the camp and made it seem as if his whole army was bivouacked there. White's men also rode around the British bivouac shouting out orders to fictitious units. White demanded the detachment's surrender, and the whole British command was taken prisoner, along with three schooners. The schooners were carrying a shipment of salt. Colonel White continued the charade by telling Lieutenant French that he must keep his men back, or, in their animosity, his men would slaughter the British. He told Lieutenant French that he would place him under three guides, which was in reality half of Colonel White's force. The three soldiers moved the 150 captured British troops off in much haste. White was placed in command of the "Georgia Navy" until after the siege of Savannah. He would die in Virginia a year later.

1164 Colonel James Williams was born in November 1740 in Hanover County, Virginia. In 1773 he moved to South Carolina and during 1775 he became a captain in the militia. In 1779 he became a lieutenant colonel and commanded the North and South Carolina militia in the battle of Stono Ferry. He commanded the Laurens District Militia in the siege of Savannah. He commanded the South Carolina militia in the battle of Musgrove's Mill on 18 August 1780. He was rewarded with a commission of brigadier general. He was killed in the battle of King's Mountain on 7 October 1780. His estate supplied one hundred fifty gallons of whiskey for militia use.

1165 Samuel Espey served with Colonel Elijah Clark and was killed in the summer of 1780.

1166 Captain George Wyche. Both of these officers had been captured by the British and then taken parole. If an officer takes parole he swears that he will not take up arms against his enemy until such a time as the parole states. This usually was when the officer was exchanged or when the officer was officially released from "captivity". If an officer broke his parole it was the same as breaking his oath or his word. He would be tried for this, either by the enemy, or by his own army. If found guilty he could be returned to captivity or if he was in enemy hands he could be sentenced to death.

with Breach of their Parole – President Lieu[t] Colonel Marion – Members One Continental & One Militia Captain of the first Brigade One Continental And One Militia Cap[t] of the Second Brigade –

The pres[t] Will Appoint where the Court will set –

Command At 2 OClock this Afternoon Col° Few [1167] Lieutenant Col° Reid [1168] & Major Motte

	Sb S R
For guard tomorrow	1 1 16
	C Sb S R&F
Command this Evening	- 1 - 1 - 2 - 10

Brigade Orders by General Huger

One Captain from Virginia Regiment & One from Col° Skirvings for the Court of Enquiry this Afternoon

---

Oct[r] 6th / P. Bayard C[n] _ S[ns] _ S[t]. Mark & Madrid

Field Officer for to morrow

Brigade Major –

Reg[t] Ord[r] A regimental court martial to set immediately to try all prisoners orderd to it, evidences to Attend

Capt[n]. Mason president

L[ts] Foissin, Ogier – Bush, & Gray Members –

~~September~~ 6th

Camp before Savannah G. O. by Gen[l]. Lincoln

Parole Bayard – C[n]. S[ns] S[t]. Mark, Madrid

field Officer for tomorrow Col° Skirving

Brigade Major Captain Mc.Intosh

Command at four OClockGuard tomorrow

Sb S R	S R
1.. 3 - 10 0 -	2 – 16

for tomorrow Adjutant Parker

For command this afternoon Lt Colo Marion Lieut Col Few & Major Smith – Captn 4 8 Sb. 16 Serjts & 200 Rank & file –

Octr. 6	C Sb S R file
Genl. Orders by Genl. Lincoln	1. 2. 3. 86 –

to be paraded immediately they take their Orders from Col Laumoy Fatigue 2d R 1 Serjt. 16

1167 Benjamin Few commander of the Upper Richmond County Militia Battalion of the Georgia Militia.

1168 Colonel James Read was an ensign in the 1st North Carolina Regiment on 4 January 1776. He was promoted to 2nd Lieutenant on 6 July 1776 and to 1st Lieutenant on 7 July 1776. He was promoted to captain on 8 July 1777. He was captured on 1 June 1778. He was exchanged and became the colonel of the 2nd North Carolina Regiment of Militia. He was taken prisoner when Charleston fell on 8 May 1780.

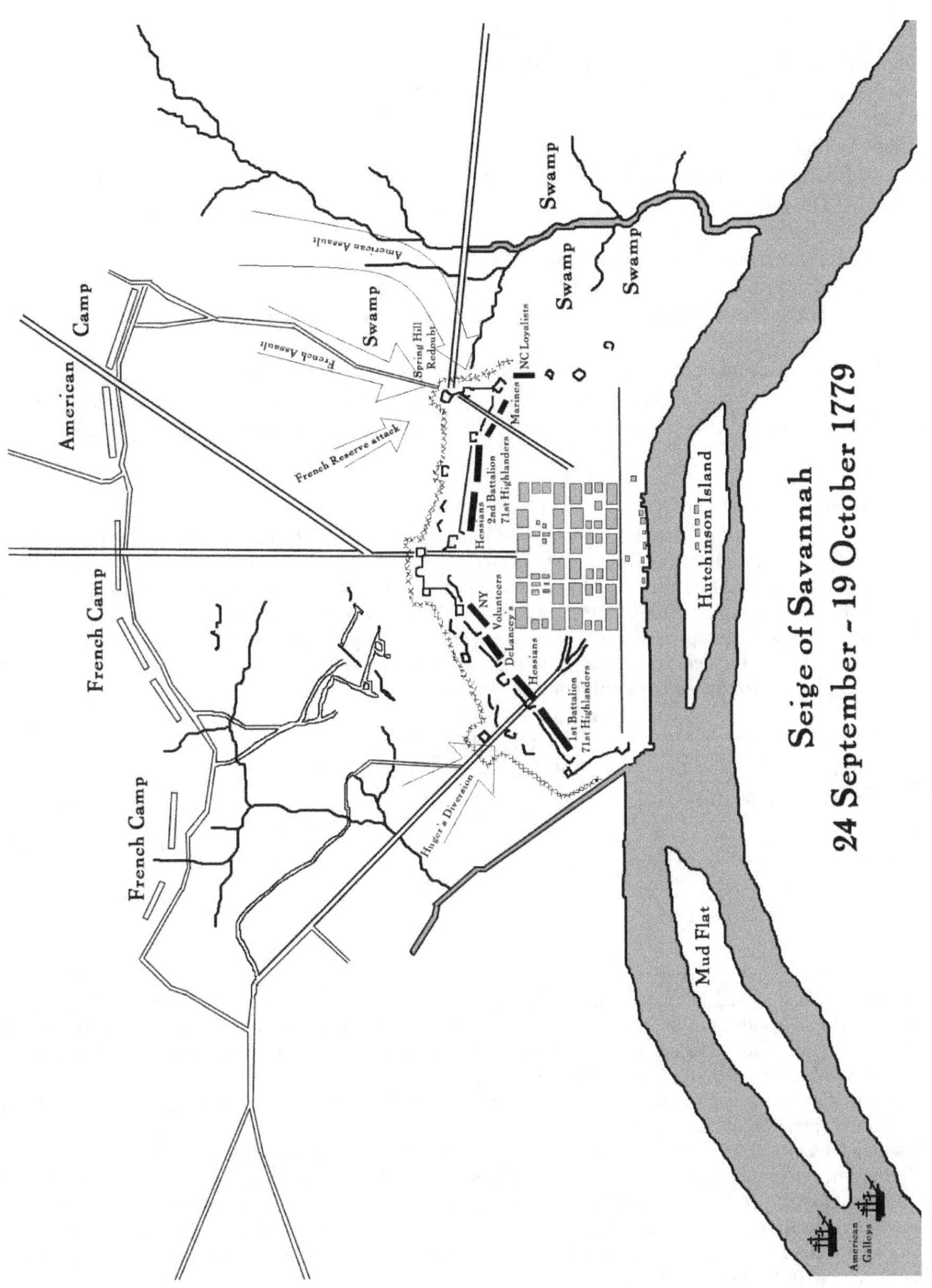
Seige of Savannah
24 September - 19 October 1779
American Camp
French Camp
French Camp
Swamp
Swamp
Swamp
Swamp
American Assault
French Assault
Spring Hill Redoubt
NC Loyalists
Marines
French Reserve attack
Hessians
2nd Battalion
71st Highlanders
NY
Volunteers
DeLancey's
Hessians
1st Battalion
71st Highlanders
Huger's Diversion
Hutchinson Island
Mud Flat
American Galleys

Oct^{r}_ 7: 1779

Parole. – Gustavius ___ C^{n} S^{ns} St John & Paris
Field officer for to Morrow Col_ Few
Brigade Major Bowie [1169]
Adjutant Welch[1170]

Two careful men are immediately to be Sent From the 2^{d} Regiment to attend the Sick at the Continental Hospital –

Guard to Morrow from 2^{d}. Regiment 1 Sb. 2 S. 16 R.
For command at four oClock – 1 Capt. 0 – 1 S. 18 R.
The Field officers for command this afternoon At Four O Clock Col. Downs [1171]

Agreeable to the Regimental court Martial that Octr 6th wich cap^{t}. Mason was President – Corporal Andrews of cap^{t} Bakers comp^{y} was Reduced to the Rank & Rec^{d}. 75 Lashes on the bare Back with a cat onine Tails for Insolent behaviour And abuse of Language to Doct^{r}. Theus

On the morning of October 8th Captain L'Enfant, with a handful of troops, tried to set fire to the abatis of felled trees in front of the British lines. Unfortunately the wood was too damp and did not catch fire. D'Estaing's engineers told him they would need at least ten more days before they could penetrate the British works. D'Estaing insisted that the Allies should launch a frontal attack. Lincoln and his officers protested this tactic. It seems that everyone was opposed to the assault, except D'Estaing. In the end Lincoln gave in to D'Estaing, and issued his men forty rounds and a spare flint. The American army began preparations for the early morning attack.[1172]

~ Octr. 8th. 1779 ~

P. Gibraltar _________ C^{n}. S^{ns} S^{t}. Luke & Josh
Field officers for to Morrow Col.
Brigade Major
Adjutant –

[1169] John Bowie had served as a captain of an independent company in 1775. On 25 February 1776 he became a captain in the 5th South Carolina Regiment. He was wounded in Savannah on 9 October 1779. During 1781 he became a major. From 25 January 1781 to 30 April 1783 he was aide de camp to General Pickens. He was in the battle of Guilford Courthouse in March 1781.

[1170] Edward Welch was commissioned a first lieutenant in the 5th South Carolina Regiment in 1777. He was promoted to captain on 20 January 1778.

[1171] Lieutenant Colonel Jonathan Downs who enlisted under Captain James Williamson in October 1775. He was in the Snow Campaign in November – December 1775 and at the first siege of Ninety-Six in 1775. He served three months. In January 1776 he was commissioned a major and was in the battles of Lyndley's Fort, Seneca, Tugaloo River and the Ring Fight. At the Ring Fight, in August of 1776, he was wounded in the abdomen and hand. He served seven months. He was taken prisoner at Hayes Station on 19 November 1781.

[1172] O'Kelley, *NBBAS, Volume One*, pg 342.

R. O. A court martial to sett Immediately for the trial of Frederick Huges & Amos Tubbs, Confined for Taking & Offering to Sell the Waggon horse of the Reg^t^. ~ Hugh Holland & Captn. Gray to Attend as Evidence [1173]
President Captain Gray Member Lieu^ts^ Martin ~~Bush~~ Warly Bush & Gray ~

There is a month missing from the orderly book, due to what happened the next day. Three small forts, or redoubts, protected the British right flank. Fifty South Carolina Loyalists, led by Lieutenant Colonel Thomas Brown, defended the most exposed one, the Spring Hill Redoubt. The British called this redoubt the Carolina Redoubt due to the defenders. Within the redoubt there was 28 dismounted cavalry of Captain Tawse's Georgia Light Dragoons and 28 soldiers of the Royal American Regiment. Sailors under the command of Captains Manley and Steel manned six cannon in the redoubt. The Spring Hill Redoubt was considered to be the weak link in the defenses. The other redoubts were held by Loyalist troops, which means the bloodiest part of the assault pitted Americans against Americans. Supporting these redoubts were 74 Grenadiers of the 60th Regiment and 37 Marines.

Right after midnight on October 9th five columns were formed for the assault against the town. Three of these columns were French, led by D'Estaing himself. The allied plan called for the French Grenadiers to rush the Spring Hill Redoubt, while the other columns attacked the forts to the British right. Two American assault columns, under Lieutenant Colonel Laurens and Brigadier General McIntosh, would support the French. The 1st and 5th South Carolina Regiments were under McIntosh, and the Light Infantry of the 2nd South Carolina Regiment and the 1st Battalion of Charles Town Militia were under Laurens.

In the camp the wounded and sick would keep the fires burning to deceive the defenders. Pulaski and his cavalry were to follow the left column of French troops at the Spring Hill Redoubt. They would arrive at the redoubt before Laurens and his Light Infantry, and they were to penetrate the lines between the Spring Hill Redoubt and the battery to the left of the redoubt. The American and French artillery would remain silent until the defenses were broken, then move to the redoubt and fire upon the whole British defenses and the city.

All of the troops were forbidden to fire except upon orders. The redoubts were to be taken silently with only the bayonet. This plan would have been effective, but unfortunately the night before Sergeant Major James Curry of the Charles Town Grenadier Company deserted and told the British the time and place of the attack. Prévost strengthened his defenses on his right flank near the Spring Hill Redoubt, and put Maitland in command there. To prevent any fratricide the Allied troops were to wear "large white cockades" in their hats and their "shirts over their coats." The British learned of this from Curry, and they dressed their troops the same way.

Unfortunately, the French arrived late at their point of departure. To further compound the problem, the French military etiquette of the day gave precedence to regiments in accordance to their seniority. The company's precedence was given to the date of commissions of their captains. One company commander insisted on marching his company down the whole column to its rightful place, with the drums and fifes playing in the early morning darkness.

1173 Hugh Holland enlisted under Captain Bernard Elliot in the 2nd South Carolina Regiment on 5 August 1775. In 1779 he was under Captain Thomas Hall at the siege of Savannah.

Spring Hill Redoubt - 9 October 1779

From the direction of the Spring Hill Redoubt the eerie wail of bagpipes drifted toward them through the heavy pre-dawn fog. The music did not come from the normal location of the 71st Highlanders, but instead came from the point of attack. D'Estaing wrote it was as if the enemy "wanted us to know their best troops were waiting for us." D'Estaing considered calling off the attack, but pride got the better of him.

D'Estaing did not wait for the columns to complete their attack formation. The drummers beat the command to charge bayonets. D'Estaing's Irishmen of the Régiment de Dillon and the Régiment Walsh surged forward with a cry of "Vive le Roi!" The red and white-coated soldiers came out of the fog and swarmed over the glacis of the Spring Hill Redoubt. The British and Loyalist troops in the fort opened up on them with a vicious crossfire of muskets and cannons.

The grenadiers cleared the abatis in front of the redoubt, and then under heavy fire, they pushed their way up the parapet. However, the vanguard jumped the ditch which lay before them and uselessly climbed up the side. The artillery in the redoubt decimated their ranks. D'Estaing was wounded in the arm by a musket just before he reached the redoubt. The fighting became intense, and the attackers were sprayed with musket fire and grapeshot. More murderous fire came from a British galley in the river. The bloody D'Estaing urged his troops forward, crying, "Advance, my brave grenadiers, kill the wretches!" However only a few hundred were able to follow.

Because D'Estaing had charged without waiting for the columns to form, the supporting column was slow in following them. By the time they arrived to reinforce the vanguard, the enemy fire had driven the grenadiers back. They were "mown down by the right battery which took them in the flank."

Despite three assaults on the fort, the French could not stand up to the British artillery. D'Estaing reluctantly ordered a retreat. As the French fell back the British troops rose up from the parapet and delivered a point-blank volley. The British artillery fired "packets of scrap iron, the blades of knives and scissors, and even chains five and six feet long."

D'Estaing was wounded for a second time in the thigh and was nearly left for dead. Naval Lieutenant Truguet saved his life by placing D'Estaing on the shoulders of two grenadiers. The grenadiers were immediately killed by grapeshot, and two more grenadiers picked their wounded commander up and carried him to safety.

The Americans arrived near the Spring Hill Redoubt at the height of the battle's confusion, just as the wounded D'Estaing tried to re-form his troops. Laurens' column and Pulaski's Legion came upon the carnage, and rushed the Spring Hill Redoubt. They cut their way across the abatis and crossed the ditch in front of the redoubt, only to be stopped by the high parapet. The Loyalists in the redoubt rained death upon the attackers. The Charles Town Militia "fled in a cowardly manner in the woods even before the action commenced", but the South Carolina Continentals held their ground. The 2nd South Carolina planted both their colors on the rampart. Lieutenant Hume was carrying one of the colors when he was shot down. Sergeant John McDonald was wounded, but still pulled the colors from the redoubt when the command came to retreat. He brought the colors off and saved them from capture.

Lieutenant John Bush was carrying the other 2nd South Carolina colors and was wounded when he planted them on the redoubt. Sergeant Jasper, the hero of Fort Sullivan, took up those colors. Lieutenant Gray was part of the color guard, and was killed trying to protect Jasper. Jasper was wounded and passed the colors back to the dying Lieutenant

Bush. The flag fell under Lieutenant Bush lying in the ditch and the British captured them.[1174]

The Carolinians were exposed to a murderous musket and artillery crossfire from the other redoubts, and they were butchered in the ditch. A British officer wrote "Their assault was as furious as ever I saw; The Ditch was choke full of their Dead." Laurens was seen standing before the British volleys, "with his arms wide extended" wanting to die with his men. When an officer tried to stop his rash move he replied, "My honour does not permit me to survive the disgrace of this day." Seeing all of this D'Estaing ordered a retreat. As the retreating troops tried to make their way back through a hollow or in the swamp, the British guns fired upon them unmercifully.

When Pulaski saw the slaughter that was happening he spurred his horse and rode off towards the fight. Pulaski's cavalry veered to the left to avoid the heavy fire from the defenses, and rode through Laurens' men. The momentum of the cavalry horses forced the Carolinians into Yamacraw Swamp. As he rode across the French lines Pulaski was shot in the groin by a swivel gun from one of the galleys. He would die on October 11th, two days later.

This was the bloodiest day of American and French troops in the whole war. Of the six hundred Carolinians who charged up the hill, two hundred and fifty did not return. Eighty men lay dead in the ditch and on the parapet, 93 were dead within the abatis on the other side of the ditch. Under a flag of truce the Americans and French buried their dead in front of the British lines.

The French had lost 183 men on the assault on Spring Hill Redoubt. This included 15 officers leading from the front. The truce lasted from 10:00 in the morning until 4:00 that afternoon. At 8:00 that night the French beat a parley to negotiate a further truce, but the British refused. Afterwards there was some artillery fired between the lines, but it was only a shadow of the days before. Several deserters left the allied army and surrendered to the British.

Lincoln lost 125 killed and 332 wounded during the whole siege. The French lost 600 killed and 411 wounded. The British only lost 40 killed and 63 wounded. Count D'Estaing had lost his taste for land warfare. On October 18th he hurried his men on board ship, then sailed back to the West Indies. Lincoln withdrew into South Carolina to consider the next course of action.

1174 The 2nd South Carolina regiment had two colors, a red ensign and a blue ensign. The blue colors were captured at Savannah and the other at the surrender of Charleston in May 1780. When the colors had visited South Carolina in 1970 for the Tricentennial of the State, Major General A.E.W. Williams said that "This marks the first occasion in British military history when a color taken following a battle has ever been returned even on loan to the place of its origin." The blue ensign was returned for good in 1992 and is on display at the South Carolina State Museum in Columbia. Up until 2006 they were the only known Southern regimental colors to exist. In 2006 the two captured colors of Buford's Virginia regiment were brought forward by the descendants of Banastre Tarleton and sold at auction.

The 2nd South Carolina Regiment lost many men during the siege of Savannah. Many of these men had served in the militia, or in the South Carolina Continental Regiments earlier in the war and had reenlisted when the British captured Savannah and threatened to invade South Carolina. They had joined to defend their homes from attack by a foreign invader. The 2nd South Carolina Regiment in Savannah consisted of about 350 men, in a regiment that normally would have had a full strength of about 500. The youngest of these soldiers was 13 years old, while the oldest was 73. During the assault the regiment had 24 men killed and 10 wounded. After the assault on the Spring Hill Redoubt the official records only mentioned wounds that were mortal, or wounds that crippled the men for life. Any wounds that were not considered severe and that the men could recover from were not listed. The total number of wounded will never be known, but with the reported numbers the regiment suffered 10% casualties during the assault on the Spring Hill Redoubt.

1st Company

Captain Adrian Proveaux
Lieutenant Josiah Kolb
Sergeant Robert Matthews (**killed**)
Sergeant Noble Barnett
Corporal John Mills
Corporal Solomon Long
Drummer Enoch Boolk
Fifer Jacob George

Privates

Jacob Benhoist
Samuel Blackford [1175]
Isaac Chinners
Samuel Cortney
Archy M. Daniel
John Caddy (**deserted**)
Edward George
John Harper [1176]
James Gaskey (**wounded**) [1177]
John Hawkins
Frederick Hughes
George Hughes
William Johnston[1178]
James McDaniel [1179]
William Norman
Lewis Patrick [1180]
John Perry
William Phillips [1181]
John Ratford [1182]
Rolly Rawlins [1183]
Alexander Stewart
Phillip Thomas
John Thompson
Amos Tubbs
David Vaughn
Walkinsheer Thompson
John White

[1175] Samuel (Lemuel) Blackford enlisted in the 2nd South Carolina Regiment on 20 February 1779.

[1176] John Harper enlisted in the 2nd South Carolina Regiment on 14 February 1779.

[1177] James Gaskey (Gaskery) enlisted in the 2nd South Carolina Regiment on 1 August 1779. He died on 27 October 1779 due to his wounds.

[1178] William Johnston enlisted in the 2nd South Carolina Regiment on 4 November 1775.

[1179] James McDaniel enlisted in the 2nd South Carolina Regiment on 1 July 1779 after the British invaded South Carolina and fought at the Battle of Stono Ferry.

[1180] Lewis Patrick enlisted in the 2nd South Carolina Regiment on 1 July 1779 after Stono Ferry.

[1181] William Phillips enlisted in the 2nd South Carolina on 1 July 1779 after Stono Ferry. One source says that he died on 6 September 1779, but he is mentioned as being in Captain Proveaux's company on 1 November 1779.

[1182] John Ratford (Radford) enlisted in the 1st South Carolina Regiment on 4 November 1775. He enlisted in the 2nd South Carolina Regiment on 1 August 1779.

[1183] Rolly Rawlins (Raulins) enlisted in the 2nd South Carolina Regiment on 1 July 1779 after Stono Ferry.

2nd Company

Captain Richard Baker
Lieutenant Alexander Hume (**killed**)
Sergeant John Roberts [1184]
Sergeant Alexander McDonald [1185]
Corporal Joseph Wilkins
Corporal Levi Brown
Fifer Silas Gibson [1186]

Privates

Andrew Adams (**wounded**) [1187]
Daniel Andrews
James Costello
Robert Clyatt [1188]
Jonathan Collins [1189]
William Connelly [1190]
Timothy Downing
John Fenwick
Lewis Domas (**killed**) [1191]
Dickerson Green [1192]
Thomas Hagartey
Samuel Horn
Joshua Morgan [1193]
Moses Mace [1194]
Adam Meek
John Richardson [1195]
James Russell [1196]
Solomon Stapleton
Nathaniel Swobb
George Taylor
Henry Taylor [1197]
William Waites
Benjamin Webster

1184 John Roberts enlisted in the 2nd South Carolina Regiment on 4 November 1775. He was promoted to corporal on 2 June 1777 and to sergeant on 14 July 1778.

1185 Alexander McDonald enlisted in the 2nd South Carolina Regiment on 1 August 1779. He was promoted to sergeant in Captain Baker's company. This is not Sergeant Major Alexander McDonald who was in the light infantry company.

1186 Silas Gibson enlisted in the 2nd South Carolina on 14 July 1778. He became a fifer on 28 October 1778.

1187 Andrew Adams enlisted in the 2nd South Carolina Regiment on 25 November 1775. On 23 December 1776 he was reprimanded and mult 14 days pay. He was under Captain John Blake. In 1779 he was under Captain Richard Baker. He was wounded and lost an arm in Savannah.

1188 Robert Clyatt (Clyett) served in Captain Benjamin Screven's militia during 1775. He enlisted in the 2nd South Carolina Regiment on 1 July 1779 after Stono Ferry.

1189 Jonathan Collins enlisted in the 2nd South Carolina Regiment on 1 July 1779 after Stono Ferry. He was taken prisoner at the surrender of Charleston in May 1780 and then kept in irons on a prison ship. He was exchanged at James's Tavern in Virginia and returned to South Carolina. He joined Marion's partisans and was with Captain John Rogers and Captain Simons. He was at the fight at Blue Savannah in September 1780 and he conducted Marion's raids on Georgetown at beginning of 1781. He was in the Battle of Eutaw Springs in September 1781. He was discharged by Marion at Bowling Green towards the end of the war.

1190 William Connelly enlisted in the 2nd South Carolina Regiment on 1 July 1779 after Stono Ferry.

1191 Lewis Domas enlisted in the 2nd South Carolina Regiment on 1 August 1779 after Stono Ferry. He was killed during the assault on the Spring Hill Redoubt on 9 October 1779.

1192 Dickerson Green enlisted in the 2nd South Carolina Regiment on 1 August 1779 after Stono Ferry.

1193 Joshua Morgan enlisted in the 2nd South Carolina Regiment on 1 August 1779 after Stono Ferry.

1194 Moses Mace was born in Maryland in 1743. He enlisted in the 3rd South Carolina Rangers on 18 July 1775 under Captain Samuel Wise. He enlisted in the 2nd South Carolina Regiment on 1 August 1779.

1195 John Richardson enlisted in the 2nd South Carolina Regiment on 1 August 1779 after Stono Ferry.

1196 James Russell was born in Ireland in 1756. He enlisted in the 3rd South Carolina Rangers in 1775 under Captain John Purvis. He was later under Captain Felix Warley. He enlisted in the 2nd South Carolina Regiment on 1 August 1779 under Captain Richard Baker after Stono Ferry. After the fall of Charleston he served in the militia of Captain Ballard and Colonel Marshall. From 15 December 1780 to 15 March 1781 he was in Marion's Partisans. He lost a horse during this service. He served 109 days in the militia in 1782.

1197 Henry Taylor served in the Salt Catcher Volunteer Company under Captain James Jones in 1775. He later enlisted in the 2nd South Carolina Regiment and was under Captain Baker.

3rd Company

Major Charles Motte (**killed**)
Captain John Hart
Lieutenant Alexander Petrie (**wounded**)
Sergeant John Barnet Taylor
Sergeant George Brewton
Corporal William Oliver (**killed**)
Fifer Peter Area

Captain Henry Gray (**wounded**)
Lieutenant Albert Roux (**wounded**)
Lieutenant Christopher Rogers
Sergeant William Murphy
Corporal William Jones [1198]
Drummer William Burbridge [1199]

Privates

Matthew Anderson [1200]
Joseph Cooper [1203]
Francis Ferrill [1205]
Oswald Hackle [1207]
Ralph Ingram
Thomas Mills
Hugh Newman (**killed**)
Nathaniel Rogers

Jonathan Burbridge [1201]
Cornelius Constantine
James Fitzsimmons
Benjamin Huggins [1208]
Samuel Kinney
Vincent Maroni (**killed**)
Robert Pinhorn
James Stanton

Thomas Burbridge [1202]
William Easton [1204]
John Godbolt (**killed**)[1206]
Humphrey Haines
George McCormack
Malcolm McFarlan [1209]
Lewis Powell [1210]
William Willis

1198 William Jones enlisted in the 2nd South Carolina Regiment on 4 November 1775 under Captain Charles Motte. He was promoted to corporal. He was demoted to private on 13 January 1778. He was promoted again back to corporal by 1779.

1199 William Burbridge (Burbage) enlisted on 19 June 1778 in the 2nd South Carolina Regiment as a drummer under Captain Motte. He remained in the service until 1783.

1200 Matthew Anderson enlisted in the 2nd South Carolina Regiment on 4 November 1775. He was promoted to corporal on 9 June 1777. He was demoted to private on 25 August 1777.

1201 Jonathan Burbridge (Burbage) enlisted in the 2nd South Carolina Regiment on 26 February 1779 under Captain Motte. He was still in the regiment in 1782.

1202 Thomas Burbridge (Burbage) enlisted in the 6th South Carolina Regiment at the age of 17. He was in Captain Richbourg's company. He fought the Cherokees while in this regiment, and he ranged into Georgia to fight against the Loyalist Daniel McGirth. He enlisted in the 2nd South Carolina Regiment under Captain Motte. He was captured when Charleston surrendered and was held as a prisoner for 14 months. After he was released he joined Marion's partisans and would be under Captain William Brown. He was later under Captain William Duke.

1203 Joseph Cooper enlisted in the 2nd South Carolina Regiment on 4 July 1779 under Captain Motte after Stono Ferry.

1204 William Easton served 7 months in the 2nd South Carolina Regiment under Captain Motte. He was in the regiment in 1780.

1205 Francis Ferrill enlisted in the 2nd South Carolina Regiment on 20 November 1775.

1206 John Godbolt enlisted in the 2nd South Carolina Regiment on 1 July 1779 under Captain Motte after Stono Ferry. He was killed on the assault on the Spring Hill Redoubt on 9 October 1779.

1207 Oswald Hackle enlisted in the 2nd South Carolina Regiment on 4 November 1775. He was under Captain Motte in 1779.

1208 Benjamin Huggins was in the 2nd South Carolina Regiment after 1 July 1779 under Captain Motte after Stono Ferry. He also served in the 3rd South Carolina Regiment as an ensign under Captain Joseph Warley. During April 1781 he became an ensign, and then in 1782 he was promoted to lieutenant. He served as an adjutant under Colonel Peter Horry in Marion's partisans.

1209 Malcolm McFarlan enlisted in the 2nd South Carolina Regiment on 4 November 1775.

1210 Lewis Powell enlisted in the 2nd South Carolina Regiment on 1 July 1779.

4th Company

Captain Peter Gray
Lieutenant James Gray (**killed**)
Sergeant James Feast
Corporal Abraham Kerslick
Corporal Samuel Brown
Drummer Lewis McClendall
Lieutenant John Wickham (**killed**)
Sergeant John McDowell
Sergeant Joseph Turner
Corporal Thomas Galloway
Drum Major John Hyrne
Fifer Frederick Lamb [1211]

Privates

John Peter Allen
Charles Bentley [1213]
John Campbell
Alexander Ferguson
Charles Hutton
William Leaton, Sr. [1218]
William Martin [1219]
William Simpson
Bartholomew Hine
Isaac Withersford
Edward Bambrick [1212]
John Bewly [1214]
Moses Childs
Joshua Hall
Thomas Jones
Robert Lance
Joseph Pain (**deserted**)
Jesse Simmons
Reuben Wales
Abraham Berlin (**deserted**)
Ambrose Bray [1215]
Richard Clark
William Hughes [1216]
Hendrick Keyler [1217]
Henry Martin
John Riley [1220]
Charles Skipper
William Winford

1211 Frederick Lamb was born around 1765 in Virginia. He ran away from home at a very young age and enlisted in the 1st South Carolina Regiment on 7 October 1777. He served in the 2nd South Carolina Regiment from 30 September 1778. He was a fifer in the regiment from 20 October 1778 until 15 November 1783. At the end of the war he was discharged at Camden, and he met Celia Bowen, who he later married. He settled in Glynn County, Georgia. He died sometime after 1805.

1212 Edward Bambrick served in the 2nd South Carolina Regiment in 1776 and in 1779.

1213 Charles Bentley enlisted in the 2nd South Carolina Regiment on 1 July 1779 after Stono Ferry.

1214 John Bewly enlisted in the 2nd South Carolina Regiment on 1 March 1779.

1215 Ambrose Bray enlisted in the 2nd South Carolina Regiment on 4 November 1775.

1216 William Hughes served in Captain Gray's company of the 2nd South Carolina Regiment in 1779. He was in Colonel Richard Winn's regiment of Sumter's partisans from in March of 1781. He served under Captain Gray at that time. He was an acting lieutenant under Captain Durham from 15 October to 20 December 1781. He became a captain under Colonel Winn on 24 December 1781. In June 1782 he was a private under Captain Joseph Hughes and Colonel Brandon.

1217 Hendrick Keyler (Kiler) enlisted in the 2nd South Carolina Regiment on 10 February 1779.

1218 William Leaton, Sr. enlisted in the 2nd South Carolina Regiment on 28 July 1777 under Captain Gray. His son, William Leaton, Jr. enlisted in the regiment on 22 September 1777.

1219 William Martin enlisted in the 2nd South Carolina Regiment on 7 October 1776.

1220 John Riley was born in 1760 in Amherst County, Virginia. He enlisted in the 2nd South Carolina Regiment on 5 February 1779 under Captain Gray. After the fall of Charleston he lived in the Camden District. He enlisted under Captain Jacob Barnett in Sumter's partisans. He was in the siege of Orangeburgh. He was in the battle of Eutaw Springs under Captain Thomas Gillian and Colonel Bratton and was wounded in the leg during the fighting. He moved to Alabama after the war.

5th Company

Captain Thomas Hall
Sergeant William Henderson
Sergeant William Harvey
Sergeant William McCullock (**killed**)
Sergeant Nehemiah Watt
Corporal John Dubose
Fifer Conrad Fitner

Lieutenant James Legare
Sergeant Reuben Minor
Sergeant John Burtell
Sergeant William Hasman
Corporal Robert Rain
Drummer Phillip Fry

Privates

Enoch Andrews [1221] (**wounded & captured**)
Thomas Bowen
John Crawford
John Caton [1222]
John Clements
Thomas Crozier (**killed**)
Thomas Davis
Henry Dishers
Thomas Faulder [1223]
James Fitzpatrick[1224]
Edward Fry
Rapes Going [1225]
Thomas Goodson
Needham Gunter[1226]
Samuel Henderson
Hugh Holland
William Lindsey[1227]
John Marlow
Daniel McFarlin[1228]
Solomon Mitchell[1229]
Samuel Moet [1230]
Dempher Oldfield [1231]
Benjamin Owens [1232]
John Proby
James Reid
Benjamin Sergenor
Francis Simpson
Thomas Welch

[1221] Enoch Andrews enlisted in the 2nd South Carolina Regiment on 1 August 1779 after Stono Ferry.

[1222] John Caton enlisted in the 2nd South Carolina Regiment on 15 December 1777.

[1223] Thomas Faulder enlisted in the 2nd South Carolina Regiment in 1775.

[1224] James Fitzpatrick enlisted in the 2nd South Carolina Regiment on 1 August 1779 after Stono Ferry. After the fall of Charleston he was a lieutenant in Colonel Myddleton's 2nd Regiment of South Carolina State Dragoons of Sumter's partisans. He also served 33 days in Marion's partisans in 1782.

[1225] Rapes Going enlisted in the 2nd South Carolina Regiment on 1 July 1779 after Stono Ferry.

[1226] Needham Gunter enlisted in the 2nd South Carolina Regiment on 1 August 1779 after Stono Ferry. After the fall of Charleston he was in Marion's partisans. He was at the skirmish at Biggin's Church. He continued under Marion until the end of the war.

[1227] William Lindsey (Linsey, Linzly) enlisted in the 2nd South Carolina Regiment on 1 August 1779 after Stono Ferry.

[1228] Daniel McFarlin enlisted in the 2nd South Carolina Regiment on 1 August 1779 after Stono Ferry.

[1229] Solomon Mitchell was born in 1759 in Granville County, North Carolina. He enlisted in the militia while living in Abbeville District in 1777. He served under Captain Francis Logan and Colonel Pickens. While in the militia he was in the battle of Stono Ferry, and the attacks on Augusta. He enlisted in the 2nd South Carolina Regiment on 1 August 1779 after Stono Ferry. He married Nancy Broton in May 1781. After the war he moved to Tennessee. He died 27 January 1839 at the age of 80.

[1230] Samuel Moet enlisted in the 2nd South Carolina Regiment on 1 July 1779 after Stono Ferry.

[1231] Dempher Oldfield enlisted in the 2nd South Carolina Regiment on 1 August 1779 after Stono Ferry.

[1232] Benjamin Owens enlisted in the 2nd South Carolina Regiment on 1 August 1779 after Stono Ferry.

6th Company

Captain Thomas Moultrie
Lieutenant Peter Foissin
Sergeant Douglas O'Neal
Corporal William Rogers
Corporal Samuel Murray
Fifer Archibald Robertson [1234]
Lieutenant John Bush (**killed**)
Sergeant Stephen Roberts
Sergeant Daniel Green
Corporal David Manly
Drummer Robert Logan [1233]

<u>Privates</u>
Abraham Baggett [1235]
Francis Bridges [1236]
Peter Deviney (**killed**)
Moses Groom
James Hain
Archibald Lamb [1243]
Edward Murphy
John Steel (**killed**)
Nicholas Barger
Robert Cox [1237]
Edmund Gainey [1238]
James Grover
James Houston [1241]
Henry McCall [1244]
William Russell [1245]
Elisha Tomplat
John Bently
John Friday
Frederick Gowin [1239]
James Grubbs [1240]
Richard Lackey [1242]
William McCallister
James McGowen (**wounded**)
Edward Wainwright

7th Company – Light Infantry Company

Captain Thomas Dunbar
Lieutenant William Capers
Sergeant Henry Webb
Sergeant Reuben Dewitt
Corporal Harris Dewitt
Drummer Jesse Martin [1246]
Lieutenant Cornelius Van Hempstead Vlieland (**killed**)
Sergeant Major Alexander McDonald
Quarter Master Sergeant William Jasper (**killed**)
Sergeant John McDonald (**wounded**)
Corporal William Manning
Corporal & Fifer James Newton [1247]

[1233] Robert Logan enlisted in the 2nd South Carolina Regiment on 10 August 1778. He became a drummer under Captain Moultrie on 17 November 1778.
[1234] Archibald Robertson enlisted in the 2nd South Carolina Regiment on 30 August 1778. On 12 January 1779 he became a fifer under Captain Moultrie.
[1235] Abraham Baggett enlisted in the 2nd South Carolina Regiment on 20 January 1779.
[1236] Francis Bridges enlisted in the 2nd South Carolina Regiment on 1 July 1779 after Stono Ferry.
[1237] Robert Cox enlisted in the 2nd South Carolina Regiment on 1 July 1779 after Stono Ferry.
[1238] Edmund Gainey enlisted in the 2nd South Carolina Regiment on 2 September 1778.
[1239] Frederick Gowin enlisted in the 2nd South Carolina Regiment on 1 August 1779 after Stono Ferry. After Savannah he was made a sergeant and he was killed during the siege of Charleston in 1780.
[1240] James Grubbs enlisted in the 2nd South Carolina Regiment on 19 October 1778.
[1241] James Houston enlisted in the 2nd South Carolina Regiment on 1 August 1779 after Stono Ferry.
[1242] Richard Lackey enlisted in the 2nd South Carolina Regiment on 12 January 1779.
[1243] Archibald Lamb enlisted in the 2nd South Carolina Regiment on 1 August 1779 after Stono Ferry.
[1244] Henry McCall was in the 2nd South Carolina Regiment at Savannah under Captain Dunbar. After the surrender of Charleston he served under Sumter and Huger. He served as a sergeant for 30 days in 1782.
[1245] William Russell enlisted in the 2nd South Carolina Regiment on 22 January 1779. He was captured at the surrender of Charleston and he was a prisoner at Haddrell's Point in October 1780.
[1246] Jesse Martin enlisted in the 2nd South Carolina Regiment on 23 August 1778. He became a drummer under Captain Dunbar on 20 October 1778.
[1247] James Newton enlisted in the 2nd South Carolina Regiment on 4 February 1779.

Privates

Barnaby Brian [1248]
John Butler [1249]
John M. Cade
William Cade [1250]
George Carrick
William Chancelly[1251]
John Chavis
James Clark
William Clarke
William Cook [1252]
Joseph Davis [1253]
John Dius
Nicholas Flinn
James Ford [1254]
John Frances
Robert Gammell
Christopher Gamond
John Hampton [1255]
Aaron Harris
Hezekiah Heath [1256]
William Henson
Kindred Hollisman
John Holmes
Joseph Hughes (**killed**)
Burrell Jones [1257]
James Jones [1258]
Daniel Jordan [1259]
James Leaton
John Martin Sharp
Shadrack McClendon (**wounded**)[1260]
John McBride
William Mimms [1261]
James Moody [1262]
Thomas Nutt
James Oliver
Thomas Oliver [1263]

[1248] Barnaby Brian (Bryan) enlisted in the 2nd South Carolina Regiment on 4 November 1779.

[1249] John Butler enlisted in the 2nd South Carolina Regiment on 1 July 1779 after Stono Ferry.

[1250] William Cade was born on 24 August 1725. He enlisted in the 2nd South Carolina Regiment in 1779. He served under Captain Moultrie, except during the siege of Savannah when he was in Captain Dunbar's Light Infantry Company. After the fall of Charleston he served 16 months in the partisans with Hezekiah Maham's Light Dragoons in 1782 and 1783. He married Elizabeth H. Smith and died sometime after 1787.

[1251] William Chancelly enlisted in the 2nd South Carolina Regiment on 11 September 1778.

[1252] William Cook enlisted in the 2nd South Carolina Regiment on 17 September 1778. After the fall of Charleston he was in Colonel Peter Horry's cavalry with Marion's partisans. He was court martialed on 4 October 1781 for being absent without leave and he received 12 lashes on his buttocks.

[1253] Joseph Davis enlisted in the 2nd South Carolina Regiment at Post's Ferry on the Great Pee Dee River. He was in the battles of Fort Moultrie, Stono Ferry, and the siege of Savannah. After the fall of Charleston he was in Marion's partisans under Colonel Horry. He was in the battles of Nelson's Ferry, Black Mingo, Blue Savannah and Coosawhatchie. He served one year as a private, two years as a sergeant in the infantry and three years as a lieutenant in the cavalry. He died on 13 November 1838.

[1254] James Ford enlisted in the 2nd South Carolina Regiment in 1775.

[1255] John Hampton enlisted in the 2nd South Carolina Regiment on 1 August 1779 after Stono Ferry.

[1256] Hezekiah Heath enlisted in the 2nd South Carolina Regiment on 23 August 1778.

[1257] Burrell Jones enlisted in the 2nd South Carolina Regiment on 1 August 1779 after Stono Ferry.

[1258] James Jones enlisted in the 2nd South Carolina Regiment on 1 August 1779 after Stono Ferry.

[1259] Daniel Jordan enlisted in the 2nd South Carolina Regiment on 6 November 1775. He was court martialed on 6 April 1778 for scaling the fences of the general hospital. He was in Dunbar's light infantry at the siege of Savannah. He was still in the service in 1782.

[1260] Shadrack McClendon was born in 1753. He enlisted for three years in the 2nd South Carolina Regiment on 31 August 1778 under Captain Mazyck. He lost his left eye in the siege of Savannah. He was captured at the fall of Charleston and held as a prisoner for 14 months, before he was able to escape. After the war he moved to Mississippi and Louisiana. He died in 1846.

[1261] William Mimms was born in 1762. He enlisted in the 2nd South Carolina Regiment on 1 August 1779 after Stono Ferry. He died in 1825.

[1262] James Moody enlisted in the 2nd South Carolina Regiment on 1 August 1779 after Stono Ferry.

[1263] Thomas Oliver enlisted in the 2nd South Carolina Regiment on 28 August 1777. He was captured at the fall of Charleston and he was a prisoner at Haddrell's Point. He was also a servant to Lieutenant Foissin while at Haddrell's Point.

Richard Richardson
John Smith
David Stuart
John Whitley
James Scurry
John Sparrow [1264]
David Wiley
William Wilkinson
Frederick Simmons
Thomas Stafford
Robert Whiley
Thomas Windsor

8th Company

Captain Richard Mason
Sergeant John Taylor [1265]
Sergeant James Clatworthy
Drummer William Crapps [1266]
Lieutenant Paul Warley
Sergeant John Davis
Corporal Thomas Kidwell
Fifer Moses Newton

Privates

William Brown
William Chaney
Hugh Derberry
Anthony Hinds
Matthew Kennedy [1270]
Jeremiah Peters [1272]
Thomas Rybold
Peter Rosman
Jacob See
John Thompson
Charles Burnham [1267]
John Conner
William Enochs
Stephen Irons [1269]
Phillip Newton [1271]
Michael Peters [1273]
Benjamin Reeves
William Ryan
Matthew Skipper
Samuel Butler
William Dalton (**killed**) [1268]
Timothy Green (**killed**)
Isaac Herring (**wounded & captured**)
William H. Jones (**deserted**)
Thomas Rawlins
Joseph Reeves
Henry Savage
Adam Smith

1264 John Sparrow became a servant to Lieutenant Dunbar while they were prisoners at Haddrell's Point after their capture at Charleston.

1265 John Taylor served in the 2nd South Carolina in 1776 in Captain Eveleigh's company. He was wounded at Fort Sullivan on 28 June 1776. He may have been the John Taylor who died of smallpox during the war.

1266 William Crapps enlisted in the 2nd South Carolina Regiment on 28 February 1778. He became a drummer in Captain Mason's company on 1 July 1779 after Stono Ferry.

1267 Charles Burnham enlisted in the 2nd South Carolina Regiment on 23 February 1779.

1268 William Dalton enlisted in the 2nd South Carolina Regiment on 4 November 1775. He was killed during the assault on the Spring Hill Redoubt at Savannah in 1779.

1269 Stephen Irons enlisted in the 2nd South Carolina Regiment on 24 February 1779.

1270 Matthew Kennedy enlisted in the 2nd South Carolina Regiment on 1 July 1779 after Stono Ferry.

1271 Phillip Newton enlisted in the 2nd South Carolina Regiment on 11 September 1777.

1272 Jeremiah Peters enlisted in the 2nd South Carolina Regiment on 8 February 1779.

1273 Michael Peters enlisted in the 2nd South Carolina Regiment on 1 July 1779 after Stono Ferry.

9th Company

Captain Daniel Mazyck
Lieutenant George Ogier
Sergeant Benjamin Stone
Corporal Marmaduke Ethridge
Corporal Samuel McMillian
Fifer David Parrish [1276]

Lieutenant John Martin
Sergeant William Wood
Sergeant Jacob Kalckoffen
Corporal William McCullough [1274]
Fifer Benjamin Booth [1275]
Drummer John Robinson

Privates

James Beard (**killed**) [1277]	Benjamin Breeler [1278]	John Breeler
Edward Brown [1279]	Blake Calcott	John Carter (**killed**) [1280]
Charles Caves	William Clary [1281]	Arthur Colson [1282]
Thomas Cowen [1283]	Abram Debraudy [1284]	John Dubose [1285]
Hugh Davis	Christopher Gallington [1286]	William Gunter (**killed**) [1287]
Jacob Heigle [1288]	William Hyde (**deserted**)	John Keith [1289]

1274 William McCullough (McCullock) enlisted in the 2nd South Carolina Regiment on 4 November 1775. On 2 February 1779 he was promoted to corporal in Captain Mayzyck's company. He was at the siege of Savannah and at the surrender in Charleston. In 1782 he served 66 days in the militia. There were four soldiers named William McCullough or McCullock in the 2nd South Carolina. All three were at the siege of Savannah and two of them died there.

1275 Benjamin Booth enlisted in the 2nd South Carolina on 12 January 1779 as a fifer under Captain Mazyck.

1276 David Parrish enlisted in the 2nd South Carolina on 31 January 1779 as a fifer under Captain Mazyck.

1277 James Beard enlisted in the 2nd South Carolina Regiment on 14 February 1776. He was killed on 9 October 1779 during the assault on the Spring Hill Redoubt.

1278 Benjamin Breeler enlisted in the 2nd South Carolina Regiment on 1 July 1779 after Stono Ferry.

1279 Edward Brown enlisted in the 2nd South Carolina Regiment on 2 November 1777.

1280 John Carter enlisted in the 2nd South Carolina Regiment on 9 February 1779.

1281 William Clary (Clay) enlisted in the 2nd South Carolina Regiment on 4 November 1775. He was accused of murder while he was in Marion's Partisans on 9 May 1782, but he was acquitted.

1282 Arthur Colson enlisted in the 2nd South Carolina Regiment on 1 July 1779 after Stono Ferry.

1283 Thomas Cowen enlisted in the 2nd South Carolina Regiment on 19 April 1777.

1284 Abram Debraudy (Debraidy, Debrawder) served in the 2nd South Carolina Regiment under Captain Mazyck in 1779.

1285 John Dubose was born in 1715. He was in the 2nd South Carolina Regiment in 1779 under Captain Mazyck. He supplied provisions for the miltia and the Continental army in 1782. He married Mary Francis DeWitt. He died some time before 1785.

1286 Christopher Gallington enlisted in the 2nd South Carolina Regiment on 1 July 1779 after Stono Ferry.

1287 William Gunter enlisted in the 2nd South Carolina Regiment on 1 July 1779 after Stono Ferry. He was killed at Savannah on 9 October 1779 during the assault on the Spring Hill Redoubt. Another source says he lived and he was in the militia as a sergeant under Captain James Crawford and Colonel Brandon from 5 May 1781 to 25 October 1783. There may have been two men named William Gunter in the 2nd South Carolina in 1779.

1288 Jacob Heigle (Heiglar) served in the 2nd South Carolina Regiment after 21 September 1778.

1289 John Keith enlisted in the 2nd South Carolina Regiment on 1 July 1779 after Stono Ferry. He was taken prisoner at the fall of Charleston. He was held 14 months, and then he was exchanged. He joined the militia and was wounded at the siege of Yorktown. He died on 10 February 1839.

Joseph Mallery [1290]
Hector McLane [1292]
Frederick Rowland [1295]
John Smith [1296]
Robert Marker
William Pawling [1293]
John Skipper
John Teague
William McCullock (**killed**) [1291]
Thomas Poston [1294]
Drury Smith
Rowland Walker

The Cost

[1290] Joseph Mallery enlisted in the 2nd South Carolina Regiment on 14 February 1779.

[1291] William McCullock (McCulloch) was killed at Savannah while serving with the 2nd South Carolina.

[1292] Hector McLane (McLean) served in the 2nd South Carolina Regiment under Captain Blake in 1778. He was in Captain Mazyck's company at the siege of Savannah.

[1293] William Pawling was born on 23 July 1765 in Georgetown, South Carolina. He enlisted in the 2nd South Carolina Regiment when he was 13 years old on 1 July 1779 after Stono Ferry. He was in the siege of Savannah under Captain Mazyck. In 1781 he was in the partisans under Colonel Reid. He also rode with Colonel Wade Hampton in Sumter's partisans. He was at the battle of Shubrick's Plantation, and he was in the battle of Eutaw Springs, all before he was 15 years old.

[1294] Thomas Poston enlisted in the 2nd South Carolina Regiment on 1 July 1779 after Stono Ferry.

[1295] Frederick Rowland enlisted in the 2nd South Carolina Regiment on 4 November 1775.

[1296] John Smith enlisted in the 1st South Carolina Regiment on 7 May 1777. He was promoted to sergeant on 2 June 1777. On 9 March 1778 he was demoted to private. He was with Dunbar's Grenadiers in the 2nd South Carolina Regiment and then was transferred to Captain Mazyck's company during the siege of Savannah. He was transferred back to the 1st South Carolina Regiment and served in Captain William Hext's company during the siege of Charleston. He served in the 1st South Carolina until 1 July 1781.

Camp at Sheldon November 9th 1779 [1297]

The Main Guard to be Augmented so as to give three More Centinels – The Picquets & out Centinels are ordered Not to suffer soldiers Waggoners or followers of the Army to carry fire Arms or Ammunition out of camp without a written permit signed by the Commanding officer

Any one that shall hereafter be found out of Camp with fire Arms will be punished for a Breach of orders – Fire Hunting is strictly forbid:

Any person who shall be convicted of a stabbing cutting or otherwise abusing the Horses in or about camp – Must Expect to be punished with Unrelenting Severity ______________

This order & the order Issued the 23d of October last to be Read to the Men in the two following Days of Parade ________________

Returns to be Made tomorrow of the Stores in the Qr Ms Commissarys & Artillery Departments

The 2d 6th & 13th Sections of the Articles of war, to be read this afternoon to the Respective Corps at this post _____________

Regimental orders 9th of November 1779

Capt John Vanderhorst is Appointed a Major in the 2d Regiment he is to be Respected & Obey'd as such

to 1st November

The pay bills to be Made by tomorrow 10 OClock in forenoon & given in to Major Vanderhorst

The Regiment to Exercise every Monday, Wednesday & Friday at 2 OClock in Afternoon all officers & soldiers of duty to Attend –

Parole	Cn Sns Vernice Valour	
		S – C – R
Detail for to Day		1 – 1 – 7
Qr. Guard _ _ _		1 – 1 – 6

After Orders 1 Gallon of Rum to be Drawn for Each Officer & One gill pr. Man

________________________________ Novr 10th ________________________________

Parole Cn Sns Wisper Way

B: Orders by Colo Beckman/ Application to the Commanding Officer for leave of absence must be signed by the Officer Commanding the Regiment or Corps to which they belong

Camp at Sheldon 11th November 1779

B Orders by Colo Beckman / It is with Pleasure the Commanding officer is Authorized to give assurance to the men of this Army that their Winter Clothing is in Hand & in great forwardness which will be sent up to Camp as soon as they are done – in the mean time it is Desired, that fences of Poles be rais'd Round the separate Line of the Incampment to be Whattled close with ever greens or other Bushes which will Shield the men from the Cold Northerly Winds the Qr. Masters & their Serjeants to see this Compleat

---

[1297] Sheldon is located about ten miles north of Beaufort. After the disastrous attack on the Spring Hill Redoubt the French left the Carolinas. General Lincoln retreated into South Carolina and was determined to salvage something from the ill fated expedition. The British still held Beaufort with a small force, but they were supported by armed galleys surrounding the defenses. For Lincoln this appeared to be another Stono Ferry. He stayed in Sheldon keeping an eye on the British and determining his next move.

The Officer of ye Day to be Releaved Dayly on the Parade at Guard mounting who is to visit all the Guards & Centinels except the Command at Port Royal ferry, at Least once During the day & once During ye Night & Specify the hour of going the Rounds in His Reports which will be maid to the
Camp at Sheldon 11th Novr. 1779
Parole Smith Cn Sn Youth York
Commanding Officer at 10 OClock
The Officer of the day will also on his Command Prevent any Riot or noise in Camp by Day or Night it is expected that the Officers of the Several Corps & Companies Keep up their Orderly Books –

	C	L	S	Sg	R
Detail for the Day	1..	– ..	1 ..	0	7
Quarter Guard	–	–	–1..	1	6

Parole America 12th CSn Amity Array
B Orders by Col° Beckman / The Salt that will be wanted for the use of the Troops at this Post is to be taken from Lt James Garveys[1298] plantation on new River which is to be accounted for by the Commissary of Purchases with Col° Garden [1299] – the Qr Master General to Provide Waggons to Transport the Same
A Court Martial to set from the Line Immediately for the trial of such Prisoners as shall come before them all Evidences to Attend
Capt. Hopkins, President[1300] from ye Second 1 Sub from the 3d 1 Sub Members

Camp at Sheldon 13 November 1779
Parole Moultrie CSn Barc Brilliant

	S ..	S ..	C ..	P
Detail of the Guard	1 ..	1 ..	1 ..	7
Quarter Guard	- ..	1 ..	1 ..	6

Head Quarter Chs town 12th Novr 1779
Genl. Orders by G Lincoln } at a General court martial whereof Col°. Horry is president Conrade Bessinger a private Soldier in the 3d S°. Carolina Battalion for Desertion & being in Arms Against his country, he is Found Guilty & Sentenced to be hanged by the Neck Untill Dead [1301]

The General Approves of the Sentence & order it to be ~~put~~ Executed on Wednesday next between the hours of nine & Eleven in the forenoon

---

[1298] James Jarvis was a lieutenant in the 5th South Carolina Regiment.
[1299] Colonel Benjamin Garden was in the Granville County Militia from 1775 to 1780.
[1300] David Hopkins was born in 1739 in Virginia. He served as a lieutenant in the 3rd South Carolina (Ranger) Regiment under Captain Robert Goodwyn from 25 June 1775. In 1779 he became a captain. He served in the militia during from 1780 to 1782. He died in 1816 at the age of 77.
[1301] Conrad Bessinger enlisted in the 3rd South Carolina (Ranger) Regiment on 9 December 1777. He was court-martialed on 12 November 1779 on the charge of desertion and taking up arms against his country and sentenced to be hanged.

At the same Court Captn Lieu^t^ Wilson[1302] & Lieu^t^ Fields [1303] both of the S^o^ Carolina Reg^t^ of Artillery were try'd for Absenting themselves from Camp & Duty on the 25. 26. 27. & 28th Days of October 1779 without Leave & found Guilty of the Charge, but the court considering the particular circumstances which induced them to over stay their leave of Absents, they think a punishment adequate to the Offence & recommend that they may be Discharged from their Arrest, & Orderd to their Duty –

the General Approves the Sentence & Expects the Officers will immediately return to their Duty
The General Court Martial is Dissolved

Transmitted by A. Dilliant B: M:

Orders by Lt. Colo. Marion Commanding the Continental Troops at Sheldon Novembr. 14th 1779 –

The Adjutant of the 3d Regiment will immediately Read the Sentence against Conrade Bessinger to him and Inform him of the Generals Order respecting it, that he may prepare himself for the Aughfull moment:

the Adjutant Quartr. Mastr. Genl. to prepare a Coffin by Wednesday 8 OC: in morning, who is also to make a return Immediately of all Stores & Intrenching tools now in his hands, that it may be transmitted to the General

---

November 14th
Parole Marion CSn. Courage Conduct

	C. S:	S: C P
Detail for the Day	1 . . - . .	1 . 1. . 6
Quarter Guard	- - . - .. - . .	.1 .. 1 . 6

15th
Parole Horrey C Sign 1st Diama 2d Dido

	S C P
Detail for the Day	1 .. 1 – 6
Quarter Guard ___	1 .. 1 .. 6

Parole Rutledge C Sn 1 England 2 Enquiry } Novr 16th 1779

Orders by Colo Garden / Candles to be served out to the Officers in the Line & Artillery the Colo Desires the Commissary & Quarter Master Genl will pay more attention to ye Orders Issued for their Departments, and that they do attend ye Order of ye Fourteenth the entreats their Returns by nine OClock this morning – The Prisoner under sentence of Death is to be executed tomorrow morning agreeable to ye Sentence of the Court Martial Between the ours of nine & Eleven oClock in the forenoon the Capt of the Day to see the Sentence put in Execution ye Negro fellow under Guard may

---

1302 James Wilson.

1303 James Fields served in the 4th South Carolina Regiment (Artillery) in 1779 and 1780. On 13 November 1779 he was court martialed with some other officers for being absent from the camp after the failed assault on the Spring Hill Redoubt in Savannah. However the court considered the "particular circumstances which induced them to over stay their leave of Absents" and recommended that he be discharged from arrest. He was promoted to captain in 1780 and was taken prisoner at the fall of Charleston.

be made use of an Executioner – The Q[r] M Gen[l] is to have a Gallows Erected for y[e] Purpose & a Coffin made to enter the Corps – The Line & y[e] Artillery to be turned out & March to y[e] Place of Execution & after the Prisoner is Executed to return to camp – The command at Port Royal Ferry to be releaved tomorrow Morning by an Equal number of Officers and Privates, who are to be releaved Weekly untill further Orders –

	S C P
Detail for the Day –	1 .. 1 .. 7
Quarter Guard - - . .	1 .. 1 .. 6

Parole Bull C S[n]. Firm Fortitude Nov[r]_ 17th – 1779

	S _ S _ C .. P
Detail for the Day	1 . . 2 . . 1 .. 14
Quarter Guard - -	.. 1 . . 1 .. 6

Orders by L[t] Col[o] Marion at Sheldon 18 Nov[r]. 1779

A court Martial from the Continental Line to Set at 10 OC: this forenoon for the tryal of such prisoners as shall be Order'd to it All Evidences to Attend this Court to Consist of a president and four members, to be taken from the 2[d], 3[d] , & 5 Reg[ts]

After Orders by L[t]. Col[o]. Marion } The Guards to be relieved tomorrow Morn by the Continentals only, the Malitia in future to do separate Duty & to be Commanded by their own Officers ~ who is requested to take the picquet at the Church

No Officer to Leave Camp without permission from the Commanding Officer

Parole ~~Gasdden~~ Garden C[n] S[ns] Continental, Militia

	S – C – R
Detail for to day	1 – 1 – 7
Quarter Guard	1 – 1 – 6

19th

Parole Savannah C[n] S[ns] Commet & pines

	S – C – R
Detail for to day	1 – 1 – 7
Quarter Guard	1 – 1 – 6

20th Nov[r] Sheldon –

Parole Gen[l]. Moultrie C[n] S[ns] Friend,

	S – C – R
Detail for to day	1 – 1 – 2
Quarter Guard	1 – 1 – 6

21st

Parole Lincoln, C. S. New England, & Boston

	S – C – R
Detail for to day	1 – 1 – 7
Quarter Guard	1 – 1 – 6

22d

Parole Genl. Howe CSns New York, & Kingsbridge

	C – S – S – C – R
Detail for to day	1 – 1 – 1 – 1 – 5
Quarter Guard	1 – 1 – 6

23d

Parole Charlestown Cn. Sns . Beaufort & Port Royal

	Sb – S – C – R
Detail for tomorrow	1 – 1 – 1 – 13
Quarter Guard	1 – 1 – 6

24th

Sheldon 24th Novr. ~ 1779 Genl. Orders by Lt Colo C. Marion ~

Commanding Officer is Induced to believe from many Guns fired in and about the Camp & the possibility of purchasing Ammunition that the men must make use of their Cartridges, the Orders of the Ammunition Given each man be Examin'd at today Roll call & if Lost or made away with not in the service that they shall be immediately Confin'd & try'd by a Court martial, that they may Suffer Accordingly – he recommends that the Exercising Officers of Corps, would have their men Exercise at least three times a week, & hope the Strictest Dicipline (so very necessary) will be Observed with the greatest punctuality

The Adjutant Q. M Genl. will Issue one weeks Soap and Candles to the troops today

Parole Genl. McIntosh, Cn Sns Pocotaligo & Prince

	C – Sb – S – C – R
Detail for tomorrow	1 – – 1 – 1 – 6
Quarter Guard	– – – 1 – 1 – 6

25th

Parole Sweden Cn. Sns Gustavus & Freedom

	Sb = S = C – R
Detail for tomorrow	1 – 1 – 1 – 6
Quarter Guard	1 – 1 – 6

26th Novr

G. Ordrs By Lt Colo C Marion at Sheldon

The horses belonging to the Waggons of the line to be picquetted every night, the Loss of any of them must be Accounted for, by the waggoner to his Commanding Officer of the Regiment or Corps he belong & Such Commanding Offr of Regiment & Corps must Account for all Losses when called on

Any Negroe purchasing arms Accoutrements or any Clothing from Soldiers will be punished immediately, this to be made known to the Malitia & any Soldiers sellg Arms Accoutrements or any Cloathing to Negroes or any other person may Expect to be punished with the greatest severity

For the day tomorrow Captn. Dunbar

Main Guard Lt.

The Weekly returns to be made & Given in every Munday at 10 OC: in forenoon

After Orders, Muster Rolls to be made out by Sunday at 10 OC: in the forenoon specifying the time of Inlistment & time of service of All the Continental troops & to be Deliverd in at that hour ~~when~~ at which time All the troops are to turn out to be mustered, by the Muster Mast[r]. Gen[l].

Parole Gen[l] Gadsden – C. S. Cha[s] Town & Bravery

	S – C – R
Detail for tomorrow	1 – 1 – 6
Quarter Guard	1 – 1 – 6

27th November

Parole Count De Estaing C[n] S[ns] Fleet & Philadelphia
For guard Cap[t] Mason & Lieu[t] Hall

~~Quarter Guard~~	S – C – R
Detail for tomorrow	1 – – 13
Quarter Guard	1 – 1 – 6

Orders by Col[o] Marion } Mustering of the Continental Troops Mentioned in Yesterday Orders & for tomorrow is put off till Monday – The Infantry to be Mustered at 10 OClock in the forenoon & Major Varniers Horse at 3 OClock in the Afternoon the same Day – [1304]

November 29th 1779

Parole Oaketu C: S[ns] Combind – Fleet

	S – C – R
for guard tomorrow	1 – 1 – 6
Quarter guard	1 – 1 – 6

Nov[r] 29th 1779

Parole Dresden, C[n] S[ns] Col[o]. Horry & Fifth
For the Day tomorrow Captain Mazyck

	Sb – S – C – R
For Guard	1 – 1 – 1 – 6
Quarter Guard	1 – 1 – 6

Novr 30th 1779

Parole Brandenburgh C[n] S[ns] France, & Spain
For the Day tomorrow Captain Mazyck

	Sb – S – C – R
For Guard	1 – 1 – 1 – 6
Quarterguard	1 – 1 – 6

[1304] Chevalier Pierre-François Vernier became the commander of the remnants of Pulaski's Legion of Horse and Foot after they were annihilated at Savannah. During the siege of Charleston Pulaski's Legion consisted of 55 troopers.

December 1st 1779

Parole Philadelphia C^{n} S^{ns} Newburn & Brunswick
For the Day tomorrow Captain Baker

	Sb	S	C	R
For Guard	1	1		7
Quarterguard		1	1	6

Decem^{r} 2^{d} 1779

Parole Portsmouth C. S^{n}. Darien & Altamaha

	S	C	R
For Guard	1	1	7
Quarter Guard	1	1	6

December 3^{d} 1779

Parole Forbay C^{n}. S^{ns} Ireland & Glasgow
For the Day tomorrow Capt Mazyck

	S	C	R
For guard tomorrow	1	1	7
Quarter guard	1	1	6

December 4th 1779

Parole Augusta C^{n}. S^{ns} Camden & Waxhaws
L^{t}. Lagare to Join & do duty in Captn Grays Company till further Order
A regimental Court martial to set immediately to try all prisoners brought before them Evidences to attend
Capt Moultrie president Lieu^{ts} Kolb & Rogers members of the Court Martial

	Sb	S	C	R
For guard tomorrow	1	1	0	6
Quarter guard		1	1	6

December 5th 1779

Parole Holland C^{n} S^{ns} Denmark & Sweden
For the Day tomorrow Captain Provause

	Sb	S	C	R
For guard tomorrow	1 =	1	1	6
Quarter guard		1	1	6

Decr 5th 1779 { Charles town the 26 – November 1779
General orders by General Lincoln

The pay rolls of the Different Regiments Are to be Immediately Made out to the first of December And hereafter they Must be Made Monthly

A delliant B, M,

NB Agreeable to sentance of Last Court Martial Jacob See of Captain Masons Company for Quitting his party & Attempting to Return to Charles Town was Sentenced to Receive & did receive One Hundred Lashes on the Bare back with a Cat a Nine tails

December 6th 1779

Parole Stockholm Cn. Sns Newhaven - & Charlotte Ville
For the day tomorrow Captain

	Sb =	S	C	R
For guard tomorrow	–	1	1	6
Quarter guard		1	1	6

December 7th 1779

A court martial of the line to set immediately to try such prisoners as shall be orderd to it Evidence to attend Capt Hopkins Presedent Detail 1 Sub from 3d 1 Sub. From 5th Members
Parole Jamaca C. S. St. Thomas & Barbados

	C	Sb	S	C	R
For guard tomorrow		1	1	1	6
Quarter Guard			1	1	6
For Command	1	0	1	0	6

Sheldon 8th Decr. 1779 Ordrs by LCol Marion

The Adjutant Q. Mastr Genl to Issue every week soap & candles to the Army, agreeable to ~~orders~~ issue positive of Congress

Any Soldier or Other persons who shall be found pulling down the Church & carrying away the Bricks may Depend on being punished in the most Severest manner

It is recommended to Commanding Officers of Battalions & Corps, to make their men make Warm ~~hutts~~ huts of Either Loggs or ~~Clay~~ Dirt & Order a number of fatigue men every day for that purpose till they have a Sufficient Number for their men

Any person Doing their Occasion within one hundred & fifty Yards of the Encampment to be Confined & to be brought to a Court martial And Suffer Accordingly

Fatigue men from Each Battalion & Corps to be Order'd to make Nessesarys Near the front of the Line

Parole Hanabal – C. S. Marius & Sylla

	Sb	S	C	R
For guard		1	0	6
Quarter guard		1	1	6

December 9th 1779

Parole Virginia C. Sn Williamsburg & Newbern

	Sb	S	C	R
For guard	1	1	1	6
Quarter guard		1	1	6

December 10th 1779

Parole Cn Sns Sword & Sheath

	Sb	S	C	R
For guard				
Quarter guard				

/ Decr 11 /

Parole Constantinople C. S^{n} Hampstead & Winden

	Sb –	S – C – R
For guard	1 –	1 – 1 – 7
Quarter guard		1 – 1 – 6

A regimental court martial to set today at 4 OC: in the forenoon to try all prisoners as shall be orderd to it Evidences to Attend

William Henderson of Captn Halls company is made a Corporal in the Light Infantry Company he is to be Obey'd as Such –

For the Court Martial President Lieu^{t}. Capers Members Lieu^{ts} Rogers & Legare

NB, Agreeable to sentence of the Above Court Arthur Coleson of Capt Mazycks Comp^{y} put under stoppages for Loosing his gun

The Col^{o} Approves the sentence –

12 Decemb^{r} 1779 Sheldon

Parole Hungary C S^{n} Venice & Madonna

	Sb –	S – C – R
For guard	1 –	1 – 1 – 7
Quart^{r} guard		1 – 1 – 6

No Soldier to go to any plantations without Leave from the Commanding Officer, all Such Soldiers who may be found one mile from Camp or in any plantation without a pass from the Command^{g} Off^{r} will be Deemed a ~~Deser~~ Deserter & will be tried by a court Martial & Suffer Accordingly

No houses in or out of Camp to be pulled down or Destroyed on any pretence whatever all persons who are found Acting contrary to this Order will be Deemed a Disobeyer of Orders & will Assured by Suffer

Decem^{r}. 13th 1779 at Sheldon

Parole Paris C S^{n} Janis & Algion

	Sb –	S – C – R
For guard	1 –	1 – 1 – 10
Quart^{r} - -		1 – 1 – 6

One Sargeant 1 Corporal & 20 privates from the Line to go Immediately & Build a logg House at Cross roads for the picquetts They will Apply to the Asst Q. M. Genl for tools Giving a receipt for them this fatigue party to be continued till the House is built

Decr 14th 1779

Parole Russia C. S^{ns} Moscow & Pultawaa

	Sb –	S – C – R
For guard		1 – 1 – 10
Quarter guard		1 – 1 – 6

December 15th 1779

Parole Purisburgh C^{n} S^{ns} Ogeechee, & Abercorn

	Sb –	S – C – R
For guard tomorrow	1 –	1 – 1 – 9
Quarter guard	- - -	1 – 1 – 6

December 16th 1779

Parole Prone C^n^. S^ns^ Cesar, & Cato

	Sb –	S –	C –	R
For guard tomorrow		1 –	1 –	10
Quarter guard	---	1 –	1 –	6

/ Decr. 17th /

Parole Mecklenburg – C S^ns^ Camden & Salisbury

	Sb –	S –	C –	R
For Guard	----	1 –	1 –	10
Quart^r^ Guard	---	1 –	1 –	6

GO. one Gallon rum to be Issued to ea Commissioned Officer of the 2nd, 3d, 4th. & 5th Reg^t^. Maj^r^. Verneer & Col^o^. Horrys Corps also one Gill rum per day to Each man belonging to Reg^ts^. & Corps, the returns to be signed by the Respective Commanding Off^rs^.

RO. A regimental court to set this forenoon to try all prisoners Order'd to it Evidences to Attend

Sheldon

Parole – Widsor – C S^ns^ Plymouth & Depthford

	Sb –	S –	C –	R
For Guard	1 –	1 –	1 –	10
Quart^r^ Guard	1 –	1 –	6	

~~GO a court martial to set to day~~

December 19th. 1779

Parole Augusteen C S^ns^ . Pensicola, & St. Marks

	Sb –	S –	C –	R
For guard tomorrow	1 –	1 –	1 –	11
Quarter guard	---	1 –	1 –	6

Decr. 20th 1779

Parole Pittsburgh C S^ns^ Virginia & Lake George

	Sb –	S –	C –	R
For guard tomorrow		1 –	1 –	10
Quarter guard	---	1 –	1 –	6

GO A court martial of the Line to set today for the trial of all prisoners Orderd to it Evidences to Attend

A sergt. & 12 men for fatigue the sergeant will call on Col^o^. for Directions

Capt Leggar Presedent – 1 Sub from the 2nd & 1 Sub from the 3d – Member

December 21st 1779

Parole North Carolina _ C S^ns^ Edenton, & Brunswick

	Sb –	S –	C –	R
For guard tomorrow	1 –	1 –	1 –	10
For Command			1 –	10
Quarter Guard	----	1 –	1 –	6

December 22d 1779

Parole Lancaster C Sns Essex, & Chester

Sb – S – C – R

For guard tomorrow --- 1 – 1 – 11

Quarter guard --- 1 – 1 – 6

December 23rd 1779

Parole Marlborough Cn Sns Newport, & Marblehead

S – C – R

For guard tomorrow 1 – 1 – 10

Quarter guard --- 1 – 1 – 6

GO. The Commandg. Offrs of the 2d, 3d, & fifth Regiment will give in a return of the (Sergts?) in their respective Regts Specifying their names [1305]

December 22nd 1779 [1306]

Parole Crownpoint Cn. Sns St. Johns Quebec

Sb – S – C – R

For guard tomorrow --- 1 – 1 – 1 – 10

Quarter guard --- 1 – 1 – 6

Genl. Orders December 6th 1779

The General directs that no French Soldier or Sailor be Enlisted into the Continental service untill he has been Previously carried before Monsieur Plombard the French Consul, and has Obtained his Certificate that he is Not in the french service ____________ [1307]

A delliant B. Major

25 Decr

Parole Edenton – C Sns York & Newport

Sb – S – C – R

For Guard tomorrow --- 1 – 1 – 11

Quarter --- 1 – 1 – 6

GO. The Commanding Officer is truly Astonished & shamed to see Soldiers Behave ~~in~~ so Disorderly at last night & this morning firing of Guns Contrary to all good order & Discipline, he Order no Guns in future to be fired in or near the Camp, all those who Disobey this Order will be taken up & immediately punished, he therefore calls on all Commissioned & NonCommissioned Officer to take

[1305] I am unable to make out the word of what they want a return of. It looks like "Sergts" but I am not sure. It may have something to do with finding out how many French are in the ranks, since after this an order was sent out telling the different regiments that they could not enlist any French soldiers.

[1306] This should read December 24th.

[1307] During the siege of Charleston two separate French units were formed for the defense of the city. One was Britigney's Volunteer Corps of Frenchmen. This Regiment was made up of non-English speaking French citizens of Charlestown. This unit may have also contained the guards to Monsieur Plombard, the French Consul to Charlestown. The second one was known as the French Company. These were D'Estaing's French troops sent to Charlestown after Savannah that were too wounded to travel. Sixty-two of the Chasseur Volontaires de Saint-Domingue escorted those wounded men to Charleston. The Chasseurs were free mulatto and Black volunteers from Haiti.

any man who shall be found firing of Guns in Camp & confine them that they may be punished for their disobedience & Guards to send out a party so that they may find them & confine them
The Sutler and all other persons are forbid to Sell any Liquor to any Soldiers on pain of having the Liquor Destroyed & themselves immediately punished – [1308]

After the French left the Georgia coast Sir Henry Clinton in New York realized he had a golden opportunity to capture Charleston and erase the defeat he had suffered there in 1776. Unfortunately one of the worst winters in the 18th century hit the American coast. The British invasion fleet was supposed to have departed on December 19th, but the weather stopped any ships from leaving. To compound the problems, one of the transports was destroyed by ice floes pushing into the harbor. Six other transports were damaged by the ice and had to have their cargoes removed to other ships. This took even more time.

Finally on December 26th, 1779 ninety British troopships, escorted by fourteen warships, sailed out of New York Harbor bound for Charlestown. The fleet had barely escaped being caught by the ice floes in New York Harbor. Unfortunately they were caught in a gale that lasted for four days. The British fleet was scattered across the Atlantic Ocean. One troopship, the *Anna*, was dismasted and then drifted for eleven weeks until landing at St. Ives on the coast of Cornwall, England.

The British fleet rendezvoused in Savannah in the early part of February. Eleven ships had come up missing in the gale. In addition to the *Anna,* the transports *Judith,* and *Russia Merchant,* and a one-masted artillery transport were lost. The *Russia Merchant* had carried most of the heavy siege artillery and ammunition needed to conquer a fortified town. The gale had pushed other ships far southeastward and the men were dangerously low on food and water. A voyage that would normally take 10 days took the British fleet five weeks to make. The Patriots looked upon this as Divine intervention, similar to how the Japanese called the storm the stopped the Mongol invasion the "Divine Wind" or Kamikaze.[1309]

Decr 26th 1779

Parole Halifax. C^n. S^ns S^t. Lawrence, & Biscay
For the Day tomorrow Captain Provause

	Sb	S	C	R
For Guard	1	1	1	10
Quarter guard	---	1	1	6

Decr 27th 1779

Parole Holland C S^ns Amsterdam, & Portugal

	S	C	R
For guard tomorrow ---	1	1	11
Quarter Guard ---	1	1	6

1308 Evidently these orders were due to excessive celebration of Christmas. The tradition of firing guns into the air during Christmas has been in the South since before the beginning of the nation, and came from an even more ancient time when men would gather around a tree on Epiphany Eve, drink cider and fire guns and blow horns in the air to drive away evil spirits.

1309 Patrick O'Kelley, *"Nothing but Blood and Slaughter" The Revolutionary War in the Carolinas, Volume Two, 1780,* (Booklocker.com, Inc 2004), pp. 34-35.

/ Dec[r] 28 1779

Parole Santee C[n]. S[ns] Gooscreek & Dorchester
For the Day tomorrow Capt[n]. Provause

	S – C – R
For guard	1 – 1 – 12
Quarter d[o].	1 – 1 – 6

GO. one Serg[t]. & 6 privates to be added to the Post Royall Guard tomorrow

Roster 16th August
Capt[n] Moultrie – 27 Oct: 1779
N Carolina – 28 March 1779
Lieu[ts]: Capers – 24 Feb[y]. 78 –
N Carolina – W[m]. Bush [1310] 1 Feb[y]. 1779
Foissin – 13 July 78
N Carolina – Tho[s]. Clark 1 Feb[y]. 79
Petrie – 23[d] Feb[y]. – 79
N Carolina ~ W[m]. Lord[1311] 12 June 79
2 L[t].- Rogers
N. Carolina – N. Williams[1312] Ensign 14 March 79
2 L[t].- Hume . . . . 23 Feb[y]. 79
~~Noland~~
2[d] L[t].- Legare – 20 Aug[t]. 79
Capt[n]. Moultrie

[1310] William Bush was an ensign in the 8th North Carolina on 10 April 1777. He was promoted to 2nd lieutenant on 15 August 1777. After the ten North Carolina regiments were reconsolidated he transferred to the 1st North Carolina Regiment on 1 June 1778. He was promoted to 1st lieutenant on 1 February 1779. He was promoted to captain in 1781 and served to the close of the war.

[1311] William Lord was the paymaster for the 1st North Carolina Regiment on 12 December 1776. He resigned on 5 March 1777. He was commissioned a lieutenant in the 10th North Carolina Regiment on 1 August 1779.

[1312] Nathaniel B. Williams was commissioned a 2nd lieutenant in the 8th North Carolina Regiment on 28 November 1776. He was promoted to a 1st lieutenant in the 10th North Carolina Regiment on 23 January 1781. He transferred to the 4th North Carolina on 6 February 1782 and served until the close of the war.

Captn. Lesesne resigned 6th August 1779 Roster Aug^t^.
Captn. Moultrie
Anderson discharged Jany. 1780 [1313]

Roster13th Aug^t^.
Capt^ns^. Moultrie
Ramsay [1314] N.C – Commandant
Dunbar ___ __ Light Infantry
Chapman N.C
Lieu^t^. 1st Hart ____ ___ Act^g^. Adj^t^.
Roux _ ___Light Infantry
Capers
Foissin
Bush N.C
Clark NC
W^m^. LordNC
2 – Humes
Rogers
Vlieland
Williams NC

229
118 Dunbar
22
3
272

Hart
Roux
Capers
Petrie
Kolb
Rogers
Bush
Vlieland

Captn Dunbar
Resigned 9th Dec^r^.

[1313] This page is extremely waterlogged and is hard to read. This sentence appears to say "Anderson discharged" but I am unable to determine who Anderson is.
[1314] Matthew Ramsay had been a captain in the 9th North Carolina Regiment on 28 November 1776. He transferred to the 4th North Carolina Regiment on 1 June 1778. He resigned in November 1781. He received a land warrant on 22 October 1783 for 59 months of Continental service.

This list is not in the orderly book, but I created it to show the

Lineage and Honors of the 2nd South Carolina Regiment of the Continental Line

6 June 1775	Activation of the 2nd South Carolina Regiment
August 1775 – January 1776	1775 Defense of Charleston, South Carolina
14 September 1775	Capture of Fort Johnson, South Carolina
11-12 November 1775	Battle of Hog Island Channel
May – July 1776	1776 Defense of Charleston, South Carolina
28 June 1776	Battle of Sullivan's Island, South Carolina
8 August 1776-15 February 1777	Florida Campaign of 1776
4 November 1776	Adopted into the Continental Army
27 February 1776	Assigned to the Southern Department
23 November 1776	Assigned to the 2nd South Carolina Brigade
February – March 1777	Relief expedition of Fort McIntosh, Georgia
January 1778 – May 1778	Privateer Campaign of 1778
26 August 1778	Assigned to the 1st South Carolina Brigade
May – June 1779	1779 Defense of Charleston, South Carolina
11-13 May 1779	Siege of Charleston Neck, South Carolina
15 June 1779	Assigned to McIntosh's Brigade
14 September 1779	Assigned to Huger's Brigade
24 September – 19 October 1779	Siege of Savannah, Georgia
January – May 1780	1780 Defense of Charleston, South Carolina
11 February 1780	Consolidated with 5th and 6th South Carolina Regiments
11 February 1780	Assigned to the South Carolina Brigade
27 February 1780	Assigned to the 2nd Virginia Brigade
29 March 1780	Skirmish at Gibbes Plantation, South Carolina
1 April – 11 May 1780	Siege of Charleston, South Carolina
12 May 1780	2nd South Carolina Regiment surrenders with the city
1 January 1783	2nd South Carolina Regiment disbanded

Lieutenant Col^{o}. Francis Marion

Roster of 2^{d}. Regiment

~~Harleston~~	______	{~~Baker~~
		{~~Gray~~
Motte	______	{ Provaux
		{ Fascine
~~Charnock~~	______	{ ~~Moultrie~~
		{ ~~Kolb~~
Lesesne	______	{Gray
		{-------
Moultrie	______	{Hart
		{-------
Mazick	______	{Roux
		{-------
Dunbar	______	{ Martin
		{ ------
Hall	______	{ Caper
		{ -------
Baker	______	{Warley
		{ -------
Provaux	______	{ _____
		{ _____
Mason	______	{ _____
Gray	______	{ _____

Roster 17 May 1779

Capt^{ns} Dunbar
Mazyck
Proveaux
Gray

L^{ts} ~~Worley~~ Roux
Foissin
Guerry
Kolb
Ogier

Morning Report

Lieu^{ts} Roux
Malitia
Foissin
~~Guerry~~ ~~Petrie~~
~~Martin~~
Kolb
Malitia
Ogier
Malitia
~~Malitia~~

Lieu^{t}. McDonald

Roster 25th May

Captn Mazyck
Vanderhorst
Dunbar
Fogartie
Proveaux
Ayers [1315]
Gray
Kamell [1316]

Morning Report

This was the last entry in the orderly book for the 2nd South Carolina Regiment. Marion remained at Sheldon until after the new year. The British would eventually surround Charleston and lay siege to it until the middle of May, when Lincoln's army, and the 2nd South Carolina, would surrender.

[1315] Captain Thomas Ayer served as a captain in the militia from 1777 to 1781

[1316] John Gamble (Gambell) was commissioned as a lieutenant in a volunteer company of militia in October 1775. He served as a captain and major in Marion's Brigade of Partisans in 1780 and 1781

Letter located in the South Carolina Historical Society that shows how many of the 2nd South Carolina Regiment soldiers began wearing moustaches, in what I think was imitating the French that they fought beside at Savannah, who they looked upon as professional soldiers:

Lt. Col. Francis Marion to Maj. Isaac Harleston (at Chs Town)

Sheldon, 26 Jany 1780

Dr Sir:

I am happy to find you are again in the 2d Regiment, tho I am sorry for Vanderhorst, who wishes to continue in the Service & have Given me Great Satisfaction the Little time he has been with me.

As some Captains from the 6th Regiment will be put in the 2nd, shall be glad you would point out 2 or 3 which you think are good, as for Subs they will be so few they can be no Choice. I think Lt. [John] Buchanan to be a good officer and wish to have him.

Majr. Vanderhorst goes to town & Leave me with one Captn. and 2 Subs, my Command here in Chief prevents me from seeing to the regimt.

Shall be oblige to You to get me all the Articles in the public store which I have a right (Except the Cloath for a coat Jacket & Breeches which I have) but no trimings. Linen I particularly am in want of.

When you come up if, you think proper to be in my mess (I have nobody but Captn Moultrie) it will be agreable to me.

I dont know When the General will grant me a Little time to transact my private affairs. I waited with patients, Expecting I should be one of those Officers who was to go to the right about. I am Disappointed & Suppose I am for the war, or a Ball. When you see me you will find I have a formidable pr of Mustasho, which all the regimt. now ware & if you have not one you will be Singular.

We find it Cold here. I don't know how it may be where you are; you may see it by my Scroll.

I am Dr Sr. with Great regard,
Yr Most Obt Servt.
Frans Marion

N. B. Our men are in Great Want of Shoes & Shirts & Blankets to Compleat them. Many of the men is without a Shirt & Shoes. I wish you woud try to get them & Send by two Waggons now in town.

Francis Marion ??

1780

Though there is no record of any entries in the 2nd South Carolina orderly books from 29 December to the surrender on 8 May, there are other orderly books that are still intact. One of these is the Order book of John Faucheraud Grimké, that is located at the South Carolina Historical Society in Charleston, South Carolina. Entries from that book that effected the 2nd South Carolina Regiment, that was not considered normal guard and fatigue duty, are as follows:

January 20th. Lieut. Capers of the 2d S°. Carolina Regiment having resigned his Commission is no longer to be considered as an Officer in the Contl. Service.

February 2d 1780. One third of the Troops are to be paraded at eight oClock every Morning for fatigue in their Regimental or Brigade parades & from thence they will march to the Rendezvous which shall be appointed by the Engineer. The fatigue parties must be fully officered & the Officers are requested to be careful that the Men do not idle away their time.

February 9th. The 2d, 3d, & 5th Regts are to go to Haddrell's Point tomorrow Morning – Major Harleston will command them.

February 11th. Whereas the Honble the Contl. Congress have resolved that the number of Battalions of Infantry of this State shall be reduced to three – the Genl. & Governor to whom the mode of Reduction & Arrangement was committed have appointed the following Field Officers.

Col°. C. C. Pinckney
Lieut. Col°. Scott
Major Pinckney } To the first Regiment

Lieut. Col°. Marion
Major Harleston } to the Second Regiment

Col°. Thompson
Lt. Col°. Henderson
Major Hyrne } to the Third Regiment

And the Genl. & the Governor having left the Appointment of the Capts to the above Field Officers, A Majority of them have agreed that 27 Capts oldest in service as Commissioned Officers shall be retained and Capts Turner, Lieuts Elliott, Hext, Lining, Gadsden, Williamson, Jackson, Lavacher, Moultrie, Mazyck, R.B. Baker, Provaux, Gray, Mason, Roux, F. Warley [1317], Smith, I.

[1317] Felix Warley was born in South Carolina in 1750. He became a lieutenant in the 3rd South Carolina Rangers on 17 June 1775. He was promoted to captain on 24 May 1776. On 12 April 1779 he was appointed as the

Darley, Goodwin, Farrar, Liddle,[1318] Shubrick, G. Warley,[1319] Buchanan, Baker [1320] of the 6th & Pollard appearing to be the oldest in the Service (except Captains Caldwell, Towles & Hennington who have not signified their Intention to continue in Service since their Releasement from Captivity & Capts St. Martin & Hampton the former of whom chooses to decline and the latter is supposed from his long Absence to have quitted the service) are therefore appointed Capts in the three retained Regts. Each Officer to continue in the Regt to which he belongs except those of the 5th & 6th who are to fill up Vacancies in the retained Regt. The Senior Officers going into the oldest – and if any of them should die or quit the service, the Vacancy so happeng shall not be filled up until each of the Field Officers in the Regt in which such vacancy may happen shall have companies, and the new arrangement thereby completely adopted.

And whereas some of the Officers above named may choose now to retire from the Service; any Vacancy happening by non acceptance shall be filled up by the Supernumerary Officers according to Seniority of Service. And the Genl. hopes if there are any who do not mean to continue in the Service during the War, the public good & Justice to those Supernumeraries who wish to be retained will prompt them to declare their intentions Immediately in which Case they will be put on the List of Supernumeraries & their places filled accordingly.[1321]

February 12th. Whereas there are many Frenchmen in Charleston who from their want of Knowledge in the English Language are incapable of rendering equal services by being incorporated with Americans – The Genl. desires that the Officers commanding Militia Companies in town will erase all such out of their Rolls, and directs that they do Duty in the Marquise de Bretangne's Corps – this to take place immediately.

The Drummers & Fifers wanting to compleat in the first So. Carolina Regt. are to be taken from the 6th & the remainder together with the Drum Majors and Drummers & Fifers of the 5th Regt. are to be incorporated with the third.

The British fleet left Savannah on February 9th and anchored off of Trench Island.[1322] On the evening of February 11th the Light Infantry and Grenadiers landed unopposed on Simmons Island.[1323] On February 12th the rest of invasion force was issued three days

---

auditor paymaster, director, clothier or barrack master general and the commissary general of provisions, prisoners and military stores. He was captured at the fall of Charleston in May 1780. He was exchanged in June 1781. He was made a brevet major on 30 September 1783. He married Anne Turquand. He died in 1814 at the age of 64.

[1318] George Lidell was born in in 1754 in Maryland. He became a sergeant in the 2nd South Carolina Regiment on 19 August 1775. He transferred to the 3rd South Carolina Regiment and was a sergeant on 24 July 1776. He was appointed a 2nd Lieutenant on 1 January 1777 and then promoted to 1st Lieutenant in 1777. He was promoted to captain on 20 December 1778. He was taken prisoner at the fall of Charleston. After the war he married Rachel Thomson. He died on 28 December 1789 at the age of 35.

[1319] George Warley was a captain in the 6th South Carolina Regiment on 19 June 1777. He transferred to the 2nd South Carolina Regiment in February 1780, when the regiments were consolidated due to losses in the siege of Savannah. He was captured at the fall of Charleston in May 1780. He was exchanged in October 1780. He became a brevet major on 30 September 1783.

[1320] Stephen Baker was a lieutenant in the 6th South Carolina Regiment.

[1321] A supernumerary officer is one who did not have any soldiers to command due to shortages.

[1322] Present day Hilton Head Island.

[1323] Present day Seabrook Island.

rations and disembarked. The artillery had to remain on board because there were no horses to pull them. On the morning of February 14th the Jägers and the 33rd Regiment set out in search of Stono Ferry.[1324]

February 13th. If there are any corps whose Alarm Posts are not yet assigned the Com[g] Officers thereof will immediately appoint them & report them to the Adj[t] Genl.

In case of an Alarm the several Corps will instantly assemble on their Alarm posts & there wait for Orders.

Orderly Hours to be 9 oclock in the morning & 5 in the Evening.

A.O. The troops are immediately to be supplied wth. fifty Round of Cartridges per Man.

As the Approach of the Enemy makes the utmost vigilance & circumspection necessary – the Taptoo will beat at ½ past 9 oclock which shall be the signal for every one to repair to their respective Homes – at ten the Countersign will be given out after which no person whatever will be permitted to walk the Streets without it.

When it became clear that Clinton's objective was Charlestown, Lincoln began to consolidate his forces. The 1st South Carolina Regiment continued to garrison Fort Moultrie and improve the defenses, with the labor of "a number of Negroes." The rest of the Continental infantry and the Charlestown Battalion of Artillery manned the trenches on Charlestown Neck. The Charlestown Militia would man the batteries in the city facing the harbor. Every man who was left was utilized, improving the defenses.

Almost four thousand American troops were stationed around Charlestown, of which 2,000 were the militiamen from the Carolinas. When Lincoln began bringing the troops into the city, the militiamen from the outlying areas refused to enter. They feared an outbreak of smallpox. Even after Rutledge threatened to confiscate any property that the militiamen owned in Charlestown, only a little over 200 responded and came into the city.

On February 10th, 1,248 North Carolina militiamen under General Alexander Lillington did finally come into town. On February 14th, Continental Marines blew up the lighthouse. Two days later the Marines were landed on James Island and they planted explosives inside Fort Johnson.[1325]

February 18th. The Non-Comd Officers & Soldiers of the 5th & 6th Balt[o] of S[o] Carolina are to be immediately incorporated with the 1st & 2d in such manner as to make those two equal.

Francis Marion was ordered by Moultrie to form a light infantry unit from the best men of the 2nd South Carolina Regiment. Marion's unit had been reduced during the assault on the Spring Hill Redoubt in Savannah, but he had been reinforced with remnants of the 6th South Carolina Regiment.

Lincoln ordered that the five South Carolina Continental regiments be consolidated into three on February 11th, due to the losses suffered at Savannah. Out of the 246 men in Marion's regiment his "best" consisted of 227. Marion's force was moved to Bacon's Bridge near Fort Dorchester to delay a British force that was marching up from Savannah.[1326] The normal load of the Continental soldier was 40 rounds per man, but Marion's men only had 25 rounds per man to hold back the British invasion.

1324 O'Kelley, *NBBAS, Volume Two*, pg 26

1325 Ditto, pg 27-28

1326 Bacon's Bridge is located south of Summerville, where route 165 crosses the Ashley River.

11 February 1780
British fleet arrives

February 27th. The Detachmts of the 2d & 3d South Carolina Regts are to do Duty as part of Colo Parker's Brigade till further Orders [1327]

March 1st, 1780. Lieuts Langford & Lieut Buchannan late of the 6th are ordered to join the 2d Regt of So Carolina

On February 26th the British occupied the destroyed Fort Johnson and began rebuilding it. On February 27th some transport ships arrived from Savannah with the grenadier companies of the 63rd and 64th Regiments and one battalion of the 71st Highlanders. The next day, while work was being done on a redoubt in Fort Johnson, the frigate *Boston* and the sloop *Ranger* both fired into the fort from an unprotected side. One shot by the *Boston* killed a gunner, and two grenadiers of the Hessian Grenadier Battalion von Graff. By March 1st the British controlled James Island. Their Light Infantry had crossed the Wappo Cut to establish itself on the mainland.[1328]

March 6th. The different Brigades & Corps are to have 50 Rounds of Cartridges per man, but as it would occasion great waste of ammunition to deliver it all out – the men are to be furnished with only 36 Rounds & the remainder lodged with the respective Quarter Masters.

On March 2nd the *Providence, Boston, Ranger, Bricole, Notre Dame* and several other galleys, fired into Fort Johnson again, with no effect. When the first British schooners appeared off the bar the American fleet ceased the shelling of Fort Johnson. The capture of the fort allowed the British to protect their ships when they crossed the bar, and send occasional harassment fire into Charlestown.

Lincoln ordered the sentries in the trenches in Charlestown to be relieved every half hour. He did not want any man to sleep and allow the British to go undetected. He also ordered St. Michael's church bell to ring every 15 minutes, so that the sentries would know when to call out to each other. When the bell rang each sentry would cry out "All's Well" to the next one, until the cry had gone around the entire cities defenses. To make sure that no one wasted any ammunition by shooting at shadows Lincoln ordered that there be no firing in the vicinity of the camp or lines. He sent a detail to kill all the dogs they found because he did not want the garrison to be alarmed by a barking dog and think that the British were attacking.

[1327] The detachment from the 2nd South Carolina were the men who were not able to be part of Marion's delaying force. Colonel Parker's brigade was the 2nd Virginia Brigade, also known as Brigadier General Charles Scott's Brigade of Virginia Continentals. After serving with Washington's army in the north the Virginia regiments had to be consolidated into other regiments. This Brigade was made up of the 1st, 2nd and 3rd Virginia Detachments. The 1st Virginia Detachment consisted of recruits still in Virginia. The 2nd Virginia Detachment consisted of soldiers and officers from the depleted Virginia Regiments. The 3rd Virginia Detachment consisted of men from the 7th Virginia Regiment and new levies from Virginia. Only the 1st and 2nd Virginia Detachments made it to Charlestown. The 3rd Virginia Detachment was under the command of Colonel Abraham Buford, and turned back to North Carolina once Charlestown fell. They would be attacked at the Waxhaws by Lieutenant Colonel Banastre Tarleton in what became known as "Buford's Massacre".

[1328] O'Kelley, *NBBAS, Volume Two*, pg 21.

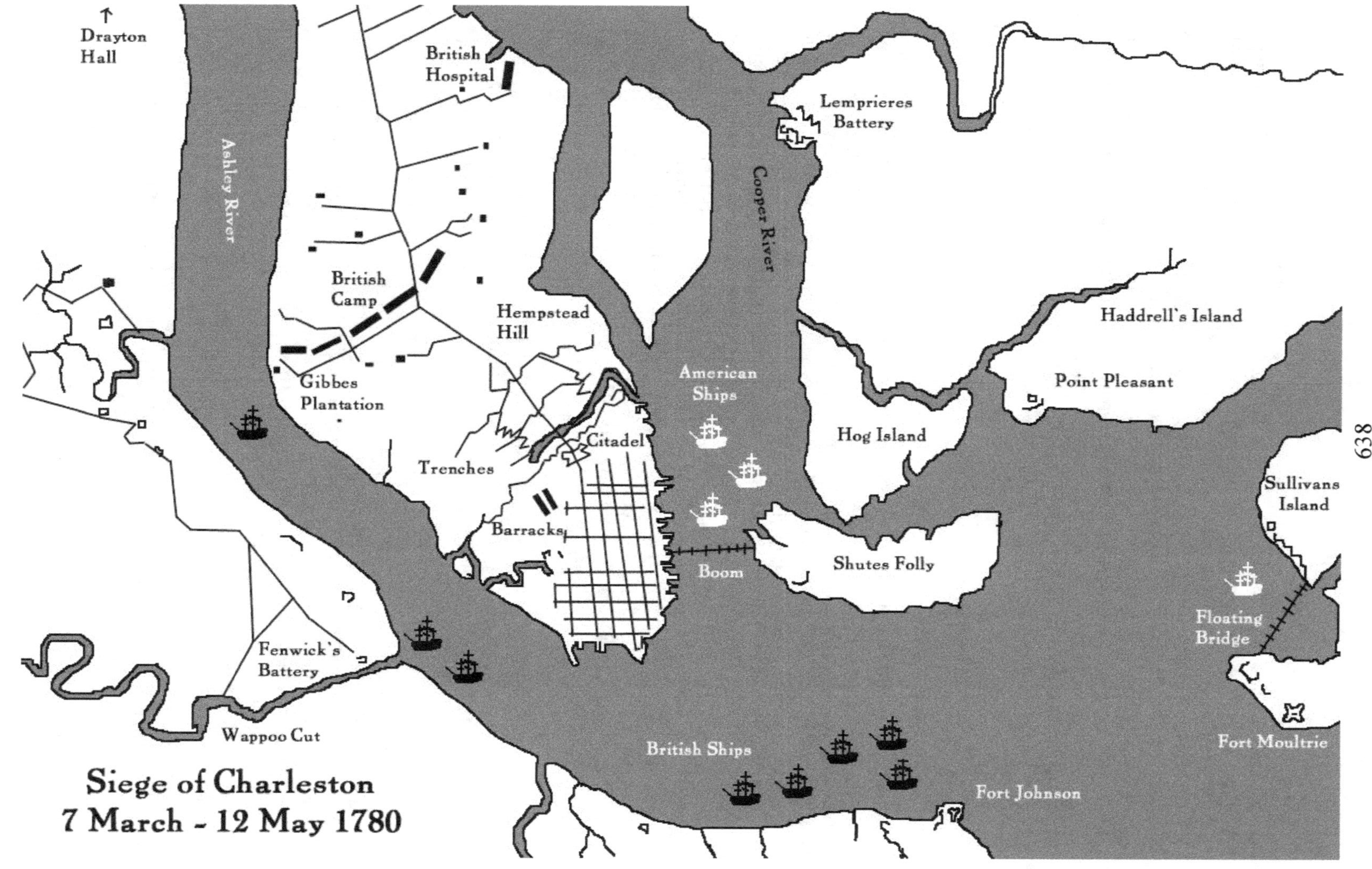
Siege of Charleston
7 March - 12 May 1780
Drayton Hall
Ashley River
British Hospital
British Camp
Gibbes Plantation
Hempstead Hill
Trenches
Barracks
Citadel
Fenwick's Battery
Wappoo Cut
American Ships
Boom
Cooper River
Lempricres Battery
Hog Island
Shutes Folly
Haddrell's Island
Point Pleasant
Sullivans Island
Floating Bridge
Fort Moultrie
British Ships
Fort Johnson

On the 3rd of March the Charlestown garrison was increased by the arrival of 600 North Carolina Continentals under Brigadier General James Hogun. The North Carolinians had marched south from Washington's army and had been in the hottest fighting there was in the north.[1329]

March 7th. As the privates of the 5th So. Carolina Regt have been transferred to the 2d – Lieut. Evans[1330] & Lieut. Frierson [1331] late of the 5th are appointed Lieuts in the 2d Regt & Lt. Buchannan & Lt. Langford late of the 6th are appointed Lieuts in the 3d Regt.

The whole Army is ordered to be at their Alarm post at the So. End of the Town tomorrow Morng. at five oclock.

On March 12th two 32-pounders, two 24-pounders and a howitzer was mounted on the platforms built at Fenwick's Point.[1332] These cannons fired on the city for the first time, but the citizens of the city didn't know it until they found two 32-pound balls at Cumming's Point and the Sugar House. The galleys *Lee* and the *Britigney* fired upon the British battery, but did not cause any damage. The American ships that were located at Sullivan's Island moved up the Cooper River for better protection.

The next day the British built a second battery near the Fenwick battery. This artillery was mounted to shell American ships and the fortifications on the opposite shore. A redoubt was built between the two batteries so that they would not be assaulted by troops on the land.

The British fleet was still outside the Bar, waiting for the conditions to improve so they could run for the harbor. Admiral Whipple had decided not to stand and fight at Five Fathom hole because he said that there would be no room to maneuver, and his ships would be shot full of holes. Lincoln ordered Whipple's fleet to construct a barrier in Rebellion Road that would slow the British ships. It was decided to let the British pass over the Bar and then take a stand near Fort Moultrie, where the guns of the ships and the fort would be able to rake the British fleet. This obstruction would consist of sunken ship hulks lashed together with chains and cables. If the British tried to remove these obstacles, they would do so under the guns of Fort Moultrie.

On the 17th of March the morale of the defenders took a blow. Lillington's North Carolina militia's term of service expired. Rutledge offered a $300 bounty and new clothes to those who would stay on an additional three months, but even with that incentive 700 of the militia headed home. Only 150 stayed in Charlestown.[1333]

1329 Ditto, pg 61.

1330 George Evans entered into the service in 1775. He was commissioned a lieutenant in the 5th South Carolina Regiment on 18 August 1779. He transferred to the 2nd South Carolina Regiment on 7 March 1780 when the regiments were consolidated due to losses suffered at Savannah. He was taken prisoner at the fall of Charleston and was exchanged in June 1781. He served in the militia of Captain Edmund Kirby in 1781 and 1782.

1331 John Frierson became a lieutenant in the 5th South Carolina Regiment on 9 March 1778. He transferred to the 2nd South Carolina Regiment on 7 March 1780 when the regiments were consolidated due to losses suffered at Savannah. He was captured at the fall of Charleston. He was exchanged in June 1781, and then served in Marion's partisans under Captain Benjamin Waring.

1332 Fenwick Point was located on Albermarle Point, where the James Island Expressway crosses the Wappoo Creek.

1333 O'Kelley, *NBBAS, Volume Two*, pp 63 - 66.

March 19th. Whereas an Act of Gen[l] Assembly of South Carolina passed the 11th Sept last for the purpose of filling up the Cont[l] Batt[s] of this State has expired – His Excellency the Governor and Privy Council have thought proper to extend the Operation of the said Act two Months from this day – Therefore every able bodied Man who shall voluntarily inlist in either of the Cont[l] Batt[s]; of this State for the Term of 21 Months, shall at the time of his Enlistment receive a Bounty of 500 Dollars and an Indent for a further Bounty of 2000 Dollars payable at the End of their faithful Service. The Indent to carry 10 per ct. interest & that payable half yearly they shall also be entitled to 100 Acres of Land & every other Advantage of pay Clothing & Rations as expressed in the said Act.

The British fleet had been stalled outside the Bar for sixteen days, waiting for the weather to cooperate. A hard rain had hit the low country on March 18th and 19th, and the wind blew against the British fleet. Finally on March 20th the wind changed its course and Admiral Arbuthnot was able to cross the Bar with his smaller warships. He had to leave the large ships of the line behind, since they would not have been able to make it over the shallow sandbar. The citizens of Charlestown did not realize that their situation just became more desperate. They still had hope that the British would be easily defeated, just like they had been in 1776 at Fort Sullivan, and when Prévost tried to take the city in 1779.[1334]

March 20th. The whole Garrison to turn out on fatigue this Afternoon, they will parade at the Horn Work.

In the center of the Patriot works was an eighteen-gun fortress known as the "old royal work" or the Hornwork, constructed of lime and seashells that had the strength of reinforced concrete. Over that tough exterior were laid palmetto logs to absorb any cannonballs. The Hornwork was three tiers of artillery, and there was a ditch that would slow any attackers. It was flanked on both sides by bastions that allowed the defenders to fire on any attackers trying to go over the walls of the Hornwork. It guarded the gate to city on King Street. The gate also had a lunette right in front of it. A Lunette was a small half moon fortification.[1335]

In front of the Hornwork, stretching from river to river was a series of redoubts and batteries that were connected by a wall, or parapet. Two main redoubts were built on the Ashley and Cooper Rivers to anchor the parapet. The Cooper redoubt was in front of the line, so that the defenders could fire upon any troops trying to assault the main works. Both these redoubts were constructed in the same way that Fort Sullivan was in 1776. Palmetto logs were laid one upon the other in two parallel rows that were sixteen feet apart. The space between the logs was filled with sand.[1336]

March 25th. The Light Companies of the 2d and 3d S° Carolina Battalions are immediately to be formed agreeable to the Regulations of the Army – Lieu[t] Langford of the 6th Reg[t] is ordered to Join the 2d Reg[t].

[1334] O'Kelley, *NBBAS, Volume Two*, pp 66 – 67.

[1335] The Horwork was located in between King and Meeting Street, and the Citadel college was constructed there in 1842. An original piece of the tabby wall is located at the Old Citadel. The new Citadel is located west of the original Citadel.

[1336] O'Kelley, *NBBAS, Volume Two*, pp 70 - 71.

British Light Infantry

March 28th. The 2d and 3d Continental Battalion of S° Carolina are to be paraded for Inspection on the left of the Horn Work tomorrow afternoon at 4 oClock

On March 27th the British anchored ninety flatboats and longboats at Wappoo Neck. These boats were used to ferry the British across the Ashley River. The next day Clinton made all the necessary preparations to cross the Ashley River. A redoubt had been built five days earlier to cover the movement across the river. Seventy-five flatboats were moved by the Patriot batteries at three in the morning using muffled oars. A few hours after arriving the Royal Navy seamen had to conduct a river crossing into enemy territory.

At 8 o'clock in the morning on March 29th, the troops at Drayton Hall entered the Ashley River.[1337] As they began some Patriot riflemen fired at them from long range, but with no effect. The British sailed up the Ashley River under the escort of an armed row galley, looking for a landing location.

After four miles a slightly raised bank was located at the house of Benjamin Fuller, with a Patriot double cavalry post. The row galley approached the house and fired on it to determine if there were any cannon there. The Patriot cavalry returned fire, but did not have any artillery. The Light Infantry and Jägers quickly disembarked on the shore, and secured a half-moon perimeter around the house. The British troops at the perimeter remained there until the rest of the soldiers had crossed the Ashley River. The only opposition to the landings was a picket of "Negroes" that they saw at a distance that was "observing rather than hindering". Lincoln could not oppose the landing because he did not have enough men.[1338]

March 29th. The Light Companies of the 2d and 3d S° Carolina Battns are to hold themselves in readiness to march at 3 oClock this Afternoon – They with the two Light Companies of the North Carolina Brigade will form a Battalion under the command of Lt. Col°. Laurens.

The Inspection of the 2d and 3d S° Carolina Battns is postponed for a few days that the Officers may prepare their Returns.

On the evening of the 29th the British army moved into camp near the Quarter House.[1339] The Ashley River protected their right flank, and the army formed a front facing three sides. The entire British army, with General Patterson's reinforcements, consisted of 10,000 men. Lieutenant Colonel Laurens with this detachment of light infantry exchanged fire with the British advance elements at Gibbe's Plantation.[1340] The British artillery returned fire, killing Captain Bowman of the North Carolina Continentals. This was an intense fight that cost the British nine killed, eleven wounded, and five missing. On that same afternoon a patrol of light infantry found five Jägers who had been missing. All were dead, killed by bayonets, and the eyes of one had been cut out. Captain Ewald thought that this might have meant the Patriots were angry at losing heavily in the previous day's action at Gibbe's Plantation.

1337 Drayton Hall still exists and is open to the public.

1338 O'Kelley, *NBBAS, Volume Two*, pp 67 - 68

1339 The Quarter House would have been located in what is now North Charleston, near the intersection of Meeting and Success streets.

1340 Gibbes Plantation was located where the Lowndes Grove Plantation is today, on St. Margaret Street in Charleston.

The British began the trenches on April 1st and the siege became a constant scene of artillery barrages, sniper attacks and the threat of a sortie from the enemy trenches. As the situation became more hopless there were many parties held to celebrate the last days of freedom by the defenders.

On March 19th Marion had attended a dinner party at a house on the corner of Orange and Tradd, hosted by Moultrie's adjutant general, Captain Alexander McQueen. In a custom of the time, McQueen locked all the doors and first floor windows so his guests could not escape.

Marion was not a drinking man. He was descended from French Huguenots, and the hardest thing he drank was vinegar. Marion decided to go to the second floor and jump out the window. Upon landing he badly fractured his ankle, which put him in the category of an officer unfit for duty. By April his injury had still not healed and he was evacuated. His injury saved him from capture and would create a South Carolina legend.

Lincoln convinced Rutledge that he needed to leave the city so that he could "keep alive the Civil authority, give confidence to the people, and throw in the necessary… supplies to the garrison." Rutledge agreed and he crossed the Cooper River around noon. As he rode into exile he looked back at his city in flames, and could hear the rolling thunder of the bombardment.

Christopher Gadsden would remain behind as lieutenant governor. Lincoln hoped that Rutledge would be able to convince the country militia to come to the relief of the city. When Rutledge left the city so did a number of invalids and wounded men. One of these men was Francis Marion.

When Marion left the city there was still a chance that General Lincoln would be able to get away and fight the British on his terms in the countryside. However the city leaders made this strategy impossible when they threatened to open the gates and let the British troops into the city if Lincoln made any moves to leave.

Where Marion went during those months is not known, but he did go into hiding, eluding the various British patrols in the area. Slowly the British were tightening the noose on Charleston and May 6th, effectively eliminating the cavalry corps in South Carolina.[1341]

Marion was still recovering from his broken ankle and he was moved from house to house. Due to his reputation as the commander of the 2nd South Carolina he was well known by both the people in South Carolina, and by the British, so his capture would be a huge victory. The people of South Carolina, both wealthy and poor, hid Marion and moved him quickly from plantation to cabin, and from dark swamp to dense canebreak when any British patrols would come near. Though no account exists of this time, Marion would have had to been a major target of Tarleton's mounted Legion.

The city of Charleston would fall to the British on May 12th, after a siege of 42 days. During the siege the Patriots lost 89 killed and 138 wounded. Far more serious to the cause were the 3,371 men captured, of whom 2,571 were Continentals. There was no more Patriot army in the South. The British had lost 99 killed and 217 wounded. They captured 311 artillery pieces, 5,916 muskets, 9,178 artillery round shot, 15 Regimental colors, 33,000 rounds of small arms ammunition, 212 hand grenades, and 376 barrels of flour. The best muskets and flints had been thrown in the harbor by the Patriots. Forty-nine ships and 120 boats were captured, including large magazines of rum, rice and indigo.

The 2nd South Carolina Regiment started the siege with 266 men who were fit for duty, the rest were sick or on furlough. By April 8th the Regiment had detached 59 men to Lauren's Light Corps, had another 67 men "on command" and only had 92 men fit for

[1341] Lenud's Ferry is located north of Jamestown where 41 crosses over the Santee River.

duty. By May 2nd the regiment's chain of command had almost ceased to exist. Marion had broken his ankle and had been sent out of town to recuperate, and Major Isaac Harleston was commanding the regiment in the place of Francis Marion. Captain Marion and Lieutenants Foissin and Legare were in the Hospital sick; Captain Roux was in Georgetown, recovering from wounds. Captains Mazyck, Baker, Proveaux, Gray and Lieutenant Kolb were all commanding batteries on the defensive line. The officers of the 2nd South Carolina who were captured at Charleston were Daniel Mazyck, John Buchanan, Richard Bohun Baker, Adrian Proveaux, Peter Gray, Peter Foissin, George Ogier, George Evans, Thomas Dunbar, and Henry Gray

As soon as Marion was able to mount a horse he rode to North Carolina with his friend and comrade in arms, Peter Horry. They rode to join a new Continental army marching south under the command of Baron De Kalb. Horry described Marion's ankle as "very crazy" and Marion still had to have his servant, Oscar, lift him from his horse.

Major General De Kalb had taken control of what remained of the Continental Southern Department's troops. On April 16th he marched from Morristown, New Jersey, with 1,400 men. These were six Maryland and Delaware Continental regiments, and the 1st Continental Artillery. With the army were their wives, children, laundresses and other camp followers. De Kalb had hoped to be reinforced by state authorities along the way, but little help was given to him. His army had only gone as far as Granville County in North Carolina, when news came of the surrender in Charlestown. De Kalb's army had no horses or wagons, and the men had to carry everything on their backs. Sick and hungry, his army marched on to Buffalo Ford on the Deep River. They made their camp there, 125 miles northeast of Camden.[1342]

Congress appointed General Horatio Gates to the Southern Department on June 13, 1780, hoping that the militia would rally to him, as the New England militia had in 1777 when Gates had captured the British army of Burgoyne. General Charles Lee warned him upon his parting, "Beware lest you exchange your Northern laurels for Southern willows."

Gates arrived in Hillsborough on July 13th and took command of the beleaguered army there. Within 72 hours of his arrival Gates had his army marching south to attack the British. Many historian think that Gates had made a great mistake by doing this, and some thought him a fool, however Gates knew that to win the fight against the British he needed the thousands of militiamen that were roaming the countryside of South Carolina, looking for a fight, under the command of Sumter and Caswell.

On July 1st Captain Ardesoif of the Royal Navy had captured Georgetown with no resistance from any Patriot forces in the area. On August 1st Cornwallis ordered Tarleton to cross the Santee River and determine the size of any possible rebel forces in that area. Tarleton found nothing, but did hear of a force of 500 men led by Major John James at Indiantown. When James learned of Tarleton's raid into Williamsburg County, he asked Gates to send an experienced commander for the resistance.

On August 4th Francis Marion rode into the camp of Gates at Deep Creek..[1343] Williams wrote, "Colonel Marion, a gentleman of South Carolina, had been with the army a few days, attended by a very few followers, distinguished by small black leather caps and the wretchedness of their attire; their number did not exceed twenty men and boys, some white, some black, and all mounted, but most of them miserably equipped; their appearance was in fact so burlesque that it was with much difficulty the diversion of the regular soldiery was restrained by the officers; and the general himself was glad of an opportunity of

[1342] Buffalo Ford is located 6 miles south of Ramseurs, on highway 22.
[1343] Deep Creek is located near Chesterfield.

detaching Colonel Marion, at his own instance, towards the interior of South Carolina, with orders to watch the motions of the enemy and furnish intelligence."

Gates had decided that Marion's comical army was of no use to him in the swamps of South Carolina, and sent them out to gather intelligence and deliver a proclamation to the citizens that he would protect them from "acts of barbarity and devastation." Gates gave Marion orders to go to the Williamsburg Township, and destroy all the boats to prevent Cornwallis from escaping his approaching army.

Marion rode off to Williamsburg District to begin the partisan warfare that would become legend. Gates and his army marched off to fight one of the most one sided battles of the war, losing to an outnumbered General Cornwallis and losing his Continental army at the Battle of Camden.

Meanwhile, Major John James had gone to the British commander in Georgetown, and asked him what was expected of the people in Williamsburg. James told Captain Ardesoif that he represented the people. Ardesoif drew his sword and stated, "I shall require unqualified submission from them; and as for you, I shall have you hanged." James parried the sword with a chair, and escaped on his horse *Thunderer*.

James told his story to the clan at King's Tree, and the word was passed to Pudding Swamp, then to Cedar Swamp, then to Thorntree Swamp and finally to Lynch's Lake. Two days later four companies formed in Williamsburg, and called on James to command them. This battalion would be the nucleus of Marion's partisans.

Marion was an officer of the Continental Line on special assignment, and had no rank in the militia. He also had no legal authority over the Williamsburg Township Militia, but this was not a court of law, and lack of authority did not deter him. He immediately took command of the Williamsburg District militia on August 10th. The men in the district knew Marion, and many of them served with him in the 2nd South Carolina Regiment.

It was said that Marion "did not talk much. He did things." He was described as "lean and swarthy. His body was well set, but his knees and ankles were badly formed, and he still limped upon one leg. His eyes were black and piercing. He was dressed in a close round-bodied crimson jacket of a coarse texture, and wore a leather cap, part of the uniform of the 2nd Regiment, with a silver crescent in front, inscribed with the words, 'Liberty or Death!!" [1344] Marion had his partisans wear white cockades to distinguish them from the Tories and he armed his men with swords fashioned from saw blades by blacksmiths in the area.

Marion's first action was to try to rescue the prisoners taken by Cornwallis at the Battle of Camden. These prisoners were marched through the Williamsburg District on the way to prison ships in Charleston. On August 25th Marion and about 150 men were able to surprise the British guards at Nelson's Ferry and rescue 147 soldiers of the Maryland and Delaware line.[1345] Ironically when 85 of the Continentals saw the ragged condition of their rescuers, they preferred to stay in captivity. Many of these men probably thought there was no hope of winning the war after the surrender at Charleston and the panicked retreat from Camden. Marion took the two dozen British prisoners and the 62 Continentals and

[1344] There are several descriptions of Marion and other officers wearing red short coats. This may have been a uniform from a militia unit they belonged to before the war began. Both the South Carolina Independent Company, and the Charlestown Infantry Regiment wore red short coats years before the war began.

[1345] There are two locations known as Nelson's Ferry. This site was also known as Great Savannah, or Sumter's Plantation, and is currently under Lake Marion, near Goat Island, south of Summerton.

marched to Port's Ferry, but by the time he arrived there all but three of the Continentals deserted him.[1346]

Two days later Major John Wemyss and the 63rd Regiment marched in Williamsburg District to put down the partisans led by Marion. Marion sent Major James to strike at the British column, but when he learned of the size of the force he rode into North Carolina with his men, to avoid capture. He stayed at Ami's Mill on Drowning Creek. [1347]

Marion had left some partisans in Williamsburg to determine the situation there. They told him of a gathering of Loyalists under the command of Micajah Ganey. Ganey had been in the 2nd South Carolina until he felt slighted by an officer, then he switched sides and became a thorn in the side of Marion. Ganey kept the partisans of Cheraw District busy, keeping them from joining Marion on his operations. With Ganey was Captain Jesse Barefield, who had also been with the 2nd South Carolina at the Battle of Fort Sullivan. He had gone to the side of the British after the fall of Charleston and formed a mounted Loyalist unit.

Marion rode from North Carolina and attacked Ganey and Barefield at Blue Savannah, scattering the former 2nd South Carolina members and their Loyalists into the swamps.[1348] Major Wemyss and his men laid waste to the Williamsburg District in retaliation. Wemyss had his men break up looms and burn the mills. He had his troops shoot milk cows and bayonet sheep. His destruction of the looms and the sheep was to deprive the people of clothing. He burned, and laid waste to fifty plantations, carried off their slaves for use as slave labor by the British, and hanged several men who opposed his actions. Wemyss burned the Presbyterian Church at Indian Town because he said all churches were "sedition shops." Wemyss's actions drove many men into the ranks of Marion's partisans, doubling his force.

After Wemyss's destructive army had marched away Marion returned to the Williamsburg District. Towards the end September Marion learned that Colonel John Coming Ball and his Loyalist militia were at Dollard's Tavern, on Black Mingo Creek.[1349] Colonel Ball's mission was to serve as an advance outpost for the recently completed British post at Georgetown, twenty miles to the east. Marion attacked Colonel Ball in a surprise attack at night, driving them from the area and capturing their horses, guns and equipment. Marion was promoted to Brigadier General in the militia due to his efforts in the low country.

In September General Cornwallis marched into North Carolina pursuing the remains of Gates's army, and attempting to rally the Loyalist forces there. To secure his march he sent Major Patrick Ferguson into the mountains to subdue the backwoodsmen. Early in September Ferguson arrived in Gilbert Town and sent a paroled prisoner into the mountains with a message for the backwoodsmen. He threatened to lay waste to their country with fire and sword, and hang their leaders should they stay with the Rebel cause. These strong words did not frighten the over-the-mountainmen, but instead unified them with the goal of tracking down Ferguson and wiping out his army at the Battle of King's Mountain. Cornwallis had to retreat back into South Carolina to a safer location to rebuild his shattered army.

[1346] Port's Ferry is located on the Great Pee Dee River, west of Britton's Neck.

[1347] Drowning Creek is the Lumber River, located near present day Fair Bluff, North Carolina.

[1348] Blue Savannah was located north of Centenary, at the intersection of 41 and 501.

[1349] The action at Black Mingo Creek happened where Highway 41 crosses over Black Mingo Creek, just south of Union Crossorads.

Towards the end of October Marion was tracking down the South Carolina Rangers under the command of Major John Harrison. Harrison had been in the Patriot South Carolina militia early in the war, but had switched sides when Savannah fell. While searching for Harrison, Marion discovered a company of Loyalist militia under the command of Lieutenant Colonel Samuel Tynes camped at Tearcoat Swamp. Marion's 150 partisans surrounded the Tories that night and attacked, killing six and wounding fourteen. The rest ran into the swamp to avoid capture. Many of the Loyalists were not so dedicated and joined Marion's partisan that night. Tynes was able to escape but Marion sent Captain William Clay Snipes to the High Hills of Santee to bring him back. Snipes was able to capture Tynes and several other Loyalist officers, and Justices of the Peace.

The British were outraged that Marion had the ability to do this in an area under their control. They sent Banastre Tarleton to find Marion's men. Marion learned of Tarleton's mission and tried to surprise him, but Tarleton had set up a trap of his own to capture Marion at Woodyard Swamp. Marion was warned of this and immediately rode away. Enraged that Marion had detected his ambush, Tarleton began burning thirty plantations and houses, from Jack's Creek to the High Hills.[1350] Tarleton was aware of Marion's skills, and each night he brought his Legion together to form a defensive perimeter.

Marion tried every trick he knew to lure Tarleton into an ambush, but the men were evenly matched, neither falling for the other's traps. The chase soon ended when Tarleton was ordered back to find Sumter's partisans . Cornwallis knew that Sumter was closer to his army and more immediate threat. Due to his actions, in the hills of the Santee to the west. Tarleton, like Major Wemyss, became an excellent recruiting officer for Marion's Partisans.

Some historians write that Marion received his nickname "The Swamp Fox" at this time from Tarleton. Supposedly when Tarleton was given a new mission of finding Sumter's force, he said, "Let us go back and we will find the Gamecock, but as for this damned old fox, the devil himself could not catch him!" However this is not mentioned in Tarleton's book or letters. The name "Swamp Fox" does not appear in any writings of the time, and first appears in 1821 in William Dobein James's book "A Sketch of the Life of BRIG. GEN. FRANCIS MARION". James is also the first one to mention that Sumter's nickname was "The Gamecock".

Once Tarleton left the area there was less of a threat to Marion, and he turned his attention to Georgetown. He needed supplies of salt, clothing and ammunition for his men and the capture of Georgetown would also be a huge blow to British morale in the region. Marion concealed his force in the swamp north of town, at a location near Alston's Plantation that became known as The Camp. He sent Peter Horry and his horsemen across White's bridge on a reconnaissance towards the Black River.[1351] He sent Captain John Melton to the Sampit road. With Melton was Marion's nephew, Lieutenant Gabriel Marion.

At White's Plantation Horry's force found Captain James "Otterskin" Lewis's Loyalists killing cattle. There was a short hot fight, then the Loyalists fled. Captain Melton's patrol was moving down the Sampit Road, when he learned of a Loyalist party camping at The Pens, another plantation of Colonel William Alston.[1352] As Melton's horsemen were passing through a dense swamp, they stumbled onto Captain Barfield and his Loyalist troop. Both sides fired at the same time. Barfield took a load of buckshot in

[1350] Jack's Creek flows from the Black Mingo Creek, and is located west of Highway 41, north of Rhems.

[1351] White's Plantation and White's Bridge is located west of Georgetown, where US 17 crosses White's Creek.

[1352] Alston's Plantation is located 1 mile northeast of Georgetown, at the end of Indigo Avenue.

his face and was killed. His brother Jesse Barefield was wounded. Gabriel Marion's horse was killed. The Loyalists seized Gabriel, and began clubbing him with their muskets until he was knocked senseless. When he was recognized as the nephew of Francis Marion, a mulatto named Sweat put a musket against Gabriel's chest and fired a load of buckshot into his heart, killing him instantly. The barrel was so close that it set his linen shirt on fire.

The next day Marion's Partisans captured Sweat. As they were crossing the swamps that night, a militia officer rode up to Sweat, put his pistol to his head and shot him dead. Marion was furious and publicly reprimanded the officer that killed him. Marion did not condone any acts that were against the rules of war, and demanded his men adhere to the strict discipline of the regular army.

After the fall of Charlestown the remaining South Carolina Continentals were placed into two battalions, and designated as the 1st South Carolina Regiment. The 1st Battalion was under the command of Colonel Charles Pinckney, and was imprisoned in Charlestown. The 2nd Battalion was with Francis Marion and became part of his partisans.

Towards the end of November a new threat marched into Williamsburg District to challenge Marion's partisans, and to show the local population that the British army was still strong. Major Robert McLeroth crossed the Santee River at Nelson's Ferry, and marched into the Williamsburg Township. He camped on the village green in Kings Tree. This was a direct challenge to Marion, because any British unit going into the Williamsburg Township was tempting fate.

The villagers sent a message to Marion in Black Mingo, but instead of mounting an attack, he hastily assembled his outnumbered partisans and slipped away. On November 22nd, McLeroth feared an attack by a larger unit, and abandoned Kings Tree.

Three weeks later Marion learned the Major McLeroth was marching two hundred recruits of the 7th Fusiliers, from Charlestown to Camden. This force of new recruits was more to the liking of Marion's small force of partisans and he immediately rode off searching for the British soldiers. Unfortunately for Marion, McLeroth was not Tarleton or Wemyss, and he did not wage war on the people of Williamsburg District. When Marion stopped at the house of a lady who had always been friends to the partisans, and asked her if she had known of any movements of McLeroth. The woman told Marion that she could not help him because McLeroth had been so honorable and gentle, that she didn't wish to see him harmed.

Marion found McLeroth on the road at Halfway Swamp.[1353] Marion divided his men into three sections and hit the British on the road from the rear with his riflemen under McCottry.[1354] The rest of his force attacked the column from the front and flank, driving them into an open field. The recruits panicked, but the veterans of McLeroth's 64th Regiment gathered them and placed them behind their formation. The fight became a long range duel of rifles, until a white flag of truce was sent out by McLeroth.

The British officer came up with an unusual challenge to Marion. He challenged 20 of Marion's best partisans to engage 20 of McLeroth's Regulars in a fight to the death. The fate of each army would be decided by these 20 men. Marion accepted. This style of

---

[1353] Halfway Swamp was located south of Rimini, about a mile south on the Old River Road.

[1354] William Robert McCottry had initially served in the volunteer militia under Captain Benjamin Screven in 1775. After the fall of Charleston he commanded a rifle company in Marion's partisans. McCottry's Riflemen may have been the most effective rifle unit at the time. Most of the stories told of the riflemen in the American Revolution tend to be more myth than fact, but this is not the case with McCottry's men. When the American reports told of their deadly aim, it was backed up by the same reports from the British. Tarleton, Wemyss and Ardesoif reported McCottry's riflemen to London, and they told of their deadly skill with the rifle.

fighting by using a small number of "champions" dated back to the biblical times, when it was believed that the fate of an army was already decided by God, so unnecessary bloodshed was not needed. A single man, or in the case of this challenge, 20 men, would be the only ones put in harm's way.

Marion chose Major John Vanderhorst to command the fighters, and then he chose the bravest and most accurate marksmen. Vanderhorst marched forward to within a hundred yards of the British, but the British shouldered their muskets and retreated. Vanderhorst and his men were confused, but gave three huzzahs. What they did not know was that McLeroth had outfoxed Marion, and they had been tricked. McLeroth had been stalling for time. While the preparations for the challenge had been going on, McLeroth had sent couriers racing for help. They met Captain Coffin and one hundred and forty mounted New York Volunteers. However, Coffin did not come to rescue them from Marion, but instead turned back and camped his men in Swift Creek.[1355]

During the night McLeroth's men built huge bonfires, they shouted and sang, then quieted down at midnight. This was a deception because in the dark of night they abandoned their baggage and slipped quietly away on the road to Singleton's Mills.[1356] Marion discovered the trick at daylight and dispatched Hugh Horry and a hundred horsemen to intercept the British. They found the British at Singleton's Mills, but after firing one volley they quickly abandoned the fight. They discovered that the family at the Mills had small pox, and this silent killer was feared by both sides.

The next day Marion shut down all river traffic on the Santee River and the Santee Road, shutting off any supplies coming from Georgetown and going to the British outposts in South Carolina. McLeroth marched near Marion's force, but did not engage them. Marion remained on the road until he received orders from a new Southern commander, Nathanael Greene, that ordered Marion to conduct intelligence gathering for Greene's army.

The British South Carolina Gazette condemned Marion's men and wrote that the "State of South Carolina no longer exists. Marion and the men who follow him are blue parties and traitors against rebellion itself, and are to be sacrificed by any regular enemy.[1357] Their violence and rapine mark their steps. The King will not always be merciful."

Towards the end of December Marion sent Horry to determine the strength of forces in Georgetown. This port was the main supply point for Williamsburg District, and for the British outposts up the Santee and Pee Dee River. Marion's main goal was to capture the town. The post at Georgetown had been reinforced with a cavalry troop of the Queen's Rangers under the command of Captain John Saunders. Horry skirmished with the troop, capturing 16 of them in a short bloody fight. Two of the prisoners captured by Horry were both men who had been in the 3rd South Carolina Regiment. They said that they had enlisted from the prison ship in Charlestown so that they would have a better chance to escape.

By the end of 1780 Marion's men commanded the countryside of Williamsburg District. Sumter's partisans did the same for central South Carolina, while General Greene threatened Cornwallis from the border of North Carolina. Cornwallis could not move without the partisans whittling down his supply lines, but he had to gain control of the

---

[1355] Swift Creek is located four miles southwest of Rembert.

[1356] Singleton's Mill was located on the southeast edge of Poinsett State Park, at Christmas Mill Lake, northwest of Pinewood.

[1357] Draper wrote that "Blue Parties" in the German Wars were those men who would go out without orders of proper officers and they were hanged whenever they were found.

Carolinas. Greene knew that if he could draw Cornwallis into North Carolina, he would be able to cut him off from his supply bases in South Carolina by the use of the partisans.

At the beginning of January Marion's men were still striking at the British in Georgetown from their base of operations on Snow's Island.[1358] British Lieutenant Colonel George Campbell made an expedition into the countryside around Georgetown, to try to destroy the rebel base. After a week of campaigning he was unsuccessful and returned to Georgetown.

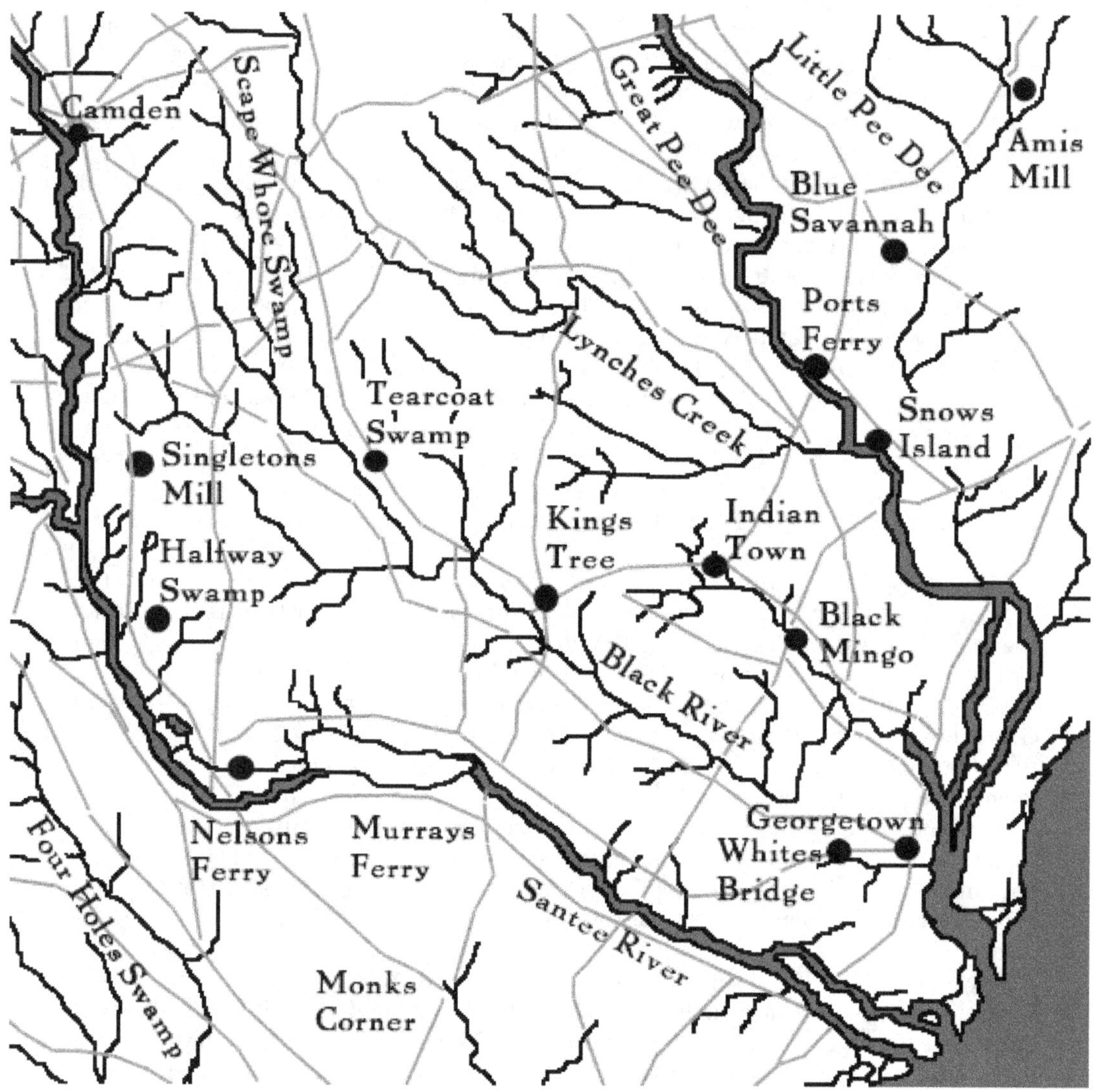

Operations August 1780 – January 1781

[1358] Snow's Island was an "island" in the swamp where Lynch's Creek, Clark's Creek and the Pee Dee River met. It is presently located somewhere at the southeast corner of Florence County, near Johnsonville, and was named for William Goddard's uncles, James and William Snow.

During this time Greene had a decisive victory that would start a chain of events into motion. Greene had detached General Daniel Morgan into South Carolina, with the instructions to try to divide Cornwallis's forces. Cornwallis sent his trusted cavalry commander, Banastre Tarleton, after Morgan's small army. In the early morning of January 17th, Tarleton fought Morgan at the Cowpens. Morgan destroyed the British force with only minimum losses, and then retreated into North Carolina. Cornwallis had no choice but to pursue after Morgan to get back the 600 prisoners captured at Cowpens. This is exactly what Greene had hoped for.

Marion received reinforcements from General Greene when Lieutenant Colonel Henry "Light Horse Harry" Lee arrived on January 23rd with his Legion. These sharp troopers were a marked contrast to Marion's poorly outfitted partisans. The two commanders learned that the troop of Queen's Rangers in Georgetown had been given permission by Cornwallis to leave South Carolina. This weakened the defenses of Georgetown, so Lee and Marion decided to strike.

The two units would wait outside Georgetown, while two companies of Lee's men would enter the fortified town by boats moving across the rice fields. Lee's amphibious troops rushed into the town and took Lieutenant Colonel Campbell without firing a shot, though one of the British officers was bayoneted to death.

Marion and Lee arrived with their cavalry, but not one British soldier had been seen. Amazingly none had tried to rescue their commander, or even appear. The British soldiers had holed up in their brick redoubt, not wanting to take on Marion's partisans. Without battering rams or scaling ladders the little fort could not be taken. Lee did not want to sacrifice his men in a direct assault. Marion agreed that it was suicide to do so, and gave the signal to withdraw.[1359]

Marion's partisan orderly book starts on 16 February 1781, two days after Captain John Postelle was able to capture 29 men of the King's American Regiment who had been occupying his father's home. Unlike the orderly book that was kept by the 2nd South Carolina Regiment, this orderly book does not have daily entries, and at times a whole week may pass by without a note. This was the life of the partisans though, and they would catch up on their paperwork after they had stopped moving.

[1359] This period of the war in the Carolinas is covered in greater detail in *Volume 2, 1780* and *Volume 3, 1781* of *Northing But Blood and Slaughter,* by the author.

This volume extends from 16 February 1781 to 15 December
1782 - &
closes with Marion's valedictory to his heroic brother soldiers the day after the evacuation of Charleston

Not Marion's
Handwriting

Francis Marion, Brig. Gen.
S.C. Militia

Note: This one contains no
F.M. handwriting

Orderly Book Commanding
16 Feb[y] 1781 –

1783

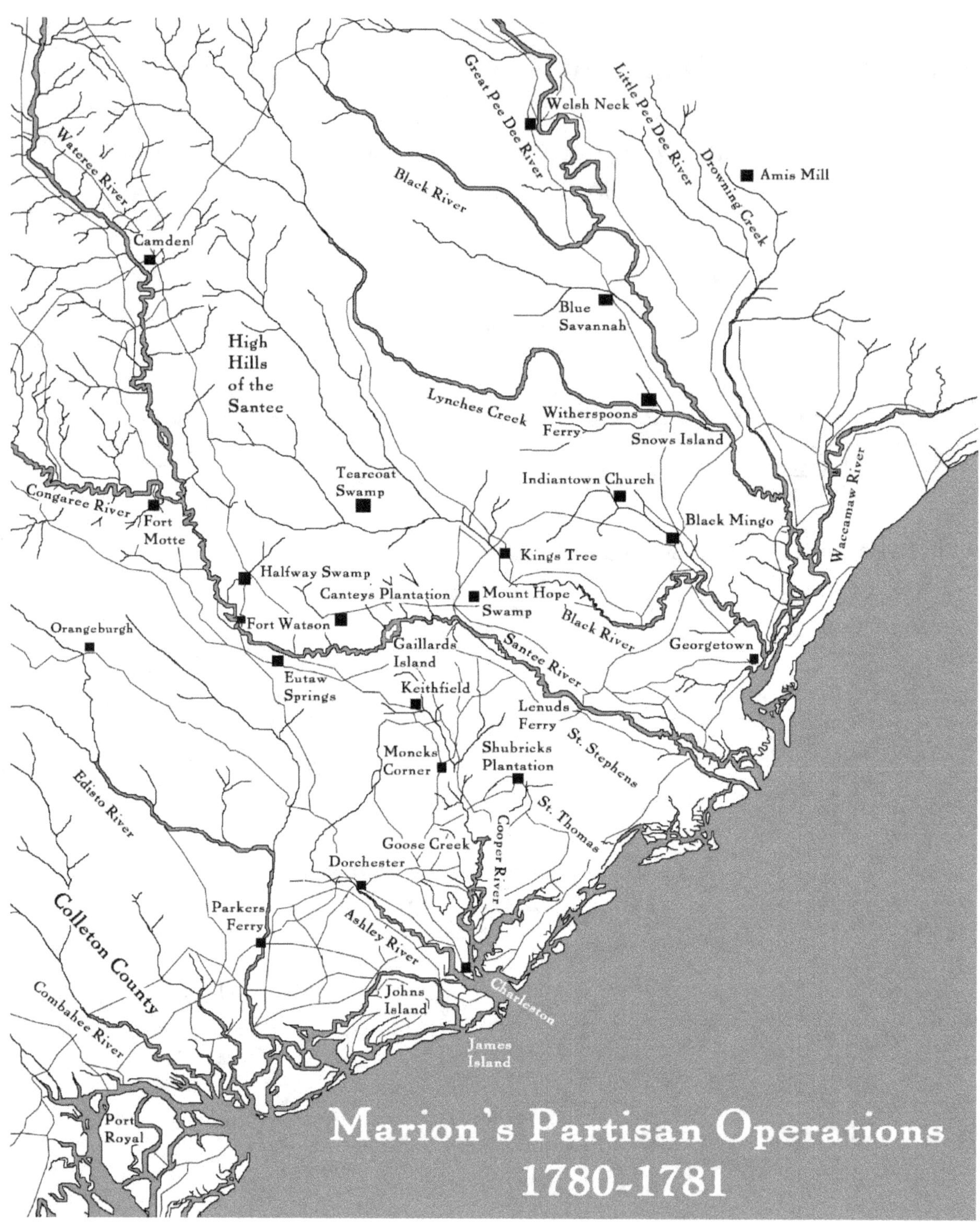
Welsh Neck
Great Pee Dee River
Little Pee Dee River
Drowning Creek
Amis Mill
Black River
Wateree River
Camden
Blue Savannah
High Hills of the Santee
Lynches Creek
Witherspoons Ferry
Snows Island
Tearcoat Swamp
Indiantown Church
Congaree River
Fort Motte
Black Mingo
Waccamaw River
Kings Tree
Halfway Swamp
Canteys Plantation
Mount Hope Swamp
Black River
Fort Watson
Orangeburgh
Gaillards Island
Santee River
Georgetown
Eutaw Springs
Keithfield
Lenuds Ferry
St. Stephens
Moncks Corner
Shubricks Plantation
Edisto River
St. Thomas
Goose Creek
Cooper River
Dorchester
Parkers Ferry
Colleton County
Ashley River
Charleston
Johns Island
Combahee River
James Island
Port Royal
Marion's Partisan Operations
1780-1781

1781

Orders 16th Febr 1781 –

It is ordered that each regiment do raise one Troop of Light horse to be Equipt with Swords & pistols & to act as kept horse to the Brigade as soon as they are compleated they are be Organized in One Corps the Command of which is Given to Col°. Peter Horry. He is Desired to procure caps & swords as soon as possible – [1360]

Captains of each Troop are to chose the men from the regiment giving a Certificate to each men of his Entering in said troop which Shall be Sufficient Discharge from the Company he formerly belong.

Each troop is to consist of forty rank & file, three Sergts & three Corperols & two trumpeters, the uniform is a Short homespun coat with cuffs & cape thus McDonalds red,[1361] Richardson Blue,[1362] Giles yellow,[1363] Kolb Green,[1364] Kershaws Black,[1365] the Colour of the Coats Blue & White lined.[1366] After a man has Entered in Sd troop he is not to Leave it without Leave of the Commanding officer of the Brigade –

Col° Horry will be given a warrant to Impress such horses as may be needed for Present Service, the General hopes that Every young man will show their Zeal for their Country by Entering in Sd troops which is so necessary at this Junction.

1360 Marion put in a requisition for all the saws in the country and had all the blacksmiths in the region make swords for the four troops of militia cavalry. He had so little ammunition that this expedient was necessary. One of the images most people have of the dragoons in the Revolutionary War is of mounted men firing carbines at the enemy, but by this time in the war, carbines were extremely hard to find and most dragoon units only had pistols and swords.

1361 Archibald McDonald was born in 1740 and had served as a colonel in the militia in the garrison of Georgetown. He died in 1785 at the age of 45.

1362 Richard Richardson, Jr. was born on 4 March 1741 in Prince Frederick Parish. He had served as a lieutenant under Captain Samuel Cantey in Lyttleton's Campaign against the Indians during 1759-1760. During 1775 he was in the Snow Campaign as a captain of militia. In 1776 he was a captain in the 2nd South Carolina Regiment. He became a major and was taken prisoner at the fall of Charleston. He was paroled and joined Marion's Partisans as a lieutenant colonel. He died in 1818 at the age of 77.

1363 Hugh Giles was born in 1750. He had been a first lieutenant in the Prince Frederick Parish Volunteers in 1775. He had been at the fall of Charleston and served as a colonel and regimental commander under Marion afterwards. He was no longer with Marion in June 1781. He died in 1802 at the age of 52.

1364 Abel Kolb had served as a colonel in the militia since 1776. On 26 April 1782 he was murdered and his house was burned by Tories near Dorchester.

1365 Joseph Kershaw was born on 17 March 1728. He was married to Sarah Mathis. He was a colonel in the militia from 1776 to 1780. He was taken prisoner at the fall of Charleston. When he was released he joined Marion's partisans. He died on 28 December 1791 at the age of 63.

1366 Clement C. Brown made the jackets and overalls. Isham Nettles made the caps.

Lt. Daniel Conyers is Appointed Captn of the Light horses in Colo McDonalds regimt. He is to be Obeyd as Such – [1367]
Captn John James is Appointed a Major in Colo McDonalds Regimt in Room of Majr Horry promotion, he is to be respected and – Obeyd as Such – [1368]

Jeffrys Creek 17th Febry 1781 – [1369]
For the Night Captn Withers [1370] & Lt Taylour [1371]

Parole Pyckens – } Countersigns
Polly –
Peggy –

Captn John Baxter is appointed to Command a Compy of Light horse in Colo. Giles Regimt. He is to be Obeyd as Such [1372]

Jeffrys Creek Febry 18th – 1781 –

For the Night Captn

Parole Quails } Countersign { Queer
Question

Bursches Plantn Febry 20th 1781 – Parole Ritchmond [1373]
CSn. Rum – Rations –
Mr William Gordon is appointed a Captn [1374] John McRae first & John Gordon Second Lieutenants in the Company formerly Captn Baxter, they are to be Obeyd as Such – [1375]

[1367] Daniel Conyers had enlisted during October 1775 in a volunteer company of militia under Captain William Fullwood. He served 582 days as a lieutenant and captain under Marion.
[1368] John James was born in Ireland on 12 April 1732. He had been at the fall of Charleston in 1780. Afterwards he served as a captain, major and lieutenant colonel under Marion. His son, William Dobein James was also in Marion's Brigade and wrote a biography on Francis Marion in 1832, titled *The Life of Marion.*
[1369] Jeffries Creek is located on the Pee Dee River, near present day Mill Branch, west of Florence.
[1370] William Withers had served as a captain in the 2nd South Carolina Regiment before the fall of Charleston.
[1371] Samuel Taylor served as a lieutenant and a captain in the light dragoons under Colonel Maham. He died on 14 January 1815.
[1372] John Baxter had served in the volunteer militia under Captain Benjamin Screven during 1775. He was promoted to major in Colonel Ervin's regiment of Marion's Brigade on 28 June 1781. He was later promoted to lieutenant colonel in Marion's Brigade. He was wounded on 17 July 1781 at Quinby Bridge. In June of 1782, he commanded the Pee Dee Regiment while it was stationed in Georgetown.
[1373] Burches Plantation and Mill is located west of Klondike on 701, where Horry, Marion and Williamsburg county come together.
[1374] William Gordon had served 150 days as a cavalryman in the Britton Neck Regiment from 15 March to 15 October 1780. He was appointed a captain in Marion's partisans on 18 February 1781. He died in 1786.
[1375] John McReeatt (McRae) served in the cavalry under Peter Horry. He was promoted to 1st Lieutenant on 18 February 1781.

M^{r}. John Thompson Green is appointed first L^{t} in Captn Baxters Company of Light horse he is to be Obeyd as Such [1376]

Mr John Parker is appointed a first L^{t} in Captn Thornly Company [1377] he is to be Ob^{d} as Such – [1378]

Burches Plant^{n} Febr^{y} 21st 1781 –

For the Night Captn

Parole Scipio } Countersign } Sampson –
Saul –

After the raid on Georgetown Marion dispatched seventy of his partisans under the Postelle brothers to destroy the British supply bases at Wadboo, Keithfield and Manigault's Ferry.[1379] This was to support a planned offensive by General Greene against Fort Ninety-Six or Camden. This was Greene's original intent, before the battle of Cowpens, to destroy the British supply bases.

Captain John Postelle and thirty-nine supernumerary officers destroyed the stores at Wadboo, consisting of fifteen hogsheads of rum, a quantity of pork, flour, rice, salt and turpentine. Supernumerary officers were "extra" officers who did not have any men to command due to their capture at Charleston. Captain Postelle next surprised the depot at Keithfield, near Monck's Corner. The partisans killed two of the British guards, wounded three, and captured 30 British soldiers, and seven wagons. He did this without the loss of a single man. He also burned "fourteen waggons loaded with soldiers cloathing and baggage, twenty hogsheads of rum, and retired with his prisoners."

Captain Postelle's brother, Major James Postelle, and forty men were ordered to Colonel Thompson's camp, but found no stores there. All had been removed a few days before. Major Postelle found the camp too heavily guarded to attack. As Major Postelle was returning to Marion's camp he heard of a great quantity of rum, sugar, salt, flour, pork, soldiers clothing and baggage at Manigault's Ferry. The British guard at Manigault's Ferry had gone after the other Postelle brother at Keithfield and had had left only four men in a redoubt of wood. This was easily taken by the partisans. Major Postelle destroyed all the stores and the redoubt, and just like his brother, he did not lose a single man.[1380]

1376 John Thompson Green first served in the 5th South Carolina Regiment, then transferred to the 1st South Carolina Regiment in February 1780. He was captured at the fall of Charleston. He broke his parole to join General Nathaniel Greene. He was promoted to first lieutenant in Baxter's company of Horry's dragoons on 18 February 1781. He remained with Marion until the end of the war.

1377 Robert Thornley served as a captain and a major during 1781 and 1782.

1378 John Parker had enlisted in the 2nd South Carolina Regiment on 4 November 1775. He was promoted to a 1st Lieutenant in Captain Thornly's company of Horry's dragoons on 18 February 1781.

1379 Wadboo and Keithfield are both under the waters at the southern end of Lake Moultrie, north of Monck's Corner. Manigault's Ferry is located under the waters of Lake Marion, just east of Santee.

1380 Patrick O'Kelley, *"Nothing but Blood and Slaughter" The Revolutionary War in the Carolinas, Volume Three, 1781*, (Booklocker.com, Inc 2005), pp 64-66.

It is ordered that each regiment do raise one Troop of Light horse to be Equipt with Swords & pistols & to act as kept horse to the Brigade as soon as they are compleated they are be Organized in One Corps the Command of which is Given to Colo. Peter Horry.
PAT OKELLEY 2006

Huges Plant^{n} Febr^{y} 22nd 1781 – [1381]

By Genl Greens Order I have the Pleasure to Return his thanks to Maj^{r} and Captn Postele for their Spirited address in Burning the Stores of the Enemy at wadboo & keithfield & Manigaults ferry.[1382] This Satisfaction is added to it hat I have to give my Thanks to Captn Postele and his officers and his men for the Spirited Behaviour in Capturing Captn DeRister one Subaltern & twenty four Granidiers with an Inferior Number without Loss –

Captain John Postelle had learned that Captain James DePeyster had occupied his father's home, and he was determined to drive the British out of the house. On the night of the 14th Postelle and his fourteen partisans crept near the kitchen of his father's house, and then waited until morning to strike.[1383] Postelle's father was not in the house at the time. When the sun came up Postelle formed his men in four ranks to make their numbers look larger than they were, and rushed the house. He demanded the surrender of DePeyster and his force.

DePeyster had with him 29 grenadiers of the King's American Regiment. De Peyster asked for some time to make up his mind. Postelle stated that he would not give five minutes, and began to set his father's house on fire. While the house was burning Postelle again demanded the surrender. This time DePeyster had his men march out, stack arms and surrender. When DePeyster surrendered his sword to Postelle he asked, "Where are your men?" Postelle told him that he only had fourteen men. DePeyster became enraged that he had surrendered to an inferior force. Postelle's mounted men quickly had the prisoners run out of the area before reinforcements could rescue them.

Parole Virtue } Vigour / Value { Countersign

Officers of the Night Captn Windham[1384] Lieu^{t} McBride – [1385]

The overall commander of the South Carolina militia was General Thomas Sumter. Sumter was considered more of a threat to the British because of the number of men he could raise at any given time. This was evidenced when Tarleton was ordered to quit pursuing Marion and instead go after Sumter. Tarleton was defeated by Sumter at the

---

[1381] Hughes Plantation was located near present day Oatland, approximately where the Lark Hill Plantation is today.

[1382] James Postelle was born in 1745. He had served as an ensign in the Gentlemen Volunteers of St. Bartholomew's Parish, and was a major under General Richardson during 1779. He was a lieutenant colonel under Joseph Kershaw. He was a brigade major to General Lachlan McIntosh when he was captured at Charleston. After being exchanged in June 1780 he joined Francis Marion. He died on 10 March 1824 at the age of 79. Captain John Postelle was the brother of James Postelle.

[1383] Postelle's house at Hasty Point (now Belle Rive) is on the Great Pee Dee River, just east of Plantersville.

[1384] Amos Windham had served as a lieutenant, captain and major in the St. David's Parish Militia in 1776. He served as a captain under Marion during 1781 and 1782.

[1385] John McBride, a former private of the 2nd South Carolina Regiment.

Battle of Blackstocks on 20 November 1780, however Sumter was severely wounded during the battle.[1386]

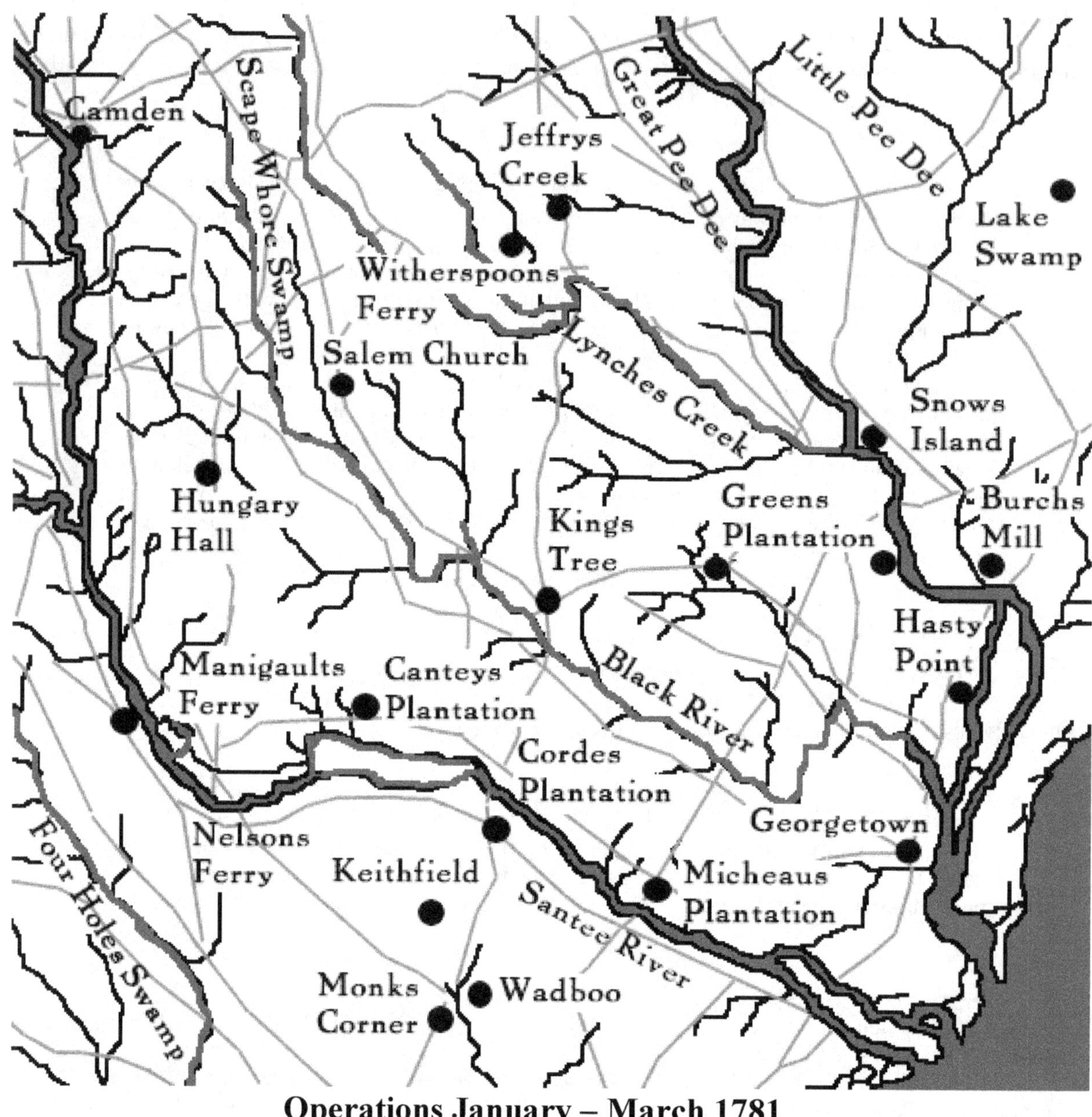

Operations January – March 1781

By the end of February Sumter had regained his health and went on the offensive against the British posts in South Carolina. His first target was Fort Granby, located where

[1386] Full details of the Battle of Blackstocks are described in "*Nothing but Blood and Slaughter, Volume Three*" by the author.

the present day town of Cayce is. Sumter believed that if he was quick enough he could sweep down the Congaree and capture the garrison by surprise. His larger plan was to quickly eliminate the remaining garrisons, and force the British into Charlestown.

Lord Rawdon learned of the siege and immediately dispatched a large relief force under the command of Lieutenant Colonel Welbore Doyle. The force consisted of 600 infantry, 100 cavalry and two pieces of artillery. Sumter asked Francis Marion to make diversionary attacks to slow Rawdon's column down, but the British were not diverted.

When Sumter withdrew his men he thought Marion would rendezvous with him, but after marching twenty miles they learned that Major McLeroth with the 64th Regiment, the New York Volunteers and a field piece had left from Camden to intercept their force. To make matters worse a detachment of Regulars and militia were approaching from Ninety-Six. Sumter's men gathered any boats along the shore of the Congaree and crossed over, taking the boats with them. Sumter quickly raced down the Congaree to attack the post at Belleville.[1387] Sumter's men would call these series of actions "Sumter's Rounds". Sumter also no longer believed he could rely on Marion for assistance.

Marion knew his men, and he did not have the hundreds of partisans that Sumter could muster. At times Marion's force consisted only of supernumerary officers without any men to command. These men would number only a few dozen. Marion realized that he did not have the men needed to march hundreds of miles away to strike at the enemy with Sumter. The few men who did ride with Marion would most likely refuse to leave the lowcountry, and their families, to go fight with another commander they did not know.[1388]

Joseph Glovers Plantation Febry 23rd 1781 – [1389]

Parole Wisdom } Wonder / Watchfull { Countersign

Officers of the Night Captn Hardon [1390] Lt Rogers – [1391]

Lemprieres Plantation Febry 25th 1781 –

Jarl Parole } Yarmouth / Yorkick { Countersign

Officers of the Night Captn Gambol [1392] Lt. King [1393]

1387 Belleville was located north of St. Matthews, where 601 crosses the Congaree River.

1388 O'Kelley, *NBBAS, Volume Three*, pp 88 – 90.

1389 Colonel Joseph Glover's Plantation was located at the head of Georgetown Creek, near Georgetown. After the raid by the Postelle brothers Marion had to move each day to a new location, setting up camp at a friendly plantation. Unfortunately the location of many of these plantations are lost to time.

1390 Edward Harden served as a private, lieutenant and captain under Marion during 1780 and 1781.

1391 John Rogers had been in the siege of Savannah and the siege of Charleston. He would be in the battle of Quinby Bridge and Eutaw Springs. He served as a private, lieutenant and captain under General Marion. He died on 9 August 1840.

1392 John Gamble, former officer of the 2nd South Carolina Regiment, served as a captain and major in Marion's Brigade during 1780 and 1781.

1393 John King served 180 days as a lieutenant in Marion's partisans.

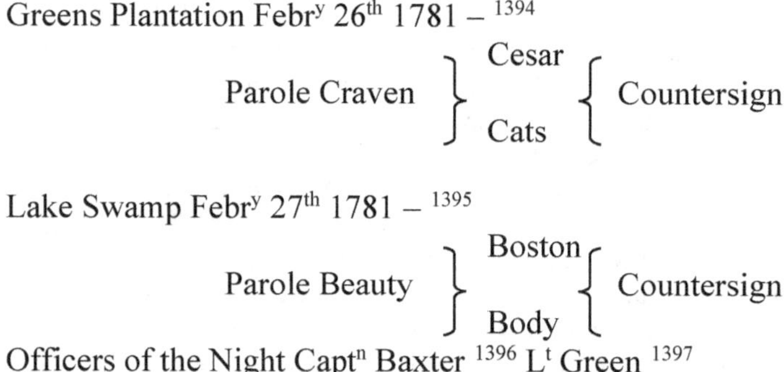
Greens Plantation Febr^y 26th 1781 – [1394]

Parole Craven } Cesar / Cats { Countersign

Lake Swamp Febr^y 27th 1781 – [1395]

Parole Beauty } Boston / Body { Countersign

Officers of the Night Captn Baxter [1396] Lt Green [1397]

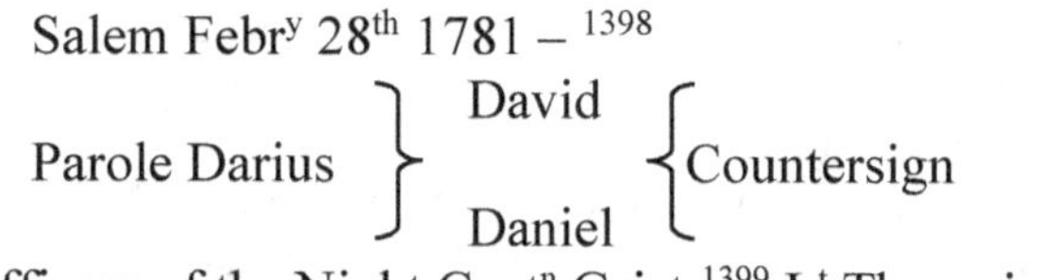
Salem Febr^y 28th 1781 – [1398]

Parole Darius } David / Daniel { Countersign

Officers of the Night Captn Grist [1399] Lt Thompion[1400]

By March the constant raids and ambushes by Marion's men were taking their toll on the ability of the British army to carry on the war in the Carolinas. Lord Rawdon decided to crush Marion's partisans and on March 1st he ordered Lieutenant Colonel John Watson Tadwell-Watson to march from Fort Watson, down the Santee and attack Marion in front. At the same time Lieutenant Colonel Doyle would march down the Pee Dee and cut off Marion's retreat. Doyle had been marching to relieve Fort Granby from Sumter's siege, but Sumter had withdrawn his forces.

Watson had with him his Provincial Light Infantry, two field pieces, Henry Richbourg's mounted Loyalist militia and 80 South Carolina Rangers. Henry Richbourg had been a captain in the 6th South Carolina Regiment, and was another South Carolinian who had taken up the King's shilling.[1401]

Camp at Salem 1st March 1781 –

Where as a Number of men go about the Country plundering the Inhabitants under pretence that they are my men, This is to inform the Publick that I will publish the names of all Such men who I know have Plundered Either on the north or South of Santee, after which it will be lawfull for any man to put them to Death where they are Not without being able to be called to account for it – for Such

---

[1394] Green's Plantation was located on the Pee Dee River, in north Georgetown County, near Port Hill Ferry.

[1395] Lake Swamp is located in between Gallivant's Ferry and Green Sea, where State Highway 19 and 917 intersect near Pleasant View.

[1396] John Baxter.

[1397] John Thompson Green.

[1398] Salem Church was located five miles south of Mayesville, where highway 527 crosses the Church branch of the Black River.

[1399] There is no additional information on Captain Grist.

[1400] There is no additional information on Lieutenant Thompion.

[1401] O'Kelley, *NBBAS, Volume Three*, pp 117 - 118.

Detestable practices, I have ever forbid & prevented as much as in my Power, and that I never sent any Party without a proper Officer and Orders in wrighting –

Mr. Benjamin Huggins is appointed Adjutant to the Corps of Light Horse, he is to be Obeyd as Such – [1402]

Clarks Plantation on Scape hore 1 March 1781- [1403]

Parole Adam } Again / Abel { Countersigns

Officers of the Night Captn Conyers and Lt McBride

Hungres Hall Mr. whites Plantations 2nd of March 1781 – [1404]

Parole Baltimore } Babylon / Bable { Countersign

Officers of the night Captn Potts & Lt. Gambol[1405]

John Kenteys March 3rd 1781 – [1406]

Parole Come } Cace / Can { Countersigns –

Captn Harden Lieutn Postell Officers of the Night [1407]

Cordes Plantation March 4th 1781 – [1408]

Parole Damage } Desert / Danger { Countersigns

Captn Black & Lt Rogers Officers of the Night -[1409]

Cordes Plantation March 5th 1781 –

Parole Ephraim } Eight / Egale { Countersigns –

1402 Benjamin Huggins, former officer of the 2nd and 3rd South Carolina Regiments.

1403 The Scape Hoar creek and swamp was "named after a woman of bad reputation who escaped in the swamp". It was located 4 miles northeast of Sumter.

1404 Hungary Hall Branch is located in Clarendon County, four miles south of Paxville, near where Highway 15 crosses Sammy Swamp.

1405 William Gamble served in the militia from 13 November 1780 to 5 January 1782.

1406 John Cantey, Jr. had served in the Cherokee Campaign of 1760 as an adjutant under his brother, Samuel Cantey. Before the war he was a major under Colonel Richardson. He was in Marion's partisans after Charleston fell. He married Margaret Richardson, then Hannah Flud, then Susannah McDonald. He died on 15 May 1786. His plantation was located east of Lake Marion, east of the dam, near Eagle Point.

1407 Jehu Postelle was born in 1749 and was the brother of the James and John Postelle. He served as a lieutenant and captain under General Marion. He died on 30 December 1797 at the age of 48.

1408 Cordes Plantation was north of Lake Moultrie, west of Pineville, on route 45.

1409 William Black was in Peter Horry's cavalry.

Officers of the Night Captn Potts L^{t}. McBride –

Parole Farmer } Faithfull { Countersign
Favour

Officers of the Night Captn Baxter L^{t}. Price [1410]

Colonel Watson's army moved down the Santee River towards a campsite below Nelson's Ferry. Marion learned of Watson's approach and he placed his men in an ambush position on the Santee road at Wiboo Swamp, between Murry's and Nelson's Ferries. [1411] When Watson came upon Marion's ambush position he was suspicious, and sent out a reconnaissance force led by Richbourg. Peter Horry did not have time to conceal his men and were seen by the British troops. Watson ordered Harrison and Richbourg to charge, but he did not see the other partisans hidden behind Horry's horsemen. Horry ordered a charge and Marion's men came on with a fury. Harrison was killed in the attack. Watson brought up his artillery and fired grapeshot at the partisans, then quickly retreated. The British charged at Marion's men with bayonets, and Marion ordered a retreat. Watson withdrew his men to Cantey's plantation to recuperate from the attack.[1412]

Coards Plantation 7th March

Parole Grantham } C Sign { Grubber
Gabeon

Captain William Benton is appointed Major to the Brigade, he is to be respected and Obey^{d} as Such[1413]
Captn Willm Capers is appointed adjutant to the Brigade, he is to be respected and Obey^{d} as Such[1414]
A Brigade court Martial to Set immediately of the tryal of Such prisoners as may be Brought to it, evidence to attend –
NB Sergeant wise for being asleep on his ~~post~~ gard, Acquitted –[1415]

[1410] Samuel Price had been in the 3rd South Carolina Regiment on 20 August 1776. He would be promoted to captain before the end of the war.
[1411] Wiboo Swamp is located on Lake Marion, near the dam, south of Jordan.
[1412] O'Kelley, *NBBAS, Volume Three*, pp 115 – 118.
[1413] Lemuel (William, Lamb) Benton served with General Marion in 1781 and 1782. He was appointed major to Marion's Brigade of Partisans on 7 March 1781. He was promoted to colonel on 28 April 1781 when Colonel Kolb was murdered.
[1414] Wiliam Capers, formerly lieutenant in the 2nd South Carolina Regiment.
[1415] Thomas Wise had served under Captain Richard Richardson prior to the surrender of Charleston. He was in the cavalry under Captain Harry Linus and Colonel Peter Horry. During a skirmish against Colonel Watson on the Santee River he lost his horse, bridal, gun and clothes.

John Port for Suffering himself to be disarmed when on centry.[1416] Sentence, to do a fortnights duty Extraordinary over this Month, approv[d]. Charles lye is appointed Capt[n] to take the Command of the Company from the lower District of the Santee, he is to be respected and Obey[d] as Such –[1417] Mack Nuggans is appointed first Lieut under Capt Gee [1418] he is to be Respected as Such[1419]

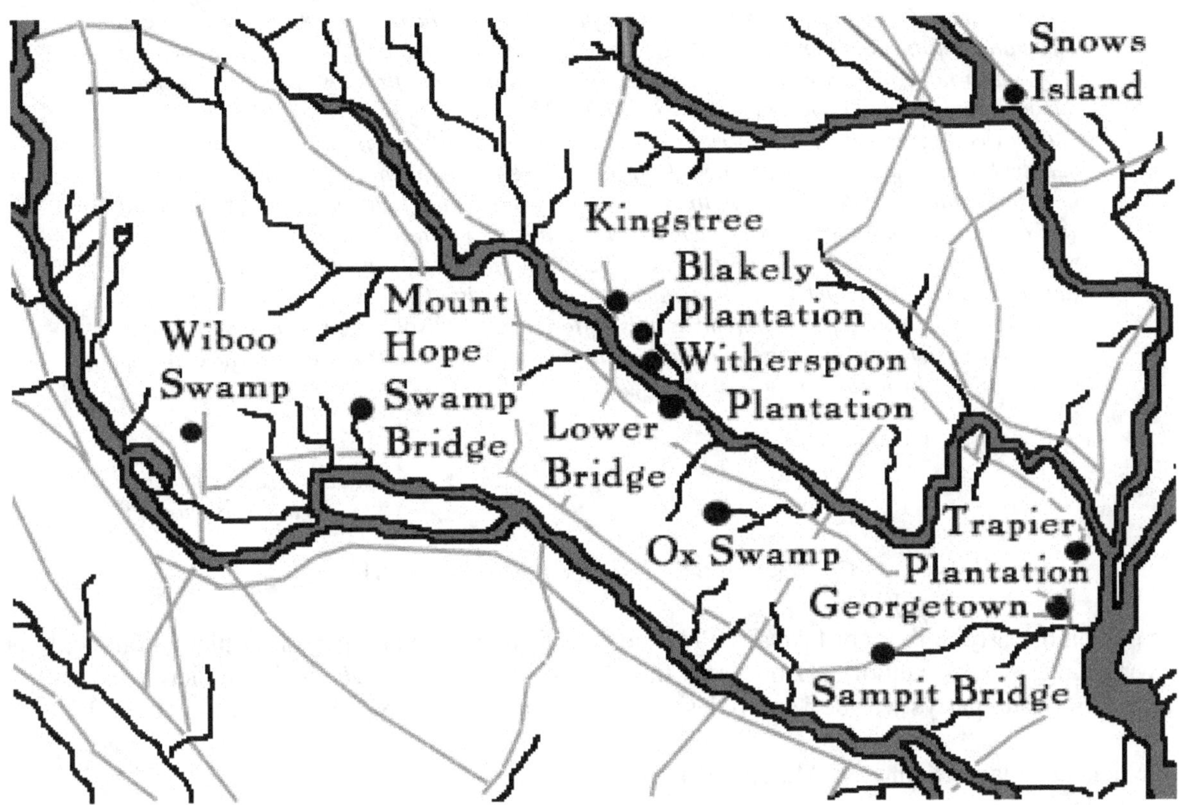

Marion's "Bridges Campaign" against Watson – March 1781

On March 7th Marion learned that his offcers who had been captured were not being treated humanely, and some of his men had been executed. He sent a letter to Colonel Balfour in Charleston:

1416 John Porter had enlisted in the 4th South Carolina Artillery Regiment on 24 June 1777 as a gunner. He reenlisted on 1 January 1780 and was in the siege of Charleston. On 7 March 1781 he was sentenced to do a fortnight's duty for allowing himself to be disarmed while on sentry. He was promoted to corporal on 15 December 1781 in Captain Lenud's Troop of Colonel Peter Horry's Cavalry Regiment. He died on 8 January 1815.

1417 Unable to find any additional information on Charles Lye.

1418 Charles Gee served in the partisans of both Sumter and Marion. He was wounded at the Battle of Eutaw Springs in September 1781.

1419 Unable to find any additional information on Mack Nuggins.

Santee, March 7, 1781.
Sir,

I sent Capt. John Postell with a flag to exchange some prisoners, which Capt. Saunders, commandant of Georgetown, had agreed to, but contrary to the law of nations, he has been seized and detained as a prisoner. As I cannot imagine that his conduct will be approved of by you, I hope orders will be immediately given to have my flag discharged, or I must immediately acquaint congress of this violation. The ill consequence of which it is now in your power to prevent. I am sorry to complain of the ill treatment my officers and men meet with from Capt. Saunders; the officers are closely employed in a small place, where they can neither stand or lie at length, nor have they more than half rations. I have treated your officers and men who have fallen into my hands in a different manner. Should these evils not be prevented in future, it will not be in my power to prevent retaliation. Lord Rawdon and Col. Watson have hanged three men of my brigade for supposed crimes, which will make as many of your men in my hands suffer. I hope this will be prevented in future, for it is my wish to act with humanity and tenderness to those unfortunate men, the chances of war may throw in my power.

I have the honour to be

Your obedient servant,
Francis Marion[1420]

Cordes Plantation Santee 8 March – 1781 –

Parole -

No person or party to press or take any Provisions or forage from any person or plantations without a wrighting ~~shxxxx~~ authority from me, those who do will be Deemed plunderers & Suffer Accordingly and partys will be Sent to destroy all such plunderers wherever they may be found –

Micheaus Plant^{n} Santee 9th March 1781 – [1421]

Parole

A Brigade court martial to Set imediatly for trial of John Kain for Plundering Negroes homes & other goods. Evidences to attend. [1422]

[1420] James, William Dobein. *A Sketch of the Life of Brig. Gen. Francis Marion, and A History of his Brigade, From its Rise in June 1780, until Disbanded in December, 1782; With Descriptions of Characters and Scenes not heretofore published*, Gould and Riley, Charleston, S. C., 1821, pp 44 – 45.

[1421] Micheaux crossroads and Plantation was located just south of present day Andrews, where the CSX railroad crosses the Santee.

[1422] John Cain was born on 22 June 1760 in North Carolina. He enlisted in 1776 while living in Union District of South Carolina. He served under General Williamson and Colonel Thomas. In 1779 he was under Captain William Farr and Colonel Thomas and was in the battle at Stono Ferry. After the fall of Charleston he was in Marion's partisans. On 9 March 1781 he was court martialed for "Plundering Negroes homes & other goods." He was sentenced to do duty constantly for three months. He was in the battle of Eutaw Springs. After the war he moved to Indiana. He died on 1 April 1835 at the age of 75.

L^{t}. Col^{o} Peter Horry prisident – Captn Nolun[1423] & Benton L^{t}. Black[1424] & Green[1425] Members
NB Sentence of the above court is that John Kains is Guilty. Court but think his guilty proceed from McDonald who went to get horses and Divided them amongst his fellows. The Horses to be Restored & he is to do three Months Constant Duty[1426]

Orders at Glovers Plant^{n} Peedee March 11th 1781 –

Captn John Gambol is appointed Major In Col^{o} Richardsons Regiment, he is to be Obey^{d} and respected as Such –

Lt James McCaully is appointed Captn in the room of Captn Gambol promoted, he is to be Obey^{d} and Respected as Such – [1427]

M^{r}. Samuel Bacot is appointed Ensign in Captn Windham Comp^{y}, he is to be Obey^{d} and respected as Such –[1428]

After skirmishing with Marion at Wiboo Swamp Colonel Watson moved to Cantey's Plantation. He let his men rest for a day and then he moved eastward, when he ran into Marion's partisans at Mount Hope Swamp.[1429] Marion had broken down the bridge at the swamp, and left Hugh Horry and McCottry's Riflemen there to prevent any crossing. Watson brought up his artillery and swept Horry's men from the opposite bank with grapeshot.

After crossing the swamp, Watson continued down the Santee road for a few miles, then turned north towards Black River, to the town of Kings Tree. After delaying Watson at the Mount Hope Swamp bridge, Marion sent Major James with 70 men to hold the Lower Bridge of the Black River.[1430] Thirty of these men were McCottry's riflemen. James and his men removed the planks from the middle span, and set fire to the stringers on the eastern end. The western bank was higher than the eastern bank, and the river was fifty yards wide. Watson attempted to sweep the defenders away with artillery like he did at Mount Hope Swamp, but the guns couldn't depress their barrels enough to fire on the east bank. The grapeshot flew harmlessly overhead. When the gunners tried to move the cannon forward to a bluff, McCottry's riflemen picked them off. Watson decided to do a frontal assault, but the attack failed and Watson remained in the field above the ford until evening. He recovered his dead and wounded after dark and moved to Witherspoon's Plantation, a mile above the ford.[1431]

1423 Unable to find any additional information on Captain Nolun.

1424 Unable to find any additional information on Lieutenant Black.

1425 James Green was at the fall of Charleston. He served as a private, sergeant major, coronet and a lieutenant in the cavalry under General Marion during 1780 to 1782.

1426 There were seven partisans that were named McDonald. These were Archibald, Charles, Henry, Francis, James, John, Martin, and William. I am unable to determine which McDonald this one was.

1427 James McCauley served 468 days as a lieutenant and captain under General Marion during 1780-1782.

1428 Samuel Bacot was born on 3 March 1745. He had been captured at Charleston on 12 May 1780, but he escaped shortly afterwards and joined Marion's Brigade. He was promoted to ensign in Captain Windham's company on 11 March 1781. He died in 1797 at the age of 52.

1429 Mount Hope Swamp is located south of present day Greeleyville, on the Santee River.

1430 The Lower Bridge of the Black River is located south of Kingstree, where 377 crosses the Black River.

1431 Witherspoon's Plantation is located south of Kingstree, on 377, and north of Rock Bluff.

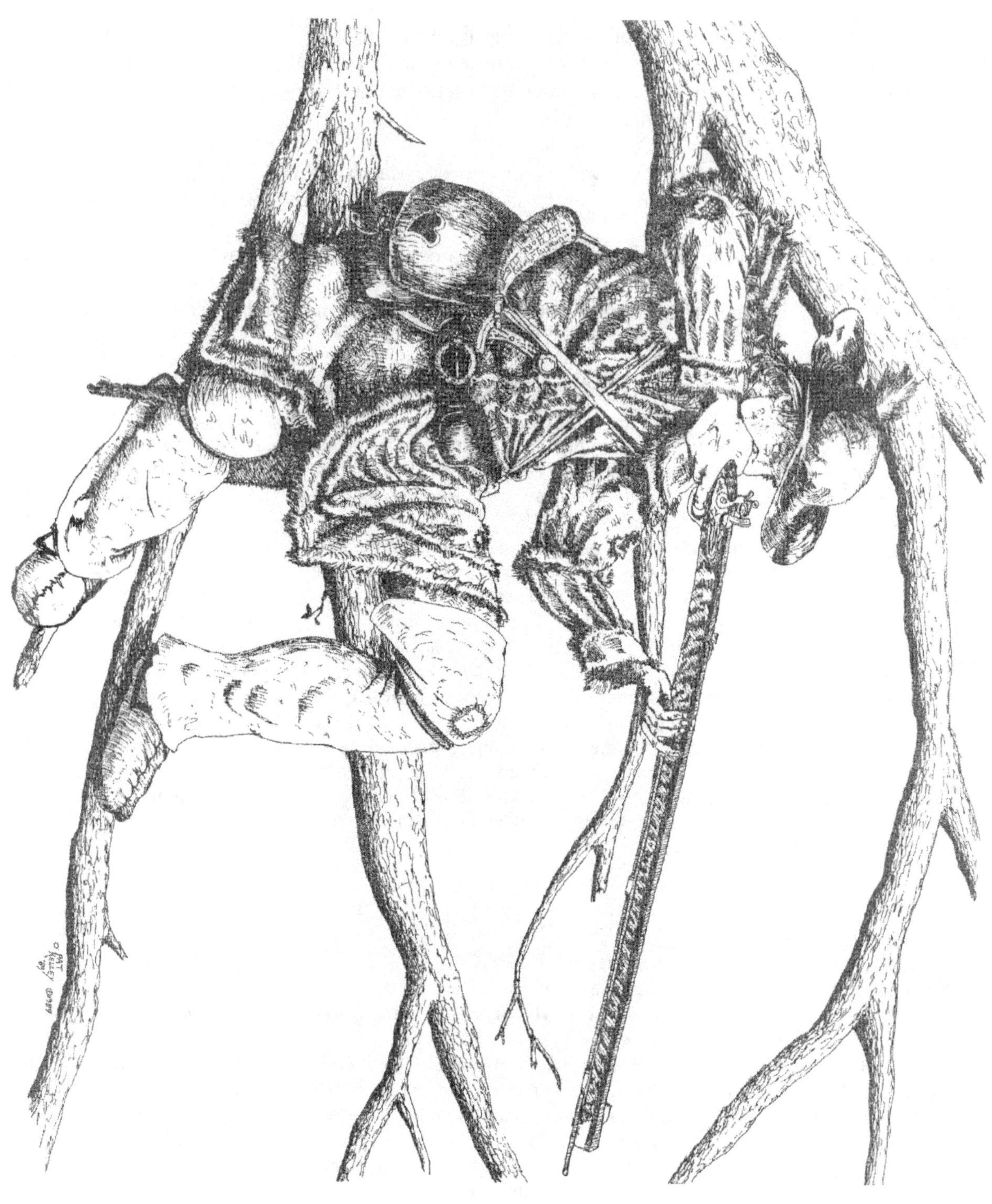

The next day Marion sent his mounted riflemen to keep harassing Watson's men. Watson sent Marion a dispatch that called the partisans "banditti and murderers." He challenged Marion to come and fight him in the open. Marion had fallen for that trick before and would not do it again. Marion's answer was to send Sergeant McDonald up into a tree. McDonald climbed into one of the large oaks that lined the avenue of Witherspoon's house and put a rifle ball into the knee of Lieutenant George Torriano, 300 yards away. Watson tried to cross the river again, but the riflemen would not let his men attempt it.

In the afternoon Watson moved half a mile further up the river and encamped on a large open field on the Blakeley Plantation.[1432] Marion's snipers continued harassing the British, keeping them in a panic and wounding eleven more soldiers. Watson sent a slave from Chavis Plantation with a message for the commandant of Georgetown asking for relief. Marion's men killed the messenger, and the letter fell into their hands. Watson remained at Blakeley Plantation under close scrutiny by Marion's men until March 27th.[1433]

Orders at Peedee 24th March 1781 –

I have thought proper for the good of the Service to appoint Hezekiah Maham Major Commandant of a Corps of light horse to be raised the S°. of Santee River, but such young men who May Incline to Enter this corps that are Inhabitants of the North of Santee Must first obtain permission from the Commanding officer of the regiment they belong too, Otherwise they cannot Enter in S^d^ Corps and for the Incoming of Such who may Enter in Either of the corps of Light Horse All horses are goods taken from the Enemy in Actual Arms Shall be the property of Such Corps or party, which may taken them, provided that such horses & goods do not belong to any Inhabitant friends to America

Cornwallis had pursued Greene into North Carolina, but was not able to catch him before he crossed the Dan River into Virginia. Now the British commander was too far from his supply lines in South Carolina, and he could not get the Loyalists in North Carolina to rise up and join his army. In the middle of March, Greene returned and deployed his army at Guilford Courthouse. Cornwallis knew he had to eliminate this threat, even though he was outnumbered two to one. Greene had over 4,000 Continentals and militia, while Cornwallis had less than 2,000. Cornwallis attacked anyway. Greene had set up his army in the same manner that Morgan did at Cowpens, with a series of defensive lines that the British would have to slug their way through. The Battle of Guilford Courthouse was one of the bloodiest of the war, and the British had to fight over four hours, but in the end they were victorious and Greene had to withdraw from the field. Though it was a victory, it was a hollow one. Cornwallis had set out in January with 3,300 men. He was now heading to Wilmington with 1,400, and these men were no longer fit to campaign. He had to march his army to Wilmington on the coast for safety.

Marion's partisans had laid siege to Watson's Loyalists at Blakeley Plantation since the 15th of March, but by the 28th Colonel Watson had enough. He sank his dead in an abandoned rock quarry to conceal their loss, loaded his wounded on wagons, and marched his troops off at the double time on the road to Georgetown.

1432 Blakeley Plantation is located ½ mile north of Witherspoon's Plantation.

1433 Full details of all of Marion's partisan fighting are described in "*Nothing but Blood and Slaughter, Volume Three*" by Patrick O'Kelley.

At Ox swamp he found the bridge destroyed and the causeway was blocked by trees dropped by Marion's men. Marion was closing fast onto his rear, so Watson sent his troops racing across 15 miles of forest and swamps to the Santee Road. Marion's men followed close behind them, firing on them from every vantage point.[1434]

Marion sent Horry's Continental horsemen to the bridge leading to Georgetown in order to throw the planks from the bridge.[1435] Horry posted Lieutenant Scott and his riflemen at the river to shoot any enemy that came within range. The Loyalists wanted to end the nightmarish pursuit and never slowed down. Upon reaching the Sampit River they formed column and plunged into the river, crossing as quickly as they could. Scott's men saw the approaching glittering bayonets, became frightened and did not fire. Marion fell upon the rear guard with a fury as they crossed the river. Watson rallied his men, but a rifleman killed his horse. He quickly mounted another horse and ordered the cannon to fire grapeshot into Horry's horsemen, driving Marion's men back. Twenty of the British were killed, and almost twice that number was wounded. Watson loaded his wounded into two wagons, and left his dead where they lay. That evening the British camped at the Trapier Plantation.[1436]

In the beginning of March Colonel Balfour sent Cornet Meritt of the Queen's Rangers with a flag of truce to find Francis Marion. Meritt was captured by Marion's Partisans and held prisoner in retaliation for the detention of Captain John Postelle. Though Marion threatened to hang any prisoner in retaliation for his men being executed, he did not harm Meritt, but put him in a jail on Snow's Island known as the "Bull Pen". Meritt was able to escape with some other prisoners and he led his men fifty miles through the swamps until he reached Georgetown. Upon his return he told Colonel Balfour about the location of Marion's hidden base.

While Marion was attacking Watson, a second British army under the command of Lieutenant Colonel Doyle marched on Marion's clandestine base at Snow's Island. When Marion learned that his base was in danger of being attacked by Doyle, he abandoned the pursuit of Watson.

Doyle burned all the structures on the island, but he knew that if Marion appeared he would be in a trap. He immediately recrossed Clark's Creek and encamped on the North Side of Witherspoon's Ferry.[1437] Marion learned of this and followed after Doyle.

Gen[l]. Orders at Burkes mill 1st Apr[l] 1781 –

All Such person who are Draughtet to Serve in my Brigade and Refuse to comply, the Commanding officer of Company will Give in their names to the Commanding officer of the Brigade and such persons May be assured that their names will be published as Enemy to the State after which their persons & property will be Seizt and authority Given to the friends of America to apply their property to their own life

1434 Ox Swamp is located on the south side of the Black River, east of Kingstree, north of Route 521.

1435 The Sampit Bridge was located where the modern day US 17 Alternate crosses the Sampit River.

1436 Trapier Plantation was located 1 ½ miles northeast of Georgetown.

1437 Witherspoon's Ferry was located just north of Johnsonville, where highway 41 crosses the Lynch's River.

Majr Lamb Benton is Appointed Lt Colo[1438] And Captn Maurice Murphey is Appointed Majr in Colo Kolb regt they are to be Obeyd & Respected as Such – [1439]

Mr Garnish Bachler is appointed Captn in troop of Light Horse in Captn Kolb [1440] regt he is to be respected & Obeyd as such – [1441]

Mr. William Lyins is appointed Lieutn [1442]& Maurice Murphy Confidential, they are to be Obeyd as such –[1443]

Marion pursued Doyle's 300 men to Witherspoon's, where they caught the panicked rear guard scuttling the ferryboat. McCottry's sharpshooters fired into the group, killing and wounding several of the Loyalists. Doyle quickly formed his men along the bank and delivered a terrific volley. Their shots cut twigs and leaves loose, but did not do any damage to Marion's men. The British struck their tents, mounted their horses and headed toward the Pee Dee.

That night Marion's men were demoralized because since they had become partisans many of them had been killed and wounded. The homes of almost a hundred of them had been burned. The Loyalist Micajah Ganey had recovered from his wounds and vowed revenge upon all the partisans, and now the military supplies on Snow's Island were gone. Marion called his men together to talk to them and try to raise their morale. When he finished his words of encouragement he had them cheering to be in his brigade. The unified partisans renewed the pursuit of Doyle's Loyalists.

Cornwallis had left Wilmington and began his march to Yorktown, Virginia. Greene did not pursue, but instead turned his attention to South Carolina. Watson learned of Greene's army entering into South Carolina, and he knew that Camden would be the primary target for Greene. He burned his baggage, wheeled his artillery into the swamp at Catfish Creek, and marched with all speed to Camden.[1444]

Marion held a council of war on whether the partisans should pursue Watson or retreat back into the swamps. His men were for retreating, but a messenger arrived with the news that Lieutenant Colonel Lee was approaching with his Legion and supplies. Lee had been ordered by Greene to rendezvous with Marion and cooperate with him to take Fort Watson.[1445]

The fort was built upon an old Indian mound, which was about forty to fifty feet high. The fort was named after Colonel Watson. On April 14th the Marion and Lee rendezvoused and the next day they began the siege of Fort Watson by cutting off British access to Scott Lake. Marion placed McCottry's riflemen to overwatch the water supply at the lake.

1438 Lemuel Benton.

1439 Maurice (or Morris) Murphy was a captain under Colonel Hicks and Colonel Powell during 1776. During 1780-1781 he served as a major under General Marion. He was promoted to lieutenant colonel under Colonel Benton on 28 April 1781.

1440 This is actually Colonel Kolb.

1441 Captain Garner Bachelor.

1442 William Lyon (Lynes) had been a corporal in the 2nd South Carolina Regiment from 1 August 1779 under Captain Charles Motte. He was promoted to lieutenant in Marion's partisans in March 1781.

1443 This is a title for a confidential secretary and may also have been an Aide de Camp. A military secretary was considered a gentleman as opposed to a clerk who would not have the same social standing.

1444 Catfish Creek is located near Pamplico, on the east side of the Pee Dee River.

1445 Fort Watson was located where Interstate 95 crosses the north side of Lake Marion and is a State Park today. The mound can still be seen and visitors can climb to the top by a wooden walkway.

Gen[l] Order at Scotts Lake 19th April 1781 –
Roll Call to be at 8 Oclock in the Morning and 11 in the Evening –
Any Person who do not attend roll call to be confined which Shall Suffer for disobedience of Orders.
Field Officers to use these orders attended to –
A morning report of Each company to be given in every morning to the Commanding Officer of the Reg[t] as soon as the roll Calls any Men who shall leave camp without Leave except for Forrage must be confin[d] and shall be severely punished –

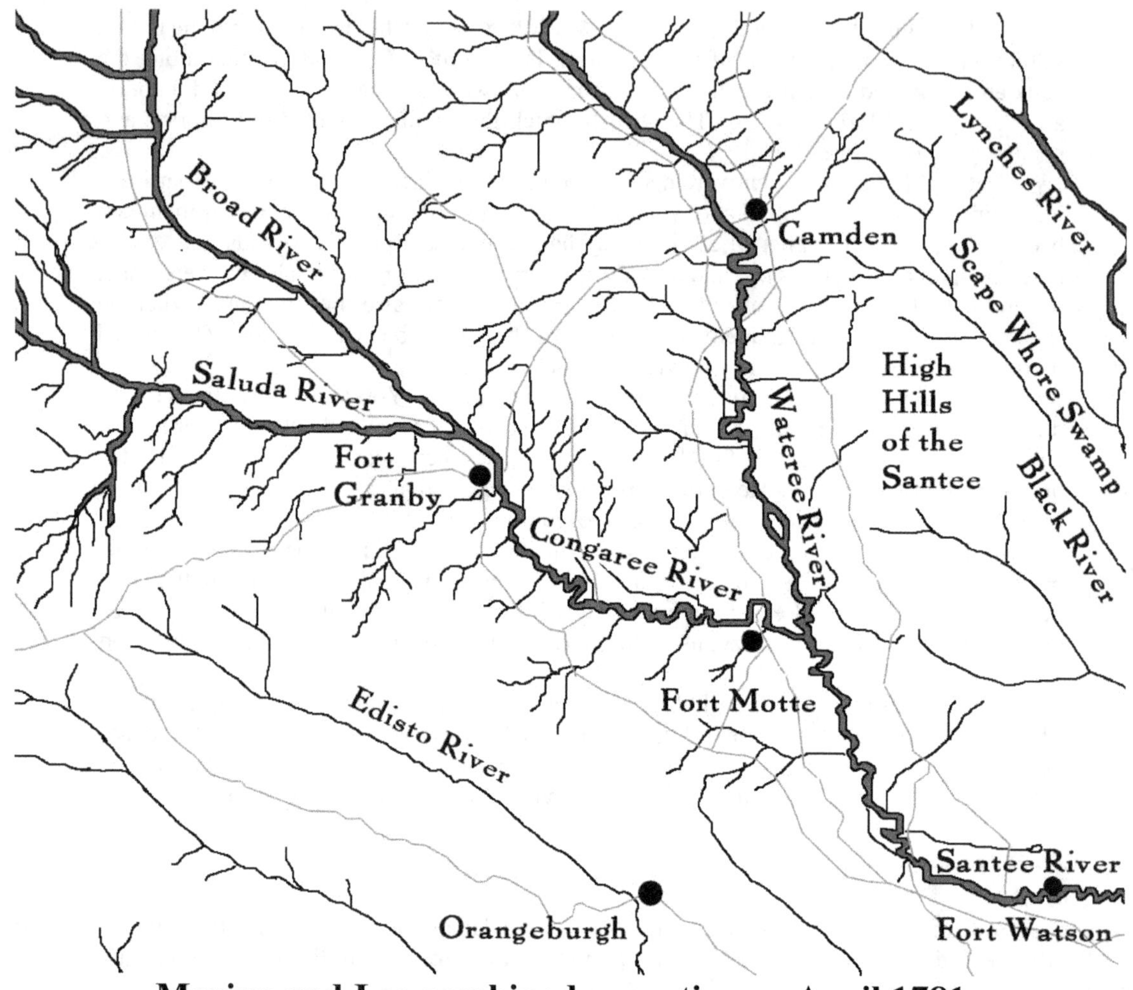

Marion and Lee combined operations – April 1781

Lieutenant James M[c]Kay, commandant of Fort Watson, was not worried by the siege. He had plenty of food and ammunition and the morale of his men was high. The steep sides of the fort and the three rings of abatis would not let Marion assault the post. All of the trees within rifle range of the fort had been cut down to remove any cover for an attacking force. McKay had his men dig a well inside the stockade for water, and then he dug a trench to the lake to fill the well.

The Patriot forces did not have the same confidence and had problems throughout the siege. There was no artillery, so there was no way to take the fort. During the siege smallpox struck Marion's camp, which demoralized the men. The militiamen not infected with smallpox began to desert. There arose bickering among his officers and men and morale plunged even further.

Scotts Lake 22nd April 1781 –

Commanding Officer of the Regt to Enther in a book all the mens names Who have Joined their agreeable to proclamation of the 10 Inst – And those who is yet obstinate and will not Serve their Country

Of their names to the Adjutant Genl Regt they may be Published & meet with such punishment as they Deserve

Any man who quit the camp without Leave & return home Will be Deemed a Deserter & his name will be published as an Enemy to the united States and Suffer as Such –

No person to go out of Camp in the Morning to forrage Until after roll call –

Major Hezekiah Maham suggested a way to take the fort without cannon. For five days they felled and notched trees out of sight of the fort. On the night of April 22nd, Maham and a squad of volunteers built a 40-foot tall, oblong tower, which was higher than the enemy's rampart. The front of the tower was reinforced with a shield of timber. At daylight on the 23rd McCottry's riflemen climbed into the crow's-nest and began firing into the fort through small loopholes in the floor of the "Maham" tower.

Lieutenant McKay and his men crawled around inside the fort, attempting to stay out of sight of the sharpshooters. With the garrison forced to stay under cover the Patriots were able to send volunteers to rush forward and begin tearing down the abatis. When McKay saw the men of the assault party ready to charge into the fort he raised the white flag. The terms of surrender were very generous. The officers were granted paroles, kept their swords, and were able to take their baggage with them to Charlestown, where they were to await regular exchange.

Scotts Lake 24th Aprl 1781 –

Genl Marion returns his thanks to Colo Lees ~~the~~ Legion and the militia for their Spirited behaviour in the Reduction of fort watson, and in particular to Major Maham for his unwarried Diligence in the Executing of a work which was the Principal occasion of the Reduction of the Fort and to Ensign

Johnson [1446] & Mr. Lee [1447] who commanded the party which made a Lodgement Near the Stockade to Lieuts McDonald [1448] & Lt Coutirice [1449] Who assisted Such Bravery can only be Equalled

After Fort Watson was destroyed Marion and Lee marched to the High Hills of the Santee, an area where Marion often rested his men. The combined army of Lee and Marion's forces also went to the High Hills of the Santee to be closer to Greene's army, and support him in the taking of Camden.

Marion sent a reconnaissance patrol towards Camden, under the command of Lieutenant McDonald. This was the famous Sergeant McDonald, who had recently been commissioned as an officer. The British had placed a company around a mill at Camden to guard it from being destroyed. The mill was used by the British to grind meal. The partisans approached the mill at night with the intention of setting fire to the building. They were ordered to not fire any weapons so they wouldn't surprise the garrison, but one of the men fired on the British sentry anyway, killing him. The British reacted by coming out of the house "swarming like bees; and alarmed the horse in Camden, whose feet roared like thunder, as they came to their relief." The partisans had to flee or be captured.

Outside of Camden the partisans came across a group of Loyalists dancing beside a house, under the command of Major William Downes. Downes had been a former officer of the Royal Irish Artillery who settled in Camden as a blacksmith, and now held a commission in the Camden District Loyalist Militia.

The partisans ordered the Loyalists to surrender. Downes did not know the number of the partisans in the darkness, so they surrendered. Once Downes saw that there were only a few of the partisans he commanded his men to fire. The Loyalist fired, then ran back into Downes' house and bolted the doors.

One of the partisans was gunned down in the yard, while the rest rushed forward. Angered by the violation of a surrender agreement they killed every man in the yard. Major Downes and his overseer, James Matthews, defended the house by firing out the windows while his wife and children loaded the weapons. After killing several of the attackers Downes decided that the threat to his family was too much, so he surrendered.

When Downes and Matthews came out of the house they were fired upon. Downes was hit by nine bullets and was killed in front of his wife and daughter. His daughter was wounded and crippled for the rest of her life. When the partisans heard the approach of the cavalry from Camden they escaped into the night.

Marion wrote of another incident at this time, possibly by the patrol led by McDonald, "A small detachment which I sent to watch the enemy's movements in Camden took at the mouth of Keneshaw's Creek, a boat laden with corn, killed 2, wounded 4, and took 6 British soldiers and 1 tory."

Marion had been apprehensive about the possibility of Colonel Watson coming to Fort Watson's rescue, but Watson was still recovering from the weeks-long battle with Marion and decided to avoid crossing his path again. Watson had marched to the Monck's Corner

---

1446 Ensign Baker Johnson served under General Marion.

1447 Lee was a volunteer in Lee's Legion. I am unable to determine his first name.

1448 This is the famous Sergeant McDonald, who was one of the Maryland Continentals that Marion rescued at Nelson's Ferry. There is not much known about him, to include his first name. He was promoted to lieutenant in 1781.

1449 This may be John Couterier, who was a captain in Colonel Richardson's militia in 1775, and then a lieutenant in the dragoons.

garrison, retrieved the troops there, and then continued on to Camden. His circuitous route took him one hundred miles out of his way and detained him until the 9th of May.

Greene had marched from North Carolina to Camden and then placed himself in a position near the town. After the fall of Fort Watson Lord Francis Rawdon, the British commander in Camden, knew that Lee and Marion could join with Greene, strengthening his army. Rawdon decided to strike at Greene, even though he was outnumbered almost two to one.

On April 25th Rawdon armed everyone in the garrison who could march, and then attacked Greene on Hobkirk's Hill. The battle only lasted fifteen minutes, but the British lost 258 men killed, wounded and missing. The Patriots lost almost the same number, 270 killed, wounded and missing. Greene had to retreat from the field, but Lord Rawdon also had enough. He gathered up his wounded and left Major John Coffin and his New York Volunteer cavalry to hold the field.

Losing did not stop Greene. He knew that as long as he was able to whittle away the British army he would eventually be successful. In a letter to Chevalier del la Luzerne, the French ambassador to the United States, he wrote "We fight, get beat, rise and fight again; the whole country is one continued scene of blood and slaughter." [1450]

Genl Orders Camp High Hill 17 Aprl 1781 – [1451]

A Genl Court Martial to Set immediately for the trial of Such prisoners as Shall be Ordered with evidence to attend. Lt Colo Peter Horry of the Light horse Lieut Colo Lee will appoint one of his officer to act as Judge Advocate and furnish the Members –

Micajah Ganey, former 2nd South Carolina officer and enemy of Francis Marion, had reassembled his Loyalists militia on Drowning Creek to return his counter-guerilla operations against Marion.[1452] Colonel Abel Kolb learned of their rendezvous and surprised them on April 26th. Ganey's men fled into the countryside.

Captain Joseph Jones of Ganey's militia gathered his men and rode to Kolb's home at Welsh Neck.[1453] While attacking Kolb's home, Jones ordered the house to be burned. In order to save his wife and children Kolb decided to surrender. As he stepped through the door with his wife, Sarah, a mulatto Loyalist named Mike Goings shot him from behind. Thomas Evans and his brother tried to run away, but the Loyalists shot them too. Jones' tories proceeded to plunder the Kolb home, seizing everything of value, and then set fire to the house and burnt it down anyway. Sarah Kolb kept the dress she wore that day for many years. The dress had holes in it from where the shot passed through it. Colonel Kolb's death outraged many in the area, driving even more volunteers into Marion's partisans.[1454]

[1450] Full details of the Battle of Hobkirk's Hill are described in "*Nothing but Blood and Slaughter, Volume Three*" by Patrick O'Kelley.
[1451] The High Hills of the Santee are located at the present day town of Sumter.
[1452] Drowning Creek is the present Day Lumber River, near Nichols.
[1453] Welsh Neck is located near present day Society Hill.
[1454] O'Kelley, *NBBAS, Volume Three*, pp 213 – 214.

Gen[l] Orders 28th April 1781 –

No person whatever to take away forage or grain from any Inhabitants of this State without properly Issued by the commissary as grain forage or other provision to be taken from Lieut Armstrong Without orders from the Commandant –

2/ any Horses as put in the wheat or other field Such Horses is to be taken and the person who put them in to be brought before the Commandant and those who may commit Companys as Ordered to receive all the public horses from their men & have them (unable to read word) & Such men who refuse to conceal them they are to take prisoners & Send them to Camp to be tried –

Marion and Lee moved to capture Fort Motte, the next major post in the line of defenses in South Carolina.[1455] Fort Motte was erected around the mansion of Mrs. Rebecca Motte on Mount Joseph Plantation. Since only a siege or cannon could reduce it, the fort became the principal depot for the convoys moving supplies up from Charlestown.

The garrison of Fort Motte was a company of the 84th Regiment under the command of Lieutenant Donald McPherson. When Marion and Lee arrived they began their siege of the fort on May 8th. Since Lee had more men than Marion, he gave him the honor of reducing the fort. Lee emplaced a 6-pounder that General Greene had sent them, so that it would rake the northern face of the enemy's defensive works. His men then dug a trench towards the fort 400 yards away.

On May 10th Lee summoned McPherson and asked if he would surrender. McPherson declined and continued to defend. He was hoping that a relief column from Camden would come to his aid. What he did not know was that after the Battle of Hobkirk Hill Rawdon did not have the men to hold the garrison. Rawdon had released the prisoners, fired the jail, burned the mills and many of the townspeople's private dwellings. On May 9th, Rawdon set fire to all the remaining supplies leaving the town a little better than a heap of ruins. He collected his sick, except those too ill to travel, and set his army retreating down the road to Charlestown. The next day Greene moved into Camden and had Sumter destroy what was left of the British redoubts while he continued the pursuit of the British.

At Fort Motte, Marion was disappointed in his militia, especially when they were compared to the well-outfitted and disciplined Continentals of Lee's Legion. He felt Greene had slighted him on matters of logistics for his men and he threatened to resign. This was because Greene had thought that Marion was not sending him the horses he had requested after Marion's attacks on the British. Lee had told Greene that Marion had plenty of horses. What neither Greene nor Lee considered was that Marion needed these horses for his men to be as effective as they were. Greene quickly apologized for any slight towards Marion, because he did not want to lose his most effective partisan. When Marion's men heard that Greene wanted their horses, many of them deserted, leaving Marion with less than two hundred men after the siege.

On May 11th, the retreating army of Lord Rawdon could be seen in the distance by the fort's defenders. Marion knew that Rawdon would be able to reach Motte's within 48 hours, so he decided upon a desperate strategy. Marion sent Lee to ask Rebecca Motte if she would let them burn her home. Mrs. Motte readily agreed. Lee gave McPherson one more chance to surrender, but seeing Rawdon's campfires on the other side of the river the

[1455] Fort Motte was located near the present day Fort Motte, just south of the Congaree Swamp National Monument.

British commander declined. Waiting until noon when the roof had become hot and dry, Lee ordered the house to be set on fire.

Weems, the writer who created many of the myths about the Revolutionary War, wrote that Mrs. Motte lent Lee and Marion a bow and "African arrows. Another account states that muskets fired the "African arrows" onto the roof. However, William Dobein James was there, and he wrote, "The house was not burnt, as is stated by historians, nor was it fired by an arrow from an African bow, as sung by the poet. -- Nathan Savage, a private in Marion's brigade, made up a ball of rosin and brimstone, to which he set fire, slung it on the roof of the house."[1456]

As the roof caught fire McPherson sent a detail aloft to rip off the burning shingles. Captain Finley fired upon them with grapeshot from the 6-pounder. When his men began jumping from the burning house McPherson raised the white flag. Marion lost two men in this action, Lieutenant Cruger and the famous Sergeant McDonald, who had been commissioned a lieutenant before he fell.

As soon as the British laid down their arms Marion sent everyone to the house to help put the fire out. Marion and Lee offered McPherson honorable terms, but friction between the Continentals and Marion's militia had become so strong that when the British marched out Lee accepted the surrender of the Regulars, and General Marion accepted the surrender of the Loyalist militia.

Mrs. Motte invited both the Patriot and the British officers to dine with her that night. The dinner was marred when one of Lee's officers, Cornet William Butler Harrison, had ordered three Loyalists to be hanged. Marion was seated at the table when McPherson received a message of the hanging. Marion leapt up from the table and stormed out of the house, arriving to find two dead Loyalists on the ground and one swinging from a noose. Marion ordered that the man be cut down. He told all of Lee's men that he was in charge and that he would kill the next man who harmed any prisoners.

A short time after Fort Motte fell Greene arrived and met Marion for the first time. He ordered Marion to move against Georgetown on the coast, and Lee was sent up the Congaree to reduce Fort Granby. Marion did move towards the coast, but he did it on the heels of Rawdon's retreating army.[1457]

S^{t}. Stephens 19th May 1781 – [1458]

Sir

I find the British has ordered all their Prisoners to go in to them agreeable to the paroles we have Ordered ~~the~~ at that Period which obliges us to Know if Such Men will obey the Summons, if not their Parole is Broken – Consequently they are liable to be Called on duty by the Americans as I am determined Every man who is a Prisoner to the British Shall go Into there and Remain until Exchanged, as they shall not Remain neuter in the State –

I may inquire your determination on the Point I wish to be as delicate as possible in Every respect But hope you will see the Necessity of the Measure

Copy of letter to Col^{o} McDonald, Major James, $^{Sen.}$ Francis Marion

[1456] Nathan Savage served under Marion from 10 July 1779 to 1 December 1782. He served as a commissary from 7 December 1778 to 28 February 1779. He served under Hugh Giles in Marion's partisans.

[1457] Full details of the Siege of Fort Motte are described in "*Nothing but Blood and Slaughter, Volume Three*" by Patrick O'Kelley.

[1458] St. Stephens Parish is the present day Francis Marion National Forest.

Patrick O'Kelley

After Camden fell General Greene marched on to Fort Ninety Six and began laying siege to the British post on May 22nd. Marion marched to Cantey's Plantation and sent out a call for his militia. He then headed for Georgetown on May 27th.

The commandant of Georgetown had been Captain Robert Gray, and he had been told by Lieutenant Colonel Balfour to evacuate his post if he should become "so press'd by the enemy as to make a retreat necessary." Marion began to lay siege by digging trenches, but the British boarded their vessels at nine o'clock that night and left the town. The British had spiked their three 9-pounders and a carronade, then knocked them off their trunnions.

Marion's men entered the town on May 28th and leveled the British works as the British ships waited outside the bar to Winyah Harbor. Marion was also able to personally replenished his wardrobe and fit himself out in a new suit of regimentals.[1459]

Before George Town 28th May 1781 –

All the guards and piquetts to keep their Arms so near as to be able to take them at an instant and without Confusion Night and Day. No Officer to leave his guard on Any pretence. Should an alarm take place all guards & Piquits must be ready to make the utmost opportunity and Not to quit their posts but on the greatest Necessity or by Orders from the Commandant or Field Officer of the day. Are to visit all Centrys ~~visited~~ every hour During the night by either the Officer of the night or patrol from their own Guard –

The first alarm the time is to turn out when it is Expected they will make the almost opposition against every attack, that may be made their Arms must be put in such places as they may take them with the greatest ease without Confusion every Officer to remain with his men during the night And dispose their men in such a Manner as to have them at Command in a Moment. Every Officer is to be answerable for ~~his~~ the Good Order of their men and the greatest attention must be paid to Every part of duty and it is hoped both Officers and Men will Cheerfully do their Duty and prevent the ill consequences which Must attend carelessness or Neglect –

No Man to be absent from Camp at any time on any Pretence whatsoever without order from the Commandant or Field Officer of the Day –

All horses must be piqueted in the rear of the Incampment at Night –

Forage will be Brought to Camp and regular Issued –

Capt Gough [1460] is appointed Confidential in room of Capt milton [1461] he is to be respected and Obeyd as Such –

Parole Washington: wine
watchful } Countersigns

G.O. by Brigd Genl Marion Camp near George Town 1781 –

Colo Peter Horry will take the Command of Geo Town And keep good order in it. He will give such orders as may be Necessary for that purpose –

[1459] O'Kelley, *NBBAS, Volume Three*, pp 267 – 268.

[1460] Richard Gough had served as a captain of the regiment of light dragoons raised by the Assembly in 1779 and was a captain in the militia under Colonel Peter Horry and General Marion during 1780 – 1782.

[1461] John Melton served under General Marion from 1780 to 1782.

He will see all the works of the Enemy demolished as soon as Possible –
Parole –
Countersigns –

Marion was summoned by Greene to help with the siege at Fort Ninety-Six, so he left a small force in the town under the command of Lieutenant Colonel Peter Horry and marched away with the captured baggage on the back of mules.

After the capture of Georgetown one of Marion's main opponents, Micajah Ganey, asked for a truce with Marion that would last for a year. During that time Ganey agreed that he would not attack Marion or his forces. Marion agreed to the truce. Marion's partisans decided that since the job in Georgetown was finished they could return to their homes. Frustrated, Marion went about gathering a new militia to harass Lord Rawdon.

G.O. George town 31st May 1781 –
I have reasons to believe there is a ~~amount~~ quantity of Arms and ammunition lodged in this Town, the property of the Enemy. You will please have a search made for them, if found they must be taken for the use of our Service –
Also a quantity of Spirituous Liquors, Dry goods & Salt Belonging to Persons gone with the Enemy which must also be taken, a particular account must be kept and Deposited in Some place untill ordered, otherwise all goods found in the possession of any Resident now in town Must prove by Oath their property, and nothing taken But from those Person mentioned above. The Flats and Boats must be ~~taken~~ Collected in one place under a Guard And no Boats allowed to cross the River without your Pass –

Colo Peter Horry –

G.O. 5 June George town 1781 –
Joseph } Parole
Job
James } Countersigns –

NB The following promotions take Place the 28 April 1781. Lt Colo Lamb Benton to be Colo Kolb, killed. Majr Maurice Murphey to be Lt Colo vice Benton. Capt T Thos [1462] to be Majr Murphey, they are to be respected and Obeyd –

On June 11th Greene learned that a relief column of 2,000 men under the command of Lord Rawdon was on the way from Charleston to Ninety-Six. A lot of Rawdon's men were fresh recruits from Ireland and were not used to the heat of the South. Greene immediately sent word to Marion and Sumter to gather their militia, get in front of Rawdon, and do everything possible to delay his march. He ordered Washington and Andrew

[1462] Tristram Thomas was born in 1752 in Maryland. He had been a first sergeant in the 3rd South Carolina Rangers on 1 July 1775 under Captain Samuel Wise and Colonel Thomson. He was at the fall of Charleston. During 1780 and 1781 he served in the militia as a captain under Colonel Hicks and Colonel Hobbs. He was promoted to major on 28 April 1781 under Colonel Benton and General Marion.

Pickens to support Marion in this mission.[1463] Sumter's partisans did strike at Rawdon's column, but he wasn't able to hit with a large force. Unfortunately Marion was still trying to muster replacement partisans and when he marched off with the few militia that he could gather, he did not find the British column in time.

When Greene learned that Rawdon was 30 miles away from Fort Ninety-Six, he decided to take the Star Fort and Fort Holmes by storm. On June 17th a heavy artillery barrage was aimed at Fort Holmes to soften it up for the coming assault. Greene attacked the fort at two locations. There was a brief, bloody encounter in the ditch around the Star Fort, but the Patriots were driven back to the trenches with heavy losses. On June 19th Greene lifted the siege and withdrew. He stopped his army about twenty miles away, and learned from a prisoner that Rawdon marched into Ninety-Six at two in the afternoon on the 21st.

G.O Nunneys ~~June~~ ferry June 22nd 1781 – [1464]

Lt Colo Marshale [1465] have been put under arrest ~~& order~~ For Disobedience of Orders & Plundering & Order to Camp –
For they all have not thought proper to Comply He is (written over word, possibly "cashiered")
He is therefore no longer to be Obeyd as an Officer in Colo Postells regiment –

Majr Fredrick Kimbull is promoted to Lt in ~~Colo~~ Colo Postells regiment, he is to be Respected & Obeyd as Such – [1466]

Capt Ths Thompson is promoted to ~~take~~ be Major in Room of Majr Kimbull – [1467]

Wilm Barnett is appointed Adjutant in Colo Postells Regt they are to be Respected & Obeyd as Such – [1468]

Many of Rawdon's relief force were troops that had recently arrived from Ireland. As they marched away from Charlestown they were "in exceeding bad order, many of them fainting a little distance from the town." Rawdon had to stop at Orangeburgh to rest his men, after "killing great numbers of them" due to the heat.

1463 Andrew Pickens was born on 19 September 1739. He married Rebecca Calhoun on 19 March 1765. He was the colonel of the Upper Ninety Six Regiment of militia and later was a brigadier general of the militia. Due to his actions in the battle of Cowpens he was awarded a silver sword from Congress. At the battle of Eutaw Springs he was wounded. He died on 17 August 1817 at the age of 78.

1464 This may be Hunnings Plantation, that was located on the Sampit River, just south of Georgetown.

1465 John Marshall served as a captain, major and lieutenant colonel in Marion's partisans under Colonels Kershaw and Postelle. He was cashiered for "disobedience of orders & plundering" in June 1781. Marion knew that to win the support of the people, his men could not loot and harm any of the farms in the area. Marshall being cashiered showed that no one was above Marion's orders, whether it was a private soldier or a high ranking officer.

1466 Frederick Kimball served as a lieutenant, captain and major in Marion's partisans. He was promoted to lieutenant colonel on 22 June 1781 under Colonel Postelle when Lieutenant Colonel Marshall was cashiered.

1467 Thomas Thompson served under General Marion's command. He was promoted to major on 22 June 1781 under Colonel Postelle.

1468 William Barnet, Jr. had served as a private under Captain John Drennan, and a lieutenant under Captain Henry White. He was appointed an adjutant under Colonel Postell in Marion's partisans on 22 June 1781.

After arriving at Ninety Six Rawdon ordered the troops to leave all gear that was not needed, including the knapsacks and blankets of the troops, and then marched out of the town on June 23rd to pursue Greene.

After a 40 mile march Rawdon had caught up to Greene's rear guard, consisting of Lee's Legion and Kirkwood's Delawares, but the British were no longer able to fight. Rawdon's soldiers had marched from Charlestown for 17 days in 100 degree heat, wearing their heavy woolen uniforms. More than fifty had died of heat exhaustion. To make matter worse Greene had dismantled all the mills along his march so that there would be no provisions for the British.

Rawdon returned to Ninety-Six and realized that he could not hold the town. He marched out on June 29th with 800 men and 60 horses. Greene ordered Lee, Kirkwood and a hundred militia under Major Alexander Ross to continue to harass Rawdon's retreat. Cruger remained behind to protect the Loyalists who were gathering their things, in preparation for the retreat from the area. On July 8th, Cruger destroyed the fort, and escorted to Charlestown any Loyalists that wished to remain under British protection. Once again Greene had not won, but the British still had to leave due to their losses.

Greene knew that if he had his partisans with him he could have been able to defeat Rawdon easily. On June 25th, Greene wrote to Marion from Sandy River and ordered him to cooperate with Sumter. Sumter was planning a raid into the lowcountry. Sumter also wrote to Marion and told him to march to Ninety Six, but Marion's men did not want to leave their homes and did not want to fight under Sumter. Marion did march to the Congaree River, but he did not have many men at all. He rode west to rendezvous with Sumter at the High Hills of the Santee.[1469]

G. O By Brigadier Genl. Marion Camp at Ancrums – [1470]
Plantation on the Congarees 28th June 1781 –

No Person to leave Camp without leave from the Commanding Officer. Return wanting to Compleat six rounds pr Man to be made out Immediately and Given to the quarter Master of Brigade, all Horses to be kept near the Camp in the day and Piquetted at Night. all Officers are to pay particular Attention to the above Order in Making their Men Comply.

The Roll to be Called morning and Evening agreeable to former Orders –

G.O By Brigadier Genl Marion Camp at Ancrums
Plantn at the Congarees 26 June 1781 –
All the Horses to be piquetted to Night the whole Brigade to March to Morrow morning at 4 OClock precisly –
The Commissary to gett the Salt & other Baggage ready to Move at the above hour –
Those Men Just come in to make return immediately for Ammunition –

1469 Full details of the Siege of Fort Ninety-Six are described in "*Nothing but Blood and Slaughter, Volume Three*" by Patrick O'Kelley.

1470 Ancrum's Plantation was located in Berkeley County.

G.O. by Brigadier Gen[l] Marion June 28 1781 –
At Farmans Plantation on the watteree River – [1471]
L[t] Col[o] John Ervin is appointed full Col[o] in the Room of Colonel Hugh Giles and is to take Command and to be Obey[d]. As Such. [1472]

Maj[r] Alexander Swinton is appointed L[t] Col[o] in the Room of Col[o] John Ervin promoted he is to take the Rank and to be Obey[d] as Such – [1473]
Capt[n] John Baxter is appointed Maj[r] in the Room of Maj[r] Alexander Swinton promoted, He is to take Rank And be Obey[d] as Such –

G.O. by Brigadier Gen[l]. Marions Camp high hills of Santee June 28th 1781 –

Whereas a Number of men Paroled by the Enemy in the district of my Brigade and it has been thought hurtful to the American service ~~this~~ It is therefore to order all Such Men to Repair to the first Enemys Port in this State and their to Remain till exchanged –

All Officers of Regiment & Companys are hereby ordered to take every paroled man who will not obey the above orders And convey them to me or the first port of the Enemy –

Camp the high Hills of Santee July 3d 1781 –
Parole –
M[r] John Hamilton is appointed Commissary of Hides To the Brigade, he is to considered as Such – [1474]

Ordered that all hides from the cattle killed for the Brigade to be deliver[d] M[r] Hamilton on his Orders –

M[r] Thomas Elliot is appointed Confidential, he is to b respected and Obey[d] as Such – [1475]
Cap[t] Anthony Ashby is appointed Brigade Major – he is to be Respected and Obey[d] as Such – [1476]

Lieutenant Colonel Watson had been broken by his campaign against Marion and he retired as the Commander of the Provincial Light Infantry. Lieutenant Colonel Stewart had taken Watson's regular regiment, the "Buffs" and the Provincial Light Infantry and

---

[1471] Furman's Plantation was south of the High Hills of the Santee, near the Wateree River, eight miles south of Rembert on US 521.

[1472] John Ervin was born on 25 March 1754 in Williamsburg County, South Carolina. He served in Marion's partisans from 1780. He was promoted to colonel on 28 June 1781 and assumed command of Colonel Giles Regiment. He died in 1810 at the age of 56.

[1473] Alexander Swinton was born in 1736. He had served in the volunteer militia under Captain Benjamin Screven in 1775. He was in the Britton Neck Regiment of Militia from 10 January 1780 to 17 January 1782. He was promoted to lieutenant colonel in Marion's partisans on 28 June 1781, under Colonel Ervin. He was wounded at Quinby Bridge on 17 July 1781. He died in 1814 at the age of 78.

[1474] John Hamilton served as a foragemaster in the militia during 1780 – 1782. He was appointed commissary of hides to Marion's brigade on 3 July 1781.

[1475] Thomas Elliott was a lieutenant in the militia and an aide de camp to General Marion from 3 July 1781 to 14 December 1782.

[1476] Anthony Ashby served 300 days in the militia as a major during 1781 and 1782. This was not the same Anthony Ashby who was in the 2nd South Carolina Regiment from 1775-1778. That Anthony Ashby was taken prisoner at Augusta on 18 September 1780 and executed by Colonel Thomas Brown.

began marching to Orangeburg to reinforce Lord Rawdon. Rawdon could not move through the High Hills of Santee without his force being decimated, so he marched south, towards Dorchester. Greene wanted him stopped before his army could reach Dorchester, so he ordered Sumter and Marion to rendezvous with Greene.

Marion's force had grown to 400 mounted men as he waited at the High Hills when he rode to intercept Stewart. Stewart was a better officer at partisan warfare than Watson had been, and did not move his men on well-traveled roads. The two corps passed by each other in the early hours of the morning. Marion learned of his mistake and sent Colonel Horry to overtake the British. Horry overtook the supply trains, but Stewart got through to Orangeburgh and joined Lord Rawdon. When Stewart's men arrived at Orangeburgh Rawdon's force had been reinforced to 1,500 men. Loyalist Colonel Cruger was also able to slip through the partisans and also join Rawdon with 1,300 more men from Ninety-Six.

On July 10th, Greene with the forces of Sumter, Marion, Washington, and Lee, took a position on the north side of Turkey Hill, four miles above Orangeburg.[1477] They prepared for battle. Greene issued ammunition and sent the women and children off with the baggage wagons. For two days his troops taunted Lord Rawdon, trying to entice him out of the well-defended garrison at Orangeburg, but Rawdon declined to do battle.

On July 12th, Greene and his generals reconnoitered the British lines. The British were defending several buildings within the town, so storming the town would be suicidal. Contented that he had offered battle, but knowing that he could not take the heavily fortified town, Greene retreated across the Santee.

Greene divided up his army by sending all of his partisan leaders to strike at the British line of posts all the way to the gates of Charlestown. With the rest of the army, Greene marched to the High Hills of the Santee to reestablish himself in force. Greene gave his cavalry and mounted militia orders to get behind Rawdon and to strike directly at Charlestown.

Since Snow's Island had been compromised Marion established a new base on the Santee River, in upper St. Stephen's Parish at Peyre's Plantation.[1478] The new camp was in a cleared cane brake, a quarter of a mile from the Santee River, and at the western end of a swamp known as Gaillard's Island.

As Sumter neared the lower country he broke into separate detachments. Lee was sent to the British fort at Dorchester, Wade Hampton was sent towards Charlestown to harass the British posts there, and Marion was sent to engage Colonel Coates at Monck's Corner.[1479] This was known as the "Raid of the Dog Days" since the weather was so hot. Rawdon ordered 500 of his men to pursue Sumter.[1480]

[1477] Turkey Hill Meeting House was located three miles northwest of Orangeburg, on Benjamin Boulevard.

[1478] Peyres Plantation is on Gaillard's Island. Gaillard's Island was located under the waters of Lake Marion, north of Wilson's Landing, where the Little River empties into the lake.

[1479] Wade Hampton was a lieutenant and paymaster in the 1st South Carolina Regiment in 1776. He was promoted to captain in 1777. In 1778 he became the paymaster of the 6th South Carolina Regiment. After the fall of Charleston he declared himself loyal to the King on 21 September 1780, but later renounced his allegiance and joined Sumter's Brigade. He was promoted to colonel in March 1781, and given the command of a brigade under Colonel Sumter. After the war he remained in the service and was promoted to major general in 1813. He married Harriett Flud in 1780, and married Mary Cantey in 1801. He died on 4 February 1835.

[1480] The term "Dog Days" had nothing to do with actual dogs on earth, but was due to the star system Sirius, the "Dog Star", rising in the sky. Greek astrology connected the days with heat, drought, sudden thunderstorms, lethargy, fever, mad dogs, and bad luck.

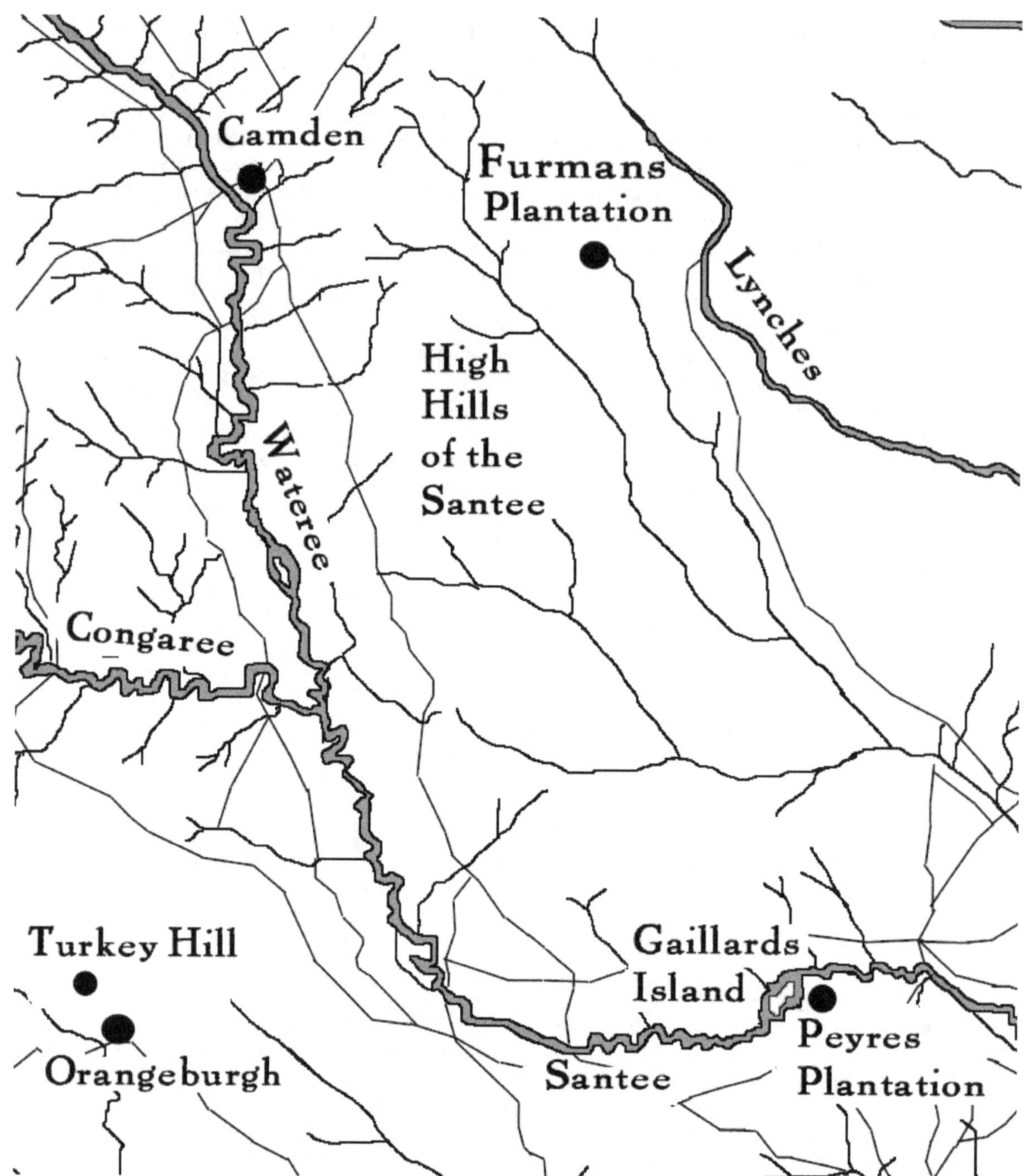

Marion attached to Greene – Summer 1781

Sumter sent the cavalry units down all the roads leading to the low country, while he came behind with the foot regiments. He did this so that all relief routes to Monck's Corner would be cut off. Lee's Legion had been sent to Dorchester, while Henry Hampton seized the bridge across Four Holes Swamp to stop any reinforcements from Orangeburgh.[1481] Colonel Maham rode down the east side of the Cooper River to burn the bridge over Wadboo Creek so that Rawdon would not be able to use it to receive supplies from Charlestown.[1482] Destruction of this bridge would also hinder Coates's movement to Charlestown. William Washington continued to shadow the movements of Rawdon's army and was not part of the fighting that would happen next.

Colonel Coates was commanding the 19th Regiment of Foot at the British outpost at Monck's Corner. The outpost consisted of a redoubt, and Biggin's Church.[1483] The building was a fortified brick church that covered the Wadboo Bridge. Eight companies of the 19th Regiment were stationed at the outpost.

When Lee and Hampton arrived after their raid on Charlestown they were expecting to find Biggin's Bridge occupied by Sumter's men, but were highly annoyed to find it in the possession of Coates's troops. The two cavalry commanders moved their men to Wadboo Bridge and let the horses run loose to feed. The British learned of this and around five o'clock in the afternoon Fraser's dragoons struck Horry's camp as they were eating dinner. Horry's men were caught by surprise, but quickly rallied and counterattacked. Colonel Edward Lacey and his mounted riflemen drove Fraser's dragoons back with minor losses.

The 19th Regiment arrived from Monck's Corner with a fieldpiece that stopped Horry's attack. Horry withdrew to Sumter's main body while Coates placed his men in and around Biggin's church. Sumter thought that Coates had marched out to meet him and on July 16th placed his 550 men into line of battle and waited. However, the skirmish with the South Carolina Royalists had not been an attack, but merely a delaying action, while Coates placed all his stores in the church then put the torch to it. At midnight on July 17th, Coates withdrew from the burning Biggins Church and moved down the Cooper River to Quinby Bridge.[1484] The burning church was a signal to Sumter that Coates was retreating, and he immediately pursued the British column. When he learned of this, Sumter left behind his artillery with Captain Singleton, so that he would not be slowed down.

Lee and Hampton led the pursuit until they came to a fork in the road at Wadboo River. Hampton followed the South Carolina Royalists, which had taken the right hand route. The Royalists had been ordered by Coates to ride to Strawberry Ferry.[1485] Hampton's pursuit was in vain because the Royalists had crossed over the river and secured the boats on the other side.

For 18 miles the British trudged under a hot July sun, constantly harassed by Lee and Maham's cavalry. About a mile north of Quinby Bridge 100 men of the 19th Regiment were overtaken. Lee's trumpeter sounded the charge and the cavalry came on at a gallop with their sabers flashing. The recruits of the 19th Regiment threw down their arms without firing a shot. Nearly all of the baggage train was captured.

[1481] Four Holes Swamp is located near present day Four Holes, south of Holly Hill.

[1482] Wadboo Bridge was located east of Monck's Corner, where route 402 crosses over Wadboo Creek.

[1483] Biggin's Church was also located east of Monck's Corner, on route 402.

[1484] Quinby Bridge is located west of Huger, where Cainhoy Road crosses over Quinby Creek.

[1485] Strawberry Ferry was located on Wambaw creek, in present day Francis Marion National Forest, near Thompsons Corner

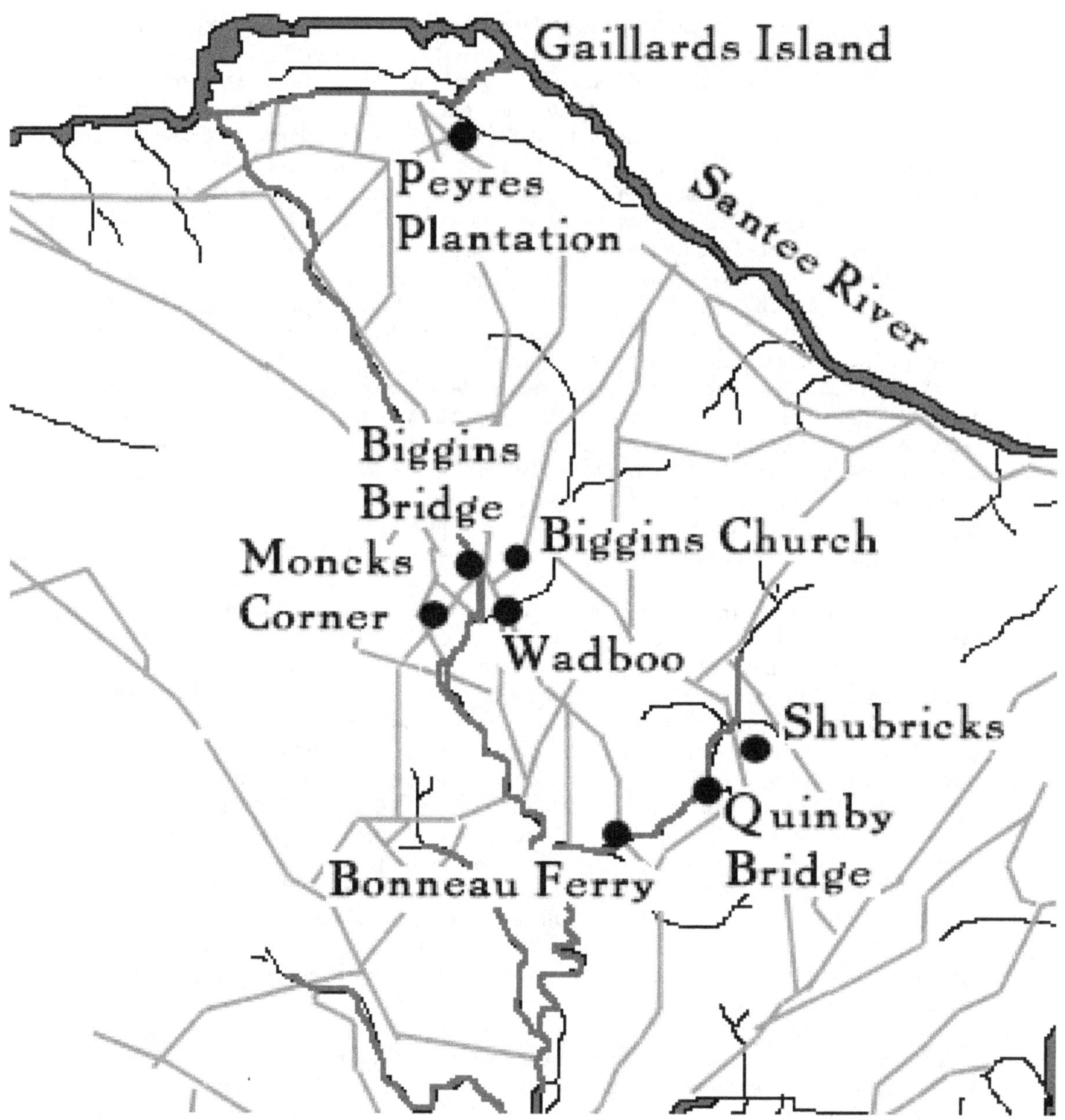

Marion, Lee and Sumter versus Coates – July 1781

When Coates arrived at Quinby Bridge he began loosening the planks to remove them. When Lee's dragoons appeared Coates placed his men into a defensive position and put a howitzer at the end of the bridge. A fatigue party of the 19th Regiment remained on the bridge and continued to remove the planks. Unfortunately the British howitzer was unable to fire due to the fatigue party on the bridge masking their fire.

Maham's dragoons were under the command of Captain Armstrong of Lee's Legion. They charged right through the fatigue party and into the howitzer, driving artillerymen from the gun. As the dragoons charged across the bridge they knocked the loosened planks in the water, leaving the infantry following behind them to cross the bridge on the stringers. The fighting around the howitzer and on the bridge was a fierce life and death struggle. Maham's horse was shot out from under him. Armstrong attacked Coates and some of his officers around a wagon, while the men of the 19th attempted to display into a line of battle. Lee and the rest of his Legion arrived and began repairing the bridge, but they were only armed with swords and were no match for the muskets of the Regulars. The only thing that saved the cavalry was that the 19th were not battle-hardened veterans, but untested recruits, unsure of what to do in the chaotic battle.

Coates decided to move his regiment across the concealment of the cornfields to find shelter in some buildings attached to the Quinby Plantation. These were owned by Captain Thomas Shubrick, who had been captured at Charlestown. At the Shubrick Plantation Coats formed his men into a square, using the buildings as part of his defense. His remaining howitzer was placed in the center. Since they did not have many bayonets Lee and Marion decided not to attack the formidable position and instead waited for Sumter to arrive with his artillery. Sumter did not arrive until 3 o'clock, allowing the British even more time to prepare their defense. He also did not bring along his artillery.

Sumter decided to divide his force into three sections and attack the plantation from different angles. Sumter placed his own brigade in the center, where they had some protection from the plantation's slave buildings. The cavalry were placed in reserve.Marion's Brigade was ordered to advance on the right, across open fields, with no cover except for a fence fifty yards in front of the plantation. Marion protested, but Sumter told him to engage immediately.

Taylor's riflemen only had seven bullets each at the beginning of the fight. Taylor with 45 men rushed up to a fence enclosure on the left of the house. Taylor's men fought until their seven bullets were gone and then they were pushed away by the 19th Regiment with the bayonet. Marion's men rushed in to help Taylor withdraw from the tight situation. During the fight, Lieutenant Bates of the Camden Company of Mounted Militia was hit by five balls and killed. Marion's men suffered heavy casualties due to Sumter's failure to bring along Singleton's artillery to support the attack. As Taylor's men withdrew their colonel was the last to retreat. Captain Baxter of Marion's force was knocked from his horse by a musket ball. Fifty of Marion's men were killed or wounded in the assault to save Taylor's men. Taylor found Sumter sitting under the shade of a tree and he told him that he would never serve with him again.

The battle lasted for three hours and only ended when it was too dark to shoot. The darkness was a welcome relief to Marion's men since they had run out of ammunition. Sumter had the men retreat three miles, and wait for Captain Singleton and the artillery piece. That night all but a hundred of Marion's militia deserted. The next day Marion and Lee left Sumter's command, resolved never to fight under Sumter again.[1486]

Sumter had to withdraw from Shubrick's plantation when Rawdon's column from Orangeburgh landed at Bonneau's Ferry four miles away.[1487] The Patriot's casualties for the battle at Quinby Bridge and Shubrick's Plantation were 30 killed and 30 wounded. The

[1486] Full details of the Battle of Quinby Bridge and Shubrick's Plantation are described in "*Nothing but Blood and Slaughter, Volume Three*" by Patrick O'Kelley.

[1487] Bonneau's Ferry is on the Cooper River, near Cordesville. It is located off of route 402, and at the end of Bonneau's Ferry road.

British casualties were six killed, 38 wounded and 100 captured. The British also lost several wagons, a load of ammunition, and the baggage of the 19th Regiment. In the baggage was a chest containing 720 guineas, which Sumter divided up among his men. As Sumter withdrew he burned two schooners at Wadboo, then he marched him men to Nelson's Ferry.[1488]

A Return of the killed and wounded at Quinbys St. Thos parish 17th July 1781 –

	Lt Collo	Majors	Capt	Lieuts			Privates
	Wounded	Wounded	Wounded	Killed	Wounded	Killed	Wounded
Colo Richardson Regt	..	..	..	..	1		
Colo McDonald Do...	..	..	1	1	..	3	5
Colo Postells Do...	..	..	..	1	..	1	
Colo Ervins Do....	1	1	..	..	1	2	1
Colo Peter Horry Do..	..	..	..	..	1		1
	1	1	1	2	3	6	13

NB
Lt Colo Swinton wounded
Major Baxter ....Do.
Capt Brown [1489] .....Do.
Lt Bates[1490] & Perry[1491] Killed
Lt Fox [1492] .... Wounded
Lt Postell .... Do.
Lt Scott [1493] .... Wounded
Wilm Cooper a private taken[1494]

Orders the 23rd July 1781 – St Stephens Cordes Plantation

It is with great Pleasure that I have this opportunity of Conveying Genl Greens Approbation of the conduct of my Men in the Action of Quinby the 17 Inst. which is as follows.

[1488] Nelson's Ferry is located north of Eutawville, and is under the waters of Lake Sumter. It is located on the borderline of Clarendon and Orangburgh counties.

[1489] John Brown had served as a captain in the 5th South Carolina Regiment. He was wounded at Shubrick's Plantation on 17 July 1781.

[1490] Henry Bates was the commander of the Camden Company of Mounted Militia. He was killed when five musketballs hit him in the battle of Shubrick's Plantation.

[1491] John Perry served as a captain in Marion's Brigade. He was killed in the battle of Shubrick's Plantation.

[1492] Joseph Fox was born in 1745 in South Carolina. He had served as a sergeant in the 3rd South Carolina Rangers under Captain Edward Richardson during 1775. On 7 May 1776 he enlisted in the 6th South Carolina Regiment as a sergeant. He was discharged on 1 June 1777. He served 93 days as a lieutenant in Marion's partisans during 1781 and 1782 and was wounded at Quinby Bridge.

[1493] Joseph James Scott had served 835 days as a lieutenant in the 3rd South Carolina Rangers under Lieutenant Colonel Hugh Horry. He was wounded at Black Mingo Swamp on 14 September 1780 and a second time at Shubrick's Plantation on 17 July 1781.

[1494] William Cooper was wounded and captured at Quinby Bridge. He was a lieutenant in the war, but must have served as a private with Marion's partisans.

The Gallantry & good conduct of your Brigade reflects the Highest honour on them. I only lament that men who Spilt their Blood in Such Noble Exertions to Serve their Country could not meet with more Deserving Success –
I beg you will Communicate my particular thanks to the Officers & men of the Respective corps. I am fully Sensible of their Merrit, and Shall take a pleasure in Doing them Justice –

After the battle at Shubrick's Plantation Captain William Ransom Davis was ordered by Sumter to go to Georgetown and seize the slaves, horses, indigo, salt and medical supplies of the Loyalist civilians. Since Sumter had threatened the Loyalists, the British decided to retaliate.

The Loyalist Privateer schooner *Peggy* was ordered by Colonel Balfour to destroy Georgetown as a supply depot. The schooner sailed over the Winyah Harbor bar and bombarded Georgetown. After the shelling had frightened everyone from the streets the Loyalists came ashore to set fire to the stores and warehouses, burning down 42 houses. As the town burned the *Peggy* shelled the streets to prevent fire fighters from putting out the flames. The town of Georgetown did not fully recover from this attack until 1830, almost 50 years later. Marion rushed supplies and men to Georgetown in the relief effort.

After the Georgetown raid Governor Rutledge outlawed the practice of retaliations and reprisals on the Loyalists. Sumter felt that this was aimed directly at him and in anger, he relinquished the command of his Brigade to Colonel William Henderson. Upon Sumter's retirement Marion became the senior Brigadier General of Militia in South Carolina.

Brigade Orders 19th Augt 1781 – S^t. Stephens Peyres Plantation

Every Person who is Draughted to Serve in the Field is to serve From the day they came in to the day of the Month following And four weeks is not considered a Month –

As many Persons have given Money to the Officer of the two Reg^ts now raising for the Cavalry, this is to acquaint them that they are not to consider themselves as Discharged until a Man is actually Inlisted in these corps, And a Certificate given to the Commanding Officer they Belong to which the Commanding Officer of Such Militia Regiments on duty are then to give a Discharge for, at least A twelve month agreeable to Genl Greens Orders, no man Whatever is of Shall be considered as Discharged till the above Terms is completed with & it is Ordered that every Man who does not Produce his Certificate from Either of the Commanding Officers of the Said Cavalry, that a Man is Inlisted as a Substitude with the name of Such Man Inlisted in it is ordered to oblidge by force to do his or Their Tour of Duty, Notwithstanding he has Given any Sum of Money whatever –

Colonel Isaac Hayne had been the commander of the Colleton County Regiment, and when Charlestown fell he went home to his plantation, as per the instructions under the Articles of Capitulation. When Sir Henry Clinton revoked all paroles Hayne signed a stipulation "to demean himself as a British subject so long as the country should be covered by the British army." When Hayne visited Charlestown he showed this paper to General Patterson, but the General refused to let him return home unless he signed a declaration of allegiance to the King. Hayne did sign the declaration, but after the Patriots had reoccupied the area of his home and plantation he believed that he was freed from the declaration.

On July 5th, Hayne had led 100 horsemen into the suburbs of Charlestown and captured Brigadier General Andrew Williamson, the man known as the "Benedict Arnold of the South." Williamson had been the leader of the Patriots in the Ninety-Six district, but had taken protection from the British after Charlestown fell. What Hayne didn't know was that Williamson was an "undercover" spy, passing information on to Colonel Laurens.

Williamson was living on a plantation at Horse Savannah on Dorchester Road, which was only seven miles from Charlestown.[1495] When Hayne captured him, he was hurried from bed without even being given the time to dress properly. The rescue of General Williamson became a priority with the British due to political reasons. The British could not allow the Patriots to discredit their protection so easily.

Major Thomas Fraser had been at Dorchester attempting to convert the entire South Carolina Royalists into a cavalry regiment, when he learned of the capture of Williamson. He took 90 of his men and pursued Hayne's raiders. After a hard ride they charged Hayne's camp at Ford's Plantation.[1496] They killed 14 of his men, wounded one, and scattered the rest. Unfortunately for Hayne, he was captured in uniform and he was bearing arms, a direct violation of the confusing parole he had been ordered to sign in Charlestown.

Hayne was brought to Charlestown for a trial and he was found guilty of violating his parole. On August 4th, he was hanged by the British for treason. Hayne had become a martyr to the Patriot cause and his name was a rallying cry for South Carolinians. Greene wrote to the British "that retaliation shall immediately take place, not on the tory militia officers, but it shall fall on the heads of regular British officers."

In August Greene ordered Marion to strike at the enemy's lines of communications to Charlestown. Marion wanted the cavalry of Maham and Horry to go with him, but they would not take orders from Marion unless they were ordered to do so by Greene. This was because Maham and Horry's troops had been placed on the Continental establishment. Greene did order them to assist Marion. Maham had 20 dragoons and Horry had 15 that needed to have swords issued to them by Marion.

Marion detached a force of men under Captain George Cooper to create a diversion near Dorchester and Monck's Corner. On August 22nd he left his dismounted men with Maham and then rode off with Horry's cavalry and two hundred hand picked men from his brigade.

After covering about 100 miles Marion joined Colonel William Harden, who had been monitoring Major Fraser's progress. William Harden was unable to take the field, but sent Major Charles Harden with eighty men to go with Marion.[1497] While Marion was camped at the Horse Shoe Colonel William Stafford[1498] and 150 more men reinforced him, increasing his army to about 400 partisans.[1499]

1495 Horse Savannah would have been located near Dorchester Avenue and Ranger Drive in North Charleston.

1496 Also known as The Horse Shoe or the Horse Neck, and is located three miles east of Walterboro.

1497 Charles Harden served 428 days in the militia as a captain and a major in 1780-1782.

1498 William Stafford served as a captain, lieutenant colonel and colonel of the Lower Granville County Regiment during 1779 to 1782.

1499 The Horse Shoe camp was located northwest of present day Jacksonboro, where 64 crosses Horse Shoe Creek.

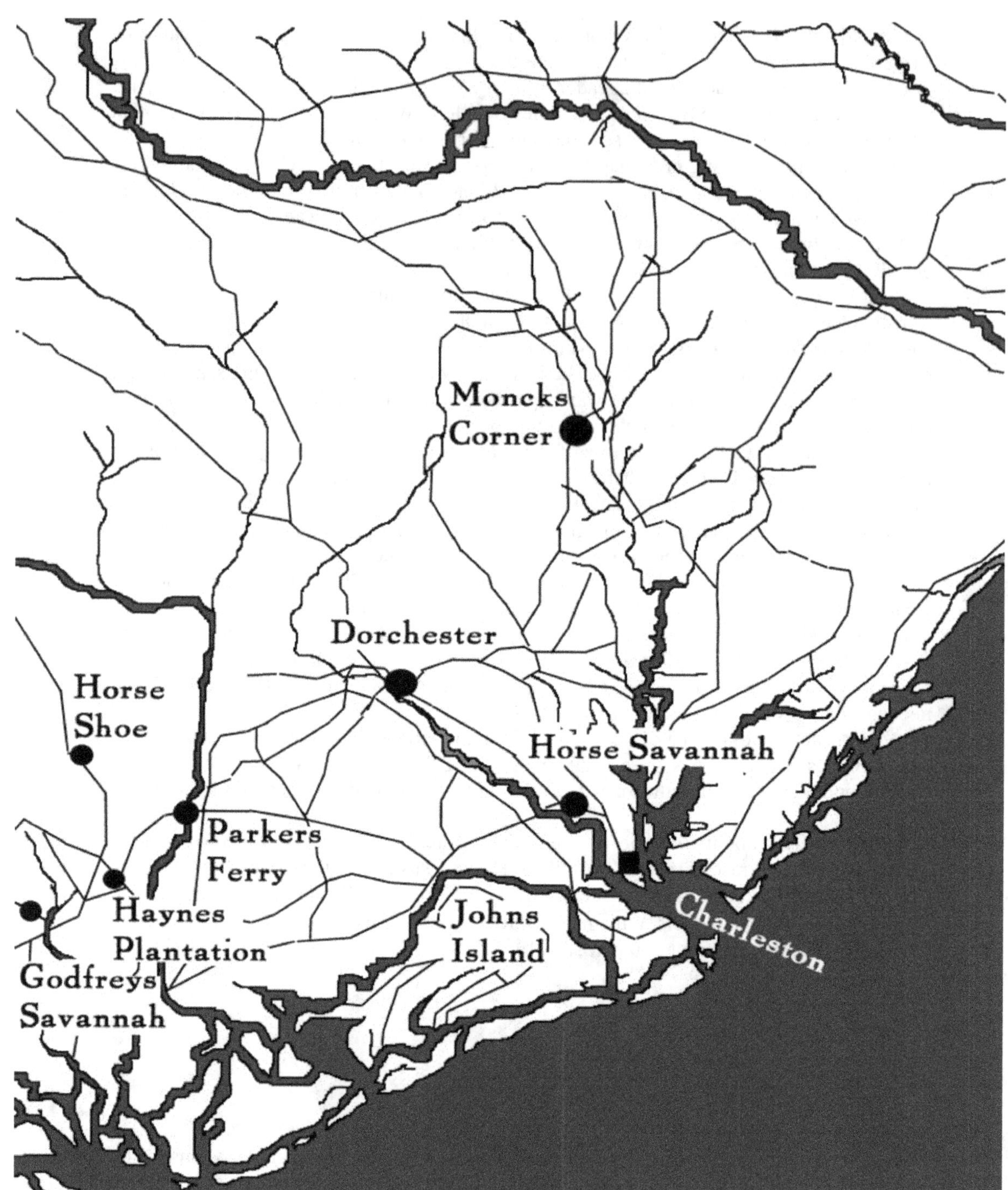

Marion Operations – Parker's Ferry

Marion sent out parties to reconnoiter British positions at Parker's Ferry and found them too strong to attack.[1500] Since he couldn't attack the ferry, Marion decided to ambush the British on the causeway leading to Parker's Ferry. That night he placed men on the causeway to observe the British movements. The partisans spotted a British patrol looking for Marion's force, and they moved off the causeway. The British were not able to find Marion's camp in the dark and instead rode on to Hyrne's Plantation.

The next morning Marion followed the British and put his men in a line of battle in the trees. A few long-range shots were fired that wounded two of the British soldiers. The British did not fall for the bait, and Marion was not able to lure them out for an attack. After two hours of sniping the partisans returned to their camp.

On the 29th the British moved out to Hayne's Plantation and Marion followed.[1501] Still looking for a fight he deliberately set up a camp five miles from theirs. The next day, Marion concealed his men in a swamp beside the causeway. He had Major Harden's 80 men move back 100 yards from the ambush line so that they could be used in reserve. Major Samuel Cooper and sixty swordsmen were told to attack the rear of the enemy after the ambush was initiated. Marion then waited for the British foraging party to appear.

At Parker's Ferry there were 100 Loyalist militia under Major "Bloody Bill" Cunningham that kept an eye on the surrounding swamps. At sunset Cunningham's pickets were walking along the road when one of the Loyalists shouted that he saw a "white feather" in the swamp. A white feather cockade was worn by Marion's men as a recognition symbol. The Loyalist shouted the password to the figure in the swamp, but there was no answer. A few shots were exchanged between Cunningham's men and Marion's partisans, then the Loyalists withdrew.

Colonel von Borck had left Hayne's Plantation in the middle of the afternoon with his infantry. He had two pieces of artillery in front of the column while Major Fraser and his mounted South Carolina Royalists were in the rear of the column. It was almost dark when Borck saw the Loyalists withdrawing from Marion's ambush. He ordered his dragoons to drive off Marion's men. The South Carolina Dragoons charged forward, but some of Marion's mounted men charged at them, so they withdrew to the main body of horsemen in the road. Fraser believed it was Harden's Regiment that was in the road, and not Marion's men and ordered the cavalry in a full gallop towards the enemy in the road. When Marion saw the cavalry in the ambush site and he gave the signal.

At a distance of 40 yards Marion's men opened fire with buckshot and the dragoons went down. Fraser rallied his men and tried to charge the partisans in the swamp, but they delivered a second volley, and then a third. There was no way to retreat and no way to attack in the swamp. Fraser had to withdraw down the causeway, down the length of the ambush. The Royalists left dead and dying men and horses on the causeway. Fraser had been badly bruised when his horse was killed and the rest of the cavalry rode over him as he lay in the road. Marion reported that 20 dragoons and 23 horses were dead on the spot.

The partisans continued to occupy the causeway for three more hours until Marion saw a large body of infantry with a fieldpiece charging towards the sound of the ambush. Marion's riflemen fired on the artillery piece, wounding and killing many of them. Marion could have easily slaughtered more of the Royalists with his rifles, but he was low on ammunition. His men had also not eaten in 24 hours, so he slipped away through the swamp.

[1500] Parker's Ferry is located north of where US 17 crosses the Edisto River, near the present day Jacksonboro.

[1501] Hayne's Plantation was located northwest of Jacksonboro on Highway 64

Marion lost one man killed, and three from Colonel Stafford's command were wounded. Marion was less than pleased with Harden and Cooper's men. He wrote that Harden's men never fired a gun, and that Cooper's cavalry had never been in sight during the fighting. Cooper had left the area and took a long circuitous ride to Charlestown, then rejoined Marion at Eutaw Springs.

The British evacuated the area and moved back to Charlestown. Marion sent a party after them and found 40 dead horses on the road. Marion estimated that Fraser had paid with 100 dragoons and 27 horses. A captured British ferryman told Marion that 50 of the British had been wounded. This loss of British cavalry would play a critical part in the upcoming battle at Eutaw Springs. Marion returned with the prisoners captured at Parker's Ferry and delivered them to Maham. He then then continued to watch the line of communication between Dorchester and Eutaw.[1502]

Orders near Jenkens Ferry 31st Augt 1781 –

Genl Marion returns his thanks to Col^{o} Stafford, Ervin and Horry & the Officers & men of their Division for their Spirited and good Behaviour in the ~~Return~~ Action of Yesterday. Had they been Assisted by the Officer of the cavalry And the night Division agreeable to Orders, he is Sertain that Very few of the Enemy would have Escaped –

A Return of the Killed & wounded in the action At Parkers ferry the 30th Augt 1781 –

Col^{o} Staffords reg^{t} three privates wounded & one Killed

The Loss of the Enemy, 18 men Killed & one negro taken

23 horses d^{o} and five wounded and Seven taken –

Maj^{r} Frazer wounded & Capt Campbell, by the best Accounts they had Eighty men wounded

---

Augt 22nd 1781 –

Mr Albert Arny Muller[1503] is appointed Brigade Major –

He is to be respected and Obey^{d} as Such –

Lieutenant Colonel Alexander Stewart had moved his British force to the hills near the Congaree and Wateree Rivers so that he could receive supplies without risking an attack. Every shipment was harassed or captured by the partisans in the lowcountry. Stewart felt secure since the hot Southern weather suspended all regular military operations, and the large rivers would make a sudden attack unlikely. Since recent the partisans had recently been successful, Stewart reinforced his supply routes. Fort Dorchester was strengthened and a force was placed at Fairlawn, near the head of the Cooper River. Stewart also adapted his supply boats with wheels, so that they could bypass areas of the river controlled by the partisans.

On August 22nd, Greene called in all his detachments, except Marion, Maham and Harden. Greene broke camp and marched north to the nearest ford near Camden, where his troops crossed the Wateree River. At Camden Greene left his sick soldiers with the campfollowers, especially those with children, to serve as nurses.

Towards the end of August the army proceeded to Howell's Ferry on the Congaree, intending to cross and pursue Stewart. The Patriot army had spent their time drilling and firing with blank rounds to insure that none of the men would fire early, which was

---

[1502] Full details of the Battle of Parker's Ferry are described in "*Nothing but Blood and Slaughter, Volume Three*" by Patrick O'Kelley.

[1503] Major Albert Arney Muller served as a brigade major under General Marion during 1781 and 1782.

normally their habit. On August 27th, Greene's men were ordered to leave their baggage and tents and only take the camp kettles and provisions. The militia was issued 20 rounds a man and the Continentals were given 30 rounds. When Greene crossed the Congaree on August 28th, Stewart hastily retreated thirty miles and took a strong position at Eutaw Springs. [1504]

Seven miles from Eutaw Springs, Marion's partisans rendezvoused with Greene at Laurens Plantation. This was a surprise to Greene because he had no idea that Marion was in the area and thought that the partisans were 20 miles away. With his reinforcements Greene decided to attack Stewart at Eutaw Springs before he had a chance to make it a permanent post. On September 5th, Greene began to creep slowly towards Eutaw Springs arriving three days later, at 4:00 in the morning on September 8th.

Believing himself secure, Stewart had sent out a rooting party of 310 men to dig sweet potatoes. The plan was for the diggers to go out around 5 o'clock in the morning, and work until it became too hot. An hour after the rooting party left, two deserters from Greene's army appeared and told Stewart that Patriots were near. Greene's slow advance had worked and he took Stewart by surprise. Not trusting the deserters, Stewart sent Major Coffin out with a detachment of 140 infantry and 50 cavalry to verify if the information was true.

At 8 o'clock in the morning Major Coffin met some of Lee's cavalry, under Captain James Armstrong, four miles from Stewart's camp. Coffin attacked what he thought was a small force, but it was actually the advance of Greene's army, consisting of Lee's Legion and Sumter's partisans, under the command of Colonel William Henderson. Armstrong's men quickly retreated to Lee's advance party with the British in pursuit.

Lee told Armstrong to place his men in plain sight across the road, while Lee and Colonel William Henderson concealed their men in some nearby woods. Coffin charged Armstrong's men in the road and rode right into the ambush. Four of Coffin's infantry were killed, and at least 40 others, including their captain, were captured. The British rooting party heard the firing and came to Coffin's aid. After some fierce fighting, Coffin's horsemen galloped back to Eutaw, spreading panic as they rode. Sixty of the British rooting party were killed, wounded, or taken prisoner.

Hearing the shooting, Greene halted his column, and let his men drink from the rum casks to allow the "animal spirit" to come out. Greene then formed his men in the familiar three-line Cowpens pattern. The units in the first line were Marion's South Carolina Brigade, Pickens South Carolina militia and Marquis Francis de Malmedy's North Carolina Dragoons. Captain-Lieutenant William Gaines was in charge of the two 3-pounders in the center of the first line.

Greene's second line consisted of Sumner's North Carolina Continentals, Campbell's Virginians, and Otho Williams Maryland Line. In the center of this line were two 6-pounders under Captain William Brown. Lee's Legion made up the right flank, and the South Carolina State Troops under Colonel Henderson was on the left flank. Washington's dragoons and Kirkwood's Delaware would make up a reserve force.

[1504] Eutaw Springs is located just south of Lake Marion, east of Eutawville.

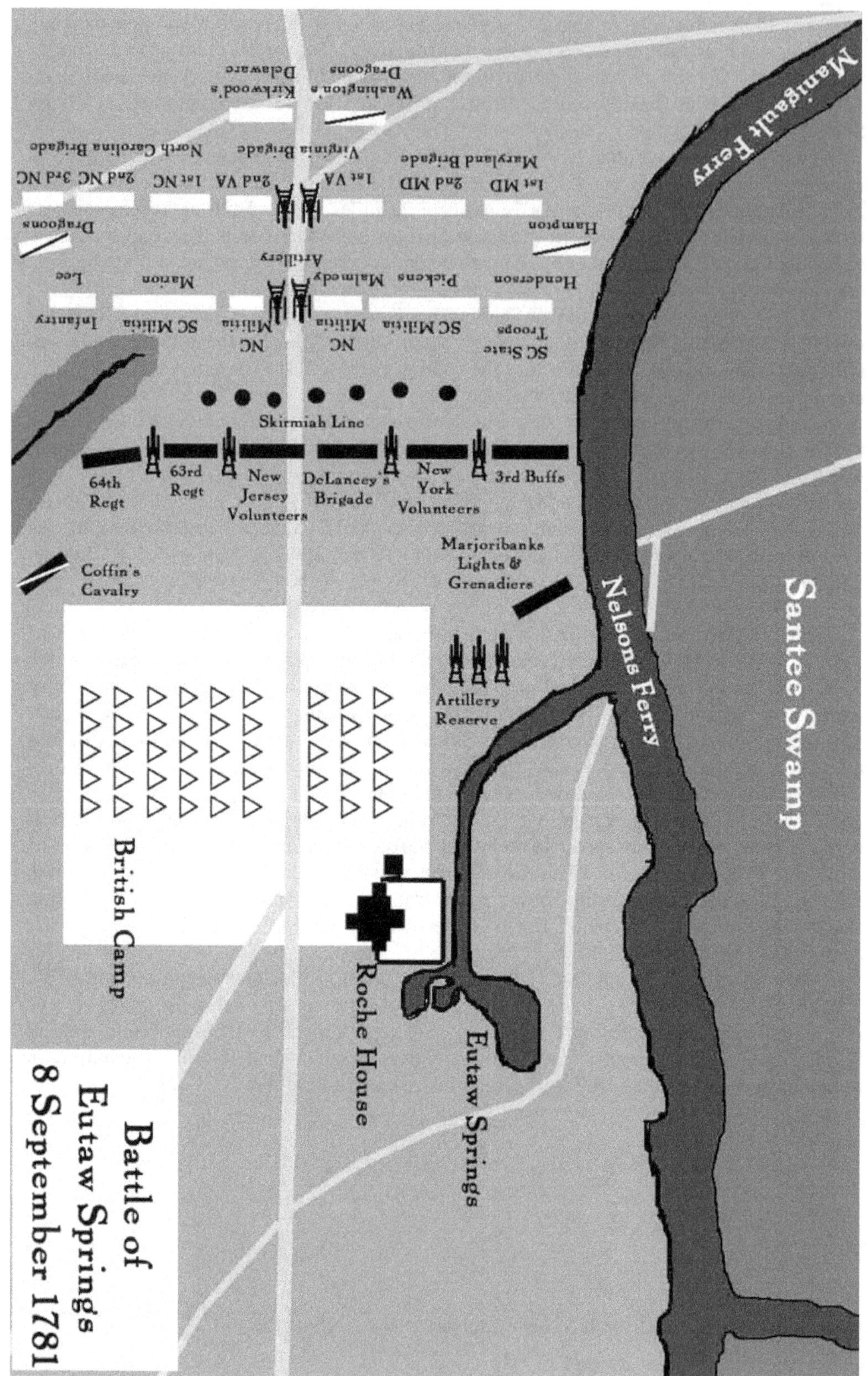
Battle of
Eutaw Springs
8 September 1781
Manigault Ferry
Santee Swamp
Nelsons Ferry
Eutaw Springs
Roche House
British Camp
Artillery
Reserve
Marjoribanks
Lights &
Grenadiers
Coffin's
Cavalry
64th
Regt
63rd
Regt
New
Jersey
Volunteers
DeLancey's
Brigade
New
York
Volunteers
3rd Buffs
Skirmish Line
SC State
Troops
SC Militia
NC
Militia
NC
Militia
SC Militia
Infantry
Henderson
Pickens
Malmedy
Marion
Lee
Artillery
Hampton
Dragoons
1st MD
2nd MD
1st VA
2nd VA
1st NC
2nd NC
3rd NC
Maryland Brigade
Virginia Brigade
North Carolina Brigade
Washington's
Dragoons
Kirkwood's
Delaware

Stewart placed a detachment of British infantry a mile forward of his camp with orders to stop Greene's army. He deployed the rest of his troops in one line across the Congaree Road, just about a hundred yards in front of their tents. The entire British line was in the woods, with the exception of an open field beside the brick building to their rear. Realizing the importance of the brick house and the outbuildings Stewart ordered them to be occupied by Major Henry Sheridan's men. Sheridan would be able to cover a withdrawal of Stewart's army if the battle turned against them.

The British skirmishers had advanced two miles in front of their camp when they bumped into Francis Marion's line. Both sides were surprised, but Marion's men drove the skirmishing parties back through the woods and then fell on the line of British Regulars. Captain Charles Gee of Marion's Brigade was commanding the front platoon when he was shot in the head. The ball passed through the cock of his hat and grazed his head. He laid there for most of the day, everyone assuming he was dead.

The British right of the line was anchored on the steep riverbank and protected by Major John Marjoribanks with 300 light infantry and grenadiers. The men there took shelter in the scrubby blackjack oaks that were impenetrable by cavalry, and almost unassailable by infantry. Stewart placed the survivors of Coffin's cavalry and infantry in reserve behind some hedges on the left of the house.

Lee tried to flank the British left, but he was driven back by the 63rd Regiment and the British cannon sweeping the road with grapeshot and canister. Gaines's two 3-pounders with 22 North Carolina Continentals as mattrosses, pushed forward for nearly a mile, until they overtook Lee. Gaines broke through the woods at 25 yards from the British line and fired a round of canister through the ranks. The British became disorganized and begun to panic while Gaines continued firing until the trunion straps on the cannons broke and disabled the guns. The British cannons that were returning fire also dismounted at the same time.

Stewart was worried about the possibility of Greene's cavalry attacking his flanks. Since Marion's militia made the initial attack Stewart was content to hold his ground and not expose his flanks. Unfortunately due to a mistake the British troops on the left advanced and the rest followed. Lee held against the 63rd Regiment while Marion held the other flank against the 3rd Regiment of Foot. The British were shocked to see Marion's militia fighting like veterans.

Greene said of Marion's men, "they would have graced the veterans of the great King of Prussia. But it was impossible that this could endure long, for those men were all this time receiving the fire of double their number. Their artillery was dismounted and disabled, and that of the enemy was vomiting destruction in their ranks." William Dobein James wrote "The British shot generally about five feet too high; but the wind blew that day favourably for Marion's marksmen, and they did great execution."

Greene responded to the new threat by ordering Sumner's North Carolina Continentals to plug the gap being made by Marion's withdrawal. Marion's militia had no more ammunition and they retired in good order leaving the fighting to the North Carolina Continentals. William Dobein James wrote, "They fired from fifteen to twenty rounds each man." William Vaughan, another one of Marion's men, wrote that "On the fire of twelve rounds we obliqued to the Right, still firing on they enemy's flank." [1505]

The North Carolina Continentals were mostly new levies, but there were many veterans in their ranks. Almost all of their officers had served in Washington's army during the battles of Brandywine, Germantown and Monmouth. The North Carolinians pushed the British back, but they had two-thirds of their men killed or wounded. The North Carolina line began to sag and yield ground to the British, that sensed victory and rushed forward. Though the North Carolinians had been given 300 muskets by Lafayette, none of them came with bayonets. The men of the 63rd and 64th Regiments of Foot broke the line of Carolinians and began charging through the gap created by the retreating North Carolina troops.

To counter this threat Greene sent in his strongest troops, the third line of Maryland and Virginia Continentals. The Marylanders fired a terrific volley into the British ranks and stopped the breakthrough. As the British fell back the North Carolina and Maryland Line combined their forces and Colonel Williams ordered them to charge after the fleeing enemy. Two British 6-pound cannon were captured as the Maryland line swept over them.

John Eager Howard's men fired a volley within forty yards of the retreating British line, while the second line continued to charge forward with trailed muskets.[1506] The officers led from the front and suffered the consequences. Colonel Howard had a musket ball break his collarbone, and Lieutenant Colonel Richard Campbell fell mortally wounded with a musket ball in his chest.[1507]

Stewart rallied his men in front of Roche's brick house, and then continued to retreat through their camp. As the Patriot army moved through the camp they noticed that food and drink lay in the British tents. The militia of Pickens, Malmedy and Marion, along with the Virginia Continentals stopped to plunder the tents. Greene was unaware that his line had fallen into confusion.

Major Coffin's cavalry was still on the field refusing to yield. Lieutenant Colonel Lee knew that the key to success was to drive this last unit off the field, so he sent for Major Joseph Eggleston to lead the Legion cavalry against Coffin's horsemen. Unfortunately Eggleston had already been committed on the other flank. Lee later wrote that this was the principal cause of Greene's defeat.

Major Sheridan poured fire from the brick house, aided by two swivel guns. Another British force fired from a palisaded garden attached to the house. Sheridan's marksmen fired on Greene's disorganized mob with deadly accuracy. When their officers lost control Greene ordered a retreat back away from the brick house. Coffin tried to push them with his cavalry, but Colonel Wade Hampton drove the British back.

[1505] William Vaughan served 143 days in the militia as a sergeant in 1781 under Captain Anderson Thomas. Towards the latter half of 1781 he served as a lieutenant under Captain Thomas. He was at Orangeburg, Four Holes and the Forks of the Edisto.

[1506] John Eager Howard was the commander of the 1st Maryland Regiment at Eutaw Springs. A musketball broke his collar bone in the battle.

[1507] Richard Campbell was the commander of the Virginia Continental Brigade during the battle.

Marjoribanks held in his natural fortress of the blackjack thicket. Washington tried to drive the British from the thicket by a frontal assault, but Marjoribanks opened up on the cavalry and swept them from their horses. All of Washington's officers except two were killed or wounded. Washington was pinned under his horse and about to be killed by a bayonet-wielding soldier, when a British officer stopped him. He was captured and moved to the rear.

Greene brought up two captured 6-pounders to fire against the house and breach the walls. The cannon were brought to within 50 yards of the house, but they were within range of the deadly effects of the British swivel gun mounted in the second floor window. All the artillerymen were shot down.

The loss of artillery and his troops looting the tents had turned against Greene. He tried to bring up his cavalry to smash the British flanks, but his cavalry had ceased to exist. Greene's men in the British camp were trapped there. Every time a head appeared outside a tent a rifle ball shattered it. Captain Thomas Polk was killed as he stood beside his brother, William, "the blood spurting and spattering on him."[1508] Pickens "was knocked from his horse by a musketball and his life only saved by his sword buckle." Greene realized that he needed to withdraw and he ordered his men to retreat. He had Hampton and his cavalry cover the retreat.

Major John Marjoribanks took that moment to emerge from his thicket and engage Hampton's cavalry. While fighting the British cavalry Hampton moved too close to the Roche house. A murderous fire fell upon Hampton's horsemen, killing or wounding a third of his force and scattering the rest. Captain John Hood had his hunting shirt pierced with seven bullets.[1509]

Marjoribanks counterattacked against the infantry, driving Greene back with the loss of the two field pieces, but Marjoribanks was wounded in the counterattack. His men wheeled the two captured guns into the fenced yard of the house. Greene rallied his troops at the edge of Roche's woods and collected his wounded, then he withdrew his exhausted army towards Burdell's Tavern.

Stewart claimed victory against Greene, but his army was destroyed. Eutaw Springs was the bloodiest battle of the Southern campaign. The battle lasted over three hours and had the highest percentage of losses sustained by any force during the war. Stewart had lost 42% of his force.

Greene had suffered about one fourth of his force being killed or wounded. The North Carolina Brigade suffered the worst losses with 154 killed or wounded. Colonel Otho Williams and Lieutenant Colonel Lee were the only Continental regiment commanders who had not been wounded. Washington, Howard, and Henderson were wounded, and Campbell had been killed. [1510]

Greene considered this battle a victory because the Patriots were able to boast that they had swept the British off the field with the bayonet, the weapon of choice of the British army.

[1508] Thomas Polk was a lieutenant in the light dragoons in Sumter's Partisans, under Captain Samuel Marin and Lieutenant Colonel William Polk. There was also a Colonel Thomas Polk, but he is a different person. William Polk was also a lieutenant in Sumter's State troops, who was killed that day at Eutaw Springs.

[1509] John Hood had been a lieutenant under Captain Adam Meek in Sumter's Partisans. When Meek was promoted to major, Hood became a captain.

[1510] Full details of the Battle of Eutaw Springs are described in "*NBBAS, Volume Three*".

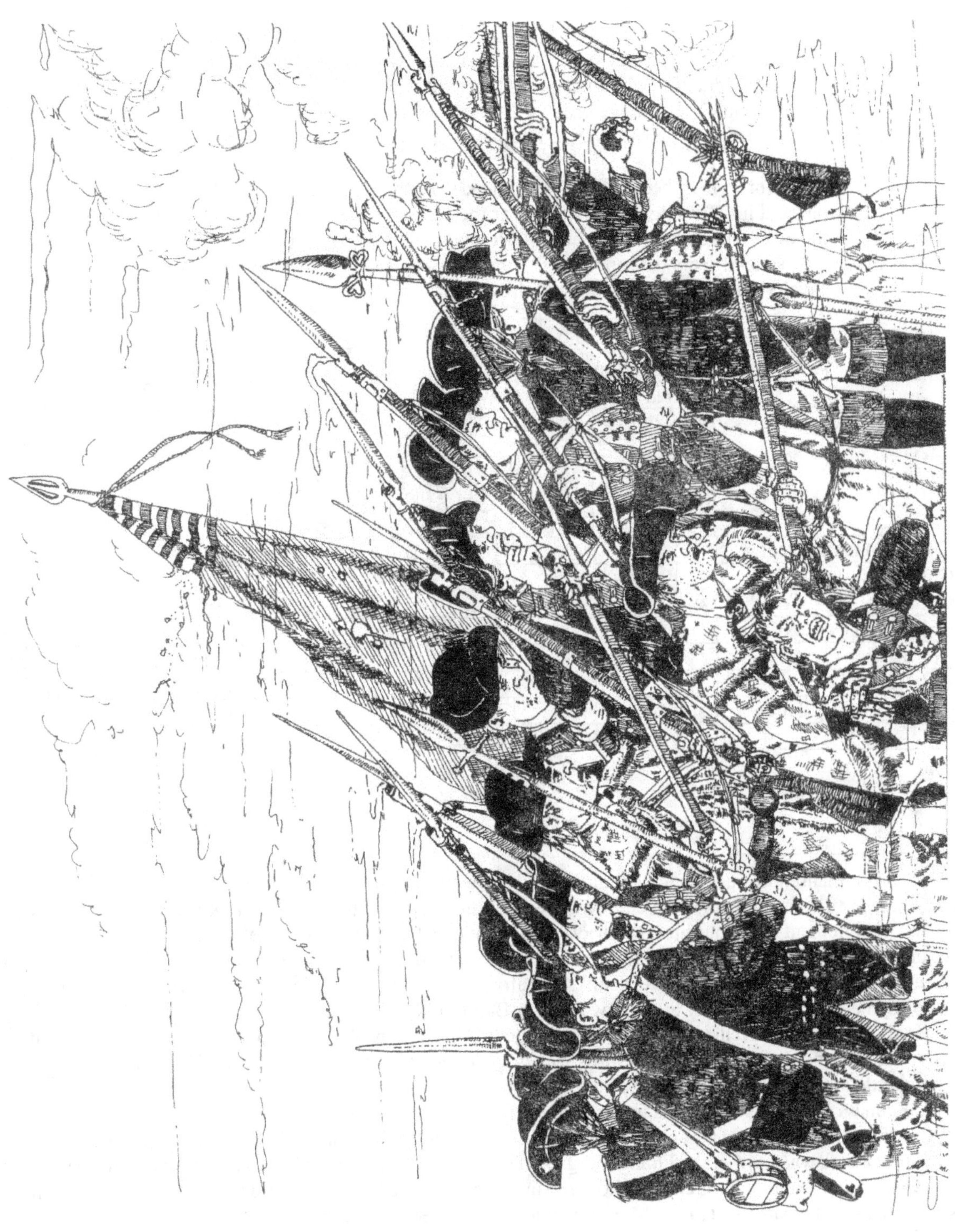

Fearing a second attack Stewart burned his stores, left his dead unburied, and left 70 wounded men under a flag of truce. He retreated to Monck's Corner. Along the way the wounded Major Marjoribanks died of a fever and was buried on the side of the road at Wantoot Plantation.[1511] Greene ordered Marion and Lee to turn Stewart's flank, but after it was determined that the British had left they were ordered to intercept Stewart before he reached Fairlawn.

Stewart force marched down the Monck's Ferry Road and had Major Archibald McArthur and his 71st Highlanders to cover the retreat. The Highlanders had not been in the battle at Eutaw Springs. Lee and Marion were able to capture the cavalry of the rear guard and several wagons containing stores and baggage. The Highlanders fired upon Lee's cavalry on the road, and Lee withdrew.

A Return of the Killed and wounded in the Action at Eutaw 8th Sept 1781

	Killed					Wounded						
	Lt Colº	Captain	Lieutenant	Sergeant	Privates	Lt Colº	Captain	Lieutenant	Sergeant	Privates	Total	
Colº McDonalds Regt	"	"	"	"	1	1	1	1	1	5	"	NB
Colº Richardsons dº	"	"	"	"	"	"	"	"	"	3	"	Lt Colº H Horry[1512] } Wounded
Colº Postells . . . . dº	"	"	"	"	"	"	"	"	"	3	"	Capt Gee . . . .
Colº Ervins . . . . . dº	"	"	1	"	1	"	"	"	"	4	"	Lt Boon [1513] . . .
Colº Bentons . . . .dº	"	"	1	"	1	"	1	"	"	2	"	Capt Peques [1514] wd
Colº Tates [1515] . . .dº.	"	"	"	"	"	"	"	"	"	1	"	Lt Holmes [1516] } Killed
Capt McCaulley Troop	"	"	"	"	"	"	"	"	"	1	"	Lt Simons[1517]
	"	"	2	"	3	2	2	1	1	20	31	

[1511] Wantoot Plantation was located about five miles west of Bonneau, and under the waters of Lake Moultrie. After the war the owner, Daniel Ravenel wrote to the English war office wanting to know if the country would pay for a monument to be put up on Marjoribanks grave. The War Office wrote back, "If England put up a monument wherever one of her heroes lay, the world would be white with stones." Daniel Ravenel's sons erected a monument to Major Majoribanks. Before the waters of Lake Moultrie covered the plantation, Marjoribanks was re-interred at the Eutaw Springs Battlefield.

[1512] Hugh Horry had served as a major and a lieutenant colonel of the Lower Craven County Regiment. He was at the fall of Charleston and afterwards was in Colonel Richardson's regiment of Marion's partisans. He was wounded at the Battle of Eutaw Springs.

[1513] Lieutenant Boon's first name is not known. It is possibly Thomas Boon.

[1514] Claudius Pegues served as a captain in Marion's Brigade and was wounded at Eutaw Springs.

[1515] Samuel Tate was born in the Orangeburg District of South Carolina. He served 115 days as a captain and 214 days as a colonel under General Thomas Sumter's during 1780. In addition he supplied corn for militia use. He was in the battle of King's Mountain and he had been wounded at the battle of Guilford Courthouse.

[1516] Lieutenant Holmes first name is not known.

[1517] Captain John Simons had been a lieutenant in the 1st South Carolina Regiment on 10 May 1776. He became a captain under Marion during 1780. He is listed in some sources as being killed at Quinby Bridge.

Camp Santee 4th October 1781 –

No man is to be Excused from Militia Duty under pretence Of working for Continental Regiments or as Drovers to the army Without a Written permission from me –

All who disobey Orders, and will not come out on duty Agreeable to their Draught, must be find agreeable to the militia Law, without Distinction –

Camp at Doughty at Santee 17th Octr 1781 – [1518]

Mr John Edwards [1519] is appointed Aid DeCamp. He is to be Respected and Obeyd as Such –

Instead of fighting the British, Marion spent the month trying to get two of his former partisans to follow his orders. The units of both Hezekiah Maham and Peter Horry had come under the Continental establishment, and due to this they no longer considered themselves under the command of Marion, who was only a general in the militia, but not in the Continental army. Marion had to threaten Maham with a court martial when he took horses and cattle from local citizens. Marion knew that he had been successful due to the trust he had with the people of South Carolina, and he did not want to have them turn against the Patriot cause for any reason. Marion had to admonish Horry because he was doing the same.

Greene and Governor Rutledge became involved, and Greene ordered Horry and Maham to place themselves under Marion. This power struggle between Marion, Maham and Horry would lead to a disaster in the coming new year.

Camp at Capt Canteys Santee 25th Octr 1781 –

A Return of Ammunition wanting to Compleat Eight Rounds pr Man in each corps to be made out and deliverd in by 8 OClock to morow Morning to the Brigade Major –

It is Conjecture by the many guns fired around the camp That the men expend their Ammunition given them in game. It is ordered that every Officer of Regt and corps do Examin their mens ammunition once a day and if any is wanted of what is given them, such men must be confined, and will be made to Suffer for every Deficiency that is not Lost unavoidably, and the Officer of the day is ordered to Sent a party to take every man who fires a Gun in or within hearing of the Camp, and confine him, who Shall suffer Immediately According to the custom of war in giving such Alarms. Every commanding Officer of Regt & Corps is charged with the ammunition of their men, and will be called to account For such men, who has made away with it, and no notice taken of them, agreeable to the order of the day –

1518 Doughty's Plantation is located where 701 crosses the Black River, and is now known as Greenfield Plantation.

1519 John Edwards was born in 1760. He had enlisted in the 1st South Carolina Regiment on 4 November 1775. He served as a captain and aide-de-camp to General Marion during 1781 and 1782. He died on 31 December 1798 at the age of 38.

One of the reasons that the Battle of Eutaw Springs is not well known is due to what happened just a month later. Almost 18,000 American and French soldiers surrounded General Cornwallis at the small town of Yorktown, Virginia in October of 1781. The allied army was led by General George Washington and Count de Rochambeau. Cornwallis had marched to Virginia to rebuild his army and strike back into the Carolinas against General Greene. Cornwallis only had 8,300 men to withstand the siege by the French and American soldiers.

After three weeks of constant bombardment, and realizing that there was no relief effort that could reach him, Cornwallis surrendered to Washington on October 19th. Though this was not the end of the war, it was the last major battle fought during the war. After the surrender of Cornwallis's army the British realized that the war with America had gone on too long and had cost too much. The real enemy was not thousands of miles away in America, but it was less than a hundred miles away in France. Negotiations began on ending the war.

John Canteys S° Santee 29th Octr 1781 – [1520]

His Excellency the Governor has been pleased to comply with Genl Greens request in Exempting the men who have Given Substitudes to Colonels Horry & Mahams Regt from Doing duty, untill further Orders all prosecutions against Such men for Default of duty, is Ordered to be with drawn and made void –

John Canteys S° Santee 3 Novr 1781 –

A field Officer to be appointed for the day who is to take the Command in Camp & see that Capt of the day do their Duty & good Order be kept in camp when any Alarm he is to See the men Drawn up with the Greatest Expedition & is to make overall report to the general of all Occurency of the day when released-

Such men as are relieved must give all their ammunition to the Men who releave them in Default of which the Officers of the Regt & Company will be made to pay for every round of Ammunition that is not given up unless they make it appear it is Expended in action or Lost unavoidable and Such men Who do not give it up, their Officers must make them pay for It –

[1520] Though Marion's force was camped at Cantey's Plantation, Horry's dragoons were dispersed to different locations. Captain Gough's Troop was quartered at Scotts, the Blacksmith the partisans used. Captain Wither's Troop was at Patterson's Plantation, and Captain Lenud's Troop was at McKnight's Plantation. I am unable to determine where these plantations were. Horry was losing his patience with his soldiers and noncommissioned officers because they would go out each night and steal food from the plantations, or raid the bee hives. The troops also let the horses run free, grazing wherever they wished. Horry ordered the horses tethered so to stop the "Shamefull and wanton conduct of the Soldiers in damaging & ruining every Plantation they Quartr at."

Be Cool and Do Mischief

John Canteys Plant[n] 6 Nov[r] 1781 –
The Governor has Ordered a Return made of all the Houses Burnt by the Enemy, with time, when, & by whom – And whom belonging, also all the men that Enemy has hanged in the District of my Brigade –
Commanding Officers of Reg[t] will Immediately as possible make Such a Return to me, that it may be transmitted to his Excellency –

Due to the losses suffered by Marion's partisans at the battle of Eutaw Springs General Greene placed a contingent of 600 militia under the command of Marion. These men were the Over-the-Mountainmen of Colonels Isaac Shelby and John Sevier. These men were not used to the swamps of the South Carolina low country, and instead were more suited to combat in the hills and mountains of the Carolinas.[1521]

Marion ordered Maham's Regiment and a detachment of Shelby's backwoodsmen to attack the redoubt at Wappetaw. When the British saw the Carolinians approaching they abandoned the post without a shot.[1522]

On November 10th, Marion received word that Cornwallis had surrendered his army at Yorktown. He held a victory ball at John Cantey's house on the north side of the Santee River, and invited the ladies of the area to attend. Marion had good reason to celebrate because with the victory in Virginia, a large number of reinforcements would be coming to South Carolina.

On November 7th, Captain Murdock MacLaine was sent with 50 men of the 84th Regiment of Foot to relieve the post at Fair Lawn Plantation.[1523] The plantation protected a landing on the Cooper River and was located just south of Monck's Corner. The post had been garrisoned by a detachment of 150 Hessians, who were being recalled because of excessive desertions.

MacLaine was concerned about how close Marion's forces were and transported 50 convalescent soldiers from the Colleton Hospital to Charlestown.[1524] This still left over 80 patients in the hospital. Marion took advantage of MacLaine's confusion and dispatched Maham to attack Fair Lawn Plantation. Maham's force consisted of 180 of his men and 200 of Shelby's frontiersmen.

On November 12th, MacLaine's scouts reported that a very large force of mounted rebels was in the area. MacLaine pleaded to Stewart for reinforcements, but received only 20 mounted militia. MacLaine decided that it was impossible to defend the hospital and the other two posts, so he chose to abandon the hospital. The only thing left in the hospital was the medical staff and the patients who could not be moved.

On the way to Fair Lawn, Maham and Shelby passed another British post and attempted to entice the British cavalry out to fight. The British refused to be lured into a trap and the Patriots moved on. However, the British horsemen followed Maham after they left.

1521 Colonel Isaac Shelby had been in Kentucky securing lands that he had marked out five years before, but when he learned of the surrender of Charlestown he returned home to serve the Patriot cause. He commanded the Over-the-Mountainmen of Tennessee at the Battles of Wofford Iron Works, Musgrove Mills, and King's Mountain. John Sevier commanded the Tennesseans of the Washington County miltia.

1522 Wappetaw Bridge was located five miles east of Wando, where the Guerin Bridge road passes over the Wando River.

1523 Fairlawn Plantation was located at Monck's Corner, at the Old Santee Canal State Park.

1524 Colleton Plantation was also located at Monck's Corner, at the Old Santee Canal State Park.

When the Patriots arrived at the Fair Lawn redoubt on the morning of the 17th, they found that the fort was too formidable to attack. Time was against Maham because of the threat of the enemy cavalry to their rear from the previous post. Shelby and Maham decided to attack one of the outbuildings, which was the hospital. Maham had Shelby's riflemen cover the redoubt at Fairlawn, while his cavalry boldly rode up to the building and demanded its surrender.

The medical defenders of the hospital offered no resistance at all. Maham captured 300 stands of arms and other stores. About 150 patients and staff were made prisoner. Eighty of the prisoners were found able to walk and were taken back to Marion's camp. The rest of the prisoners were paroled. MacLaine's garrison watched Maham's raid from the redoubt, but did nothing to interfere. The hospital was burned, but there is a dispute as to who burned it. William Moultrie said that Maham was forced to burn the house because his men were drinking too much of the British liquor supplies.

On November 25th, Shelby left Marion's force and returned to the mountains, because of what Shelby described as "a leave of absence." In all probability it was due to the impatience of the mountainmen to the style of partisan warfare waged by Marion.

Genl Orders 20th Novr 1781 –
at Peyers Plantation Santee – [1525]

Notwithstanding repeated Orders that no Persons should fire Guns in or out of Camp to give false Alarms, this is to acquaint such Persons, who continue this Practice that they Shall suffer Death whenever they are detected & Patrols Will be sent around the Camp to take such Persons when they shall immediately suffer agreeable to articles of war & Custom in all Armys, Officers of Regt are order'd not to Permit their men to go from camp without any Officer. Any Man who are found half a mile from camp without an Officer Or Written pass from one, will be confined & Suffer for Disobedience of orders – [1526]

Quashs Plantn St Thomas – [1527]
Novr 28th 1781 –

The Enemy getting great Supplies from the Parishes of St Thomas and Christ Church,[1528] it is ordered that no Persons Whatever do send any Provisions, Forage, navel Stores, or Lumber to Charles Town. Any Person who may be found Supplying the Enemy with any of those Articles will be Deem'd an Enemy And suffer accordingly. Any Persons who go to Charles Town Without my written permission will be deem'd a Spy and Suffer as such and all Persons are forbid to have any communication With the Inhabitants of the town on any pretence whatever

---

1525 Peyer's Plantation was located about two miles east of the dam on Lake Marion, in Berkeley county.

1526 Though they were conducting partisan, or guerilla warfare, the standards of the military were still enforced. Before a march Horry ordered each of his men to have their "Clothes washed, and his Arms in best Order, their Hairs Cut Short and themselves & Horse furniture in the best Order."

1527 St Thomas Parish was located at the southern end of the Francis Marion National Forest, in between Huger, Awendaw and Wando. Quash's Plantation was located two miles north of Huger, where Highway 41 crosses Turkey Creek.

1528 Christ Church Parish was a narrow strip of land, that went from Mount Pleasant, and followed the present US 17 north to Bull's Island.

Quash's Plant^n^ St Thomas Decr 5th 1781

A Genl Court Martial to be held for the trial of L^t^_ Frances Green, for disobeying the orders of his commanding Officers, Evidences to attend.[1529] Col^o^ Ritchardson President, one Capt & two Subalterns from Col^o^ McDonalds Reg^t^, one Capt & two Subalterns from Col^o^ Richardson, One Capt & two Subalterns from Col^o^ Bentons, one Capt & two Subalterns From Col^o^ Ervins Reg^t^. The Court to sit Today or early Tomorrow morning at 8 OClock –

Quash's Plant^n^ Decr 6th 1781 –

The court Martial of which Col^o^ Richardson was president Is dissolved, the Oppinion of the court is that Lt Green of Capt Postells Comp^y^ tried for Disobedience of Orders is not Guilty & is therefore Acquitted. Lt Greens Arrest is taken off & is to do Duty –

~~Quashs Plant^n^~~

The united States in Congress Assembled Octr 24th 1781 Resolved –

That the thanks of the United States in Congress Assembled be Presented to Brigade Genl Marion of the South Carolina Militia for his Wise, Gallant and Decided conduct in defending the Liberties of his Country and particularly for his prudent and intrepid Attack on a Body of the British troops on the thirtieth day of August last and for the distinguished part he took in the battle of the Eight of September –

Charles Thomson Sec^y^ –

Sir

Your distinguished exertions for the Support and defence of you Country in the hour of her distress have often claimed the attention of Congress for which And particularly for you prudence and Gallantry Exhibited on the 10th of August and 8th of Septr last – [1530]

I have the pleasure to present you their thanks, as Expressed in an act of the 29th instant, of which a copy Is inclosed –

I have the Honor to be with the Highest Respect and regard –

Sir

your most Obedient
And very humble Servt

Thos M^c^Kean president

Philadelphia
October 31st 1781 –

The Honb^l^ Genl
Marion

[1529] Francis Green was born on 26 April 1746. He served in Marion's partisans in 1781 and 1782. He died on 7 January 1825 at the age of 79.

[1530] This should read "30th of August" and it is in reference to the ambush at Parker's Ferry and the Battle of Eutaw Springs.

Gen[l] Orders 28th December 1781 –

No person whatever to obstruct or prevent the Commissary of the Brigade from taking any provision or Forrage from any Plantation. Those that take that Liberty may Depend on being brought to a Court Martial for Disobedience of orders and preventing supplys to the troops –

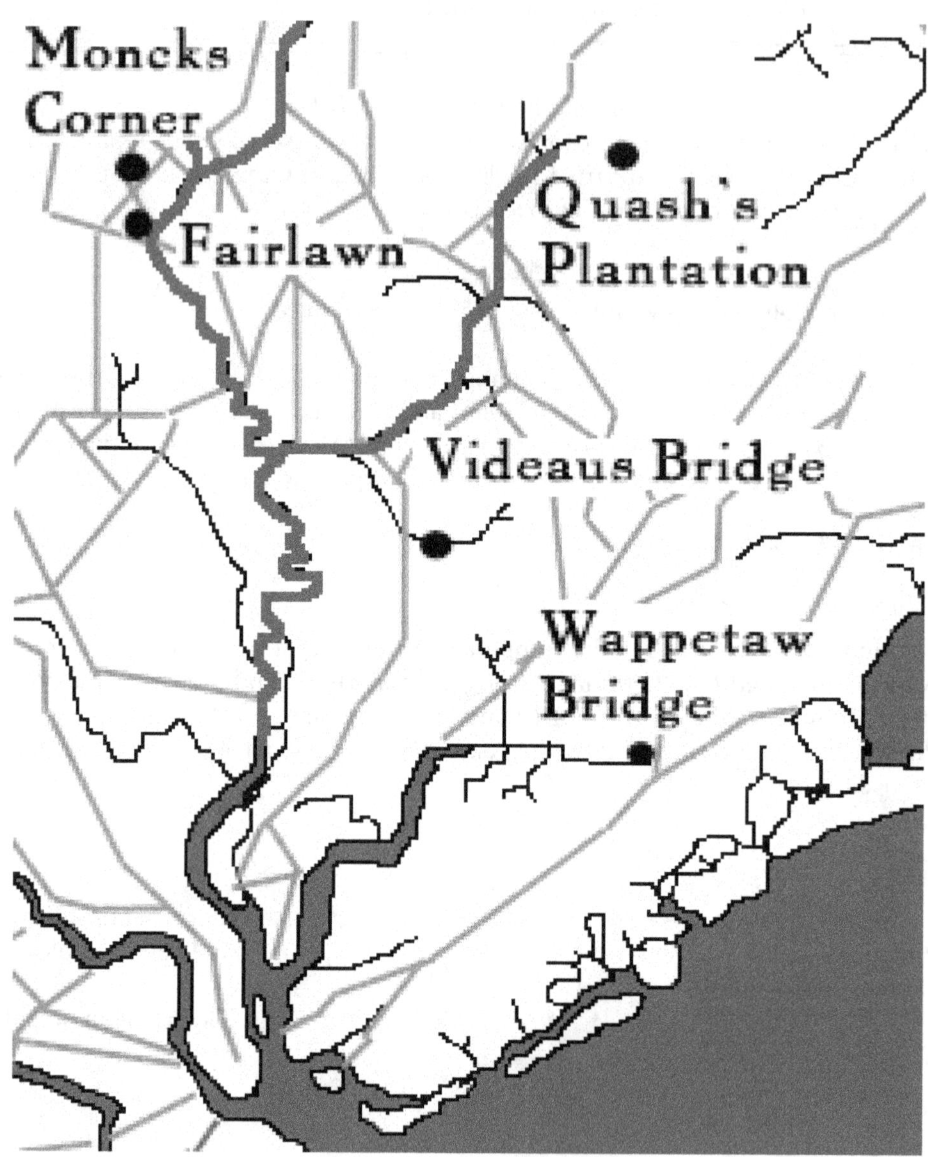

Operations November 1781 – January 1782

1782

In 1782 the British were penned up inside the city of Charlestown, surrounded by the Patriot forces of General Greene. Major General Alexander Leslie was the commandant of the city and had to slaughter 200 horses due to the lack of feed for the animals. To protect the approaches to the city Leslie had several outposts which were guarded by armed galleys. One of these galleys was in the Wando River guarding the outpost on Daniel's Island. This outpost was under surveillance by Colonel Richard Richardson and a small body of men who were based in the settlement of Cainhoy.

Leslie had received reports that Francis Marion's partisans might be vulnerable to attack and he ordered Major William Brereton to cross the river and invade St. Thomas's Parish.[1531] On January 2nd, Brereton crossed from the city to Daniel's Island with 350 infantry and cavalry and then moved up the Strawberry Road.[1532] Colonel Richardson knew that he was outnumbered so he requested for reinforcements from Marion. Marion was low in men, but he did send a detachment of new and inexperienced cavalry from Lieutenant Colonel Thomas Screven's troop, led by Captain John Carraway Smith.[1533]

Brereton marched across Daniel's Island and crossed Beresford Creek. On January 3rd, he rested his troops at Brabant Plantation, the plantation of Reverend Robert Smith. Brereton put troops on Videau's Bridge to guard the approaches leading to the plantation. Richardson had his men circle around and come at the bridge from the north.[1534]

Brereton went out on a reconnaissance alone, but soon came racing back with the New York Volunteers, the South Carolina Royalists, and the Independent Troop of Black Dragoons hot on his heels. When Richardson reached his men he wheeled his mount and ordered a charge. Captain Smith charged with his cavalry and the British quickly retreated. Unfortunately Smith's cavalry was untrained and they were soon strung out and disorganized.

At Videau's Bridge the British infantry fired a volley into the disorganized partisans and killed 22 of them. Major Coffin charged the confused partisans with fresh cavalry, causing Marion's new recruits to flee. Some of Maham's veterans, led by Major Samuel Cooper, stayed and fought, but they were either slaughtered or they ran.

In the intense fight Captain G. Sinclair Capers had taken three sword thrusts through the body.[1535] Captain Archibald Campbell of the South Carolina Royalists was killed when he attempted to escape after he had been captured. The British pursued the partisans on a six mile running gun battle before they turned back to their main force. Brereton continued

[1531] William Brereton commanded the 64th Regiment in South Carolina. The 64th Regiment was described as "This Noble Corps, the 64th were almost all Grey-headed men".

[1532] Daniel Island is located on the east side of the Cooper River, beside Charleston.

[1533] Thomas Screven served since 1775.

[1534] Videau's Bridge was located on the east side of the Cooper River, where the Cainhoy Road crosses Freshing Lead Creek, in between the towns of Red Hill and Charity.

[1535] George Sinclair Capers served as a captain in the militia and the state regulars under General Marion.

on the cattle raid, foraging as far north as Quinby Bridge before he returned to Haddrell's Point.[1536]

Genl Orders at wambaw by Genl Marion 6th January 1782 – [1537]

A general court martial to set tomorrow at the usual Hours at the presidents Quarters for the trial of John Nesmith Senr, and David Snow charged as Being Spies to the British, and Such prisoners as may be Ordered to it – [1538]

Lt James Smith of Col^{o} Mahams Corps is appointed Judge Advocate, Lt Colonel Maham President – Six Officers from Col^{o} Mahams Legion & Six Officers From Col^{o} Peter Horry's Corps Members –

Evidences are ordered to attend

Before the battle of Eutaw Springs, Greene had heard of a possible peace initiative from the British government. Greene had feared that Britain might claim civil authority over South Carolina and so could continue to rule over the region as part of a peace compromise. If a Patriot government were functioning in South Carolina, the British claims would not be justified. He wrote to Governor John Rutledge that a civil government should be "set up immediately" so that the state could be run by a "civil rather than military authority."

By 1782, Governor Rutledge decided since South Carolina was almost free of British soldiers he would reestablish the General Assembly. Greene had written to Governor Rutledge and told him that civil government needed to be "set up immediately" because the public should be ruled by the "civil rather than military authority." Greene new that with the negotiotians of a peace treaty between England and the United States, Britain could claim civil authority over the Southern states, and could possibly continue to rule over the region. However if South Carolina had a functioning government, British claims to the region would not be justified.

Each of the militia generals were sent orders to organize and supervise the elections. The people elected those men who had led the soldiers to be their representatives in the assembly that would meet at Jacksonboro. The meeting place was picked to show the people of South Carolina that there was nothing to fear from the British army anymore. Jacksonboro was 32 miles from Charleston, but Greene's army was camped at Round O, fourteen miles north of Jacksonboro.

Francis Marion had been elected to the South Carolina General Assembly as the senator for the Parish of St. John, and Berkeley County. He left Horry in charge of his brigade while he was at the general assembly in Jacksonboro. Hezekiah Maham still considered his unit totally independent and would only take orders from General Greene, not Peter Horry. On the advice of Marion, Horry moved the brigade to Wambaw Creek

---

1536 Full details of the Battle of Videau's Bridge are described in "*Nothing but Blood and Slaughter, Volume Four*" by Patrick O'Kelley.

1537 Wambaw Bridge is located eight miles north of McLellanville, where Rutledge Road crosses Wambaw Creek.

1538 Private John Nesmith, Sr. His son, John Nesmith, Jr. was a sergeant Horry's dragoons.

near the Santee River. The forage was more available there and it had better protection from the British troops.[1539]

Genl after Orders, by Genl Marion at Wambaw the 6th January 1781 –

It appears that officers ordered on the Genl Court Martial to Sit to morrow are not to be had without Lt James Smiths being a Member, who was ordered to act as Judge Advocate. It is orderd that Capt Thos Elliot, my Aid De Camp do act & he is appointed Judge Advocate to the Said court Martial –

Genl Orders by Genl Marion 7th Jany 1781

The General court Martial ordered to Sit to day is ordered to be postponed untill further orders –

Colo Horry's and Maham will come to my quarters to morrow, to Settle the rank of their respective Regiments & Officers and produce their orderly books –

A Return of the Killed and wounded in the Action of the 3d Instant Jan 8th 1782 –

	Dragoons wounded	Dragoons Killed	Dragoons Missing	Total –	Horses Missing
Colo Mahams Cavalry	..2..	..4..	..6..	..12..	..11..
	Wounded	Killed	Missing	Total	Horses Missing
Militia ___	..4..	..5..	..9..	..18..	..18..
Total	.6.	.9.	.15.	.30.	.29.

Enemy lost Capt Archd Campbell & three Dragoons Killed and Major Coffin, Captn Alexr Campbell, Officer and 12 Dragoons wounded – two Horses taken – One Dragoon prisoner –

Genl Orders by Brigadier Genl Marions at Wambaw 10th Jany 1782 –

The court which was postponed is ordered to Set to Day & proceed on the trial of the Prisoners ordered to it –

The Sentence of the above Court upon examination Report that Mr. Nesmith & Mr. Snow are not Guilty of the Charges exhibited against them, the above Court is therefore dissolved

On January 12th, Major General Anthony Wayne crossed the Savannah River with 100 of Colonel Anthony White's dragoons and a detachment of artillery. Wayne and his men had been at the siege of Yorktown, and once Cornwallis surrendered they were sent south to restore United States authority in Georgia. Wayne was joined by 300 mounted

[1539] The South Carolina Historical and Genealogical Magazine, *Letters to General Greene and Others*, (Volume XVI, Number 4, October 1915), pp. 139-140.

infantry of Sumter's Brigade under the command of Colonel Wade Hampton and 170 Georgia militia under the command of Colonel James Jackson.

Wayne aggressively attacked the British, though they had superior numbers. Due to this the British thought they were being surrounded by a much larger force. Wayne easily drove the British outposts back to Savannah, but he was not strong enough to take the city.

British Brigadier General Alured Clarke had withdrawn before Wayne leaving behind nothing of any use. He ordered a "scorched earth policy" burning the outposts and anything his soldiers could not carry back to Savannah. Wayne sent slaves along the British perimeter, enticing the soldiers there to desert and convincing them that they were outnumbered.

Orders the 25 January 1782 –

Ordered that all the men in the Berkly county and Charles Town Regiment do immediately Join the Brigade otherwise parties will be Sent Out to take them into and will be punished for their Neglect notwithstanding any order they may have Rec^{d} from ~~the~~ any Officer whatsoever –

Jacksonburgh Feb^{r} 23rd 1782 [1540]

Sir,

Inclosed is Twenty Blank Commissions which I Sent to your discretion to be filled up for the vacancies in your Brigade. Rank ought to be preserved except in cases of glaring inability should any such arise you'll first inform me before you proceed to an appointment out of the usual instances

I am Sir,
Your most Obed^{t} Servt
John Mathews [1541]

Brig^{r} Genl Marion

Jacksonburgh Feb^{r} 23rd 1782

Sir

Inclosed is a Resolution of the Legislation for supplying the Widow & Orphans of persons killed in the Service of this State and of such as have been Disabled in the Service with Provisions. I therefore must request you will take the necessary measures for carrying the same into Execution.

I am Sir,
Your most Obed^{t} Servt
John Mathews

Brig^{r} Genl
<u>Marion</u>

In the House of Representatives
Feby 7th 1782 –

1 Copy /

[1540] Jacksonburgh is Jacksonoboro.

[1541] John Mathews was born in 1744 in Charleston. In 1760 he was commissioned an ensign in the South Carolina Provincial Regiment and fought in the Cherokee War. He was a lawyer and he became a barrister in London. When he returned he sided with the Patriots from the beginning of the war. He was elected to the Assembly in 1772. He was also a captain in the Colleton County militia. In 1780 he was elected to Congress and he endorsed the Articles of Confederation. In 1782 he was elected governor of the state and served for one term. In 1784 he was elected judge of the court of chancery. In late 1784 he was elected to the State house of representatives again. He was elected judge of the court of equity in 1791 and served until 1797, when he resigned. He died in Charleston, on 17 November 1802 at the age of 58.

Resolved That his Excellency The Governor be empowered to order a Sufficient Quantity of Provisions to be purchased at the Publick Expense & distributed amongst the Widows & Orphans of such persons as have been killed in the Defence of this State & also of Such Persons who have been Disabled in the said Service ~~Service~~ as may be Deemed proper Objects of Relief –

Colonel Benjamin Thompson, a Loyalist from Massachusetts, learned that Marion was at the general assembly and that there was a breakdown in communications between Marion's two colonels. Thompson decided to attack the partisans while their guard was down. Thompson had put together a cavalry force that consisted of all the mounted units in Charlestown. Horry was on the other side of the Santee River visiting his plantation and had left Colonel Archibald McDonald in command while he was gone. Marion had told Horry that if he had to absent himself for any reason the command should go to Maham, however Maham was with Marion at the assembly.

On the morning of February 24th, Thompson set out from Daniel's Island and rode towards Marion's camp. Colonel Lemuel Benton held a position at Durant's Plantation.[1542] Major William Benison commanded the scouts in St. Thomas's and told Benton that the British were approaching his position. Benison proceeded to Colonel McDonald's headquarters and also told him of the approaching enemy. Many of the officers were eating dinner and most of the Patriot officers did not believe that the British were going to attack.

Benton was one of the few who did believe the reports and rode to Durant's plantation only to encounter the advance of Thompson's army. Major John Doyle did not wait for the rest of the cavalry force to arrive and charged Marion's men at Wambaw Bridge. Benton was wounded and his dragoons fled across the Wambaw Bridge.

The stress was too much for the old bridge and it broke under the weight of men and horses. Many of Benton's men tried swimming across and a few drowned. The rest of Marion's Brigade fell back to Mrs. Tydiman's Plantation in between Echaw and Wambaw.[1543] Thompson continued to raid the countryside and was able to capture and parole Charles Cotesworth Pinckney.[1544]

Jacksonburgh Feb^r 26th 1782

Sir/

As some little Trade has begun to flow into George Town and our necessities requiring every encouragement to be given it, and as nothing will tend more effectually to do so than a proper protection given the vessels & Goods brought in, I therefore desire you would have a party under the command of an active and Vigilant Officer stationed at And about George Town so as to be always in a situation to give the most prompt assistance when requested. I must leave to you to Judge what Number of Men will be sufficient for this Service –

I am Sir,
Your most Obed^t Servt
John Mathews

Brig^r Genl Marion

1542 Durant's Plantation was located south of Wambaw creek, in present day Francis Marion National Forest, northeast of Thompsons Corner.

1543 Tydiman's Plantation is located in the Francis Marion National Forest, three miles northwest of the Hampton Plantation State Park.

1544 Full details of Thompson's Raid and the Battle of Tydiman's Plantation are described in "*Nothing but Blood and Slaughter, Volume Four*" by Patrick O'Kelley

Jacksonburgh Feb^r^ 26th 1782

Sir/

The Legislature having requested me To have the Number of which Inhabitants in ~~town~~ This State ascertained as soon as possible I therefore Request you would take the necessary measures for this purpose and transmit the returns to me without delay distinguishing men, men Women and Children and men able to bear arms Although I could wish to have the business done speedily, yet I would not have it hurried in Such a Manner as to render the Returns inaccurate on the contrary I must desire them to be made with the greatest exactness –

I am Sir,
Your most Obed^t^ Servt
John Mathews

Brig^r^ Genl Marion

Once Marion learned of the attack against Durant's Plantation he left Jacksonboro with Colonel Maham and quickly rode back to his brigade. Marion took Maham's dragoons and covered thirty miles until he reached Mrs. Tydiman's Plantation on February 25th. At the plantation Marion set up camp while Maham continued on to his own plantation. Maham had left Captain John Carraway Smith in command of his dragoons in his absence.

Colonel Thompson had let his Loyalist infantry continue walking along the road with their stolen cattle. This gave the appearance to Marion's spies that this was merely a foraging expedition. Thompson in reality had taken his cavalry and mounted infantry and headed back towards Wambaw Bridge to strike at Marion a second time, since he knew that Marion would return after he had attacked the first time.

Both sides were startled by the appearance of the each other's cavalry, but Thompson quickly recovered and swung his cavalry into a field and formed a line of battle. Marion ordered Captain Smith to charge the forming Loyalists. As Smith bore down on the enemy he was suddenly seized with panic and dashed into the woods on the right. His cavalry followed him, veering to the left to avoid a pond in their path. This threw the whole attack into disorder. The British saw this confusion and charged. Smith's dragoons broke and fled, some attempting to swim across the Santee. The British riflemen fired into the water, killing all they could see. Lieutenant Jacob Smiser of Horry's cavalry drowned trying to cross the river.

A half a mile away Marion rallied the broken horsemen, but Thompson never followed. Marion's men lost most of their arms and many of their horses. After the battle his regiment had only 60 dragoons left and Horry's dragoons were decimated. Thompson had killed 20 of Marion's men and took another 12 prisoner. Marion reverted to the tactics that made him a successful partisan and withdrew his men into the woods. Thompson declined to attack him in the covered location. Thompson rejoined his infantry and marched back to Charlestown. Captain Smith resigned over the controversy of his cowardice after this action.

This action finally settled the dispute between Maham and Horry. Governor Mathews ordered the two decimated regiments to be combined and Maham was placed in charge of the new regiment. Horry felt slighted and resigned. Marion consoled him by placing him in command of Georgetown.

Maham took ill after this and returned to his plantation in St. Stephen's Parish. When the Loyalists learned of this they rode to capture him.[1545] Lieutenant Robins and a troop of "Bloody Bill" Cunningham's dragoons captured Maham, his doctor and another officer while they were sitting down to supper. Robins had Maham sign a parole, but Lieutenant Robins could not read or write, and did not sign it. The Loyalist lieutenant also left the parole lying on the table as they rode away. Maham believed that he was no longer under obligation to honor the parole. When Maham attempted to take the field later Marion would not let him. Marion always believed in honoring any obligation, and since Maham had signed the document he was bound by honor to obey the parole.

Greene dispatched Lieutenant Colonel John Laurens to the support of Marion's Brigade. Marion retreated across the Santee River to rebuild his forces. Thompson's raid did allow the British to forage at will from the end of February until the beginning of April.

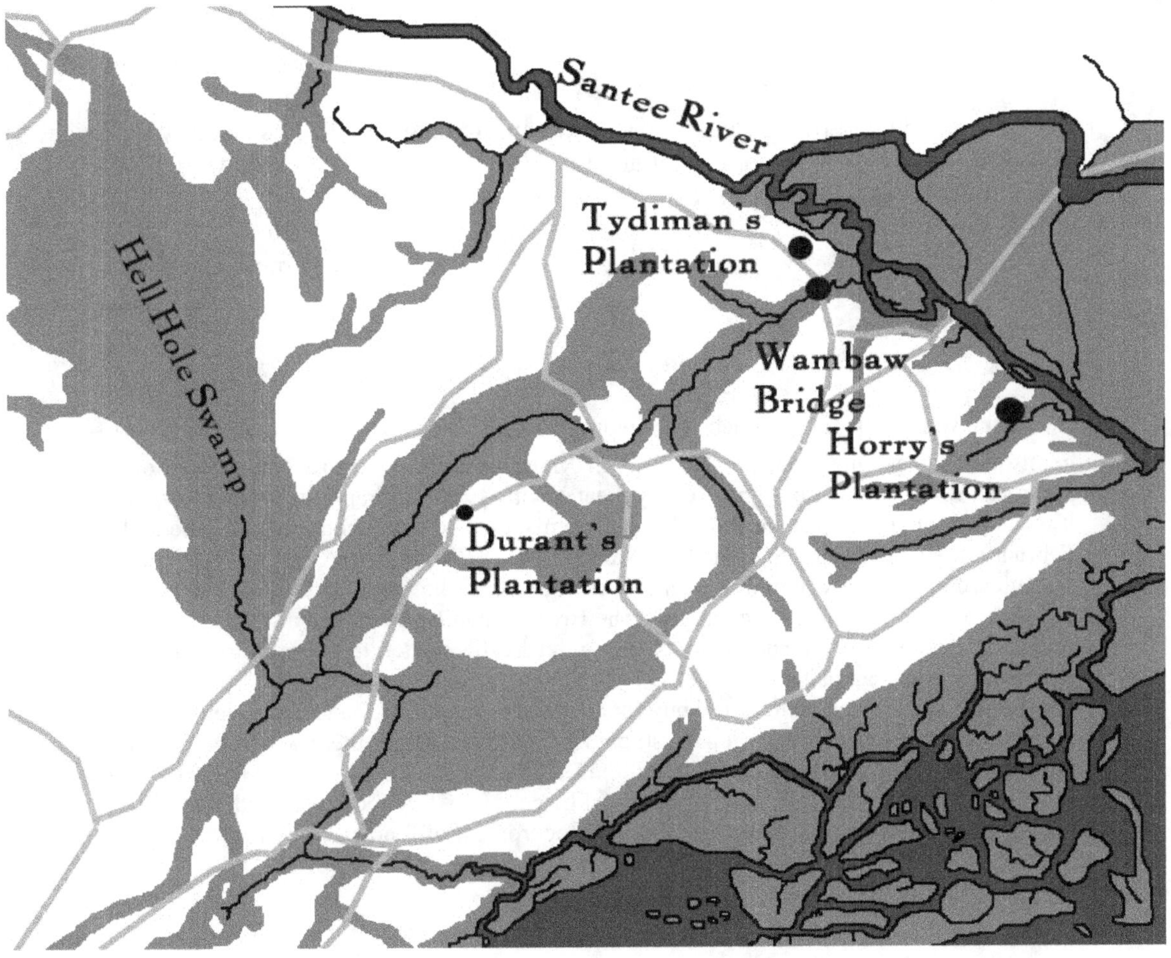

Operations February - March 1782

[1545] Maham's Plantation was located near Pineville.

Camp St Stephens Murry's ferry 1st March – 1782 [1546]

Commanding Officer of Reg^{t} will see that Every Commanding Officer of Company keep a Company Book And be very Exact that their men on Duty are recorded In that Book, the time when they Join the Brigade & when Discharged, every man when Discharged must be given a Certificate of the time he has ~~Done~~ Served in the field, to be Signed by the Officer Commanding the Company & Reg^{t}. Which Certificate must be Brought to the Brig^{d} or commandant of the Briga^{d} to Countersign & this mode is ~~done~~ to Initiate the men to their pay agreeable to law, the new Militia Law is to be Read to every relief which come in Camp & the Commanding Officer of the Regim^{t} will transmit a Copy of the Law to Different parts of their Districts, that Every man may be acquainted with the Law, which is to be & Shall be put Strictly Inforce –

Jacksonburgh Feb^{r} 13th 1782

Sir/

You will be pleased to give orders Immediately in the District of your Brigade to put a Stop to any more provisions of any kind Being Sent to Charles Town, without a permit either from Genl Green or myself

I am Sir,
Your most Obed^{t} Servt
John Mathews

Copy
Brig^{r} Genl Marion

The war for Britain had become a world war by 1781. There was fighting in every hemisphere, stretching from the Far East in Malaysia and India, to Central and South America. Britain could not fight in every theater with its full military might and knew that it would have to give up some ground in one hemisphere so that it could win in others. In February, the British Parliament had tried to enact a decree ending the war, but it failed by a single vote. The next month a decree went out to all the generals in the field to stop fighting, pending a formal peace treaty. The British intended to pursue a war in the West Indies against the French, and General Alexander Leslie needed to get food for his army for the upcoming campaign.

Leslie asked Greene if he could buy food from the South Carolina farmers, but Greene refused. Greene knew that if the British paid for gold and silver there would be no food left for his army. Greene also knew that if the British could not provision themselves, they would soon have to leave. Leslie stated that if he could not buy food he would take it, and the fighting in 1782 would center around the procurement of food and supplies.

[1546] Murry's Ferry was located north of the present day town of St. Stephens, on the Santee River. The CSX Railroad bridge is built where the ferry was located.

A Return of the Killed & Wounded & Missing of Col° Maham's Legion at Mr Tydimans 25th Febr 1782

	Killed	Wounded	Missing	Horses Missing
Cavalry	"	"	2	7
Infantry	4	1	4	"
Total	4	1	6	7

NB Lt Smiser of Col° Horrys Regt drowned in crossing the River [1547]
Militia –

Lt Martin & one man killed at Strawberry & one Wounded.[1548] 20th Febr Capt Browns party reconnoitering two men Killed. 22d Febr Lt Herds party reconnoitering the 24th in the morning two killed one Wounded – [1549]

A Return of the Killed Wounded and missing of Col° Peter Horrys Cavalry 24th Febr 1782 –

Killed	Wounded	Missing	Horses Missing
4	6	7	24

Major Benson [1550]
Thos Broughton[1551] } Volunteers } Killed
Mayes [1552] _ _

[1547] Jacob Smiser had been commissioned a lieutenant on 6 December 1781 in Horry's Cavalry Regiment under Captain William Fishburn.

[1548] John Martin had served under Colonel Richardson in Horry's dragoons and was killed at Durant's Plantation.

[1549] Charles Herd served as a private and a lieutenant in the militia from 1779 to 1783.

[1550] William Benison served as a captain in the militia and a major in the 2nd South Carolina Regiment. He was appointed a major in Horry's Dragoons on 4 November 1781. He was killed at Wambaw Creek.

[1551] Thomas Broughton served as a foragemaster, quartermaster and lieutenant under Captain Joseph Calhoun in Maham's Regiment. He was killed at Wambaw Creek.

[1552] James Mays served 29 days as a wagoner. He was killed at Wambaw Creek.

Horry's Plantation, Santee 16th March 1782 [1553]

A Gen[l] Court martial to set to day for the trial of Norman M[c]Donald,[1554] taken as a Spy & going to the Enemy –

L[t]. Sinkler Capers & Will[m] Cleland Evidence.[1555] Cap[t] Elliot, Judge Advocate –

Col[o] tho[s] Scriven President, and twelve officers From the Brigade members, Cap[t] John Palmer,[1556] Cap[t] John Norwood,[1557] Cap[t] Will[m] Bennett, Cap[t] Jacob All,[1558] Lieutenants B. Davier [1559] Ja[s] Witherspoon,[1560] Will[m] Lewis,[1561] Ja[s] Jenkiman[1562] Edw[d] Tho[s],[1563] Will[m] Skilling[1564] Pet[r] Hubbard,[1565] Henry Kuhn,[1566] Members

The Gen[l] Court which Col[o] Scriven is president are of Opinion the Prisoner Norman M[c]Donald is Guilty of the Charge & do Sentence him to death by being hanged. The Sentence is Approv[d] & S[d] M[c]Donald is Ordere[d] to be Executed on Monday 18th Inst. at 1 oClock in the forenoon before the Camp, the Officer of the day to see it Executed.

The Court is Dissolv[d] –

1553 This is Daniel Horry's Plantation on the Santee River, and is the current Hampton Plantation State Park north of McClellanville. Horry had taken parole, possibly to save his estate, and he had left for England. George Washington visited this plantation in 1791.

1554 Unable to find any information on Norman McDonald, but he was hanged for being a spy on 18 March 1782.

1555 William McLeland.

1556 John Palmer had served as a captain under General Marion since 1780.

1557 John Norwood had been a private in the St. David's Parish Volunteer Company before the fall of Charleston. When Captain Andrew Miller died at the Battle of Cowpens he became a captain. He was wounded at Pratt's Mill on 3 October 1781 when he fought "Bloody Bill" Cunningham.

1558 Unable to find any information on William Bennett or Jacob All.

1559 Unable to find any information on B. Davier. It might possibly be Lieutenant Benjamin Davis.

1560 James Witherspoon was with Marion from 1780 to 1782 and provided food and forage. He died in 1791. His son, James Witherspoon, Jr. served 55 days as a private and 30 days as an adjutant. He also served as a lieutenant during 1780 and 1781 under Colonel McDonald.

1561 This may be the same Willam Lewis who had enlisted in the 2nd South Carolina on 4 November 1775.

1562 James Jenkins had enlisted as a sergeant under Captain Thomas Ellerbee and Colonel George Hicks in February 1776. In September 1779, he was commissioned as a lieutenant. In 1781 he was under Colonel Wade Hampton and General Thomas Sumter and was in the skirmish at Fort Granby. He joined Marion and was in a skirmish near Orangeburg. He was also in a skirmish near Beaufort, in the battle of Eutaw Springs and the skirmish at Strawberry Ferry. He died on 26 August 1839.

1563 Edward Thomas was under Captain Benjamin Marion and Colonel Singleton in the Berkeley County Militia Regiment in 1775. He was a lieutenant in the militia before the fall of Charleston. He was dead prior to August 1785.

1564 William Skellin served in the 1st South Carolina Regiment under Captain Thomas Lynch during 1775.

1565 Peter Hubbard was born in 1756. He enlisted in the 3rd South Carolina Rangers on 18 July 1775 under Captain Samuel Wise and John C. Smith. He was in the battle of Sullivan's Island in June 1776. After Sullivan's Island he joined the militia under Colonel Hicks. After the war he moved to Tennessee and Illinois. He died in 1832 at the age of 76.

1566 Henry Kuhn served as a 2nd Lieutenant in Marion's partisans.

17th March 1782 –

Francis Marion
Brigd Genl Militia

Horrys Plantn Camp Santee 21st March 1782

At a Genl court Martial to Set 23d Inst. at the usual Hours and at the Presidents Quarters for the trial of Daniel [1567] and Berry [1568] of Colo Horry's Regt for desertion, And Such other prisoners as are Ordered before them. Colo Maham President, Capt withers, Lts Maxwell [1569] Lesesne [1570] & Ferguson [1571] of Horry's Corps with Eight Officers of Colo Mahams Corps, Members. Capt Elliot Judge Advocate, Evidences are ordered to attend –

Horry's Plantn Camp Santee March 23rd 1782

By the Number of Guns fired in & around the Camp it is expected that the men are making away with their ammunition. It was formerly orderd And now repeated that Commanding Officers of Regiments & Companys are to be liable for all ammunition that may be lost by their Men, not unavoidably when those Men have not been Called to an account by Bringing them to trial may Depend such Officers will be Called on for any Deficiencies their men may want of the Quantity Given them. It is again repeated that roll Call shall be at 4 OClock in the Afternoon. An Officer from each Compy to attend when their Arms & ammunition will be Examin'd. Every Man who ~~Should~~ Shall fire a Gun in or within hearing of the Camp must be confined for causing a faulse Alarm, and be tried agreeable to the late Law. The Officers of the Day are particularly order'd to see it is put in Execution All Orders respecting the Good order of the Camp. The Eldest Field Officer in Camp is to see it put in Execution & to Give orders for keeping of the Camp Clean & picqueting all Horses at a Proper place. With such other regulations as he May think necessary for the Camp Commanding Officer of Regt will order and see their men trained Once a Day constantly when the Weather will Permit it –

Lt Peter Farssen Resigned his Commission in the South Carolina 2d Regt is no Longer to be Considerd As ~~such~~ an Officer – [1572]

[1567] There were two John Daniels with Marion at this time. One was John Daniel, born in 1760 in Marboro District, South Carolina. He served under Captain Moses Pearson. After the war he married Rebeca Stephens and moved to Georgia. However, this is most likely John M. Daniel, who served in Horry's dragoons.

[1568] Thomas Berry enlisted in the 2nd South Carolina Regiment on 4 November 1775. He was in the 3rd South Carolina Rangers on 1 October 1777, and he was taken prisoner at Savannah on 29 December 1778. He was in the cavalry under Horry with Marion's partisans. On 21 March 1782 he was court martialed for desertion. He was in the 1st South Carolina Regiment (Marion's partisans) on 17 June 1782.

[1569] Josiah Maxwell was a lieutenant in Horry's Dragoons.

[1570] Thomas Lesesne had been a captain in the 2nd South Carolina before the fall of Charleston.

[1571] Artemas Ferguson was appointed a cornet in Horry's Dragoons on 24 November 1781 under Captain Henry Lenud. He was under Captain Youngblood in 1782.

[1572] Peter Foissin of the 2nd South Carolina Regiment.

Brigade Orders 3^{d} April 1782 Horrys Plant^{n}

Commanding Officers of Regiments to make Return of all their men who have Surrendered since the 27 Septr 1781, distinguishing those who have Surrendered Since the 16th of Decr last from those before that day –

This return is expected immediately that it May be transmitted to the Governor before the First of Next month –

General Orders Porcher's Plantation April 17th 1782 [1573]

As Great Irregularities have been made by the different Corps under my Command, Contrary to the Rules & Articles of war & the Militia Law _ _ It is Hereby Order'd that no Soldier shall be admitted to go in to Any Plantation for Provisions & forage with out a Commision'd Officer or Quarters Master from each Reg^{t} or Corps – all Commanding Officers are ordered to Observe the above with the Greatest strictness –

The Quarter Master of the Brigade will Issue Provisions, Forage & Potts & every other Article necessary For the troops to the quarter Master of Reg^{ts} or Corps According to their Returns

Any Soldier of any Denomination who is found taking any Article from Any Plantation wither from white or Black will be Deem'd A Marauder & Plunderer & Shall suffer immediate Death. Orders will be given to parties for that purpose. Any Soldier found Burning Fences or ~~other~~ destroying Orchards shall suffer agreeable to the Article of war – Which says If an Officer be cashier'd, If a Soldier corporate Punishment according to Sentence of a Court martial and turn'd in the Ranks, and Oblig'd to double the Duty he was Liable to do on failure of which to be turned in the Continental Service not exceeding one Year –

Genl Orders by Genl Marion

Camp St Stephens 19th April 1782 Bluford [1574]

Whereas the Quarter Master & Commissary of the Brigade had been Interrupted and abused in the Execution of his Business to the detriment of the Service, It is hereby Strictly ordered that no Person, or persons whatever do Interrupt him or prevent him in any Respect From doing his duty, agreeable to my directions or use Any abusive words to him, under pain. If an Officer to be put under Arrest and If a Private to be confin'd and tried by a Court for disobedience of Orders & Misdemeanors to the hurt of the Service

If said Quarter Master or Commissary should Neglect his Duty or act Contrary to Orders Complaint must be made the Genl Commanding the troops when he shall be Regularly tried & suffer accordingly to his Demerits –

//

[1573] Porcher's Plantation was located where the Old Santee Canal passes under Route 45, one mile east of Eadytown.

[1574] Bluford's Plantation is now the Oakland Club in Berkeley County, located in between Pineville and Eadytown, east of Lake Marion.

Genl Orders St Stephens 22d April 1782

Whereas a number of women & other persons go to Chars Town without permission Contrary to the good of our Service It is ordered that no person, or women whatever go to Chas Town, without my written pass, such as may person to go, without such pass Shall be if a woman, not suffer to Return but ~~If~~ be sent by a party to the Enemys Lines & be Obligd to Remain there, If a Man or neagro be Deemed a Spy & Suffer accordingly –

B.O. May 4th 1782 –

No person to go out of camp without a pass from one of my Aid de Camps or Brigade Major & such passes must be first signed by the Commanding Officer of the Regt he belongs to.

As it is probable the Troops will be on the present Ground for some time it is recommended that they made good convenient Hutts –

Every man must put his arms in the best order & Keep them in such Places as they may take them in the darkest night. In case of alarm they will parade In front of their Encampment without confusion & Observe the Strictest Silence.

A Return of ammunition wanting to make up Twelve rounds pr a Man to be made Immediately & deliver'd Majr Muller An adjutant for The day to be appointed who will attend Majr Muller at one OClock to Receive Orders and the Men for Duty must parade in front of the Encampment opposite the Left and Right of the two Brigades –

G.O. May 5th 1782 –

Parole Frankford } Germantown / Haddenfield { Countersigns

The Heads of Each great Department in the army viz. the quarter Master consisting of the Waggon and forage Branches, the Commissarys in both the Purchasing and Issuing Branches, the Medical Orders Clothier And Hides Department, will on future make monthly Returns of the first of Each Month of the State of Each of their Departments in their respective Branches in a Summary way of the Stores receiv'd, Issu'd & on Hand.

Genl Marion will order the Picquit at the Cross Road [1575] to be furnish'd by the Militia likewise the Necessary Picquits for the Security of the Right Flank of the Army –

G.O. May 6th 1782 –

Parole Hanover } Mackinsack / Mowerstraw { Countersigns

The Commissary of the Brigade of the Light troops & detached parties, will report to Lt Colo Hamilton [1576] the D.C. Genl an exact State of all the hides which fallen under their Charge where they are, to whom they have been Delivered, or how disposed of. This Report is wanted immediately.

[1575] The Cross Road was Monck's Corner.

[1576] Deputy Commissary General John Hamilton.

In future all the Hides are to be dispose[d] of as L[t] Hamilton shall direct. The Commissary of Hides will therefore take this order from him on this Business –

Gen[l] Orders May 7th 1782 _ Paroles – Countersigns {

The North Carolina Brigade is permitted this afternoon to fire three Rounds blank Cartrig[s] – [1577]

The pay Mast[r] of the maryland & Pensil[v] Brigade are to make immediate application to L[t] Hamilton D.C. who will Issue to them the Proportion of Linnen arriv[d] for Coates.[1578] The Commanding Officers will order them made up Agreeable to a pattern which will be furnished by the Cloathier. When made they are not to be Delivered out, but Lodged with the pay master. At the Gen[l] Court Martial of which L[t] Col[o] Adams [1579] is president Held the 3d Instant was tried L[t] Ruben Wilkinson[1580] of the North Carolina Brigade on the Charge of Behaving in a Scandalous & infamous Manner, unbecoming the Carract[r] of an Officer and Gentleman, for going in Maj[r] Rees name on the &[tc] and owners of Ferrys & furnish an Express On his way to Philadelphia with Forage &[tc] and to pass Him over the Differ[t] ferreys on this publick Account, which the Person passing as an Express was one L[t] Wilkinson on private Business the Court having matturely considered the matters are of Oppinion that L[t] Wilkinson is Guilty of a Breach of the 21 Act 14 Section of the rules and art of war, are therefore under the necessity of annexing the punishment Specify[d] Dismission from the means adopted to afford him his wished for Relief were certainly unjustifiable & unprecedented & necessity

---

1577 The camp of Greene's army was located at Pon Pon, near Jacksonboro. This may mean that Marion was camped with the main army at this time, or it may mean that he has included Greene's general orders in his orderly book, though Marion may have been still camped in St. Stephens. The North Carolina Brigade consisted of the 1st, 2nd, 3rd and 4th North Carolina Regiments under the command of Hardee Murfree. However they were a mere ghost of a brigade, with many of their men going home. In March they conducted blank fire volleys, which scared the inhabitants. They received permission to do it again here, and it was said that "they made a verry bad fire."

1578 In January 1781, the Pennsylvania Continentals had mutinied in Washington's Grand Army because they said their enlistment contracts had been violated. Soldiers who had enlisted for three years were being detained on the pretext that they had enlisted for the duration of the war. The Pennsylvania line had been reorganized into three Provisional Battalions. Once they arrived in South Carolina there was more dissent. The new grievance of the Pennsylvanians was that they were not used to the conditions of fighting in the South. General St. Clair wrote, "Can soldiers be expected to do their duty, clothed in rags and fed on rice." Two sergeants, one from Pennsylvania and one from Maryland, were approached by the British to capture Greene and give him to the British. The plan was foiled by the sergeant's wife, who told General Greene. Greene had them arrested, and hanged for mutiny. The Pennsylvania line was the only northern unit from the Continental army to fight in the Carolinas. Though Maryland and Delaware fought extensively in the South, some historians consider them to be Southern or Middle Atlantic states, and not northern regiments.

1579 Lieutenant Colonel Peter Adams, commander of the 3rd Maryland Regiment. He is referred to as lieutenant colonel commandant because a regiment is commanded by a colonel. If there is no colonel, the lieutenant colonel is now in command, and he is known as commandant.

1580 Reuben Wilkinson had enlisted in the 4th North Carolina Regiment on December 20, 1776. He retired on June 1, 1778. He then reenlisted as an ensign in the 3rd North Carolina Regiment on May 1, 1779. He was promoted to Lieutenant in 1780. He retired again on July 2, 1782, possibly due to these charges being leveled against him.

must have been more than Commonly Importunate in Urging an Officer, after Six Years Service to so far from the Customary practice of Gentl^{mn} urged in his Diffence is a prerogative of that being only who Judges by heart, the Court will not presume to Decide that point, but Leave it to operate on the Gen^{ls} Feelings & recommen'd at the Same time L^{t}. wilkinson as an object for him to exercise his Clemency on –

The Genl is very Sorry that an offic^{r} who Append to bear as good a Carract^{r} as L^{t} wilkinson should be Guilty of an act that Should bring either his honour or integrity into question, neither merit, distress or bright of Service can Justify an Offic^{r} adopting dishonorable Measures, but as the present has more the appearance of that evil intention The Genl does not confirm Sentence, However it is much to be wished that this may Prove an useful Lesson & Serve to Show the necessity of Suffering with dignity, rather than attempt relief through improper Channel –

L^{t}. Wilkinson is released from his arrest –

Genl Orders 9th May 1782 –

At the Request of Brigadier Genl Gist,[1581] M^{r}. Brown is Appointed Volunteer Aid to him & is to be accordingly Respected – [1582]

At the Genl Court Martial of which Lt Col^{o} Adams was President held 7th Inst was tried Elias Smith private in the Maryland Line for desertion & Reenlisting in Lt Col^{o} Mahams Corps of Cavalry, the Court Sentence him to Receive One Hundred Lashes & then to Join his Regiments at the same court was tried Wm Roberts, Drummer in the Maryland line for desertion & Reenlisting in the Cavalry Commanded by Col^{o} Maham The court Sentence him to Receive Twenty five lashes & then To Join his Regiment –

The Genl approved of the above mentioned Sentences & Orders the Punishment to be inflicted this Morning at Roll Call –

At the same court held the 8th Int - was tried Clare on suspicion of Murder, the court taking the case of the Prisoner under Consideration can not fix any degree of Criminality on him, but think he acted strictly up to the Line of his Duty, they therefore acquit him – [1583]

The Genl approves the Sentence and orders him to be Released & Join his Regt immediately –

In future Soldiers whose Crimes are Commensurable by a Regimental Court Martial are not to be Sent to the Provost _ _ _ _ _ _

Genl Orders May 11th 1782

The Genl Court Martial of which Lt Col^{o} Command^{t} Adams is president will convene Tomorrow Morning – 10 OClock –

A party of active Infantry furnished with Provisions for To Morrow & Next Day is to be Ready to March To morrow Morning at Sunrise, the commanding Officers will attend at Head Quarters for his Orders –

Maj^{r} Eccleston will command the party _ _ _ _ _[1584]

[1581] Brigadier General Mordecai Gist of the Maryland Line.

[1582] Unable to find any information on who Mr. Brown is, but he is most likely from Maryland.

[1583] Possibly Private William Clary who served with the 2nd South Carolina Regiment since 1775.

[1584] Major John Eccleston of the 1st Maryland Regiment.

Genl Orders May 13th 1782 –

For tomorrow { Lt Colonel Stewart – [1585]
Brig^{d} Maj^{r} Bankson – [1586]

The Dept Quart^{r} Master Genl is directed daily to inspect the environs of Camp & order all dead horses offal & filth to be Buried

The ~~Genl~~ Necessity of having the camp Perfectly kept clean in order to ~~Prevent~~ preserve the Health of the Soldiery is an object of such importance that the Genl expects the Commanding Officers of Corps will pay a strict attention thereto –

Soldiers coming off guards on future are Not to draw their load but by a special order from the Commanding Officer of their Regt –

Brig^{d} Ordrs by Genl Marion 13th May 1782

Commanding Officers of Regiments to appoint fatuge Men every day to clean the filth and Rubbish in front And rear of their Camp at least 60 yards from the Line of their Hutts [1587]

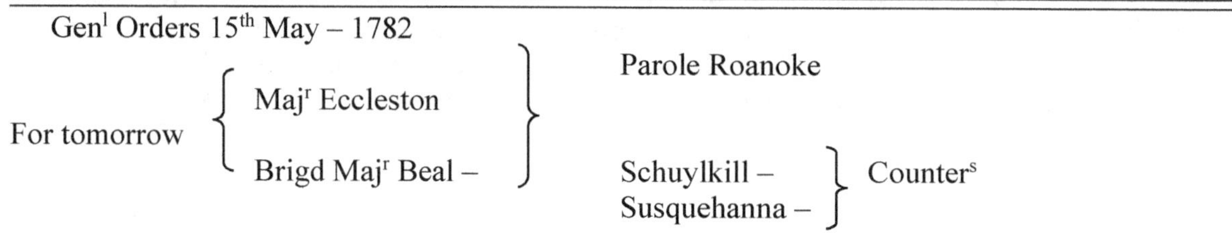

Genl Orders 15th May – 1782

For tomorrow { Maj^{r} Eccleston
Brigd Maj^{r} Beal – } Parole Roanoke

Schuylkill –
Susquehanna – } Counter^{s}

The Brigad^{e} Maj^{r} of the Light Corps is to attend at the Orderly Office twice a week on Monday and Friday for orders, The Dep^{y} Adj^{t} Genl will transmit it, the Commanding Officer of the Light Troops extracts of any particular orders which May be Issued _ _ _ _ _ _ [1588]

Genl orders 17th May 1782 –

The Genl Desires of establishing the army upon a permanent footing as its present form and composition will admit as well as for perfecting its discipline, as for introducing regularity in the Accounts & Economy in the Corps, directs that the Officers and Soldiers in the Pensilvania and Maryland lines continue in the several Corps & companies Conformable to the Present arrangement without any Alteration whatever, except such as may be ordered by the Commanding In Chief, the Secretary of War or as Genl Orders of this Army –

The Officers & Soldiers in future not actually Serving with the Army, except Such Officers and their waiters As were included in the Arrangement & have been Since furlough'd, are to be included in the Muster Roll – And the Returns of the Army, all other both Officers And Men not on duty in either of the Carolinas, or Georgia, or in the Hospitals of those States are to Immediately be Struck

---

[1585] Lieutenant Colonel John Stewart commander of the 1st Maryland Regiment.

[1586] John Bankson, the brigade major of the Maryland Continental brigade.

[1587] This should read "fatigue Men".

[1588] The Light Corps was commanded by Lieutenant Colonel John Laurens. It consisted of the Delaware Regiment's Light Infantry Company and the dismounted troops of Lee's Legion.

of the Muster Roll & a list of the rank & names of the Officers, likewise a list of the names & a description of the men to be Immediately Reported to Head Quarters & a copy of it to be Sent to the Commanding Officer Superintending at the State Rendesvous to which they Belong –

Genl Orders 18th May 1782

At the General Court martial of which Bⁿ Genl Guest is President, heald the 17th Int was tried[1589]

Lt Col^o Commandant Adams upon the following charges

1st for refusing to deliver to the G^r Guard of the Second Maryland Regiment too Soldiers of s^d Regiment two Soldiers politely & propperly required from him –

2^d In having contrary to the rules of the army Sent to the provost Guard a Soldier of the 2^d Maryland Reg^t without having Sent any Report of his name, or Crime to the Commanding Officer of the Reg^t to which he belongs & for returning a Rude & impertinent answer to Lt Col^o Stewarts Second Battalion, unbecoming an Officer or a Gentleman

The Col^o after duly considering the Evidence of Lt Col^o Adams defence are of Oppinion that he was justifiable in his Conduct & they acquit him with honour –

The General approves the Sentence of the court is Dissolved

The Genl whishes for the honor of the Army As well as for the good of service, that Officers of high Rank would treat each other with a polite attention as it tends to cultivate respect & subordination from the Example, besides which, it promotes the pleasure of society And increases the force of an Army, by uniting its interests And efforts. It is also much to be wished that all Genl Rules which Serve as governing principles, in the order of Service Should be as little deviated from as possible –

After Orders –

Mr Samuel Dwight is appointed Assistant Commissary of Issues to the Poast at George Town is accordingly to be respected – [1590]

Marion's old enemy, Major Micajah Ganey, commanded the Loyalists in the area of the Pee Dee River, near present day Marion. Ganey had made a truce with Francis Marion the previous year after Georgetown had been captured. This truce was due to expire on June 17th.

A "Scotsman from Charleston" was able to make it through Marion's lines on the pretense that he was returning to his home, but the man was sent to call any other Highlanders to arms in the Pee Dee area. He misinformed the Loyalists that the British were stronger than Marion and would be mounting an expedition soon. Within the Pee Dee the Loyalists began gathering their forces. The Scotsman who spread disinformation was caught and hanged. The governors of North and South Carolina mounted a joint expedition under Marion to stop the rising Loyalist rebellion. Marion sent three columns into the truce ground from different directions.

[1589] This should read Brigadier General Gist.

[1590] Samuel Dwight (D'Witt) enlisted in the Volunteer Company of Horse of St. Mark's Parish under Captain Mathew Singleton on 26 August 1775. He was made the assistant commissary of issue at Georgetown on 18 May 1782.

Orders at Burches Mill 3 June 1782
As a number of persons have come & submitted to the Americans – And have obtained pardon for the Offences committed against the State, it is hereby ordered that Such men shall not be molested. Those that do in any respect commit Such outrage by taking what is Called private Satisfaction will be made to Suffer Agreeable to the Laws of this State in the most Rigorous manner And it is Recommended As Christians to forgive & forget all Injuries which have been committed by such who have been Led away by our Enemies –

Camp at Burches Mill 7th June 1782 –
Col° Lamb Benton having given good & sufficient reasons For the appearance of Neglect of Duty, the Genl is Satisfied with his Conduct & Orders the Col° to resume the Command of his Regt

Colonel John Baxter led one of the columns into the truce ground and followed a large force of 500 men led Major Ganey. Baxter's men learned that the Loyalists had seized a boatload of rice on Black Lake.[1591] The Patriots were not large enough to take on Ganey's militia and Baxter requested reinforcements from Marion. He had his men fire on the Loyalists as they proceeded up the lake in canoes. Only one man was wounded in the exchange and that was Robert James, a friend of Marion.

After the brief skirmish Ganey realized that he was surrounded on three sides and he sued for an armistice. Marion invited him to cross the Pee Dee and come to conference at Burches Mill. Marion knew that if the country were to be united there would have to be forgiveness for its enemies.

The two commanders agreed that the Loyalists should restore all plundered property wherever possible, become peaceable citizens, submit to the laws of the State and sign a declaration of allegiance to South Carolina and to the United States. The treaty was good for all Loyalists except Colonel David Fanning, Major Samuel Andrews and "Bloody" Bill Cunningham. They were to receive no mercy.

Ganey's followers laid down their arms at Bowling Green. Ganey told Marion that he could not relinquish his command to Marion, but would have to do that to Colonel Balfour from whom he received his commission. Once that was done Ganey promised that he would return.[1592]

Genl Or. Camp Smiths Mill the 27th June 1782 – [1593]
Any Person who ~~may~~ Should strike, molest or In any way abuse any of the men who submit to us & have Recd protection shall be immediately taken & confind and suffer agreeable to Law with the utmost severity

On 11 July the British evacuated Savannah. The British Regular troops embarked for Charlestown and New York, while about 2,500 white and 4,000 Black civilian refugees were transferred to the Tybee and Cockspur islands. The Loyalists would end up in Florida. After the British had left Major General Anthony Wayne marched in with his Continentals

---

[1591] Black Lake was located west of Klondike, on the Pee Dee River.

[1592] Bowling Green is close to where Burches Mill was located, on the Pee Dee River.

[1593] Smith's Mill was located near Georgetown.

and occupied Savannah. The only Southern city, north of Florida, that still had any British troops was Charleston. Greene ordered Marion to patrol the area between the Santee and the Cooper River.

Camp Bleufort Santee 22d July – 1782 – [1594]
Militia –

By a Genl Militia Court Martial Held in St Stephens 21st July – 1782 of which Colo Richardsn Was President the following Persons were tried & Sentence Past which Sentences are approv'd of and Confirmed And Ordered to be Carry'd in Execution by Genl Marion

Captn Linder tried for disobedience of Orders was found Guilty & Sentence him to be Cashier'd & reduc'd to the Ranks & to serve Forty Days extraordinary immediately After being so reduc'd – [1595]

Lt Syders tried for Disobedience of Orders.[1596] The Court are of Opinion that the Charge is by no Means supported, they therefore acquit him, his Arrest is taken off & he is ordered to take commd

James Cooper tried for ~~Disobedience~~ Desertion is guilty & by the 2nd clause of the Militia Act sentence him to do double the Length of time he was draught'd For – [1597]

John Camball Charg'd with disobedience of Orders is acquitted[1598]

John Falkner tried for refusing to March When ordered is sentenc'd to serve double the Length of Time he was Draught'd for[1599]

William Beesley tried For disobedience of orders is Acquitted by the court– [1600]

William Cox charged with Neglect of Duty is Acquitted by the court – [1601]

Anthony Noon tried For Disobedience of orders is acquitted by the Court – [1602]

Thomas Hitchcock tried for Sleeping on his Post is sentenc'd to serve Twelve Months in one of the Continental Regts of this State – [1603]

1594 Bluefield Plantation was located where Lake Marion is now, near the present day Eadytown.

1595 Daniel Linder had served in the Berkeley County Militia Regiment during 1776.

1596 John Syders served 48 days in the militia.

1597 James Cooper was a special quartermaster appointed by Governor Rutledge to have charge of stores from 10 April 1779 to 9 February 1780. After the surrender of Charleston he was a private in the cavalry and was in the battle at Parker's Ferry. On 22 July 1782 he was found guilty of desertion and sentenced to serve double the length of time he was drafted for. He was dead by 1786.

1598 John Campbell served as a horseman and as infantry under Captain John Cowan from 12 May 1780 to 20 February 1783.

1599 John Faulkner was supposed to serve 30 days in Marion's Brigade in 1782 but he was found guilty of refusing to march when ordered on 22 July 1782. He was sentenced to serve double the length of time he was drafted for.

1600 William Beesley served 65 days in Marion's partisans in 1782.

1601 William Cox was born on 4 February 1763 and lived in the Marlboro District in South Carolina. He entered the service in 1779 and served numerous short terms under various officers until 1783.

1602 Unable to find any information on Anthony Noon.

1603 Thomas Hitchcock was born on 3 November 1755 and lived in the Marborough District. He enlisted in December 1780 and served in Peter Horry's cavalry. He was in the battles at Wacamaw and Goose Creek Bridge, where he was wounded. He was also at Black Mingo, Camden, and Sand Hill Creek. After his punishment he was ordered to serve in the 1st South Carolina Regiment. He deserted to the British in the spring

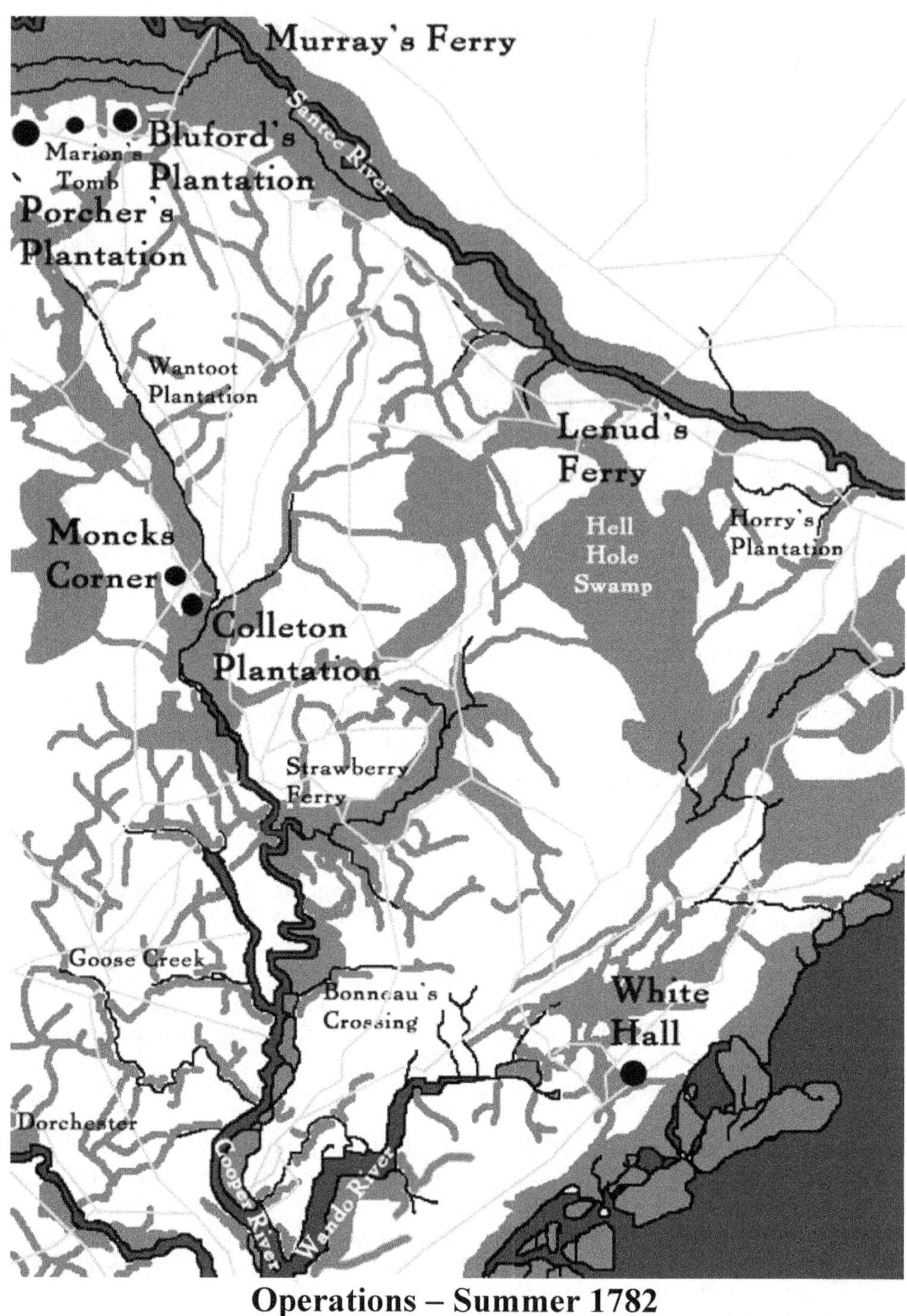

Operations – Summer 1782

of 1783, but was able to return. He married, moved to North Carolina, then to Tennessee. He died on 28 July 1844 at the age of 89.

Camp Watbo – 27th July 1782 – [1604]

At a Court of which Col° Benton was president the following Persons were try'd & the Sentence approv'd of by the General –

Thomas Beacon – Charg'd with disobedience of orders is sentenced to serve double the Length of time he was otherwise Liable to – [1605]

Nathaniel Baker, disobedience of orders, double Duty – [1606]

Adwell Adkins – D°. _________ D°. Double the length of Duty[1607]

Solomon Moody - D°. ________ D°. – acquitted –[1608]

Fredh Adkins – D°. _________ D°. his trial is postponed Not being able to attend the Summons _ [1609]

Hugh Gamble disobedience of orders – acquitted – [1610]

Michael Cobey– D°. ________ D°. – Acquitted [1611]

Alexr Robeson 50 years old Exempted duty & acquitted [1612]

Willm Barrett – disobedience of Orders, Guilty and liable to serve double the time he was Drafted for, his case being recommended as somewhat exasible to the Persenal, he hereby remits the Sentence _ [1613]

Edmund Collins,[1614] Dillard Collins [1615] & John (*unable to determine name since it is cut off in original*) to Serve double the time they were Drafted for –

Thomas Kolb – disobedience of orders, acquitted [1616]

John Chiney [1617] and Justiner [1618] disobedience of orders to serve double duty –

[1604] Though Marion moved camps often, he frequently camped at Colleton Plantation, near Monck's Corner and the Wadboo Bridge.

[1605] Unable to find any information on Thomas Beacon.

[1606] Nathaniel Baker served in the militia and had been at the fall of Charleston. On 27 July 1782 he was found guilty of disobeying orders and was sentenced to do double the length of time he was drafted for.

[1607] Ardwell Atkinson served 365 days in the militia under General Marion. On 27 July 1782 he was found guilty of disobeying orders and was sentenced to do double the length of time he was drafted for.

[1608] Solomon Moody served 67 days in the militia under General Marion.

[1609] Unable to find any information on Frederick Atkinson.

[1610] Hugh Gamble served under General Marion and provided sundries for militia and Continental use during 1781 and 1782.

[1611] Michael Cobia was born in 1756. He had served in the German Fusiliers of Charleston during 1775. He was at the siege of Savannah in 1779. He served with his brothers, Nicholas and Francis Cobia under Marion.

[1612] Alexander Robinson had served in an independent company of Rangers under Captain Robert Ellison during 1775. When he was drafted to serve in Marion's partisans he showed up for muster, but was exempted from duty since he was 50 years old.

[1613] William Barrett served in Colonel Richardson's regiment during 1780 – 1782.

[1614] Unable to find any information on Edmund Collins.

[1615] Unable to find any information on Dillard Collins. He may possibly be Daniel Collins.

[1616] Thomas Kolb had enlisted in the 5th South Carolina Regiment on 18 March 1776 as a sergeant under Captain Thomas Potts. He was discharged on 13 July 1777. He returned and was in the battle of Stono Ferry. He was a sergeant under General Marion in 1781 and 1782.

[1617] John China was born on 8 December 1764 in the Marlborough District. He served under Colonel Richard Richardson in 1781. He served under Captain James Theus and Colonel Maham in 1782. He died on 24 August 1847 at the age of 83.

[1618] Possibly Pierre Juttier who had served in the 1st South Carolina Regiment during May 1777.

Jonathan Burbridge,[1619] Thos Burbridge & Ritch Burbridge[1620] – disobedience of orders, acquitted

John Hilton [1621] & Folliver Martin,[1622] to Serve Double the time they were Otherwise liable to –

August 4th – 1782 –

Proceedings of the Consolation of Colo Maham & Majr Conyers Corps of Cavalry at Captn Lenuds[1623] Plantn [1624]

Lt Colo Commandant Hezekiah Maham Comn dated 22 Jun 1781

1st Majr Edmund Massing Hyrne ___________ 22 Do. _

2d Majr James Conyers[1625] _________________ 1st March 1782

Captains –

1	Thos Giles [1626]	}	
2	James Theus [1627]	}	Cavalry .... 22d June 1781
3	Samuel Taylor [1628]	}	.... 4th Febr – 1782
4	James Simons [1629]	}	Infantry .... 1st March – 82

[1619] All three Burbridge brothers had been with the 2nd South Carolina in Savannah.

[1620] Possibly William Burbridge, who enlisted on 19 June 1778 as a drummer in the 2nd South Carolina Regiment with his brothers, under Captain Charles Motte. He remained in service until 1783.

[1621] John Hilton was born in 1759 in South Carolina. He had served in the 3rd South Carolina Rangers under Captain Edward Richardson in 1775.

[1622] Possibly Sergeant Robert Martin, who served in the militia from 1779 to 1782. He was at the surrender of Charleston. He was in the cavalry under Colonel Maham and General Marion in 1781 and 1782.

[1623] Captain Henry Lenud was in the cavalry under Colonel Peter Horry in 1781.

[1624] This unit became known as the South Carolina State Legion.

[1625] James Conyers was a captain and commanded a cavalry troop in Marion's partisans in 1781. He was promoted to major on 1 March 1782. His cavalry was consolidated with Maham's in June 1782 and then he assumed command due to Maham being on parole. He served 362 days.

[1626] Thomas Giles served in the cavalry under Colonel Daniel Horry in 1779 and was in the siege of Savannah. After the fall of Charleston he served as a captain in Colonel Maham's cavalry and was in the battle of Eutaw Springs. He commanded 1st Cavalry troop of Maham's Dragoons in 1782.

[1627] James Theus was born in 1755. He served as a lieutenant in the light dragoons from 1779 to 1782 under Colonel Maham. He was promoted to captain on 22 June 1781. He commanded 2nd Cavalry troop of Maham's Dragoons. He died on 8 September 1806 at the age of 51.

[1628] Samuel Taylor was promoted to captain on 4 February 1782 and commanded 3rd Infantry Troop of Maham's Dragoons. He married Ann Bonneau in August 1784. He died on 14 January 1815.

[1629] James Simons was promoted to captain on 1 March 1782 and commanded the 4th Infantry Troop of Maham's Dragoons.

First Lieutenants –

No.	Name	Branch	Date
1	Samuel Cooper [1630]	}	19th Novr _ 1781
2	Jacob Barnet [1631]	Cavalry	1st March 1782
3	Henry Hyrne [1632]	}	5th d^{o}. _ _ _ _
4	William Basquin [1633]	Infantry	6 _ d^{o}. _ _ _ _

Second Lieutenants

No.	Name	Branch	Date
1st	Isaac Dubose [1634]	}	6th May _ 1782
2^{d}.	Robt Mis Campbell [1635]	Cavalry	1st March 1782
3^{d}	Jervis Henry Stevens [1636]	}	1st Augt _ 1782
4th	John Bryan [1637]	Infantry	2^{d} d^{o}. _ _ _ _

Cornets

No.	Name	Branch
1st	Mirron Winn [1638]	} . . . .
2^{d}.	Thos Dryton [1639] _	Cavalry . . . .
3^{d}.	William Postell [1640] –	Infantry

[1630] Samuel Cooper was promoted to 1st Lieutenant on 19 November 1781 and was in 1st Cavalry troop in Maham's Dragoons in 1782.

[1631] Jacob Barnett served as a lieutenant under Captain William Byers and Colonel Neel in the Cherokee War. He was a captain in the militia under Colonel Brandon when he was wounded at Rocky Mount in 1780. He served as a captain-lieutenant in the light dragoons under Lieutenant Colonel Henry Hampton and General Sumter in 1781. He was promoted to 1st Lieutenant (Continental rank) on 1 March 1782 and was in 2nd Cavalry troop in Maham's Dragoons.

[1632] Henry Hyrne served as a captain of the Cheraw Company of the Colleton County Regiment of Foot during 1775. He was a lieutenant in the 5th South Carolina Regiment until he resigned on 29 November 1779. He lost a horse at the battle of Briar Creek. He was promoted to 1st Lieutenant on 5 March 1782 and was in the 3rd Infantry troop in Maham's Dragoons.

[1633] William Basquen served in the light dragoons under Colonel Maham from 12 October 1781 to 20 January 1783. He was promoted to 1st Lieutenant on 6 March 1782 and was in the 4th Infantry troop in Maham's Dragoons.

[1634] Isaac Dubose, former 1st Lieutenant of the 2nd South Carolina Regiment in 1776.

[1635] Robert Miscampbell served 75 days as an adjutant and a lieutenant in the Marion's partisans. He was promoted to 2nd Lieutenant on 6 March 1782 and was in the 2nd Cavalry troop of Maham's Dragoons.

[1636] Jervis Henry Stevens served as an assistant deputy commissary general during 1779 and 1780. He was promoted to 2nd Lieutenant on 1 August 1782 and was in the 3rd Infantry troop of Maham's Dragoons. At sometime he was the captain of the brigantine *Richard.*

[1637] John Bryan served as a cornet under Colonel Maham. He was promoted to 2nd Lieutenant on 2 August 1782 and was in the 4th Infantry troop of Maham's Dragoons.

[1638] Minor Winn was commissioned as a cornet in the 1st Cavalry troop of Maham's Dragoons. He was later promoted to lieutenant.

[1639] Thomas Drayton was commissioned as a cornet in the 2nd Cavalry troop of Maham's Dragoons. He was later promoted to lieutenant.

[1640] William Postell was commissioned as a cornet in the 3rd Infantry troop of Maham's Dragoons. He was later promoted to captain.

Ordered that Majr Conyers take the Command of Colo Mahams State Legion & the above Arrangement to take Place, he will make the arrangement of the Non Commissd And Privates agreeable to the above By order of his Excellency Governor Mathews – [1641]

On the Consolation it appeared that Capt Samuel tyler had a Brevet dated the 11th day of Febr, it also appeared that he succeeded Capt John Carraway Smith who Commanded a Troop in Colo Mahams Corps of Cavalry Untill the 25th of said Month & resigned on the 26, the Board of Officers met this day have therefore determined that Capt Taylors Commission should be dated the 26 February 1782 –

On the Promotion of Capt Taylor a vacancy appeared In favour of Cadit Thomas Drayton whose Brevet was dated the 5th March, the Remove of Capn Taylor gave rank to Mr. Drayton

In the Arrangement of the Cavalry & Infantry Officers, the Question was whether the Eldest Officers of each rank should be return'd as Cavalry Officers or Balotted for, determined against the Ballot [1642]

4 For it 3 –

By a Genl Militia Court Martial held at Whites Bridge 4th Augt 1782 of which Lt Colo James was President, the following persons were try'd & Sentences Past which Sentenced are approv'd of & confirm'd & Order'd to be carry'd in Execution by Genl Marion –

Wilm Smart try'd for Neglect of Duty is Acquitted.[1643] Sergt Maginnes try'd for neglect of Duty is Acquitted,[1644] Alexr Anderson tried for ~~neglect~~ Disobedience of Orders is Acquited.[1645] Thos Martin Sanders tried for Disobedience of Orders is Acquitted.[1646] Thomas Hinling tried for Disobedience of Orders is Acquitted.[1647] John Goodwin chargd with disobedience of Orders is sentenc'd by the court to serve double the Length of time he was liable to.[1648] Charles Evans tried

---

[1641] Major James Conyers had to take command of Maham's dragoons after Maham had been captured by Loyalists and signed a parole. Marion would not allow him to take the field again.

[1642] Some of Marion's men were now dismounted infantry. William Dobein James wrote "The militia were now so far relieved, that, by law, they were obliged to turn out only one month in three; but were ordered, as we have mentioned above, to be dismounted, which discouraged them, and rendered their movements less rapid. The experience derived both from the history of the revolutionary and the late war, fully shows that the militia are effective only when mounted."

[1643] William Smart served 151 days in Marion's partisans.

[1644] Jeremiah McGuiness had been in the 2nd South Carolina Regiment.

[1645] Unable to find any information on Alexander Anderson.

[1646] Thomas Martin Sanders served 168 days in Marion's partisans.

[1647] Possibly Thomas Hagen who had served in the 3rd South Carolina Rangers during 1775 and was in the militia during the siege of Charleston.

[1648] John Goodwin was at the fall of Charleston and afterwards served under Colonel Peter Horry and then under Colonel Maham. On 4 August 1782 he was found guilty of insubordination and sentenced to do double the length of time he was drafted for.

for Disobedience of Orders, is Acquitted.[1649] John Hubanks try[d] for Disobedience of Orders sentenc'd to serve double the Length of time he was Draughted.[1650]
For Samuel more try'd for Disobedience of Orders, sentenc[d] to serve double the Length of time he was Draughted for.[1651] William Cockran try[d] for Disobediance of orders, to serve double the Length of time he was draughted for – [1652]
Robert Furnell try[d] for Disobediance of orders to give double the length of time he was draught[d] for _ [1653]
James cannon try[d] for disobedience of Orders to serve double the Length of time he was draughted for.[1654] Robert Hurst try[d] for disobedience of Orders to serve double the Length of time he was draughted for –[1655]
Moses Cunning try'd for disobedience of Orders to serve double the Length of time he was draughted for –[1656]

7th August 1782 – JH Edwards ADCamp

[1649] Charles Evans had served as a captain under Colonel Powell during 1775 and 1776. He served as a private in Marion's partisans for 36 days in 1782.
[1650] John Eubanks had served as an ensign in the Volunteer Company of St. David's Parish during 1775 and 1776. On 4 August 1782 he was found guilty of insubordination and sentenced to do double the length of time he was drafted for. He later was a lieutenant in Marion's Brigade.
[1651] Samuel More served in the light dragoons under Captain Philemon Waters, Lieutenant Colonel John Thomas, Jr. and General Sumter in 1781. In 1782 he served with Marion's partisans. On 4 August 1782 he was found guilty of insubordination and sentenced to do double the length of time he was drafted for.
[1652] William Cochran served under Captain Peter Burns, Colonel Wade Hampton and General Sumter. In 1782 he was with Marion's partisans. On 4 August 1782 he was found guilty of insubordination and sentenced to do double the length of time he was drafted for.
[1653] Robert Purnal (Furnell) served 48 days in Marion's partisans in 1782. On 4 August 1782 he was found guilty of insubordination and sentenced to do double the length of time he was drafted for.
[1654] James Cannon was born in 1755 in the Ninety Six District. He volunteered during December 1775 under Captain Forard Smith to guard the frontier in Smith's Station against Indians and Loyalists. In May 1777 he volunteered under Colonel McCrary and marched to St. Mary's River, Florida. He volunteered in December 1778 under Colonel Willliams. He was in Marion's partisans in 1782. On 4 August 1782 he was found guilty of insubordination and sentenced to do double the length of time he was drafted for. After the war he moved to North Carolina, Virginia and Indiana.
[1655] Robert Hurst served in the militia during 1780-1782, and provided food and forage. He was with Marion's partisans in 1782. On 4 August 1782 he was found guilty of insubordination and sentenced to do double the length of time he was drafted for.
[1656] Moses Cummings enlisted in the 6th South Carolina Regiment on 12 August 1777. He was a fifer on 20 October 1777. He was with Marion's partisans in 1782. On 4 August 1782 he was found guilty of insubordination and sentenced to do double the length of time he was drafted for.

Gen[l] Orders Geo Town 12th Aug[t] 1782 –

Forrage to be Issued only to field Officers, Quarter Masters, Commissarys & Adjutants & such Men Imploy under the Commissarys, a Drivers & Only such as are Actively on Duty, those off Duty has no Right Either to Forage or Provisions –
Except such who have a Special Order from the General –

Field Officers . . . . . . two Horses
Quat[r] Masters . . . . . two . . Ditto
Commissary . . . . two . . Ditto
Adjutants . . . . two . .Ditto
Drivers . . . . One . Ditto
Cap[tn] Command Reg[t] one horse

Francis Marion continued to send out patrols to check on British intentions as the war wound down. Captain G. Sinclair Capers of Horry's cavalry was sent into southeastern Berkeley County with 12 troopers. At Whitehall Capers discovered 26 Black Dragoons led by two Black officers, Captain March and Lieutenant Mingo. Capers charged the dragoons and defeated them, freeing three of his neighbors who were in handcuffs as prisoners. Two of Capers' men were wounded.[1657]

Gen[l] O[r] Whites Bridge 4th Aug[t] 1782 –
The Town adjutant will attend M[r] Tho[s] Mitchell at 12 OClock every day for orders.[1658]

It is recommended to the Troops in Camp to make good Hutts that will Shelter them from Rain as it is probable the Brigade Will Remain on the ground for Some time – Commanding Officers of Reg[ts] & corps to have their Mens arms clean & in good Order & inspect once a Day their Ammunition at Roll Call which must Be at 5 OClock in the Afternoon, when on Officer of a Camp must attend – Any non Commissioned Officer or Private who absents himself from Roll call Without leave must be confirmed & brought to trial When they will suffer agreeable to Law. The Quarter Master of the Brigade will Issue Provisions & Forage to the Cavalry as well as the Infantry, to him all Returns must be made – Forage to be Issued to the Field Officers for two Horses – A Return of arms wanting to be made immediately & Deliver[d] to the Brigade Majors –

A Court to sett immediately for the tryal of Defalters that will be brought to it, Evidence to Attend. L[t] Col[o] James President, four other officers from the line Members – Field Officers of Reg[t] to turn out fatigue men to sink wells at the Brow of the Hill to obtain good water and Necessary Sinks, 150 yards in front of their Reg[t]. any Persons who do their Occations within 150 yards of the line of Encampment will be severely Punished. All Bones & Spoiled meat to be removed from before their hutts. Commanding Officers of Reg[ts] to be answerable for the conduct of their Men –

[1657] Full details of the skirmish at White Hall and details about the "Black Dragoons" are described in "*Nothing but Blood and Slaughter, Volume Four*" by Patrick O'Kelley.
[1658] Thomas Mitchell had served 45 days in the militia in 1779 under Captain Birch. He was an aide de camp to his uncle Francis Marion in 1782.

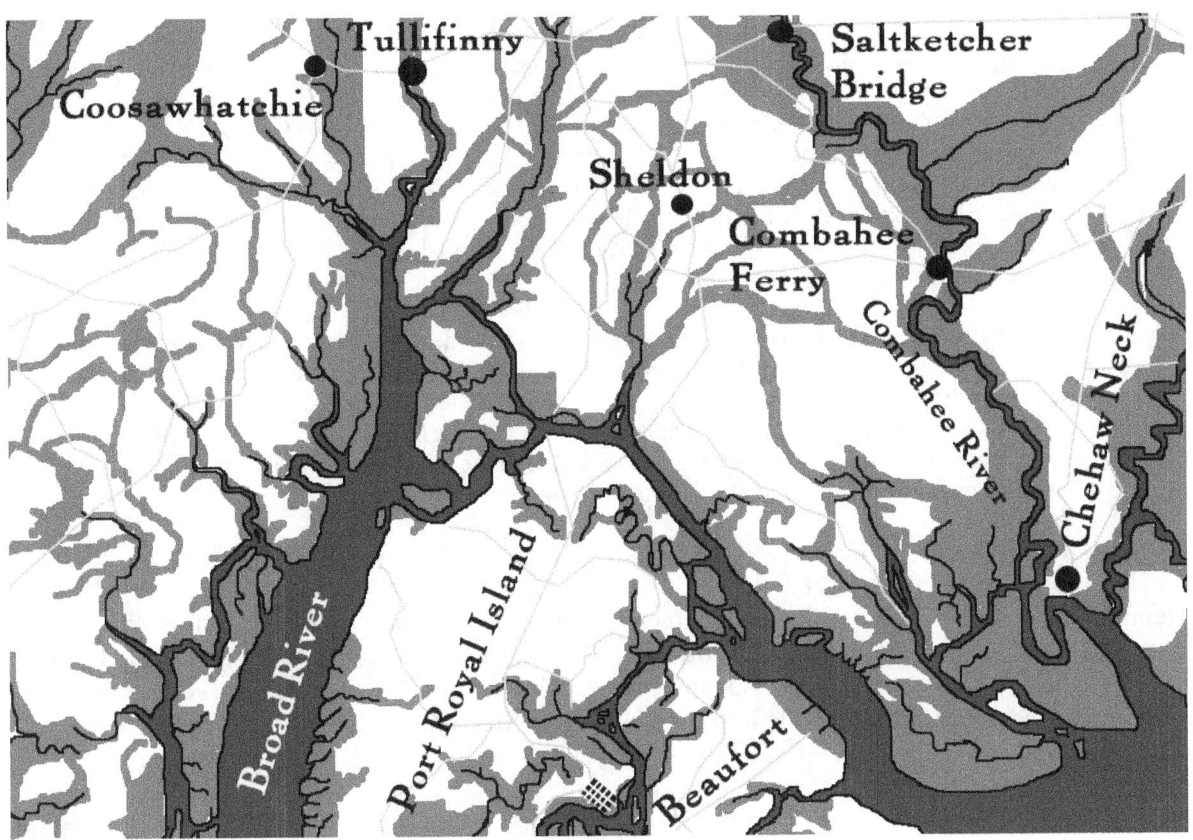

Brereton's Combahee Raid – August 1782

British General Alexander Leslie asked Greene if he could buy food from the South Carolina farmers, but Greene refused. Leslie stated that if he could not buy food he would take it. He ordered small fleets of galleys up the rivers from Charlestown to raid the plantations for food. Greene needed to stop the raiders, so he set up a howitzer on the Combahee River to prevent British ships from bringing supplies into Charlestown harbor. Leslie retaliated by sending a force to remove the artillery.

On August 25th British troops, under Major William Brereton, landed undetected near Combahee Ferry and took the howitzer Greene had set up.[1659] Greene ordered General Gist and 300 Continentals to march to the relief of the rice plantations being raided. A British covering force of 140 men had been formed to protect the rest of the expedition as it made its way to the waiting boats.

Colonel John Laurens was sick in bed near Charlestown, but when he heard about the upcoming fight he left and headed to Combahee Ferry. Gist's men couldn't cross the river and get at the British soldiers because of the enemy vessels anchored in the river. The British couldn't get to supplies on the north bank due to Gist's men. The two sides were at a stalemate.

[1659] Combahee Ferry was located seven miles east of Sheldon, where US 17 crosses the Combahee River.

Gist sent his cavalry across the Salkehatchie Bridge[1660] and he sent Laurens to Chehaw Neck where the river did a loop.[1661] On the way to Chehaw Neck Laurens stopped at the plantation of the Stock family and stayed up all night having a party with the plantation owner and his family. At 3 a.m. he roused his troops and headed towards Chehaw Neck.

Major Brereton had been informed by spies of Lauren's approach and he landed a force of infantry that would ambush the Patriots when they arrived. As Laurens approached the site he detected the British ambush. He did not retreat, but instead decided to attack the superior British force. As Laurens and his men charged, the British fired upon the Patriots. Laurens was mortally wounded and fell to the ground. Laurens had led his men in a double envelopment and half of his men were killed or wounded.

Gist saw the fight and had his men swim the river to help Laurens. Under covering fire from their vessels the British were able to withdraw, leaving Gist with nothing but a few horses. The British casualties were one man killed and seven wounded.

Gist's mission to protect the rice plantations and drive the British away was successful. The British moved down river and decided to forage in the country around Beaufort. Gist took his army and proceeded to Port Royal Ferry to block them there.[1662]

Brigade Orders, 31st Augt 1782 –
Henry Woodward,[1663] a private in Genl Baxters Regiment Being charged by his Commanding Officer with Deserting his post when a disposition of Battle was Made to Receive the Enemy, the Court of which Capt Shad Simons[1664] was President have found him guilty & deserve that he suffer the pinalty inflicted by the Second clause in the Regulations for the government of the Militia of this State. The Genl approved the Sentence –

Thos Elliot
Aid dCamp

General Leslie saw an opportunity to send another foraging party towards Monck's Corner to get fresh meat for the hospital. Leslie sent Major Thomas Fraser and his Royalists on this mission.

Fraser crossed the Cooper River and set out to surprise the guards at Biggin Bridge and Strawberry Ferry. He thought that Marion was supervising the defense of Georgetown and the removal of stores there. What he didn't know was that Marion had finished the supervision of the Georgetown defense and had returned to his post at Fair Lawn, the house of the Loyalist John Colleton, on the south side of the Wadboo River.

When Marion learned of the approaching raid, his cavalry was away patrolling down the Wadboo River looking for the British fleet. Marion's cavalry did not know that the fleet had moved up the Combahee River, where they had ambushed Laurens.

1660 The Salkehatchie bridge is located east of Yemasee, where Highway 17/21 crosses over the Combahee River.

1661 The Chehaw Neck is located about one mile from the mouth of the Combahee River, just south of Wiggins.

1662 Full details of the death of Laurens and the fight at Combahee Ferry are described in "*Nothing but Blood and Slaughter, Volume Four*" by Patrick O'Kelley.

1663 Unable to find any information on Henry Woodward.

1664 Shadrack Simons was a captain in Baxter's Regiment of Marion's Partisans. He served over 360 days in the militia during the siege of Charleston.

Marion organized a force under Captain Gavin Witherspoon and sent him to find Fraser. After Witherspoon left, Marion put his infantry into order of battle. Part of his infantry he placed on the side of a cedar lined road in an ambush position. The rest of his infantry he placed in and around the slave cabins on the plantation. He did not wholly trust his men, due to the fact that they were all "new made Whigs." These were men who saw which way the war had turned and joined the Patriot army for immunity. Among Marion's men were former Loyalist Major Micajah Ganey and 40 of his followers.

Fraser was able to capture some of Marion's pickets as he approached the houses. He detected Witherspoon in the woods and he immediately charged. Witherspoon and his troop turned back toward Fair Lawn at full gallop. As Witherspoon neared the plantation he fell behind in the ambush kill zone to let the Loyalist cavalry catch up. As he waited a Loyalist dragoon darted forward and was prepared to strike Witherspoon with his sword. Witherspoon coolly brought up his carbine and fired buckshot into the man's chest.

As Fraser's dragoons came within 30 yards of the ambush site, Marion's hidden men in the cedars shouted a cheer and fired a volley. Four of the Loyalists and five of their horses were killed and six wounded. At the sound of the volley the horses pulling Marion's ammunition wagon bolted. Five of Marion's men armed with captured broadswords went after the wagon. After a fight with Fraser's men they returned without the wagon. Fraser tried to rally his men, but they were being cut down from both sides of the road.

The Loyalists hovered near the plantation for an hour looking for an advantage, but Marion had planned too well. Unfortunately without the ammunition wagon Marion had no cartridges and was not able to continue the fight. He gave the order for his men to retreat to the Santee. This would be Marion's last fight.

Marion was later asked to strike against the British foraging parties coming out of Charlestown, but he guarded the British instead as they looked for provisions. Marion said, "My Brigade is composed of citizens, enough of whose blood has been shed already. If ordered to attack the enemy, I shall obey; but with my consent, not another life shall be lost, though the event that the enemy are on the eve of departure, so far from offering to molest, I would rather send a party to protect them."

Brigade Orders 1st Sept … 1782 –

No person to fire a Gun within Camp or within hearing of the Encampment, such as do will be Condemned to serve as a Continental Soldier during the war

Any persons leaving Camp without the Generals pass to be punished for disobedience of Orders

Brigade Orders 5th Sept 1782 –

At the Court of which Colo Baxter was President Lt Ivey [1665] of Colo Bentons Regt charged with willfull Disobediance of Orders, was found not Guilty & they Acquit him accordingly the ~~Court~~ arrest is therefore taken off & he is ordered to Join his Regiment & do Duty –

Brigadier Orders 10th Sept 1782

A Brigade Court of Enquiry to Set this day at 10 OClock in the forenoon to enquire into the Complaints of men who Say they are not fit to Bear arms, nor do duty. Major Sabb [1666] president, three Captains & 5 Subalterns from the line members, Evidence to be warned to attend

---

1665 Robert Ivy served 31 days as a lieutenant in Marion's partisans in 1782.

1666 Morgan Sabb served as a private, sergeant and major in the militia from 1779 to 1782.

At the above Court the following persons were Found incapable to do Duty in Camp, viz – John Palmer [1667] William Fountain [1668] George Right [1669] Pierce King,[1670] Thos Tant [1671] and Able Waddle [1672] – and James Wise [1673] complaining of a White Swelling, was found Fit for Duty & Recommended to be put on duty Immediately, there Determinations are approved of by the General.

Thos Elliot a DCamp

~~Of the General~~ Brigade Orders 9th Sept 1782 –
A Militia Court will this afternoon for the trial of all Prisoners that Shall be brought to it, Capt Lesesne President & four Officers from the line members, Evidence to attend. This Court to Sett at four OClock
Herron charged with Neglect of Duty & disobedience of Orders, is sentenced to do double Duty. The Sentence of the Court on Herron is approved, & his Officers are Directed to See he Performs it, or bring him to a Second trial agreeable to Law – [1674]

Brigade Orders St Stephens 14th Septr 1782
A Brigade Court to Set in Camp at 10 oClock the forenoon for the trial of Such Defalters as will be Ordered to it, all Evidences to attend Major Gambol President, three Capts & three Subalterns taken From the line members –

Brigade Orders St Stephens Septr 15th 1782 –
A Brigade Court to Set in Camp at 10 oClock this Forenoon For the trial of Such Defalters as will be Ordered to it, all Evidences to attend Major Thornly President, three Captains & three Subalterns taken from the line Members –

[1667] John Palmer had served under Captain Maurice Murphy and had been in the battles of Fort Moultrie, Parkers Ferry and Eutaw Springs. He was excused from duty in Marion's partisans on 10 September 1782 due to a "white swelling", which was tuberculus arthritis.
[1668] William Fountain served in Marion's Brigade during 1782 and 1783. He was excused from duty in Marion's partisans on 10 September 1782 due to a "white swelling".
[1669] George Wright served 220 days in the militia in 1781 and 1782. He was excused from duty in Marion's partisans on 10 September 1782 due to a "white swelling".
[1670] Unable to find any information on Pierce King.
[1671] Unable to find any information on Thomas Tant.
[1672] Abel Waddle served in the militia under Captain William Standard. He was excused from duty in Marion's partisans on 10 September 1782 due to a "white swelling".
[1673] James Wise served 43 days in the militia in 1782 and 29 days in 1783. He was excused from duty in Marion's partisans on 10 September 1782 due to a "white swelling".
[1674] Frederick Herron served in the cavalry under Colonel Peter Horry. On 9 September 1782 he was found guilty of neglect of duty and sentenced to do double the length of time he was drafted for.

Brigade Orders –
At the Court Martial of which Major Gamble was President for trying Defalters for Neglect of Duty, John Harleston [1675] & Robert Williams [1676] Sentenced to do Double Duty Nicholas Pough [1677] & David Snow [1678] Acquitted by the Said Court the General approves the Sentence –
Action at Watboo 29th Augt 1782 –

The British horse repulsed, their Loss, Capt Gillant, three privates, Seven horse Killed, one private & five horses taken,[1679] they buried before they got to fight, hope their men died and one did Died there & one Horse they had wounded, Capt Dawkins, 2 other Officers & twelve men wounded,[1680] our Loss one man Wounded & three taken, on the out picquett, one waggon & team with 300^{lb} powder, Some Lead, My tent & Baggage & one mule taken –

THE HORSE AMERICA, *throwing his Master*

[1675] This may be the same John Harleston as the lieutenant who resigned his commission in the 2nd South Carolina on 27 January 1777.

[1676] Robert Williams served 30 days in the militia during 1782. On 15 September 1782 he was found guilty of neglect of duty and sentenced to do double the length of time he was drafted for.

[1677] Nicholas Powers was born in 1756 in Cheraw, South Carolina. He enlisted while residing in Camden in 1776 as a sergeant and clerk under Captain Robert Lyle in the 3rd South Carolina Rangers. In 1779 he was in the same regiment under Captain John Hennington. He was wounded in Savannah. He was taken prisoner on 9 January 1779. After the fall of Charleston he was in the militia of Colonel Richardson. He served 36 days in the militia under General Marion in 1782.

[1678] Unable to find any information on David Snow.

[1679] This was Captain Robert Gillies of the North Carolina Dragoons.

[1680] Captain George Dawkins of the South Carolina Royalists.

At a Brigade Court Martial held the 15th Septr 82 of which Major Thornby was President – Benjm Holladay [1681] Joseph Burges [1682] Elisha Nettles [1683] John Right [1684] Henry White [1685] Nicholas Burbridge [1686] John Hilton, Wilm Barnett [1687] Samuel Little [1688] & Willm Little [1689] were sentenc'd to do double Duty which Sentence is approv'd of by the Genl –

Gl Or St Stephens 17th Septr 1782

The Scandalous & Infamous practice of Some men in the Brigade plundering the Neighbouring corn fields to Feed their Horses and Graveling and Destroying the Fields of Potatoes for two or three Miles around, oblidge one to move where they are no Inhabitants for a great distance

If Officers were careful to confine every Man whom they are bringing in or Eating Potatoes or New corn which they know they did not Come Honestly by, Such Practices would soon be Put a stop to, the General is sorry to see so great a Neglect so much to the hurt & Distress of the Inhabitants around. The Brigade to March to off Tomorrow by Sun rise –

The Quarter Master will get Carts to Convey Rice & Ammunition & have his Cattle Ready to move with the Brigade.

A Horse will be provided for the Sick And a Doctor to Attend them the State Legion to March with the Brigade ~~By the~~ [1690]

[1681] Benjamin Holliday had served as a sergeant in the 2nd South Carolina Regiment during 1777. On 15 September 1782 he was found guilty of neglect of duty and sentenced to do double the length of time he was drafted for.

[1682] Joseph Burgess served in the militia under Captain John Calhoun and Colonel Anderson from May 1781 to April 1783. On 15 September 1782 he was found guilty of neglect of duty and sentenced to do double the length of time he was drafted for.

[1683] Unable to find any information on Elisha Nettles. He is most likely related to George, Isham, Jesse, Joseph and Robert Nettles, who were with Marion in 1781 and 1782.

[1684] This may have been John Wright, who served with the 3rd South Carolina Rangers in 1775.

[1685] Henry White served from 15 April to 5 May 1780 under Colonel Robert Crawford. From 25 June to 29 August 1780 he was a horseman under Captain Hugh White. From 2 –23 October 1780 he was under Lieutenant John Tomlinson. From 1 - 25 November 1780 he was under Lieutenant Thomas Thompson. In 1782 he served under General Sumter as a private and a saddler. On 15 September 1782 he was found guilty of neglect of duty and sentenced to do double the length of time he was drafted for.

[1686] Unable to find any information on Nicholas Burbridge, but he is likely related to William Burbridge, who was in Marion's partisans.

[1687] William Barnet was born on 15 May 1759 in Lancaster County, Pennsylvania. He moved to South Carolina when he was five years old. While living in the Camden District in 1779 he volunteered under Colonel Kershaw and served in Kershaw's Company of Mounted Spies. In the spring of 1780 he served under Lieutenant James Cannada guarding the jail in Camden. In June of 1780 he served under Captain Caldwell and was in the battle of Ramsour's Mill where he was wounded. He was in the battles of Fishing Creek, Fish Dam Ford, and Camden. At various times from 1780 to 1782 he served under General Sumter and General Marion. On 15 September 1782 he was found guilty of neglect of duty and sentenced to do double the length of time he was drafted for. After the war he moved to Kentucky.

[1688] Unable to find any information on Samuel Little.

[1689] William Little served 175 days in the militia under Colonel Brandon. He served with Marion's partisan in 1782. On 15 September 1782 he was found guilty of neglect of duty and sentenced to do double the length of time he was drafted for.

[1690] The South Carolina State Legion referred to here is Maham's unit of cavalry and infantry. The Legion was a concept introduced to the 18th century by Marshal Saxe of France in 1744 when he wrote a *Treatise*

//

By a Certificate from J Barnet Aid de Camp to Gen[l] Green, Major Hyrne has risigned His Commission in the State Legion, he is not to be considered as a Maj[r] in said Corps in Future

Camp Blufort 18th Sep[t] 1782 –

The Sedition Act makes every person who goes or attempts to go to the Enemy or who gives any Intelligence or Supplies them with Provisions or any Article, Death without Benefit of Clergy – [1691]

This is to Acquaint all persons that By Virtue and Power Invested in me that I Shall take up every person who Comes under the Above Law and on proof send them to Jail to be tried by the Next Court.

All Persons who have Surrendered to me and are Liable to do duty for Six Months in the Militia agreeable to a Late Law passed at Jacksonborough are Required to Join My Brigade to perform that duty or give two Substitutes to Serve for three years or during the war in the Continental line of this State. Those who do not Comply will be taken into Custody and Delt with agreeable To the Militia Law without Distinction. Those who are Above fifty years of age must give One Substitude or Perform Six Months Duty –

<u>21st September 1782 –</u>

Watboo Sep[tr] 23rd 1782 – [1692]

A Militia Court will Set this Afternoon at 4 oClock for the Trial of all prisoners & Defalters that Shall be brought to it, Captain Allison [1693] president & four Officers from the line Members –
The above Court of which Cap[t] Robert Ellison was president Sentence Joseph Hurst [1694] who was tryd for Desertion to do Double Duty, the General approves of the Sentence –

Concerning Legions. The Legion that he proposed was to include 64 companies of foot divided into four regiments each of four battalions, plus one company of horse, one company of grenadiers, and one company of light-armed foot per regiment. Saxe later wrote *Reveries, or Memoirs Concerning the Art of War* and renamed the battalion, calling it the Century that also had artillery. Each regiment would have four centuries plus a half-century of horse and a half-century of light-armed foot. Each legion would have four regiments and two 12-pounders, and each century would have one amusette of half-pound caliber. An amusette was a large musket, known as a wall gun. In 1743, a Volontaires de Saxe was formed, containing a mix of lancers and dragoons. By the time of the Revolutionary War the tactic of having a mobile, combined arms army of cavalry, infantry (with rifles and muskets) and artillery, was being utilized by both sides.

[1691] Benefit of Clergy was a plea allowed in the colonies in lieu of punishment. Originally this plea was used in the Catholic church to exempt the clergy from being prosecuted, but by the 18th century anyone accused of a crime could use the plea. This plea was used by some of the British soldiers who were found guilty of manslaughter during the Boston Massacre. A person could only plead this once, and then would be branded, usually on the hand or thumb. If another crime was ever committed by this person, they would then be put to death. The State of North Carolina still has a version of this in their judicial system, and a person can claim "Prayer for Judgement" and not be prosecuted for many crimes.

[1692] Wadboo was the name for Marion's camp at Colleton plantation.

[1693] Robert Ellison had commanded an independent company of Rangers as a captain during 1775. He served as a private under General Marion from 1 November 1780 to 20 August 1781. He was promoted to lieutenant on 20 August 1781 and served 92 days. In 1782 he became a captain under Marion.

[1694] Joseph Hearst (Hurst) served 68 days in the militia from May 1781 to August 1782 under Captain Dawson. On 23 September 1782 he was found guilty of desertion and sentenced to do double the length of time he was drafted for. He served with his brother, John Hearst, who served 36 days from May 1781 to August 1782.

Wadboo 21st Sep[tr] 1782 –

At Ringing the great bell & firing of one Gun from the House every man must turn out and Immediately Repair to their alarm post –

It has been made Known to me that a Number of The Militia hire Substitudes to do their duty, which is Contrary to Law, every Officer in future who take such Substitudes Shall be tried for Disobediance of Orders & Suffer Accordingly & Such persons who have Substitudes Will be tried as Defalters as if he never had hired a man in his room, Commanding Officers of Regiments & Comp[ys] are therefore Ordered to bring such Defalters to trial –

Wadboo Sep[tr] 29th 1782

No Officer or Private of the State Legion or Militia to leave Camp without my Permission, any Officer or Private who disobeys this Order may Expect to be called to an Account agreeable to the Articles of war & the Militia Law of this State – [1695]

Orders at Wadboo 1 Oct[r] 1782 –

The provisions Returns of Yesterday is most Astonishing, Some of the Regiments was forty Rations more than it should be & not one return but what was more than it ought

The Quarter Master of the Brigade is Ordered Not to Issue to field Officers more than three & other Officers two Rations – no Black Servants is to be Allow'd Rations in future –

Proceedings of an Ordinary Militia Court Held at Wadboo of which Cap[t] M[c]Intosh was President,[1696] Jacob Fort was Brought before the Court & Charged with Disobediance of Orders,[1697] neglecting to carry an Express where Orderd, the Court Sentence him to serve twelve Months In one of the Continental Reg[ts] of this State. The General approves of the Sentence of the Court –

[1695] As the war was winding down it was becoming harder to get the men to do their duty, which was to keep the British surrounded in Charleston. No man wants to be the last one killed in a war, and many did not want to risk death or injury in the final days. The men also wanted to see to their crops, and bring in the harvest. Another reason that so many men were deserting was to protect their families. Since there was little food in Charleston due to the siege, bands of Loyalists would roam the countryside plundering all that they could, and exacting revenge for any who were suspected of being Whigs.

[1696] Alexander McIntosh served as a captain and major under Colonel Benton and General Marion in 1781 and 1782.

[1697] Unable to find any information on Jacob Fort.

At a Brigade Militia Court held at Wadboo – 23rd October of which Capn Amos windham was prisidet the following Sentences was pass'd –

Col° Ritchardsons Regiment

John Way [1698] . . . . Sentencd to do double duty
Caleb Gayle [1699] . . . . . . Sentencd to do double duty
Abijah Rambart [1700] . . . . Sentencd to do double duty
Joseph Winn [1701] . . . . . . Acquitted _ _ _ _
Charles Brunson [1702] . . . Acquitted _ _ _ _
William Way [1703] . . . . . . Sentencd to do double duty
Isaac Brunson[1704] . . . . . Acquitted _ _ _ _
Seth poole [1705] . . . . Sentencd to do double duty
John Holiday [1706] . . . . . . Acquitted _ _ _ _
Moses Brunson [1707] . . . . Acquitted _ _
Thomas Jones [1708] . . . . . Sentencd to do double duty

[1698] John Waugh served 66 days in the militia during 1781 under Captain Edward Martin and Colonel Winn. In 1782 he served under Captain John Turner. He was in Colonel Richardson's Regiment of Marion's partisans in 1782. On 1 October 1782 he was found guilty of desertion and sentenced to do double the length of time.

[1699] Caleb Gayle had enlisted on 26 August 1775 in the Volunteer Company of Horse of St. Marks Parish under Captain Matthew Singleton. He was in Colonel Richardson's Regiment of Marion's partisans in 1782. On 1 October 1782 he was found guilty of desertion and sentenced to do double the length of time.

[1700] Isaac Rembert was in Colonel Richardson's Regiment of Marion's partisans in 1782. On 1 October 1782 he was found guilty of desertion and sentenced to do double the length of time he was drafted for.

[1701] Joseph Winn served twelve months in the cavalry under Colonel Peter Horry. He was in Colonel Richardson's Regiment of Marion's partisans in 1782.

[1702] Charles Brunson had enlisted on 26 August 1775 in the Volunteer Company of Horse of St. Mark's Parish under Captain Matthew Singleton. He was in Colonel Richardson's Regiment of Marion's partisans in 1782.

[1703] William Way was at the fall of Charleston and afterwards served 37 days in Marion's partisans in 1782. On 1 October 1782 he was found guilty of desertion and sentenced to do double the length of time.

[1704] Isaac Brunson was born in 1745. He served in the 3rd South Carolina Rangers under Captain Edward Richardson during 1775. He was in the militia at the fall of Charleston. He was in Colonel Richardson's Regiment of Marion's partisans in 1782. He died in 1827.

[1705] Zeth Poole was born in 1755 in North Carolina. He had enlisted in the 3rd South Carolina Rangers under Captain Edward Richardson on 25 June 1775. He was in the 6th South Carolina Regiment on 7 May 1776. He was in Colonel Richardson's Regiment of Marion's partisans in 1782. On 1 October 1782 he was found guilty of desertion and sentenced to do double the length of time he was drafted for.

[1706] John Holliday was born in 1746. He had enlisted in the 4th South Carolina Artillery on 14 February 1776. He was in Colonel Richardson's Regiment of Marion's partisans in 1782.

[1707] Moses Brunson was at the fall of Charleston. He also served 71 days in the militia under General Marion in 1782.

[1708] Thomas Jones served 369 days in the militia, alternately from 1779 to 1783. On 1 October 1782 he was found guilty of desertion and sentenced to do double the length of time he was drafted for.

Captain Nelsons Company [1709]

Thomas Sumner [1710] . . . . . Sentencd to do double duty
Thomas McNight [1711] . . . . . Acquitted _ _ _ _
John McNight [1712] . . . . Acquitted _ _ _ _
Nathaniel Pijate [1713] . . . . . . Sentencd to do double duty
Benjn Cooper [1714] . . . . Sentencd to do double duty
Joseph Hanington [1715] . . . . Acquitted _ _ _ _
James Burgess [1716] . . . . Acquitted _ _ _ _

The General approves of the Sentence of the Court –

A Militia ordinary Court held at Wadboo the 28th October 1782, Order of Brig^d Genl Marion and of which Capt Davis was President,[1717] sentence Jam^s Canty to Serve three Months in one of the continental Regiments of this State for Deserting from a Command[1718]

The General Approves the said Sentence –
Thos Elliot
_ ADC _

At a Militia Court of which Capt Elliot was prisident And held by order of Genl Marion the 29th oct^r 1782 – the following prisoners were tried & sentenced as follows Viz –
Samuel Harford for Disobedience of orders acquitted[1719]
Mayberry Holmes – D^o. D^o. _ _ _ _ _ _ _ _ _ D^o. _ [1720]

[1709] John Nelson had enlisted in October 1775 in a Volunteer Company of Militia under Captain William Fullwood. He served in the militia from 1 April 1781 to 25 February 1782. During this time he was promoted to captain.

[1710] Unable to find any information on Thomas Sumner.

[1711] Thomas McKnight served in the militia as a horseman from 27 March 1780 to 23 October 1782 under Captain Thomas Parsons, John Gowen, Moses Wood, and Colonel Roebuck.

[1712] John McKnight served 142 days in the militia during 1779. He also supplied forage and provisions.

[1713] Nathaniel Pigott was at the fall of Charleston. He served 35 days in the militia in 1782. On 1 October 1782 he was found guilty of desertion and sentenced to do double the length of time he was drafted for.

[1714] Benjamin Allen Cooper had enlisted in the 6th South Carolina Regiment on 4 April 1776. On 17 December 1776 he was a corporal. He was discharged on 1 June 1777. He was in Marion's partisans in 1782. On 1 October 1782 he was found guilty of desertion and sentenced to do double the length of time.

[1715] Unable to find any information on Joseph Hanington.

[1716] Possibly John Burgess, who served in Marion's Brigade. James Burgess was in the 3rd South Carolina Rangers and was killed at Savannah on 9 October 1779.

[1717] William Ransom Davis, formerly of the 3rd South Carolina Rangers.

[1718] James Canty served in the militia under Captain John Chesnut during 1780 and was at the fall of Charleston. He was under General Marion in 1782. On 28 October 1782 he was found guilty of desertion and sentenced to serve three months in a Continental regiment.

[1719] Samuel Harper served in Marion's partisans from 1780 to 1782.

[1720] Mawbry (Mabry, Mayberry) Holmes (Helms) served 30 days in Marion's partisans in 1782 under Captain Charles Lewis.

John Holmes – to serve double the time he was Otherwise liable to –[1721]

John Avary to do double Duty – [1722]

The General approves the above sentences _ _ _ _

Thos Elliot
Aid deCamp

Wadboo November 13th … 1782

A Militia Court will sit this morning at 10 oClock for the trial of all Prisoners & Defalters, that shall be Brought to it, Major Wardin[1723] President and four Officers from the Line Members –

At the above Court of which Major Warden was Prisident, William Sabb [1724] & Ritchard ward [1725] of Colo Gressets Regiment was tried for neglect of Duty & Sentenced to double Duty. [1726] The General approves of the Sentence –

18th Novr 1782 –

A Militia Court is ordered to set this day for the trial of such Defaulters as may be Brought before it. Major John Green President, & 4 Officers from the line Members

At the above Court David Rumph [1727] John Brothers [1728] and Enor Easterling [1729] charged with disobedience of orders, are acquitted –

Joel Spell [1730] & William Williams [1731] also charged with Disobedience of Orders & not appearing agreeable to Summons are sentenced to do double Duty –

// . The General approves the above Sentences –

1721 John Holmes had been in the 2nd South Carolina Regiment.

1722 John Averite who had enlisted in the 3rd South Carolina Rangers on 1 July 1777. On 29 October 1782 he was found guilty of disobeying orders and sentenced to do double the length of time he was drafted for.

1723 Major John Warden.

1724 Possibly William Stubb who was born on 22 December 1748 in the Marlborough District of South Carolina. He enlisted in 1776 under Colonel Kolb.

1725 Richard Ward had served in the 3rd South Carolina Rangers under Captain Joseph Warley and John Hennington during 1779. On 13 November 1782 he was found guilty of neglect of duty and sentenced to do double the length of time he was drafted for.

1726 There is no further information on Colonel Grissett.

1727 David Rumph served 180 days in the Orangeburg District Independent Cavalry under Captain Jacob Rumph. He served 61 days in the militia in 1781 and served 120 days in the militia in 1782.

1728 Unable to find any information on John Brothers, Jr.

1729 Henry Easterling.

1730 Unable to find any information on Joel Spell.

1731 William Williams served 624 days from 20 May 1779 to October 1782 and was a prisoner for 391 of those days. He had been held after the fall of Charleston and after the defeat at Camden. On 18 November 1782 he was charged with disobeying orders. He did not appear before the court martial and was sentenced to do double the length of time he was drafted for.

Wadboo Novr 23rd 1782 –
A Militia Court to Set this Day from the trial of Such Defalters as may be Brought before them
Lt Col^o^ Lushington President [1732]
Two Cap^n^ & two Subs Members –

At the above Court John Coone [1733] charged with Disobedience of Orders was acquitted –
Sergeant Kean [1734] charged with Fraud – also acquitted
// The General approves the sentences –

Wadboo Novr 28th 1782 –
A Militia Court will Sit this morning at 11 oClo for the trial of all prisoners & Defalters that shall be Brought to it, Lt Col^o^ James President & four Officers from the Line members –
At the above Court Christian Rumph[1735] being charged with Sleeping on his Post was acquitted. The Genl Approves the Sentence, Orders the prisinor to be Released & to Join his Regiment –

Novr 28th _ 1782 _
Major John Vanderhorst is appointed Aid De Camp,[1736] he is to be respected & Obey^d^ as Such _ _

Nineteen months after the surrender of Charlestown British General Leslie evacuated Charlestown. He had agreed not to destroy the city if the Patriots would allow his troops to depart in safety. Upon the firing of the morning cannon they moved out of the forward works, while the Continentals of General Wayne moved in. Moultrie wrote in his memoirs, "This fourteenth day of December, 1782, ought never be forgotten by the Carolinians; it ought to be a day of festivity with them, as it was the real day of their deliverance and independence."

On September 3, 1783 the Treaty of Paris was signed and Britain's war with the United States was officially over.

[1732] Richard Lushington had served as a lieutenant and captain in the Charleston Regiment of Militia in 1775, 1778, 1780 and 1781. He was captured at Charleston and taken to St. Augustine as a prisoner. After his release he was a lieutenant colonel in Marion's partisans from 3 July to 18 December 1782.

[1733] John Coon had served 117 days in the militia under Captain John Summers from 9 January to 8 July 1779. He was in Marion's partisans in 1782.

[1734] Possibly James Kean who served in the militia in 1781 and 1782.

[1735] Christian Rumph was drafted during November 1781 in the Indianfield Company under Captain Jacob Linder, Colonel Simons and General Marion. In 1782 he was under Major Sabb and Colonel Ervin in Marion's partisans.

[1736] Major John Vanderhorst of the 2nd South Carolina Regiment.

General Orders at Wadboo Dec[r] 15th 1782 [1737]

General Marion Congratulates the troops under his Command on the Evacuation of Charles Town. This Happy Event has made it unnecessary for the country Militia being Kept any longer in the field. He therefore discharges them. The General returns his warmest thanks to the officers and ~~Soldiers~~ men who with unwaried patience & Fortitude have under gone the greatest fatigues & Hardships & with a Spirit & Bravery which must ever reflect the Highest honour on them. No Citysins in the world Have ever done more they have. He begs leave to give his Particular thanks to all the Officers & Men of the Country Militia for that Partiality to his Person & ready Obedience to all Orders for two years And an half, which will be Remember'd with Gratitude To the end of his life. He will always consider them with the Affection of a Brother & will be happy to Render them every Service in his power – he cannot doubt in the least of their Readiness to turn out Should their Country be ever again so unhappy as to be invaded by her cruel & barbarous Enemys. He wishes them a long Continuance of happiness & the Blessing of Peace –

He is requested by the Governor to return his Thanks To them in the following Words, be pleased Sir to Issue in Gen[l] Orders at the time of Discharging Your Men, my Sincere & hearty thanks in behalf of the State for their long faithful & Important Services, especially at a time, when nothing but their unabated Love for their Country & in the cause of Liberty could have supported them under the Variety of Hardships they had to Encounter

---

[1737] Marion's Farewell address was at Colleton Plantation.

Lineage of and Honors of Marion's Partisans

10 August 1780	Marion assumes command of Williamsburg District
25 August 1780	Raid at Nelson's Ferry and rescues prisoners
27 August 1780	Ambush at Kings Tree, South Carolina
4 September 1780	Raid at Blue Savannah, South Carolina
29 September 1780	Raid at Black Mingo Creek, South Carolina
25 October 1780	Raid at Tearcoat Swamp, South Carolina
15 November 1780	Raid on White's and Ashton's Plantation, Georgetown
13 December 1780	Attack at Halfway Swamp, South Carolina
14 December 1780	Ambush of ships at Nelson's Ferry, South Carolina
27 December 1780	Raid on Georgetown, South Carolina
6 January 1781	Ambush at Georgetown, South Carolina
13 January 1781	Raid on Waccamaw Neck, South Carolina
24 January 1781	Attack Wiggan's Plantation, South Carolina
25 January 1781	Raid on Georgetown, South Carolina
31 January 1781	Raid on Wadboo, Keithfield, and Manigault's Ferry
14 February 1781	Raid on Postelle's House, Georgetown
6 March 1781	Attack at Wadboo Swamp Bridge, South Carolina
13 March 1781	Attack at Mount Hope Swamp Bridge, South Carolina
14 March 1781	Attack at the Lower Bridge of the Black River
15-28 March 1781	Siege of Blakely Plantation, South Carolina
28 March 1781	Attack at Sampit Bridge, South Carolina
29 March 1781	Defense of Snow's Island, South Carolina
2 April 1781	Attack at Black River, South Carolina
3 April 1781	Attack at Witherspoon's Ferry, South Carolina
15-23 April 1781	Siege of Fort Watson, South Carolina
26 April 1781	Attack at Drowning Creek, South Carolina
27 April 1781	Attack at Hulin's Mills, South Carolina
8-12 May 1781	Siege of Fort Motte, South Carolina
28 May 1781	Capture of Georgetown, South Carolina
8-10 July 1781	Siege of Orangeburgh, South Carolina
17 July 1781	Battle of Quinby Bridge and Shubrick's Plantation
16 August 1781	Ambush on Santee River, South Carolina
31 August 1781	Ambush at Parker's Ferry, South Carolina
8 September 1781	Battle of Eutaw Springs, South Carolina
17 November 1781	Attack at Fairlawn Plantation, South Carolina
3 January 1782	Attack at Videau's Bridge, South Carolina
24 February 1782	Defense of Brabant's Plantation, South Carolina
25 February 1782	Attack at Tydiman's Plantation, South Carolina
15 March 1782	Attack at Middleton's Plantation, South Carolina
8 June 1782	Attack at Burch's Mill, South Carolina
August 1782	Attack at White House, South Carolina
29 August 1782	Defense of Wadboo, Colleton Plantation, South Carolina
15 December 1782	Marion's partisans disbanded

AFTERWORD

When Marion returned to his plantation at Pond Bluff he found it destroyed. Both British and Patriots had plundered Marion's home, and the British had burned down the house. Marion had no money, since he had not been paid for his time in the partisans, and he had to buy everything on credit to rebuild his home. He continued to represent his parish in the South Carolina Senate. When the legislation offered him the same protection against lawsuits from the war, like they did Sumter, he declined. He replied, "If I have given any occasion for complaint, I am ready to answer in property and person. If I have wronged any man, I am willing to make him restitution. If, in a single instance, in the course of my command, I have done that which I cannot fully justify, justice requires that I should suffer for it."

Marion was also vigorously opposed to the confiscation of Tory property. He sponsored laws giving the Loyalists equality and justice. In the north the Loyalist were burned, beaten and driven out of their homes. Many settled in Canada, where even after two centuries there is some hard feelings against the United States. However due to Marion, many of the Loyalists in the South remained and became citizens of the United States.

On 30 September 1783 Congress promoted Marion to the rank of colonel in the Continental line, and a year later South Carolina made Marion the commandant of Fort Johnson. He was promised an annual salary of £500, but the South Carolina Congress soon forgot about the heroes and focused on the economy. They reduced the pay to $500.

In 1786 he married his cousin, 49 year old Mary Esther Videau. She was wealthy, so he was no longer in debt. He resigned as the commandant of Fort Johnson and rebuilt a home at Pond Bluff. This was not a luxurious mansion, but was merely a simple one story house.

He was a Federalist and was on the side of George Washington. Though he was not a politician, he was a delegate to the convention that wrote the South Carolina Constitution. After South Carolina joined the union he turned down any further political office. He continued to command his militia brigade, attending the musters and the training until 1794 when the militia was reorganized.

His health began to deteriorate in 1795 when he was 63 years old. He died at his home on Pond Bluff on February 27th, with Mary Esther by his side. He is buried in the family cemetery on Gabriel Marion's plantation on Belle Isle.[1738] His tomb is marked by a slab that reads that he "lived without fear, and died without reproach."

[1738] The Francis Marion tomb and monument is located near Lake Marion is now, east of Eadytown.

ARTICLES OF WAR

Resolved, That from and after the publication of the following articles, in the respective armies of the United States, the rules and articles by which the said armies have heretofore been governed, shall be, and they are, hereby repealed.

Section I

Article 1. That every officer who shall be retained in the army of the United States, shall, at the time of his acceptance of his commission, subscribe these rules and regulations.

Art. 2. It is earnestly recommended to all officers and soldiers diligently to attend divine service: and all officers and soldiers who shall behave indecently, or irreverently, at any place of divine worship, shall, if commissioned officers, be brought before a general courtmartial, there to be publicly and severely reprimanded by the president; if non-commissioned officers or soldiers, every person so offending shall, for his first offence, forfeit &frac16th of a dollar, to be deducted out of his next pay; for the second offence, he shall not only forfeit a like sum, but be confined for twenty-four hours; and, for every like offence, shall suffer and pay in like manner; which money, so forfeited, shall be applied to the use of the sick soldiers of the troop or company to which the offender belongs.

Art. 3. Whatsoever non-commissioned officer or soldier shall use any prophane oath or execration, shall incur the penalties expressed in the foregoing article; and if a commissioned officer be thus guilty of prophane cursing or swearing, he shall forfeit and pay, for each and every such offence, two-thirds of a dollar.

Art. 4. Every chaplain who is commissioned to a regiment, company, troop, or garrison, and shall absent himself from the said regiment, company, troop, or garrison, (excepting in case of sickness or leave of absence) upon Pain of being brought to a Court Martial and punished as their judgment and the circumstances of his offence may require. shall be brought to a court-martial, and be fined not exceeding one month's pay, besides the loss of his pay during his absence, or be discharged, as the said court-martial shall judge most proper.

Section II

Art. 1. Whatsoever officer or soldier shall presume to use traiterous or disrespectful words against the authority of the United States in Congress assembled, or the legislature of any of the United States in which he may be quartered, if a commissioned officer, he shall be cashiered; if a non-commissioned officer or soldier, he shall suffer such punishment as shall be inflicted upon him by the sentence of a courtmartial.

Art. 2. Any officer or soldier who shall behave himself with contempt or disrespect towards the general, or other commander in chief of the forces of the United States, or shall speak words tending to his hurt or dishonor, shall be punished according to the nature of his offence, by the judgment of a court-martial.

Art. 3. Any officer or soldier who shall begin, excite, cause or join, in any mutiny or sedition, in the troop, company or regiment to which he belongs, or in any other troop or company in the service of the United States, or in any party, post, detachment or guard, on any pretence whatsoever, shall suffer death, or such other punishment as by a court-martial shall be inflicted.

Art. 4. Any officer, non-commissioned officer, or soldier, who, being present at any mutiny or sedition, does not use his utmost endeavor to suppress the same, or coming to the knowledge of any intended mutiny, does not, without delay, give information thereof to his commanding officer, shall be punished by a court-martial with death, or otherwise, according to the nature of the offence.

Art. 5. Any officer or soldier who shall strike his superior officer, or draw, or shall lift up any weapon, or offer any violence against him, being in the execution of his office, on any pretence whatsoever, or shall disobey any lawful command of his superior officer, shall suffer death, or such other punishment as shall, according to the nature of his offence, be inflicted upon him by the sentence of a court-martial.

Section III

Art. 1. Every non-commissioned officer and soldier, who shall inlist himself in the service of the United States, shall at the time of his so inlisting, or within six days afterwards, have the articles for the government of the forces of the United States read to him, and shall, by the officer who inlisted him, or by the commanding officer of the troop or company into which he was inlisted, be taken before the next justice of the peace, or chief magistrate of any city or town-corporate, not being an officer of the army, or, where recourse cannot be had to the civil magistrate, before the judge-advocate, and, in his presence, shall take the following oath, or affirmation, if conscientiously scrupulous about taking an oath:

I swear, or affirm, (as the case may be,) to be true to the United States of America, and to serve them honestly and faithfully against all their enemies or opposers whatsoever; and to observe and obey the orders of the Continental Congress, and the orders of the generals and officers set over me by them. Which justice or magistrate is to give the officer a certificate, signifying that the man inlisted, did take the said oath or affirmation.

Art. 2. After a non-commissioned officer or soldier shall have been duly inlisted and sworn, he shall not be dismissed the service without a discharge in writing; and no discharge, granted to him, shall be allowed of as sufficient, which is not signed by a field-officer of the regiment into which he was inlisted, or commanding officer, where no field-officer of the regiment is in the same state.

Section IV

Art. 1. Every officer commanding a regiment, troop, or company, shall, upon the notice given to him by the commissary of musters, or from one of his deputies, assemble the regiment, troop or company, under his command, in the next convenient place for their being mustered.

Art. 2. Every colonel or other field-officer commanding the regiment, troop, or company, and actually residing with it, may give furloughs to non-commissioned officers and soldiers, in such numbers, and for so long a time, as he shall judge to be most consistent with the good of the service; but, no non-commissioned officer or soldier shall, by leave of his captain, or inferior officer, commanding the troop or company (his field-officer not being present) be absent above twenty days in six months, nor shall more than two private men be absent at the same time from their troop or company, excepting some extraordinary occasion shall require it, of which occasion the field-officer, present with, and commanding the regiment, is to be the judge.

Art. 3. At every muster, the commanding officer of each regiment, troop, or company, there present, shall give to the commissary, certificates signed by himself, signifying how long such officers, who shall not appear at the said muster, have been absent, and the reason of their absence; in like manner, the commanding officer of every troop or company shall give certificates, signifying the reasons of the absence of the non-commissioned officers and private soldiers; which reasons, and time of absence, shall be inserted in the muster-rolls opposite to the names of the respective absent officers and soldiers: The said certificates shall, together with the muster-rolls, be remitted by the commissary to the Congress, as speedily as the distance of place will admit.

Art. 4. Every officer who shall be convicted before a general court-martial of having signed a false certificate, relating to the absence of either officer or private soldier, shall be cashiered.
Art. 5. Every officer who shall knowingly make a false muster of man or horse, and every officer or commissary who shall willingly sign, direct, or allow the signing of the muster-rolls, wherein such false muster is contained, shall, upon proof made thereof by two witnesses before a general court-martial, be cashiered, and shall be thereby utterly disabled to have or hold any office or employment in the service of the United States.
Art. 6. Any commissary who shall be convicted of having taken money, or any other thing, by way of gratification, on the mustering any regiment, troop, or company, or on the signing the muster rolls, shall be displaced from his office,and, moreover, forfeit all such Pay as may be due to him at the time of conviction of such offences. and shall be thereby utterly disabled to have or hold any office or employment under the United States.
Art. 7. Any officer who shall presume to muster any person as a soldier, who is, at other times, accustomed to wear a livery, or who does not actually do his duty as a soldier, shall be deemed guilty of having made a false muster, and shall suffer accordingly.

Section V

Art. 1. Every officer who shall knowingly make a false return to the Congress, or any committee thereof, to the commander in chief of the forces of the United States, or to any his superior officer authorized to call for such returns, of the state of the regiment, troop, or company, or garrison, under his command, or of arms, ammunition, clothing, or other stores thereunto belonging, shall, by a court-martial, be cashiered.
Art. 2. The commanding officer of every regiment, troop, or independent company, or garrison of the United States, shall, in the beginning of every month, remit to the commander in chief of the American forces, and to the Congress, an exact return of the state of the regiment, troop, independent company, or garrison under his command, specifying the names of the officers not then residing at their posts, and the reason for, and time of, their absence: Whoever shall be convicted of having, through neglect or design, omitted the sending such returns, shall be punished according to the nature of his crime, by the judgment of a general court-martial.

Section VI

Art. 1. All officers and soldiers, who having received pay, or having been duly inlisted in the service of the United States, shall be convicted of having deserted the same, shall suffer death, or such other punishment as by a court-martial shall be inflicted.
Art. 2. Any non-commissioned officer or soldier, who shall, without leave from his commanding officer, absent himself from his troop or company, or from any detachment with which he shall be commanded, shall, upon being convicted thereof, be punished, according to the nature of his offence, at the discretion of a court-martial.
Art. 3. No non-commissioned officer or soldier shall inlist himself in any other regiment, troop or company, without a regular discharge from the regiment, troop or company, in which he last served, on the penalty of being reputed a deserter, and suffering accordingly: And in case any officer shall, knowingly, receive and entertain such non-commissioned officer or soldier, or shall not, after his being discovered to be a deserter, immediately confine him, and give notice thereof to the corps in which he last served, he, the said officer so offending, shall, by a court-martial, be cashiered.

Art. 4. Whatsoever officer or soldier shall be convicted of having advised or persuaded any other officer or soldier to desert the service of the United States, shall suffer such punishment as shall be inflicted upon him by the sentence of a court-martial.

Section VII

Art. 1. No officer or soldier shall use any reproachful or provoking speeches or gestures to another, upon pain, if an officer, of being put in arrest; if a soldier, imprisoned, and of asking pardon of the party offended, in the presence of his commanding officer.

Art. 2. No officer or soldier shall presume to send a challenge to any other officer or soldier, to fight a duel, upon pain, if a commissioned officer, of being cashiered, if a non-commissioned officer or soldier, of suffering corporal punishment, at the discretion of a court-martial.

Art. 3. If any commissioned or non-commissioned officer commanding a guard, shall, knowingly and willingly, suffer any person whatsoever to go forth to fight a duel, he shall be punished as a challenger: And likewise all seconds, promoters, and carriers of challenges, in order to duels, shall be deemed as principals, and be punished accordingly.

Art. 4. All officers, of what condition soever, have power to part and quell all quarrels, frays, and disorders, though the persons concerned should belong to another regiment, troop or company; and either to order officers into arrest, or non-commissioned officers or soldiers to prison, till their proper superior officers shall be acquainted therewith; and whosoever shall refuse to obey such officer (though of an inferior rank) or shall draw his sword upon him, shall be punished at the discretion of a general court-martial.

Art. 5. Whatsoever officer or soldier shall upbraid another for refusing a challenge, shall himself be punished as a challenger; and all officers and soldiers are hereby discharged of any disgrace, or opinion of disadvantage, which might arise from their having refused to accept of challenges, as they will only have acted in obedience to the orders of Congress, and done their duty as good soldiers, who subject themselves to discipline.

Section VIII

Art. 1. No suttler shall be permitted to sell any kind of liquors or victuals, or to keep their houses or shops open, for the entertainment of soldiers, after nine at night, or before the beating of the reveilles, or upon Sundays, during divine service, or sermon, on the penalty of being dismissed from all future suttling.

Art. 2. All officers, soldiers and suttlers, shall have full liberty to bring into any of the forts or garrisons of the United American States, any quantity or species of provisions, eatable or drinkable, except where any contract or contracts are, or shall be entered into by Congress, or by their order, for furnishing such provisions, and with respect only to the species of provisions so contracted for.

Art. 3. All officers, commanding in the forts, barracks, or garrisons of the United States, are hereby required to see, that the persons permitted to suttle, shall supply the soldiers with good and wholesome provisions at the market price, as they shall be answerable for their neglect.

Art. 4. No officers, commanding in any of the garrisons, forts, or barracks of the United States, shall either themselves exact exorbitant prices for houses or stalls let out to suttlers, or shall connive at the like exactions in others; nor, by their own authority and for their private advantage, shall they lay any duty or imposition upon, or be interested in the sale of such victuals liquors, or other necessaries of life, which are brought into the garrison, fort, or barracks, for the use of the soldiers, on the penalty of being discharged from the service.

Section IX
Art. 1. Every officer commanding in quarters, garrisons, or on a march, shall keep good order, and, to the utmost of his power, redress all such abuses or disorders which may be committed by any officer or soldier under his command; if, upon complaint made to him of officers or soldiers beating, or otherwise ill-treating any person; of disturbing fairs or markets, or of committing any kind of riots to the disquieting of the good people of the United States; he the said commander, who shall refuse or omit to see justice done on the offender or offenders, and reparation made to the party or parties injured, as far as part of the offenders pay shall enable him or them, shall, upon proof thereof, be punished, by a general court-martial, as if he himself had committed the crimes or disorders complained of.

Section X
Art. 1. Whenever any officer or soldier shall be accused of a capital crime, or of having used violence, or committed any offence against the persons or property of the good people of any of the United American States, such as is punishable by the known laws of the land, the commanding officer and officers of every regiment, troop, or party, to which the person or persons so accused shall belong, are hereby required, upon application duly made by or in behalf of the party or parties injured, to use his utmost endeavors to deliver over such accused person or persons to the civil magistrate; and likewise to be aiding and assisting to the officers of justice in apprehending and securing the person or persons so accused, in order to bring them to a trial. If any commanding officer or officers shall wilfully neglect or shall refuse, upon the application aforesaid, to deliver over such accused person or persons to the civil magistrates, or to be aiding and assisting to the officers of justice in apprehending such person or persons, the officer or officers so offending shall be cashiered.
Art. 2. No officer shall protect any person from his creditors, on the pretence of his being a soldier, nor any non-commissioned officer or soldier who does not actually do all duties as such, and no farther than is allowed by a resolution of Congress, bearing date the 26th day of December, 1775. Any officer offending herein, being convicted thereof before a court-martial, shall be cashiered.

Section XI
Art. 1. If any officer shall think himself to be wronged by his colonel, or the commanding officer of the regiment, and shall, upon due application made to him, be refused to be redressed, he may complain to the general, commanding in chief the forces of the United States, in order to obtain justice, who is hereby required to examine into the said complaint, and, either by himself, or the board of war, to make report to Congress thereupon, in order to receive further directions.
Art. 2. If any inferior officer or soldier shall think himself wronged by his captain, or other officer commanding the troop or company to which he belongs, he is to complain thereof to the commanding officer of the regiment, who is hereby required to summon a regimental court-martial, for the doing justice to the complainant; from which regimental court-martial either party may, if he thinks himself still aggrieved, appeal to a general court-martial; but if, upon a second hearing, the appeal shall appear to be vexatious and groundless, the person so appealing shall be punished at the discretion of the said general court-martial.

Section XII

Art. 1. Whatsoever commissioned officer, store-keeper, or commissary, shall be convicted at a general court-martial of having sold (without a proper order for that purpose) embezzled, misapplied, or wilfully, or through neglect, suffered any of the provisions, forage, arms, clothing, ammunition, or other military stores belonging to the United States, to be spoiled or damaged, the said officer, store-keeper, or commissary so offending, shall, at his own charge, make good the loss or damage, shall moreover forfeit all his pay, and be dismissed from the service.

Art. 2. Whatsoever non-commissioned officer or soldier shall be convicted, at a regimental court-martial, of having sold, or designedly, or through neglect, wasted the ammunition delivered out to him to be employed in the service of the United States, shall, if a non-commissioned officer, be reduced to a private sentinel, and shall besides suffer corporal punishment in the same manner as a private sentinel so offending, at the discretion of a regimental court-martial.

Art. 3. Every non-commissioned officer or soldier who shall be convicted at a court-martial of having sold, lost or spoiled, through neglect, his horse, arms, clothes or accoutrements shall undergo such weekly stoppages (not exceeding the half of his pay) as a court-martial shall judge sufficient for repairing the loss or damage; and shall suffer imprisonment, or such other corporal punishment, as his crime shall deserve.

Art. 4. Every officer who shall be convicted at a court-martial of having embezzled or misapplied any money with which he may have been entrusted for the payment of the men under his command, or for inlisting men into the service, if a commissioned officer, shall be cashiered and compelled to refund the money, if a non-commissioned officer, shall be reduced to serve in the ranks as a private soldier, be put under stoppages until the money be made good, and suffer such corporal punishment (not extending to life or limb) as the court-martial shall think fit.

Art. 5. Every captain of a troop or company is charged with the arms, accoutrements, ammunition, clothing, or other warlike stores belonging to the troop or company under his command, which he is to be accountable for to his colonel, in case of their being lost, spoiled, or damaged, not by unavoidable accidents, or on actual service.

Section XIII

Art. 1. All non-commissioned officers and soldiers, who shall be found one mile from the camp, without leave, in writing, from their commanding officer, shall suffer such punishment as shall be inflicted upon them by the sentence of a court-martial.

Art. 2. No officer or soldier shall lie out of his quarters, garrison, or camp, without leave from his superior officer, upon penalty of being punished according to the nature of his offence, by the sentence of a court-martial.

Art. 3. Every non-commissioned officer and soldier shall retire to his quarters or tent at the beating of the retreat; in default of which he shall be punished, according to the nature of his offence, by the commanding officer.

Art. 4. No officer, non-commissioned officer, or soldier, shall fail of repairing, at the time fixed, to the place of parade of exercise, or other rendezvous appointed by his commanding officer, if not prevented by sickness, or some other evident necessity; or shall go from the said place of rendezvous, or from his guard, without leave from his commanding officer, before he shall be regularly dismissed or relieved, on the penalty of being punished according to the nature of his offence, by the sentence of a court-martial.

Art. 5. Whatever commissioned officer shall be found drunk on his guard, party, or other duty under arms, shall be cashiered for it; any non-commissioned officer or soldier so offending, shall suffer such corporal punishment as shall be inflicted by the sentence of a court-martial.
Art. 6. Whatever sentinel shall be found sleeping upon his post, or shall leave it before he shall be regularly relieved, shall suffer death, or such other punishment as shall be inflicted by the sentence of a court-martial.
Art. 7. No soldier belonging to any regiment, troop, or company, shall hire another to do his duty for him, or be excused from duty, but in case of sickness, disability, or leave of absence; and every such soldier found guilty of hiring his duty, as also the party so hired to do another's duty, shall be punished at the next regimental court-martial.
Art. 8. And every non-commissioned officer conniving at such hiring of duty as aforesaid, shall be reduced for it; and every commissioned officer, knowing and allowing of such ill-practices in the service, shall be punished by the judgment of a general court-martial.
Art. 9. Any person, belonging to the forces employed in the service of the United States, who, by discharging of fire-arms, drawing of swords, beating of drums, or by any other means whatsoever, shall occasion false alarms in camp, garrison, or quarters, shall suffer death, or such other punishment as shall be ordered by the sentence of a general court-martial.
Art. 10. Any officer or soldier who shall, without urgent necessity, or without the leave of his superior officer, quit his platoon or division, shall be punished, according to the nature of his offence, by the sentence of a court-martial.
Art. 11. No officer or soldier shall do violence to any person who brings provisions or other necessaries to the camp, garrison or quarters of the forces of the United States employed in parts out of said states, on pain of death, or such other punishment as a court-martial shall direct.
Art. 12. Whatsoever officer or soldier shall misbehave himself before the enemy, or shamefully abandon any post committed to his charge, or shall speak words inducing others to do the like, shall suffer death.
Art. 13. Whatsoever officer or soldier shall misbehave himself before the enemy, and run away, or shamefully abandon any fort, post or guard, which he or they shall be commanded to defend, or speak words inducing others to do the like; or who, after victory, shall quit his commanding officer, or post, to plunder and pillage: Every such offenders being duly convicted thereof, shall be reputed a disobeyer of military orders; and shall suffer death, or such other punishment, as, by a general court-martial, shall be inflicted on him.
Art. 14. Any person, belonging to the forces of the United States, who shall cast away his arms and ammunition, shall suffer death, or such other punishment as shall be ordered by the sentence of a general court-martial.
Art. 15. Any person, belonging to the forces of the United States, who shall make known the watch-word to any person who is not entitled to receive it according to the rules and discipline of war, or shall presume to give a parole or watch-word different from what he received, shall suffer death, or such other punishment as shall be ordered by the sentence of a general court-martial.
Art. 16. All officers and soldiers are to behave themselves orderly in quarters, and on their march; and whosoever shall commit any waste or spoil, either in walks of trees, parks, warrens, fish-ponds, houses or gardens, cornfields, enclosures or meadows, or shall maliciously destroy any property whatsoever belonging to the good people of the United States, unless by order of the then commander in chief of the forces of the said states, to annoy rebels or other enemies in arms against said states, he or they shall be found guilty of offending herein, shall (besides such penalties as they are liable

to by law) be punished according to the nature and degree of the offence, by the judgment of a regimental or general court-martial.
Art. 17. Whosoever, belonging to the forces of the United States, employed in foreign parts, shall force a safe-guard, shall suffer death.
Art. 18. Whosoever shall relieve the enemy with money, victuals, or ammunition, or shall knowingly harbour or protect an enemy, shall suffer death, or such other punishment as by a court-martial shall be inflicted.
Art. 19. Whosoever shall be convicted of holding correspondence with, or giving intelligence to the enemy, either directly or indirectly, shall suffer death, or such other punishment as by a court-martial shall be inflicted.
Art. 20. All public stores taken in the enemy's camp, towns, forts, or magazines, whether of artillery, ammunition, clothing, forage, or provisions, shall be secured for the service of the United States; for the neglect of which the commanders in chief are to be answerable.
Art. 21. If any officer or soldier shall leave his post or colors to go in search of plunder, he shall upon being convicted thereof before a general court-martial, suffer death, or such other punishment as by a court-martial shall be inflicted.
Art. 22. If any commander of any garrison, fortress, or post, shall be compelled by the officers or soldiers under his command, to give up to the enemy, or to abandon it, the commissioned officers, non-commissioned officers, or soldiers, who shall be convicted of having so offended, shall suffer death, or such other punishment as shall be inflicted upon them by the sentence of a court-martial.
Art. 23. All suttlers and retainers to a camp, and all persons whatsoever serving with the armies of the United States in the field, though no inlisted soldier, are to be subject to orders, according to the rules and discipline of war.
Art. 24. Officers having brevets, or commissions of a prior date to those of the regiment in which they now serve, may take place in courts-martial and on detachments, when composed of different corps, according to the ranks given them in their brevets or dates of their former commissions; but in the regiment, troop, or company to which such brevet officers and those who have commissions of a prior date do belong, they shall do duty and take rank both on court-martial and on detachments which shall be composed only of their own corps, according to the commissions by which they are mustered in the said corps.
Art. 25. If upon marches, guards, or in quarters, different corps shall happen to join or do duty together, the eldest officer by commission there, on duty, or in quarters, shall command the whole, and give out orders for what is needful to the service; regard being always had to the several ranks of those corps, and the posts they usually occupy.
Art. 26. And in like manner also, if any regiments, troops, or detachments of horse or foot shall happen to march with, or be encamped or quarterd with any bodies or detachments of other troops in the service of the United States, the eldest officer, without respect to corps, shall take upon him the command of the whole, and give the necessary orders to the service.

Section XIV.
Art. 1. A general court-martial in the United States shall not consist of less than thirteen commissioned officers, and the president of such court-martial shall not be the commander in chief or commandant of the garrison where the offender shall be tried, nor be under the degree of a field officer.

Art. 2. The members both of general and regimental courts-martial shall, when belonging to different corps, take the same rank which they hold in the army; but when courts-martial shall be composed of officers of one corps, they shall take their ranks according to the dates of the commissions, by which they are mustered in the said corps.
Art. 3. The judge advocate general, or some person deputed by him, shall prosecute in the name of the United States of America; and in trials of offenders by general courts-martial, administer to each member the following oaths:
"You shall well and truly try and determine, according to your evidence, the matter now before you, between the United States of America, and the prisoners to be tried. So help you God.
"You A. B. do swear, that you will duly administer justice according to the rules and articles for the better government of the forces of the United States of America, without partiality, favor, or affection; and if any doubt shall arise, which is not explained by the said articles, according to your conscience, the best of your understanding, and the custom of war in the like cases. And you do further swear, that you will not divulge the sentence of the court, until it shall be approved of by the general, or commander in chief; neither will you, upon any account, at any time whatsoever, disclose or discover the vote or opinion of any particular member of the court-martial, unless required to give evidence thereof as a witness by a court of justice, in a due course of law. So help you God."
And as soon as the said oath shall have been administered to the respective members, the president of the court shall administer to the judge-advocate, or person officiating as such, an oath in the following words:
"You A. B. do swear, that you will not, upon any account, at any time whatsoever, disclose or discover the vote or opinion of any particular member of the court-martial, unless required to give evidence thereof, as a witness, by a court of justice, in a due course of law. So help you God."
Art. 4. All the members of a court-martial are to behave with calmness and decency; and in the giving of their votes, are to begin with the youngest in commission.
Art 5. All persons who give evidence before a general court-martial, are to be examined upon oath; and no sentence of death shall be given against any offender by any general court-martial, unless two-thirds of the officers present shall concur therein.
Art. 6. All persons called to give evidence, in any cause, before a court-martial, who shall refuse to give evidence, shall be punished for such refusal, at the discretion of such court-martial: The oath to be administered in the following form, viz.
"You swear the evidence you shall give in the cause now in hearing, shall be the truth, the whole truth, and nothing but the truth. So help you God."
Art. 7. No field-officer shall be tried by any person under the degree of a captain; nor shall any proceedings or trials be carried on excepting between the hours of eight in the morning and of three in the afternoon, except in cases which require an immediate example.
Art. 8. No sentence of a general court-martial shall be put in execution, till after a report shall be made of the whole proceedings to Congress, or to the general or commander in chief of the forces of the United States, and their or his directions be signified thereupon.
Art. 9. For the more equitable decision of disputes which may arise between officers and soldiers belonging to different corps, it is hereby directed, that the courts-martial shall be equally composed of officers belonging to the corps in which the parties in question do then serve; and that the presidents shall be taken by turns, beginning with that corps which shall be eldest in rank.
Art. 10. The commissioned officers of every regiment may, by the appointment of their colonel or commanding officer, hold regimental courts-martial for the enquiring into such disputes, or criminal

matters, as may come before them, and for the inflicting corporal punishments for small offences, and shall give judgment by the majority of voices; but no sentence shall be executed till the commanding officer (not being a member of the court-martial) or the commandant of the garrison, shall have confirmed the same.

Art. 11. No regimental court-martial shall consist of less than five officers, excepting in cases where that number cannot conveniently be assembled, when three may be sufficient; who are likewise to determine upon the sentence by the majority of voices; which sentence is to be confirmed by the commanding officer of the regiment, not being a member of the court-martial.

Art. 12. Every officer commanding in any of the forts, barracks, or elsewhere, where the corps under his command consists of detachments from different regiments, or of independent companies, may assemble courts-martial for the trial of offenders in the same manner as if they were regimental, whose sentence is not to be executed till it shall be confirmed by the said commanding officer.

Art. 13. No commissioned officer shall be cashiered or dismissed from the service, excepting by an order from Congress, or by the sentence of a general court-martial; but non-commissioned officers may be discharged as private soldiers, and, by the order of the colonel of the regiment, or by the sentence of a regimental court-martial, be reduced to private sentinels.

Art. 14. No person whatever shall use menacing words, signs, or gestures, in the presence of a court-martial then sitting, or shall cause any disorder or riot, so as to disturb their proceedings, on the penalty of being punished at the discretion of the said court-martial.

Art. 15. To the end that offenders may be brought to justice, it is hereby directed, that whenever any officer or soldier shall commit a crime deserving punishment, he shall, by his commanding officer, if an officer, be put in arrest; if a non-commissioned officer or soldier, be imprisoned till he shall be either tried by a court-martial, or shall be lawfully discharged by a proper authority.

Art. 16. No officer or soldier who shall be put in arrest or imprisonment, shall continue in his confinement more than eight days, or till such time as a court-martial can be conveniently assembled.

Art. 17. No officer commanding a guard, or provost-martial, shall refuse to receive or keep any prisoner committed to his charge, by any officer belonging to the forces of the United States; which officer shall, at the same time, deliver an account in writing, signed by himself, of the crime with which the said prisoner is charged.

Art. 18. No officer commanding a guard, or provost-martial, shall presume to release any prisoner committed to his charge without proper authority for so doing; nor shall he suffer any prisoner to escape, on the penalty of being punished for it by the sentence of a court-martial.

Art. 19. Every officer or provost-martial to whose charge prisoners shall be committed, is hereby required, within twenty-four hours after such commitment, or as soon as he shall be relieved from his guard, to give in writing to the colonel of the regiment to whom the prisoner belongs (where the prisoner is confined upon the guard belonging to the said regiment, and that his offence only relates to the neglect of duty in his own corps) or to the commander in chief, their names, their crimes, and the names of the officers who committed them, on the penalty of his being punished for his disobedience or neglect, at the discretion of a court-martial.

Art. 20. And if any officer under arrest, shall leave his confinement before he is set at liberty by the officer who confined him, or by a superior power, he shall be cashiered for it.

Art. 21. Whatsoever commissioned officer shall be convicted, before a general court-martial, of behaving in a scandalous, infamous manner, such as is unbecoming the character of an officer and a gentleman, shall be discharged from the service.

Art. 22. In all cases where a commissioned officer is cashiered for cowardice, or fraud, it shall be added in the punishment, that the crime, name, place of abode, and punishment of the delinquent, be published in the newspapers, in and about the camp, and of that particular state from which the offender came, or usually resides: After which, it shall be deemed scandalous in any officer to associate with him.

Section XV

Art. 1. When any commissioned officer shall happen to die or be killed in the service of the United States, the major of the regiment, or the officer doing the major's duty in his absence, shall immediately secure all his effects, or equipage, then in camp or quarters; and shall, before the next regimental court-martial, make an inventory thereof, and forthwith transmit the same to the office of the board of war, to the end, that his executors may, after payment of his debts in quarters and interment, receive the overplus, if any be, to his or their use.

Art. 2. When any non-commissioned officer or soldier shall happen to die, or to be killed in the service of the United States, the then commanding officer of the troop or company, shall, in the presence of two other commissioned officers, take an account of whatever effects he dies possessed of, above his regimental clothing, arms, and accoutrements, and transmit the same to the office of the board at war; which said effects are to be accounted for, and paid to the representative of such deceased non-commissioned officer or soldier. And in case any of the officers, so authorized to take care of the effects of dead officers and soldiers, should, before they shall have accounted to their representatives for the same, have occasion to leave the regiment, by preferment or otherwise, they shall, before they be permitted to quit the same, deposit in the hands of the commanding officer or of the agent of the regiment, all the effects of such deceased noncommissioned officers and soldiers, in order that the same may be secured for, and paid to, their respective representatives.

Section XVI

Art. 1. All officers, conductors, gunners, matrosses, drivers, or any other persons whatsoever, receiving pay or hire in the service of the artillery of the United States, shall be governed by the aforesaid rules and articles, and shall be subject to be tried by courts-martial, in like manner with the officers and soldiers of the other troops in the service of the United States.

Art. 2. For differences arising amongst themselves, or in matters relating solely to their own corps, the courts-martial may be composed of their own officers; but where a number sufficient of such officers cannot be assembled, or in matters wherein other corps are interested, the officers of artillery shall sit in courts-martial with the officers of the other corps, taking their rank according to the dates of their respective commissions, and no otherwise.

Section XVII

Art. 1. The officers and soldiers of any troops, whether minutemen, militia, or others, being mustered and in continental pay, shall, at all times, and in all places, when joined, or acting in conjunction with the regular forces of the United States, be governed by these rules or articles of war, and shall be subject to be tried by courts-martial in like manner with the officers and soldiers in the regular forces, save only that such courts-martial shall be composed entirely of militia officers of the same provincial corps with the offender.

That such militia and minute-men as are now in service, and have, by particular contract with their respective states, engaged to be governed by particular regulations while in continental service, shall not be subject to the above articles of war.
Art. 2. For the future, all general officers and colonels, serving by commission from the authority of any particular state, shall, on all detachments, courts-martial, or other duty wherein they may be employed in conjunction with the regular forces of the United States, take rank next after all generals and colonels serving by commissions from Congress, though the commissions of such particular generals and colonels should be of elder date; and in like manner lieutenant-colonels, majors, captains, and other inferior officers, serving by commission from any particular state, shall, on all detachments, courts-martial, or other duty, wherein they may be employed in conjunction with the regular forces of the United States, have rank next after all officers of the like rank serving by commissions from Congress, though the commissions of such lieutenant-colonels, majors, captains, and other inferior officers, should be of elder date to those of the like rank from Congress.

Section XVIII
Art. 1. The aforegoing articles are to be read and published once in every two months, at the head of every regiment, troop or company, mustered, or to be mustered in the service of the United States; and are to be duly observed and exactly obeyed by all officers and soldiers who are or shall be in the said service.
Art. 2. The general, or commander in chief for the time being, shall have full power of pardoning or mitigating any of the punishments ordered to be inflicted, for any of the offences mentioned in the foregoing articles; and every offender convicted as aforesaid, by any regimental court-martial, may be pardoned, or have his punishment mitigated by the colonel, or officer commanding the regiment.
Art. 3. No person shall be sentenced to suffer death, except in the cases expressly mentioned in the foregoing articles; nor shall more than one hundred lashes be inflicted on any offender, at the discretion of a court-martial.
That every judge-advocate, or person officiating as such, at any general court-martial, do, and he is hereby required to transmit, with as much expedition as the opportunity of time and distance of place can admit, the original proceedings and sentence of such court-martial to the secretary at war, which said original proceedings and sentence shall be carefully kept and preserved in the office of said secretary, to the end that persons entitled thereto may be enabled, upon application to the said office, to obtain copies thereof. That the party tried by any general court-martial, shall be entitled to a copy of the sentence and proceedings of such court-martial, upon demand thereof made by himself, or by any other person or persons, on his behalf, whether such sentence be approved or not.
Art. 4. The field officers of each and every regiment are to appoint some suitable person belonging to such regiment, to receive all such fines as may arise within the same, for any breach of any of the foregoing articles, and shall direct the same to be carefully and properly applied to the relief of such sick, wounded, or necessitous soldiers as belong to such regiment; and such person shall account with such officer for all fines received, and the application thereof.
Art. 5. All crimes not capital, and all disorders and neglects which officers and soldiers may be guilty of, to the prejudice of good order and military discipline, though not mentioned in the above articles of war, are to be taken cognizance of by a general or regimental court-martial, according to the nature and degree of the offence, and be punished at their discretion.

Bibliography

Stoney, Samuel G. editor, *The Great Fire of 1778 Seen Through Contemporary Letters,* The South Carolina Historical Magazine, Volume 64, Number 1, January 1963

Bass, Robert D. *Swamp Fox, The life and campaigns of General Francis Marion*, Sandlapper Publishing Company, 1974

Moss, Bobby Gilmer. *Roster of the South Carolina Patriots in the American Revolution,* Genealogical Publishing Co., Inc. 1983

Grimkè, John Faucheraud. *Journal of the Campaign to the Southward, May 9th to July 14th, 1778,* The South Carolina Historical and Genealogical Magazine, Volume XII, Number 4, October 1911

O'Kelley, Patrick. *"Nothing but Blood and Slaughter" Military Operations and Order of Battle of the Revolutionary War in the Carolinas, Volume One 1771-1779,* Booklocker.com, Inc 2003

O'Kelley, Patrick. "*Nothing but Blood and Slaughter" The Revolutionary War in the Carolinas, Volume Two, 1780,* Booklocker.com, Inc 2004

O'Kelley, Patrick. *"Nothing but Blood and Slaughter" The Revolutionary War in the Carolinas, Volume Three, 1781*, Booklocker.com, Inc 2005

O'Kelley, Patrick. *"Nothing but Blood and Slaughter" The Revolutionary War in the Carolinas, Volume Four, 1782*, Booklocker.com, Inc 2006

Salley, A. S. editor. *Col. Peter Horry's Order Book*, The South Carolina Historical Magazine, Volume 35, Number 3, July 1934

Simons, Robert Bentham. *Regimental Book of Captain James Bentham, 1778-1780,* The South Carolina Historical Magazine, Volume 53, Number 1, January 1952

1

3

A

B

C

D

E

F

G

I

J

K

L

M

N

O

P

Q

R

S

T

U

V

W

Y

Connect with Blacksmith Publishing

www.ThePinelander.com

www.BlacksmithPublishing.com

www.ingramcontent.com/pod-product-compliance
Lightning Source LLC
Chambersburg PA
CBHW081003140626
46546CB00019B/3184

* 9 7 8 1 9 5 6 9 0 4 2 9 1 *